MW00675269

CONTINUUM MODELS AND DISCRETE SYSTEMS

Proceedings of the Eighth International Symposium on

CONTINUUM MODELS AND DISCRETE SYSTEMS

11 – 16 June 1995 Varna, Bulgaria

Editor

Konstantin Z Markov
Faculty of Mathematics and Informatics
University of Sofia
Bulgaria

Published by

World Scientific Publishing Co Pte Ltd
P O Box 128, Farrer Road, Singapore 912805
USA office: Suite 1B, 1060 Main Street, River Edge, NJ 07661
UK office: 57 Shelton Street, Covent Garden, London WC2H 9HE

Library of Congress Cataloging-in-Publication Data

Continuum models and discrete systems : proceedings of the eighth
international symposium : 11–16 June 1995, Varna, Bulgaria / editor,
Konstantin Z. Markov.
xviii, 662 p. ; 22.5 cm.
Includes index.
ISBN 9810225520
1. Continuum mechanics--Congresses. I. Markov, Konstantin Z.
QA808.C674 1996
531--dc20 95-47204
CIP

British Library Cataloguing-in-Publication Data
A catalogue record for this book is available from the British Library.

This book is printed on acid-free paper.

Printed in Singapore by Uto-Print

PREFACE

The Eight International Symposium on Continuum Models and Discrete Systems (CMDS8) took place in the Grand-Hotel "Varna" at the "St. Constantine" Black Sea resort near the town of Varna, Bulgaria, from 11th to 16th June, 1995. The Symposium was placed under the auspices of the *International Society for the Interaction of Mechanics and Mathematics* (ISIMM). The previous Symposia of this series were:

CMDS1 (Kielce, Poland, 1975)
CMDS2 (Mont Gabriel, Canada, 1977)
CMDS3 (Freudenstadt, FRG, 1979)
CMDS4 (Stockholm, Sweden, 1981)
CMDS5 (Nottingham, UK, 1985)
CMDS6 (Dijon, France, 1989)
CMDS7 (Paderborn, Germany, 1992)

By tradition the participants whom have been personally invited by the International Scientific Committee include the following:

B. Adams	Carnegie Mellon University, USA
K.-H. Anthony	Universität-GH-Paderborn, Germany (preceding Chairman)
G. Capriz	Consorzio Pisa Ricerche, Italy
S. Cowin	City University of New York, USA
B. U. Felderhof	RWTH-Aachen, Germany
J. Kratochvil	Czech Technical University, Czech Republic
K. Markov	"St. Kl. Ohridski" University of Sofia, Bulgaria (Chairman)
G. Maugin	Université Pierre et Marie Curie, France
N. Rivier	Université Louis Pasteur, France
K. Sobczyk	Polish Academy of Sciences, Poland
A. Vakulenko	St. Petersburg State University, Russia
J. Willis	University of Cambridge, United Kingdom
Xu Bingye	Tsinghua University, P. R. China

The Chairman has been supported by the Local Committee including his colleagues A. Baltov (Bulgarian Academy of Sciences) and I. Mihovsky (University of Sofia).

As with the earlier Symposia of the CMDS series, the purpose of CMDS8 was to bring together scientists with different backgrounds, actively working on continuum theories of discrete mechanical and thermodynamical systems in the fields of mathematics, theoretical and applied mechanics, physics, material science and engineering. The main objective did not lie, however, in reporting specific results as such, but rather in joining together different languages, questions and methods developed in the respective disciplines and to stimulate thus a broad interdisciplinary research. Judging from the lively discussions, the friendly, unofficial and stimulating atmosphere, both inside and outside Conference rooms, this goal was achieved. Moreover, the excellent conditions offered by the Grand-Hotel "Varna", the pleasant surroundings of the "St. Constantine" resort, the good weather and warm sea also contributed toward the success of the meeting.

Six broad topics, which have been traditional for the CMDS-series, have been selected for the Symposium by the International Scientific Committee:

1. Thermodynamics, transport theory and statistical mechanics, in the context of continuum modeling of discrete systems.
2. Continuum mechanics of complex fluids and deformable solids with microstructure.
3. Continuum theory of living structures.
4. Defect dynamics, synergetics, solitons, coherent structures.
5. Dislocations and plasticity.
6. Fundamentals of fracture mechanics.

Within these topics, 15 General Lectures and 56 Research Communications were delivered. All of them, and especially the General Lectures, were carefully "nominated" by the International Scientific Committee, so as to illustrate the newest trends, ideas and results. Moreover, the Lecturers have made special efforts to render their presentations understandable to an audience of scientists who are not narrow specialists. In this way they have all decisively contributed toward the main CMDS objectives — the fruitful and stimulating interdisciplinary exchange of ideas and results.

Almost all of the lectures, delivered at the Symposium, are recorded in this volume. Also, the editor took the liberty to include two or three texts which have been handed to him prior to or during the meeting and which seemed really worthy and interesting, but whose authors were not able to attend due to unforeseen reasons.

In the proceedings volume such as the present one, the authors usually record their original results in a more or less concise form that reflects both the limited time of their presentations and the restrictions imposed by the limited bulk of the volume. Several authors have however expressed this time, before or during the meeting, a special desire to be permitted, if possible, a longer than the average length for their manuscripts, since they felt that their ideas needed a bit more detailed exposition. After CMDS8 it became clear that some of the contributors would not present their texts, hence there was some space left within the page limit, fixed by the Publisher, and the editor took the liberty to include these longer texts. But this was only one of the reasons for doing so. Far more important was editor's deep conviction that these manuscripts indeed concern new approaches of considerable interdisciplinary interests, inherently tangled with the CMDS topics, and hence the reader should be made available to a more detailed record as soon as possible.

Finally, it is a pleasure to express sincere gratitude to the members of the International Scientific Committee and of the Local Committee for their constant help, valuable advices and support in organizing CMDS8. Cordial thanks go to all Lecturers for their good presentations and careful preparation of the manuscripts.

It is with great pleasure and gratitude that we also acknowledge the support of several Bulgarian institutions and firms, namely, the National Fund of Scientific Research at the Bulgarian Ministry of Science, Education and Technology (under Grants MM26-91 and MM416-94), "Cyril and Methodius" Foundation, "Evrika" Foundation, Union of Bulgarian Mathematicians, Union of Scientists in Bulgaria, "Antibiotic" Co. (Razgrad) and "Abritus" Co. (Razgrad). The support of Directorate General for Science, Research and Development at the Commission of the European Union, Brussels, for the East and Central European participants, is gratefully acknowledged as well.

We are looking forward to CMDS9 which should tentatively take place in Istanbul in 1998.

K. Z. Markov
Sofia, October 1995

HENRYK ZORSKI AND 20 YEARS OF CMDS SYMPOSIA: A DEDICATION

Professor Dr **Henryk Zorski** (Warsaw, October, 1995)

The first International Symposium on **Continuum Models and Discrete Systems (CMDS1)** was organized by Henryk Zorski, together with Ronald Rivlin, in Jodlowy Dwor near Kielce, Poland, in 1975. This symposium was a result of the well realized necessity to bring together specialists from different branches of exact sciences who, in the root, do one and the same: model the behaviour of various real systems with complicated discrete microstructure by means of macroscopical continuum models. The importance *per se* of such a modelling in any specific field and problem is, of course, beyond all doubt. But, on a higher interdisciplinary level, it is even more desirable to recognize the common background and the general principles that lie behind the continuum modelling independently of the origin of the specific problems at hand that may spread from biological structures and material science to rigorous theorems of homogenization in pure mathematics.

The success of the first meeting, the friendly, unofficial and stimulating atmosphere and the vivid interest of many of the most actively working specialists from all over the world led naturally to the organization of its followers, thus turning CMDS1 into a series of symposia. After Jodlowy Dwor, Mont Gabriel (1977) followed, then Freudenstadt (1979), Stockholm (1981), Nottingham (1985), Dijon (1989), Paderborn (1992), and now, twenty years later, St. Constantine. Henryk Zorski actively participated in the preparation of all these meetings: he co-chaired the first one with Ronald Rivlin, chaired the second and was a member of the International Scientific Committees until CMDS6 in Dijon, where he resigned as such, but continued his participation. He delivered two memorable General Lectures, namely "Direct Continuum Model of a Solid" in Nottingham (1985), and "Nonlinear Waves in Structured (Dipolar) Media" in Paderborn (1992). Equally important was his personal presence at all the CMDS Symposia with his sharp observations, remarks, useful advice, his jokes and keen sense of humour. And now, twenty years after CMDS1, it is clear that Henryk Zorski not only initiated and shaped the CMDS Symposia, but also played a central and decisive role in establishing them as a well recognized and noted international tradition. Just to express at least a part of our affection and gratitude to Henryk Zorski for everything he has done and is doing for the CMDS Symposia, the International CMDS-community dedicates, with great pleasure, both CMDS8 and this volume to him.

CONTENTS

---

*GL — General Lecture, RC — Research Communication
†Names underlined — Lecturers

PART II.

CONTINUUM MECHANICS OF COMPLEX FLUIDS AND DEFORMABLE SOLIDS WITH MICROSTRUCTURE

PART III.

CONTINUUM THEORY OF LIVING STRUCTURES

PART IV.

DEFECT DYNAMICS, SYNERGETICS, SOLITONS, COHERENT STRUCTURES

PART V.

DISLOCATIONS AND PLASTICITY

PART VI.

FUNDAMENTALS OF FRACTURE MECHANICS

PART I

THERMODYNAMICS, TRANSPORT THEORY, STATISTICAL MECHANICS

Continuum Models and Discrete Systems
Proceedings of 8th International Symposium, June 11-16, 1995, Varna, Bulgaria, ed. K.Z. Markov
©World Scientific Publishing Company, 1996, pp. 2–9

SELF-ORGANIZED CRITICALITY IN STATISTICAL MECHANICS

J. G. BRANKOV
Laboratory of Statistical Mechanics and Thermodynamics,
Institute of Mechanics, Bulgarian Academy of Sciences, Acad. G. Bonchev Str., bl. 4,
BG-1113 Sofia, Bulgaria

Abstract. The theory of self-organized criticality (SOC) aims at explaining the appearance of fractal (self-similar or self-affine) spatial and temporal structures in various nonlinear dissipative systems with many interacting degrees of freedom. One of its most interesting applications is to earthquakes. There is a simple cellular automaton model on a lattice, known as the Abelian sandpile model (ASM), which exhibits essential features of SOC. The problem of exact evaluation of the pair correlation functions in this model is still unsolved. Here we develop an approach to its solution which is based on the construction of one-to-one mappings of the configurations of the ASM, defined on a finite connected multigraph, onto the spanning trees in an associated multigraph.

1. The Concept of Self-Organized Criticality

A new era in natural sciences has started by shifting the focus of investigations from "reduction of complexity to simplicity" to "developing of complexity out of simplicity" [1]. The recent theory of "self-organized criticality" (SOC) [2] aims at explaining how simple local rules can lead to a great variety of natural phenomena organized on all spatial and/or temporal scales, such as fractal structures and $1/f$-noise. Their mathematical description is based on the concepts of scale-invariance, self-similarity and self-affinity [3]. Here is an example which is believed to exhibit essential features of SOC.

Earthquakes [4-7]. The number N of observed earthquakes of magnitude greater than M is given by the empirical Gutenberg-Richter law: $\log N = a - bM$, valid typically for $2 < M < 8$, with almost universal constants a and b. The lack of characteristic time scale is demonstrated by the Omori law describing the decay with time t of the number of aftershocks: $n(t) \propto t^{-p}$ with $p \approx 1$. The lack of characteristic spatial scale is manifested by the "fractal" self-similarity of the set of earthquake epicenters. It has been suggested that the concept of SOC is relevant for understanding the underlying processes of tectonic plate motion and release of accumulated strain energy. It is generally assumed that the dynamics of earthquakes is due to a stick-slip sliding of the continental plates along faults. The fundamental idea is that the whole history of previous earthquakes has organized the crust and its array of faults in a

SOC state [5]. Computer simulations of simple mechanical models, largely based on the instability of the stick-slip motion under nonlinear friction, have supplied evidences in support of the Gutenberg-Richter law, the existence of $1/f$ noise in the spectrum of the time gaps between large earthquakes, etc.

2. The Abelian Sandpile Model

Bak, Tang and Wiesenfeld [2] were the first to suggest several sandpile models, including the Abelian sandpile model which was found later to admit some analytical solutions [8-11]. The attractor of its nonlinear (threshold) diffusive dynamics is a state of SOC which is characterized by power-law correlations in space and time. In the simplest version of the model, the dynamic variables are the integer heights $h_i \in \{1,2,3,4\}$ of the sandpile at each vertex v_i, $i = 1,2,\ldots,N$, of a finite square lattice $\mathcal{L}$. The boundary conditions are specified by connecting with edges some (or all) of the vertices at the perimeter $\partial\mathcal{L}$ of $\mathcal{L}$ to a vertex g in the exterior face of $\mathcal{L}$. One step of the dynamics is defined by the addition of a grain of sand at a randomly chosen vertex v_i which increases h_i by unity. If the height h_j at any vertex v_j exceeds the critical value $h_{cj} = \deg(v_j)$, where $\deg(v_j)$ is the number of edges incident with the vertex v_j, then one unit of sand slides down along each of the edges connecting that vertex with its neighbours; the grains of sand sliding down to the "ground" vertex g leave the system. The process continues until a stable configuration is reached in which all $h_j \leq h_{cj}$. Then another grain of sand is added, and so on. The effect of sand sliding upon addition at v_i can be expressed as $h_j \to h_j - \Delta_{i,j}$ for all $j = 1,2,\ldots,N$, where $\mathbf{\Delta} = \{\Delta_{i,j}\}$ is (up to a minus sign) the discrete Laplacian matrix for the square lattice $\mathcal{L}$: $\Delta_{i,i} = \deg(v_i)$, $\Delta_{i,j} = -1$, if v_i and v_j are neighbours, and $\Delta_{i,j} = 0$ otherwise. Note that for v_i away from $\partial\mathcal{L}$, $\Delta_{i,i} = 4$. The standard boundary value problems are: the Dirichlet boundary conditions, when $\Delta_{i,i} = 4$ for $v_i \in \partial\mathcal{L}$, corresponding to the case when the sand is allowed to leave the system through $\partial\mathcal{L}$ (open boundary); the Neumann boundary conditions, when $\Delta_{ii} = 3$ for $v_i \in \partial\mathcal{L}$, if v_i is not a corner, and $\Delta_{ii} = 2$, if v_i is a corner, corresponding to the case when the sand cannot leave the system through $\partial\mathcal{L}$ (closed boundary).

Thus, the dynamics of the sandpile is a Markoff process on the set of stable configurations. A stable configuration is recurrent if it has nonzero weight in the invariant probability measure for the process. There are exactly $\det(\mathbf{\Delta})$ recurrent configurations and all of them have equal probability [8].

A plausible link between the sandpile avalanches, which may accompany the transition from one recurrent configuration to the next, and earthquakes is provided by letting the sandpile heights correspond to the local strain or energy accumulated at the boundary between the moving plates; then the avalanche size corresponds to the energy released by an earthquake.

3. The Mathematical Problem

The two-point correlation functions for the different local heights are one of the most important spatial characteristics of the sandpile. In a properly defined infinite-lattice limit, their long-distance asymptotic form is expected to obey a power-law. Actually, as pointed out by Priezzhev [11], that has been proved only for the correlations between unit heights [9]. In general, the problem of determining the power-law exponents remains open and one of the ways to tackle it is by mapping it onto problems one knows more about. Thus, Majumdar and Dhar [10] found a one-to-one mapping of the recurrent configurations of the ASM, with open boundaries at the whole perimeter of the lattice $\mathcal{L}$, onto the spanning trees in a lattice with boundary vertices $\partial\mathcal{L}$ connected to an additional vertex g ("the ground") in the exterior face.

The present work is a part of an attempt to approach the solution of the problem by following the chain of isomorphisms: exactly solved dimer model on a square lattice $\rightarrow$ spanning trees on a related planar multigraph $\rightarrow$ sandpile configurations on its subgraph. The first isomorphism has been established in [12]. Our aim here is to define precise one-to-one mappings of the set of spanning trees $\mathcal{T}$ in a finite connected multigraph G' onto the set of configurations $\mathcal{C}$ of the ASM on the multigraph $G = G' \setminus g$, where g is an arbitrary vertex of G'. The vertex g determines the open boundaries of the sandpile: the sand can leave the system through the neigbours of g in G'.

Here are some basic definitions and notations. A graph $G = (V, E)$ is defined by the set of vertices V and the set of edges E incident with some pairs of vertices. Let I be the incidence relation between edges and vertices: for each $e \in E$, $I(e) = \{v, v'\} \subseteq V$, where $\{v, v'\}$ is an unordered pair of vertices, called the ends of e; since we do not consider graphs with loops, v and v' must be distinct vertices. A subgraph of a graph $G = (V, E)$ is a graph $G' = (V', E')$ such that $V' \subseteq V$, $E' \subseteq E$ and $I'(e) = I(e)$ for all $e \in E'$; the incidence relation I' is the restriction of I to the domain $E' \subseteq E$ and will be denoted simply by I in the remainder. The subgraph with $V' = V \setminus W$, where $W \subset V$, and $E' = \{e \in E : I(e) \subseteq V \setminus W\}$ is called the section graph induced by the set $V \setminus W$, to be denoted by $G \setminus W$. A path $p(v, v')$ between two vertices v and v' in a graph G is a finite sequence of alternating vertices and edges of G: $v = v_0, e_1, v_1, e_2, \ldots, v_{n-1}, e_n, v_n = v'$, such that $I(e_k) = \{v_{k-1}, v_k\}$ for all $k = 1, 2, \ldots, n$. A polygon in a graph G is a subgraph consisting of the vertices and edges of a closed path in G with all vertices distinct except the first and the last (which coincide). A tree T in a graph G is a connected nonempty subgraph of G containing no polygon. A spanning subgraph G' of G is a subgraph with $V' = V$. A spanning tree in a graph G is a spanning subgraph of G which is a tree. If the definition of a graph is extended to include multiple edges, i.e., more than one edge incident with the same pair of vertices, the resulting object is called multigraph.

Let $G' = (V', E')$ be a finite connected multigraph and let $\mathcal{T}$ be the set of all spanning trees $T = (V', E^T)$ in G'. Choose a vertex $g \in V'$ and let $\text{dist}\,(v, g|T)$ denote the distance between the vertices $v \in V = V' \setminus g$ and g along the edges of the tree T (the so called chemical distance between v and g). Let $E(v) = \{e \in E' : I(e) = \{v, v'\}, v' \in V'\}$ denote the set of all edges incident with a vertex $v \in V'$ in the multigraph G', and let $N(v)$ be the neighbourhood of v in G', i.e., $N(v) = \{v' \in V' : \{v, v'\} \in E'\}$. The local heights h_v of the sandpile configuration $\{h_v, v \in V\}$, defined below as the map of T, depend on the spanning tree T through the following subsets of $N(v)$:

$$N_{\geq}(v) \;=\; \{v' \in N(v) : \text{dist}\,(v', g|T) \geq \text{dist}\,(v, g|T)\}\,,$$

$$N_{-1}(v) \;=\; \{v' \in N(v) : \mathrm{dist}\,(v', g|T) = \mathrm{dist}\,(v, g|T) - 1\}. \tag{1}$$

Actually, when G' is a multigraph, the local sandpile height at $v \in V$ depends on the following subsets of the set of all edges incident with v:

$$\begin{aligned} E_{\geq}(v) &= \{e \in E(v) : I(e) = \{v, v'\}, v' \in N_{\geq}(v)\} \\ E_{-1}(v) &= \{e \in E(v) : I(e) = \{v, v'\}, v' \in N_{-1}(v)\}. \end{aligned} \tag{2}$$

Since there is a unique path along the edges of a tree T from each $v \in V$ to g, the set $E_{-1}(v) \cap E^T$ contains exactly one edge.

Definition 1. A local ordering ϕ_v of the set $E(v)$ of edges incident with a vertex $v \in V$ in the finite multigraph G is called any one-to-one mapping

$$\phi_v : E(v) \to \{1, \ldots, |E(v)|\}. \tag{3}$$

Here and in the remainder $|S|$ denotes the cardinality of the set S. The inverse mapping is denoted by $\phi_v^{-1} : \{1, \ldots, |E(v)|\} \to E(v)$.

Definition 2. Local ordering $\phi_v(.|E')$ of any nonempty subset of edges $E'(v) \subset E(v)$, induced by the ordering ϕ_v of $E(v)$, is called the one-to-one mapping

$$\phi_v(\cdot|E') : E'(v) \to \{1, \ldots, |E'(v)|\}, \tag{4}$$

which obeys the condition $\phi_v(e_1|E') < \phi_v(e_2|E')$ if and only if $\phi_v(e_1) < \phi_v(e_2)$ for all $e_1 \neq e_2$ from E'. The inverse mapping is denoted by $\phi_v^{-1}(.|E') : \{1, \ldots, |E'(v)|\} \to E'(v)$.

Remark. Let $T = (V', E^T)$ be any spanning tree in the multigraph $G' = (V', E')$ and let $\Phi = \{\phi_v, v \in V\}$ be any set of local orderings of the set of edges $E(v)$ of G' incident with a vertex $v \in V = V' \setminus g$. The edge set E^T of T can be written in the form

$$E^T = \bigcup_{t=1}^{n(T)} \left\{ \bigcup_{v \in A_t} \phi_v^{-1}(k_v|E_{-1}(v)) \right\}, \tag{5}$$

where $k_v \in \{1, \ldots, |E_{-1}(v)|\}$,

$$A_t = \{v \in V : \mathrm{dist}\,(v, g|T) = t\},\; t = 0, 1, ..., n(T), \tag{6}$$

and

$$n(T) = \max_{v \in V} \mathrm{dist}\,(v, g|T) \geq 1. \tag{7}$$

Obviously, $A_0 = \{g\}$, and

$$\bigcup_{\tau=1}^{n(T)} A_\tau = V. \tag{8}$$

The representation given in Eq. (5) follows in an obvious way from Eq. (8) and the definition of a spanning tree.

4. Mapping of Recurrent Sandpile Configurations into Spanning Trees

Here we suggest a formalized version of the mapping found by Majumdar and Dhar [10], valid for any connected and finite multigraph. It is based on the burning algorithm [8] which allows one to determine whether or not a given stable configuration of the sandpile is recurrent. We consider the ASM on a multigraph $G = (V, E)$ supplied with the following boundary conditions: a subset of vertices $\partial V \subseteq V$, called open boundaries, are connected by one or several edges each to an additional vertex g in the external face of G, thus forming the extended multigraph $G' = (V', E')$ with $V' = V \cup \{g\}$. The application of the burning algorithm to any stable sandpile configuration C on $G = G' \backslash g$ generates a finite sequence of subgraphs of $G: G_1 = G \supset G_2 \supset \cdots \supset G_n \supset G_{n+1}$. The configuration C is recurrent if and only if G_{n+1} is the empty graph. The restriction $C(G_t)$ of a configuration C to a nonempty subgraph $G_t = (V_t, E_t)$ of G is called forbidden if $h_v \leq \deg(v|G_t)$ for all $v \in V_t$, where $\deg(v|G_t)$ denotes the number of edges incident with v in the subgraph G_t. The sequence $\{G_t\}$, $t = 1, \ldots, n+1$, is obtained by applying the following rules.

1. At step $t = 0$ the ground vertex g is deleted from the multigraph G' and the sandpile configuration C on $G = G' \setminus g = (V, E)$ is considered.

2. Define recurrently $G_{t+1} = G_t \setminus B_t$ with initial conditions $G_{t=0} = G'$ and $B_{t=0} = \{g\}$. At each step $t = 1, \ldots$, check whether the set $B_t = \{v \in V_t : h_v > \deg(v|G_t)\}$ is empty or not.

If $B_t = \emptyset$ but the (multi)graph G_t is nonempty, then the test ends with the result that the configuration C contains the forbidden subconfiguration $C(G_t)$ and, therefore, C is not recurrent.

If $B_t \neq \emptyset$ and the (multi)graph G_t is nonempty, then consider the restriction of C onto

$$G_{t+1} = G_t \setminus B_t = G' \setminus \{\bigcup_{\tau=0}^{t} B_\tau\}. \tag{9}$$

Since G' is a finite multigraph, the iteration with respect to t ends at some final step $t = n$ with the result $G_{n+1} = \emptyset$ which implies that C is recurrent.

The above algorithm can be used to construct different mappings of the set $\mathcal{C}$ of recurrent sandpile configurations on G into the set $\mathcal{T}$ of spanning trees in G'. Each mapping is specified by the set of arbitrary local orderings of the edges incident with each vertex $v \in V$.

Definition 3. For any set of local orderings $\Phi = \{\phi_v, v \in V\}$, let $\mathcal{P}_\Phi : \mathcal{C} \to \mathcal{T}$ be the mapping of the set $\mathcal{C}$ of recurrent configurations C of the ASM into the set $\mathcal{T}$ of spanning subgraphs T of the multigraph G' defined for all $C \in \mathcal{C}$ by $\mathcal{P}_\Phi(C) = T \equiv (V', E^T(C))$, where

$$E^T(C) = \bigcup_{t=1}^{n} \{\cup_{v \in B_t} \phi_v^{-1}(h_v - \deg(v|G_t)|E^b_{-1}(v))\}. \tag{10}$$

Here $G_1 = G \supset G_2 \supset \cdots \supset G_n$ is the sequence of nonempty subgraphs of G' generated by the application of the burning algorithm to C; the set of vertices deleted at step t from $G_t = (V_t, E_t)$ is given by

$$B_t = \{v \in V_t : h_v > \deg(v|G_t)\}, \; t = 1, \ldots, n; \tag{11}$$

$B_0 = \{g\}$ and for each $v \in B_t$, $1 \leq t \leq n$,

$$E^b_{-1}(v) = \{e \in E(v) : I(e) = \{v, v'\}, v' \in B_{t-1}\}. \tag{12}$$

Proposition 1. For each $C \in \mathcal{C}$, the subgraph $\mathcal{P}_\phi(C) = T \equiv (V', E^T(C))$ is a spanning tree in the multigraph $G' = (V', E')$.

Proof. Since $B_{t+1} \subseteq V_{t+1} = V_t \setminus B_t$, then $B_t \cap B_{t'} = \emptyset$ for all $t \neq t'$. Due to $V = \cup_{t=1}^n B_t$, each vertex $v \in V = V' \setminus g$ belongs to exactly one set B_t with some $t \in \{1, ..., n\}$. By the definition of T, each $v_t \in B_t$ is connected by exactly one edge $e_t \in E^T(C)$ to a vertex $v_{t-1} \in B_{t-1}$. Therefore, there exists a unique path $v_t = v, e_t, v_{t-1}, e_{t-1}, \ldots, v_1, e_1, v_0 = g$ from each vertex v to g along the edges of T.

5. Mapping of Spanning Trees into Recurrent Sandpile Configurations

The mapping presented here is again a formalized and generalized to multigraphs version of the correspondence between spanning trees and recurrent configurations of ASM established by Majumdar and Dhar [10] with the aid of the burning algorithm [8]. We will use that algorithm to prove that the image of a spanning tree is indeed a recurrent sandpile configuration.

Definition 4. For any set of local orderings $\Phi = \{\phi_v, v \in V\}$, let $\mathcal{M}_\Phi : \mathcal{T} \to \mathcal{C}$ be the mapping of the set $\mathcal{T}$ of spanning trees T in the multigraph G' into the set $\mathcal{C}$ of configurations C of the ASM (on $G = G' \setminus g$, with open boundaries at the set of vertices $N(g)$), defined by $\mathcal{M}_\Phi(T) = C = \{h_v, v \in V\}$ with

$$h_v = |E_\geq(v)| + |\{e \in E_{-1}(v) : \phi_v(e) \leq \phi_v(E_{-1}(v) \cap E^T)\}| \tag{13}$$

for all $v \in V$.

Proposition 2. The configuration $C = \mathcal{M}_\Phi(T)$ is stable and recurrent for all $T \in \mathcal{T}$.

Proof. Since $|E_\geq(v)| + |E_{-1}(v)| \leq \deg(v|G') = h_{cv}$ and

$$|\{e \in E_{-1}(v) : \phi_v(e) \leq \phi_v(E_{-1}(v) \cap E^T)\}| \leq |E_{-1}(v)|, \tag{14}$$

it follows that $1 \leq h_v \leq h_{cv}$, i.e., C is stable. Set $G^a_0 = G'$ and define

$$G^a_{t+1} = G^a_t \setminus A_t = G' \setminus \{\cup_{\tau=0}^t A_\tau\}, \quad t = 0, 1, \ldots, n(T), \tag{15}$$

where the sets A_t are defined in Eq. (6) and the integer n(T) is given by Eq. (7). Let $G^a_t = (V^a_t, E^a_t)$; obviously, $G^a_1 = G = (V, E)$. Since $A_t \subseteq V^a_t$ is nonempty at $t = 0, 1, \ldots, n(T)$, for each $v \in A_t$ one has $|E_\geq(v)| = \deg(v|G^a_t)$ and hence

$$h_v \geq |E_\geq(v)| + 1 = \deg(v|G^a_t) + 1 > \deg(v|G^a_t). \tag{16}$$

Therefore $A_t \subseteq B_t$ for all $t = 0, 1, \ldots, n(T)$. Due to Eq. (8), the graph $G^a_{n(T)+1}$ is empty, i.e., the configuration C burns out along the sequence of subgraphs (15).

Proposition 3. For any spanning tree $T \in \mathcal{T}$, the sequence of subgraphs G_t, $t = 1, \ldots, n$, generated by the application of the burning algorithm to the configuration

$C = \mathcal{M}_\Phi(T)$, coincides with the sequence of subgraphs G_t^a, $t = 1, \dots, n(T)$, defined by Eqs. (6), (7) and (15). In particular, $n(T) = n$ and $A_t = B_t$ for all $t = 0, 1, \dots, n$.

Proof. From $V = \bigcup_{t=1}^n B_t$ and $A_t \subseteq B_t$ for all $t = 1, \dots, n(T)$, see the proof of proposition 2, one concludes that $n \geq n(T)$. Suppose that $n < n(T)$. Then, by definition, $B_t = \emptyset$ for all $t > n$ but $A_n \subseteq B_n = \emptyset$ is a contradiction due to $A_n \neq \emptyset$, $n < n(T)$. Thus $n = n(T)$. Suppose now that $A_t \neq B_t$ for some $t = 1, \dots, n$. Then there exists a vertex $v \in B_t \setminus A_t$ belonging to some $A_{t'} \subseteq B_{t'}$ with $t' \neq t$, which contradicts the fact that $B_t \cap B_{t'} = \emptyset$ for all $t' \neq t$, see the proof of proposition 1. Therefore, $A_t = B_t$, and since $A_0 = B_0 = \{g\}$, $G_0^a = G_0 = G'$, it follows that $G_t^a = G_t$ for all $t = 1, \dots, n$.

Lemma 1. For any set of local orderings $\Phi = \{\phi_v, v \in V\}$, $\mathcal{M}_\Phi(\mathcal{P}_\Phi(C)) = C$ for all $C \in \mathcal{C}$.

Proof. Let $C = \{h_v,\ v \in V\}$ be any recurrent sandpile configuration and let G_t, $t = 1, \dots, n$, be the sequence of nonempty subgraphs generated by the application of the burning algorithm to C. By definition, the spanning tree $\mathcal{P}_\Phi(C) = (V', E^T(C))$ has the edge set given by Eq. (10), where the sets B_t, $t = 1, \dots, n$, defined in Eq. (11), are nonempty. Let $\mathcal{M}_\Phi(\mathcal{P}_\Phi(C)) = \{h'_v, v \in V\}$ be the corresponding sandpile configuration with local heights

$$h'_v = |E_{\geq}(v)| + |\{e \in E_{-1}(v) : \phi_v(e) \leq \phi_v(E_{-1}(v) \cap E^T(C))\}|, \tag{17}$$

where the sets of edges $E_{\geq}(v)$ and $E_{-1}(v)$ are defined with respect to the distance from the neighbours of v to g along the edges of the spanning tree $T = \mathcal{P}_\Phi(C)$. We shall prove that $h'_v = h_v$ for all $v \in V$. From proposition 3 it follows that the configuration $C' = \mathcal{M}_\Phi(T)$ burns out along the sequence $G_t = G_t^a$, $t = 1, \dots, n$, see Eq. (15), where $n = n(T)$, see Eq. (7), and $B_t = A_t$, see Eq. (6). Hence,

$$\begin{aligned} E_{-1}(v) &= E^b_{-1}(v), \quad \deg(v|G_t) = |E_{\geq}(v)|, \\ e^T(v) &\equiv E_{-1}(v) \cap E^T(C) = E^b_{-1}(v) \cap E^T(C), \quad \forall v \in V. \end{aligned} \tag{18}$$

Then, by using the definition of $E^T(C)$, see Eq. (10), one obtains

$$e^T(v) = \phi_v^{-1}(h_v - |E_{\geq}(v)||E_{-1}(v)). \tag{19}$$

Since $e^T(v) \in E_{-1}(v) \subseteq E(v)$,

$$\begin{aligned} &|\{e \in\ E_{-1}(v) :\ \phi_v(e) < \phi_v(e^T(v))\}| \\ &= |\{e \in\ E_{-1}(v) :\ \phi_v(e|E_{-1}(v)) < \phi_v(e^T(v)|E_{-1}(v))\}| \\ &= h_v -\ |E_{\geq}(v)|. \end{aligned} \tag{20}$$

Thus, Eqs. (17) and (20) imply that $h'_v = h_v$ for all $v \in V$.

Lemma 2. The mapping $\mathcal{P}_\Phi : \mathcal{C} \to \mathcal{T}$ is injection.

Proof. Let $C \neq C'$. Suppose that $\mathcal{P}_\Phi(C) = \mathcal{P}_\Phi(C')$. Then $\mathcal{M}_\Phi(\mathcal{P}_\Phi(C)) = \mathcal{M}_\Phi(\mathcal{P}_\Phi(C'))$ contradicts the fact that $\mathcal{M}_\Phi(\mathcal{P}_\Phi(C)) = C$ and $\mathcal{M}_\Phi(\mathcal{P}_\Phi(C')) = C'$ are different due to Lemma 1.

Lemma 3. The mapping $\mathcal{P}_\Phi : \mathcal{C} \to \mathcal{T}$ is surjection.

Proof. We shall prove that each $T \in \mathcal{T}$ is the image under $\mathcal{P}_\Phi$ of $C = \mathcal{M}_\Phi(T) \in \mathcal{C}$, i.e., that $\mathcal{P}_\Phi(\mathcal{M}_\Phi(T)) = T$ for each $T \in \mathcal{T}$.

Let the edge set E^T of the tree T be given by Eq. (5) with some integers $k_v \in \{1, \ldots, |E_{-1}(v)|\}$, $\forall v \in V$. Consider the image of T, $\mathcal{M}_\Phi(T) = C$, which is a recurrent configuration $C \in \mathcal{C}$ with local heights given by Eq. (13). Since the sets $E_{\geq}(v)$ and $E_{-1}(v)$ are defined with respect to dist $(., g|T)$, for each $v \in V$ we have $\phi_v(E_{-1}(v) \cap E^T) = k_v$ and, therefore, $h_v = |E_{\geq}(v)| + k_v$. As we have already proved, see proposition 1, the image of the recurrent configuration $\mathcal{M}_\Phi(T)$ under the mapping $\mathcal{P}_\Phi$ is a spanning tree, say $\tilde{T} = (V', \tilde{E}^T)$. According to definition 3, its edge set is given by Eq. (10), where n, B_t, G_t are obtained by applying the burning algorithm to $C = \mathcal{M}_\Phi(T)$. Due to proposition 3 and Eqs. (18), we can write

$$\tilde{E}^T = \cup_{t=1}^{n} \left\{ \cup_{v \in A_t} \phi_v^{-1}(h_v - |E_{\geq}(v)|) | E_{-1}(v)) \right\}. \tag{21}$$

The above derived expression $h_v = |E_{\geq}(v)| + k_v$ completes the proof that $\tilde{E}^T = E^T$.

Our main result can be summarized in the following

Theorem. For any set of local orderings $\Phi = \{\phi_v, v \in V\}$, the mapping $\mathcal{P}_\Phi : \mathcal{C} \to \mathcal{T}$ of the set $\mathcal{C}$ of recurrent configurations of the ASM onto the set $\mathcal{T}$ of spanning trees of the multigraph G' is a bijection. Its inverse mapping is $\mathcal{M}_\Phi : \mathcal{T} \to \mathcal{C}$.

The proof follows from Lemmas 2 and 3.

The apparently nonlocal definition of the local heights of the ASM, see definition 4, first found in a different form by Priezzhev [11], makes the analytical evaluation of the height probabilities and the height-height correlation function a challenging problem. One may hope some simplifications to occur in the infinite-lattice limit.

Acknowledgements. The support of the Bulgarian National Foundation for Scientific Research under Grant No MM405/94 is gratefully acknowledged.

References

1. P. W. Anderson, *Physics Today* **44** (1991) 9.
2. P. Bak, C. Tang and K. Wiesenfeld, *Phys. Rev. A* **38** (1988) 364.
3. B. B. Mandelbrot, *The Fractal Geometry of Nature*, Freeman, San Francisco, 1982.
4. P. Bak and C. Tang, *J. Geophys. Res.* **94** (1989) 15635.
5. A. Sornette and D. Sornette, *Europhys. Lett.* **9** (1989) 197.
6. J. M. Carlson and J. S. Langer, *Phys. Rev. A* **40** (1989) 6470.
7. K. Ito and M. Matsuzaki, *J. Geophys. Res.* **95** (1990) 6853.
8. D. Dhar, *Phys. Rev. Lett.* **64** (1990) 1613.
9. S. N. Majumdar and D. Dhar, *J. Phys. A* **24** (1991) L357.
10. S. N. Majumdar and D. Dhar, *Physica A* **185** (1992) 129.
11. V. B. Priezzhev, *J. Stat. Phys.* **74** (1994) 955.
12. J. G. Brankov and E. G. Litov, *J. Theor. Appl. Mech.* **2** (1993) 46.

Continuum Models and Discrete Systems
Proceedings of 8th International Symposium, June 11-16, 1995, Varna, Bulgaria, ed. K.Z. Markov
© World Scientific Publishing Company, 1996, pp. 10–14

CRITICAL MODELLING ASPECTS: GRANULAR FLOWS, LAYERING IN SMECTICS

G. CAPRIZ
Dipartimento di Matematica dell'Università di Pisa, Pisa, Italy

and

G. MULLENGER
Department of Civil Engineering, University of Canterbury, Christchurch, New Zealand

1. Introduction

The concept of microstructure is protean: almost each Author uses a (slight or radical) variant. The proposal in [1] is to consider each body element as a Lagrangian system with coordinates of place and order parameters as Lagrangian coordinates; it is then implied that the element maintains its identity through any process and that the microstructure is material.

Many contrary examples could be provided where one or both of those assumptions fail; utterly at variance are cases where the microstructure is purely local or purely kinematic. Two instances, in a sense intermediate, are considered below: granular materials (during fast flow) and smectic liquid crystals (under quasi-static conditions). In the first instance the local complexity of the flow needs to be put in evidence; in both instances there may be permeation and hence a partial loss of identity of the element; in the second instance a local microstructure (the layering of molecules) is superposed to the material one (the nematic structure).

In [2] an anholonomic constraint is assumed to apply for the field of relative microvelocities, which, though it be the simplest imaginable, nevertheless allows us to evidence at least some essential gaits.

2. Local Kinematics

To account, within a slightly enlarged classical continuum scheme, for the possibly discordant motion of agitation of the molecules around a place x it is suggested in [2]:

i) to consider, at each place x and time t, together with the usual mass-weighted average v of the molecular velocities over a "small", "spherical" neighbourhood, also partial averages v^+, v^- over "hemispherical" neighbourhoods, bounded by the plane of unit normal n;

ii) to accept for the difference $w = v^+ - v^-$, odd in n, a linear estimate

$$w(x,t;n) = A(x,t)n\,,$$

where A is the second order "tensor of agitation".

Then the permeation rate α through the plane with normal n at x can be expressed in terms of A and n

$$\alpha(x,t;n) = -A \cdot (n \otimes n)$$

and so can the slip rate (I, identity tensor)

$$s = (I - n \otimes n)An,.$$

The tensor A by itself characterizes some important aspects of the kinetic behaviour around x. For instance, the average permeation rate

$$\eta = \frac{1}{4\pi}\int_{\mathcal{S}} \alpha\, d\mathcal{S}$$

($\mathcal{S}$, the unit sphere) is given by $\frac{1}{3}\operatorname{tr} A$, because

$$\frac{1}{4\pi}\int_{\mathcal{S}} n \otimes n\, d\mathcal{S} = \frac{1}{3}I$$

The average Reynolds' tensor is

$$H = \frac{1}{4\pi}\int_{\mathcal{S}} w \otimes w\, d\mathcal{S} = \frac{1}{3}AA^T\,;$$

the average "granular temperature"

$$\theta = \frac{1}{8\pi}\int_{\mathcal{S}} w^2\, d\mathcal{S} = \frac{1}{6}A \cdot A = \frac{1}{2}\operatorname{tr} H\,,$$

and the average squared slip-rate

$$\sigma^2 = \frac{1}{4\pi}\int_{\mathcal{S}} s^2\, d\mathcal{S} = \frac{4}{15}A \cdot A - \frac{1}{15}A \cdot A^T - \frac{1}{15}(\operatorname{tr} A)^2 = \frac{8}{5}\theta - \frac{3}{5}\eta^2 - \frac{1}{15}A \cdot A^T\,.$$

The last result uses the identity

$$\frac{1}{4\pi}\int_{\mathcal{S}} n_i n_j n_h n_k\, d\mathcal{S} = \frac{1}{15}\left(\delta_{ij}\delta_{hk} + \delta_{ih}\delta_{jk} + \delta_{ik}\delta_{jh}\right)\,,$$

or

$$\frac{1}{4\pi}\int_S n\otimes n\otimes n\otimes n\, dS = \frac{1}{15}\left(I\otimes I + \mathsf{I} + \mathsf{T}\right),$$

where I, T are the identity and the transposing fourth-order operators.

3. Granular Flows

In the study of fast flows of granular materials it is usual to adjoin to the classical fields (density ρ, velocity v, Cauchy stress T) also the Reynolds' tensor H, the corresponding hyperstress S and stirring external actions S (or at least the scalar θ and a kind of "granular" heat flux). An extra balance equation is also required; it must be either derived from molecular dynamics via a process akin to one suggested by Grad to come to his 13-moment theory (see [3] or surmised from developments in elementary dynamics for systems of mass-points. In the latter case one starts by defining, with obvious notation,

$$\widehat{H} = \frac{1}{m}\sum m^{(i)}\left(v^{(i)} - v_G\right)\otimes\left(v^{(i)} - v_G\right)$$

and finds that

$$\dot{\widehat{H}} = 2\,\mathrm{sym}\,\widehat{S},$$

where the "stir" $\widehat{S}$ is the tensor moment

$$\widehat{S} = \sum\left(v^{(i)} - v_G\right)\otimes f^{(i)}.$$

Proceeding to the scheme of the continuum one is led to a balance equation

$$\frac{1}{2}\rho\dot{H} = \mathrm{sym}\left(-\,\mathrm{div}\,\mathsf{S} + (\mathrm{grad}\,v)T + S\right), \tag{3.1}$$

or, simply,

$$\frac{1}{2}\rho\dot{\theta} = -\,\mathrm{tr}(\mathrm{div}\,\mathsf{S}) + (\mathrm{grad}\,v)\cdot T + \mathrm{tr}\,S. \tag{3.2}$$

Thus, the new balance equation is nearer to some in "extended thermodynamics" than to the more common equations in continuum mechanics.

Details are given in [4] together with a comparison with the results in [3] and other papers of Jenkins.

4. Permeation in Smectic Liquid Crystals

Nematic liquid crystals are almost archetypal as continua with microstructure: the order parameters characterize the direction of the long molecules which prevails locally. Correspondingly a microstructural balance equation obtains, where the director-stress is usually written as a sum of a term which derives from Frank's potential and of a linearly viscous term.

In the dynamics of smectics (see ch. 8 of [5]), when one wants to allow for permeation, a further kinematic variable enters the picture: the permeation velocity

$$p = \frac{1}{\delta^2}\left(v \cdot d + \frac{\partial \omega}{\partial \tau}\right) d\,, \tag{4.1}$$

where ω is a function of place and time such that

$$\omega(x, \tau) = \xi$$

(ξ, a constant within an appropriate interval) represents the surfaces bounding the lamellae of the smectic;

$$d = \operatorname{grad} \omega$$

and $\delta = |d|$ is construed as a measure of the number of lamellae per unit length.

A uniform additive change in ξ translates all lamellae with respect to a reference: thus $-\frac{\partial \omega}{\partial \tau} d$ coincides with the quantity denoted by $\frac{\partial u}{\partial \tau}$ in [5], at least when the latter is parallel to d.

The extra balance equation (Eq. $(8.17)_c$ in [5]) reads

$$-\frac{p}{\lambda} + \frac{\mu}{\lambda}\frac{\operatorname{grad} \tilde{\theta}}{\tilde{\theta}} + \operatorname{div} T^* = 0\,, \tag{4.2}$$

where λ and μ are two constants, $\tilde{\theta}$ is absolute temperature, T^* is the contribution to T due to the energy connected with the shifting of molecules between layers (see (8.4) of [5]).

With regard to (4.2) the following remarks are in order:

i) No inertia term appears in it, as, in most circumstances, inertia effects are minute.

ii) The first term expresses Darcy's law of filtration and the constant λ^{-1} is the resistivity.

iii) The second term accounts for Soret's thermomechanical effect.

The question may be asked whether there be any link between (4.2) and (3.1) (or (3.2)), all grave differences notwithstanding. We argue that a link may be guessed, if A is taken as

$$A = -\frac{1}{\delta} p \otimes d\,,$$

so that

$$H = \frac{1}{3} p \otimes p \quad \text{and} \quad \theta = \frac{1}{6} p^2\,;$$

in fact scalar multiplication by p reduces (4.2) to

$$0 = -\operatorname{tr} \operatorname{div}(p \otimes T^*) + (\operatorname{grad} p) \cdot T^* + \operatorname{tr}\left(\frac{1}{\lambda} p \otimes \left(p - \frac{\mu}{\tilde{\theta}} \operatorname{grad} \tilde{\theta}\right)\right). \tag{4.3}$$

Remembering the remark (i) above, obvious identifications ensue by comparison of (4.3) with (3.2). Perhaps a link is thus also established with procedures, common in theories of liquid crystals, where balance of energies rather than balance of momenta is invoked to help in explaining phenomena.

Acknowledgements. The visit of George Mullenger to the Dipartimento di Matematica of the University of Pisa was partially supported by the Gruppo Nazionale di Fisica Matematica.

References

1. G. Capriz, *Continua with Microstructure*, Springer Tracts in Natural Philosophy, **35**, 1989.
2. G. Capriz and P. Trebeschi, *Bull. Tech. Univ. Instanbul* **47**(3) (1994) 1.
3. J. T. Jenkins and M. W. Richman, *Arch. Rat. Mech. Anal.* **87** (1985) 355.
4. G. Capriz and G. Mullenger, Extended continuum mechanics for the study of granular flows, to appear in *Rend. Acc. Lincei, Matematica*.
5. P. G. de Gennes, *The Physics of Liquid Crystals*, 2nd edition, Clarendon Press, 1993.

Continuum Models and Discrete Systems
Proceedings of 8th International Symposium, June 11-16, 1995, Varna, Bulgaria, ed. K.Z. Markov
© World Scientific Publishing Company, 1996, pp. 15–20

HYDRODYNAMIC INTERACTIONS

B. CICHOCKI [1]
Institute of Theoretical Physics, Warsaw University,
Hoza 69, PL-00 618 Warsaw, Poland

Abstract. Recently developed theory of the low-Reynolds-number hydrodynamic interactions of spherical particles immersed in a fluid is discussed. The numerical scheme for efficient calculations of the friction and mobility matrices for many spheres is presented. Both the infinite fluid case and the periodic boundary conditions are considered.

1. Introduction

It is easy to understand qualitatively what hydrodynamic interactions are. In a system composed of solid particles immersed in a viscous fluid a motion of each particle causes a flow pattern which in turn affects motions of other particles. A proper account of hydrodynamic interactions is required in understanding phenomena in wide variety of systems such as suspensions, colloids, polymers and proteins [1,2]. However a quantitative description of the interactions is not an easy task.

In many situations it is sufficient to consider the so-called creeping flows only, i.e. stationary low-Reynolds-number flows of incompressible fluid. In this case there are three fundamental difficulties in calculations of the hydrodynamic interactions. First, the interactions have long range. The velocity field caused by a single particle decays as the inverse power of distance. Second, the hydrodynamic interactions have many-body character. The complete flow pattern for many particles has the nature of a multiple scattering series. Third, the friction functions diverge at short distances due to strong velocity gradient in the fluid set up by near particles in relative motion (lubrication effects).

During the last years, important progress has been made in obtaining an efficient numerical scheme for the calculation of hydrodynamic interactions between many spheres. In 1987, Durlofsky *et al.* [3] proposed a scheme based on the idea of combining the long-range behavior of the mobility matrix with short-range lubrication effects in a pair superposition approximation for the friction matrix. Later Ladd [4] performed more systematic calculations, taking into account not only forces, torques, and stresslets as in [3], but also higher order multipole moments. Recently, Cichocki

[1] Also at: Institute of Fundamental Technological Research, Polish Academy of Sciences, Świętokrzyska 21, PL-00 049 Warsaw, Poland.

et al. [5] proposed a scheme that may be regarded as an improvement on the work of Durlofsky *et al.* and of Ladd. A general structure of their scheme is discussed here.

2. Friction and Mobility Matrices

We consider a system of N spheres immersed in an incompressible fluid of shear viscosity η. The configuration of their centers is described by $\mathbf{X} = (\mathbf{R}_1, \ldots, \mathbf{R}_N)$. The fluid velocity $\mathbf{v}$ and the pressure p are assumed to satisfy the Stokes equations [1]

$$\eta \nabla^2 \mathbf{v} - \nabla p = 0, \qquad \nabla \cdot \mathbf{v} = 0. \tag{1}$$

In addition we assume stick boundary conditions on the surfaces of the spheres. The fluid may be infinite or the periodic boundary conditions are added on walls of a container. The latter is very important for computer simulation algorithms. We consider a situation in which the particles are moving with given translational velocities $\mathbf{U}_1, \ldots, \mathbf{U}_N$ and rotational velocities $\boldsymbol{\Omega}_1, \ldots, \boldsymbol{\Omega}_N$ and the fluid motion is due entirely to the motion of the spheres. Then the forces $\mathbf{F}_1, \ldots, \mathbf{F}_N$ and torques $\mathbf{T}_1, \ldots, \mathbf{T}_N$ applied to the particles are related linearly to the velocities, i.e.

$$\mathsf{F} = \boldsymbol{\zeta} \cdot \mathsf{U}, \tag{2}$$

where the $6N$-dimensional vector F comprises both forces and torques and the vector U translational and rotational velocities. The configuration-dependent friction matrix $\boldsymbol{\zeta}$ and its inverse $\boldsymbol{\mu}$, called mobility matrix, are the basic objects in the theory of low-Reynolds number hydrodynamic interactions [2]. The main goal of the theory is to develop an efficient scheme for the numerical calculation of these matrices.

To describe the method of Ref. 5 we consider the more general problem of spheres moving with velocities U immersed in an incident flow $\mathbf{v}_0(\mathbf{r})$. Let $\mathbf{f}(\mathbf{r}; i)$ be the force density exerted by particle i on the fluid. In the case of stick boundary conditions, this force density may be regarded as localized on the particle surfaces. Then the fluid velocity outside the spheres is given by [6]

$$\mathbf{v}(\mathbf{r}) - \mathbf{v}_0(\mathbf{r}) = \int d^3\mathbf{r}' \, \mathbf{T}(\mathbf{r} - \mathbf{r}') \cdot \mathbf{f}(\mathbf{r}'), \tag{3}$$

where

$$\mathbf{f}(\mathbf{r}) = \sum_i \mathbf{f}(\mathbf{r}; i) \tag{4}$$

and

$$\mathbf{T}(\mathbf{r}) = \frac{1}{8\pi\eta} \frac{\mathbf{1} + \hat{\mathbf{r}}\hat{\mathbf{r}}}{r} \tag{5}$$

with $\hat{\mathbf{r}} = \mathbf{r}/r$. The tensor $\mathbf{T}(\mathbf{r})$ is called the Oseen tensor. In the case of periodic boundary conditions this tensor should be replaced by the Hasimoto tensor [7].

To satisfy the stick boundary conditions, the relation

$$\mathbf{v}(\mathbf{r}) = \mathbf{U}_k + \boldsymbol{\Omega}_k \times (\mathbf{r} - \mathbf{R}_k) \tag{6}$$

must hold for $\mathbf{r} \in S_k$, $k = 1, \ldots, N$, where S_k denotes the surface of particle k. Defining the velocity field $\mathbf{u}(\mathbf{r})$ as

$$\mathbf{u}(\mathbf{r}) = \sum_k [\mathbf{U}_k + \boldsymbol{\Omega}_k \times (\mathbf{r} - \mathbf{R}_k)]\, \Theta(\mathbf{r} - \mathbf{R}_k), \tag{7}$$

where the step function $\Theta(\mathbf{r} - \mathbf{R}_k)$ localizes the velocity field to the inside of particle k, we see that Eq. (3) together with Eq. (6) is an integral equation relating $\mathbf{f}(\mathbf{r})$ to $\mathbf{u}(\mathbf{r}) - \mathbf{v}_0(\mathbf{r})$. The formal solution of this equation defines the friction operator $\mathbf{Z}$. In an abbreviated form, one can write

$$\mathbf{f} = \mathbf{Z}\,[\mathbf{u} - \mathbf{v}_0]\,. \tag{8}$$

The kernel of the operator $\mathbf{Z}$ is localized on the surfaces of the particles.

Now we write the right-hand side of Eq. (3) for $\mathbf{r} \in S_i$ as a sum of two types of contributions. Those from the force density located on the surfaces of particles $j \neq i$ we write as

$$\sum_{j \neq i} \int d^3\mathbf{r}'\, \mathbf{T}(\mathbf{r} - \mathbf{r}') \cdot \mathbf{f}(\mathbf{r}'; j) = \sum_{j \neq i} \mathbf{G}(ij)\mathbf{f}(j), \tag{9}$$

which defines the Green operator $\mathbf{G}$. The contribution from $\mathbf{f}(i)$ is given by

$$\int d^3\mathbf{r}'\, \mathbf{T}(\mathbf{r} - \mathbf{r}') \cdot \mathbf{f}(\mathbf{r}'; i) = \mathbf{Z}_0^{-1}(i)\mathbf{f}(i), \tag{10}$$

where the operator $\mathbf{Z}_0(i)$ is the one-particle friction operator, i.e. it is defined by Eq. (8) for a single sphere. The friction operator $\mathbf{Z}$ for the N-particle system is then given by the formal expression

$$\mathbf{Z} = \mathbf{Z}_0\,[\mathbf{1} + \mathbf{G}\mathbf{Z}_0]^{-1}\,. \tag{11}$$

This expression has previously been derived using the multiple scattering expansion [8]. The friction matrix $\boldsymbol{\zeta}$ can be calculated from Eq. (8) by setting $\mathbf{v}_0 = 0$ and taking into account that the forces $\mathbf{F}_i$ and torques $\mathbf{T}_i$ are related to the first two moments of $\mathbf{f}(\mathbf{r}, i)$.

The separation of the right-hand side of Eq. (3) into Eqs. (9) and (10) is not merely a cosmetic one, but has a deeper reason. First, the relation (11) is valid not only for stick boundary conditions, but can be used for arbitrary boundary conditions by changing only the single-particle operator $\mathbf{Z}_0$. This is usually an easy task since it requires only the solution of a one-body problem. In the general case, e.g. for permeable spheres, the localization of the force density and the kernels of the operators $\mathbf{Z}$, $\mathbf{Z}_0$ and $\mathbf{G}$ must be extended to the volume of the particles. Second, this separation allows the construction of a similar scheme for the direct calculation of the mobility matrix $\boldsymbol{\mu}$, which otherwise can be obtained only by an explicit inversion of the friction matrix. This advantage is missed in schemes proposed by other authors, in which the above separation is not done.

3. Multipole Expansion and Lubrication Corrections

To be able to perform the inversion operation in Eq. (11), we use a multipole expansion of the force densities. The Cartesian force multipole tensor of rank $p + 1$

for a single sphere j is defined by

$$\mathbf{f}^{(p+1)}(j) = \frac{1}{p!} \int d^3\mathbf{r}\ (\mathbf{r} - \mathbf{R}_j)^p\, \mathbf{f}(\mathbf{r}; j) \tag{12}$$

and the multipole tensor for the velocity field $\mathbf{u} - \mathbf{v}_0$ around sphere j by

$$\mathbf{c}^{(p+1)}(j) = \frac{1}{p!} \nabla^p \left[\mathbf{u}(\mathbf{r}) - \mathbf{v}_0(\mathbf{r})\right]_{\mathbf{r}=\mathbf{R}_j}. \tag{13}$$

Due to the spherical symmetry of the particles it is advantageous to express the multipole moments in terms of irreducible tensors. In our case it is sufficient to specify a set of irreducible multipole moments $\mathbf{f}_{l\sigma}$ and $\mathbf{c}_{l\sigma}$, $l = 1, 2, \ldots$, which for given l and σ have $2l + 1$ independent components [9]. The index σ takes the three values $0, 1, 2$, which correspond to the superscripts S, T, P (S standing for symmetric, T for tangential, and P for pressure) used in earlier publications.

The tensor $\mathbf{f}_{10}(j)$ is equal to the total force $\mathbf{F}_j$, and $2\mathbf{f}_{11}(j)$ is equal to the torque $\mathbf{T}_j$. For the velocity field, we have $\mathbf{c}_{10}(j) = \mathbf{U}_j - \mathbf{v}_0(\mathbf{R}_j)$ and $\mathbf{c}_{11}(j) = 2\boldsymbol{\Omega}_j - \nabla \times \mathbf{v}_0(\mathbf{r})\Big|_{\mathbf{r}=\mathbf{R}_j}$. In the multipole notation the operators $\mathbf{Z}_0$ and $\mathbf{G}$ become matrices. The friction matrix can then be written as

$$\boldsymbol{\zeta} = \mathcal{P}\mathbf{Z}_0 \left[\mathbf{1} + \mathbf{G}\mathbf{Z}_0\right]^{-1} \mathcal{P}, \tag{14}$$

where $\mathcal{P}$ is the projection operator on the subspace defined by the first two moments i.e. with $l = 1$ and σ=0,1.

In practical numerical calculations the infinite-dimensional matrices must be truncated to finite ones. We truncate the matrices $\mathbf{Z}_0$ and $\mathbf{G}$ in such a way that only the elements with $l \leq L$ are taken into account. The friction matrix calculated in this approximation is denoted by $\boldsymbol{\zeta}_L$. With this truncation the number of multipole components per particle is $n_L = 3L(L + 2)$. However, by a simple resummation the P-multipoles can be eliminated and the number of components reduces to $2L(L+2)$. It has been shown by Briggs *et al.* [10] that for stick boundary conditions $\boldsymbol{\zeta}_L$ is a positive definite matrix.

The Green matrix elements describing the interactions between two multipoles $l\sigma$ and $l'\sigma'$ at distance R has the distance dependence $R^{-l-l'-\sigma-\sigma'+1}$. From that it is clear that the matrix $\boldsymbol{\zeta}_L$ describes the long-range effects correctly even for relatively small L. It was shown in [5] that a relatively small L is also sufficient to describe the collective motion of a group of spheres, even when they touch each other. However, it is important to have $L \geq 3$. To obtain a good approximation for large sets of spheres, all Green matrix components decaying as $1/R$, $1/R^2$, and $1/R^3$ must be taken into account, since for such elements the range of the interactions is infinite and their relative contribution to the friction matrix grows as the number of particles is increased. The minimal truncation in which all such elements are included is $L = 3$.

Finally, one must consider the relative motion of spheres at short distances, where lubrication effects dominate the friction, and multipole components of very high order are required for an accurate description. Here Cichocki *et al.* [5] follow the procedure of Durlofsky *et al.* [3] by assuming that the lubrication effects are well described by the superposition approximation for the friction matrix, i.e.

$$\boldsymbol{\zeta}_{\text{sup}} = \sum_{i<j} \boldsymbol{\zeta}(ij), \tag{15}$$

where $\boldsymbol{\zeta}(ij)$ is the pair matrix for particles i and j, which can be calculated with high accuracy [2,11]. The results of the multipole expansion and the short-range contributions are finally combined in the form

$$\bar{\boldsymbol{\zeta}}_L = \boldsymbol{\zeta}_L + \boldsymbol{\zeta}_{\text{sup}} - \boldsymbol{\zeta}_{\text{sup},L} \tag{16}$$

with

$$\boldsymbol{\zeta}_{\text{sup},L} = \sum_{i<j} \boldsymbol{\zeta}_L(ij), \tag{17}$$

where $\boldsymbol{\zeta}_L(ij)$ is the pair matrix calculated with truncation $l \leq L$.

The examples considered in [5] have shown that the scheme with $L = 3$ leads to an accurate approximation for the friction matrix. The observed accuracy is of the order of 1%.

We note that a transformation of the operator $\mathbf{Z}_0$ has been used in [5] to construct a method for the direct calculation for the mobility matrix, similar to Eq. (16). This follows from the following relation equivalent to Eq. (14)

$$\boldsymbol{\mu} = \boldsymbol{\mu}_0 \left[\mathbf{1} + \mathbf{Z}_0\mathbf{G}\hat{\mathbf{Z}}_0\mathbf{Z}_0^{-1}\right]^{-1} [\mathbf{1} + \mathbf{Z}_0\mathbf{G}]\,\mathcal{P}, \tag{18}$$

where

$$\hat{\mathbf{Z}}_0 = \mathbf{Z}_0 - \mathbf{Z}_0\boldsymbol{\mu}_0\mathbf{Z}_0 \tag{19}$$

and $\boldsymbol{\mu}_0$ is mobility matrix valid for large separation of spheres.

With other schemes, the mobility matrix must be calculated from the friction matrix by an explicit inversion, which is an expensive numerical operation.

4. Concluding Remarks

We have discussed an efficient scheme for numerical calculation of the low-Reynolds-number hydrodynamic interactions between many spheres. The nature of the scheme is such that in principle any desired accuracy can be attained. In practice we get satisfactory results by truncation at relatively low multiple order with account of lubrication effects in pair superposition. In numerical implementation the irreducible Cartesian multipoles tensors are used which allows to avoid calculations with complex numbers and leads to a faster convergence compared to Ladd's expansion. The implementation has been done for an infinite fluid case and for systems with periodic boundary conditions.

The scheme has been applied to solve various problems. Examples are calculations of the mobilities of rods consisting of many touching, rigidly connected spheres [5] and Stokes drag coefficients for almost spherical conglomerates of particles [6]. The largest conglomerate considered in Ref. 6 consisted of 167 spheres. The numerical results agree very well with the experimental data.

References

1. J. Happel and H. Brenner, *Low Reynolds Number Hydrodynamics*, Noordhoff, Leyden, 1973.
2. S. Kim and S. J. Karrila, *Microhydrodynamics*, Butterworth-Heinemann, Boston, 1991.
3. L. Durlofsky, J. F. Brady and G. Bossis, *J. Fluid Mech.* **180** (1987) 21.
4. A. J. C. Ladd, *J. Chem. Phys.* **88** (1988) 5051; **90** (1989) 1149; **93** (1990) 3484.
5. B. Cichocki, B. U. Felderhof, K. Hinsen, E. Wajnryb and J. Bławzdziewicz, *J. Chem. Phys.* **100** (1994) 3780.
6. P. Mazur and W. van Saarloos, *Physica A* **115** (1982) 21.
7. B. Cichocki and B. U. Felderhof, *Physica A* **159** (1989) 19.
8. B. U. Felderhof, *Physica A* **151** (1988) 1.
9. R. Schmitz, *Physica A* **102** (1980) 161.
10. K. Briggs, E. R. Smith, I. K. Snook and W. van Megen, *Phys. Lett. A* **154** (1991) 149.
11. B. Cichocki, B. U. Felderhof and R. Schmitz, *PhysicoChemical Hydrodynamics* **10** (1988) 383.
12. B. Cichocki and K. Hinsen, *Phys. Fluids* **7** (1995) 285.

Continuum Models and Discrete Systems
Proceedings of 8th International Symposium, June 11-16, 1995, Varna, Bulgaria, ed. K.Z. Markov
© World Scientific Publishing Company, 1996, pp. 21–29

A THERMOMECHANICAL APPROACH OF MARTENSITE PHASE TRANSITION DURING PSEUDOELASTIC TRANSFORMATION OF SHAPE MEMORY ALLOYS

A. CHRYSOCHOOS and M. LÖBEL
Laboratoire de Mécanique et Génie Civil,
URA 1214 CNRS, Université Montpellier II, Place E. Bataillon,
F-34095 Montpellier Cédex 05, France

Abstract. Quasi-static and uniaxial thermomechanical tests were performed using calorimetric and infrared techniques on a *CuZnAl* shape memory alloy. The experiments evidence the main influence of the temperature variations induced by the deformation process on the stress-strain curves. These variations are essentially due to the latent heat of phase change and are of the same order of magnitude as the transition domain "width" in the stress-temperature plane. The analysis of the associated energy balance also shows a very weak intrinsic (i.e. mechanical) dissipated energy in the sense that it remains very small compared to the deformation work or to the latent heat. On the basis of these experimental data, a behavioural model that assumes both non-dissipative and anisothermal phase change under stress is evoked.

1. Introduction

It is now well known that the pseudoelastic behaviour of shape memory alloys (SMA) is due to austenite-martensite phase transitions. From a mechanical point of view, these phase transitions give rise in certain conditions to hysteresis loops of the stress-strain curves during load-unload cycles at constant room temperature.

The literature dealing on the SMA phenomenology gives several possible interpretations of that pseudoelastic hysteresis ([1,2] *et al.*), whose area is related to internal energy variation of the material. This energy variation can be interpreted in terms of dissipated energy due to internal friction, of stored energy associated with inter and intragranular deformation incompatibilities, of thermomechanical couplings like phase change under stress.

In this paper, after a brief recall of the thermodynamic framework used to interpret the experiments, a theoretical analysis of the energy balance associated with pseudoelastic behaviour during a load-unload thermodynamic cycle is proposed. It shows that the energy corresponding to the hysteresis may come a priori from the thermoelastic and phase change effects and from the dissipation developed during the transformation. This theoretical analysis is correlated with experiments performed on setups using calorimetric and infrared thermography techniques. The experimental observations lead us to consider the solid-solid phase change as a non-dissipative (in the sense of intrinsic dissipation) and, above all, as an anisothermal thermodynamic process.

Finally the consequences of such results on the modelling are briefly evoked. One-dimensional numerical exercises emphasize the role played by temperature variations in the existence of the pseudoelastic hysteresis.

2. Thermomechanical Framework

Concepts and results of Classical Irreversible Thermodynamics are used [3], [4]. At each instant t of a quasi-static process, a thermodynamic equilibrium state of a homogeneous volume element is characterized by a set of $n+1$ state variables $(\alpha_0, \alpha_1, \ldots, \alpha_n)$.

Let us take T ($T = \alpha_0$) as the absolute temperature, ε ($\varepsilon = \alpha_1$) as the strain tensor and α_j ($j = 2, \ldots, n$) as a vector α composed of $n-1$ internal variables completing the description of the thermodynamic state. Here, these α_j's are assimilated to the volume proportions of martensite variants. For the considered SMA, at zero stress, we classically suppose that: $\alpha_j = 0$ for $j = 2, \ldots, n$, if $T \geq A_f$ (austenite finish), $\sum_{j=2}^{n} \alpha_j = 1$, if $T \leq M_f$ (martensite finish), and $0 \leq \alpha_j \leq 1$, if $M_f \leq T \leq A_f$ (possible coexistence of the two phases). The difference $A_f - M_f$ is termed transition domain "width".

If ψ and s denote respectively the specific Helmholtz free energy and the specific entropy, the Clausius-Duhem inequality that comes from the local form of the Second Principle of Thermodynamics, can thus be written as:

$$d = \sigma : \dot{\varepsilon} - \rho \frac{\partial \psi}{\partial \varepsilon} : \dot{\varepsilon} - \rho \frac{\partial \psi}{\partial \alpha} \cdot \dot{\alpha} - \frac{q}{T} \cdot \mathrm{grad} T \geq 0, \tag{2.1}$$

where σ is the Cauchy stress tensor, ρ the mass density, q the heat influx vector and d the dissipation. The dot stands for the time derivative.

Classically, the intrinsic (mechanical) dissipation d_1 and the thermal dissipation d_2 are supposed to be separately positive and are defined by:

$$d_1 = \sigma : \dot{\varepsilon} - \rho \frac{\partial \psi}{\partial \varepsilon} : \dot{\varepsilon} - \rho \frac{\partial \psi}{\partial \alpha} \cdot \dot{\alpha} \geq 0, \quad d_2 = -\frac{q}{T} \cdot \mathrm{grad}\, T \geq 0. \tag{2.2}$$

In the experimental conditions of the tests presented here [5], the heat equation can be simplified as:

$$\rho C_\alpha \dot{\theta} + \mathrm{div}\, q = d_1 + \rho T \frac{\partial^2 \psi}{\partial T \partial \varepsilon} : \dot{\varepsilon} + \rho T \frac{\partial^2 \psi}{\partial T \partial \alpha} \cdot \dot{\alpha} = w'_{\mathrm{ch}}\,, \tag{2.3}$$

where θ is the temperature variation ($\theta = T - T_0$, with T_0 the equilibrium absolute temperature). The specific heat capacity C_α and the isotropic conduction coefficient

k ($q = -k \operatorname{grad} T$) are assumed constants. The term w'_{ch} denotes the volume heat source composed of the intrinsic dissipation and of crossing terms corresponding to thermoelastic effects and to the latent heat rate.

For a load-unload cycle, experimental observations show that strain and stress go back to their initial values. Because of thermal diffusion, the temperature returns to the equilibrium temperature T_0 that is chosen greater than A_f to ensure initial and final austenitic states. As a consequence, each load-unload cycle is then considered as a thermodynamic cycle. Integrating the local form of the First Principle on the duration $\mathcal{C}$ of such a cycle gives an energy balance that yields a definition of the hysteresis area A_h:

$$A_h = \int_{\mathcal{C}} \sigma : \dot{\varepsilon} \, d\tau = \int_{\mathcal{C}} \operatorname{div} q \, d\tau, \tag{2.4}$$

where the volume external heat supply has been neglected. The entropy balance on a same cycle $\mathcal{C}$ can be written as:

$$\int_{\mathcal{C}} \rho \dot{s} \, d\tau = \int_{\mathcal{C}} \rho C_\alpha \frac{\dot{T}}{T} \, d\tau - \int_{\mathcal{C}} \rho \frac{\partial^2 \psi}{\partial T \partial \varepsilon} : \dot{\varepsilon} \, d\tau - \int_{\mathcal{C}} \rho \frac{\partial^2 \psi}{\partial T \partial \alpha} \cdot \dot{\alpha} \, d\tau. \tag{2.5}$$

The entropy being a state function, the left hand of Eq. (2.5) equals zero for all thermodynamic cycles; the first integral of the right hand vanishing, the sum of the two others then equals zero. Thus, with Eq. (2.3) intergrated along the cycle $\mathcal{C}$ and with Eq. (2.4) we get:

$$A_h = \int_{\mathcal{C}} d_1 \, d\tau + \int_{\mathcal{C}} \rho \theta \frac{\partial^2 \psi}{\partial T \partial \varepsilon} : \dot{\varepsilon} \, d\tau + \int_{\mathcal{C}} \rho \theta \frac{\partial^2 \psi}{\partial T \partial \alpha} \cdot \dot{\alpha} \, d,\tau = \int_{\mathcal{C}} w'_{\mathrm{ch}} \, d\tau. \tag{2.6}$$

Eq. (2.6) shows that the hysteresis area is given both by intrinsic dissipation and thermomechanical couplings. If $d_1 = 0$, the presence of A_h is only due to the couplings that give thermal dissipation as long as the process is neither adiabatic ($q = 0$) nor isotherm ($T = T_0$). Nevertheless, note that conversely a thermal dissipation could not give hysteresis without thermomechanical coupling mechanism.

To estimate the relative importance of the intrinsic dissipative phenomena, we propose to introduce the ratio R_T defined by:

$$0 \leq R_T = \frac{A_h}{2L} \leq 1 \quad \text{with} \quad 2L = \int_{\mathcal{C}} |w'_{\mathrm{ch}}| \, d\tau. \tag{2.7}$$

Its construction implies the following result [6]: the more the dissipation is preponderant, the closer R_T is to 1; conversely, the more R_T is closer to 0, the more the dissipation can be energetically negligible.

3. Experimental Results

3.1. Experimental Arrangements

The original feature of these experimental setups is the recording and the use of the temperature field on the sample surface to evaluate the amount of heat evolved during the tests. Both setups and data processing are fully described in [5,7].

3.2. Tests and Results

Quasi-static uniaxial tests were performed on polycrystalline CuZnAl samples. Various loading conditions were defined; the influences of parameters like loading paths, loading rates, room temperatures (imposed by an environment chamber), were considered. In what follows, the stress-strain curve is shown in each case with the corresponding thermal and energetic responses.

The deformation work W_m and the evolved heat W_{ch} were estimated for an equivalent volume V_0 corresponding to the gauge length of the sample:

$$W_m = V_0 \int_0^t \sigma : \dot{\varepsilon}\, d\tau, \quad W_{\text{ch}} = V_0 \int_0^t w'_{\text{ch}}\, d\tau. \tag{3.1}$$

Loads were strain controlled and unloads stress controlled to ensure zero stress during the thermal return to the equilibrium state.

3.2.1. Load-unload Tension Cycles with Increasing Strain Amplitudes

A load-unload test with increasing strain amplitude (0.5%, 1%, 1.5%) is presented in Fig. 1. During the experiment, the room temperature was held constant at $30^o C$ (greater than A_f) and the absolute value of the strain rate was less than $10^{-3} s^{-1}$.

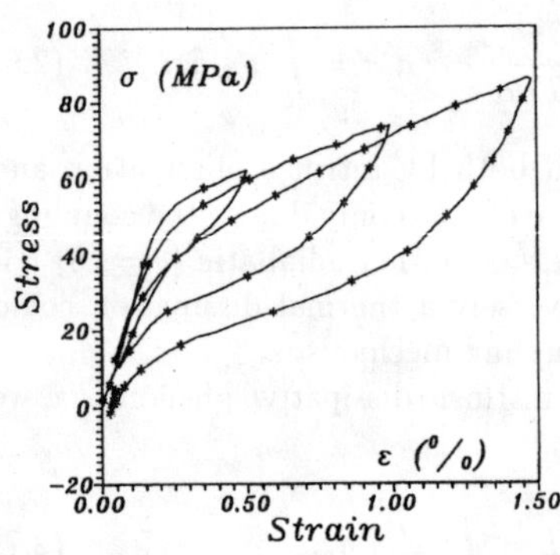

Fig. 1. Hysteresis loops increasing.

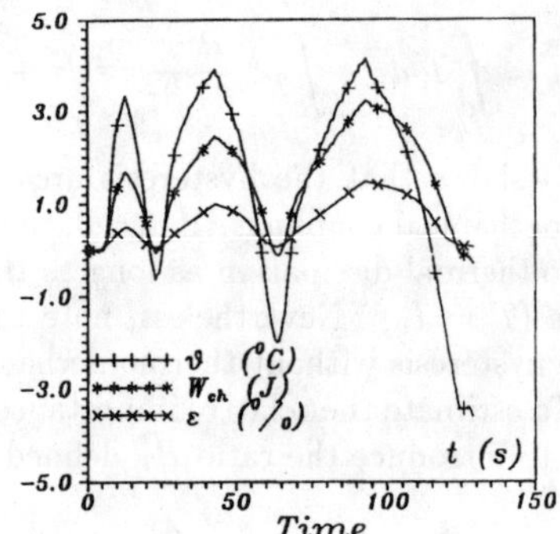

Fig. 2. Thermal and energetic responses.

In Fig. 2, we note that the amplitude of θ (around 7 ^oC) is not negligible in comparison with the "width" in temperature of the transition domain (around 13 ^oC). We also note the non-monotonic evolution of W_{ch} and its weakness indeed, at the end of each hysteresis loop. These features characterize non-dissipative endo-exothermic reactions. If the dissipative phenomena were preponderant, the evolved heat would increase in a monotonic way.

3.2.2. Influence of the Strain Rate

A significant influence of the strain rate on the stress-strain curves can be observed in Fig. 3. The increase of the stress level may be attributed to the thermo-mechanical couplings (stress effect due to the sample temperature variations induced by phase change ... under stress !).

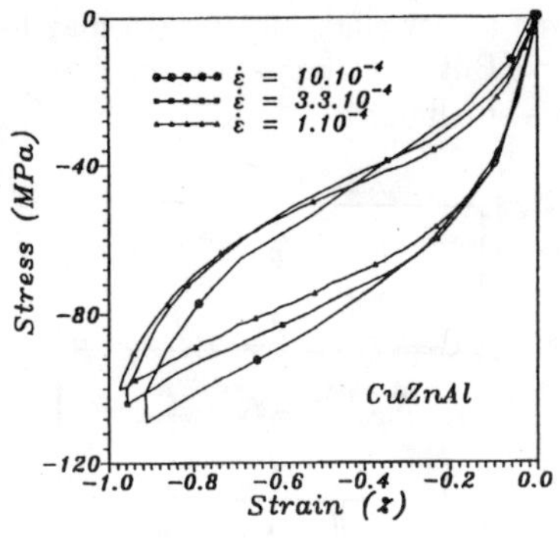

Fig. 3. Influence of the strain rate.

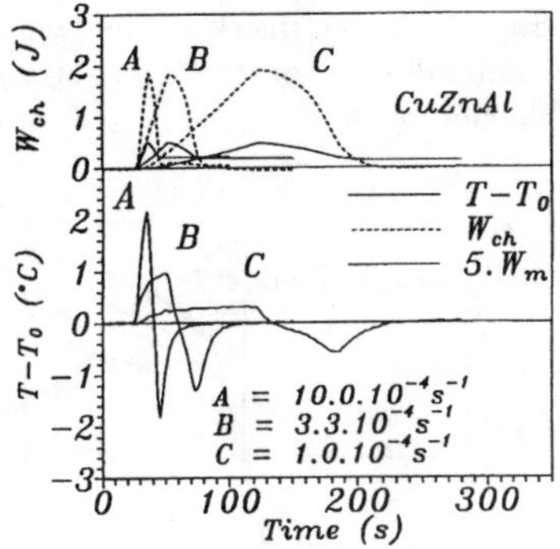

Fig. 4. Thermal and energetic responses.

Because of heat diffusion, the maximal temperature amplitude increases with the strain rate while the evolved heat level remains constant, see the curves A, B and C in Fig. 4. This could not be the case if the stress increase were related to viscous effects. Note also that $\mathrm{Sup}(W_{\mathrm{ch}})$ is ten times greater than $\mathrm{Sup}(W_m)$. Here, $\mathrm{Sup}(W)$ terms the maximal value reached by W during the test at the end of the loading path for a load-unload cycle.

3.2.3. Influence of the Room Temperature during Tension-Compression Tests

To emphasize the strong influence of temperature on mechanical behaviour, several tests were done with very close room temperatures, see Fig. 5. The greater T_0 is than M_s (martensite start) the longer the elastic path is to reach the transition domain. The T_0 values were chosen so that the transition domain boundary was reached before the elastic yield stress.

During these tension-compression or compression-tension tests, we also noticed that the "loading order" has no significant influence. Nevertheless, a slight but systematic asymmetry was observed between the hysteresis obtained in compression and that obtained in tension, see Fig. 5. This asymmetry is also detectable in the thermal and energetic responses, see Fig. 6. Everything happens as if, for one per cent of strain amplitude, the phase change starts later and develops more intensively in compression than in tension.

3.2.4. R_T and A_h Estimations

For all the above-mentioned experiments, the ratios R_T are less than 0.02. Thus, the amounts of intrinsic dissipated energy remain very small indeed in comparison with the ones of latent heat of phase change. Because of Eq. (2.6), this result shows that temperature variations induced by the transformation play an important role in the creation of hysteresis; however, as the energy A_h is itself small (1.5 % to 4 % of Sup(W_{ch})), we insist on the fact that even a very slight dissipation may have a significant influence on the stress-strain curves. But this dissipation which is always possible cannot be detected here because of its smallness.

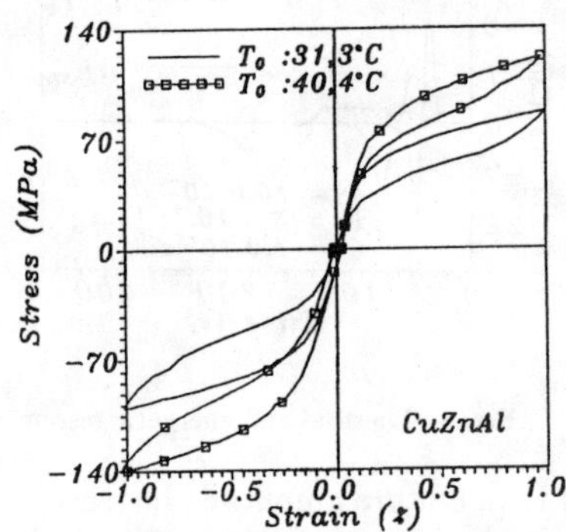

Fig. 5. Cyclic tests at different T_0.

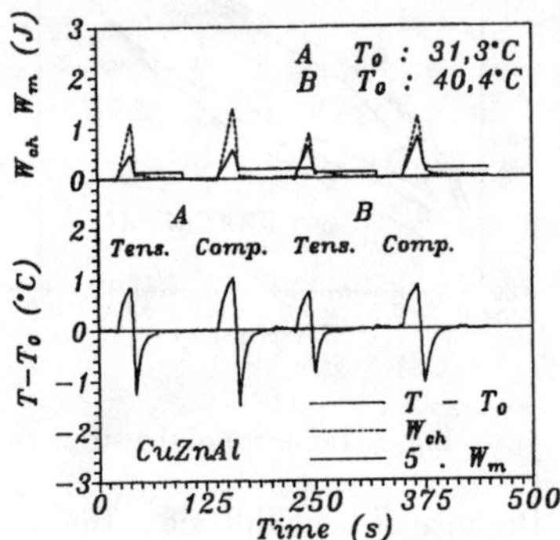

Fig. 6. Thermal and energetic responses.

4. Behavioural Constitutive Equations

In the modelling evoked below, the phase change will be considered as a non-dissipative and anisothermal phenomenon to point out the role of the thermomechanical couplings. The Generalized Standard Materials formalism [8] has been chosen and the small-perturbation hypothesis has been adopted. The suggested model takes into account coupled thermoelasticity and considers two self-accommodating martensite variants (x_1, x_2). The constitutive equations are divided into two groups deriving from two potentials. The thermodynamic potential gives the state laws while the dissipation potential gives the complementary laws. The state variables are T, ε, x_1 and x_2.

4.1. Thermodynamic and Dissipation Potentials

As done by several authors [9], the pseudoelastic free energy is decomposed into an "elastic" part ψ_e and a "chemical" part ψ_{ch} related to the phase change:

$$\psi_e = \frac{1}{2\rho}\left[E:(\varepsilon-\lambda\theta I - x_i\beta^i):(\varepsilon-\lambda\theta I - x_i\beta^i) - \left(\frac{\rho C_\alpha}{T_0} + E:\lambda I:\lambda I\right)\theta^2\right], \quad (4.1)$$

where E is the isotropic elasticity modulus tensor, λ is the coefficient of thermal expansion, I is the identity tensor and β^i, $i = 1,2$, are the constant orientation tensors of both variants. In turn,

$$\psi_{\text{ch}} = \frac{\mathcal{D}}{\rho}\left[(A_f - M_f)(x_1^2 + x_2^2) + (T - A_f)(x_1 + x_2)\right] + I_{\mathcal{V}}(x_1, x_2), \qquad (4.2)$$

where $\mathcal{D}$, A_f and M_f are phenomenological constants. As brilliantly done by Frémond [10], an indicator function has been added up to Eq. (4.2) to input in the potential properties the bounded evolutions of the x_is. The convex set $\mathcal{V}$ is here defined by:

$$\mathcal{V} = \left\{(x_1, x_2) \in I\!R^2 \ / \ 0 \leq x_i \leq 1, \ i = 1,2 \ \text{ and } \ 0 \leq x_1 + x_2 \leq 1\right\}. \qquad (4.3)$$

To translate the experimentally undetected intrinsic dissipation, the mechanical dissipation potential is chosen identically equal to zero. From both potentials, the constitutive equations can then be deduced, see [11] for a detailed presentation.

4.2. Transition Domain and Kinetics of Phase Change

We just focus here on the consequences of a zero dissipation on the phase change formulation. With such an hypothesis, the thermodynamic forces associated with variants are naturally equal to zero. During the phase change, this property yields the following equations for $i = 1,2$:

$$\beta^i : \sigma = \mathcal{D}\left[2(A_f - M_f)x_i + (T - A_f)\right], \qquad (4.4)$$

$$\beta^i : \dot{\sigma} = \mathcal{D}\left[2(A_f - M_f)\dot{x}_i + \dot{T}\right]. \qquad (4.5)$$

The relations (4.4) define the manifold of the thermodynamic states where phase change can take place. Eqs. (4.5) define the kinetics of phase change in the transition domain. Referring to classical first order phase change formalism, Eqs. (4.4) can be interpreted as an extension of transition lines defining the phase diagram while Eqs. (4.5) generalize the Clapeyron equation. Let us however stress the fact that Eqs. (4.4) and (4.5) are derived from dissipation properties and not from a non convex thermodynamic potential in a Reversible Thermodynamics framework.

4.3. Illustrating Example of Numerical Simulation

Finally the influence of thermomechanical couplings on the stress-strain curves are pictured through 1D numerical exercises. In Fig. 7, experimental loading conditions close to those of Fig. 1 have been computed. The corresponding latent heat and temperature variations are plotted in Fig. 8.

5. Concluding Comments

Thermomechanical tests were performed on a `CuZnAl` SMA for various loading conditions. The experiments evidence couplings existing between the temperature variations induced by the phase change under stress and the mechanical response. Associated energy balances show the weakness of the intrinsic dissipated energy and the

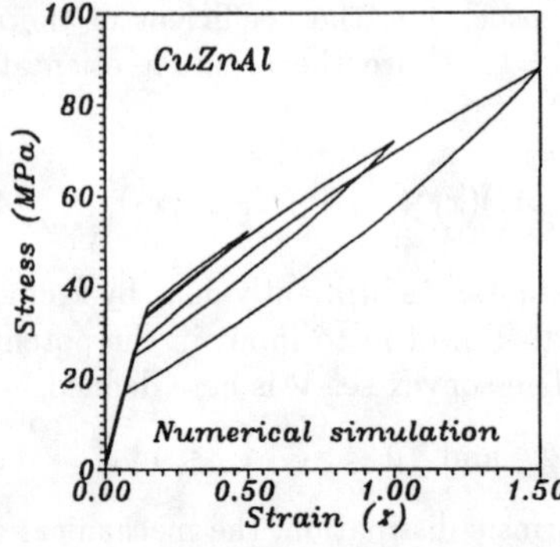

Fig. 7. Hysterisis loops increasing.

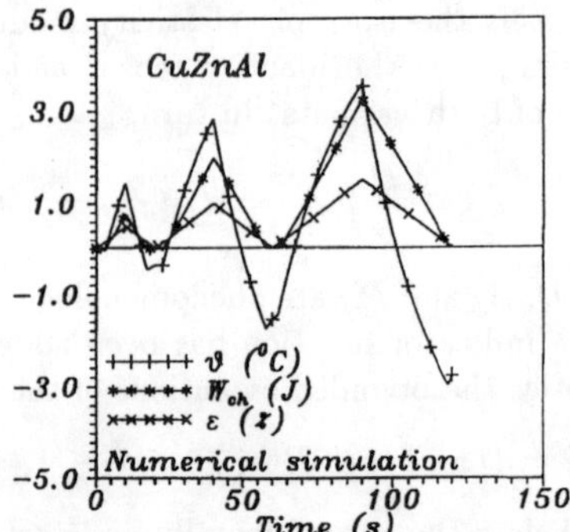

Fig. 8. Thermal and energetic responses.

great role played by the thermomechanical couplings in the creation of hysteresis. The constitutive equations of the kinetics of phase change are derived here both from a convex form of the free energy and from dissipation properties. To check the physical basis of our approach, numerical tests were also computed using a simple but heuristic model. This model predicts consistent mechanical hysteresis loops qualitatively in good agreement with the experiments. Because of thermomechanical couplings, hysteresis does exist without any intrinsic dissipation.

References

1. J. Van Humbeek, *J. de Phys.* **C5 42** (1981) 1007.
2. J. Ortin and A. Planes, *Acta. Met.* **36** (1988) 1873.
3. H. B. Callen, *Thermodynamics and an Introduction to Thermostatistics*, John Wiley, 1985.
4. D. Jou, J. Casas-Vazquez and G. Lebon, *Extended Irreversible Thermodynamics*, Springer-Verlag, 1993.
5. A. Chrysochoos, J. C. Dupré, *An infrared setup for continuum thermomechanics*, QIRT, Quantitative Infrared Thermography, Ed. Europ. Thermique and Industrie, 1992, p. 129.
6. A. Chrysochoos, M. Löbel and O. Maisonneuve, *C. R. Acad. Sci. Paris* 2b **T320** (1995) 217.
7. M. Löbel, *Caractérisation thermomécanique d'alliages à mémoire de forme de type NiTi et CuZnAl*, Thesis, Université Montpellier II, France, 1995.
8. B. Halphen, Q. S. Nguyen, *J. de Mécanique* **14** (1975) 39.
9. E. Patoor, A. Eberhart, M. Berveiller, *Acta Met.* 38 (1987) 2779.

10. M. Frémond, *C. R. Acad. Sci. Paris* 2 (1987)(7) 239.

11. R. Peyroux, A. Chrysochoos, C. Licht and M. Löbel, *Phenomenological constitutive equations for numerical simulations of SMA structures. Effects of thermomechanical couplings*, Mecamat 95, Int. Seminar, 16-19 May 1995, France.

Continuum Models and Discrete Systems
Proceedings of 8th International Symposium, June 11-16, 1995, Varna, Bulgaria, ed. K.Z. Markov
© World Scientific Publishing Company, 1996, p. 30

HAMILTONIAN DESCRIPTION OF DISSIPATIVE PROCESSES

R. KOTOWSKI
Institute of Fundamental Technological Research,
Polish Academy of Sciences, Świętokrzyska 21, PL-00 049 Warsaw, Poland

Abstract

There are many approaches to describe the irreversible processes by the field theoretic methods. One of them, which can be classified as the doubling of fields variational principle, belongs to K.-H. Anthony. This approach exploits the full power of the Lagrange formalism. There are obtained not only the Euler-Lagrange equations, but also the constitutive relations and the conservation laws by applying the Noether theorem, as well as the stability conditions of the system by making use of the accessory variational problem. The paper presents the attempt to represent the irreversible processes in the Hamilton formalism. There are proposed the Hamiltonians for the classical harmonic oscillator with friction and for the heat conduction in rigid bodies. These Hamiltonians are obtained from the Lagrangians given by other authors. The obtained results are discussed.

Continuum Models and Discrete Systems
Proceedings of 8th International Symposium, June 11-16, 1995, Varna, Bulgaria, ed. K.Z. Markov
© World Scientific Publishing Company, 1996, pp. 31–38

MICROMECHANICS-BASED CONSTITUTIVE MODEL OF SINGLE CRYSTALLINE SHAPE MEMORY ALLOYS WITH LOADING RATE EFFECT

C. LEXCELLENT, B. C. GOO
Laboratoire de Mécanique Appliquée R.C., Université de Franche-Comté, Besançon, 25030, France.

and

Q. P. SUN*, J. BERNARDINI**
** Departement of Engineering Mechanics, Tsinghua University, Beijing 100084, P. R. China.*
*** Laboratoire de Métallurgie, Université d'Aix Marseille III, Marseille, 13397, France.*

Abstract. A micromechanics constitutive model of shape memory alloys allows us to describe single crystal behavior with hysteresis loops. The elaboration and characterization of CuZnAl single crystals permit to obtain pseudoelastic stress-strain curves for two different applied stress rates. This stress rate effect is taken into account by integration of heat equation.

1. Introduction

Thermoelastic martensitic transformations are responsible for the various behavior of shape memory alloys and some kinds of zirconia ceramics. According to the material science and physical metallurgy, thermoelastic martensitic transformations contain the following three elementary processes : the forward transformation (p→m, i.e. the transformation from parent phase to martensite), the reverse transformation (m→p) and the reorientation (m→m) between different kinds of martensite variants. The corresponding changes in microstructure lead to various macroscopic phenomena such as pseudoelasticity, one-way shape memory effect, ferroelasticity, rubber-like elasticity and two-way shape memory effect etc.. With the increasing application of shape memory alloys, structural ceramics and intelligent material systems and structures, the research on the constitutive relations of these materials attracts strong interest from solid mechanics and a lot of work has been done. It includes :

(1) Application of the Landau-Devonshire phenomenological theory to shape memory alloys and the derivation of one-dimensional stress-strain relations at different temperatures (Falk [1], Müller and Xu [2]).

(2) Phenomenological thermodynamics models of transformation plasticity (Raniecki and Lexcellent [3], Tanaka et al. [4]).

(3) Application of the continuum theory of thermoelasticity in study of the constitutive law and material instability during phase transformation (Abeyaratne and Knowles [5]).
(4) Micromechanics constitutive theory of transformation plasticity (Patoor et al. [6], Sun and Hwang [7]).

The purpose of the present paper is to perform a micromechanics-based study on the hysteresis phenomenon. In the first part, we present the general frame of the micromechanics constitutive model for transformation and reorientation of martensite variants. In the second part, elaboration and characterization of CuZnAl single crystals are presented. Modelling of these experiments by integrating heat equation is given in the third part.

2. The Micromechanics Constitutive Model[8]

Let us choose the volumic fraction of each variant $f_i\,(i = 1,\ldots,N)$ as internal variables of a single crystal. The Helmholtz free energy expression is:

$$\Phi = \Phi(T, E, f_1, f_2, \ldots, f_N) = W^m + \Delta G_{chem} + W_{surf} \tag{1}$$

Mechanical energy W^m is composed of two terms :

$$W^m = W^{ext} + W^* \tag{2}$$

W^{ext} represents the energy due to external loading and W^* due to internal stress resulting from incompatibilities of deformations between a variant and its matrix. ΔG_{chem} represents the latent heat contribution and W_{surf} the surface interaction energy between parent phase and variants and also between variants themselves.

2.1. Kinematical Relations of Transformation Plasticity

Starting from the crystallographic theory of martensitic transformation (see Wechsler et al[9]), we can calculate the transformation strain $\boldsymbol{\varepsilon}^p_s$ of sth variant :

$$\boldsymbol{\varepsilon}^p_s = g\mathbf{R}_s = \frac{1}{2}g(\mathbf{e}_s \otimes \mathbf{n}_s + \mathbf{n}_s \otimes \mathbf{e}_s) \tag{3}$$

where $\mathbf{R}_s$ is called the orientation tensor of the martensite variant s, $\mathbf{e}_s$ is the displacement direction of the invariant plane of the sth variant, $\mathbf{n}_s$ is the corresponding normal of the invariant plane and g a material constant. We can derive the relation between the microscopic strain $\boldsymbol{\varepsilon}$ and macroscopic strain $\mathbf{E}$ (here, strains are assumed to be small and we neglect the rotation components) :

$$\mathbf{E} = <\boldsymbol{\varepsilon}^e>_v + <\boldsymbol{\varepsilon}^p>_v = \mathbf{E}^e + \mathbf{E}^p = \mathbf{M}:\boldsymbol{\Sigma} + \mathbf{E}^p \tag{4}$$

where $\mathbf{M}$ is the elastic compliance tensor.
According to the crystallographic theory of martensitic transformation, we further have :

$$\mathbf{E}^p = <\boldsymbol{\varepsilon}^p> = \sum_{s=1}^{N} f_s \boldsymbol{\varepsilon}^p_s = g\sum_{s=1}^{N} f_s \mathbf{R}_s \tag{5}$$

2.2. Free Energy of The Constitutive Element

By using micromechanics self-consistent approach [7], the transformation-induced internal stress $\bar{\sigma}_s$ in the sth kind of inclusions (an inclusion is approximated by an oblate spheroid with principal radii $a_1 = a_2$, $a_3 / a_1 = \rho < 1$, a_3 parallel to $\mathbf{n}_s$) can be expressed by :

$$\bar{\sigma}_s = \mathbf{L}:(\mathbf{S}_s - \mathbf{I}):\varepsilon_s^p - \mathbf{L}:\sum_{t=1}^{N}(\mathbf{S}_t - \mathbf{I}):\varepsilon_t^p f_t \tag{6}$$

where $\mathbf{L} = \mathbf{M}^{-1}$ if $\mathbf{M}$ is not singular and $\mathbf{S}$ is the Eshelby tensor determined by elastic constants and by orientation and ratio ρ of the inclusion.

By Eq.(6) the elastic strain energy W^* produced by internal stress in a unit volume of the element can be calculated (Mura [10]) as :

$$W^* = -\frac{1}{2V}\sum_{s=1}^{N}\int_{V_M}\bar{\sigma}_s:\varepsilon_s^p dV = \sum_{s=1}^{N} f_s W_s - \sum_{s=1}^{N}\sum_{t=1}^{N} f_s f_t W_{st} \tag{7}$$

with

$$W_s = -\frac{1}{2}\varepsilon_s^p:(\mathbf{S}_s - \mathbf{I}):\varepsilon_s^p, \quad W_{st} = -\frac{1}{2}\varepsilon_s^p:(\mathbf{S}_s - \mathbf{I}):\varepsilon_t^p \tag{8}$$

The total elastic strain energy W^m per unit volume of the constitutive element can be expressed by

$$W^m = \frac{1}{2}E^e:L:E^e + W^* \tag{9}$$

The p-m interface-surface energy can be reasonably expressed[2,3] as :

$$W_{surf} = Af(1-f) = A\left(\sum_{s=1}^{N} f_s\right)\left(1 - \sum_{s=1}^{N} f_s\right) \tag{10}$$

The total change in the chemical free energy ΔG_{chem} is :

$$\Delta G_{chem}(T) = (\Delta G)f = C^*(T - T_0)\left(\sum_{s=1}^{N} f_s\right) \tag{11}$$

with $T_0 = \frac{1}{2}(M_s^0 + A_f^0)$.

Now, we can formulate the Helmholtz free energy Φ :

$$\Phi(E,T,f_1,\ldots,f_N) = W^m + W_{surf} + \Delta G_{chem}$$
$$= \frac{1}{2}\left(\mathbf{E} - \sum_{s=1}^{N}\varepsilon_s^p f_s\right):\mathbf{L}:\left(\mathbf{E} - \sum_{s=1}^{N}\varepsilon_s^p f_s\right)$$
$$+\sum_{s=1}^{N} f_s W_s - \sum_{s=1}^{N}\sum_{t=1}^{N} f_s f_t W_{st} + A\left(\sum_{s=1}^{N} f_s\right)\left(1 - \sum_{s=1}^{N} f_s\right) + C^*(T - T_0)\sum_{s=1}^{N} f_s \tag{12}$$

Hence, the macroscopic stress Σ and the driving force F_s associated with the variant f_s can be obtained :

$$\Sigma = \frac{\partial \Phi}{\partial E} \quad ; \quad F_s = -\frac{\partial \Phi}{\partial f_s} \tag{13}$$

During the thermodynamic processes involving both transformation and reorientation, the second law of thermodynamics requires

$$-\dot{\Phi}|_{E,T} = \sum_{s=1}^{N} F_s \dot{f}_s = \dot{W}_d \geq 0 \tag{14}$$

where $\dot{W}_d$ is the energy dissipation rate.

The volume fraction rate of the sth kind of variants can be expressed as :

$$\dot{f}_s = \dot{f}_{s0} + \dot{f}_{s1} + \ldots + \dot{f}_{ss-1} + \dot{f}_{ss+1} + \dot{f}_{sN} \, (s = 1, \ldots, N) \tag{15}$$

where $\dot{f}_{s0}$ correspond to p↔m and $\dot{f}_{st}(t = 1, s-1, s+1, \ldots, N)$ represent the reorientation process between s and t variant.

The choice of the expression $\dot{W}_d$ provides m↔p phase transition or m→m reorientation criteria and the evolution law of f_s by the Eq. (14) [8].

At last, in a classical way, heat equation is obtained without internal heat source :

$$\rho c \dot{T} - k\Delta T = \sum_{s=1}^{N} F_s \dot{f}_s + T\left[\frac{\partial \Sigma}{\partial T} : \dot{\mathbf{E}} - \sum_{s=1}^{N} \frac{\partial F_s}{\partial T} \dot{f}_s\right] \tag{16}$$

with $\dfrac{\partial \Sigma}{\partial T} = 0, \quad \dfrac{\partial F_s}{\partial T} = -C^{*}.$

$$\rho c \dot{T} - k\Delta T = \sum_{s=1}^{N} F_s \dot{f}_s + C^{*} T \left(\sum_{s=1}^{N} \dot{f}_s\right) \tag{17}$$

where ρ, c and k are the density, the specific heat and the coefficient of thermal conductivity respectively. Δ denotes the Laplace operator.

3. Elaboration, Characterization and Thermomechanical Behavior of CuZnAl Single Crystals [11]

CuZnAl S.M.A. were grown as single crystals by the Bridgman technique with a final shape directly suitable for thermomechanical tests (cylinders with taper heads: 25mm gauge length, 4mm in diameter). The classical transformation temperatures were checked by Differential Scanning Calorimetry and resistivity. The orientation, crystal structure and mechanical structural effect (dislocations, antiphase boundaries...) have been investigated by x-ray diffraction, transmission and scanning electron microscopy.

For several single crystals, isothermal loading-unloading tensile tests were performed (i). in the pseudoelastic range $10°C < (T - A_s^0) < 50°C$, (ii). for two applied stress rates (0.1 and 1 MPa s^{-1}) between stress free state (Σ=0) and sufficient elongation to obtain the complete phase transformation (Fig. 1).

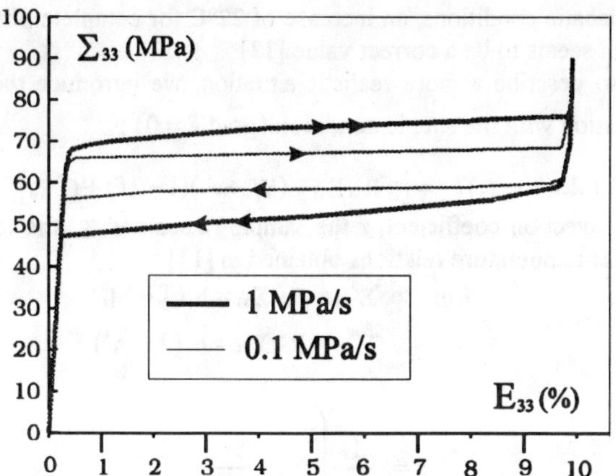

Figure 1. Experimental stress-strain curves of tension test.

For CuZnAl alloys, 24 martensitic variants (six groups of four self-accommodated variants) can be created. Directions of the habit plane normal **n** and transformation directions **e** are of {2 11 12} type. The observed plateau for (p→m) or (m→p) is correlated to the growth and the propagation of only one variant of martensite (the variant having the largest Schmid factor is chosen). Sample orientation measurements permit to obtain **R** and g=0.206 (Fig. 1). Fig. 1 shows that the hysteresis width increases with the applied stress rate. This fact was also observed by Otsuka et al. [12] on CuAlNi single crystals.

4. Modelling CuZnAl single Crystal Behavior

With the assumption of only one active variant, the thermodynamical force associated with f reads in case of tension or compression test :

$$F(\Sigma_{33}, f, T) = g R_{33} \Sigma_{33} - (W_s + A)(1-2f) - C^*(T - T_0) \tag{18}$$

We suppose that when applied stress rate is very slow, forward and reverse phase transformation take place at constant stress Σ_{AM}^{ap} and Σ_{MA}^{ap} respectively.

The measurements in Fig. 1 of yield stress for forward and reverse transformations Σ_{AM}^{ap} and Σ_{MA}^{ap} respectively permit to obtain $(W_s + A)$ and C^* by the following conditions [3] :

$$\begin{cases} F(\Sigma_{AM}^{ap}, f = 0, T) = 0 \\ F(\Sigma_{MA}^{ap}, f = 1, T) = 0 \end{cases} \tag{19}$$

The heat equation is written for any point M inside the body for tension or compression test:

$$\begin{aligned} \rho c \dot{T} - k\Delta T &= F\dot{f} + C^* T \dot{f} \\ &= [g R_{33} \Sigma_{33} - (W_s + A)(1-2f) + C^* T_0]\dot{f} \end{aligned} \tag{20}$$

In the case of adiabatic conditions, an increase of 22°C for complete phase transformation is predicted which seems to be a correct value [13].

In order to describe a more realistic situation, we introduce the convective term into the heat equation with the simple condition ($\overrightarrow{\text{grad}}\, T = 0$) :

$$\rho c\dot{T} + 2h(T - T_a)/r = \left[g R_{33}\Sigma_{33} - (W_s + A)(1 - 2f) + C^* T_0\right]\dot{f} \tag{21}$$

where h is the convection coefficient, r the sample radius and T_a the temperature of test room. Using stress-temperature relations obtained in [11] :

$$\begin{cases} p \to m & : \Sigma^0_{AM} = R_{33}\Sigma_{33} = b_1(T - M^0_s) \\ m \to p & : \Sigma^0_{MA} = R_{33}\Sigma_{33} = b_2(T - A^0_s) \end{cases} \tag{22}$$

and

$$\dot{f} = \frac{1}{g R_{33}}\left(\dot{E}_{33} - \frac{\dot{\Sigma}_{33}}{E}\right) \tag{23}$$

We can obtain the stress-strain relation from Eq. (21), (22) and (23) by numerical integration.

Pure shear test can be simulated by the same procedure with Σ_{12} and R_{12}. For modelling the curves given in Fig. 1 and pure shear test, the material parameters identified are given in table 1.

Table 1. identified material parameters.

M^0_s (°K)	A^0_s (°K)	T_0 (°K)	T_a (°K)	C^* (MJ/°Km³)	b_1 (MPa/°K)	b_2 (MPa/°K)	h (W/°Km²)
302	301	307	335	0.21	0.91	0.82	160

The calculated **R** values from the sample orientations measured by the back reflection method of Laüe are R_{33}=0.459 for tension, R_{33}=-0.369 for compression and R_{12}=0.427 for pure shearing.

We have to note that the test temperature was controlled by the flux of warm air but we did not measure the air rate. Here the used value of h corresponds to that measured by Gosse [14]when air rate is 0.03 ms^{-1} .

The predictions given in Fig. 2 and Fig. 3 explain very well the increase of hysteresis width and hardening with the applied stress rates.

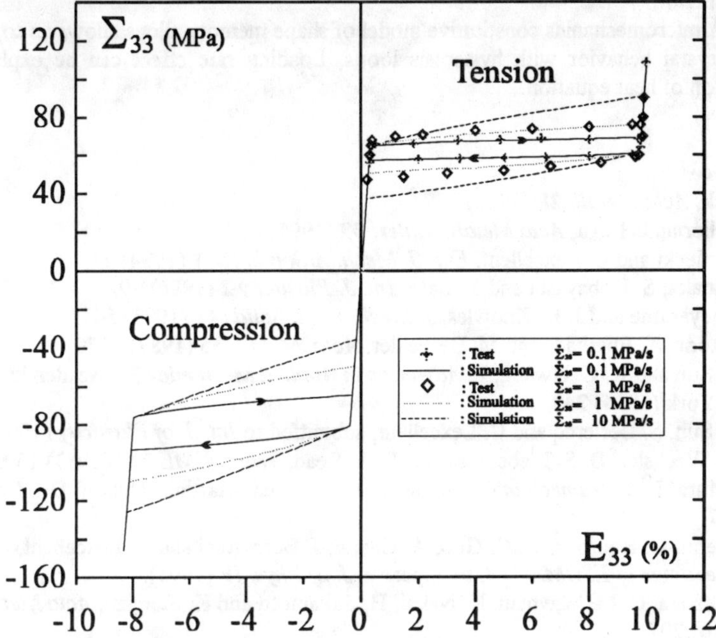

Figure 2. Simulated and experimental stress-strain curves of tension-compression.

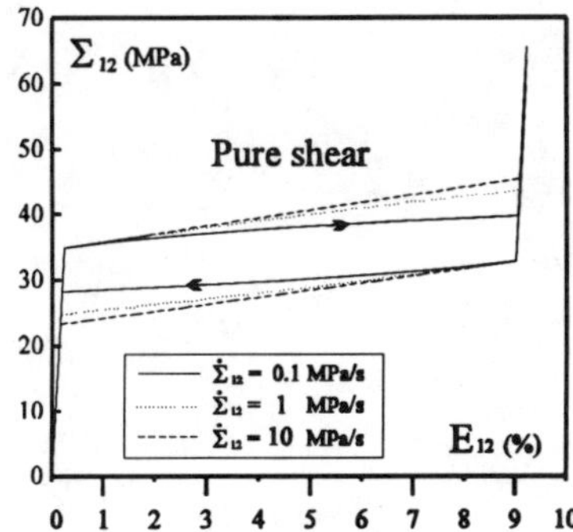

Figure 3. Simulated stress-strain curves of pure shear.

5. Conclusion

A micromechanics constitutive model of shape memory alloys allows us to describe single crystal behavior with hysteresis loops. Loading rate effect can be explained by integration of heat equation.

References

1. F. Falk, *Acta Metall.* **28** (1980) 1773.
2. I. Müller and H. Xu, *Acta Metall. Mater.* **39** (1991) 263.
3. B. Raniecki and C. Lexcellent, *Eur. J. Mech. A/Solids* **13, 1** (1994) 21.
4. K. Tanaka, S. Kobayashi and Y. Sato, *Int. J. Plasticity* **2** (1986) 59.
5. R. Abeyaratne and J. K. Knowles, *J. Mech. Phys. Solids* **41** (1993) 541.
6. E. Patoor, A. Eberhadt and M. Berveiller, *Acta. Metall.* **35** (1987) 2779.
7. Q. P. Sun and K. C. Hwang, In *Advances in Applied Mechanics* **31**, Academic Press, New York (1994) 249.
8. Q. P. Sun, S. Leclercq and C. Lexcellent, submitted to *Int. J. of Plasticity* (1995).
9. M. S. Wechsler, D. S. Lieberman and T. A. Read, *Trans AIME* **197** (1953) 1503.
10. T. Mura, In *Micromechanics of defects in solids*, ed. Martinus Nijhoff, Dordrecht, 1987.
11. P. Perin, G. Bourbon, B. C. Goo, A. Charai, J. Bernardini and C. Lexcellent, *Proceedings of ESOMAT 94*, to appear in *J. of Phys. IV* (1995).
12. K. Otsuka, C. M. Wayman, K. Nakai, H. Sakamoto and K. Shimizu, *Acta Metall.* **24** (1976) 207.
13. P. G. McCormick, S. Miyazaki and Y. Liu, *Proceedings of ICOMAT 92*, ed. Monterey Institute (1993) 999.
14. J. Gosse, *Thèse*, Ecole Nationale Supérieure de Mécanique et d'Aérotechnique et Institut de Recherches du Centre-ouest, Poitiers, France, 1955.

Continuum Models and Discrete Systems
Proceedings of 8th International Symposium, June 11-16, 1995, Varna, Bulgaria, ed. K.Z. Markov

INCLUSION OF INTRINSIC SPIN OF DISCRETE IONS IN A CONTINUUM THEORY OF CRYSTALLINE MAGNETISM

D. F. NELSON and B. CHEN
Department of Physics, Worcester Polytechnic Institute,
Worcester, MA 01609, USA

Abstract. Rather recent work in quantum field theory has shown that a quantum fermionic excitation, such as arising from a half-integral intrinsic spin, must be represented in the classical limit by an anticommuting, but otherwise classical, quantity. Thus, anticommuting or Grassmann algebra must be introduced into classical physics. Nonrelativistic interactions of a spin $\frac{1}{2}$ particle were shown to be characterized by attributing to the particle a three-vector of anticommuting components (a Grassmann G_3 vector) with the spin being proportional to the vector product of the G_3 vector with itself. This is readily incorporated into our Lagrangian theory of dielectric crystals which begins from a particle point of view. Each particle is endowed with a mass, a charge, and a G_3 vector. A long-wavelength or continuum limit then yields a classical or macroscopic Lagrangian containing all long-wavelength modes of motion (acoustic, optic, electromagnetic, and spin) and their interactions to all orders of nonlinearity in a crystal of any symmetry, anisotropy, and structural complexity. Equations of motion of all the modes are found and the conservation laws of energy, momentum, and angular momentum are presented.

1. Introduction

Since its discovery some seventy years ago, intrinsic spin has been regarded as somehow a nonclassical quantity. This belief has persisted in spite of the fact that many spin interactions such as occur in magnetooptics, magnetoacoustics, and spin waves are long-wavelength or continuum interactions and so can be viewed as classical or macroscopic phenomena. The reason is that a classical description of intrinsic spin applicable to these phenomena has eluded physicists.

The reason for this is apparent when attempting a Lagrangian formulation. If some sort of a kinetic energy is associated with spin, then the Lagrange equation inevitably becomes a second-order differential equation (a Newton's equation). However, the spin and its associated magnetization are known to obey a first-order differential equation, the spin precession equation. Thus, spin has no kinetic energy, a fact well-known also from the form of a quantum mechanical Hamiltonian. The puzzle has been, then, what is the dynamical term in the Lagrangian that produces a first-order dynamical equation of spin?

The answer to this puzzle has emerged rather recently from quantum field theory studies by Berezin and Marinov [1] and Casalbuoni [2]. Their work shows that the classical limit of a coherent wave representation of a fermionic operator (such as represents a half-integral spin excitation) is an anticommuting, but otherwise classical, variable. Their work also demonstrates the reverse connection, that is, that the quantization of the canonical formalism of a particle characterized by a G_3 vector (and a mass and charge) produces the nonrelativistic spin $\frac{1}{2}$ quantum theory of the Pauli-Schrodinger equation. This canonical formalism produces a classical Lagrangian of a spin $\frac{1}{2}$ particle including the needed dynamical term.

Previously we have constructed a very general continuum Lagrangian-based theory of dielectric crystals [3]. The Lagrangian is formulated at the microscopic, discete particle point of view where we are confident of the physics input. The long-wavelength or continuum limit is taken in a manner so as to preserve all dynamical modes: acoustic waves, optic mode vibrations, electromagnetic waves, and, with the new addition, spin waves. The approach is general enough to contain the interactions of these modes with themselves and each other to all orders of nonlinearity and to apply to a crystal of arbitrary symmetry, anisotropy, and structural complexity. With various spin sublattices now included the theory can be applied to ferromagnetic, antiferromagnetic, and ferrimagnetic crystals. Note that this has been made possible only by allowing the admission of anticommuting Grassmann algebra into the domain of "classical physics".

We believe that the Lagrangian approach with its generality and economy of formulation and its guaranteed self-consistency of results and with the Grassmann algebra formulation of spin offers advantages over other approaches based on the virtual work principle [4,5], on postulation of the forms of the conservation laws [6,7], or on the virtual power principle [8]. It also has a more intrinsic origin than a previous rather perplexing Lagrangian formulation [9].

2. Grassmann Algebra and Classical Spin Variables

The anticommuting Grassmann variable of G_3 algebra, a real 3-vector ξ, obeys

$$\xi_i\xi_j + \xi_j\xi_i = 0 \quad (i = j \text{ or } i \neq j; i, j = 1, 2, 3). \tag{1}$$

The Lagrangian of a particle having internal degrees of freedom characterized by the Grassmann variable ξ was shown [1,2] to be

$$L = \frac{i}{2}\xi \cdot \dot{\xi} + \frac{1}{2}m\dot{\mathbf{q}}^2 - V_1(\mathbf{q}) - \xi_i\xi_j V_{ij}(\mathbf{q}), \tag{2}$$

where m is its mass, $\mathbf{q}$ is its spatial position, i is the imaginary unit, and V_1 and V_{ij} are potential functions. The quadratic form in ξ follows from the need for L to be even in its commutativity with ξ and the fact that all higher even products of ξ can be reduced to a quadratic combination. Note that the first term is linear in the time derivative and proportional to the imaginary unit. It is shown below that this term disappears from the energy density or Hamiltonian. These characteristics show that the term is not a kinetic energy, but it does play an essential role in producing the dynamical equation of ξ. For these reasons we call it the kinetic Lagrangian.

Quantization of the system by the Dirac procedure shows that ξ itself does not represent spin. However, the quantity

$$\mathbf{S} = -\frac{i}{2}\xi \times \xi \tag{3}$$

quantizes to

$$\mathbf{S} = \frac{\hbar}{2}\sigma, \tag{4}$$

thus showing that $\mathbf{S}$ of Eq. (3) is the classical representation of the spin.

Since the Lagrangian of Eq. (2) must be invariant under spatial rotations, the tensor V_{ij} must be antisymmetric, $V_{ij} = (i/2)\,\epsilon_{ijk}V_k$. This allows rewriting the Lagrangian as

$$L = \frac{i}{2}\xi \cdot \dot{\xi} + \frac{1}{2}m\dot{\mathbf{q}}^2 - V_1(\mathbf{q}) + \mathbf{S}\cdot\mathbf{V}(\mathbf{q})\,. \tag{5}$$

The kinetic Lagrangian cannot be expressed only in terms of the spin $\mathbf{S}$. Thus, ξ remains the Lagrangian variable even though $\mathbf{S}$ is more physically interpretable.

The Lagrange equation for ξ yields the equation of motion of ξ to be

$$\dot{\xi} = \xi \times \mathbf{V}. \tag{6}$$

Note that this differential equation is first-order in time and of a form that conserves the magnitude of ξ. Since the spin $\mathbf{S}$ is more physically interpretable than ξ, we can use the time derivative of Eq. (3) and Eq. (6) to obtain

$$\dot{\mathbf{S}} = \mathbf{S} \times \mathbf{V}. \tag{7}$$

Interestingly the dynamic equation for $\mathbf{S}$ has the same form as that for ξ and so the same properties as just mentioned for it. It is the desired spin precession equation.

3. Lagrangian Density

We consider a closed system characterized by a Lagrangian L that is the sum of a matter Lagrangian L_M, an electromagnetic field Lagrangian L_F, and a field-matter interaction Lagrangian L_I, $L = L_M + L_I + L_F$. Since interactions in crystals can be very complex, we believe it is important to begin at a microscopic, discrete particle formulation, where we are confident of the physics, and to proceed to a continuum formulation by taking a long-wavelength limit. This leads to the Lagrangian becoming an integral over a Lagrangian density which is the quantity entering the Lagrange equations. The density may be either in the ordinary spatial frame or in the material frame defined presently. The former is the most natural for calculating the electromagnetic equations while the latter is most natural for calculating the matter equations. Of course, transformation between the systems can be done whenever it is necessary or instructive. For a more extensive description of the construction of the Lagrangian see [3].

3.1. Matter Lagrangian

Consider a primitive unit cell of a crystal containing N particles labeled by a lowercase Greek letter α. These particles should include all the ions and, depending on the problem under study, one or two bonding electrons. Each particle has a fixed mass m^α and a fixed charge e^α. Its position is $\mathbf{x}^{\alpha n}$ where n is a three-component index labeling the primitive unit cell. The particle is also assigned a three-component Grassmann vector $\xi^{\alpha n}$ to describe its internal (spin) degrees of freedom.

The particles reside in a vacuum and are subject to short-range bonding forces and to long-range electromagnetic body forces. The short-range forces are described by a potential energy $V(\mathbf{x}^{\alpha n}, \xi^{\alpha n})$ in the matter Lagrangian. The long range electromagnetic forces are described by the interaction Lagrangian.

The discrete particle matter Lagrangian can be written as

$$L_M = \frac{1}{2}\sum_{\alpha,n}\left[m^\alpha(\dot{\mathbf{x}}^{\alpha n})^2 + i\xi^{\alpha n}\cdot\dot{\xi}^{\alpha n}\right] - V(\mathbf{x}^{\alpha n}, \xi^{\alpha n}). \tag{8}$$

The continuum limit of the discrete Lagrangian can be done by replacing the discrete index n by a continuous variable $\mathbf{X}$. By not affecting the index α this replacement retains all modes of motion of the discrete lattice and all the symmetry and anisotropy of the crystal. It also fullfills the same function of labeling the matter as the discrete n did. Therefore we call $\mathbf{X}$ the continuum material coordinate vector. All the mechanical and internal coordinates can be replaced by their continuum counterparts,

$$\mathbf{x}^{\alpha n}(t) \to \mathbf{x}^{\alpha}(\mathbf{X},t), \quad \xi^{\alpha n}(t) \to \xi^{T\alpha}(\mathbf{X},t), \quad \mathbf{s}^{\alpha n}(t) \to \mathbf{s}^{T\alpha}(\mathbf{X},t) \equiv \frac{i}{2\Omega_0}\xi^{T\alpha}\times\xi^{T\alpha}, \tag{9}$$

where T stands for total to indicate that the variable consists of a spontaneous value and a variation from that value and Ω_0 is the volume of a primitive unit cell. The continuum limit causes sums over the cell index n to become integrals over the continuous material coordinate $\mathbf{X}$, that is,

$$\sum_n F(\mathbf{x}^{\alpha n}(t), \xi^{\alpha n}(t)) \to \frac{1}{\Omega_0}\int F\left(\mathbf{x}^{\alpha}(\mathbf{X},t), \xi^{T\alpha}(\mathbf{X},t)\right) dV, \tag{10}$$

where $dV \equiv dX_1 dX_2 dX_3$. To account for intercellular forces the continuum limit (10) must be improved when applied to the stored energy by adding derivatives of the independent variables in the functional argument. For the long wavelength limit first derivatives suffice [3].

The continuum limit yields a continuum matter Lagrangian in the material frame

$$\mathcal{L}_{MM} = \frac{1}{2}\rho^0(\dot{\mathbf{x}})^2 + \frac{1}{2}\sum_{\nu=1}^{N-1} m^\nu\left(\dot{\mathbf{y}}^{T\nu}\right)^2 + \frac{i}{2\Omega_0}\sum_{\alpha=1}^{N}\xi^{T\alpha}\cdot\dot{\xi}^{T\alpha} - \rho^0\Sigma\left(\mathbf{y}^{T\nu}, \mathbf{y}^{T\nu}_{,A}, \xi^{T\alpha}, \xi^{T\alpha}_{,A}\right). \tag{11}$$

Linear combinations of the coordinates $\mathbf{x}^{\alpha}$ have been introduced that are the continuum center-of-mass position $\mathbf{x}$ and a set of dispacement-invariant internal vibration coordinates $\mathbf{y}^{T\nu}$ [3]. The undeformed mass density is ρ^0 and the mass density of the ν-sublattice is m^{ν} and $\mathbf{X}, t$ are the independent variables. The notation $_{,A}$ means $\partial/\partial X_A$.

The stored energy $\rho^0\Sigma$ can be expanded in a series provided the expansion variables are rotationally invariant measures of its variables. Such a series can be truncated at a few terms for low-order effects only if the new measures vanish in the natural state (the perfect crystal without external influences). Further, the Grassmann variables must occur quadratically, as pointed out for Eq. (2), and so can be replaced by spin functions in $\rho^0\Sigma$. These several requirements are met by the set of rotational invariants

$$\Pi_A^{\nu} = X_{A,i}\left(\delta_{iB}Y_B^{\nu} + y_i^{\nu}\right) - Y_A^{\nu}, \quad \Pi_{AB}^{\nu} = y_{i,A}^{\nu} X_{B,i}, \tag{12}$$

$$\Gamma_A^{\alpha} = X_{A,i}\left(\delta_{iB}S_B^{\alpha} + s_i^{\alpha}\right) - S_A^{\alpha}, \quad \Gamma_{AB}^{\alpha} = s_{i,A}^{\alpha} X_{B,i}, \tag{13}$$

$$E_{AB} = \left(x_{i,A}x_{i,B} - \delta_{AB}\right)/2, \tag{14}$$

$$y_i^{T\nu} = \delta_{iB}Y_B^{\nu} + y_i^{\nu}, \quad s_i^{T\alpha} = \delta_{iB}S_B^{\alpha} + s_i^{\alpha}, \tag{15}$$

and where each of these spins is defined in terms of the corresponding Grassmann variable as in Eq. (9) and the capitalized quantities are the spontaneous values. The term linear in Γ_A^{α} exists when a spontaneous magnetic field is present and corresponds to the energy of that field. Of all the other terms in the expansion of spin variables the term quadratic in Γ_{AB}^{α} is dominant and represents the spin exchange energy.

3.2. Interaction Lagrangian

The discrete particle field-matter interaction Lagrangian is

$$L_I = \sum_{\alpha,n}\left[e^{\alpha}\left(\dot{\mathbf{x}}^{\alpha n}(t)\cdot\mathbf{A}\left(\mathbf{x}^{\alpha n},t\right) - \Phi\left(\mathbf{x}^{\alpha n},t\right)\right) + \mu^{\alpha}\mathbf{s}^{\alpha n}\left(\mathbf{x}^{\alpha n},t\right)\cdot\mathbf{B}\left(\mathbf{x}^{\alpha n},t\right)\right], \tag{16}$$

where $\mu^{\alpha} \equiv ge^{\alpha}/m^{\alpha}$ is the magnetic moment, g is the gyromagnetic ratio, and $\mathbf{A}$ and Φ are the vector and scalar potentials of the electromagnetic fields. The form chosen for the spin-magnetic field interaction guarantees that after quantization it produces the well known nonrelativistic spin $\frac{1}{2}$ theory.

In the continuum limit the interaction Lagrangian density arising from charge in the spatial frame $\mathcal{L}_{IS}^{q}$ (volume element $dv = dz_1dz_2dz_3 \equiv JdV$, where J is the Jacobian between the spatial and material coordinates) is

$$\mathcal{L}_{IS}^{q} = \mathbf{j}^c\cdot\mathbf{A} - q\Phi = J\mathcal{L}_{IM}^{q}, \tag{17}$$

$$q\left(\mathbf{z},t\right) \equiv \sum_{\alpha} q^{\alpha}\int\delta\left(\mathbf{z} - \mathbf{x}^{\alpha}\left(\mathbf{X},t\right)\right)dV, \tag{18}$$

$$\mathbf{j}^c\left(\mathbf{z},t\right) \equiv \sum_{\alpha} q^{\alpha}\int\dot{\mathbf{x}}^{\alpha}\left(\mathbf{X},t\right)\delta\left(\mathbf{z} - \mathbf{x}^{\alpha}\left(\mathbf{X},t\right)\right)dV. \tag{19}$$

For a dielectric (D) a multipole expansion of these yields

$$q^D(\mathbf{z},t) = -\nabla\cdot\mathbf{P} + \nabla\nabla : \mathbf{Q}, \tag{20}$$

$$\mathbf{j}^D(\mathbf{z},t) = \frac{\partial \mathbf{P}}{\partial t} + \nabla\times(\mathbf{P}\times\dot{\mathbf{x}}) - \frac{\partial}{\partial t}(\nabla\cdot\mathbf{Q}) - \nabla\times[\nabla\cdot(\mathbf{Q}\times\dot{\mathbf{x}})] + \nabla\times\mathbf{M}^c, \tag{21}$$

where the polarization $\mathbf{P}$, the tensor quadrupolarization $\mathbf{Q}$ and the magnetization $\mathbf{M}^c$ from the motion of bound charge are given by

$$\mathbf{P} = \sum_\nu \frac{q^\nu \mathbf{y}^{T\nu}}{J}, \quad \mathbf{Q} = \frac{1}{2}\sum_{\mu\nu} \frac{q^{\mu\nu}\mathbf{y}^{T\mu}\mathbf{y}^{T\nu}}{J}, \quad \mathbf{M}^c = \frac{1}{2}\sum_{\mu\nu}\frac{q^{\mu\nu}\mathbf{y}^{T\mu}\times\dot{\mathbf{y}}^{T\nu}}{J}. \tag{22}$$

Here q^ν and $q^{\mu\nu}$ are charges defined [3] in terms of the primitive charges q^α. The continuum limit of the spin interaction Lagrangian is

$$\mathcal{L}^s_{IS} = \sum_\alpha \int \delta(\mathbf{z} - \mathbf{x}^\alpha(\mathbf{X},t))\,\gamma^\alpha \mathbf{s}^\alpha(\mathbf{x}^\alpha(\mathbf{X},t),t)\cdot\mathbf{B}(\mathbf{z},t)\,dV = \mathbf{M}^s\cdot\mathbf{B}, \tag{23}$$

where the spin magnetization is given by

$$\mathbf{M}^s(\mathbf{z},t) = \sum_\alpha \gamma^\alpha \int \mathbf{s}^\alpha(\mathbf{x}^\alpha(\mathbf{X},t),t)\,\delta(\mathbf{z}-\mathbf{x}^\alpha(\mathbf{X},t))\,dV = \sum_\alpha J^{-1}\gamma^\alpha\mathbf{s}^\alpha(\mathbf{z},t), \tag{24}$$

the last form being the first term of a Taylor series expansion of the integrand. The spin interaction Lagrangian Eq. (23) is equivalent to

$$\mathcal{L}^s_{IS} = \mathbf{j}^s\cdot\mathbf{A} = J\mathcal{L}^s_{IM}, \quad \mathbf{j}^s = \nabla\times\mathbf{M}^s, \tag{25}$$

since the difference is only a perfect derivative which cannot affect the equations of motion. Thus the entire spatial-frame interaction Lagrangian is

$$\mathcal{L}_{IS} = \mathbf{j}\cdot\mathbf{A} - q^D\Phi = J\mathcal{L}_{IM}, \quad \mathbf{j} \equiv \mathbf{j}^D + \mathbf{j}^s = \nabla\times\mathbf{M}, \tag{26}$$

where $\mathbf{j}$ is the total bound current and $\mathbf{M}$ is the total magnetization $\mathbf{M}\equiv\mathbf{M}^c+\mathbf{M}^s$.

3.3. Electromagnetic Field Lagrangian

The field Lagrangian is known directly as a density in the spatial frame

$$\mathcal{L}_{FS} = \frac{1}{2}\left(\epsilon_0\mathbf{E}^2 - \frac{1}{\mu_0}\mathbf{B}^2\right), \tag{27}$$

where the electromagnetic potentials are related to the fields by

$$\mathbf{E} = -\nabla\Phi - \frac{\partial\mathbf{A}}{\partial t}, \qquad \mathbf{B} = \nabla\times\mathbf{A} \tag{28}$$

and are the Lagrangian variables of the electromagnetic field.

4. Dynamical Equations

The matter equations are most naturally found in the material frame using $\mathbf{s}^\alpha$, $\mathbf{y}^{T\nu}$ ($\nu = 1, 2, \ldots, N-1$), and ξ^α ($\alpha \leq N$) as Lagrangian variables and $\mathbf{X}$, t as independent variables. Only the matter Lagrangian density, Eq. (11), and the interaction Lagrangian density, Eq. (26), are needed since the field Lagrangian density does not contribute.

The electromagnetic equations are most naturally found in the spatial frame using $\mathbf{A}$ and Φ as Lagrangian variables and $\mathbf{z}$, t as the independent variables (with $\mathbf{z} = \mathbf{x}$ inside matter). Only the field Lagrangian density, Eq. (27), and the interaction Lagrangian density, Eq. (26), are needed.

4.1. Maxwell Equations

The Lagrange equations for Φ and $\mathbf{A}$ yield, respectively,

$$\epsilon_0 \nabla \cdot \mathbf{E} = q^D, \qquad \frac{1}{\mu_0} \nabla \times \mathbf{B} - \epsilon_0 \frac{\partial \mathbf{E}}{\partial t} = \mathbf{j}. \tag{29}$$

These electromagnetic equations are in Maxwell-Lorentz form, that is, with only $\mathbf{E}$ and $\mathbf{B}$ fields present. They can be converted to the Maxwell form by defining the electric displacement field $\mathbf{D}$ and the magnetic intensity $\mathbf{H}$ by

$$\mathbf{D} \equiv \epsilon_0 \mathbf{E} + \mathbf{P} - \nabla \cdot \mathbf{Q}, \quad \mathbf{H} \equiv \mathbf{B}/\mu_0 - \mathbf{P} \times \dot{\mathbf{x}} - \mathbf{M} + \nabla \cdot (\mathbf{Q} \times \dot{\mathbf{x}})\,. \tag{30}$$

We then have the Maxwell equations for a dielectric

$$\nabla \times \mathbf{H} - \partial \mathbf{D}/\partial t = 0, \quad \nabla \cdot \mathbf{D} = 0, \quad \nabla \times \mathbf{E} + \partial \mathbf{B}/\partial t = 0, \quad \nabla \cdot \mathbf{B} = 0, \tag{31}$$

where the last two equations are direct consequences of Eqs. (28) and so arise as ancillary conditions in the Lagrangian approach.

4.2. Internal Motions

The Lagrange equations for the internal coordinates $\mathbf{y}^{T\nu}$ expressed in the material frame after some rearrangement become

$$\begin{aligned} m^\nu \ddot{y}_i^{T\nu} &= q^\nu \mathcal{E}_i + \sum_\mu q^{\mu\nu} y_j^{T\mu} \mathcal{E}_{i,j} + \sum_\mu q^{\mu\nu} \epsilon_{ijk} \dot{y}_j^{T\mu} B_k \\ &- \sum_\mu q^{\mu\nu} \epsilon_{ikl} y_j^{T\mu} \dot{x}_{k,j} B_l - \frac{\partial \rho^0 \Sigma}{\partial y_i^{T\nu}} + \frac{d}{dX_A} \frac{\partial \rho^0 \Sigma}{\partial y_{i,A}^{T\nu}}, \end{aligned} \tag{32}$$

where $\mathcal{E} \equiv \mathbf{E} + \dot{\mathbf{x}} \times \mathbf{B}$. Linear combinations of the internal coordinates are the optic mode normal coordinates and play an important role in optical properties.

4.3. Continuum Center-of-Mass Motion

The Lagrange equation for the continuum center-of-mass coordinate $\mathbf{x}$ becomes after considerable processing [3]

$$\rho \ddot{\mathbf{x}} = q\mathbf{E} + \mathbf{j} \times \mathbf{B} + \nabla \cdot \mathbf{t}^E, \tag{33}$$

where the D has been dropped from q and the divergence acts on the second tensor subscript of the elastic stress tensor, which is given by

$$\begin{aligned} t^E_{il} &= \mathcal{E}_i P_l + \epsilon_{ijk}\left(\frac{\partial Q_{lj}}{\partial t} + (Q_{lj}\dot{x}_m)_{,m} + \epsilon_{ljm}M_m\right)B_k \\ &- 2\epsilon_{ijk}Q_{lm}\dot{x}_{j,m}B_k + 2Q_{lm}\mathcal{E}_{i,m} - (Q_{lm}\mathcal{E}_i)_{,m} + J^{-1}\frac{\partial \rho^0\Sigma}{\partial x_{i,A}}x_{l,A} \end{aligned} \quad (34)$$

and $\rho = J^{-1}\rho^0$. Though the inertial term naturally involves material frame time derivatives ($\mathbf{X}$ held constant), the applied force terms are converted here to a spatial description to exhibit the fact that body forces can be expressed in the Lorentz force form with all remaining forces expressed as contact forces from stresses. This form is also most convenient for deducing the conservation laws. The elastic stress tensor has been shown [3,10] to be asymmetric.

4.4. Spin Sublattice Motion

The Lagrange equation for $\xi^{T\alpha}$ can be converted to the dynamic equation for spin $\mathbf{s}^{T\alpha}$ in the manner used for Eq. (7) with the result

$$\frac{d\mathbf{s}^{T\alpha}}{dt} = \mu^\alpha \mathbf{s}^{T\alpha} \times {}^{eff}\mathbf{B}^\alpha, \quad {}^{eff}B^\alpha_k \equiv B_k - \frac{1}{\mu^\alpha}\left(\frac{\partial \rho^0\Sigma}{\partial s^{T\alpha}_k} - \frac{d}{dX_A}\frac{\partial \rho^0\Sigma}{\partial s^{T\alpha}_{k,A}}\right). \quad (35)$$

This is the desired first-order spin precession equation. Notice that in this theory we obtain a dynamical equation for each sublattice spin degree of freedom, not a precession equation for a saturated magnetization.

5. Conservation Laws

The conservation laws can be found by combining the equations of motion or by use of Noether's theorem. Different insights result from the two methods. Here we simply state the result obtainable from either method.

5.1. Momentum Conservation

The spatial-frame momentum conservation law for the system is

$$\frac{\partial}{\partial t}(\rho\dot{\mathbf{x}} + \epsilon_0 \mathbf{E}\times\mathbf{B})_i - \frac{\partial}{\partial z_l}\left(t^E_{il} + m_{il} - \rho\dot{x}_i\dot{x}_l\right) = 0, \quad (36)$$

where the electromagnetic stress tensor is

$$m_{il} = \epsilon_0 E_i E_l + \frac{1}{\mu_0}B_i B_l - \frac{1}{2}\left(\epsilon_0 E_k E_k + \frac{1}{\mu_0}B_k B_k\right)\delta_{il}. \quad (37)$$

This is identical to that found in [11] since spin does not contribute to momentum.

5.2. Angular Momentum Conservation

The angular momentum conservation law for the system here considered expressed in the spatial frame is

$$\frac{\partial}{\partial t}\left(\mathbf{x}\times\mathbf{g}+\mathbf{l}^{i}+\mathbf{l}^{s}\right)_i+\frac{\partial}{\partial z_l}\Big[-\epsilon_{ijk}x_j t^T_{kl}+\left(l^i_i+l^s_i\right)\dot{x}_l$$
$$-\epsilon_{ijk}\frac{1}{J}\left(\sum_\nu \frac{\partial\rho^0\Sigma}{\partial y^{T\nu}_{k,A}}y^{T\nu}_j+\sum_\alpha\frac{\partial\rho^0\Sigma}{\partial s^{T\alpha}_{k,A}}s^{T\alpha}_j\right)x_{l,A}\Big] = 0, \qquad (38)$$

where the angular momentum densities from internal vibrations and spin are

$$\mathbf{l}^i=\sum_\nu \rho^\nu \mathbf{y}^{T\nu}\times\dot{\mathbf{y}}^{T\nu}, \qquad \mathbf{l}^s=J^{-1}\sum_\alpha \mathbf{s}^{T\alpha} \qquad (39)$$

and where $t^T_{kl}\equiv t^E_{kl}+m_{kl}-\rho\dot{x}_k\dot{x}_l$ is the total stress tensor and $\rho^\nu=J^{-1}m^\nu$. Here spin plays an important role and in a functionally expected manner.

5.3. Energy Conservation

The spatial-frame energy conservation law for the system is

$$\frac{\partial}{\partial t}\left(\frac{\rho\dot{\mathbf{x}}^2}{2}+\sum_\nu\frac{\rho^\nu}{2}\left(\dot{\mathbf{y}}^{T\nu}_i\right)^2+\rho\Sigma-\mathbf{M}^s\cdot\mathbf{B}+\frac{\epsilon_0\mathbf{E}^2}{2}+\frac{\mathbf{B}^2}{2\mu_0}\right)$$
$$+\frac{\partial}{\partial z_j}\Bigg[\left(\frac{\rho\dot{\mathbf{x}}^2}{2}+\sum_\nu\frac{\rho^\nu}{2}\left(\dot{\mathbf{y}}^{T\nu}_i\right)^2+\rho\Sigma-\mathbf{M}^s\cdot\mathbf{B}\right)\dot{x}_j-\left(t^E_{ij}-M_iB_j-\mathbf{M}\cdot\mathbf{B}\,\delta_{ij}\right)\dot{x}_i$$
$$+\left(\mathbf{E}\times(\mathbf{B}/\mu_0-\mathbf{M})\right)_j+\left(\dot{x}_{i,k}Q_{kj}-\frac{\partial Q_{ji}}{\partial t}-\left(Q_{ji}\dot{x}_k\right)_{,k}\right)\mathcal{E}_i$$
$$-\frac{1}{J}\left(\sum_\nu\frac{\partial\rho^0\Sigma}{y^{T\nu}_{i,A}}\dot{y}^{T\nu}_i+\sum_\alpha\frac{\partial\rho^0\Sigma}{\partial s^{T\alpha}_{i,A}}\dot{s}^{T\alpha}_i\right)x_{j,A}\Bigg]=0. \qquad (40)$$

Note that the two types of magnetization (from charge motion and from spin) enter this conservation law differently even though each enters as a part of the interaction Lagrangian. Normally, as with the magnetization from charge motion, interaction Lagrangian terms disappear from the energy density. However, the spin magnetization appears here as if it had originated as a (negative) potential energy. Also note that the kinetic Lagrangian has disappeared from the energy density showing, as remarked earlier, that it is not a kinetic energy.

6. Conclusions

We showed how a classical Lagrangian theory of ferromagnetic, antiferromagnetic, or ferrimagnetic crystals can be constructed using the anticommuting Grassmann G_3 algebra to represent internal degrees of freedom of discrete particles in a crystal lattice. We also showed that this procedure fits in naturally with the long-wavelength average used to obtain a macroscopic, continuum theory that retains all of the degrees of freedom of the spin sublattices as well as of the internal optic modes of vibration. This theory is the proper expression of long-wavelength dynamics of intrinsic spin $\frac{1}{2}$ theory. Since nonrelativistic quantum mechanical spin theory has the same general form for any spin value, we surmise that this theory can apply to other spin values by an adjustment of the gyromagnetic ratio factor.

This theory shows that the origin of magnetization, whether from intrinsic spin or from the motion of bound charge, determines how its energy density or flow enters the energy conservation law. Experimental tests of this would be valuable. Because of the fundamental Grassmann algebra formulation of spin and the inclusion of all optic modes this theory should be uniquely qualified for magnetooptic applications.

Perhaps the overarching lesson of this work is not that intrinsic spin is a non-classical quantity but that classical physics has been wrongly restricted to commuting algebra and thus rendered incapable of accountimg for fermionic excitations.

Acknowledgments. Support of this work under National Science Foundation grant DMR-9315907 is gratefully acknowledged.

References

1. F. A. Berezin and M. S. Marinov, *JETP Lett.* **21** (1975) 320; *Ann. Phys.* **104** (1977) 336.
2. R. Casalbuoni, *Nuovo Cimento* **33** (1976) 115 and 389.
3. D. F. Nelson, *Electric, Optic, and Acoustic Interactions in Dielectrics*, Wiley, New York, 1979. Though out of print, paperback copies are available from the author.
4. W. F. Brown, Jr., *Magnetostatic Principles in Ferromagnetism*, North-Holland, Amsterdam, 1962; *Micromagnetics*, Interscience, New York, 1963; *Magnetoelastic Interactions*, Springer-Verlag, New York, 1966.
5. H. F. Tiersten, *J. Math. Phys.* **5** (1964) 1298 and **6** (1965) 779.
6. H. F. Tiersten and C. F. Tsai, *J. Math. Phys.* **13** (1972) 361.
7. G. A. Maugin and A. C. Eringen, *J. Math. Phys.* **13** (1972) 143 and 1334.
8. G. A. Maugin, *J. Math. Phys.* **17**(1976) 1727 and 1739.
9. C. F. Valenti and M. Lax, *Phys. Rev.* **B16** (1977) 4936.
10. D. F. Nelson, *Phys. Rev. Lett.* **60** (1988) 608.
11. D. F. Nelson, *Phys. Rev.* **A44** (1991) 3985.

Continuum Models and Discrete Systems
Proceedings of 8th International Symposium, June 11-16, 1995, Varna, Bulgaria, ed. K.Z. Markov
©World Scientific Publishing Company, 1996, pp. 49–54

SELF-CONSISTENT APPROACH TO MARTENSITIC ALLOYS

E. PATOOR, A. EBERHARDT, M. BERVEILLER
Laboratoire de Physique et Mécanique des Matériaux (URA CNRS 1215)
Institut Supérieur de Génie Mécanique et Productique, Université de Metz,
Ile du Saulcy, 57045 Metz , France.

Abstract : We propose a micromechanical modelling of the shape memory behavior in polycrystalline alloys. Global behavior is derived using a scale transition method. Local constitutive equation is obtained at the grain level from a kinematical description of the physical strain mechanisms and the definition of a local thermodynamical potential. Volume fractions of variants of martensite are chosen as internal variables. Good agreement is obtained between computation and experimental observations performed on Cu-based Shape Memory alloys for superelasticity.

1. Introduction

Thermomechanical properties of metals undergoing solid phase transformations are a growing field of interest related with the use of advanced materials such as TRIP Steels and Shape Memory Alloys. Macroscopic thermodynamic framework does not succeed to model accurately these behaviors. We propose to deal with a scale transition method issued from the field theory (integral equation of behavior, self-consistent approximation) developed for thermoelastic composite materials and elastoplastic metals.

Thermoelastic martensitic transformation at the origin of these behaviors produces an elementary strain mechanism. In shape memory alloys it is reasonable to assume this strain is the only inelastic one. In consequence study of this typical behavior constitute a good basis for further development dealing with more complex material like TRIP Steels. Taking account granular structure of the material and existence of several variants of martensite [1], the proposed approach is based on a kinematical description of the transformation strain mechanisms and a definition of local constitutive equations at the single crystal level. Volume fractions of the different variants of martensite are chosen as internal variables. Constitutive equations are defined in the single crystal case and applied to a polycrystalline material using a homogenization method. Results obtained by this way are in good agreement with experimental observations performed on Cu-based shape memory alloys. Such approach gives information on the microstructure evolution during the loading process and can be successfully applied to complex loading conditions.

2. Kinematical Description of the Transformation for a Single Crystal

In Shape Memory Alloys, the total strain field is composed by thermal and elastic components and by a large inelastic reversible transformation strain denoted by ε^T (one may consider that no plasticity occurs during a thermoelastic transformation). In the infinitesimal deformation framework these contributions act in an additive way. Assuming the elastic compliance tensor M is isotropic and uniform through the material and regarding the thermal expansion factor and the temperature field are also uniform, the global transformation strain E^T is defined over a crystal of volume V as :

$$E_{ij}^T = \frac{1}{V}\int_V \varepsilon_{ij}^T(r)\, dV \tag{2.1}$$

Assuming the transformation strain keeps a constant value inside each variant of martensite, the transformation strain field $\varepsilon^T(r)$ is considered as piecewise uniform.

$$\varepsilon_{ij}^T(r) = \sum_n \varepsilon_{ij}^n\, \theta^n(r) \quad \text{with} \quad \theta^n(r) = 1 \text{ if } r \in V^n \ (= 0 \text{ if } r \notin V^n) \tag{2.2}$$

where ε^n denotes the transformation strain inside a variant of volume V^n. At this microstructural level, ε^n is defined from measurement of the lattice parameters in both phases using a crystallographical theory [2]. Introducing the volume fraction ($f^n = V^n/V$) for each variant, the global transformation strain E^T is defined as :

$$E_{ij}^T = \frac{1}{V}\int_V \sum_n \varepsilon_{ij}^n\, \theta^n(r)\, dV = \sum_n \varepsilon_{ij}^n\, f^n \tag{2.3}$$

with some physical limitations exerted on f^n :

$$f^n \geq 0 \quad \text{and} \quad \sum_n f^n \leq 1 \tag{2.4}$$

Evolution with regard to the loading conditions of the transformation strain defined in Eq. (2.3) is obtained from a thermodynamical study.

3. Thermodynamical Aspects

Let us consider a reference volume V of parent phase such as given surface forces are large enough to transform a part V_M of this domain into martensite. This process involves elastic energy W_{elas}, chemical energy ΔG_{ch} and interfacial energy W_{int}. They are function of control parameters (applied stress Σ and temperature T) and a set of internal variables $\{\varepsilon^T(r)\}$ related to the transformation strain field. Complementary free energy Ψ is defined by :

$$\Psi(\Sigma_{ij}, T, \{\varepsilon^T(r)\}) = - [\ \Delta G_{ch} + W_{elas} + W_{int} - \Sigma_{ij}\ E_{ij}\ ] \tag{3.1}$$

Chemical energy depends on the temperature ; a linear approximation around the thermodynamical equilibrium temperature T_0 is commonly used [3].

$$\Delta G_{ch}\,(T) = B\,(T\text{-}T_0) \sum_n f^n \tag{3.2}$$

Oblate shape of martensite plates leads surface energy negligible in regard of the elastic one. Incompatibilities in the transformation strain field produce an internal stress field $\tau(r)$ that gives the following expression for the elastic energy :

$$W_{elas} = \frac{1}{2} \Sigma_{ij}\ M_{ijkl} \Sigma_{kl} - \frac{1}{2V} \int_V \tau_{ij}\,(r)\ \varepsilon^T_{ij}\,(r)\ dV \tag{3.3}$$

Micromechanical analysis [4] shown that contribution related with the internal stress field in Eq. (3.3) can be represented using an interaction matrix H^{nm}.

$$E_{int} = - \frac{1}{2V} \int_V \tau_{ij}(r)\,\varepsilon^T_{ij}(r)\ dV = \frac{1}{2} \sum_{n,m} H^{nm}\ f^n\ f^m \tag{3.4}$$

In fine variables f^n describe the evolution of the microstructure in crystal and are used as internal variables. According to Eqs. (3.2), (3.3) and (3.4) Eq. (3.1) is now expressed by:

$$\Psi(\Sigma_{ij},T,\ f^n) = \frac{1}{2} \Sigma_{ij}\ M_{ijkl}\ \Sigma_{kl} + \Sigma_{ij} \sum_n \varepsilon^n_{ij}\ f^n - B(T\text{-}T_0) \sum_n f^n - \frac{1}{2} \sum_{n,m} H^{nm}\ f^n f^m \tag{3.5}$$

To account with kinematical constraints defined in Eq. (2.4) Lagrange multipliers λ_0 and λ_n are introduced to define a Lagrangian functional $\mathcal{L}(\Sigma_{ij},T,\ f^n)$.

$$\mathcal{L}(\Sigma_{ij},T,\ f^n) = \Psi(\Sigma_{ij},T,\ f^n) - \lambda_0 [\sum_n f^n - 1] - \sum_n \lambda_n [-\ f^n\ - 0] \tag{3.6}$$

Thermodynamical forces are then obtained like

$$F^n = \frac{\partial \mathcal{L}}{\partial f^n} = \frac{\partial \Psi}{\partial f^n} - \lambda_0 + \lambda_n = \Sigma_{ij}\ \varepsilon^n_{ij} - B(T\text{-}T_0) - \sum_m H^{nm}\ f^m\ - \lambda_0 + \lambda_n \tag{3.7}$$

Despite the thermoelastic character of the martensitic transformation an hysteretic behavior is observed in Shape Memory Alloys. This phenomenon is related to dissipation process occurring during the transformation.

4. Constitutive Equations in the Single Crystal Case

The superelastic behavior is determined considering driving force must reach a critical value F_c to produce growing or shrinkage of a variant. This value, related to the microstructural state of the material, is assumed to be a positive material constant acting in the same way on each variant. Conditions to obtain a transformation are established like :

$$\dot{f}^n \neq 0 \qquad \text{if} \qquad F^n = F_c \qquad \text{and} \quad \dot{F}_c = \dot{F}^n = 0 \tag{4.1}$$

Assuming there are no coupling effect for dissipation on each variant second law of thermodynamics and energy balance give different expressions for direct and reverse transformation on each variant.

$$F^n \dot{f}^n = F_c \dot{f}^n \quad \text{(direct)} \qquad F^n \dot{f}^n = - F_c \dot{f}^n \quad \text{(reverse)} \tag{4.2}$$

From Eqs. (3.7) and (4.2) local transformation criteria are defined in both cases.

$$\Sigma_{ij} \varepsilon^n_{ij} - B(T-T_0) - \sum_m H^{nm} f^m - \lambda_0 + \lambda_n = {}^+_- F_c \tag{4.3}$$

The existence of kinematical constraints given by Eq. (2.4) impose to Lagrange multipliers involved in Eq. (4.3) to have a positive or null value, these give n+1 additional relations. If criteria (4.3) and constraints (2.4) are satisfied, the evolution of $\dot{f}^n$ is obtained using the consistency condition (4.1). Setting these quantities into the time derivative of the Eq. (2.3) defines a constitutive equation at the crystal level.

$$\dot{E}^T_{ij} = \sum_n \varepsilon^n_{ij} \sum_m (H^{nm})^{-1} [\, \varepsilon^m_{kl} \dot{\Sigma}_{kl} - B \dot{T} \,] \tag{4.4}$$

Eqs. (2.3), (2.4), (4.3) and (4.4) characterize the transformation plasticity for single crystal. These relations are used as local relationships in a polycrystalline model.

5. Self-Consistent Approximation

In polycrystalline materials, due to the existence of an internal stress field associated to the granular structure, an accurate determination of a thermodynamical potential is impossible. This difficulty can be solved using a homogenization method. Let us consider a homogeneous reference medium characterized by a set of uniform tangent modulus (L^o and M^o) and subject to a uniform strain E^o. Using the Green tensor method a localization expression between local and overall strain is defined on an integral form :

$$\dot{\varepsilon}_{mn}(r) = \dot{E}^o_{mn} + \int_V \Gamma^o_{mnij}(r-r') \, [\delta l^o_{ijkl}(r') \, \dot{\varepsilon}_{kl}(r') - \delta m^o_{ij}(r') \, \dot{T}] \, dV' \tag{5.1}$$

where Γ^o denotes the modified Green tensor and $\delta l(r)$ and $\delta m(r)$ the deviation parts from L^o and M^o of the local tangent moduli defined in Eq. (4.4). In this work, the self-consistent approximation is chosen to solve Eq. (5.1) [5]. This framework considers the effective overall medium as the homogeneous reference one and assumes tangent local modulii are piecewise uniform. Such approximations transform Eq. (5.1) into a concentration equation and allow to compute the effective behavior of the polycrystal [6].

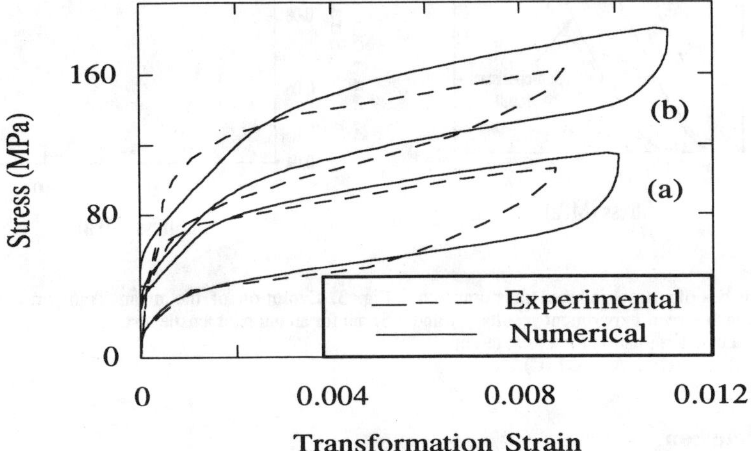

Fig. 1. Comparison between experimental results [7] and self-consistent determination (M_s = -97°C, A_f = -91 °C, T_a = -80°C, T_b = -70°C).

6. Results

Numerical applications need to characterize both the polycrystalline structure and the martensitic transformation. The granular aspect is described using spherical grains having orientation randomly chosen to induce no particular texture effect. Computation is applied to the superelastic behavior of Cu-Zn-Al Shape memory alloy. In these alloys, 24 variants of martensite forming six self-accommodated groups, are observed. WLR theory allows to define the transformation strain ε^n associated to these variants from the lattice parameters in both phases. Parameter B in Eq. (3.2) can be experimentally obtain from tensile test performed on single crystal. Critical force F_c used in Eq. (4.3) is deduced from measurement of the transformation temperatures in the austenitic phase (M_s and A_f). In these alloys micromechanical determination of the interaction matrix gives to two kinds of terms, weak one (μ/1000) for self accommodated variants and strong one (μ/150) for the other (μ denotes the elastic shear modulus). Result obtained in that way are successfully compared with experimental superelastic tensile test performed by P. Vacher [7] (figure 1).

Evolutions of some macroscopic variables are also computed as illustrated here for the amount of martensite (figure 2) and the mean transformation strain (figure 3). Two global parameters that are often used as internal variables in phenomenological models.

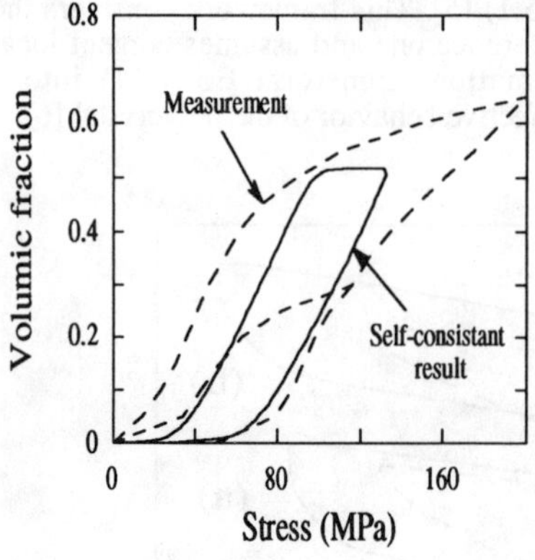

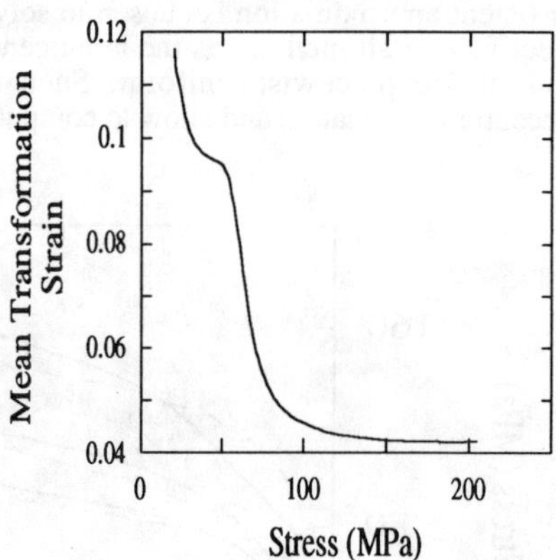

Fig. 2. Kinetics of a stress-induced transformation. Comparison between experiment results [7] and self-consistent determination for tensile test at T = -80°C (M_s = -97°C A_f = -91 °C).

Fig. 3. Evolution of the mean Transformation Strain for an uniaxial tensile test.

7. Conclusion

The micromechanical approach presented in this paper is in very good agreement with experimental observations. It is worth to notice that such agreement is obtained without any fitting parameter. The input data used in this framework are experimentally determined. The overall behavior is completely determined from the evolution of microstructural parameters (volumic fraction of variant of martensite) with respect to the loading conditions. Large predictive capacity is shown by this approach, for instance the dissymmetry experimentally observed between tensile and compressive test is well described. Extension of such result for other loading conditions allows determine a macroscopic transformation criterion on first importance for engineering applications.

References.

1. E. Patoor, A. Eberhardt, M. Berveiller, *Pitman Research Notes in Math. Series*, **296** (1993) 38.
2. M.S. Wechsler, D.S. Lieberman, T.A. Read, *Trans. AIME*, **197** (1953) 1503.
3. J. Ortin and A. Planes, *Acta Metall.*, **36** (1988) 1873.
4. O. Fassi Fehri, A. Hihi, M. Berveiller, *Scripta Met.*, **21** (1987) 771.
5. P. Lipinski and M. Berveiller, *Int.J. of Plasticity*, **5** (1989) 149.
6. E. Patoor, A. Eberhardt, M. Berveiller, *Proc. ASME WAM '94*, Chicago, IL (USA), AMD- Vol. 189 / PVD- Vol. 292, pp. 23-37 (1994).
7. P. Vacher and C. Lexcellent, *Procs. ICM 6, Kyoto*, Japan, 231 (1991).

Continuum Models and Discrete Systems
Proceedings of 8th International Symposium, June 11-16, 1995, Varna, Bulgaria, ed. K.Z. Markov
© World Scientific Publishing Company, 1996, pp. 55–72

MECHANICS AND TRANSPORT OF QUASIPARTICLES IN DEFORMABLE LATTICE STRUCTURES

D. I. PUSHKAROV
Institute of Solid State Physics, Bulgarian Academy of Sciences,
BG-1784 Sofia, Bulgaria

Abstract. Mechanics equations of Hamiltonian form for quasiparticles with arbitrary dispersion law in lattice structures subjected to time-varying deformations are deduced and transformations to replace Galilean ones are derived. A kinetic equation of Boltzmann type valid in the whole Brillouin zone is obtained. This equation contains an essentially new term which cannot be obtained in the previous (linear) theories and is responsible for inertial effects. An exact (in the quasiparticle approach) full selfconsistent set of nonlinear equations which consists of the elasticity theory equation, kinetic equation and Maxwell's equations is deduced. Some instructive applications are presented.

1. Introduction

The impressive success of the solid-state physics in the last half a century is to a great extent related to the quasiparticle approach. The main idea is that low energy excitations in solids can be considered as a gas of quasiparticles. Quasiparticles are not directly related to the structural units (e.g. atoms) of the body. Each type of them (magnons, phonons, etc.) corresponds to a given type of motion or interaction.

The main characteristic of the system—its energy spectrum (or *the quasiparticle dispersion law* $\varepsilon(\mathbf{k})$) depends on some conserved quantities (quantum numbers) the most important being the quasimomentum $\mathbf{k}$. The later appears due to the translational symmetry of the crystal lattice just in the same way as momentum in the classical mechanics appears due to homogeneity of space.

Quasiparticle approach is not a property only of the solid-state theory. It may be successfully applied to every system in which *elementary carriers of a phenomenon* do not coincide with the structural elements of the substance. Well known examples are theories of superfluidity, Fermi-liquid, etc. There is a fundamental difference, however, between continuum media on one hand, where momentum is well defined, and crystals, on the other hand, where momentum is a bad quantum number and the states are classified according to the values of the quasimomentum. As a consequence of the lattice periodicity all physical quantities (energy, momentum, mass-flux, etc.) are periodic functions of $\mathbf{k}$.

In such a way the quantum mechanical consideration replaces the real particles and the periodic lattice potential by new objects—quasiparticles with complicated dispersion law. This creates, however, *new problems*—one has to develop for quasiparticles *new mechanics, new scattering theory, new statistics, new kinetics, etc.* The crystal lattice becomes a privileged coordinate frame and therefore a problem arises about *the relationship between quasiparticle characteristics in different coordinate systems.* It is worth noting that, in virtue of the privileged role of the lattice, *there are no Galilean transformations for quasiparticles and hence quasiparticle* mechanics is not Galilean. The transformation laws are of fundamental importance not only for mechanics equations. Note that *all conservation laws* as well as *all fundamental physical equations* (including variational principles) exist only in a laboratory system while the co-moving frame set upon the lattice, in any real crystal, is not only deformed but moves noninertially with respect to it (e.g. due to lattice site vibrations, elastic waves etc.). On the other hand, all macroscopic quantities, such like densities of energy, momentum, etc., must obey Galilean transformations, independently of the fact that their origin in a microscopic level are not Galilean periodic functions.

2. Local Lattice Approach

As we pointed out in Introduction there are two main problems when dealing with quasiparticles. They both are created by the privileged role of the lattice. The first problem is put by the absence of *Galilean transformations.* The second one is due to the fact that every real crystal lattice is always deformed and, hence, quasimomentum and dispersion law are not well defined. The deformation length is, however, much larger than the size of the region within which the dispersion law is established. These two completely different scales allow us to introduce a local dispersion law $\varepsilon(\mathbf{k}, u_{ik})$ as a function of the small deformation tensor components $u_{ik}(\mathbf{r})$. In a linear approximation with respect to u_{ik} this yields

$$\varepsilon(\mathbf{k}, u_{ik}) = \varepsilon_0(\mathbf{k}) + \lambda_{ik} u_{ik}\,, \tag{1}$$

$$u_{ik} = \frac{1}{2}\left(\frac{\partial u_i}{\partial x_j} + \frac{\partial u_j}{\partial x_i}\right), \tag{2}$$

where $\lambda_{ij}(\mathbf{k})$ are the deformation potential constants and $\mathbf{u}$ is the deformation vector.

To avoid any misunderstanding, note that Eq. (1) is written in the co-moving frame attached to the shifted lattice sites. As a rule, this system moves noninertially with respect to the laboratory frame set upon the initial undeformed lattice. So the problem is how to transform quasiparticle characteristics from the co-moving frame to the laboratory one.

In order to take into account that the co-moving frame is not inertial, Landau has proposed (cf. the footnote in Ref. 2) to add in Eq. (1) the term $-m\dot{\mathbf{u}}\dfrac{\partial\varepsilon}{\partial\mathbf{k}}$, where $\dot{\mathbf{u}}$ is the local frame velocity. This yields:

$$\varepsilon(\mathbf{k}, u_{ij}) = \varepsilon_0(\mathbf{k}) + \lambda_{ij}(\mathbf{k}) u_{ij} - m\dot{\mathbf{u}}\frac{\partial\varepsilon}{\partial\mathbf{k}}\,. \tag{3}$$

This relation has been used by many authors. However, it cannot be well-grounded because of its internal inconsistency. In fact, if $\dot{\mathbf{u}} = \text{const}$, then Eq. (3) leads to the wrong conclusion that energy in an inertial frame may depend on the frame velocity. Note that Eq. (3) leads to the same confusion for the energy density which as a macroscopic quantity must strictly be in agreement with the Galilean principle.

A generalization of expression (1) can be made using a local dispersion law of the form $\varepsilon = \varepsilon(\mathbf{k}, g_{ij})$, where g_{ij} is the metrical tensor (cf. bellow) related to the full deformation tensor

$$w_{ij} = \frac{1}{2}\left(\frac{\partial u_i}{\partial x_j} + \frac{\partial u_j}{\partial x_i} + \frac{\partial u_k}{\partial x_i}\frac{\partial u_k}{\partial x_j}\right) \tag{4}$$

by the equality

$$w_{ij} = \frac{1}{2}(g_{ij} - g_{ij}^0) \tag{5}$$

and g_{ij}^0 corresponds to the undeformed lattice.

3. Notations

As it has been mentioned in the previous section, all physical quantities must be periodic functions of the quasimomentum and hence all fundamental equations have to be consistent with this condition. If we are going to use canonically conjugated variables, then we must take into account that quasimomentum has to be introduced as conjugate to some "discrete" coordinate taking into account the lattice periodicity.

The most convenient way to do this is to introduce a local frame using the primitive translation vectors of the crystal lattice $\mathbf{a}_\alpha(\mathbf{r}, t)$ ($\alpha = 1, 2, 3$) at a given point $\mathbf{r}$ at a given time t. In such a description each lattice site is determined by three integer numbers (discrete coordinates) N^α equal to the numbers of steps in the lattice in units $\mathbf{a}_\alpha$. Then the differential coordinates $d\mathbf{r}$, which is a physical infinitesimal (i.e. large in comparison with the lattice constant but small in comparison with the distance over which the properties of the lattice vary substantially) at a given time can be written in the form

$$d\mathbf{r} = \mathbf{a}_\alpha dN^\alpha. \tag{6}$$

The evolution of the primitive translation vectors $\mathbf{a}_\alpha$ in time can be obtained from plain geometrical considerations and may be expressed by the following equation:

$$\dot{\mathbf{a}}_\alpha + (\dot{\mathbf{u}}\nabla)\mathbf{a}_\alpha - (\mathbf{a}_\alpha\nabla)\dot{\mathbf{u}} = 0. \tag{7}$$

Eq. (7) describes deformations which do not break or cross crystalline lines with equal α. Actually, the condition of $\mathbf{a}_\alpha$-vector line conservation,

$$[\dot{\mathbf{a}}_\alpha + (\dot{\mathbf{u}}\nabla)\mathbf{a}_\alpha - (\mathbf{a}_\alpha\nabla)\dot{\mathbf{u}}] \times \mathbf{a}_\alpha = 0, \quad \alpha = 1, 2, 3, \tag{8}$$

is automatically satisfied.

Denote by $\mathbf{a}^\alpha(\mathbf{r}, t)$ the reciprocal lattice vectors, which satisfy the relations

$$\mathbf{a}_\alpha \mathbf{a}^\beta = \delta_\alpha^\beta, \quad a_{\alpha i} a_k^\alpha = \delta_{ik}. \tag{9}$$

Then it follows from (6) and (7) that

$$\mathbf{a}_\alpha = \frac{\partial \mathbf{r}}{\partial N^\alpha}, \quad \mathbf{a}^\alpha = \frac{\partial N^\alpha}{\partial \mathbf{r}} = \nabla N^\alpha. \tag{10}$$

Multiplying (7) by $\mathbf{a}^\alpha$ and using (9) and (10) yields the evolution equation for the reciprocal lattice vectors:

$$\dot{\mathbf{a}}^\alpha + \nabla(\mathbf{a}^\alpha \dot{\mathbf{u}}) = 0. \tag{11}$$

In the same way the deformation velocity $\dot{\mathbf{u}}$ may be expressed by the derivatives of N^α:

$$\dot{\mathbf{u}} = -\mathbf{a}_\alpha \dot{N}^\alpha. \tag{12}$$

Using Eqs. (10)–(12) yields

$$dN^\alpha = \nabla N^\alpha d\mathbf{r} + \dot{N}^\alpha dt = \mathbf{a}^\alpha d\mathbf{r} - \mathbf{a}^\alpha \dot{\mathbf{u}}\, dt \tag{13}$$

and hence

$$d\mathbf{r} = \mathbf{a}_\alpha dN^\alpha + \dot{\mathbf{u}}\, dt. \tag{14}$$

Eqs. (10) and (11) show that three functions $N^\alpha(\mathbf{r}, t)$ completely determine the configuration and velocity of the lattice.

In the same notation the metrical tensors in the real and reciprocal space can be written in the form:

$$g_{\alpha\beta} = \mathbf{a}_\alpha \mathbf{a}_\beta \quad g^{\alpha\beta} = \mathbf{a}^\alpha \mathbf{a}^\beta. \tag{15}$$

Then the lattice cell volume is equal to $g^{1/2}$, where $g = \det \mid g_{\alpha\beta} \mid$.

4. Deformation Tensor

Deformation tensor components w_{ik} may be obtained using the relation (6). The squared distance ds^2 is

$$ds^2 = g_{\alpha\beta} dN^\alpha dN^\beta. \tag{16}$$

The interval between the same points before deformation is

$$ds_0^2 = \overset{0}{g}_{\alpha\beta} dN^\alpha dN^\beta. \tag{17}$$

Hence,

$$ds^2 - ds_0^2 = 2 w_{\alpha\beta} dN^\alpha dN^\beta, \tag{18}$$

where

$$w_{\alpha\beta} = \frac{1}{2}\left(g_{\alpha\beta} - \overset{0}{g}_{\alpha\beta}\right) \tag{19}$$

plays the role of a deformation tensor in our notation.

Due to the invariance of the interval we have also the identity

$$w_{\alpha\beta} dN^\alpha dN^\beta = w_{ik} dx^i dx^k \tag{20}$$

and, therefore (cf. (10)),

$$w_{ik} = w_{\alpha\beta} a_i^{\alpha} a_k^{\beta}. \tag{21}$$

Let us denote by u_{ik} the tensor of small deformations (2) used in the linear theory of elasticity, where $\mathbf{u}$ is the deformation vector $\mathbf{u} = \mathbf{r} - \mathbf{r}_0$. Obviously, the quantities $\partial u_i / \partial x_k$ coincide with the matrix elements α_{ik} which describe the coordinate transformations

$$x_i = \overset{\circ}{x}_i + \alpha_{ik} \overset{\circ}{x}_k \tag{22}$$

and hence, the transformations of lattice vectors a_α:

$$a_{\alpha i} = \overset{\circ}{a}_{\alpha i} + \alpha_{ik} \overset{\circ}{a}_{\alpha k} \,. \tag{23}$$

To express u_{ik} and $\mathbf{u}$ in terms of $\mathbf{a}_\alpha$ and N^α let us write

$$\mathbf{a}_\alpha = \overset{\circ}{\mathbf{a}}_\alpha + \delta \mathbf{a}_\alpha \tag{24}$$

$$\mathbf{a}_\beta = \overset{\circ}{\mathbf{a}}_\beta + \delta \mathbf{a}_\beta . \tag{25}$$

Multiplying (24) and (25) and taking into account relations (9) yield in a linear approximation with respect to $\delta \mathbf{a}^\alpha$ and $\delta \mathbf{a}_\alpha$:

$$\delta a_{\alpha i} = -(\overset{\circ}{a}_{\beta i}\, \delta a_k^{\beta})\, \overset{\circ}{a}_{\alpha k} \,. \tag{26}$$

On the other hand the quantities N^α in a deformed lattice can be written in the form

$$N^\alpha = N_0^\alpha - w^\alpha , \tag{27}$$

where $N_0^\alpha = \overset{\circ}{\mathbf{a}}{}^{\alpha} \cdot \mathbf{r}$ and $\overset{\circ}{\mathbf{a}}{}^{\alpha}$ are the reciprocal lattice vectors of the undeformed crystal and w^α are the deviations of quantities N^α with respect to their values in the undeformed lattice. Hence [4,5]

$$\mathbf{a}_\alpha = \nabla N^\alpha = \overset{\circ}{\mathbf{a}}_\alpha - \nabla w^\alpha \tag{28}$$

and

$$\delta a_i^\alpha = \frac{\partial w^\alpha}{\partial x_i} . \tag{29}$$

Comparing (23), (26) and (29) yields

$$\alpha_{ik} = \overset{\circ}{a}_{\alpha i} \frac{\partial w^\alpha}{\partial x_k} = \frac{\partial u_i}{\partial x_k} \tag{30}$$

and therefore $\mathbf{u} = \overset{\circ}{\mathbf{a}}_\alpha \, w^\alpha$ coincides with the deformation vector. Substituting (28) into (21) one obtains

$$u_{ik} = \overset{\circ}{w}_{\alpha\beta} \overset{\circ}{a}{}_i^{\alpha} \overset{\circ}{a}{}_k^{\beta} , \tag{31}$$

where $\overset{\circ}{w}_{\alpha\beta}$ are obtained as a linear approximation of (19) with respect to small deviations $\delta\mathbf{a}_\alpha$ and $\delta\mathbf{a}^\alpha$:

$$\overset{\circ}{w}_{\alpha\beta} = \frac{1}{2}\left(\overset{\circ}{\mathbf{a}}_\alpha\, \delta\mathbf{a}_\beta + \overset{\circ}{a}_\beta\, \delta\mathbf{a}_\alpha\right) = \frac{1}{2}\left(\frac{\partial w_\alpha}{\partial N^\beta} + \frac{\partial w_\beta}{\partial N^\alpha}\right) \tag{32}$$

and $w_\alpha = \overset{\circ}{g}_{\alpha\beta}\, w^\beta$ are the covariant components of $\mathbf{w}$.

In the same way one may obtain the full deformation tensor [1]:

$$w_{\alpha\beta} = \frac{1}{2}\left(\overset{\circ}{\mathbf{a}}_\alpha \frac{\partial \mathbf{u}}{\partial N^\beta} + \overset{\circ}{\mathbf{a}}_\beta \frac{\partial \mathbf{u}}{\partial N^\alpha} + \frac{\partial \mathbf{u}}{\partial N^\alpha}\frac{\partial \mathbf{u}}{\partial N^\beta}\right). \tag{33}$$

From here on we shall not use these cumbersome formulae. They are only listed here in order to show the relationship between the two kinds of notations. An advantage of our notation is the possibility to avoid the description by means of deformation tensor components in their explicit form.

5. Quasiparticle Mechanics

In the co-moving lattice frame (C-system) the dispersion law $\varepsilon(\mathbf{k}, g^{\alpha\beta})$ coincides with both the energy and the Hamiltonian of the quasiparticle. Hence, the Hamiltonian equations in C-system have the form:

$$\dot{\mathbf{r}}' = \frac{\partial \varepsilon}{\partial \mathbf{k}}, \quad \dot{\mathbf{k}} = -\frac{\partial \varepsilon}{\partial \mathbf{r}'}; \tag{34}$$

$\mathbf{r}'$ are the coordinates in the C-system.

We are interested in the equations of motion in the laboratory system (L-system)

$$\dot{\mathbf{r}} = \frac{\partial H}{\partial \mathbf{p}}, \quad \dot{\mathbf{p}} = -\frac{\partial H}{\partial \mathbf{r}}, \tag{35}$$

where the Hamiltonian function $H(\mathbf{p}, \mathbf{r}, t)$ as well as the quantity $\mathbf{p}$ are to be determined. Our aim is to express the unknown $H(\mathbf{p}, \mathbf{r}, t)$ and $\mathbf{p}$ in terms of the known $\varepsilon(\mathbf{k}, g^{\alpha\beta})$ and $\mathbf{k}$. It is worth noting that Hamiltonian function is, in general, not unique, and hence the sense of the quantity $\mathbf{p}$ depends on the choice of that function. In addition to that the Hamiltonian in our case depends on time, and therefore does not coincide with the energy. Finally, $H(\mathbf{p}, \mathbf{r}, t)$ may not be a periodic function with respect to $\mathbf{p}$ because the Hamiltonian is not a physical quantity and the only condition required is to produce correct equations of motion.

Before we determine the Hamiltonian $H(\mathbf{p}, \mathbf{r}, t)$ and the energy E of an excitation in L-system, we note the following. In the theory which we are constructing, we assume that the quantities $\mathbf{a}_\alpha$ and $\mathbf{v}$ ($\mathbf{v} \equiv \dot{\mathbf{u}}$ is introduced for convenience) vary slowly in space and time, and we are expanding in derivatives of these quantities. In the expressions for $H(\mathbf{p}, \mathbf{r}, t)$ and E it is then sufficient to retain only terms of lower order. In this approximation, H and E are the same as their values in a lattice with constant but otherwise arbitrary vectors $\mathbf{a}_\alpha$ and $\mathbf{v}$, i.e. in a crystal which is uniformly deformed (periodic) and in uniform motion.

Let us first consider an electron in the co-moving frame. The wave function of an electron which belongs to a definite energy band and which is executing a semiclassical motion must be treated as a function of the discrete coordinates N^α. It has the form:

$$\psi(N^\alpha, t) \propto \exp\left\{\frac{i}{\hbar} S_0(N^\alpha, t)\right\}, \tag{36}$$

where S_0 is the classical action. In the immobile periodic lattice the Hamiltonian, H_0, is known to be equal to the energy $\varepsilon = \varepsilon(k_\alpha, g^{\alpha\beta})$ which is a periodic function of the components of the *invariant quasimomentum* k_α with periods $2\pi\hbar$ and which depends on the invariant characteristics of the unit cell, determined by the quantities $g^{\alpha\beta}$. Hence, the Hamiltonian and the new invariant quasimomentum which is conjugate to N^α can be obtained as derivatives of the action:

$$\left(\frac{\partial S_0}{\partial t}\right)_{N^\alpha} = -\varepsilon(k_\alpha, g^{\alpha\beta}), \quad \left(\frac{\partial S_0}{\partial N^\alpha}\right)_t = k_\alpha \,. \tag{37}$$

At $\mathbf{v} \neq 0$ the action S can be found by means of the well known laws by which the wave functions of a real particle (e.g. an electron) transform under Galilean transformations [6]:

$$S = S_0 + m\mathbf{v}\mathbf{r} - mv^2/2, \tag{38}$$

where m is the mass of a free particle.

The new variable $\mathbf{p}$ canonically conjugate to the coordinates $\mathbf{r}$ can be obtained differentiating S [1,3-5]:

$$\mathbf{p} = \left(\frac{\partial S}{\partial \mathbf{r}}\right)_t = k_\alpha \nabla N^\alpha + m\mathbf{v} = \mathbf{a}^\alpha k_\alpha + m\mathbf{v} \tag{39}$$

and thus

$$k_\alpha = \mathbf{a}^\alpha(\mathbf{p} - m\mathbf{v}). \tag{40}$$

The Hamiltonian is now obtained differentiating S with respect to t [1,3-5]:

$$H(\mathbf{p}, \mathbf{r}, t) = -\left(\frac{\partial S}{\partial t}\right)_r = -k_\alpha \dot{N}^\alpha + \varepsilon + \frac{mv^2}{2} = \varepsilon + \mathbf{p}\mathbf{v} - \frac{mv^2}{2}, \tag{41}$$

where $\varepsilon = \varepsilon(\mathbf{a}_\alpha(\mathbf{p} - m\mathbf{v}), g^{\alpha\beta})$ is a periodic function of $\mathbf{p}$ with periods $2\pi\hbar\mathbf{a}^\alpha$ determined by the local values of the reciprocal-lattice vectors. This periodicity is the reason to call $\mathbf{p}$ quasimomentum of the quasiparticle in L-system.

The energy E of the quasiparticle in the L-system is given by the Galilean transformation

$$E = mv^2/2 + \mathbf{p}_0\mathbf{v} + \varepsilon, \tag{42}$$

where $\mathbf{p}_0 = m\partial\varepsilon/\partial\mathbf{p}$ is the average value of the momentum (mass flow) in the system with $\mathbf{v} = 0$.

Thus

$$E = \varepsilon + m\mathbf{v}\frac{\partial\varepsilon}{\partial\mathbf{p}} + \frac{mv^2}{2}. \tag{43}$$

A comparison of (41) and (43) yields:

$$H(\mathbf{p},\mathbf{r},t) = E(\mathbf{p},\mathbf{r},t) + \left(\mathbf{p} - m\frac{\partial\varepsilon}{\partial\mathbf{p}}\right)\mathbf{v} - mv^2. \tag{44}$$

Hence, the Hamilton equations of motion in L-system are

$$\dot{\mathbf{r}} = \frac{\partial H}{\partial\mathbf{p}}, \quad \dot{\mathbf{p}} = -\frac{\partial H}{\partial\mathbf{r}}. \tag{45}$$

The transformation formulae (40)-(42) are to replace the Galilean transformations.

To avoid any misunderstanding note that in the previous works [2,7-9] the following relations have been obtained:

$$\mathbf{k} = \mathbf{p} + \frac{\partial}{\partial\mathbf{r}'}(\mathbf{u}\mathbf{p}), \quad \tilde{\varepsilon}(\mathbf{k},\mathbf{r}',t) = \tilde{H}(\mathbf{p},\mathbf{r},t) - \dot{\mathbf{u}}\mathbf{p}. \tag{46}$$

It can be shown that the Hamiltonian defined in (46) is inconsistent with the kinetic equation. The transformation (46) does not contain the bare mass of the electron and hence cannot describe inertial effects like the Stewart-Tolman effect in metals. This difficulty has been overcome in [2,7-9] by somewhat artificial methods: by transforming to a noninertial co-moving coordinate frame and by essentially postulating the form of the noninertial terms in the electron Hamiltonian (cf. the discussion following Eq. (3)).

Having the Hamiltonian and the relationship between quasimomenta $\mathbf{p}$ and k_α *we actually have constructed the quasiparticle mechanics in deformable solids.* The only condition which must be satisfied for its applicability is that the properties of the lattice should be slowly varying functions of coordinates and time at the scale of interatomic distances and atomic times.

A linearization of the Hamiltonian (44) by means of the formulae (27)–(29) yields

$$H = \varepsilon_0(\mathbf{p}) + \left(\lambda_{ik}(\mathbf{p}) + p_i\frac{\partial\varepsilon_0}{\partial p_k}\right)\frac{\partial u_k}{\partial x_i} + \dot{\mathbf{u}}\left(\mathbf{p} - m\frac{\partial\varepsilon_0}{\partial\mathbf{p}}\right). \tag{47}$$

Here $\varepsilon(\mathbf{p})$ and $\lambda_{ik}(\mathbf{p})$ are the values of the function and of its derivatives with respect to u_{ik} at $u_{ik} = 0$, respectively.

Expression (47) was suggested by Landau and then used by many authors (cf. [7-9] and the literature cited there). We see that the term $-m\partial\varepsilon/\partial\mathbf{p}$ proposed by Landau leads to a correct result in a linear approximation if expression (3) is formally considered as a Hamiltonian. However, this is in conflict with the basic conception that the dispersion law in the co-moving frame coincides both with energy and with Hamiltonian. This conception was essentially used when writing Eqs. (34). Note also that the coincidence mentioned is valid only as concerned the form of the Hamiltonian. The transformation laws for the quasimomentum are different. Due to this chance, however, the results obtained by previously are correct in a linear approximation.

6. Kinetic Equation

The kinetic equation for the quasiparticle distribution function $f(\mathbf{p},\mathbf{r},t)$ in L-system may be written in the usual form:

$$\frac{\partial f}{\partial t}+\frac{\partial f}{\partial \mathbf{r}}\frac{\partial H}{\partial \mathbf{p}}-\frac{\partial f}{\partial \mathbf{p}}\left(\frac{\partial H}{\partial \mathbf{r}}-\mathbf{F}\right)=\hat{I}f, \tag{48}$$

where $\hat{I}$ is the collision operator. Note, that this equation becomes well defined for quasiparticles only after obtaining the Hamiltonian (41) and the Hamiltonian equations.

Let us show that kinetic equation (48) is valid in the whole Brillouin zone. This means that it is compatible with the periodicity condition for the partition function $f(\mathbf{p},\mathbf{r},t)=f(\mathbf{p}+2\pi\hbar\mathbf{a}^{\alpha}(\mathbf{r},t),\mathbf{r},t)$. This is not obvious because Eq. (48) contains aperiodic terms $\left(\dfrac{\partial f}{\partial \mathbf{p}}\dfrac{\partial H}{\partial \mathbf{r}}\right)$. Of course, one may check the compatibility of the kinetic equation with the periodicity condition by a direct substitution of (41) into (48). Then all the additional terms cancel each other owing to the evolution equation (11) and one sees that $f(\mathbf{p}+2\pi\hbar\mathbf{a}^{\alpha}(\mathbf{r},t),\mathbf{r},t)$ satisfies the same equation as $f(\mathbf{p},\mathbf{r},t)$. Instead, we shall write the kinetic equation in a form which does not contain aperiodic terms at all. This can be done if the Hamiltonian (41) is written as a function of k_α:

$$H=\varepsilon(k_\alpha,g^{\alpha\beta})+k_\alpha\mathbf{a}^{\alpha}\dot{\mathbf{u}}+\frac{1}{2}m\dot{\mathbf{u}}^2\,. \tag{49}$$

Then the kinetic equation for the distribution function $f(k_\alpha,\mathbf{r},t)$ takes the form

$$\frac{Df}{Dt}+\mathbf{a}_\alpha\frac{\partial\varepsilon}{\partial k_\alpha}\left(\frac{\partial f}{\partial \mathbf{r}}\right)_{k_\alpha}-\mathbf{a}_\alpha\frac{\partial f}{\partial k_\alpha}\left\{m\frac{D\dot{\mathbf{u}}}{Dt}+\left(\frac{\partial\varepsilon}{\partial \mathbf{r}}\right)_{k_\alpha}-m\frac{\partial\varepsilon}{\partial k_\beta}\left[\mathbf{a}_\beta\,\mathrm{rot}\,\dot{\mathbf{u}}\right]-\mathbf{F}\right\}=\hat{I}f, \tag{50}$$

where $\dfrac{D}{Dt}=\dfrac{\partial}{\partial t}+(\dot{\mathbf{u}}\nabla)$ and all quantities are differentiated with respect to the coordinates and the time at constant k_α .

The term $m\dfrac{D\dot{\mathbf{u}}}{Dt}$ takes into account noninertial properties of the local frame. This is the term which is responsible, e.g., for the Stewart-Tolman effect in metals. It is seen that the inertial force is proportional to the bare mass of the particle.

The term

$$m\frac{\partial\varepsilon}{\partial k_\beta}\left[\mathbf{a}_\beta\,\mathrm{rot}\,\dot{\mathbf{u}}\right] \tag{51}$$

is of *essentially new kind* and cannot be obtained in linear theories. It is proportional to the bare mass m and, hence, is also responsible for noninertial effects. An effect owing to this term will be considered in Sec. 8.

It follows from the form of the Hamiltonian that the velocity $\frac{\partial H}{\partial \mathbf{p}}$ satisfies the periodicity condition too. Hence, external forces $\mathbf{F}$ which depend on the velocity and its derivatives are also permissible. A force of this kind is, e.g. the Lorentz force:

$$\mathbf{F} = -e\mathbf{E} - \frac{e}{c}\frac{\partial \varepsilon}{\partial k_\alpha}[\mathbf{a}_\alpha \mathbf{B}] - \frac{e}{c}[\dot{\mathbf{u}}\mathbf{B}], \tag{52}$$

where $\mathbf{E}$ and $\mathbf{B}$ are the strengths of the electric and magnetic fields, respectively. Substituting (52) into (50) yields the kinetic equation for charged quasiparticles with a charge $-e$:

$$\frac{Df}{Dt} + \mathbf{a}_\alpha \left(\frac{\partial f}{\partial \mathbf{r}}\right)_{k_\alpha} \frac{\partial \varepsilon}{\partial k_\alpha} - \frac{\partial f}{\partial k_\alpha}\left\{\mathbf{a}_\alpha \left(\frac{\partial \varepsilon}{\partial \mathbf{r}}\right)_{k_\alpha} + \mathbf{a}_\alpha e\mathbf{E}' + \frac{e}{c}\frac{\partial \varepsilon}{\partial k_\beta}[\mathbf{a}_\alpha \mathbf{a}_\beta]\,\mathbf{B}'\right\} = \hat{I}f, \tag{53}$$

where

$$e\mathbf{E}' = e\mathbf{E} + m\frac{D\dot{\mathbf{u}}}{Dt} + \frac{e}{c}[\dot{\mathbf{u}}\mathbf{B}], \quad \mathbf{B}' = \mathbf{B} - \frac{mc}{e}\operatorname{rot}\dot{\mathbf{u}}\,.$$

Further it will be necessary to integrate physical quantities over the Brillouin zone, to transform such integrals by parts as well as to perform differentiation with respect to time and space. However, the zone boundaries are moving under the time-varying deformation and are dependent not only on the deformation in a given instant but also on the velocity of the lattice. As a result the integration over the Brillouin zone does not commutate with the differentiation with respect to $\mathbf{r}$ and t. Due to this non-commutativity some fluxes appear through the zone boundaries. This effect can be of great importance in non-equilibrium systems as well as at high temperatures compared to the energy band width. In such cases typical of defecton gas in quantum crystals as well as in metals with open Fermi surfaces the distribution function values on the boundaries are non-vanishing and the non-commutativity may not be neglected. This inconvenience can be eliminated by introducing the renormalised distribution function

$$\varphi = f\sqrt{g}. \tag{54}$$

The kinetic equation for $\varphi(k_\alpha, \mathbf{r}, t)$ has the form

$$\begin{aligned}\dot{\varphi} \;+\; \operatorname{div}\left\{\left(\dot{\mathbf{u}} + \mathbf{a}_\alpha \frac{\partial \varepsilon}{\partial k_\alpha}\right)\varphi\right\} - \frac{\partial}{\partial k_\alpha}\left\{\varphi \mathbf{a}_\alpha \left(\nabla \varepsilon + m\frac{D\dot{\mathbf{u}}}{Dt}\right)\right. \\ \left. -\; m\varphi \frac{\partial \varepsilon}{\partial k_\beta}[\mathbf{a}_\alpha \mathbf{a}_\beta]\operatorname{rot}\dot{\mathbf{u}} - \mathbf{a}_\alpha \mathbf{F}\varphi\right\} = \hat{I}\varphi.\end{aligned} \tag{55}$$

In this notation the differentiation with respect to t and $\mathbf{r}$ is carried out at constant k_α and hence commutates with $\int d^3k_\alpha$. Results obtained by such a procedure can easily be rewritten in the previously adopted variables by the following substitutions

$$\begin{aligned}\langle f\ldots\rangle &\equiv \int \frac{d^3p}{(2\pi\hbar)^3} f(\mathbf{p},\mathbf{r},t)\ldots = \frac{1}{\sqrt{g}}\int \frac{d^3k_\alpha}{(2\pi\hbar)^3} f(\mathbf{p},\mathbf{r},t)\ldots \\ &= \int \frac{d^3k_\alpha}{(2\pi\hbar)^3}\varphi\ldots \equiv \langle\langle\varphi\rangle\rangle\,, \end{aligned} \tag{56}$$

$$\int d^3r\ldots = \int d^3N^\alpha g^{1/2}\ldots \tag{57}$$

It follows from the above considerations that correct results may be obtained also directly from Eq. (48) or (50), if all fluxes through the Brillouin zone boundaries appearing as a result of integrating by parts formally equal zero in spite of the non-periodic form of the integrand.

As an illustration, let us calculate the time derivative of the energy

$$\frac{\partial}{\partial t}\langle\varepsilon f\rangle = \langle\langle\dot\varepsilon\varphi\rangle\rangle + \langle\langle\varepsilon\dot\varphi\rangle\rangle,$$

where

$$\dot\varepsilon = \left(\frac{\partial\varepsilon}{\partial t}\right)_{k_\alpha} = \frac{1}{2}\lambda_{\alpha\beta}\dot g^{\alpha\beta} = -\lambda_{\alpha\beta}a_i^\alpha a_k^\beta\frac{\partial\dot u_i}{\partial x_k} - \dot u(\nabla\varepsilon)_{k_\alpha}\,, \tag{58}$$

$$\lambda_{\alpha\beta} = 2\frac{\partial\varepsilon}{\partial g^{\alpha\beta}} = \lambda_{\beta\alpha} \tag{59}$$

and $\dot\varphi$ is given by (55).

Consequently,

$$\begin{aligned}&\frac{\partial}{\partial t}\langle\varepsilon f\rangle = \\ &= \langle\langle\varepsilon\hat I\varphi\rangle\rangle - \operatorname{div}\left(\dot{\mathbf{u}}\langle\langle\varepsilon\varphi\rangle\rangle + \mathbf{a}_{\alpha\beta}\left\langle\left\langle\varphi\varepsilon\frac{\partial\varepsilon}{\partial k_\alpha}\right\rangle\right\rangle\right) + \mathbf{a}_\alpha\left\langle\left\langle\varphi\mathbf{F}\frac{\partial\varepsilon}{\partial k_\alpha}\right\rangle\right\rangle \\ &- m\ddot{\mathbf{u}}\mathbf{a}_\alpha\left\langle\left\langle\frac{\partial\varepsilon}{\partial k_\alpha}\varphi\right\rangle\right\rangle - \frac{\partial u_i}{\partial x_k}\left(m\dot u_k a_{\alpha i}\left\langle\left\langle\frac{\partial\varepsilon}{\partial k_\alpha}\varphi\right\rangle\right\rangle + a_i^\alpha a_k^\beta\langle\langle\varphi\lambda_{\alpha\beta}\rangle\rangle\right) \\ &= \langle\varepsilon\hat I f\rangle - \operatorname{div}\left(\dot{\mathbf{u}}\langle\varepsilon f\rangle + \left\langle\varepsilon\frac{\partial\varepsilon}{\partial\mathbf{p}}f\right\rangle\right) + \left\langle\mathbf{F}\frac{\partial\varepsilon}{\partial\mathbf{p}}f\right\rangle - m\ddot{\mathbf{u}}\left\langle\frac{\partial\varepsilon}{\partial\mathbf{p}}f\right\rangle \\ &- \frac{\partial\dot u_i}{\partial x_k}\left(m\dot u_k\left\langle f\frac{\partial\varepsilon}{\partial p_i}\right\rangle + a_i^\alpha a_k^\beta\langle\lambda_{\alpha\beta}f\rangle\right).\end{aligned} \tag{60}$$

7. Conservation Laws and Dynamics Equations for Charged Quasiparticles

In this Section we shall derive a full selfconsistent set of dynamical equations from conservation laws following a standard procedure. The continuity equation for quasiparticles follows directly from the kinetic equation and has the form:

$$m\dot n + \operatorname{div}\mathbf{J}_0 = 0, \quad n = \langle\langle\varphi\rangle\rangle = \langle f\rangle, \tag{61}$$

$$\mathbf{J}_0 = mn\dot{\mathbf{u}} + \mathbf{j}_0\,, \quad \mathbf{j}_0 = m\left\langle \frac{\partial \varepsilon}{\partial \mathbf{p}} f \right\rangle. \tag{62}$$

The total mass current is

$$\mathbf{J} = \mathbf{J}_0 + \mathbf{j} = \rho\dot{\mathbf{u}} + \mathbf{j}_0, \tag{63}$$

where

$$\rho = \rho_0 + mn \tag{64}$$

is the density of the crystal. The quantities ρ and J satisfy the mass continuity equation

$$\dot{\rho} + \operatorname{div}\mathbf{J} = 0. \tag{65}$$

The full momentum $\mathbf{J}'$ is a sum of $\mathbf{J}$ and the electromagnetic field momentum $\mathbf{G}$:

$$\mathbf{J}' = \mathbf{J} + \mathbf{G}. \tag{66}$$

Note, that in this case *the full momentum does not coincide with the mass flow!*

Our aim is to determine momentum and energy fluxes Π_{ik} and Q in such a way as to satisfy the continuity equation (65), the momentum conservation law

$$\dot{J}_i' + \nabla_k \Pi_{ik} = 0, \tag{67}$$

the energy conservation law

$$\dot{E} + \operatorname{div}\mathbf{Q} = 0 \tag{68}$$

and Maxwell's equations

$$\operatorname{rot}\mathbf{E} = -\frac{1}{c}\frac{\partial \mathbf{B}}{\partial t}, \quad \operatorname{rot}\mathbf{B} = \frac{4\pi}{c}\mathbf{j}_e + \frac{1}{c}\frac{\partial \hat{\varepsilon}\mathbf{E}}{\partial t}, \tag{69}$$

$$\operatorname{div}(\hat{\varepsilon}\mathbf{E}) = 4\pi(q - en), \quad \operatorname{div}\mathbf{B} = 0, \quad \mathbf{j}_e = \frac{e}{m}\mathbf{j}_0, \quad e > 0. \tag{70}$$

The energy in L-system is given by the expression

$$E = -\frac{1}{2}\rho_0\dot{\mathbf{u}}^2 + E_0(g^{\alpha\beta}) + \langle\langle E\varphi\rangle\rangle + W, \tag{71}$$

where $E_0(g^{\alpha\beta})$ is the strain energy in C-system, and

$$W = \frac{\mathbf{E}^2 + B^2}{8\pi}. \tag{72}$$

The time derivative of the energy (71) is then

$$\begin{aligned}
\dot{E} &= \rho\dot{\mathbf{u}}\ddot{\mathbf{u}} + \frac{1}{2}\dot{\rho}\dot{u}^2 + \sigma_{\alpha\beta}a_i^\alpha a_k^\beta \frac{\partial \dot{u}_i}{\partial x_k} - \dot{\mathbf{u}}\nabla E_0 + \frac{\partial}{\partial t}\langle\langle\varepsilon\varphi\rangle\rangle + m\ddot{\mathbf{u}}\mathbf{a}_\alpha\left\langle\left\langle\varphi\frac{\partial\varepsilon}{\partial k_\alpha}\right\rangle\right\rangle \\
&+ m\dot{\mathbf{u}}\dot{\mathbf{a}}_\alpha\left\langle\left\langle\varphi\frac{\partial\varepsilon}{\partial k_\alpha}\right\rangle\right\rangle + m\dot{\mathbf{u}}\mathbf{a}_\alpha\left\langle\left\langle\varphi\frac{\partial\dot{\varepsilon}}{\partial k_\alpha}\right\rangle\right\rangle + m\dot{\mathbf{u}}\mathbf{a}_\alpha\left\langle\left\langle\dot{\varphi}\frac{\partial\varepsilon}{\partial k_\alpha}\right\rangle\right\rangle + \dot{W}.
\end{aligned} \tag{73}$$

On the other hand the time derivative of the momentum (66) gives

$$0 = -\mathbf{J}' + \rho\ddot{\mathbf{u}} + m\dot{\mathbf{a}}_\alpha \left\langle\left\langle \frac{\partial \varepsilon}{\partial k_\alpha}\varphi \right\rangle\right\rangle + m\mathbf{a}_\alpha \left\langle \frac{\partial \varepsilon}{\partial k_\alpha}\dot{\varphi} \right\rangle + m\mathbf{a}_\alpha \left\langle\left\langle \frac{\partial \dot{\varepsilon}}{\partial k_\alpha}\varphi \right\rangle\right\rangle + \dot{\mathbf{G}}. \quad (74)$$

Multiplying (74) by $-\dot{\mathbf{u}}$ and adding the result to the right side of (73) yield

$$\begin{aligned} \dot{E} &= \dot{\mathbf{u}}\dot{\mathbf{J}}' - \frac{1}{2}\dot{\rho}\dot{\mathbf{u}}^2 + \sigma_{\alpha\beta}a_i^\alpha a_k^\beta \frac{\partial \dot{u}_i}{\partial x_k} - m\ddot{\mathbf{u}}\mathbf{a}_\alpha \left\langle\left\langle \varphi \frac{\partial \varepsilon}{\partial k_\alpha} \right\rangle\right\rangle - \dot{\mathbf{u}}\nabla E_0 \\ &+ \dot{W} - \dot{\mathbf{u}}\dot{\mathbf{G}} + \frac{\partial}{\partial t}\langle\langle \varepsilon\varphi \rangle\rangle. \end{aligned} \quad (75)$$

The last term is given by the expression (60). Substituting the time derivatives $\mathbf{J}'$ and $\mathbf{r}$ by means of (65) and (67) one obtains after cumbersome calculations [1,4,5]:

$$\begin{aligned} \dot{E} &+ \nabla_k \left\{ \frac{1}{2}\rho \dot{u}^2 \dot{u}_k + \dot{u}_i \left(\Pi_{ik} - \rho \dot{u}_i \dot{u}_k + E_0 \delta_{ik} + \langle\langle \varepsilon f \rangle\rangle \delta_{ik} \right) \right. \\ &- \left. \frac{1}{2} m \dot{u}^2 \left\langle \frac{\partial \varepsilon}{\partial p_k} f \right\rangle + \left\langle \varepsilon \frac{\partial \varepsilon}{\partial p_k} f \right\rangle \right\} \\ &= \frac{\partial \dot{u}_i}{\partial x_k} \left\{ \Pi_{ik} - \rho \dot{u}_i \dot{u}_k + \sigma_{\alpha\beta} a_i^\alpha a_k^\beta - \langle \lambda_{\alpha\beta} f \rangle a_i^\alpha a_k^\beta + E_0 \delta_{ik} \right. \\ &\left. - \dot{u}_i j_{0i} - \dot{u}_k j_{0i} \right\} + \langle \varepsilon \hat{I} f \rangle + \left\langle \mathbf{F} \frac{\partial \varepsilon}{\partial \mathbf{p}} f \right\rangle + \dot{W} - \dot{\mathbf{u}}\dot{\mathbf{G}}. \end{aligned} \quad (76)$$

If the operator $\hat{I}$ represents elastic collisions of quasiparticles with lattice defects or with each other, we would have $\langle \varepsilon \hat{I} f \rangle = 0$. When collisions with phonons occur, the electron energy is not conserved and phonons must be taken into consideration. All equations above, including the kinetic equation, actually apply to any quasiparticles in crystals, in particular, to phonons. To take phonons into account, it is sufficient to substitute the sum of the corresponding integrals of the electron and phonon distribution functions. It must be kept in mind, of course, that for phonons $m = e = 0$. The term containing the collision integral in this case has to be replaced by the sum $\langle \varepsilon \hat{I} f \rangle + \langle \varepsilon_{ph} \hat{I}_{ph} f_{ph} \rangle$, which vanishes by virtue of the conservations of the total energy of the quasiparticles.

The last three terms in (76) describe the change of the field energy, field momentum and the effect of external forces. They depend on the concrete type of interaction. In the case of electrons in the electromagnetic field, where $\mathbf{F}$ is the Lorentz force (52), W is given by (72), and the field momentum is

$$\mathbf{G} = \frac{1}{4\pi c}[\mathbf{EB}]. \quad (77)$$

Let us consider here the case of electrons in metals. Then, the displacement current density has to be neglected by virtue of the inequality $\omega \ll \sigma \mid \hat{\varepsilon} \mid^{-1}$ (ω being the field frequency). This inequality is always satisfied due to great conductivity σ.

An account of the displacement current would be beyond the precision [10]. Due to the same reason the electromagnetic field energy in metals is taken to be $W = B^2/8\pi$. Hence, its time derivative is

$$\dot{W} = -\operatorname{div}\mathbf{S} - \mathbf{j}_e\mathbf{E}, \tag{78}$$

where $\mathbf{S} = c/4\pi[\mathbf{EB}]$ is Poynting flux vector.

Integral containing the Lorentz force in (76) may be rewritten in the form

$$\langle \mathbf{F}\frac{\partial\varepsilon}{\partial\mathbf{p}}f\rangle = \mathbf{j}_e\mathbf{E} + \frac{\dot{\mathbf{u}}}{c}[\mathbf{B}\mathbf{j}_e]. \tag{79}$$

Substituting here $\mathbf{j}_e$ from (69) one obtains finally that the contribution from the field terms has the form

$$\dot{W} + \left\langle \mathbf{F}\frac{\partial\varepsilon}{\partial\mathbf{p}}f \right\rangle = -\nabla_k(S_k + \dot{u}_i t_{ik}) + t_{ik}\frac{\partial\dot{u}_i}{\partial x_k}, \tag{80}$$

where

$$t_{ik} = \frac{1}{4\pi}\left\{B_iB_k - \frac{1}{2}B^2\delta_{ik}\right\} \tag{81}$$

is the Maxwell tensor of the magnetic field.

Comparing (76) and (68) we can determine the unknown fluxes [1,4]:

$$Q_i = E\dot{u}_i + \left\langle \varepsilon\frac{\partial H}{\partial p_i}f \right\rangle - \frac{1}{2}\dot{u}^2 J_i + (\Pi_{ik} + t_{ik})\dot{u}_k + S_i\,, \tag{82}$$

$$\begin{aligned} \Pi_{ik} &= -(\sigma_{\alpha\beta} - \langle\lambda_{\alpha\beta}f\rangle + E_0 g_{\alpha\beta})a_i^\alpha a_k^\beta - \frac{m}{e}(\dot{u}_i j_{ek} + \dot{u}_k j_{ei}) + \rho\dot{u}_i\dot{u}_k - t_{ik} \\ &= -(\sigma_{\alpha\beta} - \langle\lambda_{\alpha\beta}f\rangle + E_0 g_{\alpha\beta})a_i^\alpha a_k^\beta + \rho_0\dot{u}_i\dot{u}_k + m\left\langle f\frac{\partial H}{\partial p_i}\frac{\partial H}{\partial p_k}\right\rangle \\ &- m\left\langle f\frac{\partial\varepsilon}{\partial p_i}\frac{\partial\varepsilon}{\partial p_k}\right\rangle - t_{ik}\,. \end{aligned} \tag{83}$$

The equation of the elasticity theory takes then the form:

$$\frac{\partial}{\partial t}(\rho\dot{u}_i) = -\frac{\partial\Pi_{ik}}{\partial x_k} - \frac{m}{e}\frac{\partial j_{ei}}{\partial t}. \tag{84}$$

This equation contains an additional term—the time derivative of the current in the co-moving system. This term exists also in the previous theories, but only as a linear approximation.

To obtain the full system describing the dynamics in metals the quasineutrality condition $q = en$ has to be added. The latter can be written in the form

$$(Z/M)\rho = \langle f\rangle, \tag{85}$$

where Z and M are the total charge and mass of the ions in a unit cell. Transforming to an integral over the invariant quasimomentum k_α we find (cf. (56)):

$$\langle f\rangle = g^{-1/2}\langle\langle f\rangle\rangle, \quad \langle\langle f\rangle\rangle = 2(2\pi\hbar)^{-3}\int f(k_\alpha,\mathbf{r},t)d^3k_\alpha = Z = \text{const.} \tag{86}$$

As it follows from Eqs. (69)–(70), the continuity equation for the charge density is

$$\dot{q} - e\dot{n} + \operatorname{div}\mathbf{j}_e = 0. \tag{87}$$

Hence, the quasineutrality condition is equivalent to the equation

$$\operatorname{div}\mathbf{j}_e = \operatorname{div}\mathbf{j}_o = \operatorname{div}\mathbf{j} = 0. \tag{88}$$

Finally, the complete system of dynamics equations of the metal consists of kinetic equation, Maxwell equations, momentum conservation equation (84) and the quasineutrality condition (86) or (88).

8. Defecton Diffusion in Rotating Quantum Crystals

As an application in which the essentially new term (51) is important let us consider defecton diffusion in rotating quantum crystal [1]. As is well known [11-13], defects in quantum crystals like solid helium turn into quasiparticles (defectons) which move in the phonon gas like heavy particles in a gas of light particles. This circumstance may be used when calculating the collision integral $\hat{I}f$. Following [15] one has

$$\hat{I}f = \int\{w(\mathbf{p}+\mathbf{q},\mathbf{q})f(t,\mathbf{p}+\mathbf{q}) - w(\mathbf{p},\mathbf{q})f(t,\mathbf{p})\}d^3q, \tag{89}$$

where $w(\mathbf{p},\mathbf{q})$ is the collision probability for a process during which the defecton quasimomentum changes its value $\mathbf{p}\rightarrow\mathbf{p}-\mathbf{q}$. Since the scattering may be considered as quasielastic, the integrand in (89) may be decomposed in powers of q/p. This yields

$$\hat{I}f = \frac{\partial}{\partial p_i}\left\{A_i f + \frac{\partial f}{\partial p_k}B_{ik}\right\}, \tag{90}$$

where

$$A_i = \int q_i w(\mathbf{p},\mathbf{q})\,d^3q + \frac{\partial B_{ik}}{\partial p_k}, \quad B_{ik} = \frac{1}{2}\int q_i q_k w(\mathbf{p},\mathbf{q})\,d^3q. \tag{91}$$

Since the collision integral equals zero for the equilibrium function f_0 it follows that

$$A_i f_0 = -\frac{\partial f_0}{\partial p_k}B_{ik}\,.$$

Let us introduce the deviation of the partition function from its equilibrium value:

$$f = f_0\left(1+\frac{\chi}{T}\right) \tag{92}$$

(where χ is not supposed to be small). Then,

$$\hat{I}f = \frac{\partial}{\partial p_i}\left(\frac{B_{ik}}{T} f_0 \frac{\partial \chi}{\partial p_k}\right). \tag{93}$$

This is the collision integral typical of the Fokker-Plank equation. The quantity B_{ik} is known as diffusion coefficient in **k**-space [14]. It has been calculated for defectons in a crystal with cubic symmetry elsewhere [1,14,15] and may be written in the form:

$$B_{ik} = B_0 \delta_{ik}\,, \; B_0 = \beta s a^{-3}\left(\frac{T}{\theta_p}\right)^9, \tag{94}$$

where β is a constant and $\theta_p = \hbar s/2a$ (s, a being the sound velocity and the lattice constant respectively).

If the deviation from the equilibrium is small, Eq. (93) may be rewritten in the form

$$\hat{I}\chi = \frac{B_{\alpha\beta}}{T}\frac{\partial \varepsilon}{\partial k_\alpha}\frac{\partial \chi}{\partial k_\beta}\,, \quad B_{\alpha\beta} = B_{ik} a_{\alpha i} a_{\beta k}\,. \tag{95}$$

Consider the defecton current

$$\mathbf{j} = \left\langle \frac{\partial \varepsilon}{\partial \mathbf{k}} f \right\rangle = -\mathbf{a}_\alpha \left\langle f_0 \frac{\partial \chi}{\partial k_\alpha}\right\rangle. \tag{96}$$

Obviously, we need for its calculation only the odd part of the distribution function (under the inversion $k_\alpha \to -k_\alpha$).

It may be obtained from the equation

$$\frac{D\chi}{Dt} - \frac{\partial \varepsilon}{\partial k_\alpha}\mathbf{a}_\alpha \nabla \mu = \frac{\partial \varepsilon}{\partial k_\alpha}\frac{\partial \chi}{\partial k_\beta}\left(\frac{1}{T}B_{\alpha\beta} + m[\mathbf{a}_\alpha \mathbf{a}_\beta]\,\mathrm{rot}\,\dot{\mathbf{u}}\right), \tag{97}$$

where the collision integral is taken from (95). Thus, the term $mT\,\mathrm{rot}\,\dot{\mathbf{u}} \cdot [\mathbf{a}_\alpha \mathbf{a}_\beta]$ is added to the diffusion coefficient matrix $B_{\alpha\beta}$. In the case when the crystal rotates with constant velocity ω we have $\mathrm{rot}\,\dot{\mathbf{u}} = 2\omega$. Denote

$$T^2 D^{-1}_{\alpha\beta} = B_{\alpha\beta} + 2mT \cdot \omega[\mathbf{a}_\alpha \mathbf{a}_\beta] \tag{98}$$

and substitute $\partial\chi/\partial k_\alpha$ obtained from (97) into (90), taking into account also that $\nabla n = \frac{n}{T}\nabla \mu$. This yields

$$j_\alpha = D_{\alpha\beta} a_{\beta i} \nabla_i n \quad \text{or} \quad j_i = a_{\alpha i} D_{\alpha\beta} a_{\beta k} \nabla_k n. \tag{99}$$

Finally, the Cartesian components of the diffusion coefficient can be written in the form

$$D_{xx} = D_{yy} = \frac{as}{4\beta}\frac{\kappa^2 t^9}{\kappa^2 t^{16} + \omega^2}$$

$$D_{xy} = -D_{yx} = -\frac{as}{4\beta}\frac{\kappa \omega t}{\kappa^2 t^{16} + \omega^2}\,, \quad D_{zz} = \frac{as}{4\beta} t^{-7}, \tag{100}$$

where $\kappa = \beta\hbar/ma^2$ and $t = T/\theta_p$.

Thus we see that the temperature dependence of the diffusion coefficient can be changed continuously from $D \sim T^{-7}$ (when $\omega = 0$) to $D \sim T^9$ (when $\omega \gg \kappa t^8$) only by variation of the velocity. The critical value for vacancion diffusion in solid helium is $\omega_c = \kappa t^8 \approx 10^2\,\mathrm{s}^{-1}$, which means about 15-20 cycles/s. On the other hand the diffusion along the axis of rotation does not vary. In this way a situation can occur when the increasing temperature leads to increasing radial diffusion and decreasing diffusion along the axis.

Note that the diffusion along z-axis does not depend on the bare mass of the quasiparticle in contrast to D_{xx} and D_{yx}. Hence, having the ratios

$$\frac{D_{xx}}{D_{yx}} = \frac{\kappa}{\omega} t^8 , \qquad \frac{D_{zz}}{D_{xx}} = 1 + \left(\frac{\omega}{\kappa t^8}\right)^2 , \tag{101}$$

one can determine the bare mass of the diffusible particle or complex of particles (vacancy, bi-vacancy, vacancy + impurity, etc.).

References

1. D. I. Pushkarov, *Quasiparticle Theory of Defects in Solids*, World Scientific, Singapore-New Jersey-London-Hong Kong, 1990.
2. V. M. Kontorovich, *Usp. Fiz. Nauk* **142**(2) (1984) 265. (in Russian.)
3. D. I. Pushkarov, *J. Phys. C: Solid-State Phys.* **19** (1986) 6873; Preprint P17-85-224, Joint Inst. Nucl. Res., Dubna, USSR, 1985.
4. A. F. Andreev and D. I. Pushkarov, *Zh. Eksp. Teor. Fiz.* **89** 1883 (1985) 1883. (in Russian.) (Engl. translation: *Sov. Phys. JETP* **62** 1087 (1985).
5. D. I. Pushkarov, *Defectons in Crystals*, JINR, Dubna, USSR, 1987. (in Russian.)
6. L. D. Landau and E. M. Lifshitz, *Quantum Mechanics*, Nauka, Moscow, (problem in §17), 1974. (in Russian.)
7. V. M. Kontorovich, *Zh. Eksp. Teor. Fiz.* **45** (1963) 1638. (in Russian.)
8. V. M. Kontorovich, *Zh. Eksp. Teor. Fiz.* **59** 2117 (1970) 2117. (in Russian.)
9. V. M. Kontorovich, In *Elektroni Provodimosti*, eds. M. I. Kaganov and V. S. Edelman, Moscow, Nauka, 1985. (in Russian.)
10. L. D. Landau and E. M. Lifshitz, *Electrodynamics of Continuous Media*, Nauka, Moscow §65, 1978. (in Russian.)
11. D. I. Pushkarov, *Quasiparticle Approach in Quantum Theory of Solids*, DrSc Thesis, Dubna, USSR, 1986. (in Russian.)
12. L. D. Landau and E. M. Lifshitz, *Theory of Elasticity*, Nauka, Moscow, §2, 1965. (in Russian.)
13. D. I. Pushkarov, Preprint P17-85-224, Joint Inst. Nucl. Res., Dubna, 1985. (in Russian.)

14. D. I. Pushkarov, *JETP* **59** (1970) 1755. (in Russian.)
15. D. I. Pushkarov, *JETP* **68** (1975) 1471. (in Russian.)
16. E. M. Lifshitz and L. P. Pitaevski, *Fizicheskaja Kinetika*, Nauka, Moscow, 1979. (in Russian.)

Continuum Models and Discrete Systems
Proceedings of 8th International Symposium, June 11-16, 1995, Varna, Bulgaria, ed. K.Z. Markov

VISCOUS FLOW, THERMAL EXPANSION AND RELAXATION OF AMORPHOUS METALLIC ALLOYS: THEORY AND EXPERIMENT

K. RUSSEW and L. STOJANOVA
Institute for Metal Science, Bulgarian Academy of Sciences,
67 Shipchenski Prohod Blvd., BG-1574 Sofia, Bulgaria

Abstract. A brief survey of contemporary theoretical approaches and experimental techniques used for studying the viscous flow behaviour of glassy metallic alloys is presented together with recent authors' contributions to the problem under consideration. It is shown that the Free Volume Theory of viscous flow in amorphous materials is self-consistent and free of internal contradictions by treating the experimental results of nonisothermal viscous flow and thermal expansion of glassy metals. Relaxation phenomena in the high temperature range near the glass transition temperature T_g are determined by the annihilation and/or creation of free volume in the amorphous structure.

1. Introduction

Both the process of production of glassy metallic alloys and the subsequent formation of their useful properties during relaxation annealing depend critically on the viscosity features of the glassy material under consideration. Schematics of the viscosity temperature dependence of rapidly quenched glassy metals is shown in Fig. 1a. During the rapid quench from the melt viscosity η follows the line of quasiequilibrium viscosity until the temperature T_g of glass transition is reached. At lower temperatures, due to the decreased atomic mobility, the viscosity of deeply undercooled melt deviates from quasiequilibrium viscosity temperature dependence and follows an Arrhenian-like temperature dependence of glassy state. This is why the as-quenched state of a metallic glass is very far from equilibrium. Upon annealing for a given time structural relaxation occurs which is connected with subtle changes of the atomic structure towards metastable equilibrium structure. It is believed [1,2,5-9] that these changes are mainly connected with annihilation (at $T < T_g$) or production (at $T > T_g$) of free volume. Here T denotes the annealing temperature. This structural relaxation is reflected in a most sensitive way by the viscosity η and thermal expansion [4] of glassy alloys. Under isothermal annealing practically all glassy alloy exhibit a linear increase of viscosity with annealing time [1,2].

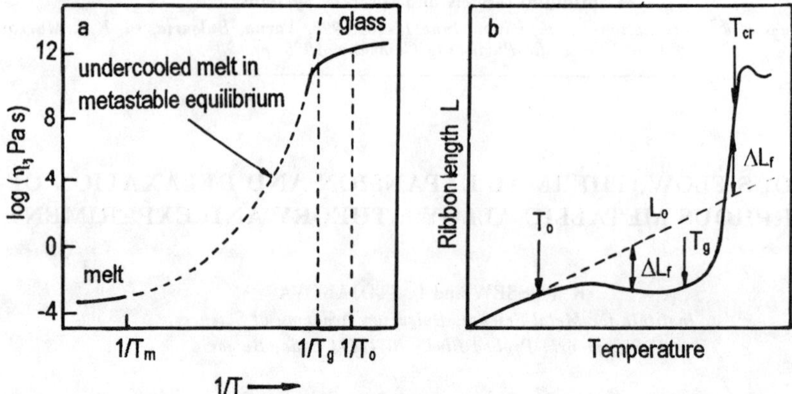

Fig. 1: a) Schematic of viscosity temperature dependence of an arbitrary rapidly quenched glassy alloy; b) Schematic of thermal expansion features of a glassy alloy heated at a constant heating rate q.

It should be expected that the viscosity will reach a constant value when the glass has reached metastable equilibrium at the temperature of isothermal annealing. However, only limited number of isothermal studies of metallic glasses in which equilibrium viscosity has been clearly reached prior to crystallization are known [1,3,5,6-9]. It has been recently shown [7] that studying the isothermal viscous flow behaviour of $Pd_{40}Ni_{40}P_{20}$ glassy alloy near its quasiequilibrium, it is possible to analyse the kinetics of viscosity approach towards its quasiequilibrium value in terms of the free volume theory [10,11]. The most general temperature dependence of the viscosity η can be represented as [6,12]

$$\eta = \frac{\eta_0 T}{c_f} \exp\left(\frac{Q_\eta}{kT}\right). \tag{1}$$

Here Q_η is the activation energy for viscous flow, η_0 is a pre-exponential factor, and c_f is the concentration of the so-called "flow defects" responsible for viscous flow taking place under applied shear stress τ. According to the free volume theory the amount of the free volume in the quasiequilibrium state is given by

$$v_{f,\mathrm{eq}} = x_{\mathrm{eq}}(T)\gamma v^* = \frac{(T - T_0)}{B}\gamma v^*, \tag{2}$$

where B and T_0 are two model parameters which correspond to the Vogel-Fulcher-Tammann (VFT) constant and to the ideal glass transition temperature of the VFT-empirical equation, respectively, γ is a geometrical overlap factor between 0.5 and 1, and v^* is the so-called "critical free volume", necessary to form a flow defect. The reduced free volume x is defined as $x = v_f/\gamma v^*$. The relation between c and x is given by [6]

$$c_f = \exp\left(-\frac{1}{x}\right). \tag{3}$$

Combining Eqs. (1), (2) and (3), the so-called "hybrid" temperature dependence of the quasiequilibrium viscosity η_{eq} is obtained

$$\eta_{eq} = \eta_0 T \exp\left(\frac{Q_\eta}{kT}\right) \exp\left(\frac{B}{T - T_0}\right). \tag{4}$$

In the cases when the glassy alloy has not reached the quasiequilibrium state c_f changes along with increasing annealing time following a bimolecular kinetics [1,6]. It has been shown [7] that the most suitable differential equation describing the bimolecular kinetics of the c_f-change with time (due to defect annihilation or defect production), which properly reflects the variation of η_{eq} by sudden changes of the annealing temperature, is

$$\frac{dc_f}{dt} = -k_r c_f (c_f - c_{f,e}), \tag{5}$$

where $k_r = \nu_r \exp(-Q_r/kT)$ is the rate constant of relaxation with ν_r denoting the jump frequency, Q_r the activation energy of relaxation, and $c_{f,e}$ the equilibrium flow defect concentration at the temperature of isothermal annealing. P. Duine *et al.* [7] have theoretically derived the time dependence of η approach towards equilibrium at different temperatures of isothermal annealing near equilibrium and fitting the measured elongation data to the theoretical prediction derived they have obtained the values of the needed seven fitting parameters, namely η_0, Q_η, ν_r, Q_r, B, T_0 and $c_{f,0}$ by using a nonlinear multiparameter regression analysis. It is clear that the annihilation or production of free volume in a given glassy alloy must be directly reflected by an "anomalous" negative or positive deviation $\Delta L_f(T)$ with respect to the extrapolated low temperature ($T \ll Tg$) linear temperature dependence $L_0(T)$ of thermal elongation in the vicinity of T_g. This phenomenon has been directly observed by us [4] for the case of $Pd_{82}Si_{18}$ glassy alloy. The nonisothermal viscosity measurements of glassy alloys under constant heating rate conditions provide the possibility to extend the theoretical and isothermal experimental approach for the case of nonisothermal viscosity measurements on practically all known glassy alloys exhibiting glass transition. The aim of this study is to perform a theoretical consideration both of the viscosity temperature dependence of glassy alloys experimentally obtained at a constant heating rate under nonisothermal conditions, and of thermal expansion anomalies of glassy alloys based on the concepts of free volume theory [10,11] in order to prove them experimentally with $Pd_{82}Si_{18}$ and $Fe_{25}Zr_{75}$ glassy alloys as examples.

2. Theoretical Considerations

As shown in our previous paper [13], at constant heating rate q Eq. (5), describing the flow defect concentration change with time under isothermal conditions, has to be transformed in order to describe the change of c_f with temperature. Taking into account that $(dc_f/dt) = (dc_f/dT)q$, the following differential equation of Bernoulli's type is obtained

$$\frac{dc_f}{dT} + P(T)c_f = c_f^2 Q(T) \tag{6}$$

with

$$P(T) = -\frac{\nu_r}{q}\exp\left(-\frac{Q_r}{kT} - \frac{B}{T-T_0}\right) \quad \text{and} \quad Q(T) = -\frac{\nu_r}{q}\exp\left(-\frac{Q_r}{kT}\right).$$

Its analytical solution is

$$c_f^{-1} = \left[c_{f,0}^{-1} - \int_{T_0}^{T} Q(x)\exp\left(-\int_{T_0}^{T} P(x)\,dx\right)dx\right]\exp\left(\int_{T_0}^{T} P(x)\,dx\right) \tag{7}$$

with the ideal glass transition temperature T_0 used as the starting temperature of calculating. By combining Eq. (7) with Eq. (1) it is possible to represent the measured temperature dependence of η at constant q in the temperature range around T_g. The fitting parameters can be obtained by using a nonlinear multiparameter least square regression analysis of the experimentally obtained viscosity temperature dependence.

Considering the anomalous deviations of thermal expansion from the low temperature range simple linear relation between the ribbon elongation $L_0(T)$ and the temperature T by heating the specimen at a constant heating rate q, the scheme, shown in Fig. 1b could be useful. The low temperature linear dependence of ribbon length $L_0(T)$ on the temperature is given by

$$L_0(T) = L_0(T_b)\left[1 + \alpha_l^0(T - T_b)\right], \tag{8}$$

where T_b denotes the starting temperature of heating, $L_0(T_b))$ is the initial length of the glassy alloy ribbon, α_l^0 is the coefficient of thermal expansion in the low temperature range $T_b \div T_0$, and T_0 is the ideal glass transition temperature. The deviation $\Delta L_f(T)$ of the real temperature dependence $L_f(T)$ of ribbon length from $L_0(T)$ can be represented as

$$\Delta L_f(T) = L_f(T) - L_0(T). \tag{9}$$

Taking into account that the reduced free volume x is related to the actual average atomic free volume v_f as $x = v_f/\gamma v^*$, its changes along with increasing temperature can be related to the length changes of the glassy alloy studied as follows

$$x(T) - x_0 = \frac{L_f^3(T) - L_0^3(T)}{\gamma v^* N}, \tag{10}$$

where N is the number of atoms in a cube of edge length $L_0(T_b)$ at $T = T_b$ of the glassy alloy under consideration. Obviously, N can be represented as

$$N = \frac{L_0^3(T_b)}{V_{\text{mol}}} N_A \tag{11}$$

with N_A denoting the Avogadro number, and V_{mol}—he molar volume of the glassy alloy studied. Taking into account the definition of $\Delta L_f(T)$ given in Eq. (9) and neglecting all terms containing $\Delta L_f^2(T)$ and $\Delta L_f^3(T)$ when expanding Eq. (10) into a power series, one obtains

$$x(T) - x_0 \cong 3L_0^2(T)\Delta L_f(T)\frac{V_{\mathrm{mol}}}{\gamma v^* N_A L_0^3(T_b)}. \tag{12}$$

Taking into account the relation between reduced free volume and the flow defect concentration c_f, see Eq. (3),

$$3L_0^2(T)\Delta L_f(T)\frac{V_{\mathrm{mol}}}{\gamma v^* N_A L_0^3(T_b)} = \frac{1}{\ln c_{f,0}} - \frac{1}{\ln c_f(T)} \tag{13}$$

with $c_f(T)$ as defined in Eq. (7). Keeping in mind that at $T < T_0$ - $x(T) = x_0$, the following temperature dependence of the experimentally observed length $L_f(T)$ of the glassy ribbon heated at a heating rate q is obtained

a) $T < T_0$:

$$L - f(T) = L_0(T) = L_0(T_b)\left[1 + \alpha_l^0(T - T_b)\right]; \tag{14}$$

b) $T > T_0$:

$$L_f(t) = L_0(T_b)[1 + \alpha_l^0(T - T_b)] + \frac{\gamma v^* N_A L_0(T_b)}{3V_{\mathrm{mol}}[1 + \alpha_l^0(T - T_b)]^2}$$

$$\times \left[\frac{1}{\ln c_{f,0}} + \frac{1}{\ln\left[c_{f,0}^{-1} - \int_{T_0}^{T} Q(x)\exp\left(-\int_{T_0}^{T} P(x)\,dx\right)dx\right] + \int_{T_0}^{T} P(x)\,dx}\right] \tag{15}$$

with $Q(T)$ and $P(T)$ as defined by Eq. (7). In a first approximation the coefficient α_l^0 of thermal expansion can be interpreted as temperature independent. It becomes obvious, that Eq. (7) appears to be a key equation, with the aid of which it is possible to examine whether the inter-relation between the nonisothermal viscosity and thermal expansion measurements of a definite glassy alloy respectively, really exists or not. Eq. (7) makes it possible as well, to prove the self-consistence of the free volume theory [10,11] itself.

3. Experimental Approach

Viscosity measurements of $Pd_{82}Si_{18}$ and $Fe_{25}Zr_{75}$ were carried out at q = 20 Kmin^{-1} using a Perkin Elmer TMS-2, and at 10Kmin^{-1} using Hereaus TMA 500 silica glass dilatometers [14-16], respectively. Scheme of the experimental equipment used is shown in Fig. 2.

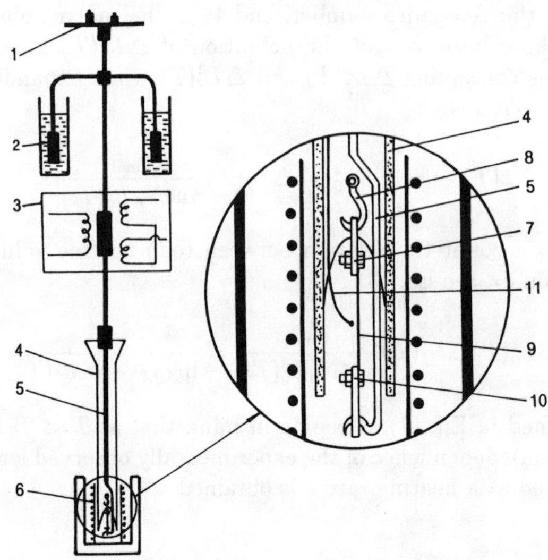

Fig. 2: Schematic of a Perkin Elmer TMS-2 analyzing unit using an assembly for creep measurements. 1 – weight tray; 2 – lifting float; 3 – LVDT; 4 – supporting silica glass tube; 5 – silica glass probe with a hook at the end applying load and connecting moving grip to LVDT; 6 – furnace; 7 – stationary grip of Invar alloy; 8 – stationary silica glass hook; 9 – specimen; 10 – movable grip of Invar alloy; 11 – thermocouple.

The temperature accuracy was ± 2 K. The accuracy of elongation measurements was better than ± 0.5μm. The apparent viscosity was calculated as a function of temperature using a modified Newtonian equation for the case of nonisothermal viscosity measurements as described elsewhere [14]. The temperature dependence of ribbon length $L_f(T)$ at the same heating rates was also monitored with the aid of Perkin Elmer TMS-2, or Hereaus TMA 500 devices, respectively.

4. Experimental Results and Discussion

Fig. 3a and Fig. 4a show the experimentally obtained temperature dependencies of the viscosity of $Pd_{82}Si_{18}$ and $Fe_{25}Zr_{75}$ glassy alloys studied together with the best fit curves obtained on the basis of Eq. (1) and Eq. (7) for heating rates of 20 or 10 K min^{-1}, respectively. The values of the fitting parameters are as follows:

$Pd_{82}Si_{18}$: $\eta_0 = 4.99 \cdot 10^{-20}$ Pa s K^{-1}, $Q_\eta = 190.9 \cdot 10^3$ J mol^{-1}, $\nu_r = 6.8 \cdot 10^{19}$ s^{-1}, $Q_r = 120 \cdot 10^3$ J mol^{-1}, $B = 2130$ K, $T_0 = 557$ K, $c_{f,0} = 2.54 \cdot 10^{-11}$.
$Fe_{25}Zr_{75}$: $\eta_0 = 5.91 \cdot 10^{-20}$ Pa s K^{-1}, $Q_\eta = 154.1 \cdot 10^3$ J mol^{-1}, $\nu_r = 7.01 \cdot 10^{19}$ s^{-1}, $Q_r = 89.1 \cdot 10^3$ J mol^{-1}, $B = 5749$ K, $T_0 = 460$ K, $c_{f,0} = 3.22 \cdot 10^{-14}$.

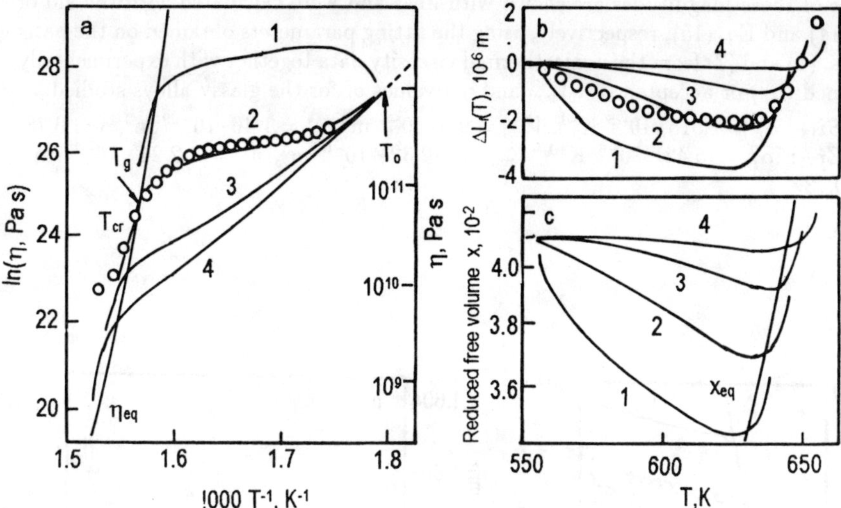

Fig. 3. $Pd_{82}Si_{18}$ glassy alloy: a) Calculated and experimental viscosity-temperature dependencies at heating rates 2,20, 200 and 2000 K min^{-1} (curves 1 to 4), respectively; b) Experimental and calculated thermal elongation anomalies ΔL_f under the same heating conditions; c) Temperature dependencies of reduced free volume x.

Fig. 3a and Fig. 4a show also the calculated viscosity temperature dependencies for heating rates of 2, 20, 200 and 2000, and 1, 10, 100, and 1000 K min^{-1}, respectively, together with the curves of quasiequilibrium viscosity η_{eq} calculated according to Eq. (4). At low heating rates, e.g., 2 K min^{-1}, an apparent increase of viscosity and its passing over a maximum along with increasing temperature is observed. This is due to the relaxation or annealing out of flow defects (annihilation of free volume) as predicted by the theoretical model used. This effect of relaxation becomes more and more concealed along with increasing the heating rate. Dealing with the problem of the definition of the so-called glass transition temperature T_g it becomes clear from Fig. 2a and Fig. 3a, that T_g corresponds to the cross-point of the nonisothermal (nonequilibrium) viscosity curves 1 to 4, calculated according to Eq. (1) combined with Eq. (7), and the quasiequilibrium viscosity curve, calculated with the aid of Eq. (4). The experimental results of the thermal expansion measurements of the glassy alloy studied at heating rates of 20 or 10 K min^{-1} are shown in Fig. 3b and Fig. 4b, respectively.

They exhibit all characteristic features schematically shown in Fig. 1b and theoretically considered in Section 2 of this article. The experimentally obtained values of $\Delta L_f(T)$ or $L_f(T)$ are shown with open circles, the theoretically expected values of these magnitudes are shown with lines and were calculated with the aid of Eq. (14) and Eq. (15), respectively, using the fitting parameters obtained on the basis of Eqs. (1) and (7) from the nonisothermal viscosity data together with experimentally obtained (α_l^0) or accepted (v^*, V_{mol} and γ) values of for the glassy alloys studied: $Pd_{82}Si_{18}$: $\alpha_l^0 = 1.515 \cdot 10^{-5}$ K^{-1}, $V_{mol} = 9.5 \cdot 10^{-6}$ m^3, $v^* = 1.46 \cdot 10^{-29}$ m^3, $\gamma = 0.78$. $Fe_{25}Zr_{75}$: $\alpha_l^0 = 5.42 \cdot 10^{-6}$ K^{-1}, $V_{mol} = 12.32 \cdot 10^{-6}$ m^3, $v^* = -2.24 \cdot 10^{-29}$ m^3, $\gamma = 0.78$.

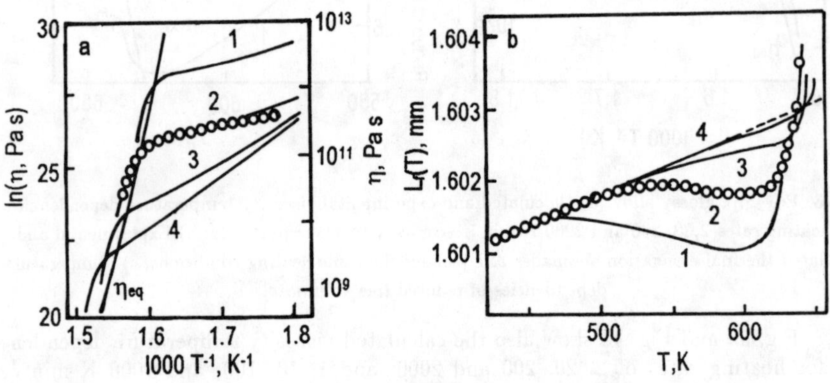

Fig. 4: $Fe_{25}Zr_{75}$ glassy alloy: a) Calculated and experimental viscosity-temperature dependencies at heating rates of 1, 10, 100 and 1000 K min^{-1} (curves 1 to 4), respectively; b) Experimental and calculated thermal elongation under the same heating conditions.

It is obvious that the coincidence of calculated and experimental thermal expansion data is almost perfect. Fig. 3c shows the calculated on the basis of Eq. (3) temperature dependence of the reduced free volume x at different heating rates from 2 K min^{-1} to 2000 K min^{-1}. It becomes obvious, that inter-relation between the free volume change, viscous flow and thermal expansion behaviour of $Pd_{82}Si_{18}$ and $Fe_{25}Zr_{75}$ glassy alloy really exists.

5. Conclusions

1. Nonisothermal constant heating rate viscosity and thermal expansion measurements of glassy alloys provide the possibility for theoretical interpretation of the experimental results on the basis of the free volume theory.

2. More specified definition of T_g of glassy alloys undergoing heating after rapid quenching is proposed. T_g is the temperature at which the trend of relaxation phenomena changes from free volume annihilation to free volume production. Glass transition temperature depends in a definite way on the rate of heating.

Acknowledgements. The financial support of the Bulgarian National Foundation for Scientific Research (Contract Th411/94) is gratefully acknowledged.

References

1. S. S. Tsao and F. Spaepen, *Acta Met.* **33** (1985) 881.
2. A. van den Beukel, E. Huizer, A. Mulder and S. van den Zwaag, *Acta Met.* **34** (1986) 483.
3. H. S. Chen and D. Turnbull, *J. Chem. Phys.* **48** (1968) 2560.
4. K. Russew, L. Stojanova and F. Sommer, *Internat. J. Rapid Solidification* **8** (1995) 279.
5. C. A. Volkert and F. Spaepen, *Acta Met.* **37** (1990) 1355.
6. A. van den Beukel and J. Sietsma, *Acta Metall. Mater.* **38** (1990) 383.
7. P. A. Duine, J. Sietsma and A. van den Beukel, *Acta Metall. Mater.* 1992 **40** (1992) 743.
8. G. W. Koebrugge, J. Sietsma and A. van den Beukel, *Acta Metall. Mater.* **40** (1992) 753.
9. P. Tuinstra, P. Duine and J. Sietsma, *J. Non-Cryst. Solids* **156-158** (1992) 519.
10. M. H. Cohen and D. Turnbull, *J. Chem. Phys.* **31** (1959) 1164.
11. D. Turnbull and M. H. Cohen, *J. Chem. Phys.* **53** (1970) 3038.
12. F. Spaepen, In *Physics of Defects, Les Houches Lectures XXXV*, ed. R. Balian *et al.*, North-Holland, Amsterdam, (1984), p. 737.
13. K. Russew, B. J. Zappel and F. Sommer, *Scripta Metall. Mater.* **32** (1995) 271.
14. K. Russew and L. Stojanova, *Mater. Sci. Eng.* **A123** (1990) 59.
15. K. Russew and L. Stojanova, *Mater. Letters* **17** (1993) 199.
16. K. Russew, L. Stojanova and F. Sommer, In *Handbook of Advanced Materials Testing*, ed. N. P. Cheremissinoff, Marcel Dekker Publishers, Inc., New Jersey, USA, 1994, Chapter 36, p. 715.

Continuum Models and Discrete Systems
Proceedings of 8th International Symposium, June 11-16, 1995, Varna, Bulgaria, ed. K.Z. Markov

MATERIAL FORCES ACTING ON PHASE-TRANSITION FRONTS IN THERMOELASTICITY

C. TRIMARCO
Istituto Matematiche Applicate "U. Dini," Università di Pisa, Via Bonanno 25/B, I-56126 Pisa, Italy

and

G. A. MAUGIN
Laboratoire de Modélisation en Mécanique, Université Pierre et Marie Curie, Tour 66, 4 place Jussieu, Case 162, F-75252 Paris Cédex 05, France

Abstract. A solid-solid phase transition occurs across a progressing surface front. We regard the front as a source of inhomogeneity. Then we appeal to the balance law of *pseudomomentum*. Hence, we find the dynamical force of inhomogeneity, a vector quantity which governs the transformation. Furthermore, the notion of *full coherency* is here introduced along with the currently adopted notion of coherency.

1. Introduction

Bragg's X-ray spectroscopy not only reveals the existence of a crystalline lattice but also addresses to the electric nature of it. There is a general agreement in that "crystal lattice sites" account for stable equilibrium positions with respect to the internal electrostatic fields. Unfortunately, the electrostatic forces are non-local forces and severe difficulties would arise in a global macroscopic approach [1]. In a Continuum model one rather disregards the electrical nature of the forces of cohesion, while retains the internal geometrical structure whenever the specific phenomenon addresses to it [2-4]. Still however, a possible inhomogeneous macroscopic response might not be completely ascribed to the enrichment of the geometrical structure of the continuum model. Hence, additional *intrinsic* forces due to the possible inhomogeneities should be accounted having in mind the electrical nature of the underlying microscopic world. This is the case for a material exhibits two different crystalline structures. The two structures may coexist in many circumstances and the shift from one to the other may occur. As those structures energetically differ one from the other, thermodynamics could be the proper framework for describing the phenomenon. Nonetheless, metallurgists retain that the classical thermodynamical description is a too reductive scheme. They rather suggest to account for a mechanical transformation, the *self-strain*, which locally converts the two structures one into the other [4,5]. Espousing the viewpoint of metallurgists one introduces an auxiliary, generally nonglobal, reference configuration, the *crystal reference* [6,7]. The access to it is ruled by the *local deformation* of the lattice which accounts for the *self-strain* [4,5,9]. Such a local deformation, which is possibly not integrable, does not correspond to that of the material point from a prescribed reference configuration. It rather describes a *rearrangement* of points in the reference configuration. Such *rearrangements* occur independently of the effective motion and deformation and are thus irrespective of the current configuration. We will denote by $\delta_x X$ the rearrangement of points which are evaluated at fixed spatial positions x. The material velocity V can be introduced by parametrizing $\delta_x X$ in time. Should there exist a Lagrangian density of the fields, then one may invoke the invariance of the Lagrangian with respect to an elementary "translation" $\overline{\delta_x X}$. Such an invariance would account for the homogeneity of the material space and lead to the dynamical balance law of pseudomomentum [8-10]. The pseudomomentum, in turn, introduces the configurational forces whose related surface densities are represented by the Eshelby stress [6-12]. Although the phenomenon is possibly dissipative, we will still invoke the validity of the balance of pseudomomentum, provided that a suitable source of *intrinsic* material forces is taken into account. Here we examine the case of a surface Σ separating the two coexisting structures, or phases, in a solid. The temperature is assumed to be continuous across Σ (homothermal front). The solid is thermoelastic in each phase, although with different forms of the free energy densities. Basing on these assumptions we regard the surface Σ as a possible source of inhomogeneities. As we wish to preserve the continuity of the crystal lattice across Σ we assume that $\delta_x X$ is continuous across it. Such a condition is equivalent to the currently accepted notion of coherency, although we will refer to it as the *geometrical coherency* [4,5,9,12-14]. The surface front may progress while deforming. As we wish to preserve the continuity of the lattice sites *also in time* we propose an additional condition: the continuity of V across Σ. This additional condition characterizes what we call the *full coherency*.

2. Coherence and Full Coherence

Assume that $\mathcal{B}_R$ and $\mathcal{B}$ represent the reference and the actual configuration of a solid material, respectively. $\mathcal{B}_R \subset \mathcal{M}^3$, $\mathcal{M}^3 \equiv$ material manifold . Let Σ be a surface which divides $\mathcal{B}_R$ into two open sets $\mathcal{B}_R^+$ and $\mathcal{B}_R^-$ so that $\Sigma \equiv \overline{\mathcal{B}_R^+} \cap \overline{\mathcal{B}_R^-}$ and $\mathcal{B}_R^+ \cup \mathcal{B}_R^- \equiv \mathcal{B}_R$. If $X \in \mathcal{B}_R$ and $x \in \mathcal{B}$, the mapping $X = k^{-1}(x,t)$ represent the inverse motion. We also assume that $\mathcal{S}(t) = k^{-1}(\Sigma(t))$ and $k^{-1} \in \mathcal{C}^3(\mathcal{B}-\mathcal{S}, \mathcal{B}_R-\Sigma)$. Furthermore, the derivatives $V \equiv \dot{X}\big|_x$ and $F^{-1} \equiv$ grad X may suffer a finite jump across Σ. Σ and $\mathcal{S}$ are endowed with regular unit normal N and n, respectively. N points from $\mathcal{B}_R^-$ to $\mathcal{B}_R^+$ and n from $\mathcal{B}^-$ to $\mathcal{B}^+$. The square brackets $[\cdot]$ denote the finite jump across the singular surfaces of the quantity enclosed. The brackets $\langle\cdot\rangle$

denote its average value at the front. The surface Σ and $\mathcal{S}$ can be represented by

$$F(Y(t),t)=0$$

and

$$f(y(t),t)=0\,, \tag{2.1}$$

respectively, where $Y\in\mathcal{B}_R$ and $y\in\mathcal{B}$. We also assume that

$$Y(t)=k^{-1}(x,t)|_{x=y}=k^{-1}(y(t),t)$$

and

$$y=k(Y(t),t)\,. \tag{2.2}$$

By differentiating the expressions (2.2) one is lead to

$$\overline{V}=\langle F^{-1}\rangle u+\langle V\rangle$$

and

$$u=\langle F\rangle\overline{V}+\langle v\rangle\,, \tag{2.3}$$

where $u\equiv\dot{y}$ and $\overline{V}\equiv\dot{Y}$ are both continuous [15, pp. 493-529]. The following identity

$$f(y(t),t)=\hat{f}(y(Y(t),t),t) \tag{2.4}$$

is assumed to hold true. By differentiating it we can write

$$u_n=\langle v_n\rangle+n\cdot\langle F\rangle\bar{V}\,, \tag{2.5}$$

where $u_n\equiv u\cdot n$ and $v_n\equiv v\cdot n$. u_n is known as the speed of displacement while $\overline{V_N}$ is the speed of propagation. It is also worth reminding that $d\Sigma\equiv\langle J^{-1}\gamma\rangle\,d\mathcal{S}$, where $J^{-1}\equiv\det F^{-1}$, $\gamma^2\equiv N\cdot C^{-1}N$ and that $C^{-1}=F^{-1}(F^{-1})^T$. Accordingly, $\langle\gamma F^T n\rangle = N$. Assume now that the differential identity $V+F^{-1}v=0$ [7-9] holds true across Σ so that

$$\langle V\rangle+\langle F^{-1}v\rangle=0\,. \tag{2.6}$$

From all the above expressions there follows that:

$$[V]=-[F^{-1}v]\equiv-[F^{-1}]u\,,$$

and

$$[V_N]=-[F^{-1}N]u=-[\gamma]u_n\,, \tag{2.7}$$

where $V_N\equiv V\cdot N$.

Basing on the lemma of Hadamard [15] one establishes the following geometrical compatibility conditions to be satisfied across Σ

$$[F^{-1}]=[F^{-1}n]\otimes n\,,\quad [V]=-[F^{-1}n]u_n\,. \tag{2.8}$$

Such conditions also define the current notion of coherency which we rather propose to denote the *geometrical coherency*. As the notion of coherency expresses the continuity of the crystalline structure one is inclined to require the continuity of the rearrangement of the crystal sites *also in time*. Basing on this idea one parametrizes $\delta_x X$ in time and writes

$$[V] = 0. \tag{2.9}$$

By contrast, the physical velocity v which is associated to the actual motion can be discontinuous, as the transformation is assumed to be a first order phase transition. The *geometrical coherency* condition provides no prescriptions for the speed of propagation of the front. Differently, the *full coherency* condition leads to

$$u_n = 0, \tag{2.10}$$

namely to a "stationary" shock-like progressing front of transformation [9].

Notice that the weaker requirement of continuity of V_N entails the continuity of V. It is also worth noticing that V_N in general differs from the speed of propagation $\overline{V_N}$ of the front [9,13-16]. Nonetheless, *full coherency* infers $\overline{V_N} = V_N$, although $\overline{V}$ and V still differ one from the other. V_N, in turn, is related to v_n as follows:

$$V_N = -\langle \gamma v_n \rangle \,. \tag{2.11}$$

3. Dynamics and Thermodynamics

3.1. Balance of Pseudomomentum

We assume that the material is thermoelastic in the two phases which are separated by the progressing front of transformation Σ. We also assume that the material is isothermally loaded by external tractions (stress induced transformation). The jump of the temperature θ across Σ can be neglected (homothermal front) as it is assumed to relax in a much shorter time scale with respect to propagation. Under the assumption that no chemical reactions occur during the transformation, thermodynamics suggests that the appropriate potential should be the Gibbs potential or free enthalpy, having in mind the continuity of the tractions and of the temperature across Σ. Nonetheless, we will refer to the Helmholtz free energy per unit mass ψ as we wish to highlight the role of the gradient of deformation as independent variable in the dynamical case. $\psi = \overline{\psi}(F, \theta)$ suffers a finite jump across Σ. The jump of ψ is due both to the discontinuity of its argument F and to the different forms it might have in the two phases. Having introduced the mass density ρ and the energy density $w = \rho\psi$ per unit volume of the current configuration, we also introduce the Eshelby stress tensor b [7,9] as

$$b = J(w_{F^{-1}})(F^{-1})^T \tag{3.1}$$

which is currently written in the following equivalent form [6,7,11,12]:

$$b^* = WI - F^T T_R \,, \tag{3.2}$$

where $W = wJ$ and $T_R \equiv \rho_0 \psi_F$ is the first Piola-Kirchhoff stress tensor. The balance of pseudomomentum will read [9]

$$\mathrm{Div}\, b^* + \mathcal{F}^{th} = \dot{\mathcal{P}}\Big|_X \quad \text{in } \mathcal{B}_R \,, \tag{3.3}$$

where $\mathcal{P} \equiv \rho_0 CV$ denotes the pseudomomentum. Pseudomomentum and configurational dynamical forces across the front (coherent or not) are balanced by the inhomogeneity force f_Σ^{in} [9]. Accordigly, we write

$$[b^*]N + [\mathcal{P} \otimes \overline{V}] + f_\Sigma^{in} = 0 \text{ at } \Sigma \tag{3.4}$$

where

$$b^* = b - (\rho_0 v^2/2)I \,, \quad \rho_0 = \rho J \,, \quad \mathcal{F}^{th} = -\rho_0 \psi_\theta \mathrm{Grad}\, \theta \,.$$

Notice that no mathematical relationship occurs between $\mathcal{F}^{th}$ and f_Σ^{in}. From the physical viewpoint they both represent two sources of inhomogeneity forces. Nonetheless, the form of $\mathcal{F}^{th}$ stems from the condition of thermoelasticity. In contrast, f_Σ^{in} is *a priori* unknown. By developing Eq. (3.4) we find the following result

$$-\rho_0 \mathcal{H} N = f_\Sigma^{in} \,, \tag{3.5}$$

where

$$\mathcal{H} = [\psi - \langle \psi_F \rangle \cdot F] \,, \tag{3.6}$$

having taken into account the balance of mass and the balance of momentum across Σ.

Eq. (3.5) and the expression (3.6) deserve some comments. First we notice that f_Σ^{in} balances a resulting configurational force $-\rho_0 \mathcal{H} N$. Despite the fact that we are dealing in *full dynamics* the quantity $\mathcal{H}$ is devoid of inertial terms and of explicit kinematical quantities. Should f_Σ^{in} vanish, Eq. (3.5) reduces to

$$\mathcal{H} = 0 \,. \tag{3.7}$$

Eq. (3.7), in turn, addresses to a meaningful interpretation in the isothermal stress-strain response diagram. In fact, it establishes the continuity of the Gibbs potential at equilibrium. The Gibbs-Maxwell rule of "equal areas" stems from it in the one-dimensional case, provided that the possible discontinuity of ψ_F can only be ascribed to that of its argument F (no inhomogeneities occur). Such an interpretation can be reasonably extended to the dynamical case. Vanishing of $\mathcal{H}$ in dynamics *generalizes* the rule of "equal areas", although the continuity of ψ_F is invalidated. In this respect, the "connection" between the two thermoelastic branches of the isothermal stress-strain diagram $(\rho_0\psi_F N, FN)$ would be a *sloped straight segment* [9,14,17]. The value of the slope is provided by the following formula

$$\overline{V_N}^2 = [\psi_F]\cdot[F]/[F]\cdot[F] \tag{3.8}$$

which, in turn, stems from the balance of momentum across Σ. Based on the formula (3.8) one easily realizes that $\overline{V_N}^2$ represents the value of the mentioned slope in the one-dimensional case and that the slope has to be positive. The aforementioned "connection" recalls the "Hugoniot shock-like connection" in the adiabatic stress-strain response diagram, although we are concerned with isothermal conditions. For this reason we suggest to call $\rho_0\mathcal{H}N$ the *Hugoniot-Gibbs configurational force*.

Should ψ and ψ_F have different forms across Σ, inhomogeneities would have to be accounted for. In this respect additional internal structures may be envisaged in the description and introduced therein [2,5,6-9]. This could be the case of the *self-strain* conjecture: a geometrical structure useful in most of the solid-solid phase transitions [4-9]. Basing on this conjecture, ψ could possibly depend on the *self-strain* and $\mathcal{H}$ would provide an explicit expression for the *force of inhomogeneity* at Σ.

3.2. Power of f_Σ^{in} and Thermodynamics

The inner product $f_\Sigma^{in}\cdot\overline{V}$ defines the power of f_Σ^{in}. With reference to the formulae (2.3) and (2.5) we write

$$f_\Sigma^{in}\cdot\overline{V} = f_\Sigma^{in}\cdot\langle F^{-1}\rangle u + f_\Sigma^{in}\cdot\langle V\rangle = -\rho_0\mathcal{H}(V_N - \langle\gamma\rangle u_n) \tag{3.9}$$

which reduces to

$$f_\Sigma^{in}\cdot\overline{V} = f_\Sigma^{in}\cdot V = -\rho_0\mathcal{H}V_N \equiv \rho_0\mathcal{H}\langle\gamma v_n\rangle \tag{3.10}$$

in the case of *full coherency*.

Basing on the balance of energy and on the possible entropy production σ_Σ across the front we are able to find that [9]

$$-\rho_0\mathcal{H}\overline{V}_N = \sigma_\Sigma\theta \tag{3.11}$$

and that

$$-\rho_0\theta[\psi_\theta] - [q_R\cdot N] = \sigma_\Sigma\theta\,, \tag{3.12}$$

where $-\psi_\theta \equiv \eta$, and q_R represent the entropy density per unit mass and the heat flux respectively. Eqs. (3.10) and (3.11) lead to the conclusion that the power of f_Σ^{in} is its dissipative, according to the general statement of thermodynamics:

$$\sigma_\Sigma\theta \geq 0\,. \tag{3.13}$$

Acknowledgements. Work supported by the Italian GNFM of CNR and by the French CNRS (URA 229).

References

1. C. Kittel, *Introduction to Solid State Physics*, J. Wiley and Sons, N. Y., 1986.
2. E. Kröner, *Int. J. Solids Structures* **29** (1992) 1849.
3. J. Ericksen, *Introduction to Thermodynamic of Solids*, Chapman and Hall, London, 1991.
4. Z. Nishiyama, *Martensitic Transformation*, Acad. Press, N. Y., 1978.
5. L. Roitburd, *Solid State Physics*, **33**, Acad. Press, N. Y., 1978.
6. M. Epstein and G. A. Maugin, *C. R. Acad. Sci. Paris* **II-320** (1995) 63.
7. G. A. Maugin, *Material Inhomogeneities in Elasticity*, Chapman and Hall, London, 1993.
8. G. A. Maugin and C. Trimarco, *Acta Mechanica* **94** (1992) 1.
9. G. A. Maugin and C. Trimarco, *Meccanica* (1995) (forthcoming).
10. D. F. Nelson, *Mechanical Modellings of New Electromagnetic Materials*, Proc. IUTAM Symposium, Stockholm, April 2-6, 1990, ed. R.K.T. Hsieh, Elsevier, Amsterdam, 1990, p. 171.
11. J. D. Eshelby, *J. Elasticity* **5** (1975) 321.
12. G. Herrmann, *Int. J. Solids Structures* **18** (1982) 319.
13. M. E. Gurtin, *Arch. Rat. Mech. Anal.* **123** (1993) 305.
14. R. Aberayatne and J. K. Knowles, *Arch Rat. Mech. Anal.* **114** (1991) 119.
15. C. A. Truesdell and R. Toupin, *Principles of Classical Mechanics and Field Theory*, **III-1**, ed. S. Flügge, Springer-Verlag, Berlin, 1960.
16. C. A. Truesdell and W. Noll, *The Nonlinear Field Theory of Mechanics. Handbuch der Physik*, **III-3**, ed. S. Flügge, Springer Verlag, Berlin, 1965.
17. L. M. Truskinovski, *P.M.M. USSR, J. Appl. Math. Mec.* **51** (1987) 777.

Continuum Models and Discrete Systems
Proceedings of 8th International Symposium, June 11-16, 1995, Varna, Bulgaria, ed. K.Z. Markov

DETERMINISTIC CHAOS IN SYSTEMS IN EQUILIBRIUM

H. ZORSKI
Institute of Fundamental Technological Research,
Polish Academy of Sciences, Świętokrzyska 21, PL-00 049 Warsaw, Poland

Abstract

The purpose of this lecture is to present some simple examples of static systems in which there occurs deterministic chaos. In a number of such systems the equations of equilibrium reduce to the so called standard (Chirikov-Taylor) dynamical system.

The examples include:

i) a long polymer (e.g. peptide) chain with or without external dipole field;

ii) the Frenkel-Kontorowa model of dislocation in a crystal.

Some generalizations of the above systems are discussed.

PART II

CONTINUUM MECHANICS OF COMPLEX FLUIDS AND DEFORMABLE SOLIDS WITH MICROSTRUCTURE

Continuum Models and Discrete Systems
Proceedings of 8th International Symposium, June 11-16, 1995, Varna, Bulgaria, ed. K.Z. Markov
©World Scientific Publishing Company, 1996, pp. 92–99

BOUNDS FOR THE EFFECTIVE PROPERTIES OF A NONLINEAR TWO-PHASE COMPOSITE DIELECTRIC

K. E. BARRETT and D. R. S. TALBOT
School of Mathematical and Information Sciences,
Coventry University, Coventry CV1 5FB, United Kingdom

Abstract. Until recently, it has only been possible to obtain at most one new bound using the extension of the Hashin-Shtrikman methodology to nonlinear problems. However generalized Hashin-Shtrikman bounds for any nonlinear composite, which is characterized by a convex potential function, can now be constructed by using a nonlinear comparison medium and trial fields having the property of bounded mean oscillation. This enables the size of the penalty incurred by using a nonlinear, rather than a linear, comparison medium to be controlled. This work has recently been extended by using a nonlinear comparison composite which exhibits linear behaviour in each phase up to some value of the field and is nonlinear thereafter. This allows both upper and lower generalized Beran-type bounds to be constructed. The new construction is described and comparisons are made with bounds for a composite spheres assemblage obtained from finite element calculations.

1. Introduction

This paper is concerned with bounding the effective energy density function of a two-phase dielectric composite. Its constitutive behaviour is described by an energy function W, given by

$$W(\mathbf{E},\mathbf{x}) = \sum_{r=1}^{2} W_r(\mathbf{E}) f_r(\mathbf{x}), \tag{1}$$

with

$$W_r(\mathbf{E}) = \frac{D_0 E_0^{(r)}}{n+1} \left(\frac{|\mathbf{E}|}{E_0^{(r)}} \right)^{n+1} . \tag{2}$$

In Eq. (2) $\mathbf{E}$ is the electric field, f_r is the characteristic function of phase r, D_0 and $E_0^{(r)}$ are constants and the exponent n, which may not be an integer, is greater than or equal to 1. One approach that has been used to bound the overall energy function of such composites is the generalization to nonlinear problems of the variational principles of Hashin and Shtrikman [3,4], which was initiated by Willis [13] and developed by Talbot and Willis [9]. The nonlinear Hashin-Shtrikman principles involve the introduction of a "comparison medium". Ideally this medium is taken to be linear, so that its energy function is quadratic in $\mathbf{E}$. However the method requires that at large field values the energy function of the comparison medium grows slower than that of any of the phases to obtain a lower bound and faster than that of any of the phases to obtain an upper bound. Clearly for a composite whose energy function is given by (1) and (2) it is possible to obtain a lower bound in this way, but no upper bound can be found using a linear comparison medium, unless $n = 1$.

In a recent paper Talbot and Willis [11] have developed an approach which uses a *nonlinear* comparison medium, whose response is linear up to some finite value of the electric field and is nonlinear thereafter. The nonlinear part introduces a penalty associated with the region over which the electric fields may be large. Although this term cannot be found exactly, the trial fields used have the property of "bounded mean oscillation" (John and Nirenberg [5]) and this fact is employed to bound the penalty term.

In [11] the comparison medium was taken to be homogeneous. A related approach, originated by Ponte Castañeda [7] and incorporated into the Hashin-Shtrikman structure by Talbot and Willis [10], uses a linear comparison composite. The approach can be used directly to find a lower bound for composites such as those with energy functions given by (1) and (2), and the bound that is obtained involves a bound on the energy of the linear comparison composite. In [7] and [10] the Hashin-Shtrikman bound for a linear composite was used, although it is possible to incorporate higher order information about the microstructure by using, for example, the bounds of Beran [1]. In [8] Ponte Castañeda used the Beran bounds to obtain lower bounds on the overall energy of a class of nonlinear composites, which includes those discussed here.

In this work a nonlinear comparison composite is used to obtain an upper bound for a composite with energy function given by (1) and (2). The bound is constructed using a generalization of the approach in [11] and allows three-point information about the microstructure to be incorporated through a geometric parameter introduced by Milton [6]. The bound applies, in particular, to the composite spheres assemblage of Hashin [2]. For this composite the geometric parameter is known explicitly and the bound is compared with one derived from finite element calculations. The calculation of the new bounds is only summarized here, a full account will be published elsewhere [12].

2. Variational Principles

An infinite periodic medium is constructed by selecting a cube Q of composite and repeating it periodically. The origin is assumed to lie in Q and units of length are chosen so that Q has unit volume. The problem is to bound the mean energy of the composite $\widetilde{W}$, defined by

$$\widetilde{W}(\overline{\mathbf{E}}) = \inf_{\mathbf{E}\in K} \int_Q W(\mathbf{E}, \mathbf{x})\, \mathrm{d}\mathbf{x}, \tag{3}$$

where

$$K = \left\{ \mathbf{E} : \mathbf{E} \text{ is Q} - \text{periodic}, \mathbf{E} = -\nabla\phi, \int_Q \mathbf{E}(\mathbf{x})\, d\mathbf{x} = \overline{\mathbf{E}} \right\} \tag{4}$$

and W is given by (1) and (2). An alternative description follows from the variational principle dual to (3):

$$\widetilde{W}^*(\overline{\mathbf{D}}) = \inf_{\mathbf{D}} \int_Q W^*(\mathbf{D}, \mathbf{x})\, d\mathbf{x}, \tag{5}$$

where $\widetilde{W}^*$ is the convex dual of $\widetilde{W}$, W^* is the dual, with respect to the variable $\mathbf{E}$, of W and $\mathbf{D}$ is divergence free in a weak sense with prescribed mean value $\overline{\mathbf{D}}$. Elementary upper and lower bounds for $\widetilde{W}$ follow directly by using the trial fields $\mathbf{E} = \overline{\mathbf{E}}$ in (3), $\mathbf{D} = \overline{\mathbf{D}}$ in (5) to get

$$\{\overline{W^*}\}^*(\overline{\mathbf{E}}) \leq \widetilde{W}(\overline{\mathbf{E}}) \leq \overline{W}(\overline{\mathbf{E}}), \tag{6}$$

where the lower bound follows by duality.

Now introduce a comparison composite, which has the same microgeometry as the nonlinear composite and energy function $\widehat{W}(\mathbf{E}, \mathbf{x})$. It follows from the definition

$$(W - \widehat{W})^*(\mathbf{P}, \mathbf{x}) = \sup_{\mathbf{E}} \left\{ \mathbf{P} \cdot \mathbf{E} - W(\mathbf{E}, \mathbf{x}) + \widehat{W}(\mathbf{E}, \mathbf{x}) \right\}, \tag{7}$$

that

$$W(\mathbf{E}, \mathbf{x}) \geq \mathbf{P} \cdot \mathbf{E} + \widehat{W}(\mathbf{E}, \mathbf{x}) - (W - \widehat{W})^*(\mathbf{P}, \mathbf{x}) \tag{8}$$

for any $\mathbf{E}$ and $\mathbf{P}$. Replacing W in (3) by the right side of (8) now gives

$$\widetilde{W}(\overline{\mathbf{E}}) \geq \inf_{\mathbf{E} \in K} \int_Q \left[\mathbf{P} \cdot \mathbf{E} + \widehat{W}(\mathbf{E}, \mathbf{x}) - (W - \widehat{W})^*(\mathbf{P}, \mathbf{x}) \right] d\mathbf{x}, \tag{9}$$

for any $\mathbf{P}$. An upper bound is found by defining

$$(W - \widehat{W})_*(\mathbf{P}, \mathbf{x}) = \inf_{\mathbf{E}} \left\{ \mathbf{P} \cdot \mathbf{E} - W(\mathbf{E}, \mathbf{x}) + \widehat{W}(\mathbf{E}, \mathbf{x}) \right\}, \tag{10}$$

and reasoning similar to that leading to (9) gives

$$\widetilde{W}(\overline{\mathbf{E}}) \leq \inf_{\mathbf{E} \in K} \int_Q \left[\mathbf{P} \cdot \mathbf{E} + \widehat{W}(\mathbf{E}, \mathbf{x}) - (W - \widehat{W})_*(\mathbf{P}, \mathbf{x}) \right] d\mathbf{x}, \tag{11}$$

for any field $\mathbf{P}$. The bounds (9) and (11) are the generalization to nonlinear problems of the variational principles of Hashin and Shtrikman. The bounds are non-trivial provided the right sides of (7) and (10) are finite. Thus for (9) to be finite $W - \widehat{W}$ should tend to infinity faster than any linear function as $|\mathbf{E}| \to \infty$ for all $\mathbf{x}$ and some $\mathbf{P}$, while for (11) to be finite the same statement should be true for $\widehat{W} - W$.

To obtain an upper bound from (11), let

$$\widehat{W}(\mathbf{E},\mathbf{x}) = \sum_{r=1}^{2} \widehat{W}_r(\mathbf{E}) f_r(\mathbf{x}) = \frac{1}{2}\varepsilon(\mathbf{x})|\mathbf{E}|^2 + \sum_{r=1}^{2} [W_r(|\mathbf{E}|) - W_r(\lambda)]\, H(|\mathbf{E}| - \lambda) f_r(\mathbf{x}), \tag{12}$$

where

$$\varepsilon(\mathbf{x}) = \varepsilon_1 f_1(\mathbf{x}) + \varepsilon_2 f_2(\mathbf{x}), \tag{13}$$

H denotes the Heaviside step function and the constants $\lambda > 0$, ε_1 and ε_2 are to be chosen. It follows that

$$\begin{aligned}\widetilde{W}(\overline{\mathbf{E}}) \le \int_Q \left[\frac{1}{2}\varepsilon(\mathbf{x})|\mathbf{E}|^2 + \mathbf{P}\cdot\mathbf{E} - (W - \widehat{W})_*(\mathbf{P},\mathbf{x})\right] \mathrm{d}\mathbf{x} \\ + \sum_{r=1}^{2} \int_{S_\lambda} [W_r(|\mathbf{E}|) - W_r(\lambda)]\, f_r(\mathbf{x})\, \mathrm{d}\mathbf{x},\end{aligned} \tag{14}$$

where S_λ is given by

$$S_\lambda = \{\mathbf{x} \in Q : |\mathbf{E}(\mathbf{x})| > \lambda\}, \tag{15}$$

for whichever trial field $\mathbf{E}$ is chosen. The last term in (14) is the penalty imposed by the nonlinear part of $\widehat{W}$. To obtain a lower bound, $\widehat{W}$ can be taken as $\frac{1}{2}\varepsilon(\mathbf{x})|\mathbf{E}|^2$, with ε given by (13), in (7) and (9). The lower bound of Ponte Castañeda then follows by choosing $\mathbf{P} = \mathbf{0}$, substituting any lower bound for the overall energy of the linear composite having dielectric constant $\varepsilon(\mathbf{x})$ and maximizing with respect to the constants ε_1 and ε_2.

3. The Two-phase Composite

An upper bound is now developed for a geometrically isotropic composite which incorporates three-point information about the microstructure through the geometric parameter introduced by Milton [6]. A lower bound of this type for a composite whose energy is given by (1) and (2) has been given by Ponte Castañeda [8]. In order to find an upper bound from (14), first let

$$\mathbf{P}(\mathbf{x}) = \sum_{r=1}^{2} \mathbf{P}_r f_r(\mathbf{x}), \tag{16}$$

where $\mathbf{P}_1$ and $\mathbf{P}_2$ are constant vectors. Next, the trial field $\mathbf{E}$ is taken to have the form

$$\mathbf{E}(\mathbf{x}) = \overline{\mathbf{E}} - (\widehat{\mathbf{\Gamma}} * (f_1\mathbf{u}))(\mathbf{x}), \tag{17}$$

where $*$ denotes convolution over Q, $\mathbf{u}$ is a constant vector and the tensor function $\widehat{\mathbf{\Gamma}}$ is related to the Q-periodic Green's function of a material with dielectric constant numerically equal to unity. Then, substituting the trial field (17) into (14) gives

$$\frac{1}{2}\bar{\varepsilon}|\overline{\mathbf{E}}|^2 + \frac{1}{2}\int_Q [\bar{\varepsilon} f_1 \mathbf{u}\cdot\widehat{\mathbf{\Gamma}} * (f_1\mathbf{u}) - 2(\varepsilon_1-\varepsilon_2) f_1 \overline{\mathbf{E}}\cdot\widehat{\mathbf{\Gamma}} * (f_1\mathbf{u})$$
$$+(\varepsilon_1-\varepsilon_2)(f_1-c_1)\widehat{\mathbf{\Gamma}} * (f_1\mathbf{u})\cdot\widehat{\mathbf{\Gamma}} * (f_1\mathbf{u})$$
$$+2\mathbf{P}\cdot\overline{\mathbf{E}} + 2f_1(\mathbf{P}_1-\mathbf{P}_2)\cdot\widehat{\mathbf{\Gamma}} * (f_1\mathbf{u})]\,\mathrm{d}\mathbf{x}, \tag{18}$$

for the integral of the first two terms, where the identity $\widehat{\mathbf{\Gamma}} * \widehat{\mathbf{\Gamma}} = \widehat{\mathbf{\Gamma}}$ and the fact that $\widehat{\mathbf{\Gamma}}$ has zero mean value have been used. The last term in (14) was shown in [11] to be bounded by

$$B\int_\lambda^\infty \exp(-\alpha s)\,\mathrm{d}\chi(s), \tag{19}$$

where

$$\chi(s) = \max_r\{W_r(s) - W_r(\lambda)\}. \tag{20}$$

The constants B and α depend on $\mathbf{u}$ and are given in [11].

Next, if the microgeometry is statistically isotropic and has so fine a scale that the shape of Q has no effect, it follows that

$$\int_Q f_1(\mathbf{x})(\widehat{\mathbf{\Gamma}} * f_1)(\mathbf{x})\,\mathrm{d}\mathbf{x} = \frac{c_1c_2}{3}\mathbf{I} \tag{21}$$

and the integral involving three points can be written in terms of the geometrical parameter ζ_1 introduced by Milton [6]. The upper bound can now be written

$$\begin{aligned}\widetilde{W}(\overline{\mathbf{E}}) \le\ & \tfrac{1}{2}\ \left\{\varepsilon_1\left[c_1|\overline{\mathbf{E}} - \frac{c_2}{3}\mathbf{u}|^2 + \frac{2c_1c_2\zeta_1}{9}|\mathbf{u}|^2\right] + \varepsilon_2\left[c_2|\overline{\mathbf{E}} + \frac{c_1}{3}\mathbf{u}|^2 + \frac{2c_1c_2\zeta_2}{9}|\mathbf{u}|^2\right]\right\}\\ & +\ c_1\left\{\mathbf{P}_1\cdot\left[\overline{\mathbf{E}} + \frac{c_2}{3}\mathbf{u}\right] - (W_1 - \widehat{W}_1)_*(\mathbf{P}_1)\right\}\\ & +\ c_2\left\{\mathbf{P}_2\cdot\left[\overline{\mathbf{E}} - \frac{c_1}{3}\mathbf{u}\right] - (W_2 - \widehat{W}_2)_*(\mathbf{P}_2)\right\}\\ & +\ B\int_\lambda^\infty \exp(-\alpha s)\,\mathrm{d}\chi(s),\end{aligned} \tag{22}$$

where $\zeta_2 = 1 - \zeta_1$. The best bound follows by minimizing the right side of (22) with respect to λ, ε_1, ε_2, $\mathbf{P}_1$, $\mathbf{P}_2$ and $\mathbf{u}$. The minimizations with respect to $\mathbf{P}_1$ and $\mathbf{P}_2$ are straightforward and the result is

$$\begin{aligned}\widetilde{W}(\overline{\mathbf{E}}) \le\ & \tfrac{1}{2}\ \left\{\varepsilon_1\left[c_1|\overline{\mathbf{E}} - \frac{c_2}{3}\mathbf{u}|^2 + \frac{2c_1c_2\zeta_1}{9}|\mathbf{u}|^2\right] + \varepsilon_2\left[c_2|\overline{\mathbf{E}} + \frac{c_1}{3}\mathbf{u}|^2 + \frac{2c_1c_2\zeta_2}{9}|\mathbf{u}|^2\right]\right\}\\ & -\ c_1(\widehat{W}_1 - W_1)^{**}(|\overline{\mathbf{E}} + \frac{c_2}{3}\mathbf{u}|) - c_2(\widehat{W}_2 - W_2)^{**}(|\overline{\mathbf{E}} - \frac{c_1}{3}\mathbf{u}|)\\ & +\ B\int_\lambda^\infty \exp(-\alpha s)\,\mathrm{d}\chi(s),\end{aligned} \tag{23}$$

where the fact that $(W - \widehat{W})_{**} = -(\widehat{W} - W)^{**}$ has been used. The remaining minimizations are described elsewhere [12].

4. Results for a Composite Spheres Assemblage

First, the penalty term in (23) is a function of $\mathbf{u}$ only. It follows that $\mathbf{u}$ and $\overline{\mathbf{E}}$ are aligned and may be replaced by scalar quantities $\overline{E}$ and u. Also, the overall energy $\widetilde{W}$ is given by

$$\widetilde{W}(\overline{E}) = \frac{D_0 \widetilde{E_0}}{n+1} \left(\frac{\overline{E}}{\widetilde{E_0}} \right)^{n+1} \tag{24}$$

and it follows that a lower bound for $\widetilde{W}$ produces an upper bound for $\widetilde{E_0}$ and conversely an upper bound for $\widetilde{W}$ produces a lower bound for $\widetilde{E_0}$.

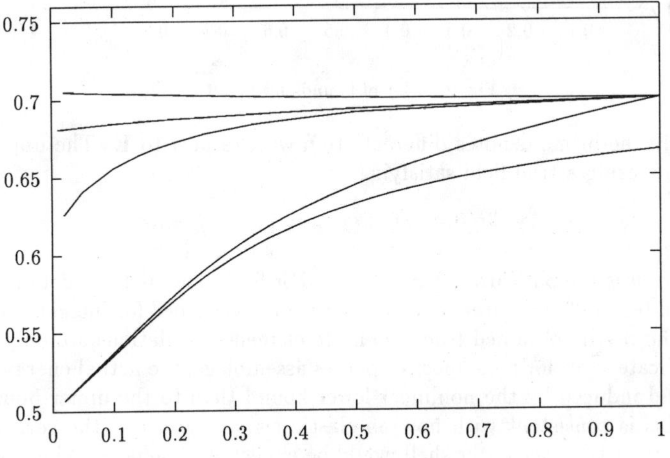

Fig. 1. A set of bounds for $\zeta_1 = 0$.

Results have been obtained for $E_0^{(2)}/E_0^{(1)} = 2$ and $c_1 = c_2 = 0.5$. In this case the coefficient of $|E|^{n+1}$ in the expression for W_1 is larger than in that for W_2. Now, for a composite spheres assemblage, when the outer shell is phase 2, $\zeta_1 = 0$ and when the outer shell is phase 1, $\zeta_1 = 1$. For such a composite, a set of bounds for $\widetilde{E_0}/E_0^{(2)}$ plotted against $1/n$ is shown in Figure 1 for $\zeta_1 = 0$ and in Figure 2 for $\zeta_1 = 1$. In each figure the outer pair of lines show the bounds induced by the simple bounds (6). The next pair are the lower bound induced by (23) and the upper bound calculated from the lower bound for $\widetilde{W}$ given in [8]. The innermost pair are upper and lower bounds obtained from finite element calculations. With C denoting a composite sphere, the lower bound was generated using a trial field satisfying

$$\operatorname{div}(W'(\mathbf{E})) = 0, \quad \mathbf{E} = -\nabla\phi, \ \mathbf{x} \in C, \quad \mathbf{E} = \overline{\mathbf{E}} \cdot \mathbf{x}, \ \mathbf{x} \in \partial C, \tag{25}$$

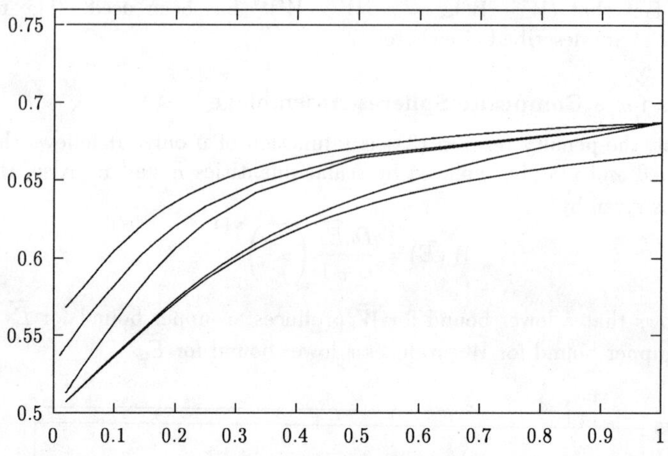

Fig. 2. A set of bounds for ζ_1 =1.

in (2), where the prime denotes differentiation with respect to $\mathbf{E}$. The upper bound was found by using a trial field satisfying

$$\mathrm{div}\mathbf{D} = 0,\ \mathbf{x} \in C, \quad \mathbf{D} \cdot \mathbf{n} = \overline{\mathbf{D}} \cdot \mathbf{n},\ \mathbf{x} \in \partial C, \tag{26}$$

in (5), where $\mathbf{n}$ is the outward normal to ∂C. For both bounds the calculations were expensive in terms of computer time and were only performed for integer values of n. Although the results obtained from the finite element calculations are only bounds, they do indicate that, for a composite spheres assemblage, the actual energy is closer to the bound induced by the nonlinear lower bound than to the upper bound when $\zeta_1 = 0$. This is consistent with how an elastic composite where the core sphere is very much stiffer than the outer shell would be expected to behave. When $\zeta_1 = 1$ the actual energy appears, again, to be closer to the bound induced by the nonlinear lower bound, at least for small values of n. However, in this case, the bounds produced by the methods described here and in [8] are themselves closer than when $\zeta_1 = 0$. In conclusion, it is worth noting that the bound generated by (23) is valid for any composite and there is certainly scope for improving this bound. Also, as remarked earlier, the finite element calculations required a significant amount of computer time and that in itself provides some justification for developing bounds such as (23) which are much easier to implement numerically.

References

1. M. Beran, *Nuovo Cim.* **38** (1965) 771.
2. Z. Hashin, *J. Appl. Mech.* **29** (1962) 143.
3. Z. Hashin and S. Shtrikman, *J. Mech. Phys. Solids* **10** (1962) 343.
4. Z. Hashin and S. Shtrikman, *J. Mech. Phys. Solids* **11** (1963) 127.
5. F. John and L. Nirenberg, *Commun. Pure Appl. Math.* **14** (1961) 415.
6. G. W. Milton, *Phys. Rev. Lett.* **46** (1981) 542.
7. P. Ponte Castañeda, *J. Mech. Phys. Solids* **39** (1991) 45.
8. P. Ponte Castañeda, *Phil. Trans. R. Soc. Lond.* **A340** (1992) 531–567.
9. D. R. S. Talbot and J. R. Willis, *IMA J. Appl. Math.* **35** (1985) 39.
10. D. R. S. Talbot and J. R. Willis, *Int. J. Solids Structures* **29** (1992) 1981.
11. D. R. S. Talbot and J. R. Willis, *Proc. R. Soc. Lond.* **A447** (1994) 365.
12. D. R. S. Talbot and J. R. Willis, *In preparation.*
13. J. R. Willis, *J. Appl. Mech.* **50** (1983) 1202.

Continuum Models and Discrete Systems
Proceedings of 8th International Symposium, June 11-16, 1995, Varna, Bulgaria, ed. K.Z. Markov
© World Scientific Publishing Company, 1996, pp. 100–110

HOMOGENIZATION PROBLEM FOR BULK DIFFUSIONAL CREEP

V. BERDICHEVSKY
Mechanical Engineering, Wayne State University,
Detroit, MI 48202, USA
and
B. SHOYKHET
Reliance Electric, Cleveland, OH 44117, USA

Abstract. The problem of homogenization for bulk diffusional creep in polycrystals with periodic microstructure is considered. The cell problem is derived in linear and nonlinear cases. Constitutive equations of macroscopic theory are obtained in an explicit form for the linear case. The stationary cell problem is posed for secondary creep. The stationary cell problem has a variational form, and the corresponding variational principle is formulated. Some results of numerical simulations for honeycomb structure are presented.

1. Introduction

We have been studying the problem of homogenization for creep in polycrystals where the leading mechanism of creep is the bulk diffusion of vacancies. The foundation of the theory of diffusional creep was laid down by Nabarro [1], Herring [2], Coble [3] and Lifshitz [4]. We have been following a nonlinear version of the theory presented in [5]. The statement of the homogenization problem is the following. Consider a polycrystalline body V consisting of a very large number of grains. The characteristic grain size ε is much smaller then the characteristic size L of region V. The body is loaded by some surface forces. Mechanical deformations inside the grains are governed by the equations of diffusional creep which are presented below. If $\varepsilon \ll L$, it is natural to expect that deformation of the body can be described by some displacement fields which are practically constant on the distances of order ε, i.e. by "averaged" displacements. The problem of homogenization concerns the derivation of the equations for averaged (or macro) displacements and the study of microdeformations of the grains. The problem of homogenization is an asymptotical problem with a small parameter ε/L. We are going to consider homogenization problem for polycrystals with periodic microstructure. The experience gained in various homogenization problems shows that macroequations for bodies with periodic and random structures are similar (see, for example, [6]).

The system of equations of bulk diffusional creep is as follows. The "elastic part" of the system of equations consists of equilibrium equations for the stress tensor σ^{ij}

$$\frac{\partial \sigma^{ij}}{\partial x^j} = 0, \tag{1.1}$$

and the constitutive equations

$$\sigma^{ij} = \frac{\partial F}{\partial \varepsilon_{ij}^{(e)}}, \tag{1.2}$$

where free energy F is a given function of elastic strains $\varepsilon_{ij}^{(e)}$, vacancy concentration c and temperature T. Elastic strains are defined as the difference of the total stains, determined by the displacement vector w_i and the plastic strains ε_{ij}^{p}. We consider "geometrically linear" theory, therefore

$$\varepsilon_{ij}^{(e)} = \frac{1}{2}\left(\frac{\partial w_i}{\partial x_j} + \frac{\partial w_j}{\partial x_i}\right) - \varepsilon_{ij}^{(p)}. \tag{1.3}$$

Small Latin indices run values 1, 2, 3 and correspond to the projections on the axes of a Cartesian frame. Temperature T is assumed to be maintained constant. In "geometrically linear" creep the deformation of the grain boundaries can be neglected in the calculation of stresses and strains and the nonlinearity is kept in the constitutive equations only.

To close the system of equations one has to formulate the evolution law for plastic strains $\varepsilon_{ij}^{(p)}$. The major feature of bulk diffusional creep is the consistency of plastic deformations: there exists a vector of plastic displacements $w_i^{(p)}$ such that plastic deformations $\varepsilon_{ij}^{(p)}$ are

$$\varepsilon_{ij}^{(p)} = \frac{1}{2}\left(\frac{\partial w_i^{(p)}}{\partial x_j} + \frac{\partial w_j^{(p)}}{\partial x_i}\right). \tag{1.4}$$

Therefore, the usual equations of elasticity should be complimented by three equations for plastic displacements $w_i^{(p)}$. These are the equations which relate the flux of vacancies and the corresponding "irreversible" flux of material

$$\dot{w}_i^{(p)} = D_i^j \frac{\partial}{\partial x^j}\frac{\partial F}{\partial c}, \tag{1.5}$$

where D_i^j is the diffusivity tensor.

System of equations (1.1)–(1.5) is closed by the diffusion equation for vacancy concentration

$$\frac{\partial c}{\partial t} = \frac{\partial}{\partial x^i} D^{ij} \frac{\partial}{\partial x^j}\frac{\partial F}{\partial c}. \tag{1.6}$$

The boundary conditions on the grain boundaries have the form ($\sigma_{nn} = \sigma^{ij} n_i n_j$, where n_i are the components of unit normal vector at the boundary, $\boldsymbol{\sigma}_\alpha$

are the tangent projections of the vector $\sigma_i^j n_j$, vectors are denoted by bold letters, Greek indices run values 1, 2; $[A]$ stands for the difference of values of A on two sides of the grain boundary Σ)

$$[w_n] = 0, \tag{1.7}$$

$$\boldsymbol{\sigma}_\alpha = 0 \quad \text{on each side of} \quad \Sigma, \tag{1.8}$$

$$[\sigma_{nn}] = 0, \tag{1.9}$$

$$\sigma_{nn} = \frac{\partial F}{\partial c} \quad \text{on each side of} \quad \Sigma. \tag{1.10}$$

Eq. (1.7) expresses the continuity of the normal component of the total displacement at the grain boundary. The tangent component of the displacements can be discontinuous: we assume that the process of stress relaxation at the boundary occurs much faster than the bulk diffusion. Therefore, Eq. (1.8) is accepted. The condition (1.9) of continuity of the normal stresses is reciprocal to (1.7). Eq. (1.10) controls the creation of vacancies at grain boundaries.

We do not formulate the boundary conditions on the polycrystal boundary ∂V because they do not affect the averaged equations and the cell equations.

In the linear case, free energy density is a quadratic function of $\varepsilon_{ij}^{(e)}$ and c

$$F = \tfrac{1}{2} A^{ijkl} \varepsilon_{ij}^{(e)} \varepsilon_{kl}^{(e)} + \tfrac{1}{2} A (c - c_0)^2, \tag{1.11}$$

where A^{ijkl} are the elastic moduli, c_0 is an equilibrium value of vacancy concentration. The material constant A can be found from elementary statistical consideration: $A = \rho_0 kT / m c_0$, m is the mass of one atom, ρ_0 is the mass density and k is the Boltzman constant.

In the following sections we formulate the results of the study of the homogenization problem for the system (1.1)–(1.10) under the natural assumption on the convexity and the smoothness of F. Before presenting the results, some description of a periodic structure is to be done.

We assume that the grains coincide with the cells of a periodic structure. Let ε be a characteristic size of the grain. We introduce dimensionless coordinates

$$y_i = x_i / \varepsilon \tag{1.12}$$

and define the periodic structure in the space of dimensionless coordinates $\boldsymbol{y}$ by the unit cell ω and the translation symmetry group G. The characteristic size of the unit cell is equal to unity. For $\boldsymbol{l} \in G$ we denote by $\omega(\boldsymbol{l})$ the image of the unit cell ω under the translation $\boldsymbol{l}$. It is assumed that different cells $\omega(\boldsymbol{l})$ may have in common the boundary points only, and that the union of the cells covers the whole y-space. It can be shown that the boundary $\partial\omega$ of the unit cell is comprised of the set of pairs of surfaces (or lines in $2D$ case) $S_1, S_1', S_2, S_2', \ldots, S_r, S_r'$, such that for every surface S_α, there exists a translation $\boldsymbol{l}_\alpha \in G$, mapping S_α onto S_α'. For hexagonal structure this notation is explained on Fig. 1. Obviously, the translation $-\boldsymbol{l}_\alpha$ maps S_α' onto S_α. Thus the periodic structure induces the certain mapping of the cell boundary

$\partial\,\omega \Longleftrightarrow \partial\,\omega$, which will be used for the formulation of the boundary conditions. For every point $\boldsymbol{y} \in \partial\,\omega$ we denote by $\boldsymbol{l}\,(\boldsymbol{y})$ the corresponding translation vector. The points $\boldsymbol{y}$ and $\boldsymbol{y}' = \boldsymbol{y} + \boldsymbol{l}\,(\boldsymbol{y})$ will be referred to as the corresponding points. Note that $\boldsymbol{l}\,(\boldsymbol{y})$ is constant within each surfaces S_α, S'_α.

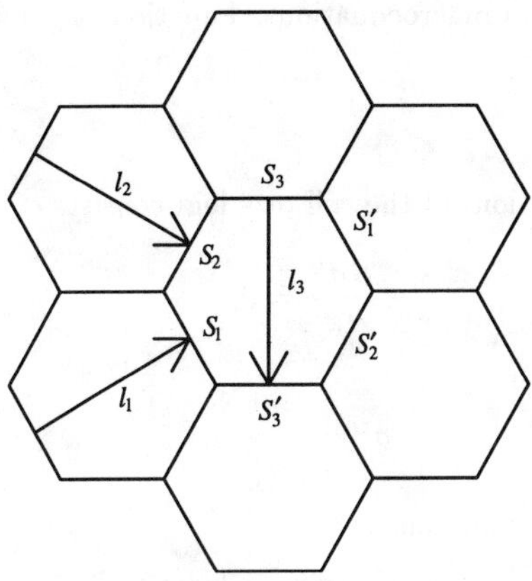

Fig. 1. Hexagonal structure. The translation vectors shown by arrows map the corresponding parts of the cell boundaries.

The unit normal $\boldsymbol{n}$ to the cell boundary is assumed to be directed outward the cell, therefore at the corresponding points $\boldsymbol{y}$ and $\boldsymbol{y}'$ we have

$$\boldsymbol{n}\,(\boldsymbol{y}) + \boldsymbol{n}\,(\boldsymbol{y}') = 0. \tag{1.13}$$

Let $f\,(\boldsymbol{y})$ be an arbitrary function, which is continuous within each grain, but may be discontinuous at the grain boundaries. Function $f\,(\boldsymbol{y})$ is called periodic if

$$f\,(\boldsymbol{y} + \boldsymbol{l}) = f\,(\boldsymbol{y}) \quad \text{for any} \quad \boldsymbol{y} \in \partial\omega \quad \text{and for any} \quad \boldsymbol{l} \in G. \tag{1.14}$$

If function $f\,(\boldsymbol{y})$ is known within the unit cell, it can be extended to the whole space by the formula (1.14).

2. Asymptotic Expansions

For $\varepsilon \to 0$ the first two terms of asymptotical expansion of the displacement field has the form

$$w_i\,(\boldsymbol{x}, t) = \bar{w}_i\,(\boldsymbol{x}, t) + \varepsilon\,\psi_i\,(\boldsymbol{x}, \boldsymbol{y}, t)\,, \tag{2.1}$$

where the "averaged" displacements do not depend on ε, $y_i = x_i/\varepsilon$ and ψ_i are some periodic functions of $\boldsymbol{y}$.

Asymptotical expansion of diffusion vacancy concentration starts with a periodic function of the fast variables

$$c = c(\boldsymbol{x}, \boldsymbol{y}, t). \tag{2.2}$$

Functions $\psi_i(\boldsymbol{x}, \boldsymbol{y}, t)$ and $c_i(\boldsymbol{x}, \boldsymbol{y}, t)$ determine microfields while functions $\bar{w}_i$ are the required functions in macroequations. Functions ψ_i and c can be found from the following cell problem.

3. Cell Problem

The system of equations of the cell problem consists of equilibrium equations

$$\frac{\partial \sigma^{ij}}{\partial x^j} = 0, \tag{3.1}$$

constitutive equations

$$\sigma^{ij} = \frac{\partial F}{\partial \varepsilon_{ij}^{(e)}}, \tag{3.2}$$

strain-displacement equations

$$\varepsilon_{ij}^{(e)} = \bar{\varepsilon}_{ij} + \frac{1}{2}\left(\frac{\partial \psi_i}{\partial y_j} + \frac{\partial \psi_j}{\partial y_i}\right) - \varepsilon_{ij}^{(p)}, \tag{3.3}$$

the law of evolution of plastic strains

$$\frac{\partial \varepsilon_{ij}^{(p)}}{\partial t} = \frac{1}{2\varepsilon^2}\left(\frac{\partial}{\partial y^i} D_j^k \frac{\partial}{\partial y^k}\frac{\partial F}{\partial c} + \frac{\partial}{\partial y^j} D_i^k \frac{\partial}{\partial y^k}\frac{\partial F}{\partial c}\right) \tag{3.4}$$

and the diffusion equation

$$\frac{\partial c}{\partial t} = \frac{1}{\varepsilon^2}\frac{\partial}{\partial y^i} D^{ij} \frac{\partial}{\partial y^j}\frac{\partial F}{\partial c}. \tag{3.5}$$

Here $\bar{\varepsilon}_{ij}$ are the macrostrains which are considered as some parameters in the cell problem, $\bar{\varepsilon}_{ij}$ do not depend on y-coordinates but might be some functions of time.

Note that the factor ε^2 in Eqs. (3.4) and (3.5) can be deleted by scaling of time, and the Eqs. (3.1)–(3.5) do not contain ε after scaling.

We search for the solution ψ_i, c of Eqs. (3.1)–(3.5) which is periodic in y-variables and obey the following boundary conditions

$$\sigma_{nn} = \frac{\partial F}{\partial c}, \quad \boldsymbol{y} \in \partial\omega, \tag{3.6}$$

$$\boldsymbol{\sigma}_\alpha = \mathbf{0}, \quad \boldsymbol{y} \in \partial\omega, \tag{3.7}$$

$$\psi_n\Big|_{\boldsymbol{y}} + \psi_n\Big|_{\boldsymbol{y}+\boldsymbol{l}(\boldsymbol{y})} = 0, \quad c\Big|_{\boldsymbol{y}} = c\Big|_{\boldsymbol{y}+\boldsymbol{l}(\boldsymbol{y})}, \quad \boldsymbol{y} \in \partial\omega. \tag{3.8}$$

If parameters $\bar{\varepsilon}_{ij}$ are given functions of time t, the system has, probably, a unique solution. Uniqueness can be proved in the linear case.

4. Macroequations

Let us define macrostresses as averaged values of stresses

$$\bar{\sigma}^{ij} = \frac{1}{|\omega|} \int_{\omega} \sigma^{ij} \, d^3 x, \tag{4.1}$$

where $|\omega|$ is the cell volume. Then $\bar{\sigma}^{ij}$ is a functional of $\bar{\varepsilon}^{ij}$ which can be found from the cell problem:

$$\bar{\sigma}_{ij} = \text{Functional}\left(\bar{\varepsilon}_{ij}\right). \tag{4.2}$$

It can be shown that macrostresses obey the equilibrium equation

$$\frac{\partial \bar{\sigma}^{ij}}{\partial x^j} = 0, \tag{4.3}$$

and macrostrains relate to macrodisplacements by the usual formula

$$\bar{\varepsilon}_{ij} = \frac{1}{2} \left(\frac{\partial \bar{w}_i}{\partial x_j} + \frac{\partial \bar{w}_j}{\partial x_i} \right). \tag{4.4}$$

Eqs. (4.3), (4.4), (4.2) and (3.6) form a closed system of equations for macrodisplacements.

5. Macroequations in Linear Case

From now on we consider the linear case when free energy is given by (1.11). It can be shown, that the system (3.1)–(3.8) satisfies the Boltzman superposition principle, and, hence, the stress-strain relation (4.2) has the form

$$\begin{aligned} \bar{\sigma}^{ij} &= \int_0^t Q^{ijkl}(t-\xi)\, \dot{\bar{\varepsilon}}_{kl}(\xi)\, d\xi \\ &= Q^{ijkl}(0)\, \bar{\varepsilon}_{kl} + \int_0^t \frac{\partial Q^{ijkl}(t-\xi)}{\partial t}\, \bar{\varepsilon}_{kl}(\xi)\, d\xi. \end{aligned} \tag{5.1}$$

Here the function $Q^{ijkl}(t)$ is the macrostress component $\bar{\sigma}^{ij}(t)$ caused by the application of the strain $\bar{\varepsilon}_{kl}(t) = 1$ for $t > 0$, $\bar{\varepsilon}_{kl}(t) = 0$ for $t \leq 0$, while all the other strain components are equal to zero. Tensor $Q^{ijkl}(0)$ is the elastic moduli tensor of polycrystal. The inversion of the relation (5.1) renders

$$\begin{aligned} \bar{\varepsilon}^{ij} &= \int_0^t R_{ijkl}(t-\xi)\, \dot{\bar{\sigma}}^{kl} d\xi \\ &= R_{ijkl}(0)\, \bar{\sigma}^{kl} + \int_0^t \frac{\partial R_{ijkl}(t-\xi)}{\partial t}\, \bar{\sigma}^{kl}(\xi)\, d\xi. \end{aligned} \tag{5.2}$$

The function $R_{ijkl}(t)$ has the sense of the macrostrain component $\bar{\varepsilon}_{ij}(t)$ caused by the constant macrostress $\bar{\sigma}^{kl}(t) = 1$ for $t > 0$, $\bar{\sigma}^{kl}(t) = 0$ for $t \le 0$, while all the other stress components are zeros. The values $R_{ijkl}(0)$ are the components of the tensor of the elastic compliance of the polycrystal. Due to that it is convenient to decompose $R_{ijkl}(t)$ into the sum

$$R_{ijkl}(t) = R_{ijkl}(0) + K_{ijkl}(t)\,, \quad K_{ijkl}(0) = 0, \tag{5.3}$$

where the functions $K_{ijkl}(t)$ are the creep macrostrains. The creep curves $K_{ijkl}(t)$ have a square root asymptotics for small values of t:

$$K_{ijkl}(t) \sim t^{1/2}, \quad \frac{\partial K_{ijkl}(t)}{\partial t} \sim t^{-1/2}.$$

Hence, at the initial moment, the creep rate tends to the infinity as $t^{-1/2}$.

An important feature of the constitutive equations (5.1), (5.2) is that these equations are not local: there is a memory of the prehistory of the process. This means that local theories of primary creep are not adequate at least in the case of bulk diffusional creep.

6. Secondary Creep

Generally speaking, the macroscopic constitutive equations are given by the integral operators (5.1) or (5.2). However, for "slow" loading processes and developed creep it is possible to use as an approximation the creep law

$$\dot{\bar{\varepsilon}}_{ij} = E_{ijkl}\,\bar{\sigma}'^{kl}, \qquad \bar{\sigma}'^{kl} = \bar{\sigma}^{kl} - \delta^{kl}\,\bar{\sigma}^{ss}/3\,, \tag{6.1}$$

or

$$\bar{\sigma}^{ij} = e^{ijkl}\,\dot{\bar{\varepsilon}}_{kl}\,, \tag{6.2}$$

$$\dot{\bar{\varepsilon}}_{kk} = 0\,. \tag{6.3}$$

The macrocharacteristics of the secondary creep E_{ijkl} are the limits of the creep rates $\dot{K}_{ijkl}(t)$ when $t \to \infty$. The fact that the creep rates tend to some constants when $t \to \infty$ constitutes the basis of the approximation (6.1)–(6.3). Tensor e^{ijkl} is the inverse tensor to E_{ijkl}.

It turns out that for the secondary creep the creep macrocharacteristics and the microfields may be found from the following variational principle.

Let $\dot{\bar{\varepsilon}}_{ij}$ be an arbitrary constant creep rates, satisfying the incompressibility condition (6.3). Denote by $J(C)$ the following functional of function $C(y)$

$$J(C) = \frac{1}{2}\int_{\omega} d^{ij}\,\frac{\partial C}{\partial y^i}\,\frac{\partial C}{\partial y^j}\,d^3y + \frac{1}{2}\int_{\partial\omega} \dot{\bar{\varepsilon}}_{ij} n^i l^j C\,d^2y\,. \tag{6.4}$$

Here $d^{ij} = AD^{ij}/\varepsilon^2$.

Consider the minimization problem

$$J(C) \to \min_{C} . \tag{6.5}$$

Minimum is sought on the set of all functions C obeying the constraints

$$C(\boldsymbol{y}) = C(\boldsymbol{y} + \boldsymbol{l}(\boldsymbol{y})), \quad \boldsymbol{y} \in \partial\omega, \tag{6.6}$$

$$\int_{\partial\omega} C \cdot \boldsymbol{l} \times \boldsymbol{n} \, d^2 y = \boldsymbol{0}. \tag{6.7}$$

After the solution of the variational problem (6.4)–(6.7) is found, the deviator of macrostresses is defined by the formula

$$\bar{\sigma}'^{ij} = \frac{A}{2|\omega|} \left(\int_{\partial\omega} C n^i l^j \, d^2 y - \delta^{ij} \int_{\partial\omega} C n^s l^s \, d^2 y/3 \right). \tag{6.8}$$

The solution C of the problem (6.4)–(6.7) depends linearly on the parameters $\dot{\bar{\varepsilon}}_{ij}$. Hence plugging this solution in (6.8) one obtains macrostresses in terms of creep velocities $\dot{\bar{\varepsilon}}_{ij}$, i.e. the relation (6.2), and the macrocharacteristics e^{ijkl}.

So, the solution of the variational problem (6.4)–(6.8) gives us macroequations of creep and the microfield of vacancy concentration. We have to also determine microstresses. To this end we note that the solution $C(\boldsymbol{y})$ of the variational problem is determined up to an arbitrary constant C_0. We fix this constant by the condition

$$\int_{\partial\omega} (C(\boldsymbol{y}) - C_0) n^s l^s \, d^2 y = 0. \tag{6.9}$$

Then the microstresses at the cell boundary are determined by formulas (3.6) and (3.7) which for the linear case takes the form

$$\sigma_{nn}(\boldsymbol{y}) = A(C(\boldsymbol{y}) - C_0), \quad \boldsymbol{\sigma}_\alpha = 0, \quad \boldsymbol{y} \in \partial\omega. \tag{6.10}$$

Now we may consider the elasticity problem for the cell with the boundary conditions (6.10). The necessary condition for solubility of this problem (total external force and moment are zeros) is satisfied due to the choice of the additive constant C_0 made through Eq. (6.9).

Denote by $\sigma^{*ij}(\boldsymbol{y})$ the solution of the elasticity problem. The microstresses of the nonstationary problem (3.1)–(3.8) tends to the sum $\sigma^{*ij}(\boldsymbol{y}) + \delta^{ij}\bar{\sigma}^{ss}/3$.

Remark 1. For some cell shapes, the relations (6.7) may be automatically satisfied. In particular, it is true if for any $\boldsymbol{y} \in \partial\omega$ the normal vector $\boldsymbol{n}(\boldsymbol{y})$ to the cell boundary is parallel to the translation vector $\boldsymbol{l}(\boldsymbol{y})$. The regular $2D$ hexagonal structure used in the next Section for numerical examples, possesses this property. It can be shown, that in this case the solution of the cell problem (3.1)–(3.8) is determined up to a rotation of the cell as a rigid body. It does not influence the computation of constants in the constitutive equations (6.1) and (6.2).

Remark 2. Let us normalize the diffusivity tensor in (3.5):

$$D^{ij} = D\tilde{D}^{ij}, \tag{6.10}$$

where D is some characteristic value of tensor D^{ij}. Analysis of physical dimensions shows that creep rate constants can be represented as follows:

$$E_{ijkl} = \tilde{E}_{ijkl}\,\frac{D}{\varepsilon^2}, \tag{6.11}$$

where the dimensionless constants $\tilde{E}_{ijkl}$ depend on the constants $\tilde{D}^{ij}$ and the unit cell shape only. An important consequence is that the creep rates in secondary creep do not depend on the elastic properties.

Remark 3. The characteristic time of approaching the steady creep under the constant load is ε^2/DA.

7. Numerical Results

The numerical modeling were done for the two-dimensional regular hexagonal structure, shown in Fig. 1. For definiteness, it was assumed that grains are isotropic, and hence only four physical constants are needed: Young modulus E, Poison ratio ν, the constant A in (1.11), diffusivity constant D in (6.11) (with $\tilde{D}^{ij} = \delta^{ij}$), and the grain size ε.

Secondary Creep Rates. In creep, the periodic hexagonal structure behaves isotropically. Thus, the creep law (6.1) contains just one macrocharacteristic, viscosity μ:

$$\bar{\sigma}'^{ij} = \mu\dot{\bar{\varepsilon}}^{ij}. \tag{7.1}$$

The dimension analysis of the cell problem shows that μ depends on the grain size ε and the diffusivity coefficient D only

$$\mu = a\frac{\varepsilon^2}{D}, \tag{7.2}$$

where a is some constant.Numerical simulations give the following value of the constant a for hexagonal structure

$$a = 0.344\,. \tag{7.3}$$

The formulae (7.2) and (7.3) inspire an assumption that the similar relation between macro- and micro-characteristics takes place for the random structure as well, where ε is the averaged grain size and D is the characteristic diffusion coefficient for monocrystals, while the coefficient a is of the order of unity.

Microdeformation. Distribution of creep velocity over the cell in the regime of secondary creep is shown in Fig. 2. The orientation of shear stress applied is given at the right top of Fig. 2. It can be seen that there are three pairs of opposite cell sides with different properties. Material departs from one pair of sides and arrives at the other pair of sides. The remaining two sides consist of two pieces: material leaves one piece and arrives at the other one.

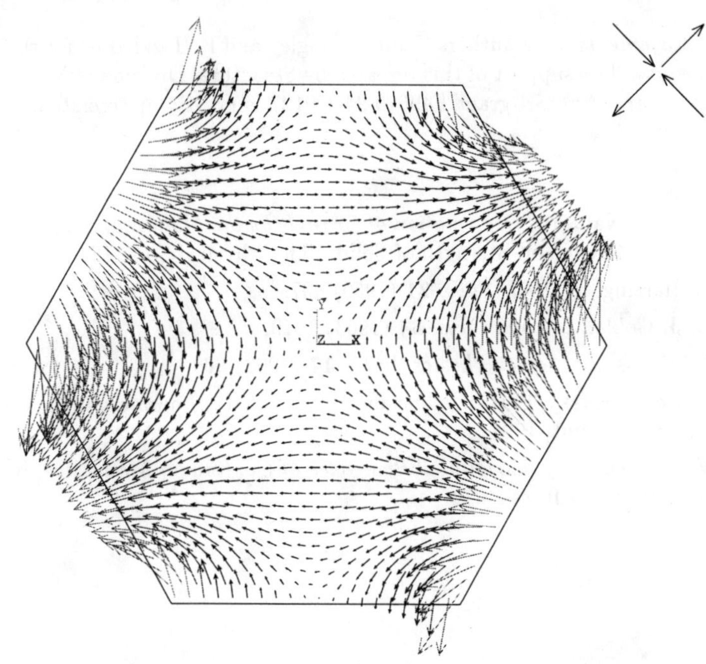

Fig. 2. Creep velocity distribution during the secondary creep.

Conclusions

Three interesting outcomes of this study seem worthy noting.

First, the constitutive macroequations of diffusional creep turn out to be non-local. It is not seen how to eliminate the non-locality by introducing additional internal variables. Probably, the elimination of the non-locality on the macroscale is impossible in principle. Since this seems to be the case, a search for the adequate local constitutive equations for the diffusional creep is unlikely to be successful.

Second, there is an intrinsic material time $\theta = tDA/\varepsilon^2$. Strain-time dependence (for constant stresses) is universal for intrinsic time in the sense that it does not depend on the material and on the temperature (temperature dependence penetrates through the material constants D and A).

Third, as the variational principle shows, the creep rates do not depend on the elastic properties in secondary creep: only diffusion constants, the grain size and the grain geometry are important. Formula (7.2) is an example of such a dependency.

The discussion of these issues in full detail will be presented in another paper.

Acknowledgements. The authors thank R. Bagley and P. Hazzledine for the discussion of the results. The support of this research by Structural Division of Wright-Paterson Laboratory and AFOSR grant F49620-94-1-0127 is greatly appreciated.

References

1. F. R. N. Nabarro, *Deformation of Crystals by the Motion of Single Ions*, Report of a Conference on Strength of Solids, The Physical Soc., 1948, p. 75.
2. C. Herring, *J. Appl. Phys.* **21** (1950) 437.
3. R. J. Coble, *J. Appl. Phys.* **34** (1963) 1679.
4. I. M. Lifshitz, *Soviet Physics JETP* **17** (1963) 909.
5. V. Berdichevsky, P. Hazzledine and B. Shoykhet, *Micromechanics of Diffusional Creep*, Preprint, 1994.
6. V. Berdichevsky, *Variational Principles of Continuum Mechanics*, Nauka, Moscow, 1983. (in Russian.)

Continuum Models and Discrete Systems
Proceedings of 8th International Symposium, June 11-16, 1995, Varna, Bulgaria, ed. K.Z. Markov

NONLINEAR BEHAVIOR IN A COMPOSITE MEDIUM: THE IMPORTANCE OF RESONANCES AND PERCOLATION

D. J. BERGMAN
School of Physics and Astronomy
Raymond and Beverly Sackler Faculty of Exact Sciences,
Tel Aviv University, Tel Aviv 69978, Israel

Abstract. The nonlinear response of a composite medium to an external field can be greatly enhanced if the system is near a percolation threshold p_c or near a strong, isolated resonance. The latter situation, which can be achieved in the electric properties of metal-dielectric composites, can result in bistability of the dielectric response even at low field amplitudes. The former situation is realizable in electric properties of metal-insulator composites, as well as in elastic properties of an elastic solid with a random distribution either of voids or of rigid inclusions. The enhancement of nonlinear response near p_c can be described by a scaling theory which makes some non-trivial predictions regarding the critical behavior. Calculational methods are also described. For nonlinear behavior near p_c these include simple approximations as well as numerical simulations of random discrete network models. Calculational methods for nonlinear behavior near a sharp resonance include a "zero virtual work" principle along with an appropriate ansatz for the trial field.

1. Introduction

The phenomenon of weakly nonlinear response, like the following relation between the electric field $\mathbf{E}$ and the current density $\mathbf{J}$ in an electrical conductor

$$\mathbf{J} = \sigma\mathbf{E} + b|\mathbf{E}|^2\mathbf{E}, \quad b|\mathbf{E}|^2 \ll \sigma, \tag{1}$$

is quite common. However, in order to be observable and lead to interesting new modes of physical behavior, the nonlinear term should not be too small. Besides increasing the magnitude of the applied or volume averaged field $\langle\mathbf{E}\rangle$, another way to increase the relative importance of that term is to mix together different materials so as to form a composite medium in such a way that the local electric field becomes extremely non-uniform. In order to analyze the properties of such a medium, where each component has a $\mathbf{J}$ vs. $\mathbf{E}$ relation like (1) but with different values of σ, b, it is natural to adopt a perturbation approach. In this way, Stroud and Hui showed that the volume averaged fields $\langle\mathbf{J}\rangle$, $\langle\mathbf{E}\rangle$, are related as follows [1]

$$\langle \mathbf{J} \rangle = \sigma_e \langle \mathbf{E} \rangle + b_e |\langle \mathbf{E} \rangle|^2 \langle \mathbf{E} \rangle, \tag{2}$$

$$\sigma_e \equiv \frac{1}{V} \int dV \sigma(\mathbf{r}) \frac{|\mathbf{E}(\mathbf{r})|^2}{|\langle \mathbf{E} \rangle|^2} + \mathcal{O}(b^2), \tag{3}$$

$$b_e \equiv \frac{1}{V} \int dV\, b(\mathbf{r}) \frac{|\mathbf{E}(\mathbf{r})|^4}{|\langle \mathbf{E} \rangle|^4} + \mathcal{O}(b^2), \tag{4}$$

where $\sigma(\mathbf{r})$, $b(\mathbf{r})$ are the position dependent local values of σ, b and $\mathbf{E}(\mathbf{r})$ is the local field *in linear approximation.* This means that $\mathbf{E}(\mathbf{r})$ is found by assuming $b(\mathbf{r}) = 0$ everywhere. From these results one expects that b_e will be more sensitive than σ_e to abnormally large local values of $\mathbf{E}$.

Such values of $\mathbf{E}$ arise when a good conductor with coefficients σ_M, b_M and a poor conductor with coefficients σ_I, b_I are mixed in a random fashion, with volume fractions p_M, $p_I \equiv 1 - p_M$ such that p_M is close to the percolation threshold p_c, namely $|\Delta p| \ll 1$, where $\Delta p \equiv p_M - p_c$, and $|\sigma_I/\sigma_M| \ll 1$. In Section 2 we discuss the enhancement of b_e which can result in such a percolating composite medium.

Another situation where $\mathbf{E}(\mathbf{r})$ can be greatly amplified in comparison with $\langle \mathbf{E} \rangle$ is when the system is in the vicinity of a sharp, isolated quasi-static resonance, also known as a "surface plasmon resonance". Such resonances have been predicted to appear in metal-dielectric composites that are either a very dilute collection of similarly shaped metallic inclusions embedded in a dielectric host or else have an accurately periodic microstructure [2,3]. These resonances can only occur when the applied or average field $\langle \mathbf{E} \rangle$ is a monochromatic ac field with a frequency in the range where the dielectric permittivity of the metal component is mostly real and negative, with only a small imaginary part. In Section 3 we argue that, near such a resonance, perturbation theory must be used carefully even when the nonlinearity is weak. When that is done, not only does the nonlinear behavior get enhanced, but bistable behavior can also occur: Different solutions $\mathbf{E}(\mathbf{r})$ can be found that correspond to the same value of $\langle \mathbf{E} \rangle$.

2. Two-Component Percolating Conductors

Because a continuum percolating system is too difficult to treat with any precision, even for a strictly linear composite conductor, therefore numerical calculations are invariably done using discrete network models. Such calculations, along with approximate real-space-renormalization-group transformations, have lead to the conclusion that the contributions to (4) from each component have simple scaling forms [4,5]

$$b_e \cong b_M |\Delta p|^{2t-\kappa} \mathcal{F}_M(z) + b_I |\Delta p|^{-2s-\kappa'} \mathcal{F}_I(z), \tag{5}$$

$$z \equiv \frac{\sigma_I/\sigma_M}{|\Delta p|^{t+s}}. \tag{6}$$

Here $\mathcal{F}_M(z)$, $\mathcal{F}_I(z)$ are scaling functions and t, s, κ, κ' are critical exponents. The precise forms of these functions and the precise values of the exponents depend on the system dimensionality but not on the detailed microstructure. In three dimensions, these exponents have the following values

$$t \cong 2.0 \text{ [6]}, \quad s \cong 0.75 \text{ [7]}, \quad \kappa \cong 1.56 \text{ [4, 5-footnote No. 19]}, \quad \kappa' \cong 0.68 \text{ [4]}. \tag{7}$$

There are three asymptotic regimes where the scaling functions have simple forms [5]

$$\mathcal{F}_M(z) \propto \begin{cases} \text{const}, & z \ll 1, \ \Delta p > 0 \ \text{ Regime I}, \\ z^4, & z \ll 1, \ \Delta p < 0 \ \text{ Regime II}, \\ z^{\frac{2t-\kappa}{t+s}}, & z \gg 1, \ \text{any } \Delta p \ \text{ Regime III}; \end{cases} \tag{8}$$

$$\mathcal{F}_I(z) \propto \begin{cases} \text{const}, & \text{I}, \\ \text{const}, & \text{II}, \\ z^{-\frac{2s+\kappa'}{t+s}}, & \text{III}. \end{cases} \tag{9}$$

From these forms, the following behavior follows for b_e near the percolation threshold

$$b_e \approx \begin{cases} b_M \Delta p^{2t-\kappa} + b_I \Delta p^{-2s-\kappa'}, & \text{I} \\ b_M \left(\frac{\sigma_I}{\sigma_M}\right)^4 |\Delta p|^{-2t-4s-\kappa} + b_I |\Delta p|^{-2s-\kappa'}, & \text{II} \\ b_M \left(\frac{\sigma_I}{\sigma_M}\right)^{\frac{2t-\kappa}{t+s}} + b_I \left(\frac{\sigma_I}{\sigma_M}\right)^{-\frac{2s+\kappa'}{t+s}}, & \text{III}. \end{cases} \tag{10}$$

In these expressions, constant coefficients of order 1 have been omitted, since their values are not known at present. Here and henceforth that qualification should always be understood to apply whenever the symbol $\approx$ is used.

Careful consideration of (10) shows that (a) either of the two terms may dominate in any of the three regimes, thus either the good conductor or the poor conductor can make a larger contribution to b_e; (b) b_e need not be a monotonic function of Δp in the vicinity of p_c, even when σ_I/σ_M and b_I/b_M are both very small. Indeed, it is possible for b_e to exhibit a maximum at some small value of Δp, with a peak value that is greater than both b_I and b_M. That this is possible has been verified by some numerical simulations of random-resistor-networks [5]. It still remains to be tested by experiments for a real, continuum composite.

The scaling analysis is very useful when Δp and σ_I/σ_M are both very close to 0—all other approximations fail in that case, which is a singular or critical point of the transport problem in a composite medium. Away from that point simple approximations have been developed, based upon (3) and (4), which can give quite accurate results for σ_e and b_e. The best of these seems to be the one based on the assumption that $\mathbf{E}(\mathbf{r})$ is approximately constant inside each component. This leads to [8]

$$b_e \cong \sum_{i=I,M} \frac{b_i}{p_i} \frac{\partial \sigma_e}{\partial \sigma_i} \left| \frac{\partial \sigma_e}{\partial \sigma_i} \right| .$$

In this expression one can substitute either the exact result for the derivatives $\partial\sigma_e(\sigma_I,\sigma_M)/\partial\sigma_i$, or else some simple approximation, obtained for example from the Clausius-Mossotti approximation or the Bruggeman self consistent effective medium approximation for $\sigma_e(\sigma_I,\sigma_M)$.

3. Quasi-Static Resonances

These resonances appear as poles of the bulk effective dielectric permittivity ϵ_e of a two-component, metal-dielectric composite, when viewed as a function of the complex permittivity ratio ϵ_1/ϵ_2 of the two components. This can be exhibited as follows:

$$s \equiv \left(1-\frac{\epsilon_1}{\epsilon_2}\right), \quad F(s) \equiv 1-\frac{\epsilon_e}{\epsilon_2} = \sum_n \frac{F_n}{s-s_n}, \quad 0 \le s_n < 1, \quad 0 < F_n < 1.$$

The spectrum of poles s_n is dense if the composite medium is disordered and has an infinite volume. By contrast, the spectrum is discrete if the microstructure is either a dilute collection of similarly shaped inclusions ϵ_1 inside the host ϵ_2, or if the microstructure has spatial periodicity [2,3].

In the vicinity of an isolated pole s_n, a naive application of perturbation theory to the discussion of weak nonlinearity misses some important consequences. In fact, if the field independent local permittivity $\epsilon(\mathbf{r})$ is such that ϵ_e is near such a pole, then even a small additive nonlinear correction $\epsilon(\mathbf{r}) \to \epsilon(\mathbf{r}) + b(\mathbf{r})|\mathbf{E}(\mathbf{r})|^2$ can have a drastic effect. E.g., it can change the local value of $|\mathbf{E}(\mathbf{r})|$ from a decreasing to an increasing function of the value of $|\langle\mathbf{E}\rangle|$. Technically, the reason for this is that there are now two small parameters in the system, namely $b|\mathbf{E}|^2$ and $s-s_n$, which compete against each other. The existence of more than one solution for $\mathbf{E}(\mathbf{r})$ is evident in simple solvable microgeometries like parallel slabs [9], and also in more complicated systems with a laminated microstructure [10]. Other cases can be discussed by using the "zero virtual work" principle

$$\int dV \mathbf{D}(\mathbf{r}) \cdot \delta\mathbf{E}^*(\mathbf{r}) = 0, \tag{11}$$

where $\delta\mathbf{E} = -\nabla\delta\phi$ and $\delta\phi$ is an arbitrary variation away from the solution of

$$0 = \nabla\cdot\mathbf{D} = -\nabla\cdot[(\epsilon(\mathbf{r}) + b(\mathbf{r})|\nabla\phi|^2)\nabla\phi].$$

The weakness of the nonlinearity, namely $b|\mathbf{E}|^2 \ll \epsilon$ everywhere, is exploited in order to use as trial function for $\phi(\mathbf{r})$ a form that is similar to the solution $\phi_l(\mathbf{r})$ of the linear problem

$$\nabla\cdot(\epsilon\nabla\phi_l) = 0.$$

Writing

$$\phi_l = \phi_0 + \phi_1,$$

where $\phi_0(\mathbf{r})$ is a linear function such that $-\nabla\phi_0 = \langle\mathbf{E}\rangle$ and $\phi_1(\mathbf{r})$ represents the field distortions due to the inhomogeneous microstructure, the trial function and its variation are taken to be

$$\phi = \phi_0 + A\phi_1, \quad \delta\phi = \delta A\phi_1. \tag{12}$$

Here A is a complex parameter whose value is determined by solving (11). In the vicinity of a sharp, isolated resonance s_n, the potential field ϕ_1 is approximately proportional to the solution of a particular eigenvalue problem, namely

$$\begin{aligned} s_n\nabla^2\phi_n &= \nabla\cdot(\theta_1\nabla\phi_n), \\ \phi_n &= 0 \quad \text{at the boundary,} \end{aligned}$$

where $\theta_1(\mathbf{r})$ is the characteristic or indicator function of the ϵ_1 subvolume. If this function is known, then (11) reduces to a cubic equation for $|A|^2$ which can have either one or three real, positive solutions when the physical parameters are in the appropriate ranges [11]. The schematic shape of the solution(s) for $|A|^2$ vs. $\langle\mathbf{E}\rangle$ is shown in Fig. 1.

Evidently, there is a region of $\langle\mathbf{E}\rangle$ values for which three different solutions are found. At the moment, we do not have a good criterion for deciding which of these solutions the system actually chooses to be in. In particular, when the $\mathbf{D}$ vs. $\mathbf{E}$ relation is both nonlinear and complex, there does not exist an energy-like functional of ϕ whose value must have an extremum and whose first variation is the integral that appears in (11).

The main features of this problem can be seen from the simple example of an isolated metallic sphere, with field independent complex permittivity ϵ_m, embedded in an infinite dielectric host with field dependent permittivity $\epsilon_d(\mathbf{E}) = \epsilon_0 + b_0|\mathbf{E}|^2$, subject to an external uniform field $\mathbf{E}_0$ [9]. Assuming $b_0|\mathbf{E}|^2 \ll \epsilon_0$ everywhere, it was found that the parameter $|A|^2$ approximately satisfies the following cubic equation [12]

$$|\epsilon_m - \epsilon_0|^2 = \left|\epsilon_m + 2\epsilon_0 + \frac{8}{5}b_0|\mathbf{E}_0|^2\frac{|\epsilon_m-\epsilon_0|^2}{|\epsilon_m+2\epsilon_0|^2}|A|^2\right|^2 \frac{|\epsilon_m-\epsilon_0|^2}{|\epsilon_m+2\epsilon_0|^2}|A|^2.$$

Using the definitions

$$\begin{aligned} t &\equiv |A^2|\frac{\frac{8}{5}b_0|\mathbf{E}_0|^2|\epsilon_m-\epsilon_0|^2}{|\epsilon_m+2\epsilon_0|^3} > 0, \\ \mu &\equiv -\frac{\mathrm{Re}(\epsilon_m+2\epsilon_0)}{|\epsilon_m+2\epsilon_0|}, \quad |\mu| < 1, \\ \alpha &\equiv \frac{\frac{8}{5}b_0|\mathbf{E}_0|^2|\epsilon_m-\epsilon_0|^2}{|\epsilon_m+2\epsilon_0|^3}, \end{aligned}$$

this is transformed into the following cubic equation for t

$$f(t) \equiv t^3 - 2\mu t^2 + t = \alpha. \tag{13}$$

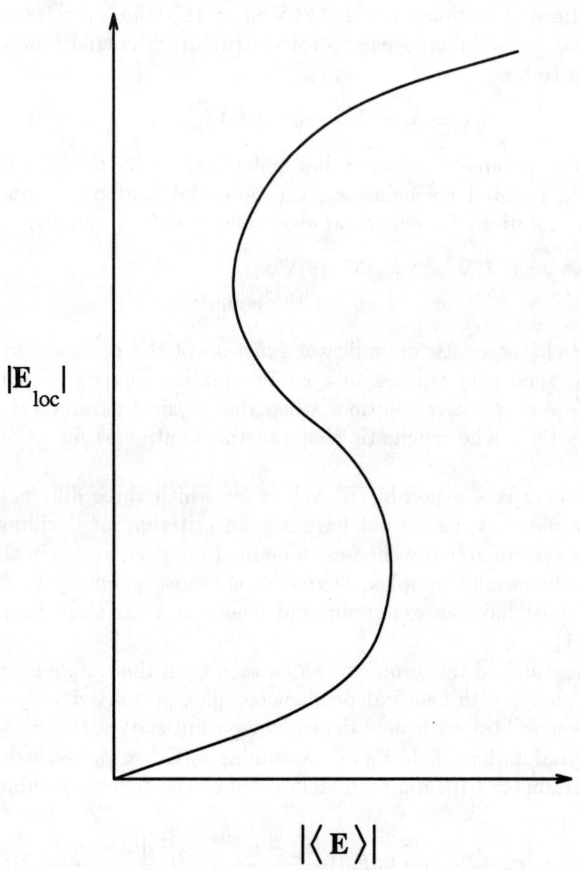

Fig. 1. Schematic plot of the local field magnitude $|\mathbf{E}_{\text{loc}}(\mathbf{r})|$, for any $\mathbf{r}$, or the variational parameter $|A|^2$, vs. the average or applied field magnitude $|\langle\mathbf{E}\rangle|$ in the bistable regime.

Eq. (13) always has at least one real, positive solution. It has three such solutions whenever

$$\frac{\sqrt{3}}{2} \leq \mu, \quad \alpha_+ \leq \alpha \leq \alpha_-,$$

where α_- (α_+) is the local maximum (minimum) value of $f(t)$. Because $\alpha_\pm$ are usually of order 1, therefore these solutions, as well as $|A|^2$, are also of order 1. In the vicinity of the single resonance of this system, which occurs when $\epsilon_m + 2\epsilon_0 = 0$, we thus find that

$$b_0|\mathbf{E}_0|^2 = \mathcal{O}(|\epsilon_m + 2\epsilon_0|^3) \ll 1.$$

Moreover, even the maximum value $\mathbf{E}_{max}$ of the *local field* $\mathbf{E}(\mathbf{r})$, which occurs just outside the poles of the spherical inclusion, satisfies

$$b_0|\mathbf{E}_{max}|^2 \approx 4b_0|\mathbf{E}_0|^2 \frac{|\epsilon_m - \epsilon_0|^2}{|\epsilon_m + 2\epsilon_0|^2}|A|^2 = \mathcal{O}(|\epsilon_m + 2\epsilon_0|) \ll 1.$$

This means that the non-unique solutions appear while the nonlinear effects are weak, thereby justifying our use of the trial function (12).

The fact that bistable behavior can occur near a sharp resonance even when the nonlinearity is everywhere weak is associated with a drastic lowering of the threshold intensity of applied field $\mathbf{E}_0$ needed for the observation of this behavior [9–11]. This is especially true in the case of the more complicated, three component laminar structures considered in Ref. 10.

Acknowledgements. Useful remarks made by O. Levy contributed to the final form of this article. Partial support for this research was provided by grants from the US-Israel Binational Science Foundation and the Israel Science Foundation.

References and Footnotes

1. D. Stroud and P. M. Hui, *Phys. Rev.* **B37** (1988) 8719.
2. D. J. Bergman, *J. Phys. C* **12** (1979) 4947.
3. D. J. Bergman, in *Les Méthodes de l'Homogénéisation: Théorie et Applications en Physique*, Préface de R. Dautray, Editions Eyrolles, Paris, 1985, p. 1.
4. R. R. Tremblay, G. Albinet and A.-M. S. Tremblay, *Phys. Rev. B* **45** (1992) 755.
5. O. Levy and D. J. Bergman, *Phys. Rev. B* **50** (1994) 3652.
6. B. Derrida, D. Stauffer, H. J. Herrmann and J. Vannimenus, *J. Phys. (Paris) Lett.* **44** (1983) L701.
7. H. J. Herrmann, B. Derrida and J. Vannimenus, *Phys. Rev. B* **30** (1984) 4080.
8. X. C. Zeng, D. J. Bergman, P. M. Hui and D. Stroud, *Phys. Rev. B* **38** (1988) 10970.
9. D. J. Bergman, O. Levy and D. Stroud, *Phys. Rev.* **B49** (1994) 129.
10. O. Levy and D. J. Bergman, *Physica A* **207** (1994) 157.
11. R. Levy-Nathansohn and D. J. Bergman, *J. Appl. Phys.* **77** (1994) 4263.
12. Note that the parameter A used here is related to the parameter B in Eq. (20) of Ref. 9 by

$$B = A\frac{\epsilon_m - \epsilon_0}{\epsilon_m + 2\epsilon_0}.$$

Continuum Models and Discrete Systems
Proceedings of 8th International Symposium, June 11-16, 1995, Varna, Bulgaria, ed. K.Z. Markov

ON THE DEFECTS IN NEMATIC LIQUID CRYSTALS

P. BISCARI
Dipartimento di Matematica, Politecnico di Milano,
Via Bonardi 9, I-20133 Milano, Italy

Abstract. A generalized topological theory of defects is applied to the study of nematic liquid crystals, to obtain a complete classification of surface defects and an accurate description of the topological onset of biaxial states.

1. Introduction

The classical topological theory of defects classifies all kind of singularities of microstructured materials [1,2,3], but it can be extended to recognize those situations in which the order parameter is forced by topological reasons to assume certain special values in the material [4,5]: let $\mathcal{R}$ be the order parameter manifold of a structured material and $\mathcal{N}$ any subset of $\mathcal{R}$. If $\mathcal{R}$ happens to be topologically trivial, then the topological structure of $\mathcal{R}\setminus\mathcal{N}$ determines the points of lines of the body where the order parameter is forced to attain its values in $\mathcal{N}$ [6].

Uniaxial nematic liquid crystals can be forced to have isotropic point or line defects [7,8]; those defects can be avoided only when the material may become biaxial in a neighborhood of the putative defects [9,10,11,12]. In the third section we give a more detailed information about the fading of isotropic defects (we refer to [6] for all details that will be omitted here). Suppose that initially we are dealing with a uniaxial material with positive order parameter s, and an isotropic defect somewhere inside it. When the isotropic points disappear, the following structure arises: a biaxial region surrounds the former defect, which is replaced by uniaxial points where the order parameter s changes its sign. Furthermore, the biaxial zone mentioned above includes a surface, which encloses the new uniaxial points, where one of the eigenvalues of $\boldsymbol{Q}$ vanishes, so that $\det \boldsymbol{Q} = 0$.

In the fourth section, we assume that surfaces induce in the order tensor the additional symmetry described in [13] and we give an exhaustive classification of line and point defects that may occur on a surface bounding a nematic liquid crystal. In addition, by studying the relative homotopy groups, we can decide whether a defect can leave the surface, and whether it is the trace on the surface of a bulk defect (in this section we refer to [5]).

2. Microstructure of Nematic Liquid Crystals

The microscopic order parameter of a nematic liquid crystal is [14] a symmetric traceless second order tensor $\boldsymbol{Q}$ that, in the bulk, takes values in the set

$$\mathcal{Q} := \left\{ \boldsymbol{Q} = \sum_{i=1}^{3} \lambda_i \, \boldsymbol{e}_i \otimes \boldsymbol{e}_i : \quad \sum_{i=1}^{3} \lambda_i = 0, \ \lambda_i \geq -\frac{1}{3} \right\}. \tag{2.1}$$

$\mathcal{Q}$ can be decomposed into the following three disjoint subsets:

(i) $\mathcal{B} := \{\boldsymbol{Q} \in \mathcal{Q} : \lambda_i \neq \lambda_j \ \forall i \neq j\}$, the class of biaxial distributions;

(ii) $\mathcal{U}^* := \{\boldsymbol{Q} \in \mathcal{Q} : \lambda_i = \lambda_j \neq \lambda_k \ for \ i \neq j \neq k\}$, the uniaxial distributions;

(iii) the singleton $\{\boldsymbol{0}\} \subset \mathcal{Q}$, which describes the isotropic distribution.

On the surface, a submanifold of $\mathcal{Q}$ seems to be fit to represent the allowed states [13]. If we call $\boldsymbol{\nu}$ the normal to the surface, it is given by:

$$\mathcal{C} := \{\boldsymbol{Q} \in \mathcal{Q} : \boldsymbol{Q}\boldsymbol{\nu} = \lambda_\nu \boldsymbol{\nu}\}. \tag{2.2}$$

In the bulk, we call *defect* any point of the body where the degree of symmetry of $\boldsymbol{Q}$ increases, and so the degree of order decreases. One can then study two different types of nematic defects: uniaxial defects in biaxial states and isotropic defects in uniaxial states.

The situation becomes more complex when we wish to study *surface defects*. As described in [5], there are three different kinds of surface defects: the first class contains the surface defects that are the restriction of bulk defects; the second, those that cannot be removed from the surface (they have a neighborhood that does not contain any further defect point); the third, those that can relax in the bulk (they can be removed from the surface by local surgery, although they are topologically stable).

3. Fading of Isotropic Defects

Let $\mathcal{Q}$, $\mathcal{B}$, and $\mathcal{U}$ be defined as above. Furthermore, let $\mathcal{U}^*_+$ and $\mathcal{U}^*_-$ be the connected components of $\mathcal{U}^*$ in which the scalar order parameter s attains respectively positive and negative values. Then the following Theorem stands [6]:

Theorem.

(i) $\mathcal{U}^*_+ \cup \mathcal{B}$ (respectively $\mathcal{U}^*_- \cup \mathcal{B}$) is topologically equivalent to $\mathcal{U}^*_+$ (resp. $\mathcal{U}^*_-$). It then follows that the same isotropic defects that can be found in a uniaxial nematic with only positive or only negative values of s do survive if we let it to become biaxial; the isotropic points cannot be then substituted just by biaxial distributions: they can only be removed by replacing them with uniaxial distributions with the opposite (with respect to the original situation) sign for s.

(ii) *If we relax isotropic defects of $\mathcal{U}_+^*$ or $\mathcal{U}_-^*$ in $\mathcal{U}^* \cup \mathcal{B}$ it will always appear a surface of biaxial tensors with null determinant surrounding the uniaxial points that substitute the original defect.*

Proof. We give here only the main ideas of the proof, that has been constructed in [6]. Part (i) of the Theorem follows from the fact that $\mathcal{U}_+^*$ is a deformation retract of $\mathcal{U}_+^* \cup \mathcal{B}$, so that they share the same homotopy groups. The equivalence between $\mathcal{U}_-^*$ and $\mathcal{U}_-^* \cup \mathcal{B}$ can be constructed exactly in the same way.

A generic $\boldsymbol{Q} \in \mathcal{U}_+^* \cup \mathcal{B}$ can be written as

$$\boldsymbol{Q} = \sum_{i=1}^{3} \lambda_i \, \boldsymbol{e}_i \otimes \boldsymbol{e}_i \,, \tag{3.1}$$

where either $\lambda_1 > \lambda_2 > \lambda_3$ (if $\boldsymbol{Q} \in \mathcal{B}$) or $\lambda_1 > 0, \lambda_2 = \lambda_3 < 0$ (if $\boldsymbol{Q} \in \mathcal{U}_+^*$). In any case, let $\boldsymbol{Q}_f\,(\boldsymbol{Q})$ be the tensor

$$\boldsymbol{Q}_f\,(\boldsymbol{Q}) := \lambda_1 \, \boldsymbol{e}_1 \otimes \boldsymbol{e}_1 + \frac{\lambda_2 + \lambda_3}{2} \left(\boldsymbol{e}_2 \otimes \boldsymbol{e}_2 + \boldsymbol{e}_3 \otimes \boldsymbol{e}_3\right); \tag{3.2}$$

note that $\boldsymbol{Q}_f\,(\boldsymbol{Q}) = \boldsymbol{Q}$ if, and only if, $\boldsymbol{Q} \in \mathcal{U}_+^*$. A retraction of $\mathcal{B} \cup \mathcal{U}_+^*$ onto $\mathcal{U}_+^*$ is then

$$\boldsymbol{F}(\boldsymbol{Q}, t) := (1 - t)\, \boldsymbol{Q} + t\, \boldsymbol{Q}_f\,(\boldsymbol{Q}) \,. \tag{3.3}$$

The second part of the Theorem can be proved by considering the continuity of the function $\det \boldsymbol{Q}$, which attains positive values far away from the defect, and a negative value at the uniaxial point(s) which has(have) replaced the defect. ■

4. Classification of Surface Defects

In this final section, we classify the *defects* (intended as points where the order tensor $\boldsymbol{Q}$ increases its symmetry with respect to its neighborhood) that can be forced on the surface of a nematic liquid crystal. We refer here to [5] for the statements and the proofs of all the theorems that yield to this classification.

4.1. Isotropic Defects on the Surface of Uniaxial Nematics

Let us suppose first that our system, for internal reasons, is not able to attain biaxial states. Then, in its uniaxial surface we can find the following kind of isotropic *defects*.

$\mathcal{C}_{\mathcal{U}^*}$, the uniaxial component of $\mathcal{C}$ where $\boldsymbol{Q}$ is supposed to take its surface values, has four path-connected components [13]: two with director along $\boldsymbol{\nu}$ (one with positive s and one with negative s), and the other two with director orthogonal to $\boldsymbol{\nu}$ and opposite sign for s.

Isotropic line defects can border upon regions with different sign of s; in this case they are the trace of a wall defect in the bulk. If the sign of s is the same on both sides of the line, but the director is parallel to $\boldsymbol{\nu}$ on one side and perpendicular to it on the other, the line cannot leave the surface. Finally, if it separates two uniaxial domains with the same sign of s and the same tilt angle, we have that: every line defect can be removed if the tilt angle is equal to $\frac{\pi}{2}$, but it exists a stable line defect if the tilt angle is equal to zero (making a half turn in the bulk around this type of line the director makes an odd number of half turns).

To classify point defects on the surface, we must again consider separately regions in which the director is parallel and orthogonal to $\boldsymbol{\nu}$. Inside a domain where the director is parallel to $\boldsymbol{\nu}$, point defects can be classified by an integer topological charge; however, they are not genuine surface defects, since they can always leave the surface to create a bulk point defect of the same topological charge. Furthermore, no point defect of this type has a tail in the volume.

Finally, in a domain where the director is perpendicular to $\boldsymbol{\nu}$, a point defect has to be assigned two topological charges: a surface charge N_S and a bulk charge N_B. If $N_S \neq 0$, the defect cannot be removed from the surface; furthermore, if N_S is odd, the point defect has a tail in the bulk.

4.2. Uniaxial Defects on the Surface of Biaxial Nematics

Let us now allow for biaxial states in our system: *defects* must now be intended as uniaxial points in a biaxial neighborhood.

Every surface line defect in a biaxial nematic has two topological charges associated with it. The first of them, N_S, indicates whether the line is able to travel in the bulk or not. All lines that cannot leave the surface ($N_S \neq 0$) are composed by uniaxial distributions with director orthogonal to the surface.

The second topological charge, N_B, shows that there are two different types of lines for every value of N_S. In fact, following the continuous variation of the eigenvector $\boldsymbol{\nu}$ around any of those lines, we find that around the non-trivial line $\boldsymbol{\nu}$ reverses its sign, while around the trivial line it does not so.

Topologically stable point defects can be classified by an integer charge N_S; none of them can relax in the bulk. They can be of four different types, three of which have non-trivial tails in the bulk: if we write $N_S = 4m + n$ with $m \in \mathbf{Z}$ and $n \in \{0, 1, 2, 3\}$, one of the following possibilities arises (we recall that the line defects in the bulk, that compose the tails of the surface point defects we are studying, can be classified by a member of the group of quaternion unities so that they can be of order one (if they are trivial), order two, or order four [15,16]):

(i) $n = 1$ or $n = 3$: the point defect has a tail of order four in the bulk;
(ii) $n = 2$: the point defect has a tail of order two;
(iii) $n = 0$: the point defect has no tail (but it can be completely removed only if $N_S = 0$).

Acknowledgements. I thank G. Guidone Peroli and T. J. Sluckin, with whom I have obtained the results presented here.

References

1. N. D. Mermin, *Rev. Mod. Phys.* **51** (1979) 591.
2. G. Toulouse and M. Kleman, *J. Physique Lettres* **37** (1976) L149.
3. R. Kutka, H.-R. Trebin and M Kiemes, *J. Phys. France* **50** (1989) 861.
4. H.-R. Trebin, *Adv. Phys.* **31** (1982) 195.
5. P. Biscari and G. Guidone Peroli, *A Hierarchy of Defects in Biaxial Nematics*, 1995, to appear.
6. P. Biscari, G. Guidone Peroli and T. J. Sluckin, 1995, to appear.
7. G. E. Volovik, *JETP Lett.* **28** (1978) 59.
8. G. E. Volovik and O. D. Lavrentovich, *Sov. Phys. JETP* **58** (1983) 1159.
9. I. F. Lyuksyutov, *Sov. Phys. JETP* **48** (1978) 178.
10. N. Schopohl and T. J. Sluckin, *Phys. Rev. Lett.* **59** (1987) 2582.
11. E. Penzenstadler and H.-R. Trebin, *J. Phys. France* **50** (1989) 1027.
12. C. Chiccoli, P. Pasini, F. Semeria, T. J. Sluckin and C. Zannoni, *J. Physique II* **5** (1995) 427.
13. P. Biscari, G. Capriz and E. G. Virga, *Liquid Crystals* **16** (1994) 479.
14. P. G. de Gennes, *The Physics of Liquid Crystals*, Clarendon Press, Oxford, 1974.
15. V. Poenaru and G. Toulouse, *J. Physique* **38** (1977) 887.
16. T. De'Neve, M. Kleman and P. Navard, *J. Physique II* **2** (1992) 187.

Continuum Models and Discrete Systems
Proceedings of 8th International Symposium, June 11-16, 1995, Varna, Bulgaria, ed. K.Z. Markov
© World Scientific Publishing Company, 1996, pp. 123–131

MORPHOLOGICALLY REPRESENTATIVE PATTERN-BASED MODELLING IN ELASTICITY

M. BORNERT
Laboratoire de Mécanique des Solides,
École Polytechnique, F-91128 Palaiseau Cedex, France

Abstract. The concept of "Morphologically Representative Pattern", more convenient for the description of some primary morphological characteristics of complex microstructures, has already been used for the derivation of Voigt-Reuss or Hashin-Shtrikman-type bounds of the overall elastic properties. Here it is used for the definition of pattern-based self-consistent estimates. These are derived from the pattern-based Hashin-Shtrikman-type bounds in the same way as the classical self-consistent scheme is derived from the classical Hashin-Shtrikman bounds. The derivation of the model requires the resolution of several composite inclusion problems, whose properties are discussed and used to establish some general properties and physical interpretations of the proposed scheme. Because of the absence of closed form solutions, a numerical FEM procedure is used, which allows for instance to derive effective properties of anisotropic matrix/inclusion composites.

1. Introduction

The theoretical prediction of the effective elastic response of random multiphase materials has been the subject of intensive study during the last decades. The main difficulty is due to the fact that only partial information on the phase distribution is available. At variance with more classical homogenization approaches which make use of point-correlation functions of the elastic moduli of the constituent phases to describe this information [9], the recently proposed [2,10] *"Morphologically Representative Pattern"* (MRP in the following)–based approach combines a deterministic description of small but finite composite patterns, with a statistical representation of the spatial distribution of the centers of these patterns in a representative volume element Ω. The concept of "phase" is then generalized to the one of "morphological phase," which is the domain occupied by all the realizations in the material of a given pattern. More precisely, Ω is described as a set of disjointed subdomains that map it completely. Some of these domains are assumed identical so that they can be grouped into families, called MRP's. Each MRP λ is characterized by the domain D_λ which gives its extension and by the known local distribution of elastic moduli $C_\lambda(x)$, $\forall x \in D_\lambda$. Each point y in Ω belongs to one and only one member of one, say

λ, of the MRP's used to describe the microstructure, centered at a point X_λ^i: there exists $x \in D_\lambda$ such that the elastic moduli at y are given by: $C(y) = C_\lambda(X_\lambda^i + x)$, and all neighbours y' of y, such that there exists $x' \in D_\lambda$ such that $y' = X_\lambda^i + x'$, satisfy $C(y') = C_\lambda(X_\lambda^i + x')$. The microstructure is completely defined when all positions X_λ^i of the MRP's members are known.

But generally only statistical information on the spatial distribution are available. When used in the classical theorems of potential and complementary energy, this information allows to derive rigorous bounds for the effective elastic moduli. When nothing particular is known on this space distribution, *"MRP-based Voigt* or *Reuss bounds"* can be derived with the statically admissible stress fields or kinematically admissible strain fields generated by the computation of the stress/strain fields that spread out in the MRP's when the same homogeneous stress or strain boundary conditions are applied at their boundaries. Let $\sigma_\lambda^{\mathrm{V}}(x)$ and $\varepsilon_\lambda^{\mathrm{V}}(x)$ be the stress and strain fields in the MRP λ when homogeneous strain conditions are considered. In the following, c_λ will be the volume fraction of the morphological phase λ in Ω and $\langle f_\lambda \rangle_\lambda$ will stand for the average of quantity f_λ over D_λ, so that the average of f over Ω will be $\langle f \rangle = \sum_\lambda c_\lambda \langle f_\lambda \rangle_\lambda$. The tensor $C_{\mathrm{MRP}}^{\mathrm{V}}$ such that $\langle \sigma^{\mathrm{V}} \rangle = C_{\mathrm{MRP}}^{\mathrm{V}} \langle \varepsilon^{\mathrm{V}} \rangle$, for any applied load, is the MRP-based Voigt bound. It is an upper bound (in the sense of the associated quadratic forms) for the tensor of effective elastic moduli. Note that in this case $\langle \varepsilon_\lambda^{\mathrm{V}} \rangle_\lambda$ is equal to the macroscopic applied strain E. The MRP-based Reuss bound $C_{\mathrm{MRP}}^{\mathrm{R}}$ is obtained in a similar way when a homogeneous stress condition Σ is used; then $\langle \sigma_\lambda \rangle_\lambda = \Sigma$.

Narrower bounds can be derived when a little bit more is known about the centers' spatial distribution, thanks to the Hashin-Shtrikman variational principle [6], in which admissible trial fields are generated by means of arbitrary "polarization fields" p that act on a "reference medium" C^0. The position of the centers can be characterized by the functions $\Psi_{\lambda\mu}$ such that $\Psi_{\lambda\mu}(u)\, du$ is the probability to find a couple (i,j) such that $(X_\lambda^i - X_\mu^j)$ belongs to the volume du around u, X_λ^i and X_μ^j being centers of members of the MRP's λ and μ respectively. Explicit *"MRP-based Hashin-Shtrikman type bounds"* can be derived in case of an *"ellipsoidal pattern distribution"*, which is such that $\Psi_{\lambda\mu}(u) = \psi_{\lambda\mu}(\| B \cdot u \|)$, where B is a second order positive definite symmetric tensor [2]. Such an assumption generalizes the concept of isotropic distribution of patterns which is obtained when B is the unit tensor. Note that it requires that the outer boundaries of all patterns are *ellipsoids*, characterized by the tensor $B^T \cdot B$, that is *with same aspect ratios and same orientation*. In this case, the explicit derivation of the MRP-based Hashin-Shtrikman bounds leads to the resolution of several auxiliary composite inclusion problems: one has to compute the average stress $\langle \sigma_\lambda^0 \rangle_\lambda$ and strain $\langle \varepsilon_\lambda^0 \rangle_\lambda$ that spread out in the (ellipsoidally shaped) patterns when they are embedded in an infinite medium C^0 and subjected to homogeneous strain E^0 (or stress $\Sigma^0 = C^0 : E^0$) conditions at infinity. More precisely, the elastic energy $W(E)$ in Ω when subjected to the macroscopic strain E can be computed as $W(E) = \mathsf{HS}^0(E,p) + \Delta\mathsf{W}^0(E,p)$, where HS^0 is the Hashin-Shtrikman functional and $\Delta\mathsf{W}^0$ is an integral over Ω that involves $C(x) - C^0$. When p is chosen in an optimal way $\mathsf{HS}^0(E,p)$ is equal to $\dfrac{|\,\Omega\,|}{2} E : C^{\mathrm{HS}\,0}_{\mathrm{MRP}} : E$ with $C^{\mathrm{HS}\,0}_{\mathrm{MRP}}$ such that $\langle\sigma^0\rangle = C^{\mathrm{HS}\,0}_{\mathrm{MRP}}\,\langle\varepsilon^0\rangle$ for all applied loads E^0 ($|\,F\,|$ stands for the volume of domain F). Note that $\langle\varepsilon^0\rangle$ is the macroscopic load E but is not equal to E^0. When C^0 is chosen such that $C(x)-C^0$ is positive (respectively negative), then $C^{\mathrm{HS}\,0}_{\mathrm{MRP}}$ is a MRP-based Hashin-Shtrikman lower (respectively upper) bound. The tensors $C^{\mathrm{HS}\,0}_{\mathrm{MRP}}$, $C^{\mathrm{V}}_{\mathrm{MRP}}$ and $C^{\mathrm{R}}_{\mathrm{MRP}}$ are positive definite and exhibit all the required symmetries for tensors of elastic moduli.

It is useful to define the pattern based averaging procedure, which transforms any field f on Ω into local fields f^M_λ on D_λ:

$$f^M_\lambda(x) = \frac{1}{N_\lambda}\sum_{i=1}^{N_\lambda} f(X^i_\lambda + x), \quad \forall x \in D_\lambda, \quad \forall\lambda, \tag{1.1}$$

where the sum is taken over the N_λ members of the pattern λ. The local fields computed in the patterns in the auxiliary inclusion problems have in fact a clear significance: they are the *pattern-based averages* of the *trial* stress and strain fields generated in Ω by the optimal polarization field and used in the variational theorems.

A particular case, referred to as the *"punctual approach"*, is obtained when the patterns reduce to points, simply characterized by the mechanical phase to which they belong. The MRP-based bounds are then identical to the classical Voigt/Reuss bounds, or classical Hashin-Shtrikman bounds in case of an isotropic pattern distribution, and the pattern-based averaging procedure reduces to the classical phase averaging. The derivation of the classical Hashin-Shtrikman bounds requires to compute the homogeneous strains that spread out in spherical homogeneous inclusions of phase s embedded in the infinite medium C^0 and which are the *averages over phase* s of the *trial* strain field generated by the Hashin-Shtrikman variational procedure.

The classical self-consistent scheme can be considered as an extension of the Hashin-Shtrikman bounds: in this model one assumes that the *average over phase* s of the *effective* local stress/strain field in Ω is given by the homogeneous field in a spherical inclusion of phase s embedded in an infinite reference medium that has the overall properties of the composite. We extend here this definition of self-consistency to the MRP-based analysis: the *pattern-based averages* of the *effective* stress and strain fields in the heterogeneous medium are given by the (non-uniform) fields that appear in the patterns, when they are embedded in an infinite medium with the effective properties and subjected to the same homogeneous conditions at infinity. With this assumption, the tensor of effective elastic moduli, $C^{\mathrm{SC}}_{\mathrm{MRP}}$, is such that, for any strain E^0 imposed at infinity,

$$\left\langle\sigma^{\mathrm{SC}}\right\rangle = C^{\mathrm{SC}}_{\mathrm{MRP}}\left\langle\varepsilon^{\mathrm{SC}}\right\rangle, \tag{1.2}$$

where $\sigma^{\mathrm{SC}}_\lambda(x)$ and $\varepsilon^{\mathrm{SC}}_\lambda(x)$ are the fields generated in D_λ, when C^0 is equal to $C^{\mathrm{SC}}_{\mathrm{MRP}}$, which is thus determined by an implicit relation. Note that the macroscopic stress $\left\langle\sigma^{\mathrm{SC}}\right\rangle$ and strain $\left\langle\varepsilon^{\mathrm{SC}}\right\rangle$ might *a priori* differ from Σ^0 and E^0.

In the punctual approach, the self-consistent estimates have been shown to correspond to materials with phases distributed according to a "perfect disorder" property [8]. One could consider that the MRP-based self-consistent scheme corresponds to materials with pattern centers arranged according to a "perfect ellipsoidal disorder", but no systematic theory is available. We shall in fact only consider the self-consistent result as an estimate of the effective properties, which might be optimal in the sense that it might minimize the complementary term ΔW^0, since the positive contributions might counterbalance the negative ones. Another interpretation will be given further. Since only the case of patterns with identical ellipsoidal shapes has a clear statistical significance for the rigorous derivation of Hashin-Shtrikman bounds, we also limit the definition of the MRP-based self-consistent scheme to such patterns. It will be shown that in this case, the model exhibits some interesting properties.

2. Properties of the MRP-based Self-Consistent Scheme

2.1. Energetical Definition of Self-Consistency

Consider first the problem of pattern λ embedded in the infinite medium ω with moduli C^0, subjected to the homogeneous strain E^0 at infinity. The strain field $\varepsilon^0_\lambda(x)$ can be written as $E^0 + \varepsilon'^0_\lambda(x)$, where the field $\varepsilon'^0_\lambda(x)$ vanishes for x far away from D_λ. Because of the linearity of the problem, one can write $\varepsilon^0_\lambda(x) = A^0_\lambda(x) : E^0$, where $A^0_\lambda(x)$ is a fourth order tensor that does not depend on E^0.

A basic result due to Eshelby [3] shows that

$$\Delta U_\lambda(C^0, E^0) = \frac{\mid D_\lambda \mid}{2} E^0 : \left\langle (C_\lambda - C^0) : \varepsilon^0_\lambda \right\rangle_\lambda$$

is the difference between the elastic energy (in fact infinite) stored in this system and the energy (also infinite) that would be stored in it if C_λ was homogeneous and equal to C^0. If these terms are summed over all patterns λ, weighted by their respective volume fractions, one gets

$$2\Delta U(C^0, E^0) = 2 \sum_\lambda \frac{c_\lambda}{\mid D_\lambda \mid} \Delta U_\lambda(C^0, E^0) = E^0 : \left[\langle \sigma^0 \rangle - C^0 \langle \varepsilon^0 \rangle \right],$$

so that, clearly, the adopted definition for self-consistency is equivalent to the condition $\Delta U(C^{SC}_{MRP}, E^0) = 0, \forall E^0$.

This generalizes the definition adopted by Christensen and Lo for their three-phase model [3]: the average of the energies stored in the composite inclusion problems, weighted by their volume fraction, is equal to the energy stored in the homogeneous medium with properties C^0. The physical interpretation is that in a material that satisfies the self-consistent hypothesis, the pattern to pattern interaction energies vanish on average; each pattern interacts only with the overall material.

So the three-phase model appears as a particular case of the wide class of here described self-consistent models: it corresponds to Hashin's "Composite Sphere Assemblage" [5], which can be described by self-similar patterns of different sizes, made of a spherical core of material 1 embedded in a spherical shell of material 2, with radii compatible with the macroscopic volume fraction; the absence of characteristic length ensures that the auxiliary composite inclusion problems are all identical.

2.2. Integral Formulation of the Composite Inclusion Problem in Case of Ellipsoids

The outer shape of the domain D_λ is now assumed ellipsoidal. The strain field $\varepsilon^0_\lambda(x)$ can formally be computed by means of the modified Green tensor Γ^0 associated with an infinite medium of elastic moduli C^0. Classically, one has:

$$\varepsilon^0_\lambda(x) + \int_{D_\lambda} \Gamma^0(x-y) : \left[C_\lambda(y) - C^0\right] : \varepsilon^0_\lambda(y)\, dy = E^0, \tag{2.1}$$

so that, after application of Fubini's relation, the average of the strain over D_λ is:

$$\left\langle \varepsilon^0_\lambda \right\rangle_\lambda + \frac{1}{\mid D_\lambda \mid} \int_{D_\lambda} \left[\int_{D_\lambda} \Gamma^0(x-y)\, dx \right] : \left[C_\lambda(y) - C^0\right] : \varepsilon^0_\lambda(y)\, dy = E^0. \tag{2.2}$$

Eshelby's well known property [4] ensures that the integral $\int_{D_\lambda} \Gamma^0(x-y)\, dx$ does not depend on y for $y \in D_\lambda$ [7]. Let $\overline{\Gamma^0}^{D_\lambda}$ be this constant tensor; it depends only on the orientation and aspect ratios of D_λ, not on its size. It can be shown that $\overline{\Gamma^0}^{D_\lambda}$ and $\overline{\Gamma^0}^{D_\lambda -1} - C^0$ are symmetric and positive definite tensors. Eq. (2.2) becomes then:

$$\left\langle \varepsilon^0_\lambda \right\rangle_\lambda + \overline{\Gamma^0}^{D_\lambda} : \left[\left\langle C_\lambda : \varepsilon^0_\lambda \right\rangle_\lambda - C^0 : \left\langle \varepsilon^0_\lambda \right\rangle_\lambda \right] = E^0. \tag{2.3}$$

The patterns are now assumed to have *identical aspect ratios and orientations*, but arbitrary sizes and contents, so that the tensors $\overline{\Gamma^0}^{D_\lambda}$ are all identical and equal to $\overline{\Gamma^0}^{D}$. After summation, one gets then:

$$\left[I + \overline{\Gamma^0}^{D} : (C^1 - C^0) \right] : \langle A^0 \rangle = I, \tag{2.4}$$

where C^1, such that $C^1 : \langle \varepsilon^0 \rangle = \langle \sigma^0 \rangle$, $\forall E^0$, depends on C^0. The adopted definition of self-consistency gives $C^1 = C^0$. Eq. (2.4) is equivalent to the condition $\left\langle A^{SC} \right\rangle = I$, that is $\left\langle \varepsilon^{SC} \right\rangle = E^0$, $\forall E^0$, or equivalently, to the condition $\left\langle \sigma^{SC} \right\rangle = \Sigma^0$, $\forall \Sigma^0$. This property generalizes the one already established for a single composite inclusion [7].

2.3. Existence of C^1

When C^0 is greater or lower than any local tensor in the material, it has already been proven that C^1 exists and is positive definite since it is a Hashin-Shtrikman bound for the effective behaviour. In the general case, the linearity of the inclusion problem ensures only that the average stress and strain are linked by a relation like $A : \langle \varepsilon^0 \rangle + B : \langle \sigma^0 \rangle = 0$. The existence of C^1 depends on the inversibility of B. In other words, $\langle \varepsilon^0 \rangle = 0$ must induce $\langle \sigma^0 \rangle = 0$.

This property can be established in some particular cases, by means of the following variational principle, that characterizes the fields $\varepsilon'^0_\lambda(x)$ and which is no more than a rewriting of the classical theorem of potential energy:

$$2\Delta U_\lambda(C^0, E^0) = \mathrm{Min}_{\varepsilon^\star_\lambda} \left[\int_\omega \varepsilon^\star_\lambda : C^0_\lambda : \varepsilon^\star_\lambda dx + E^0 : \int_{D_\lambda} (C_\lambda - C^0) : (E^0 + \varepsilon^\star_\lambda)\, dx \right.$$

$$+ E^0 : \int_{D_\lambda} (C_\lambda - C^0) : \varepsilon_\lambda^\star dx \Big] . \tag{2.5}$$

The minimum is searched over all kinematically admissible strain fields that vanish at infinity ($C_\lambda^0(x)$ is equal to $C_\lambda(x)$ for x in D_λ and C^0 elsewhere).

The solution $\varepsilon\prime_\lambda^{\mathrm{V}}(x) = \varepsilon_\lambda^{\mathrm{V}}(x) - E^0$ associated to the pattern-based Voigt bound is an admissible strain field. Assuming $\langle \varepsilon^0 \rangle = 0$, which leads to $2\Delta U(C^0, E^0) = E^0 : \overline{\Gamma^0}^{D-1} : E^0$, and using the definition of $C_{\mathrm{MRP}}^{\mathrm{V}}$, one gets, after summation for all patterns: $E^0 : [C_{\mathrm{MRP}}^{\mathrm{V}} - C^0 - \overline{\Gamma^0}^{D-1}] : E^0 \geq 0$. The properties of $\overline{\Gamma^0}^D$ ensure then that, if $C^0 - C_{\mathrm{MRP}}^{\mathrm{V}}/2$ is positive definite, E^0 vanishes and consequently $\langle \sigma^0 \rangle$ too. This proves that C^1 exists when C^0 is greater than $C_{\mathrm{MRP}}^{\mathrm{V}}/2$.

Considering stresses and complementary energy, it can be similarly shown that $\langle \sigma^0 \rangle = 0$ induces $\langle \varepsilon^0 \rangle = 0$, so that a tensor S^2 such that $S^2 : \langle \sigma^0 \rangle = \langle \varepsilon^0 \rangle$, $\forall E^0$ exists, when C^0 is lower than $2C_{\mathrm{MRP}}^{\mathrm{R}}$. So, for not too strongly heterogeneous materials, that is, such that $2C_{\mathrm{MRP}}^{\mathrm{R}} > C_{\mathrm{MRP}}^{\mathrm{V}}/2$, C^1 exists and is inversible, at least for $2C_{\mathrm{MRP}}^{\mathrm{R}} > C^0 > C_{\mathrm{MRP}}^{\mathrm{V}}/2$. In practice, the tensor C^1 could be computed in all tested cases; without a general demonstration of its existence, we shall assume it.

2.4. Symmetry of C^1

Let $\sigma_{\lambda i}^0(x)$ and $\varepsilon_{\lambda i}^0(x)$ be the mechanical stress and strain fields generated in the composite inclusion problems when two arbitrary homogeneous stress or strain conditions $\Sigma_i = C^0 : E_i$, $i = 1, 2$, are applied at infinity. Since $\sigma_{\lambda i}^0(x) - \Sigma_i$ and $\varepsilon_{\lambda i}^0(x) - E_i$, $i = 1, 2$, are self-equilibrated and kinematically admissible with 0 displacement at infinity respectively, one has:

$$\int_\omega [\sigma_{\lambda 1}^0(x) - \Sigma_1] : [\varepsilon_{\lambda 2}^0(x) - E_2]\, dx = \int_\omega [\sigma_{\lambda 2}^0(x) - \Sigma_2] : [\varepsilon_{\lambda 1}^0(x) - E_1]\, dx = 0. \tag{2.6}$$

The integrals can be decomposed onto the domains D_λ and $\omega \backslash D_\lambda$. Using the symmetry of the tensors C^0 and $C_\lambda(x)$, one rewrites the first equalities as, after summation over all patterns: $E_1 : (C^0 - C^1) : \langle A^0 \rangle : E_2 = E_2 : (C^0 - C^1) : \langle A^0 \rangle : E_1$, $\forall E_1, E_2$: $(C^1 - C^0) : \langle A^0 \rangle$ is symmetric. In case of ellipsoids, the combination with Eq. (2.4) proves the symmetry of $[(C^1 - C^0)^{-1} + \overline{\Gamma^0}^D]^{-1}$. Thus C^1 is symmetric for any choice of C^0.

2.5. Indications on Existence and Uniqueness of $C_{\mathrm{MRP}}^{\mathrm{SC}}$

The previous properties of the composite inclusion problems suggest that the fixed point algorithm could be used to determine $C_{\mathrm{MRP}}^{\mathrm{SC}}$, namely, one chooses an arbitrary initial value for C^0, and computes C^{m+1} as the value of C^1 when C^m is taken as reference medium. Two additional properties suggest that this algorithm should converge, even if no complete general demonstration is yet available.

Bounding of $C_{\mathrm{MRP}}^{\mathrm{SC}}$: The use of $\varepsilon\prime_\lambda^{\mathrm{V}}(x)$ in the variational formulation of the inclusion problems leads to $2\Delta U(C^0, E^0) \leq E^0 : [C_{\mathrm{MRP}}^{\mathrm{V}} - C^0] : E^0$, so that, if C^0 is greater than $C_{\mathrm{MRP}}^{\mathrm{V}}$, ΔU is negative. This proves that, if $C_{\mathrm{MRP}}^{\mathrm{SC}}$ exists, it is lower than $C^{\rm V}_{\rm MRP}$. A similar application of the theorem of complementary energy allows to establish a similar relation with the MRP-based Reuss bound so that:

$$C^{\rm R}_{\rm MRP} \leq C^{\rm SC}_{\rm MRP} \leq C^{\rm V}_{\rm MRP}. \tag{2.7}$$

Relation between ΔU and $C^1 - C^0$: Eq. (2.4), valid in case of ellipsoids, allows to rewrite $\Delta U(C^0, E^0)$ as $\Delta U(C^0, E^0) = \frac{1}{2}E^0 : [(C^1 - C^0)^{-1} + \overline{\Gamma^0}^D]^{-1} : E^0$. Taking benefit of the properties of $\overline{\Gamma^0}^D$, one can then establish that:

$$\Delta U(C^0, E^0) > 0\,, \quad \forall E^0 \Leftrightarrow C^1 > C^0\,. \tag{2.8}$$

Both properties ensure that in case C^m would be too large or too small, C^{m+1} would be less large or less small. Of course, this does not establish the convergence of the algorithm. In all the effectively tested cases, an initial choice for C^0 greater than $C^{\rm SC}_{\rm MRP}$ (for instance $C^0 = C^{\rm V}_{\rm MRP}$) leads to a decreasing sequence C^m and, similarly, an initial choice for C^0 lower than $C^{\rm SC}_{\rm MRP}$ generates an increasing one; both converge to the same limit. The general demonstration of this property is still an open question, but the following result could be a possible beginning of the answer. It is obtained when the fields ε'^m_λ, solutions of the problems with reference medium C^m, are used as admissible fields for the problems with reference medium C^{m+1}:

$$2\Delta U(C^{m+1}, E^0) - E^0 : [C^{m+1} - C^m] : [\overline{\Gamma^m}^{D-1} + C^{m+1} - C^m]^{-1} : [C^{m+1} - C^m] : E^0$$

$$\leq \sum_\lambda \frac{c_\lambda}{\mid D_\lambda \mid} \int_{\omega \backslash D_\lambda} \varepsilon'^m_\lambda : [C^{m+1} - C^m] : \varepsilon'^m_\lambda \, dx\,. \tag{2.9}$$

It demonstrates for instance that if $\Delta U(C^m)$ is negative, so is the first member of the inequality. Unfortunately this is not sufficient to prove that $\Delta U(C^{m+1})$ is negative as well, since the second term of this member is also negative.

3. Explicit Derivation

3.1. Numerical Procedure

The effective computation of $C^{\rm SC}_{\rm MRP}$ is performed with the iterative procedure that has been described above. The first problem is to compute the average stresses and strains in the composite inclusion problems with reference medium C^m. Since analytical solutions are only available in very specific cases, we use a FEM based numerical procedure. The approximations related to the discretization as well as to the impossibility to mesh to whole space ω have been discussed elsewhere [1].

The second problem is then to compute C^{m+1} as a function of C^m. In the most general anisotropic case, the computation of all the 21 components of C^{m+1} requires to compute $\langle \sigma^m \rangle$ for six applied strain tensors E^0 which generate six linearly independent average strains $\langle \varepsilon^m \rangle$. The components of C^{m+1} are then determined by a linear system of equations. When the symmetry of the material ensures particular properties of the effective medium, this number may be less than six. For instance, a single load is enough in the isotropic case or for the cubic symmetry. In the case of transverse isotropy, two computations are sufficient to determine the five constants that characterize such a material. More practical details on these aspects can be found elsewhere [1].

In practice, since only average stress and strain values are required to compute C^{m+1}, only rough meshes are required. Moreover it appears that the iterative procedure converges quite rapidly: three to five iterations, depending on the patterns in consideration and the expected accuracy, are usually sufficient. The method can thus be developed on any up-to-date workstation, with any classical FEM code.

3.2. Application to Anisotropic Matrix/Inclusion Composites

Composites with spherical, ellipsoidal and cubic inclusions, isotropically or "ellipsoidally" distributed in a matrix have been modelled by such MRP-based self-consistent schemes. The effect on the overall behaviour of microstructural parameters like inclusion shapes, aspect ratios, concentration fluctuations, or inclusion distributions could be investigated. Several results can be found elsewhere [1,2]. Here we just give as an example of application a result related to a composite with aligned cubic inclusions, isotropically distributed in a matrix. The patterns are all self-similar and made of a cubic inclusion of material 1 embedded in a sphere of material 2. Both constituent phases are assumed to be isotropic, with bulk moduli $K_1 = 20, K_2 = 3$ and shear moduli $\mu_1 = 10$, $\mu_2 = 1$. The overall behaviour exhibits then cubic symmetry, whose behaviour under shear is characterized by two moduli. An example of mesh is given in Fig. 1a and the evolution of both shear moduli are given as a function of the inclusion volume fraction on Fig. 1b. Both are in fact greater than the modulus obtained with the three phase model, that is, for spherical inclusions.

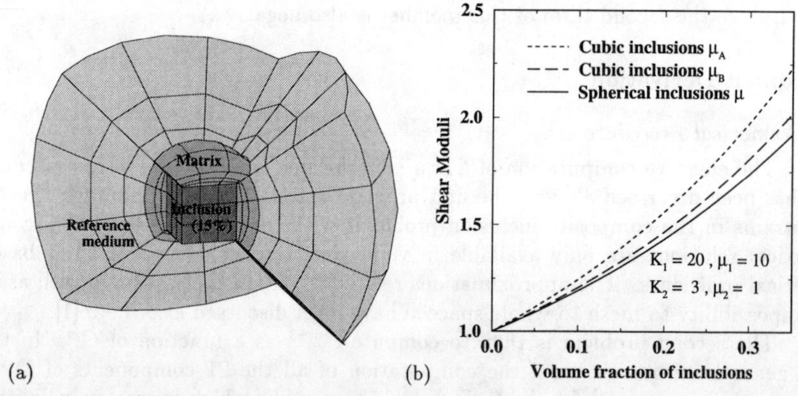

Fig. 1. Results for a composite with aligned cubic inclusions.

References

1. M. Bornert, Submitted to *Comp. Mat. Sci.* (1995).
2. M. Bornert, C. Stolz and A. Zaoui, Submitted to *J. Mech. Phys. Solids* (1995).
3. R. M. Christensen and K. H. Lo, *J. Mech. Phys. Solids* **27** (1979) 315.
4. J. D. Eshelby, *Proc. R. Soc. Lond.* **A421** (1957) 376.
5. Z. Hashin, *J. Appl. Mech.* **29** (1962) 143.
6. Z. Hashin and S. Shtrikman, *J. Mech. Phys. Solids* **11** (1963) 127.
7. E. Hervé and A. Zaoui, *Eur. J. Mech. A/Solids* **9** (1990) 505.
8. E. Kröner, *J. Mech. Phys. Solids* **25** (1977) 137.
9. C. Stolz and A. Zaoui, *C. R. Acad. Sci.* **312** (1991) 143.

Continuum Models and Discrete Systems
Proceedings of 8th International Symposium, June 11-16, 1995, Varna, Bulgaria, ed. K.Z. Markov
©World Scientific Publishing Company, 1996, pp. 132–139

INCOMPRESSIBLE FIBRE-REINFORCED COMPOSITES: EFFECTIVE ELASTOPLASTIC OR VISCOELASTIC BEHAVIOUR

L. BOURGEOIS, E. HERVÉ and Y. ROUGIER
Laboratoire de Mécanique des Solides,
École Polytechnique, F-91128 Palaiseau Cedex, France

Abstract. The prediction of the overall behaviour of fibre-reinforced materials has to take into account the specific morphology of such materials when one phase is continuous and the other consists of separate fibres. The so-called three-phase model appears adequate enough to deal with such fibre-reinforced composites. In this paper we consider a two-phase transversely isotropic material whose phases are isotropic and incompressible. In the first part, the three-phase model is extended to the nonlinear case by means of a deformation theory of plasticity using secant moduli and based on a new formulation of the original model. In the second part, a comparison of the classical self-consistent predictions (adequate for intricate morphologies) and the generalized self-consistent ones (adequate for composite-like morphologies) is performed in the case of non-ageing linear viscoelasticity: the relaxation spectra of fibre-reinforced with Maxwellian constituents exhibit strong differences according to the choice of the model, which reflects the importance of taking into account the matrix connectedness.

1. Introduction

The study of the overall mechanical behaviour of fibre-reinforced composites has always been of technological interest. A lot of works have been concerned with the theoretical determination of their effective elastic moduli. Hill [1] has proved that only three independent constants are necessary to describe the elastic behaviour of two-phase fibre composites. Hill [1], Hashin [2], Hashin and Rosen [5] and Walpole [3] have bounded these elastic constants. Self-consistent estimates were derived by Hill [4] and Walpole [3] and Christensen and Lo [6]. But all these works cannot be used to study the viscoelastic behaviour of composites with a polymeric matrix such as epoxy, or, the elastoplastic behaviour of composites with a metal matrix such as aluminium. Our aim is here to extend the field of possible application of these models in order to promote new ways of modelling the effective viscoelastic and elastoplastic behaviour of fibre-reinforced materials and to be able to prove the importance of taking a better account of the actual phase space distribution of such materials. The starting point of the present work is the classical (C.S.C.S) and generalized self-consistent schemes (G.S.C.S).

2. Elastic Behaviour

Let 1 axis denote the axis of symmetry and (x_2, x_3) define the transverse plane. Let (μ^1, μ^2, c_1, c_2) be respectively the shear modulus of phase (1) and phase (2) and their volume fractions and let $(E_{11}^{\text{hom}}, \nu_{12}^{\text{hom}}, \mu_{12}^{\text{hom}}, \mu_{23}^{\text{hom}}, K_{23}^{\text{hom}})$ be respectively the effective longitudinal Young modulus and Poisson's ratio under longitudinal load, the effective longitudinal shear modulus, transverse shear modulus and plane-strain bulk modulus. By assuming the incompressibility of each phase it is clear that the composite material (here transversely isotropic) is incompressible (K_{23}^{hom} is infinite and $\nu_{12}^{\text{hom}} = \frac{1}{2}$) and is defined by three independent constants $(E_{11}^{\text{hom}}, \mu_{12}^{\text{hom}}, \mu_{23}^{\text{hom}})$. All the following calculations are based on Hervé and Zaoui's results [7] which gives the elastic strain and stress fields in an infinite medium constituted of an n-layered transversely isotropic cylindrical inclusion, surrounded by a transversely isotropic cylindrical matrix and defines the G.S.C.S to estimate the overall elastic moduli of heterogeneous fibre-reinforced materials. The case $n = 2$ gives the local strain and stress fields, the effective elastic moduli for the G.S.C.S, and, the association of two cases with $n = 1$, yields all these results for the C.S.C.S. If phase 1 denotes the fibre and phase 2 denotes the matrix, we get then all the following results:

2.1. Normal Tensile Test

Let the specimens (with $n = 2$ or $n = 1$) be subjected to the following conditions on their terminal sections:

$$u_1^0 = \epsilon x_1, \quad T_2^0 = 0, \quad T_3^0 = 0 \tag{2.1}$$

and, on the lateral surface to:

$$u_1^0 = \epsilon x_1, \quad u_2^0 = -\frac{1}{2}\varepsilon x_2, \quad u_3^0 = -\frac{1}{2}\varepsilon x_3 \,. \tag{2.2}$$

In that particular case of sollicitation, the C.S.C.S and the G.S.C.S coincide with the rules of mixture and give identical results:

$$\langle \underline{\underline{\varepsilon}}^{(1)} \rangle = \langle \underline{\underline{\varepsilon}}^{(2)} \rangle = -\frac{1}{2}\varepsilon[\vec{e}_2 \otimes \vec{e}_2] + \vec{e}_3 \otimes \vec{e}_3] + \varepsilon \vec{e}_1 \otimes \vec{e}_1 = \underline{\underline{\varepsilon}}^0 , \tag{2.3}$$

$$\underline{\underline{\sigma}} = E_{11}^{\text{hom}} \varepsilon [\vec{e}_1 \otimes \vec{e}_1], \quad E_{11}^{\text{hom}} = 3c_1\mu^1 + 3c_2\mu^2 = c_1 E^1 + c_2 E^2 \,, \tag{2.4}$$

where $\langle f^{(i)} \rangle$ stands for the average of f over the phase (i).

2.2. Anti-plane Longitudinal Shear Mode

Let the specimens be subjected to the following conditions on their ends:

$$T_1^0 = 0, \quad u_2^0 = 0, \quad u_3^0 = 0 \tag{2.5}$$

and to the following condition on the lateral surface:

$$\underline{u}^0 = \underline{\underline{\varepsilon}}^0 \underline{x} - \frac{\gamma}{2} \vec{e}_3 \wedge \underline{x} \quad \text{with} \quad \underline{\underline{\varepsilon}}^0 = \frac{\gamma}{2}[\vec{e}_1 \otimes \vec{e}_2 + \vec{e}_2 \otimes \vec{e}_1] \,. \tag{2.6}$$

The global stress tensor is expressed by:

$$\underline{\underline{\sigma}} = \mu_{12}^{\text{hom}}\gamma[\vec{e}_1 \otimes \vec{e}_2 + \vec{e}_2 \otimes \vec{e}_1]\,. \tag{2.7}$$

The C.S.C.S leads to:

$$\langle \underline{\underline{\varepsilon}}^{(1)} \rangle = \frac{2\mu_{12}^{\text{hom}}}{\mu_{12}^{\text{hom}} + \mu^1}\underline{\underline{\varepsilon}}^0, \quad \langle \underline{\underline{\varepsilon}}^{(2)} \rangle = \frac{2\mu_{12}^{\text{hom}}}{\mu_{12}^{\text{hom}} + \mu^2}\underline{\underline{\varepsilon}}^0, \tag{2.8}$$

$$\mu_{12}^{\text{hom}} = -[(c_1 - \frac{1}{2})\mu_1 + (c_2 - \frac{1}{2})\mu_2] + \sqrt{[(c_1 - \frac{1}{2})\mu_1 + (c_2 - \frac{1}{2})\mu_2)]^2 + \mu_1\mu_2} \tag{2.9}$$

and the G.S.C.S to:

$$\langle \underline{\underline{\varepsilon}}^{(1)} \rangle = \frac{2\mu^2}{\mu^2 + \mu^1 + c_1(\mu^2 - \mu^1)}\underline{\underline{\varepsilon}}^0, \quad \langle \underline{\underline{\varepsilon}}^{(2)} \rangle = \frac{\mu^1 + \mu^2}{\mu^2 + \mu^1 + c_1(\mu^2 - \mu^1)}\underline{\underline{\varepsilon}}^0, \tag{2.10}$$

$$\mu_{12}^{\text{hom}} = \mu^2 \frac{c_2\mu^2 + (1 + c_1)\mu^1}{c_2\mu^1 + (1 + c_1)\mu^2}\,. \tag{2.11}$$

2.3. In-plane Transverse Shear Mode

Let the specimens be subjected to the following conditions on their ends:

$$u_1^0 = 0, \quad T_2^0 = 0, \quad T_3^0 = 0 \tag{2.12}$$

and, on the lateral surface to:

$$\underline{u}^0 = \underline{\underline{\varepsilon}}^0\underline{x} \quad \text{with} \quad \underline{\underline{\varepsilon}}^0 = \gamma[\vec{e}_2 \otimes \vec{e}_2 - \vec{e}_3 \otimes \vec{e}_3]\,. \tag{2.13}$$

Using the C.S.C.S the average strain and stress fields over each phase are the same as in the case of the longitudinal shear mode except that μ_{23}^{hom} stands in place of μ_{12}^{hom}. The two moduli have identical expressions.

$$\underline{\underline{\sigma}} = 2\mu_{23}^{\text{hom}}\gamma[\vec{e}_2 \otimes \vec{e}_2 - \vec{e}_3 \otimes \vec{e}_3]\,. \tag{2.14}$$

In the case of the G.S.C.S the average strains over each phase are given by:

$$\langle \underline{\underline{\varepsilon}}^1 \rangle = \frac{\dfrac{aa_h}{c} + c^2 bb_h}{D}\underline{\underline{\varepsilon}}^0 \quad \langle \underline{\underline{\varepsilon}}^2 \rangle = \frac{a}{2}\,\frac{\dfrac{aa_h}{c} + (3 - 3c + c^2)bb_h}{D}\underline{\underline{\varepsilon}}^0 \tag{2.15}$$

with

$$D = \frac{1}{4}\left[a_1a_2 + (4c - 3c^2)b_1b_2\right]\left[\frac{a_1a_2}{c} + (4c^2 - 3c)b_1b_2\right] + 3(c - c^2)^2 b_1^2 b_2^2\,,$$

$$\frac{\mu_{23}^{\text{hom}}}{\mu^2} = \frac{2c(1-c)(1-2c)ab + \sqrt{(a^2 + c^4b^2)^2 - 4a^2b^2c^2(3 - 6c + 4c^2)}}{a^2 + c^4b^2 + 2abc(2 - 3c + 2c^2)}\,, \tag{2.16}$$

$$c = c_1, \quad a = 1 + \frac{\mu^1}{\mu^2}, \quad b = 1 - \frac{\mu^1}{\mu^2}, \quad a_h = 1 + \frac{\mu^2}{\mu_{23}^{\text{hom}}}, \quad b_h = 1 - \frac{\mu^2}{\mu_{23}^{\text{hom}}}. \tag{2.17}$$

3. Elastoplastic Behaviour

The G.S.C.S and C.S.C.S have already been extended in an approximate way to non linear case for isotropic behaviour ([8] and [9]). This extension can be performed in the case of transversely isotropic behaviour when the domain of elasticity obeys Hill's criterion and when the loading is proportional. The local strain is then related to the local stress by:

$$\underline{\underline{e}} = \underline{\underline{\underline{\underline{S}}}}^{\text{sc}}(E_{11}^{\text{hom}}, \mu_{12}^{\text{hom}}, \mu_{23}^{\text{hom}}) : \underline{\underline{s}}, \tag{2.18}$$

where $\underline{\underline{e}}$ and $\underline{\underline{s}}$ denote respectively the deviatoric part of the strain and the stress field and where, $\underline{\underline{\underline{\underline{S}}}}^{\text{sc}}$ is the extension of the tensor of elastic compliances to these particular nonlinear cases, and which respects the transversely isotropic symmetry of the material. In the case of the three previously defined loadings, $E_{11}^{\text{sc}}, \mu_{12}^{\text{sc}}, \mu_{23}^{\text{sc}}$ are defined at each step by means of secant moduli. In the results given in Section 2, $\mu_1^{\text{sc}}(\varepsilon_{\text{eq}}^1)$ and $\mu_2^{\text{sc}}(\varepsilon_{\text{eq}}^2)$ are now used instead of μ^1 and μ^2 and can be identified in uniaxial tensile tests. The equivalent strains $\varepsilon_{\text{eq}}^r$ are defined by:

$$\varepsilon_{\text{eq}}^r = \sqrt{\frac{2}{3}\langle\underline{\underline{\varepsilon}}\rangle_r : \langle\underline{\underline{\varepsilon}}\rangle_r} = \varepsilon_r \text{ with } r \in [1,2]. \tag{2.19}$$

For the three kinds of loading, similar relations are to be solved. Let us consider, for instance, the case of the longitudinal shear mode: if ε denotes:

$$\varepsilon = \sqrt{\frac{2}{3}\underline{\underline{\varepsilon}}^0 : \underline{\underline{\varepsilon}}^0} \tag{2.20}$$

and if C_i are defined by:

$$\varepsilon_i = C_i(\mu_1^{\text{sc}}, \mu_2^{\text{sc}})\varepsilon \tag{2.21}$$

$\mu_{12}^{\text{sc}} = f(\mu_1^{\text{sc}}, \mu_2^{\text{sc}})$ can be determined by means of the following relation:

$$c_1 C_1 + c_2 C_2 = 1 \tag{2.22}$$

and so from (2.21), ε and γ are known ($\gamma = \sqrt{3}\varepsilon$). In turn, τ is given by (see Fig. 1):

$$\tau = 2\mu_{12}(\gamma)\gamma. \tag{2.23}$$

In order to compare our method with experimental results, we have considered a reinforced-fibre composite under an uniaxial transverse loading ($\sigma\vec{e}_2 \otimes \vec{e}_2$). By splitting it into basic loadings, the average strain and stress fields can be determined in each phase and used in order to get $\langle\varepsilon_{22}\rangle$ given by:

$$\langle\varepsilon_{22}\rangle = \frac{\sigma}{4}\left(\frac{1}{\mu_{23}^{\text{sc}}} + \frac{1}{E_{11}^{\text{sc}}}\right). \tag{2.24}$$

The numerical procedure stops when the value of $\varepsilon^1_{\text{eq}}$ and $\varepsilon^2_{\text{eq}}$ leads to the same result for σ. In Fig. 1b the results of the G.S.C.S and of the C.S.C.S are compared with experimental measurements [10]. These comparisons show that the influence of the microstructure morphology on the global curve is weak. This result can be explained by the quite similar hardening rates of each phase. A rather good agreement is found between experimental results and our calculations.

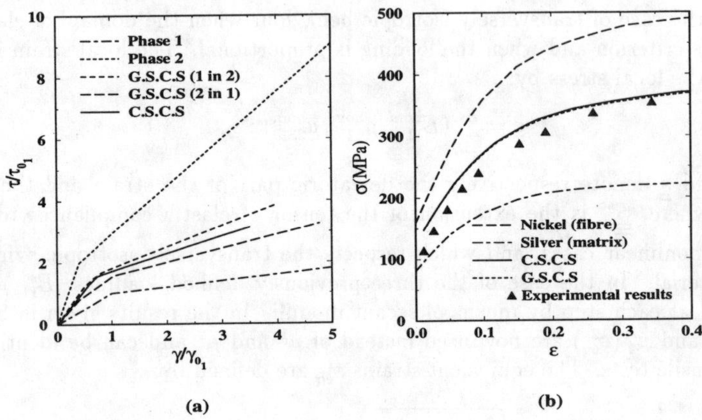

Fig. 1. Comparison of the C.S.C.S and the G.S.C.S predictions for: (a) theoretial longitudinal shear loading with c=0.5, (b) with experimental [10] results under uniaxial transverse loading for c=58.4%

4. Viscoelastic behaviour

A basic phenomenon of the viscoelasticity of inhomogeneous materials is the so-called "long-range memory effect" related to the delayed mechanical interactions between the constituents. In the linear case, it is well known [11] that this effect is responsible, for instance, for the fact that the overall behaviour of an aggregate of Maxwellian constituents is not Maxwellian.

For the sake of simplicity, in view of the derivation of closed form results for the relaxation functions $E^{\text{eff}}_{11}(t), \mu^{\text{eff}}_{12}(t)$ and $\mu^{\text{eff}}_{23}(t)$, consider a two-phase isotropic material whose constituents have a Maxwellian behaviour, governed by the equations:

$$\frac{d\underline{\underline{e}}}{dt} = a_i \underline{\underline{s}} + b_i \frac{d\underline{\underline{s}}}{dt} \quad \text{with } i = 1, 2, \tag{2.25}$$

where $\underline{\underline{e}}$ and $\underline{\underline{s}}$ are the local strain and stress deviators and a_i and b_i materials constants of the constituent (i). Through the Laplace transform (2.25) can be written:

$$\underline{\underline{s}}^L(p) = 2p\mu_i^L(p)\underline{\underline{\varepsilon}}^L(p), \qquad \mu_i^L(p) = \frac{1}{2b_i(p+\frac{1}{T_i})}, \tag{2.26}$$

where p is the complex variable, $f^L(p)$ the Laplace transform of $f(t)$, $T_i = b_i/a_i$ the relaxation time of phase (i) and $\mu_i(t)$ its shear relaxation function. A convenient representation is the spectral one, which allows to write a relaxation function $\mu(t)$ as:

$$\mu(t) = \int_0^\infty g(\tau)e^{-\frac{t}{\tau}}\,d\tau\,, \tag{2.27}$$

where $g(\tau)$ defines one of the relaxation spectrum of the materials (which reduces to a single line for a Maxwell body). A direct comparison of these spectra for the C.S.C.S and the G.S.C.S has already illustrated the influence of morphology (especially of the phase connectedness) in the case of isotropic behaviour [12]. Let us do the same for transversely isotropic overall behaviour.

The correspondence principle states that the Laplace transform of each modulus is linked to $\mu_1^L(p)$ and $\mu_2^L(p)$ through the same equation that the one which links this modulus, μ^1 and μ^2 in the elastic problem. So the same equations (2.4), (2.11) and (2.16) hold for $E_{11}^{\mathrm{effL}}(p), \mu_{12}^{\mathrm{effL}}(p)$ and $\mu_{23}^{\mathrm{effL}}(p)$ by replacing μ^i by $\mu_i^L(p)$. The Laplace inverse functions can be derived by using direct inversion or through the formula:

$$f(t) = \frac{1}{2i\pi}\int_\Delta f^L(p)e^{pt}\,dp\,, \tag{2.28}$$

where the vertical axis Δ has to leave on its left all the critical points of $f^L(p)$. By using the theorem of residues, Jordan's lemma and adequate cuts on the real negative axis, one finds [13]: for the C.S.C.S and for the G.S.C.S

$$E_{11}(t) = \frac{3}{2}\left(\frac{c_1}{b_1}e^{-\frac{t}{T_1}} + \frac{c_2}{b_2}e^{-\frac{t}{T_2}}\right) \tag{2.29}$$

and for the C.S.C.S

$$\mu(t) = \mu_{12}(t) = \mu_{23}(t) = \frac{\langle 2c_1-1\rangle^+}{2b_1}e^{-\frac{t}{T_1}} + \frac{\langle 2c_2-1\rangle^+}{2b_2}e^{-\frac{t}{T_2}}$$

$$+\,\frac{k}{\pi}\frac{T_1T_2}{\sqrt{\tau_1\tau_2}}\int_{\tau_1}^{\tau_2}\frac{\sqrt{(\tau-\tau_1)(\tau_2-\tau)}}{\tau(\tau-T_1)(T_2-\tau)}e^{-\frac{t}{\tau}}\,d\tau\,, \tag{2.30}$$

with

$$c = c_1, \quad f(x) = \frac{T_1(1-b_2x/b_1)}{(1-b_2xT_1/b_1T2)}, \quad k = \frac{|c_1-\frac{1}{2}|}{2b_2}\frac{\frac{1}{T_1}-\frac{1}{T_2}}{\sqrt{\left(\frac{1}{\tau_1}-\frac{1}{T_2}\right)\left(\frac{1}{\tau_2}-\frac{1}{T_2}\right)}},$$

$$\tau_1 = f\left(\frac{c^2 - c - \frac{1}{4} + \sqrt{c(1-c)}}{(c-\frac{1}{2})^2}\right); \quad \tau_1 = f\left(\frac{c^2 - c - \frac{1}{4} - \sqrt{c(1-c)}}{(c-\frac{1}{2})^2}\right).$$

For the G.S.C.S

$$\mu_{12}(t) = \frac{1-c_1}{1+c_1}\frac{1}{2b_2}e^{-\frac{t}{T_2}} + \frac{2c_1}{1+c_1}\frac{1}{b_1+b_2-c_1(b_2-b_1)}e^{-\frac{t}{T'}}, \tag{2.31}$$

with

$$T' = \frac{b_1+b_2-c_1(b_2-b_1)}{\frac{b_1}{T_1}+\frac{b_2}{T_2}-c_1\left(\frac{b_2}{T_2}-\frac{b_1}{T_1}\right)},$$

$$\mu_{23}(t) = K_1 e^{-\frac{t}{T_2}} + \frac{K}{\pi}\frac{T_2\Theta_1\Theta_2}{\sqrt{\tau_1'\tau_2'\tau_3'\tau_4'}}\int_{\tau_1'}^{\tau_2'}\frac{\sqrt{(\tau-\tau_1')(\tau_2'-\tau)(\tau-\tau_3)(\tau-\tau_4')}}{\tau(T_2-\tau)(\tau-\Theta_1)(\Theta_2-\tau)}e^{-\frac{t}{\tau}}d\tau$$

$$+\frac{K}{\pi}\frac{T_2\Theta_1\Theta_2}{\sqrt{\tau_1'\tau_2'\tau_3'\tau_4'}}\int_{\tau_3'}^{\tau_4'}\frac{\sqrt{(\tau-\tau_1')(\tau-\tau_2')(\tau-\tau'_3)(\tau_4'-\tau)}}{\tau(T_2-\tau)(\tau-\Theta_1)(\tau-\Theta_2)}e^{-\frac{t}{\tau}}d\tau + K_2 e^{-\frac{t}{\Theta_1}} + K_3 e^{-\frac{t}{\Theta_2}},$$

with

$$\rho_1 = -A - B, \quad \rho_2 = A - B, \quad \rho_3 = B - A, \rho_1 = A + B,$$
$$A = \sqrt{3}c(1-c), \quad B = c\sqrt{3(1-c)^2+c^2},$$

$$r_1' = -C - D, \quad r_2' = D - C, \quad C = c(2-3c+2c^2), \quad D = 2c(1-c)\sqrt{1-c+c^2},$$

$$K = \frac{1}{2b_2}\frac{\sqrt{\prod_{i=1}^{4}[b_1+b_2-\rho_i(b_1-b_2)]}}{\prod_{i=1}^{2}[b_1+b_2-r_i'(b_1-b_2)]}, \qquad g(r) = \frac{b_1+b_2-r(b_1-b_2)}{\frac{b_1}{T_1}+\frac{b_2}{T_2}-r\left(\frac{b_1}{T_1}-\frac{b_2}{T_2}\right)},$$

$$\theta_1 = g(r_1'), \quad \theta_2 = g(r_2'), \quad \tau_1' = g(\rho_1), \quad \tau_2' = g(\rho_2), \quad \tau_3' = g(\rho_3), \quad \tau_4' = g(\rho_4),$$

$$K_1 = \frac{1}{2b_2}\frac{1}{(1-r_1')(1-r_2')}(1-c)\left[2c(1-2c)+\sqrt{(1+c)^2(1+c^2)^2-12c^2}\right],$$

$$K_2 = \frac{1}{b_1(1-r_1')+b_2(1+r_1')}\frac{1}{(1-r_1')(r_2'-r_1')}4c(1-c)\langle 2c-1\rangle^+(-r_1'),$$

$$K_3 = \frac{1}{b_1(1-r'_2)+b_2(1+r_2')}\frac{1}{(1-r_2')(r_2'-r_1')}4c(1-c)\langle 1-2c\rangle^+(-r_2').$$

Fig. 2 shows the influence of the morphology on the relaxation spectra.

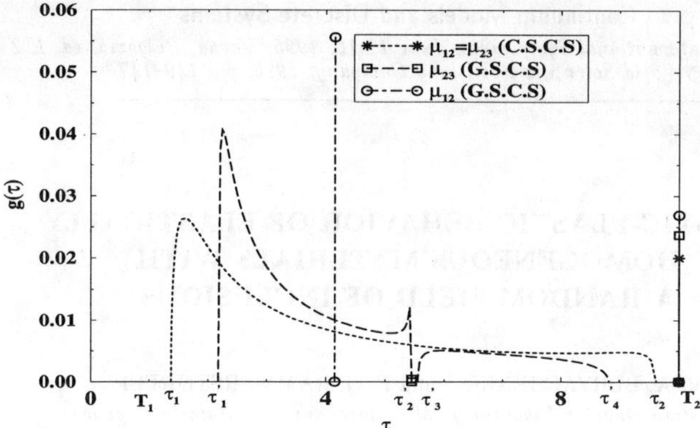

Fig. 2. Comparison of the relaxation spectra derived by the C.S.C.S and the G.S.C.S for c=0.3, $a_1 = a_2 = 1$, T_1=1 and T_2=10.

References

1. R. Hill, *J. Mech. Phys. Solids* **12** (1964) 199.
2. Z. Hashin, *J. Mech. Phys. Solids* **13** (1965) 119.
3. L. J. Walpole, *J. Mech. Phys. Solids* **17** (1969) 235.
4. R. Hill, *J. Mech. Phys. Solids* **13** (1965) 189.
5. Z. Hashin and B. W. Rosen, *J. Appl. Mech* **31** (1964) 223.
6. R. M. Christensen and K. H. Lo, *J. Mech. Phys. Solids* **27** (1979) 315.
7. E. Hervé and A. Zaoui, *Int. J. Eng. Sci* **33** (1995) 1419.
8. M. Berveiller and A. Zaoui, *Res. Mech. Let.* **1** (1981) 119.
9. E. Hervé and A. Zaoui, *Eur. J. Mech. A/Solids* **9** (1990) 505.
10. Ch. Dietrich, M.H. Poech, H.F. Fischmeister and S. Schmauder, *Comp. Mat. Sci.* **1** (1993) 195.
11. P. Suquet, In *Homogenization Techniques for Composite Media*, eds. E. Sanchez Palencia and A. Zaoui, 1985, p. 193.
12. Y. Rougier, C. Stolz and A. Zaoui, *C. R. Acad. Sci. Paris* **316** (1993) 1517.
13. L. Bourgeois, *Private Communication* (1994).

Continuum Models and Discrete Systems
Proceedings of 8th International Symposium, June 11-16, 1995, Varna, Bulgaria, ed. K.Z. Markov
©World Scientific Publishing Company, 1996, pp. 140–147

ELASTIC-PLASTIC BEHAVIOR OF ELASTICALLY HOMOGENEOUS MATERIALS WITH A RANDOM FIELD OF INCLUSIONS

V. A. BURYACHENKO [1] and F. G. RAMMERSTORFER
Christian Doppler Laboratory "Micromechanics of Materials" at the Institute of Lightweight Structures and Aerospace Engineering, Vienna Technical University, Gusshausstrasse 27-29/317, A-1040 Vienna, Austria

Abstract. A two–phase material is considered, which consists of a homogeneous elastic-plastic matrix containing a homogeneous statistically uniform random set of ellipsoidal elastic-plastic inclusions. The elastic properties of the matrix and the inclusions are the same, but the so-called "stress-free strains," i.e. the strain contributions due to temperature loading, phase transformations, and the plastic strains, fluctuate. A general theory of the yielding for arbitrary loading (by the macroscopic stress state and by the temperature) is employed. The realization of an incremental plasticity scheme is based on averaging over each component of the nonlinear yield criterion. Usually averaged stresses are used inside each component for this purpose. In distinction to this usual practice physically consistent assumptions about the dependence of these functions on the component's values of the second stress moments are applied. The application of the proposed theory to the prediction of the thermomechanical deformation behavior of a model material is shown.

1. Introduction

For random structure composites a broad class of inelastic constitutive relations for mechanical properties of composite components can be studied by the transformation field analysis proposed by Dvorak [1], which, for sufficiently fine discretization, is an exact method for deterministic microstructures. Since methods of averaged strains only allow the estimation of averaged stresses in the components, their use for linearizing functions describing nonlinear effects, e.g. strength predictions, plastic deformation, might lead to crude estimates.

In the present paper the investigation of the thermo-elastic-plastic behavior is based on the classical J_2-flow theory. However, in each step a linearization of nonlinear functions (as, e.g., the yield criterion) is applied which is based on physically consistent averaging of the phase values of the second. In this way the method of

[1]Permanent address: Department of Mathematics, Moscow Institute of Chemical Engineering, 107884 Moscow, Russia.

integral equations (contrary to the variational method [2]) is used under the simplified assumption of elastically homogeneous materials with random fields of inclusions. The approach described here has been applied for simulating the monotonic and cyclic thermo-mechanical behavior of a multiphase model material (austenite-ferrite in different topologies). Onset of plastic deformations, overall hardening behavior, shakedown and ratchetting are studied and presented at the end of the paper.

2. Description of the Mechanical Properties and the Geometrical Structure of the Components

We assume that the properties of the composite medium with, generally speaking, anisotropic components are described by the theory of small elastic-plastic strains. Additive decomposition of the total strain tensor $\boldsymbol{\sigma}$ is used in each component of the composite, i.e. $\boldsymbol{\varepsilon} = \boldsymbol{\varepsilon}^e + \boldsymbol{\varepsilon}^t + \boldsymbol{\varepsilon}^p$, with the elastic strains $\boldsymbol{\varepsilon}^e$, the thermal strains $\boldsymbol{\varepsilon}^t$, and plastic strains $\boldsymbol{\varepsilon}^p$, $\boldsymbol{\varepsilon}^t = \mathbf{m}\,\theta$, where $\mathbf{m}$ is the tensor of linear thermal expansion coefficients and θ is the temperature change from the reference value to the current temperature.

The local constitutive equation $\boldsymbol{\sigma}(x)$ is given in the form $\boldsymbol{\sigma}(x) = \mathbf{L}\boldsymbol{\varepsilon}^e(\mathbf{x})$, where $\mathbf{L} \equiv$ const is the fourth-order elasticity tensor. For isotropic materials it is given by $\mathbf{L} = (3k, 2\mu) \equiv 3k\mathbf{N}_1 + 2\mu\mathbf{N}_2$, $\mathbf{N}_{1ijkl} = \delta_{ij}\delta_{kl}/3$, $\mathbf{N}_{2ijkl} = (\delta_{ik}\delta_{jl} + \delta_{il}\delta_{jk} - 2N_{1ijkl})/2$; k and μ are the bulk and shear modulus, respectively. The local strain and stress tensors satisfy the linearized strain-displacement relations and the equilibrium equation, respectively. The traction boundary condition at the surface ∂w of the sample domain w is given. Common notations for tensor products have been employed: $\mathbf{L}\boldsymbol{\varepsilon} = L_{ijkl}\varepsilon_{kl}$, $\boldsymbol{\sigma} : \boldsymbol{\varepsilon} = \sigma_{ij}\varepsilon_{ij}$, $(\boldsymbol{\sigma} \otimes \boldsymbol{\sigma})_{ijkl} = \sigma_{ij}\sigma_{kl}$, $\mathbf{L} : \mathbf{J} = L_{ijkl}J_{ijkl}$.

Regarding the constitutive equations for the elastic-plastic materials of the phases, we use the so-called J_2-flow theory with combined isotropic-kinematic hardening

$$f \equiv \tau - F(\gamma, \theta) = 0, \quad F(\gamma,\theta) = \tau_0(\theta) + h(\theta)\gamma^{n(\theta)},$$

$$\tau = \left(\frac{3}{2} s^a_{ij} s^a_{ij}\right)^{1/2}, \quad d\gamma = \left(\frac{2}{3} d\varepsilon^p_{ij} d\varepsilon^p_{ij}\right)^{1/2}, \quad \mathbf{s}^a = \mathbf{N}_2(\sigma - \mathbf{a}^p), \tag{2.1}$$

where τ_0 is the initial yield stress, h and n are hardening parameters, $\mathbf{s}^a$ is the stress deviator; $\mathbf{a}^p$ is a symmetric second-order tensor corresponding to the "back-stresses". Ziegler's assumption $d\mathbf{a}^p = d\gamma A \mathbf{s}^a$, $A \equiv A(\gamma)$ is used as well.

A mesodomain w with a characteristic function W containing a set $X = (V_i, \mathbf{x}_i, \omega_i)$, $i = 1, 2, \ldots$, of ellipsoids v_i with characteristic functions V_i, centers x_i that form a Poisson set, semi-axes a_i^j, $j = 1, 2, 3$, and an aggregate of Euler angles ω_i is considered. It is assumed that the inclusions can be grouped into the component $v^{(1)}$ with identical mechanical and geometrical properties. In the matrix $v^{(0)} = w \backslash v^{(1)}$, the tensor $\mathbf{g}(\mathbf{x}) = \mathbf{g}^{(0)}$, ($\mathbf{g} = \boldsymbol{\varepsilon}^t$, $\boldsymbol{\varepsilon}^p$, τ_0, h, n) is assumed to be constant; in the component $v^{(1)}$, $\mathbf{g}(\mathbf{x}) = \mathbf{g}^{(0)} + \mathbf{g}_1(\mathbf{x}) = \mathbf{g}^{(0)} + \mathbf{g}_1^{(1)}$; $\mathbf{g}_1(\mathbf{x}) =$ const at $\mathbf{x} \in v^{(1)}$. The upper index of the material's properties tensor put in parentheses shows the number of the respective components. The subscript 1 denotes a jump of the corresponding quantity (e.g. of the material tensor). We assume that the phases are perfectly bonded.

It is assumed that all the random quantities under consideration are described by statistically homogeneous ergodic random fields and, thereby, the ensemble averaging could be replaced by volume averaging

$$\langle(\cdot)\rangle = \overline{w}^{-1}\int(\cdot)W(\mathbf{x})\,\mathbf{dx}, \quad \langle(\cdot)\rangle^{(k)} = [\overline{v}^{(k)}]^{-1}\int(\cdot)V^{(k)}(\mathbf{x})\,\mathbf{dx}, \tag{2.2}$$

$k = 0, 1$. The overbar indicates the measure of the respective region, e.g., $\overline{v} \equiv \operatorname{mes} v$. For the averages over an individual inclusion $v_i \in v^{(1)}$ one has $\langle(\cdot)\rangle_i = \langle(\cdot)\rangle^{(1)}$. We also introduce a conditional probability density $\varphi(v_m \,|\, \mathbf{x}_m; \mathbf{x}_1)$, which is the probability density to find the m-th inclusion in the domain v_m with the center $\mathbf{x}_m$ provided the location of the rest of the inclusions, $\mathbf{x}_1 \neq \mathbf{x}_m$, is fixed; $n^{(1)}$ is the number density of the component $v^{(1)}$; $c^{(k)}$ is the volume fraction of the component $v^{(k)}$.

3. First and Second Stress Moments Inside the Components

Let us define $\boldsymbol{\beta} \equiv \boldsymbol{\varepsilon}^t + \boldsymbol{\varepsilon}^p$ and substitute the local equations for the material state in the equilibrium equation. We obtain

$$\boldsymbol{\sigma}(x) = \boldsymbol{\sigma}_0 + \int \boldsymbol{\Gamma}(\mathbf{x}-\mathbf{y})\left[\boldsymbol{\beta}_1(\mathbf{y}) - \langle\boldsymbol{\beta}_1\rangle\right]\mathbf{dy}, \tag{3.1}$$

where the integral operator kernel $\boldsymbol{\Gamma}(\mathbf{x}-\mathbf{y}) = -\mathbf{L}^{(0)}\left[\mathbf{I}\delta(\mathbf{x}-\mathbf{y}) + \nabla\nabla\mathbf{G}(\mathbf{x}-\mathbf{y})\mathbf{L}^{(0)}\right]$ is defined by the Green tensor $\mathbf{G}$ of the Lamé's equation of a homogeneous medium with an elasticity tensor $\mathbf{L}^{(0)}$; $\delta(\mathbf{x})$ is the delta-function, $\mathbf{I}$ is the fourth order unit tensor and $\boldsymbol{\sigma}_0 \equiv \langle\boldsymbol{\sigma}\rangle$ is the external loading of the mesodomain. Let v_i be located arbitrary inclusion v_i and average (3.1) over a realization ensemble of surrounding particles under the assumption $\left\langle V_q(\mathbf{y})\boldsymbol{\beta}_1^{(q)}(\mathbf{y}) \,|\, \mathbf{x}_q; \mathbf{x}_i \right\rangle = \mathbf{r}_1(\boldsymbol{\beta}_1^{(q)}, \rho)$, where $\rho \equiv |\,\mathbf{a}_i^{-1}(\mathbf{x}_q - \mathbf{x}_i)\,|$. Here the dependence of the function $\mathbf{r}_1$ on the geometrical parameters of the inclusions v_i is indicated by scalar values ρ; $\mathbf{a}_i^{-1}$ denotes a matrix of affine transformation which transfers the ellipsoid v_i into a unit ball.

Then for two-component composites Eq. (3.1) can be recast into the simpler form

$$\langle\boldsymbol{\sigma}\rangle_i = \boldsymbol{\sigma}_0 + \mathbf{C}^{(i)}\boldsymbol{\beta}_1^{(1)}, \quad \mathbf{C}^{(i)} = (-1)^i(1 - c^{(i)})\mathbf{Q}_1, \tag{3.2}$$

$i = 0, 1$, with $\mathbf{Q}_1 = -\langle\boldsymbol{\Gamma}(\mathbf{x}-\mathbf{y})\rangle_1 = \text{const}$, $\mathbf{x}, \mathbf{y} \in v_1$.

To specify the second moment of stresses in the component $v^{(i)}$, $i = 0, 1$, it is necessary to take the tensor product of both sides of Eq. (3.1) with $\boldsymbol{\sigma}(\mathbf{x})$, $\mathbf{x} \in v^{(i)}$. By averaging the so obtained equation over the realization ensemble and taking into account only binary interaction of the inclusions we will find the average second invariant of the deviator stresses inside the components (no summing over i, $i = 0, 1$):

$$\langle\boldsymbol{\sigma}\mathbf{N}_2\boldsymbol{\sigma}\rangle_i - \langle\boldsymbol{\sigma}\rangle_i \mathbf{N}_2 \langle\boldsymbol{\sigma}\rangle_i = \mathbf{B}^{(i)}(\langle\boldsymbol{\sigma}\rangle_i - \boldsymbol{\sigma}_0) \otimes (\langle\sigma\rangle_i - \boldsymbol{\sigma}_0), \tag{3.3}$$

where

$$\mathbf{B}^{(i)} = \mathbf{N}_2 : \widetilde{\mathbf{C}}^{(i)} \left[\left(\mathbf{C}^{(i)}\right)^{-1} \otimes \left(\mathbf{C}^{(i)}\right)^{-1} \right],$$

$$\widetilde{\mathbf{C}}^{(0)} = \int [\mathbf{T}_p(\mathbf{x}_0 - \mathbf{x}_p)\overline{v}_p] \otimes [\mathbf{T}_p(\mathbf{x}_0 - \mathbf{x}_p)\overline{v}_p]\, \varphi(v_p \,|\, \mathbf{x}_p; \mathbf{x}_0)\, \mathbf{dx}_p,$$

$$\widetilde{\mathbf{C}}^{(1)} = \overline{v}_1^{-1} \int\!\!\int [\mathbf{T}_p(\mathbf{x} - \mathbf{x}_p)\overline{v}_p] \otimes [\mathbf{T}_p(\mathbf{x} - \mathbf{x}_p)\overline{v}_p]\; V_1(\mathbf{x})\varphi(v_p \,|\, \mathbf{x}_p; \mathbf{x}_1)\, \mathbf{dx}\mathbf{dx}_p,$$

$$\mathbf{T}_q(\mathbf{x} - \mathbf{x}_q) = \overline{v}_q^{-1} \int \mathbf{\Gamma}(\mathbf{x} - \mathbf{y}) V_q(\mathbf{y})\, \mathbf{dy}.$$

The omitted terms in Eq. (3.3) have a second order of smallness with respect to $c^{(1)}$ for dilute inclusion concentration $c^{(1)} \ll 1$.

4. Onset of Yielding

Elastic-plastic problems are inherently nonlinear. The commonly used linearization in solving such plasticity problems assumes homogeneity of the stress or strain fields, respectively, when computing the material tensors which describe the nonlinear effects [1] (no i-summing):

$$\tau^{(i)} = \tau_0^{(i)}, \quad \tau^{(i)} \equiv \left(\frac{3}{2} \langle \sigma \rangle_i \mathbf{N}_2 \langle \sigma \rangle_i \right)^{1/2}. \tag{4.1}$$

The hypotheses (4.1), typical for mean field approaches in mechanics of composites, in a certain sense are not fully consistent. This can be demonstrated for example by consideration of an elastic-plastic multiphase material which is macroscopically isotropic, e.g., spherical inclusions in an homogeneous isotropic elastic-plastic matrix the behavior of which is described by the von Mises J_2-flow theory. In this case, under pure thermal loading with $\langle \varepsilon_{1ij}^t \rangle = \langle \varepsilon_{10}^t \rangle \delta_{ij} \neq 0$, $\sigma_0 = 0$, we obtain $\langle \mathbf{s} \rangle_0 \equiv 0$, irrespectively of the microstructure of the inclusion phase and of the method of calculation of $\langle \sigma \rangle_0$ (for example by the formula (3.2) or any other formula). This would mean that the considered material has a purely elastic deformation even for large values of ε_{10}^t, which contradicts the experimental observations.

The above mentioned inconsistency can be overcome if averaging according to Eq. (3.3) is used, leading to

$$\widetilde{\tau}^{(i)} = \tau_0^{(i)}, \quad \widetilde{\tau}^{(i)} \equiv \left(\frac{3}{2} \langle \boldsymbol{\sigma} \mathbf{N}_2 \boldsymbol{\sigma} \rangle_i \right)^{1/2}, \quad i = 0, 1, \tag{4.2}$$

for onset of yielding in the component i. Note that, according to Hölder's integral inequality, we will obtain $\widetilde{\tau}^{(i)} \geq \tau^{(i)} \geq \langle \tau \rangle_i$. The composite yielding starts when $f^* \equiv \max\left(\widetilde{\tau}^{(i)}/\tau_0^{(i)}\right) - 1 = 0$, $i = 0, 1$.

5. Elastic-Plastic Deformations

Let us first consider situations in which at the beginning of the process $\boldsymbol{\sigma}_0 = 0$ and $f^* < 0$. Under monotonically increased external active loading elastic-plastic deformations will appear, when $f^* = 0$ and $[(\partial f^*/\partial\sigma_0) : d\sigma_0 + (\partial f^*/\partial\theta)d\theta] > 0$. Taking into account the estimation for the yield surface (4.2), the increments of the homogeneous plastic strains in a single component $v^{(s)}$

$$d\varepsilon^{p(s)} = d\lambda^{(s)} \frac{\partial \widetilde{f}^{(s)}}{\partial \langle\sigma\rangle_s} \tag{5.1}$$

are defined by means of the homogeneous functions

$$\widetilde{f}^{(s)} \equiv \widetilde{\tau}^{(s)} - F^{(s)}(\gamma^{(s)}, \theta) = 0, \quad \widetilde{\tau}^{(s)} = \left(\frac{3}{2}\langle \mathbf{s}^a : \mathbf{s}^a\rangle_s\right)^{1/2}. \tag{5.2}$$

We assume homogeneous plastic strains $\varepsilon^{p(s)}$ and the increments of the hardening parameters inside the component $v^{(s)}$ are

$$d\mathbf{a}^{p(s)} = d\gamma^{(s)} A(\gamma^{(s)}) \langle \mathbf{s}^{a(s)}\rangle_s, \quad A(\gamma^{(s)}) = \frac{H^{(s)}}{F^{(s)}(\gamma^{(s)}, \theta)}, \tag{5.3}$$

$$H^{(s)} = \frac{\partial \widetilde{\tau}^{a(s)}}{\partial \gamma^{(s)}}, \quad d\widetilde{\tau}^{a(s)} \equiv \left(\frac{3}{2}\langle d\mathbf{s}^a : d\mathbf{s}^a\rangle_s\right)^{1/2}, \quad d\gamma^{(s)} = \left(\frac{2}{3} d\varepsilon^{p(s)} : d\varepsilon^{p(s)}\right)^{1/2}, \tag{5.4}$$

where $H^{(s)}$ is the current plastic tangent modulus. The estimation (3.3) can be used for the evaluation of the second moment

$$\langle d\mathbf{s}^a : d\mathbf{s}^a\rangle_s \equiv \langle d\mathbf{s} : d\mathbf{s}\rangle_s - 2\langle d\mathbf{s}\rangle_s : \mathbf{a}^{(s)} + \mathbf{a}^{p(s)} : \mathbf{a}^{p(s)}.$$

For two-phase materials, taking Eq. (3.3) into account, we obtain from Eq. (5.2)

$$\widetilde{\tau}^{(s)} = \left[\frac{3}{2}\langle \mathbf{s}^a\rangle_s : \langle \mathbf{s}^a\rangle_s + \frac{3}{2}\mathbf{B}^{(s)} : \left(\left(\langle\sigma\rangle_s - \sigma_0\right) \otimes \left(\langle\sigma\rangle_s - \sigma_0\right)\right)\right]^{1/2}. \tag{5.5}$$

For consistency of the plastic deformation process, the following condition should hold

$$d\widetilde{f}^{(s)} \equiv \frac{\partial \widetilde{f}^{(s)}}{\partial \langle\boldsymbol{\sigma}\rangle_s} : d\langle\boldsymbol{\sigma}\rangle_s + \frac{\partial \widetilde{f}^{(s)}}{\partial \boldsymbol{\sigma}_0} : d\boldsymbol{\sigma}_0 + \frac{\partial \widetilde{f}^{(s)}}{\partial \mathbf{a}^{p(s)}} : d\mathbf{a}^{p(s)} + \frac{\partial \widetilde{f}^{(s)}}{\partial \gamma^{(s)}} d\gamma^{(s)} + \frac{\partial \widetilde{f}^{(s)}}{\partial \theta} d\theta = 0.$$

At first we will calculate some partial derivatives in the last equation and $d\langle\sigma\rangle_s$, $d\gamma^{(s)}$, $d\mathbf{a}^{p(s)}$. This leads to the relation $d\lambda^{(s)} = b_s/\beta_{ss}$ for the proportionality factor (no summing over s, $s = 0, 1$), where

$$b_s = -\left(\frac{3}{2}\langle \mathbf{s}^a\rangle_s : d\boldsymbol{\sigma}_0 + \frac{\partial \widetilde{f}^{(s)}}{\partial \langle\boldsymbol{\sigma}\rangle_s} \mathbf{C}^{(s)} d\boldsymbol{\varepsilon}_1^{t(1)} - \frac{\partial F^{(s)}}{\partial \theta} d\theta\right),$$

$$\beta_{ss} = \frac{(-1)^s \partial \widetilde{f}^{(s)}}{\partial \langle\boldsymbol{\sigma}\rangle_s} \mathbf{C}^{(s)} \frac{\partial \widetilde{f}^{(s)}}{\partial \langle\boldsymbol{\sigma}\rangle_s} - \left[\left(\frac{\widetilde{\tau}^{(s)}}{F^{(s)}}\right)^2 H^{(s)} + \frac{\partial F^{(s)}}{\partial \gamma^{(s)}}\right] \left[\frac{2}{3} \frac{\partial \widetilde{f}^{(s)}}{\partial \langle\boldsymbol{\sigma}\rangle_s} \mathbf{N}_2 \frac{\partial \widetilde{f}^{(s)}}{\partial \langle\boldsymbol{\sigma}\rangle_s}\right]^{1/2}.$$

Let us consider the deformation process when for each component $v^{(i)}$, $i = 0, 1$, the yield conditions are fulfilled

$$\widetilde{f}^{(i)} = 0, \quad \frac{\partial \widetilde{f}^{(i)}}{\partial \langle \boldsymbol{\sigma} \rangle_i} : d \langle \boldsymbol{\sigma} \rangle_i + \frac{\partial \widetilde{f}^{(i)}}{\partial \theta} d\theta > 0. \tag{5.6}$$

Then

$$d \langle \boldsymbol{\sigma} \rangle_s = d\boldsymbol{\sigma}_0 + \mathbf{C}^{(s)} \left(d\boldsymbol{\varepsilon}_1^{t(1)} + d\lambda^{(1)} \frac{\partial \widetilde{f}^{(1)}}{\partial \langle \boldsymbol{\sigma} \rangle_1} - d\lambda^{(0)} \frac{\partial \widetilde{f}^{(0)}}{\partial \langle \boldsymbol{\sigma} \rangle_0} \right) \tag{5.7}$$

and it this way we will obtain the system of two linear equations for the proportionality factors $d\lambda^{(i)}$, $i = 0, 1$, which can be easily solved.

After numerical integration of the system (5.1) we will find the plastic strains inside each component. The overall plastic strains ε^{po} and the average stresses are defined by the relations $\boldsymbol{\varepsilon}^{po} = \langle \boldsymbol{\varepsilon}^p \rangle$, $\langle \boldsymbol{\varepsilon} \rangle = \mathbf{M}\boldsymbol{\sigma}_0 + \langle \boldsymbol{\varepsilon}^t \rangle + \boldsymbol{\varepsilon}^{po}$.

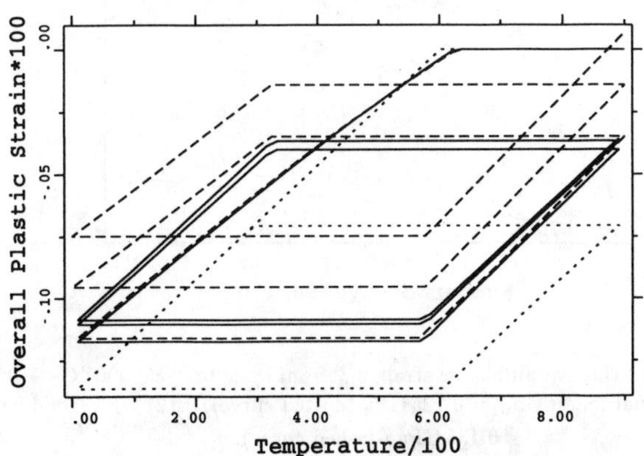

Fig. 1. The overall axial plastic strain ε_{11}^{po} as a function of current temperature during the first three temperature cycles 900 °C → 0 °C → 900 °C at fixed σ_{01}=0 (dotted curve), 0.125 GPa (solid curve), 0.135 GPa (dashed curve).

6. Numerical Results

As an example, consider an elastically isotropic composite with spherical inclusions and the step functions $\varphi(v_j \,|\, \mathbf{x}_j; \mathbf{x}_i) = H(|\,\mathbf{x}_j - \mathbf{x}_i\,| - 2a)n^{(1)}$, $\varphi(v_j \,|\, \mathbf{x}_j; \mathbf{x}_0) = H(|\,\mathbf{x}_j - \mathbf{x}_0\,| - a)n^{(1)}$ are applied; here H is the Heaviside step function. The obtained criterion for onset of yielding (4.2) of the component $v^{(i)}$ is circumscribed by an elliptic curve in the dimensionless coordinates $X^{(i)} \equiv \tau/\tau_0^{(i)}$, $Y^{(i)} \equiv \eta/\tau_0^{(i)}$, where $\tau = (3\langle \mathbf{s} \rangle : \langle \mathbf{s} \rangle/2)^{1/2}$, $\eta \equiv -3Q^k \varepsilon_{10}^t c^{1/2}$ and $3Q^k$ represents the bulk component of the tensor $\mathbf{Q}_1 = (3Q^k, 2Q^\mu)$, $\boldsymbol{\varepsilon}_1^t = \varepsilon_{10}^t \mathbf{I}^\delta$, $(\mathbf{I}^\delta)_{ij} = \delta_{ij}$. The elliptic semi-axes are determined by η and (3.3).

For example, assume that the composite represents a ferritic matrix containing identical spherical austenite inclusions with mechanical parameters of a duplex steel [3] ($c = 0.5$, $\tau^{(0)} = 0.32\,\text{GPa}$, $\tau^{(1)} = 0.2\,\text{GPa}$). The thermal strain is defined by the temperature difference $(T - T^{\max})$ between the stress-free reference temperature $T^{\max} = 900°C$ (from which the cooldown starts) and a current temperature T. Then for $\sigma_0 = 0$ the onset of yielding in the matrix takes place at the temperature $T = 593°C$ for $c = 0.5$ which is higher than that at which the inclusions would show the first onset of yielding if $\tau_0^{(0)}$ would have been significantly higher, $T = 351°C$.

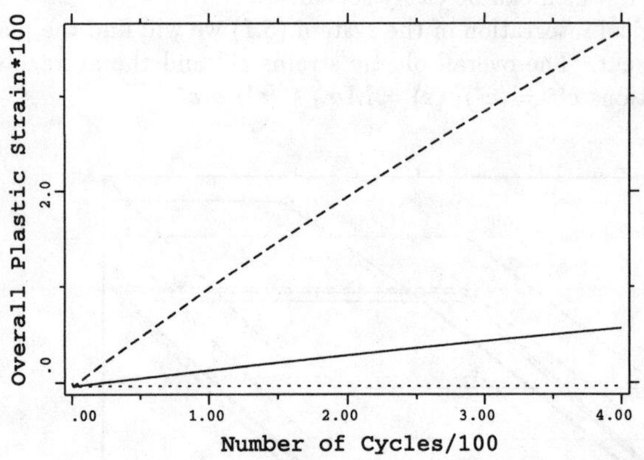

Fig. 2. Accumulation of the overall plastic strain ε_{01}^{po} from cycle to cycle 900 °C → 0 °C → 900 °C at fixed external loading σ_{01}= 0.125 GPa (dotted curve), 0.127 GPa (solid curve), 0.135 GPa (dashed curve).

Just for the sake of theoretical interest, let us consider the components with temperature independent parameters τ_0, h, $n =$ const and small kinematic hardening $H^{(i)} = 0.1\tau^{(i)}$, $h^{(i)} = 0$, $i = 0, 1$. Let the composite be loaded by a constant uniaxial external stress $\sigma_{0ij} = \sigma_{01}\delta_{i1}\delta_{j1}$ and by temperature cycles with one cycle described by $T = T^{\min} + (T^{\max} - T^{\min})(t-1)[H(t-1) - H(1-t)]$, $t \in (0, 2)$, $T^{\min} = 0°C$. According to Figs. 1 and 2 for $\sigma_{01} \leq 0.126\,\text{GPa}$ plastic shakedown takes place. For obtaining a ratchetting effect, which appears for $\sigma_{01} \geq 0.127\,\text{GPa}$), it is necessary that not only one but also the other component show plastic deformations.

For the case of temperature dependent material data for yielding we assume that the parameters $\tau(\theta)$, $h(\theta)$ decrease $1/\lambda^{(i)}$–times (as a quadratic function) under heating of the materials from $T^{\min} = 0°C$ to $T^{\max} = 900°C$. We consider fixed external uniaxial tension $\sigma_{0ij} = \sigma_{01}\delta_{i1}\delta_{j1}$, $\sigma_{01} = 0.04\,\text{GPa}$, equal softening of the components $\lambda^{(0)} = \lambda^{(1)}$ under heating of the composite and only kinematic hardening $H^{(i)} = 0.1\tau_0^{(i)}$. For small softening of the components ($\lambda^{(0)} = \lambda^{(1)} \leq 3.5$) the deformation process is within the plastic shakedown domain (see Fig. 3). For significant softening of the components ($\lambda^{(0)} = \lambda^{(1)} \geq 4$) the process jumps into the ratchetting domain.

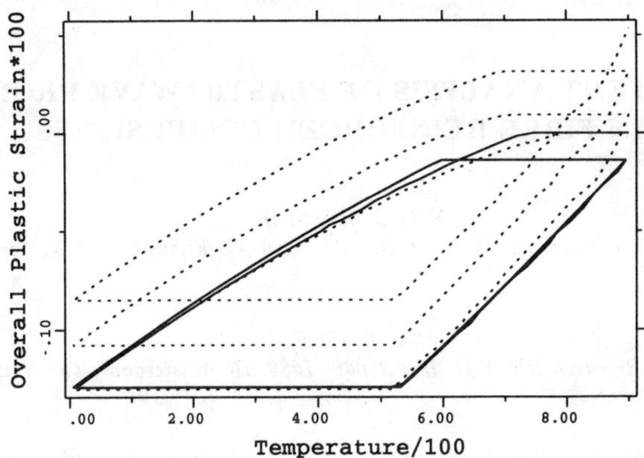

Fig. 3. The overall plastic strain ε_{11}^{po} as a function of current temperature during three temperature cycles 900 °C → 0 °C → 900 °C at fixed external loading σ_{01}=0.04 GPa. $\lambda^{(0)} = \lambda^{(1)} = 3$ (solid curve) and $\lambda^{(0)} = \lambda^{(1)} = 5$ (dotted curve), respectively.

These results were obtained by the assumption (5.1) which might be too approximative in the particular case. An extension of this theory, taking into account the inhomogeneity of the plastic strain field in the matrix, might change the results for cyclic thermal behavior and is currently under development.

Acknowledgments. Parts of this work are related to the COST 512 action, financially supported by the *Osterreichisches Bundesministerium für Wissenschaft, Forschung und Kunst* (under grant GZ 49.935/3–II/4/94).

References

1. G. J. Dvorak, *Proc. R. Soc. Lond.* **437** (1992) 311.
2. V. A. Buryachenko, W. Z. Kreher, *J. Mech. Phys. Solids* **43** (1995) 1105.
3. E. Werner, T. Siegmund and F. D. Fischer, *Comput. Mater. Sci.* **3** (1994) 279.

Continuum Models and Discrete Systems
Proceedings of 8th International Symposium, June 11-16, 1995, Varna, Bulgaria, ed. K.Z. Markov
©World Scientific Publishing Company, 1996, pp. 148–155

SELF-CONSISTENT ANALYSIS OF ELASTIC WAVE PROPAGATION IN FIBRE-REINFORCED COMPOSITES

P. G. J. BUSSINK
KSEPL, Shell Research BV, P.O. Box 60, 2280 AB Rijswijk, The Netherlands

and

P. L. ISKE
KSLA, Shell Research BV, P.O. Box 38000, 1030 BN Amsterdam, The Netherlands

Abstract. We study the propagation of harmonic elastic waves in fibre reinforced composites taking a twofold approach, based on two extensions to the elastodynamic self-consistent scheme originated by Sabina and Willis [5]. In one view, we use the extension by Smyshlyaev, Willis and Sabina [6] and model our composite as an isotropic matrix in which spheroidal inclusions of varying aspect ratio are embedded. In the other view we regard the fibres as aligned infinitely long cylinders, and restrict ourselves to elastic waves that have both their propagation direction and their polarization vector in a plane perpendicular to the axis of alignment. This is really a two-dimensional problem and to tackle it we derive the self-consistent equations for *two-dimensional* elastodynamics. We compare the respective predictions for the effective elastodynamic properties, varying the aspect ratio of the spheroidal fibres from 1 to large values to study the convergence to the system with infinitely long cylinders.

1. Introduction

Consider a system of uniformly distributed, infinitely long, parallel cylindrical fibres embedded in a matrix material, both phases homogeneous and isotropic, of which we want to know the effective elastodynamic properties. The scattering of elastic waves from parallel cylinders was studied by Varadan *et al.* [9], who applied quasi-crystalline approximation to shear waves polarized in a plane *parallel* to the cylinder axes.

In this paper we restrict ourselves to the *in-plane transversal* properties of such a system, that is, we focus on elastic waves having both their propagation direction and their polarization vector in a cross-sectional plane lying normal to the axes of the cylinders. We propose to model these properties by looking at the effective elastodynamic properties of a *two-dimensional* composite consisting of a circular inclusions embedded in a two-dimensional sheet of matrix material. To this end, we will derive the two-dimensional analogue of an effective-medium model due to Sabina and Willis [5].

2. The Self-consistent Scheme of Sabina and Willis

The theory of Sabina and Willis gives an approximate description of the propagation of harmonic elastic waves in a matrix-inclusion composite in which the inclusions are randomly distributed.

The composite is approximated by an "effective" homogeneous material, whose elastic properties are calculated using the self-consistent scheme, which was applied to elastostatics by Hill [3] and Budiansky [1]. The total wave propagating in the composite, resulting from the incident wave and the waves scattered by the inclusions, is approximated by an effective harmonic wave propagating in the effective medium.

The basic ingredient of the calculations is the description of the scattering of a harmonic wave by a single inclusion embedded in a homogeneous matrix, as treated by Willis [10]. A simplifying approximation is made for this single-scattering problem, valid for long wavelengths down to about the particle diameter. Self-consistency means here that this homogeneous material is taken to be the effective medium and the incident wave is set equal to the effective wave. For the effective material properties coupled implicit equations follow, which can be generalized straightforwardly to systems with more than one type of inclusion.

More specific, our composite material consists of a matrix with tensor of elasticity $\boldsymbol{C}^{\mathrm{mat}}$ and mass density ρ^{mat}, and inclusions with elastic moduli $\boldsymbol{C}^{\mathrm{inc}}$ and density ρ^{inc}. The inclusion phase occupies a volume fraction ϕ of the composite. All inclusions are identical in shape and orientation, and their positions are randomly distributed over the composite, lacking any geometric correlation. Stress $\boldsymbol{\sigma}$ and momentum density $\boldsymbol{p}$ are related to the usual infinitesimal strain tensor $\boldsymbol{e}$ and the particle velocity $\boldsymbol{v}$ through the material parameters in the constitutive relations:

$$\boldsymbol{\sigma} = \boldsymbol{C}\boldsymbol{e}\,, \tag{1}$$

$$\boldsymbol{p} = \rho\boldsymbol{v}\,. \tag{2}$$

Displacement fields will be restricted to polarized harmonic plane waves:

$$\boldsymbol{u}(\boldsymbol{x},t) \propto \boldsymbol{d}\exp i(\boldsymbol{k}\cdot\boldsymbol{x}-\omega t)\,, \tag{3}$$

where $\boldsymbol{d}$ is a unit polarization vector, $\boldsymbol{k}$ the wave vector, and ω the angular frequency. By analogy to Eqs. (1) and (2), the effective stiffness tensor $\boldsymbol{C}^{\mathrm{eff}}$ and the effective density tensor $\boldsymbol{\rho}^{\mathrm{eff}}$ are introduced such that, by definition,

$$\langle\boldsymbol{\sigma}\rangle = \boldsymbol{C}^{\mathrm{eff}}\langle\boldsymbol{e}\rangle\,, \tag{4}$$

$$\langle\boldsymbol{p}\rangle = \boldsymbol{\rho}^{\mathrm{eff}}\langle\dot{\boldsymbol{u}}\rangle\,, \tag{5}$$

the brackets indicating ensemble averaging over a representative volume in the random medium. The self-consistent equations for the effective-medium properties $\boldsymbol{C}^{\mathrm{eff}}$ and $\boldsymbol{\rho}^{\mathrm{eff}}$ now read:

$$\boldsymbol{C}^{\mathrm{eff}} = \boldsymbol{C}^{\mathrm{mat}} + \phi h(\boldsymbol{k})h(-\boldsymbol{k})(\boldsymbol{C}^{\mathrm{mat}}-\boldsymbol{C}^{\mathrm{inc}})\left[\boldsymbol{I}^{(4)}+\bar{\boldsymbol{S}}_x(\boldsymbol{C}^{\mathrm{inc}}-\boldsymbol{C}^{\mathrm{eff}})\right]^{-1}, \tag{6}$$

$$\boldsymbol{\rho}^{\mathrm{eff}} = \rho^{\mathrm{mat}} + \phi h(\boldsymbol{k})h(-\boldsymbol{k})(\rho^{\mathrm{mat}}-\rho^{\mathrm{inc}})\left[\boldsymbol{I}^{(2)}+\bar{\boldsymbol{M}}_t(\rho^{\mathrm{inc}}-\boldsymbol{\rho}^{\mathrm{eff}})\right]^{-1}. \tag{7}$$

We briefly discuss the quantities appearing in the self-consistent equations. The *structure factor* h is defined as the mean of the effective wave over an inclusion:

$$h(\boldsymbol{k}) = \frac{1}{V}\int_V d\boldsymbol{x}\,\exp(i\boldsymbol{k}\cdot\boldsymbol{x})\,, \tag{8}$$

where the wave vector $\boldsymbol{k}$ follows from the solution of the equation of motion for a harmonic plane wave in the effective medium. The tensor $\boldsymbol{I}^{(n)}$ denotes the unit tensor of rank n. The quantities $\bar{\boldsymbol{S}}_x$ and $\bar{\boldsymbol{M}}_t$ denote averages (over an inclusion volume) of convolution operators, the kernels of which are space and time derivatives of the Green tensor G^{eff} corresponding to the effective medium. Explicitly:

$$(\bar{\boldsymbol{S}}_x)_{ijmn} = -\frac{1}{V}\int_V d\boldsymbol{x}\int_V d\boldsymbol{x}'\left(\partial_j\partial_n G^{\text{eff}}_{im}\right)(\boldsymbol{x}-\boldsymbol{x}')\Big|_{((ij)(mn))}\,, \tag{9}$$

$$(\bar{\boldsymbol{M}}_t)_{ij} = -\frac{1}{V}\omega^2\int_V d\boldsymbol{x}\int_V d\boldsymbol{x}'G^{\text{eff}}_{ij}(\boldsymbol{x}-\boldsymbol{x}')\,. \tag{10}$$

$\bar{\boldsymbol{S}}_x$ is called the stress depolarization tensor; it has the dimension of a compliance (an inverse stiffness tensor). The suffix bracket (pq) indicates symmetrization under permutation of p and q. Its threefold use above serves to ensure that $\bar{\boldsymbol{S}}_x$ has the symmetry properties required for an elastic tensor. $\bar{\boldsymbol{M}}_t$ is the momentum depolarization tensor; its dimension is that of inverse mass-density. Since h, $\bar{\boldsymbol{S}}_x$ and $\bar{\boldsymbol{M}}_t$, on the right-hand side of the self-consistent equations, are evaluated with the effective-medium parameters of the left-hand side, the self-consistent equations are both *coupled* and *implicit*.

This general elastodynamic self-consistent scheme was worked out for three-dimensional matrix-inclusion composites by Sabina and Willis in [5] for isotropic materials and spherical inclusions, and in [6] for transversely isotropic materials and aligned spheroidal inclusions. This latter work was further generalized for random spheroidal inclusions [7] and for cracks [8]. In what follows, we will apply the self-consistent scheme to elasticity in two dimensions for isotropic materials and circular inclusions, following a more complete derivation in [2].

3. The Self-Consistent Scheme in Two Dimensions

3.1. Two-Dimensional Elastodynamics

First, we consider the propagation of harmonic waves in a homogeneous and isotropic 2D matrix, characterized by its stiffness tensor $\boldsymbol{C}$ and mass density ρ.

In what follows we will denote any isotropic elasticity tensor as (κ, μ), where κ and μ are the 2D bulk and shear modulus, respectively. A harmonic wave is polarized either parallel (P-wave) or orthogonal (S-wave) to its propagation direction. The corresponding wave speeds are

$$c_p = \sqrt{\frac{\kappa+\mu}{\rho}}\,, \tag{11}$$

$$c_s = \sqrt{\frac{\mu}{\rho}}. \tag{12}$$

The wave number k obeys the *dispersion relation*

$$k_{p,s} = \frac{\omega}{c_{p,s}}. \tag{13}$$

The time-reduced elastodynamic Green's tensor for our two-dimensional medium is now given by

$$G_{ij}(r) = \frac{1}{\rho c_s^2}\delta_{ij}\Phi_{k_s}(r) - \frac{1}{\rho\omega^2}\partial_i\partial_j\left[\Phi_{k_p}(r) - \Phi_{k_s}(r)\right] \tag{14}$$

with

$$\Phi_k(r) = \frac{i}{4}H_0^{(1)}(kr), \tag{15}$$

where $H_0^{(1)}$ is the Hankel function of the first kind of order zero.

3.2. Explicit Form of the Self-Consistent Equations

Because we are dealing with an isotropic system, the structure factor h is a function of the wavenumber $k \equiv |\boldsymbol{k}|$ only. It is easily calculated as

$$\begin{aligned} h(k) &\equiv \frac{1}{V}\int_V d\boldsymbol{x}\exp(i\boldsymbol{k}\cdot\boldsymbol{x}) \\ &= \frac{2}{ka}J_1(ka), \end{aligned} \tag{16}$$

where J_1 is the Bessel function of the first kind of order one, and a is the inclusion radius. Since in our simple medium, made up of circular disks in an isotropic matrix, the depolarization tensors are isotropic as well, we can write them in terms of just three scalars κ_S, μ_S and ρ_M:

$$\bar{\boldsymbol{S}}_x = (\kappa_S, \mu_S), \tag{17}$$

$$\bar{\boldsymbol{M}}_t = \rho_M. \tag{18}$$

These moduli are derived in [2] as

$$\kappa_S = \frac{1}{4}\frac{I_p+1}{\rho c_p^2}, \tag{19}$$

$$\mu_S = \frac{1}{8}\left(\frac{I_p+1}{\rho c_p^2} + \frac{I_s+1}{\rho c_s^2}\right), \tag{20}$$

$$\rho_M = -\frac{1}{2}\frac{I_p+I_s}{\rho}, \tag{21}$$

where $I_{p,s} \equiv I(k_{p,s})$ is given by

$$\begin{aligned} I(k) &= \frac{1}{2} i\pi (ka)^2 \sum_{n=0}^{\infty} \frac{(-1)^n (2n+1)!}{(n+2)(n!(n+1)!)^2} \left(\frac{ka}{2}\right)^{2n} \\ &\times \left\{1 + \frac{2i}{\pi}\left[\psi(2n+2) - \psi(n+2) - \psi(n+1) + \right.\right. \\ &+ \left.\left. \ln\left(\frac{ka}{2}\right) - \frac{1}{2n+4}\right]\right\}. \end{aligned} \tag{22}$$

Here, $\psi(z) \equiv \Gamma'(z)/\Gamma(z)$ denotes the logarithmic derivative of the Gamma function. For integer arguments it can be evaluated as

$$\psi(n) = -\gamma + \sum_{m=1}^{n-1} \frac{1}{m} \tag{23}$$

with γ denoting Euler's constant: $\gamma = 0.57721566\ldots$.

Writing out the self-consistent equations (6) and (7) yields the following coupled scalar equations for the effective isotropic moduli:

$$\kappa^{\text{eff}} = \kappa^{\text{mat}} + \phi \frac{h^2(k)(\kappa^{\text{inc}} - \kappa^{\text{mat}})}{1 + 4\kappa_S(\kappa^{\text{inc}} - \kappa^{\text{eff}})}, \tag{24}$$

$$\mu^{\text{eff}} = \mu^{\text{mat}} + \phi \frac{h^2(k)(\mu^{\text{inc}} - \mu^{\text{mat}})}{1 + 4\mu_S(\mu^{\text{inc}} - \mu^{\text{eff}})}, \tag{25}$$

$$\rho^{\text{eff}} = \rho^{\text{mat}} + \phi \frac{h^2(k)(\rho^{\text{inc}} - \rho^{\text{mat}})}{1 + \rho_M(\rho^{\text{inc}} - \rho^{\text{eff}})}. \tag{26}$$

4. Results

To obtain numerical results, we used the material properties listed in Table 1.

TABLE 1.
Material properties of matrix (an epoxy) and inclusions (lead).

material	c_p (km/s)	c_s	ρ (g/cc)	κ (GPa)	μ
matrix	2.64	1.20	1.20	6.03	1.73
inclusions	2.21	0.86	11.3	46.8	8.36

We solve the self-consistent equations by iteration. In general, the effective constants (and hence all derived quantities) take on complex values and are frequency-dependent, reflecting two phenomena inherent to wave scattering: attenuation and

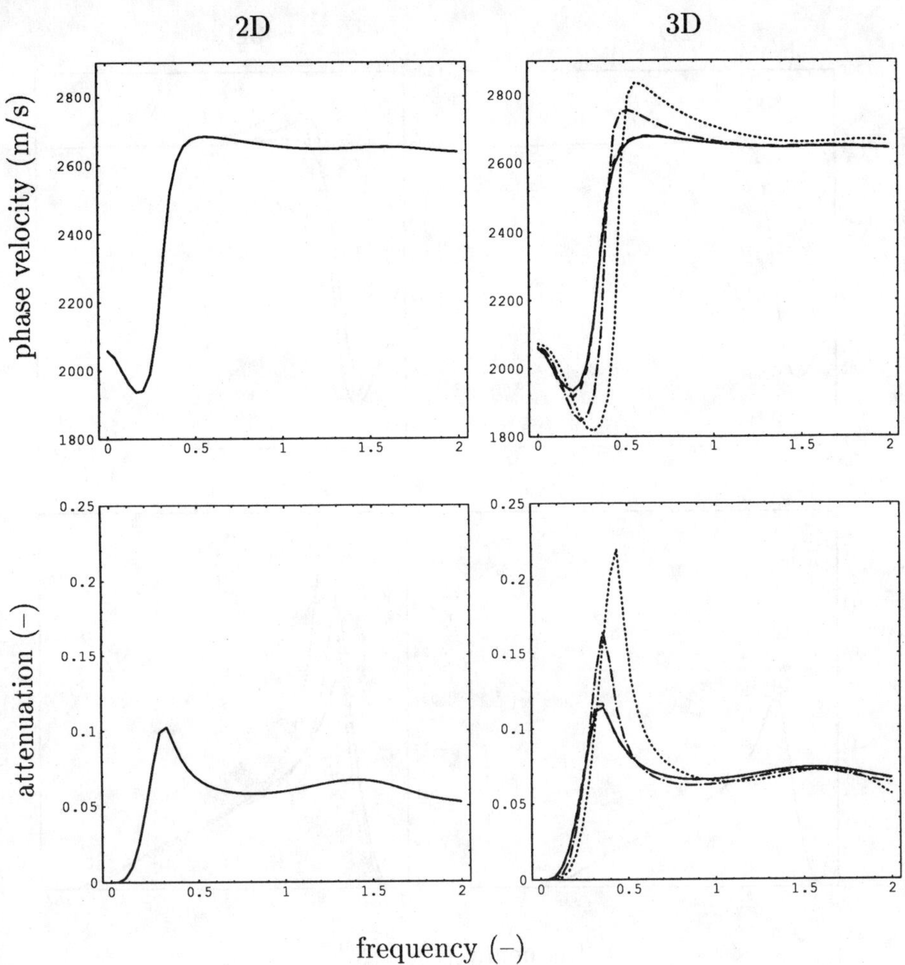

Fig. 1. Wave speed and attenuation versus dimensionless frequency for *P*-waves. On the left, the medium is a two-dimensional epoxy sheet with circular lead inclusions at a volume fraction of 10 %; on the right, the medium is a three-dimensional epoxy matrix with aligned, spheroidal lead inclusions at a volume fraction of 10 %, and an aspect ratio of either 1 (dotted line), 2 (dot-dashed), 5 (dashed) and 10 (solid).

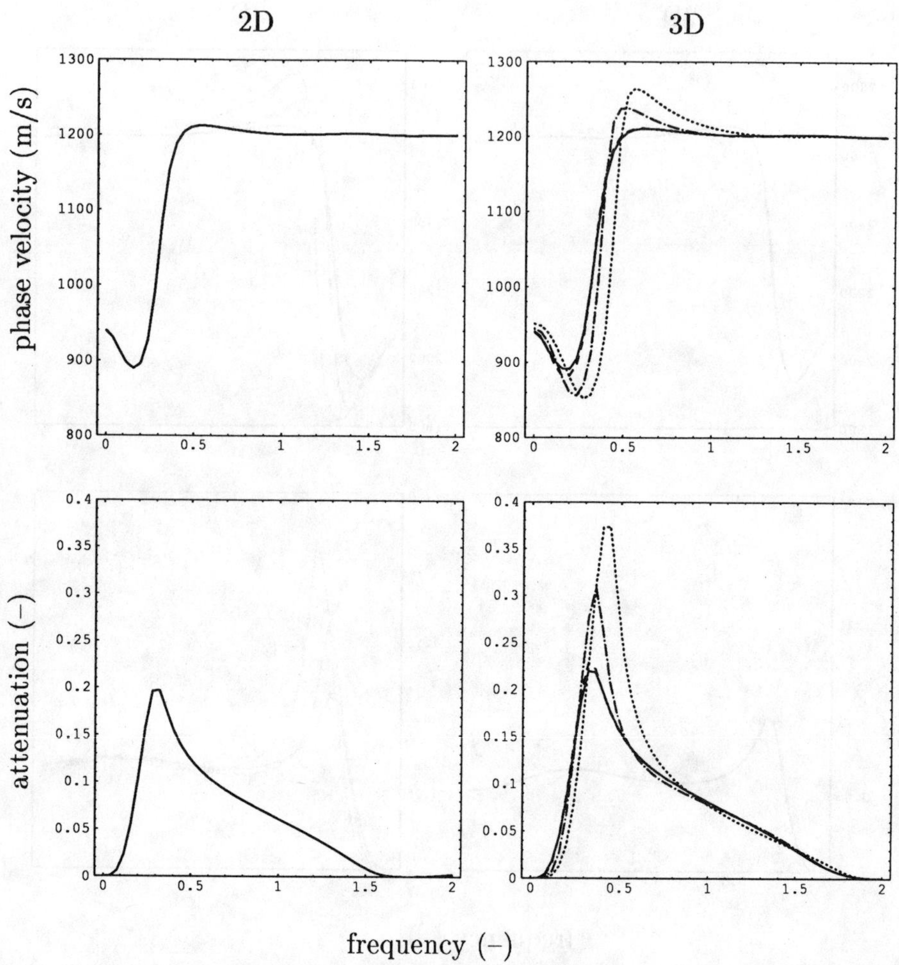

Fig. 2. Wave speed and attenuation versus dimensionless frequency for S-waves. On the left, the medium is a two-dimensional epoxy sheet with circular lead inclusions at a volume fraction of 10 %; on the right, the medium is a three-dimensional epoxy matrix with aligned, spheroidal lead inclusions at a volume fraction of 10 %, and an aspect ratio of either 1 (dotted line), 2 (dot-dashed), 5 (dashed) and 10 (solid).

dispersion. The real part of the wavenumber k corresponds to the phase velocity $c \equiv \omega/\mathrm{Re(k)}$ of the elastic wave. The imaginary part corresponds to the attenuation of the coherent wave, for which we will adopt the quantity $\mathrm{Im(k)} \cdot a$ as a measure.

We plot phase velocity and attenuation for our 2D medium with circular inclusions, as well as for a 3D composite with spheroidal inclusions of varying aspect ratio. Fig. 1 shows the results for P-waves and Fig. 2 for S-waves, where we of course took the in-plane S-waves in the 3D case. The horizontal axes display the dimensionless frequency $\omega a/c_p^{\mathbf{mat}}$, where $c_p^{\mathbf{mat}}$ is the (constant and real) phase velocity of a P-wave propagating in pure matrix material.

Looking at the velocity curves first, we observe a definite resonance and also see that the 3D curves converge to the 2D one when the aspect ratio increases. Note that in the 3D case the change in velocity around the resonance grows smaller when the aspect ratio increases and that there is hardly any difference between the curves corresponding to aspect ratios of 5 and greater. In all cases the velocities rapidly converge to the one in pure matrix material with increasing frequency.

The attenuation graphs show that the attenuation vanishes far more rapidly for S-waves than for P-waves, and that the 3D results converge slower to the 2D result than it was the case for the phase velocities.

Acknowledgements. It is our pleasure to thank Prof. J. R. Willis (University of Cambridge, UK) for many stimulating discussions. We would also like to thank Shell Internationale Research Maatschappij and Shell Internationale Petroleum Maatschappij for their permission to publish this work.

References

1. B. Budiansky, *J. Mech. Phys. Solids* **13** (1965) 223.
2. P. G. J. Bussink, P. L. Iske, J. Oortwijn and G.L.M.M. Verbist, *J. Mech. Phys. Solids*, to appear, 1995.
3. R. Hill, *J. Mech. Phys. Solids* **13** (1965) 213.
4. L. D. Landau and E. M. Lifshitz, In *Theory of Elasticity*, Pergamon Press Ltd., London, 1959.
5. F. J. Sabina and J. R. Willis, *Wave Motion* **10** (1988) 127.
6. F. J. Sabina, V. P. Smyshlyaev and J. R. Willis, *J. Mech. Phys. Solids* **41** (1993) 1573.
7. V. P. Smyshlyaev, J. R. Willis and F. J. Sabina, *J. Mech. Phys. Solids* **41** (1993) 1589.
8. V. P. Smyshlyaev, J. R. Willis and F. J. Sabina, *J. Mech. Phys. Solids* **41** (1993) 1809.
9. V. K. Varadan, V. V. Varadan and Y-H. Pao, *J. Acoust. Soc. Am.* **63** (1978) 1310.
10. J. R. Willis, *J. Mech. Phys. Solids* **28** (1980) 287.

Continuum Models and Discrete Systems
Proceedings of 8th International Symposium, June 11-16, 1995, Varna, Bulgaria, ed. K.Z. Markov

DIFFERENTIAL GEOMETRY OF CONTINUA WITH MICROSTRUCTURE

M. de LEÓN
Instituto de Matemáticas y Física Fundamental,
Consejo Superior de Investigaciones Científicas,
Serrano 123, 28006 Madrid, Spain

and

M. EPSTEIN
Department of Mechanical Engineering,
University of Calgary, 2500 University Drive NW,
T2N 1N4 Calgary, Alberta, Canada

Abstract. A geometrical theory for media with microstructure is proposed in the framework of vector bundles associated with principal bundles. The homogeneity is characterized by means of three connections.

1. Introduction

A medium with microstructure can be viewed as a simple medium each point of which is itself a deformable continuum [1,7,10,11]. In geometrical terms, we have a body manifold $\mathcal{B}$ and a fiber bundle $\pi : \mathcal{E} \longrightarrow \mathcal{B}$ over $\mathcal{B}$. The underlying simple body $\mathcal{B}$ and each fiber of $\mathcal{E}$ are deformable. For the sake of simplicity we only consider vector bundles. For instance, the geometrical model of a rod $\mathcal{B}^1$ (respectively a shell $\mathcal{B}^2$) is an embedding of $\mathcal{B}^1$ (resp. $\mathcal{B}^2$) into R^3 together with its normal bundle.

In order to determine the deformation of each fiber it is necessary to know how a basis of it is deformed. Thus, we are led to consider the principal bundle $\mathcal{FE}$ of frames of $\mathcal{E}$ and the configurations would be their embeddings into $R^3 \times Gl(m, R)$, where $n \leq 3$, with $\dim \mathcal{B} = n$ and rank $\mathcal{E} = m$. The material response depends of the 1-jet of the deformation. We define the notion of uniformity and isotropy group in terms of jets. If the body $\pi : \mathcal{E} \longrightarrow \mathcal{B}$ enjoys smooth uniformity we can characterize the homogeneity in terms of three connections: one linear connection on $\mathcal{B}$ and two connections in the vector bundle $\pi : \mathcal{E} \longrightarrow \mathcal{B}$.

The results recover the cases of materials of higher grade [3,4,6] and generalized Cosserat media [5].

2. Media with Microstructure: Configurations, Uniformity, Material Symmetry

The geometrical model of a medium with microstructure is a vector bundle $\pi : \mathcal{E} \longrightarrow \mathcal{B}$ over the body manifold $\mathcal{B}$. We assume that:

— $\mathcal{B}$ is n-dimensional;

— the rank of $\mathcal{E}$ is m, that is, the typical fiber is R^m.

In other words, an m-dimensional vector space is attached to each point $X \in \mathcal{B}$: the fiber $\mathcal{E}_X = \pi^{-1}(X)$ at X.

Next, we shall give a geometric picture for the deformations. On the one hand, the underlying body $\mathcal{B}$ is deformed and, on the other hand, the fiber at each point of $\mathcal{B}$ is also deformed. To describe the latter deformation, we only need to know how a basis of the vector space $\mathcal{E}_X$ is deformed. This fact allows us to take the principal bundle of references of $\pi : \mathcal{E} \longrightarrow \mathcal{B}$.

For each point $X \in \mathcal{B}$ we denote by $\mathcal{F}_X\mathcal{E}$ the set of ordered bases of the vector space $\mathcal{E}_X$ which can be described as follows:

$$\mathcal{F}_X\mathcal{E} = \{p : R^m \longrightarrow \mathcal{E}_X | p \text{ is a linear isomorphism}\} . \tag{2.1}$$

We set $\mathcal{F}\mathcal{E} = \cup_{X \in \mathcal{B}} \mathcal{F}_X\mathcal{E}$. If we denote by $\tilde{\pi} : \mathcal{F}\mathcal{E} \longrightarrow \mathcal{B}$ the canonical projection, we have that $\tilde{\pi} : \mathcal{F}\mathcal{E} \longrightarrow \mathcal{B}$ is a principal bundle over $\mathcal{B}$ with structure group $Gl(m, R)$. In fact, $\mathcal{E}$ is the associated vector bundle with $\mathcal{F}\mathcal{E}$ when one considers the natural action of $Gl(m, R)$ on R^m [8].

Definition 1. A configuration of $\pi : \mathcal{E} \longrightarrow \mathcal{B}$ is a principal bundle embedding $\tilde{\Phi} : \mathcal{F}\mathcal{B} \longrightarrow R^3 \times Gl(m, R)$ over an embedding $\Phi : \mathcal{B} \longrightarrow R^3$ such that $\tilde{\Phi}$ induces the identity between the structure groups.

A deformation is a change of configuration, that is, given two configurations $\tilde{\Phi}_i : \mathcal{F}\mathcal{E} \longrightarrow R^3 \times Gl(m, R), i = 1, 2,$

$$\tilde{\kappa} = \tilde{\Phi}_2 \circ \tilde{\Phi}_1^{-1} , \tag{2.2}$$

which is a principal bundle isomorphism from $\tilde{\Phi}_1(\mathcal{F}\mathcal{B})$ into $\tilde{\Phi}_2(\mathcal{F}\mathcal{B})$ inducing the identity on the structure groups and covering the diffeomorphism $\kappa = \Phi_2 \circ \Phi_1^{-1} : \Phi_1(\mathcal{B}) \longrightarrow \Phi_2(\mathcal{B})$.

We assume that the material response is completely characterized by a scalar function which depends on the first derivative of the deformation. The constitutive equation is:

$$W = W(j^1_{\tilde{X}}\ \tilde{\kappa}, j^1_X\ \kappa) , \ X = \pi(\tilde{X}) , \tag{2.3}$$

or, for simplicity,

$$W = W(j^1_{\tilde{X}}\ \tilde{\kappa}) . \tag{2.4}$$

From now on, we choose a reference configuration $(\tilde{\Phi}_0, \Phi_0)$, and make the obvious identifications: $\mathcal{B} = \Phi_0(\mathcal{B})$, $\mathcal{FE} = \tilde{\Phi}_0(\mathcal{FE})$.

Definition 2. A medium with microstructure $\pi : \mathcal{E} \longrightarrow \mathcal{B}$ is said to be uniform if for every pair of points $X, Y \in \mathcal{B}$ there exists a local isomorphism of principal bundles $(\tilde{\Phi}, \Phi)$ (inducing the identity between the structure groups) such that $\Phi(Y) = X$ and

$$W(j^1_{\tilde{\Phi}(\tilde{Y})}\ \tilde{\kappa} \circ j^1_{\tilde{Y}}\ \tilde{\Phi}) = W(j^1_{\tilde{\Phi}(\tilde{Y})}\ \tilde{\kappa})\ , \ \forall \tilde{Y} \in \tilde{\pi}^{-1}(Y), \forall j^1_{\tilde{\Phi}(\tilde{Y})}\ \tilde{\kappa}\ , \tag{2.5}$$

where $\circ$ denotes the composition of jets.

A material symmetry at a point $X \in \mathcal{FE}$ is a 1-jet $j^1_{\tilde{X}}\ \tilde{\Phi}$ as in Definition 2 such that $\tilde{\pi}(\tilde{\Phi}(\tilde{X})) = \tilde{\pi}(\tilde{X}) = X$. From (2.5) we deduce that the collection of all material symmetries at X form a group which is called the isotropy group at X.

Take the trivial principal bundle $R^n \times Gl(m, R)$ and denote by e_1 the element $(0, 1)$, where 1 is the identity matrix in $Gl(m, R)$. A 1-jet $j^1_{e_1}\ \tilde{\Phi}$ such that $\tilde{\Phi}(e_1) = \tilde{X}_0$ will be called a reference crystal at the point X_0 (or $\tilde{X}_0$), where $X_0 = \tilde{\pi}(\tilde{X}_0)$. Notice that $\Phi(0) = X_0$.

3. Non-holonomic Frames

A non-holonomic frame of $\pi : \mathcal{E} \longrightarrow \mathcal{B}$ at a point $X \in \mathcal{B}$ is a 1-jet $j^1_{e_1}\ \tilde{\Phi}$ of a local principal bundle isomorphism $\tilde{\Phi} : R^n \times Gl(m, R) \longrightarrow \mathcal{FE}$, where $\tilde{\Phi}$ induces the identity between the structure groups, and $\tilde{\pi}(\tilde{\Phi}(e_1)) = X$. The collection of all non-holonomic frames at all the points of $\mathcal{B}$ is denoted by $\bar{F}\mathcal{E}$ and we have $\bar{F}\mathcal{E} \subset F(\mathcal{FE})$, where $F(\mathcal{FE})$ denotes the linear frame bundle of $\mathcal{FE}$.

Take bundle coordinates (r^i, R^a_b), in $R^n \times Gl(m, R)$, $1 \le i, j, k, \ldots \le n$, and $1 \le a, b, c, \ldots \le m$. We have

$$\tilde{\Phi}(r^i, R^a_b) = (\Phi^i(r), \Phi^a_c(r) R^c_b)\ , \tag{3.1}$$

from which we get

$$j^1_{e_1}\ \tilde{\Phi} = (\Phi^i(0), \Phi^a_b(0), \frac{\partial \Phi^i}{\partial r^j}(0), 0, \frac{\partial \Phi^a_b}{\partial r^j}(0), \Phi^a_c(0)\delta^b_d)\ . \tag{3.2}$$

We shall use the following coordinates:

$$\begin{aligned} \mathcal{B} &: (x^i)\ , \\ \mathcal{E} &: (x^i, \xi^a)\ , \\ \mathcal{FE} &: (x^i, X^a_b)\ , \\ F(\mathcal{FE}) &: (x^i, X^a_b; x^i{}_{,j}, x^i{}_{,b}, X^a_{b,\ j}, X^a_{b,\ cd})\ . \end{aligned} \tag{3.3}$$

With these notations the coordinates of $j^1_{e_1}\ \tilde{\Phi}$ are $((x^i, X^a_b; x^i{}_{,j}, 0, X^a_{b,\ j}, X^a_c \delta^b_d)$, or, for simplicity $(x^i, X^a_b; x^i{}_{,j}, X^a_{b,\ j})$. We deduce that $\bar{F}\mathcal{E}$ is a $(n + m^2)(n + 1)$-dimensional submanifold of $F(\mathcal{FE})$.

Furthermore, if we consider the elements $j^1_{e_1}\ \tilde{\Phi}$ from $R^n \times Gl(m,R)$ into itself such that $\Phi(0) = 0$, we obtain a Lie group denoted by $\bar{G}(n,m)$ whose elements are of the form $(A,B,C) = (A^a_b, B^i_j, C^a_{bj})$ and whose multiplication law is given by the following formula obtained by applying the chain rule:

$$(A_1,B_1,C_1)(A_2,B_2,C_2) = ((A_1)^a_c(A_2)^c_b, (B_1)^i_k(B_2)^k_j, (C_1)^a_{ck}(B_2)^k_i(A_2)^c_b + (A_1)^a_c(C_2)^c_{bi})\ . \tag{3.4}$$

A simple computation shows that $\bar{\pi} : \bar{F}\mathcal{E} \longrightarrow \mathcal{B}$, where $\bar{\pi}$ is the canonical projection, is a principal bundle over $\mathcal{B}$ and with structure group $\bar{G}(n,m)$. $\bar{F}\mathcal{E}$ will be called the non-holonomic frame bundle of $\mathcal{E}$.

4. Uniform Media with Microstructure

Assume that the medium with microstructure $\pi : \mathcal{E} \longrightarrow \mathcal{B}$ is smoothly uniform. This means that for every two points $X, Y \in \mathcal{B}$ there exists an element $j^1_{\tilde{X}}\ \tilde{\Phi}$ with $\tilde{\Phi}(\tilde{X}) = \tilde{Y}$ and $\tilde{\pi}(\tilde{X}) = X$, $\tilde{\pi}(\tilde{Y}) = Y$, in such a way that this choice depends smoothly on X and Y. Choose a point X_0 in $\mathcal{B}$ and a reference crystal $j^1_{e_1}\ \tilde{\Psi}$ at X_0. A non-holonomic frame $\bar{\mathcal{P}}(X)$ at X is obtained by composition of jets:

$$\bar{\mathcal{P}}(X) = j^1_{\mathcal{P}(X)}\ \tilde{\Phi} \circ j^1_{e_1}\ \tilde{\Psi}\ . \tag{4.1}$$

Therefore, we obtain a smooth global section $\bar{\mathcal{P}} : \mathcal{B} \longrightarrow \bar{F}\mathcal{E}$ of $\bar{\pi} : \bar{F}\mathcal{E} \longrightarrow \mathcal{B}$. If we put $\mathcal{P}(X) = \tilde{\Phi}(\tilde{\Psi}(e_1))$ we obtain a global section $\mathcal{P}$ of $\tilde{\pi} : \mathcal{FE} \longrightarrow \mathcal{B}$.

Given a material symmetry $j^1_{\mathcal{P}(X_0)}\ \tilde{\Phi}$ at X_0 we obtain a 1-jet $j^1_{\tilde{\Phi}(\mathcal{P}(X_0))}\ \tilde{\Psi}^{-1} \circ j^1_{\mathcal{P}(X_0)}\ \tilde{\Phi} \circ j^1_{e_1}\ \tilde{\Psi}$, which is in fact an element of the Lie group $\bar{G}(n,m)$. Thus, we have obtained a Lie subgroup $\bar{G}$ of $\bar{G}(n,m)$, and we can extend the global section $\bar{\mathcal{P}}$ to get a $\bar{G}$-reduction $\bar{\mathcal{P}}\bar{G}$ of the non-holonomic frame bundle $\bar{F}\mathcal{E}$.

Notice that the reference crystal $j^1_{e_1}\ \tilde{\Psi}$ induces an ordinary crystal reference $j^1_0\ \Psi$ at X_0. If we compose the induced 1-jets $j^1_{X_0}\ \Phi$ with $j^1_0\ \Psi$, we obtain a linear frame at the point X which will be denoted by $\mathcal{Q}(X)$. In other words, we obtain a linear parallelism $\mathcal{Q} : \mathcal{B} \longrightarrow \mathcal{FB}$. If we compose the induced 1-jets $j^1_{X_0}\ \Phi$ from the material symmetries at X_0 with $j^1_0\ \Psi$ we obtain a Lie subgroup G of the general linear group $Gl(n,R)$ and, by prolongation of $\mathcal{Q}$, a G-structure $\mathcal{Q}G$ on $\mathcal{B}$.

In local coordinates, we have

$$\begin{aligned} \bar{\mathcal{P}}(x^i) &= (x^i, P^a_b(x), Q^i_j(x), R^a_{bj}(x))\ , \\ \mathcal{P}(x^i) &= (x^i, P^a_b(x))\ , \\ \mathcal{Q}(x^i) &= (x^i, Q^i_j(x))\ . \end{aligned} \tag{4.2}$$

Given a global uniformity field $\bar{\mathcal{P}}$ we obtain three connections as follows:

— $\mathcal{P}$ induces a connection Γ_1 in the principal bundle $\tilde{\pi} : \mathcal{FE} \longrightarrow \mathcal{B}$ by defining the horizontal subspaces to be $H_{\mathcal{P}(X)} = d\mathcal{P}(X)(T_X\mathcal{B})$, $\forall X \in \mathcal{B}$, and then translating these subspaces by the action of the group $Gl(m,R)$. The Christoffel components of Γ_1 are:

$$(\Gamma_1)^a_{jb} = -(\mathcal{P}^{-1})^c_b \frac{\partial \mathcal{P}^a_c}{\partial x^j} \,. \tag{4.3}$$

— $\mathcal{Q}$ induces a linear connection Γ_2 on $\mathcal{B}$ with Christoffel components

$$(\Gamma_2)^i_{jk} = -(\mathcal{Q}^{-1})^l_k \frac{\partial \mathcal{Q}^i_l}{\partial x^j} \,. \tag{4.4}$$

In fact, Γ_2 is the linear connection induced from the linear parallelism $\mathcal{Q}$.

— $\bar{\mathcal{P}}$ induces a second connection Γ_3 in the principal bundle $\tilde{\pi} : \mathcal{FE} \longrightarrow \mathcal{B}$ by defining:

$$H_{\mathcal{P}(X)} = d\psi(0)(T_0 R^n) \,, \tag{4.5}$$

where $\psi(r) = \tilde{\Psi}(r,1)$ and $\bar{\mathcal{P}}(X) = j^1_{e_1}\,\tilde{\Psi}$. Then, we transport the horizontal subspaces to an arbitrary point by the action of the structure group. The Christoffel components of Γ_3 are:

$$(\Gamma_3)^a_{jb} = -\mathcal{R}^a_{ck}(\mathcal{Q}^{-1})^k_j(\mathcal{P}^{-1})^c_b \,. \tag{4.6}$$

5. Homogeneous Media with Microstructure

In what follows we shall consider the case of media with microstructure with a trivial isotropy group.

We say that $\pi : \mathcal{E} \longrightarrow \mathcal{B}$ is homogeneous (with respect to the chosen crystal reference) if there exists a global configuration $(\tilde{\kappa}, \kappa)$ such that:

1. $\kappa : \mathcal{B} \longrightarrow R^3$ is an embedding into R^n, i.e. $\kappa(\mathcal{B}) \subset R^n$; and
2. $\bar{\mathcal{P}} = \tilde{\kappa}^{-1}$ is a uniformity field.

More precisely, for each $X \in \mathcal{B}$, let $\tilde{\mathcal{A}}_X : R^n \times Gl(m,R) \longrightarrow \mathcal{FE}$ be the bundle isomorphism defined by

$$\tilde{\mathcal{A}}_X(r,R) = \tilde{\kappa}^{-1}(r + \kappa(X), R) \,. \tag{5.1}$$

Then $\pi : \mathcal{E} \longrightarrow \mathcal{B}$ is homogeneous if $\bar{\mathcal{P}}(X) = j^1_{e_1}\,\tilde{\mathcal{A}}_X$ is a uniformity field. The notion of local homogeneity is the obvious one.

In that case, there exist local coordinates (x^i) in R^n such that

$$\bar{\mathcal{P}}(x^i) = (x^i, \mathcal{P}^a_b(x), 1, \frac{\partial \mathcal{P}^a_b}{\partial x^j}) \,. \tag{5.2}$$

From the Christoffel components of the three connections Γ_1, Γ_2, and Γ_3 we deduce the following result which characterizes geometrically the homogeneity of a medium with structure.

Theorem 1. A medium with microstructure $\pi : \mathcal{E} \longrightarrow \mathcal{B}$ is locally homogeneous with respect to a chosen reference crystal if and only if Γ_2 has no torsion and $\Gamma_1 = \Gamma_2$.

Notice that a connection Γ in the principal bundle $\tilde{\pi} : \mathcal{FE} \longrightarrow \mathcal{B}$ induces a connection ∇ in the vector bundle $\pi : \mathcal{E} \longrightarrow \mathcal{B}$, where ∇ denotes the covariant derivative of sections of $\mathcal{E}$ with respec to vector fields on $\mathcal{B}$

$$\nabla : \mathfrak{X}(\mathcal{B}) \times Sect\ (\mathcal{E}) \longrightarrow Sect\ (\mathcal{E})\ . \tag{5.3}$$

The two connections Γ_1 and Γ_3 induce two derivation laws ∇_1 and ∇_3 which define a linear mapping $D : \mathfrak{X}(\mathcal{B}) \times Sect\ (\mathcal{E}) \longrightarrow Sect\ (\mathcal{E})$, $(X, \sigma) \rightsquigarrow D(X, \sigma) = (\nabla_1)_X \sigma - (\nabla_3)_X \sigma$; D is called the difference of them. If T_2 denotes the torsion tensor of the linear connection Γ_2, then Theorem 1 may be stated as follows.

Theorem 2. A medium with microstructure $\pi : \mathcal{E} \longrightarrow \mathcal{B}$ is locally homogeneous with respect to a chosen reference crystal if and only if T_2 and D simultaneously vanish.

Next, we shall show what happens if we change of reference configuration and/or reference crystal.

Change of configuration

If we perform a change of configuration $(\kappa(x^i), \kappa_b^a(x^i))$ we obtain a new field of uniformities given by

$$\bar{\mathcal{P}}'(x^i) = (x^i, \kappa_c^a \mathcal{P}_b^c, \frac{\partial \kappa^i}{\partial x^k} \mathcal{Q}_j^k, \frac{\partial \kappa_c^a}{\partial x^k} \mathcal{Q}_j^k \mathcal{P}_b^c + \kappa_c^a \mathcal{R}_{bj}^c)\ , \tag{5.4}$$

which shows that the homogeneity does not depend on the choice of reference configuration. Taking into account Eq. (5.4), we deduce that if $\pi : \mathcal{E} \longrightarrow \mathcal{B}$ is homogeneous, we can make a change of configuration $(x^i, X_b^a) \rightsquigarrow (x^i, X_c^a (\mathcal{P}^{-1})_b^c)$ in such a way that the new field of uniformities reads as $\bar{\mathcal{P}}'(x^i) = (1, 1, 0)$.

Change of reference crystal

Suppose that we perform a change of reference crystal to another one. Notice that a change of crystal is given by an element of the group $\bar{G}(n, m)$, say (A, B, C). Thus, we obtain

$$\bar{\mathcal{P}}'(x^i) = (x^i, \mathcal{P}_c^a A_b^c, \mathcal{Q}_k^i B_j^k, \mathcal{P}_c^a C_{bj}^c + \mathcal{R}_{ck}^a A_b^c B_j^k)\ . \tag{5.5}$$

Since $\mathcal{P}$ is a global section of $\mathcal{FE}$, then it determines m linearly independent global sections of $\mathcal{E}$, $\mathcal{P} = (\mathcal{P}_1, \ldots, \mathcal{P}_m)$. On the other hand, the linear parallelism $\mathcal{Q}$ determines n linearly independent vector fields $(\mathcal{Q}_1, \ldots, \mathcal{Q}_n)$ on $\mathcal{B}$. Define nm global sections of $\mathcal{E}$ as follows:

$$D_{ia} = D(Q_i, P_a)\ ,\ 1 \le i \le n\ ,\ 1 \le a \le m\ . \tag{5.6}$$

Assume that we may choose a field of uniformities $\bar{\mathcal{P}}$ such that T_2 vanishes. In order to check the homogeneity with respect to a given reference crystal we have to compute the difference D. If D is not zero, can we perform a change of crystal such that the new D' vanishes identically? (Notice that the torsion tensor will remain unchanged). If we compute the new difference D', after a change (A, B, C) of reference crystal, we deduce from a straightforward calculation that

$$D' = 0 \Longleftrightarrow D_{ia} = \sigma^b_{ia} P_b \ , \text{ where } \sigma^b_{ia} = -C^b_{ck}(B^{-1})^k_i(A^{-1})^c_a \Longleftrightarrow (\nabla_1)D_{ia} = 0 \ . \quad (5.7)$$

From (5.7) we deduce the following.

Theorem 3. A medium with microstructure $\pi : \mathcal{E} \longrightarrow \mathcal{B}$ is locally homogeneous if and only if T_2 and $(\nabla_1)D_{ia}$ simultaneously vanish.

6. Examples

6.1. Generalized Cosserat Media

In this case, we take $n = 3$ and $\pi : \mathcal{E} \longrightarrow \mathcal{B}$ is the tangent bundle of $\mathcal{B}$. Thus $\mathcal{FE} = F\mathcal{B}$ is the linear frame bundle of $\mathcal{B}$. The homogeneity is characterized by three linear connections on $\mathcal{B}$. Theorems 1, 2 and 3 recover the results contained in [3], [4], [5] and [6].

6.2. Rods

In this case, we take $n = 1$, and $\pi : \mathcal{E} \longrightarrow \mathcal{B}$ is the normal bundle of $\mathcal{B}$. Here $(\mathcal{P}_1(X), \mathcal{P}_2(X))$ is a basis of the complement of $T_X\mathcal{B}$ into $T_X R^3$ with respect to the Euclidean metric. We have

$$\bar{\mathcal{P}}(x) = (x, \mathcal{P}^a_b(x), \mathcal{Q}(x), \mathcal{R}^a_{b,1}(x)) \ , \quad (6.1)$$

where x stands for an arbitrary parametrization of $\mathcal{B}$. The Christoffel components of the three connections are the following:

$$(\Gamma_1)^a_{1b} = -(\mathcal{P}^{-1})^c_b \frac{\partial \mathcal{P}^a_c}{\partial x} \ , \ (\Gamma_2)^1_{11} = -(\mathcal{Q}^{-1}) \frac{\partial \mathcal{Q}}{\partial x} \ , \ (\Gamma_3)^a_{1b} = -\mathcal{R}^a_{c1}(\mathcal{Q}^{-1})(\mathcal{P}^{-1})^c_b \ , \quad (6.2)$$

from which we deduces that T_2 always vanishes and the rod is locally homogeneous (with respect to a given reference crystal) if and only if there exists a change of parameter $x \rightsquigarrow \bar{x}$ such that

$$\mathcal{R}^a_{b1} = \frac{\partial \mathcal{P}^a_b}{\partial \bar{x}} \ . \quad (6.3)$$

6.3. Shells

In this case, we take $n = 2$, and $\pi : \mathcal{E} \longrightarrow \mathcal{B}$ is the normal bundle of $\mathcal{B}$. Here $\mathcal{P}(X)$ is a non-zero vector orthogonal to the tangent space $T_X R^3$ with respect to the Euclidean metric. We have

$$\bar{\mathcal{P}}(x^1, x^2) = (x^1, x^2, \mathcal{P}(x), \mathcal{Q}^i_j(x), \mathcal{R}^1_{1,j}(x)) \ , \quad (6.4)$$

where (x^1, x^2) denote an arbitrary parametrization of $\mathcal{B}$. The Christoffel components of the three connections are the following:

$$(\Gamma_1)^1_{j1} = -(\mathcal{P}^{-1})\frac{\partial \mathcal{P}}{\partial x^j}\ ,\ (\Gamma_2)^i_{jk} = -(\mathcal{Q}^{-1})^l_k\frac{\partial \mathcal{Q}^i_l}{\partial x^j}\ ,\ (\Gamma_3)^1_{j1} = -\mathcal{R}^1_{1j}(\mathcal{Q}^{-1})^k_j(\mathcal{P}^{-1})\ , \quad (6.5)$$

from which we deduce that $\mathcal{E}$ is locally homogeneous (with respect to a given reference crystal) if and only if

$$T_2 = 0\ ,\ \mathcal{R}^1_{1k}(\mathcal{Q}^{-1})^k_j = \frac{\partial \mathcal{P}}{\partial x^j}\ . \quad (6.6)$$

Remark. It should be remarked that the homogeneity conditions for rods and shells obtained are not the only ones possible, but only those arising naturally from the geometric model proposed (cf. [2,12]).

Acknowledgements. This work has been partially supported through grants of the National Science and Engineering Research Council of Canada and DGICYT-Spain, Proyecto PB91-0142.

References

1. G. Capriz, *Continua with Microstructure*, Springer Tracts in Natural Philosophy, 35, Berlin, 1989.
2. H. Cohen and M. Epstein, *Acta Mechanica* **47** (1983) 207.
3. M. de León and M. Epstein, *Reports on Mathematical Physics* **33**(3)(1993) 419.
4. M. de León and M. Epstein, *C. R. Acad. Sci. Paris* **319** Sér. I (1994) 615.
5. M. Epstein and M. de León, In *Proceedings Colloquium on Differential Geometry*, July 25-30, 1994, Debrecen, Hungary.
6. M. Epstein and M. de León, The geometry of uniformity in second-grade elasticity, *Acta Mechanica*, (1995), in press.
7. A. C. Eringen and Ch. B. Kafadar, In *Continuum Physics*, vol. IV, Part I, ed. A. C. Eringen, Academic Press, New York, 1976, p. 1.
8. S. Kobayashi and K. Nomizu, *Foundations of Differential Geometry*, vol. I, Interscience Publishers, New York, 1963.
9. G. Maugin, *Acta Mechanica* **35** (1980) 1.
10. C. Truesdell and W. Noll, *The Non-Linear Field Theories of Mechanics*, Handbuch der Physik, Vol. III/3, Springer, Berlin, 1965.
11. R. A. Toupin, *Arch. Rat. Mech. Anal.* **17** (1964) 85.
12. C. C. Wang, *Arch. Rat. Mech. Anal.* **47** (1972) 343.

Continuum Models and Discrete Systems
Proceedings of 8th International Symposium, June 11-16, 1995, Varna, Bulgaria, ed. K.Z. Markov
©World Scientific Publishing Company, 1996, pp. 164–171

STRUCTURED STRESS FIELDS

G. DEL PIERO
Istituto di Ingegneria
Università di Ferrara, via Saragat, 44100 Ferrara, Italy

Abstract. The concept of a structured stress field is based on the double-field description, given in [4], of the deformation of continuous bodies that undergo changes in internal structure. Each deformation field is supposed to determine a stress field: one, governed by a constitutive equation of the type encountered in classical continuum mechanics, is the stress due to the deformation of the continuum. The other one is the stress due to the change in internal structure, and is determined by a limit procedure involving the cohesive forces arising at the surfaces of discontinuity for the displacement.

1. Introduction

The mathematical description of the effects induced in a continuous body by a change in the internal structure at the microscopic level has formed the object of extensive research, involving many different models of internal structure. Examples are provided by the continuum theories of dislocations, of crystalline defects, of damage, as well as by plasticity and by theories of porous media. The name *continua with microstructure* is currently used to denote collectively a large class of such models, including continua with voids, continua with spin, polar and micropolar continua [2].

A step towards a unifying theory comes from a method recently developed in [4], where a particular class of deformations describing a change in internal structure is introduced. These deformations, called *structured deformations*, are characterized by two tensor fields, the *macroscopic deformation tensor* and the *deformation tensor without microfracture*.

The purpose of the present communication is to supplement the concept of a structured deformation with an appropriate notion of stress. Although the ideas exposed here are still at a preliminary stage, I am glad of taking advantage of this occasion to fix the present state of development of the research.

2. Simple deformations and structured deformations

A *simple deformation* of a region Ω is defined as a pair (κ, f), where κ is a surface-like subset of Ω, and f is a deformation from $\Omega\backslash\kappa$, in the sense of classical continuum mechanics [4, Sec.3]. The function f is allowed to undergo jumps across κ. These jumps are supposed to describe macroscopic cracks of the body, so that κ may be regarded as the site of the macroscopic cracks created by the simple deformation.

The idea developed in [4] is that of using sequences of simple deformations to produce, in the limit, more complicated objects, appropriate to describe both macroscopic cracks and changes in the internal structure of the body. Technically, a *structured deformation* is defined as a triple (κ,g,G), where (κ,g) is a simple deformation of Ω and G is a continuous tensor field on $\Omega\backslash\kappa$ [4, Sec.5]. It has been proved in [4] that every structured deformation (κ,g,G) admits a *determining sequence*, i.e., that there exists a sequence $m\mapsto(\kappa_m, f_m)$ of simple deformations such that the regions κ_m converge to κ, in the sense that for each point x in κ there is a number $m(x)$ such that x belongs to all sets κ_m with $m>m(x)$, and the functions f_m and their gradients ∇f_m converge to g and G, respectively, in the sense of L^∞-convergence. Examples discussed in [4] show how the difference ∇g-G can be related to changes in internal structure occurring in some specific classes of continua.

In the following, I will call ∇g and G the *macroscopic deformation tensor* and the *deformation tensor of the continuum*, respectively. The last term is preferred to the name *deformation tensor without microfracture* used in [4], because the reference to fracture was related with the original, less general purpose of that paper. The name *microfracture* will be replaced here by *change in the internal structure*. A possible alternative is the term *microdisarrangement* proposed in [10].

3. The stress arising in a simple deformation

If the body is made of an elastic material with response function $\mathcal{K}$, the Cauchy stress T in a simple deformation (κ,f) from Ω is determined by the deformation gradient ∇f

$$T = \mathcal{K}(\nabla f) \tag{3.1}$$

A referential description of the stress is provided by the Piola-Kirchhoff tensor

$$T_R = (\det \nabla f)\, \mathcal{K}(\nabla f)\, \nabla f^{-T} =: \mathcal{H}_R(\nabla f)\,. \tag{3.2}$$

If RU denotes the polar decomposition of ∇f, the principle of material indifference requires that

$$\mathcal{K}(RU) = R\,\mathcal{K}(U)\,R^T, \quad \mathcal{H}_R(RU) = R\,\mathcal{H}_R(U)\,. \tag{3.3}$$

Thus, both $\mathcal{K}$ and $\mathcal{H}_R$ are determined by their restrictions to the symmetric tensors. Moreover, the balance of angular momentum requires that $\mathcal{K}(U)$ and $\mathcal{H}_R(U)U$ be symmetric tensors.

Denote by $[f](x)$ the jump of f at the point x of κ. I assume that a jump induces surface tractions at the two sides of κ and that, in the referential description, the tractions t_R exerted at the two sides are opposite to each other, so that they do not affect the balance of linear momentum of the body. It may be assumed that the traction at x depends on the jump of f and on the orientation n_R of the normal to κ at x:

$$t_R(x) = h_R([f](x), n_R(x)). \tag{3.4}$$

This dependence can be specialized in different ways, each one reflecting a different model of physical behaviour. Indeed, the direction n_R may be either arbitrary or fixed, the second possibility denoting the presence of a specific surface of discontinuity for the displacement, as in the case of crystal defects. Moreover, the direction of the jump may be either related to n_R or free. For the purposes of the present paper, I consider a constitutive equation of the type

$$t_R = q_R(|[f]|)\, Q_R n_R \,, \tag{3.5}$$

where q_R is a scalar function and Q_R is a given orthogonal tensor. For example, Q_R equal to the identity means that the vector t_R is normal to κ, and Q_R equal to a rotation of $\pi/2$ means that t_R is tangent to κ.

Equation (3.5) is in line with the model of the *cohesive fracture* proposed by Barenblatt [1] and developed in the domain of Fracture Mechanics [3]. It is important to point out that, according to this model, a non-zero *cohesive force* t_R is transmitted across the crack when the crack is closed. For the constitutive equation (3.5), this implies that the constant $q_R(0)$ is not zero, except for the special class of continua which do not support tension. With this exception, the presence of a non-empty region κ in a simple deformation (κ,f) reveals the presence of a field of cohesive forces across κ, even if there is no discontinuity at κ.

4. The stress associated with a structured deformation

Consider a structured deformation (κ,g,G) from Ω and a determining sequence $m \mapsto (\kappa_m, f_m)$. With each term in the sequence, one can associate the Piola-Kirchhoff stress field $T_{Rm} = \mathcal{H}_R(\nabla f_m)$ over $\Omega \setminus \kappa_m$ and the surface tractions $t_{Rm} = h_R([f_m], n_{Rm})$ over κ_m. If the response function $\mathcal{H}_R$ is continuous, the convergence of $m \mapsto \nabla f_m$ to G implies the convergence of $m \mapsto T_{Rm}$ to the field

$$T_{R\infty} := \mathcal{H}_R(G)\,. \tag{4.1}$$

Similarly, if the function h_R is continuous, the sequence $m \mapsto t_{Rm}$ is expected to converge to $t_{R\infty} := h_R([g], n_R)$ over κ. But this is not the only possibility for the cohesive forces. It has been shown in [4] that the sets κ_m may diffuse across the body, originating what was called there the *fractured zone*, and that in the present context will be denoted as the *zone with changes in internal structure*. This is a closed subset of Ω, which includes the set of all points of Ω at which $\nabla g \neq G$ [4, Sec.4]. In what follows, I consider for simplicity only the second type of convergence for the cohesive forces; namely, I assume that, for the given structured deformation, κ is the empty set and the zone of Ω at which $\nabla g \neq G$ has non-zero volume. It has been shown in [5, Sec.4] that, for every region Π in Ω,

$$\lim_{m\to\infty} \int_{\Pi\cap\kappa_m} [f_m] \otimes n_{Rm}\, dA = \int_{\Pi} (\nabla g - G)\, dV \,, \tag{4.2}$$

with n_{Rm} the unit normal to κ_m. This result characterizes the tensor $\nabla g - G$ as a *volume density of the deformation due to the change in internal structure*. Consider the virtual work done by the cohesive forces t_{Rm} in the simple deformation (κ_m, f_m):

$$\int_{\Pi \cap \kappa_m} t_{Rm} \bullet [f_m] \, dA \ . \tag{4.3}$$

When $m \to \infty$, the sets κ_m converge to the empty set and the jumps $[f_m]$ converge uniformly to zero [4, Sec.4]. This does not imply, however, that the virtual work does converge to zero; it will be assumed here that for sufficiently small values of $\nabla g\text{-}G$ the limit is represented by a volume integral

$$\lim_{m \to \infty} \int_{\Pi \cap \kappa_m} t_{Rm} \bullet [f_m] \, dA \ = \int_{\Pi} T_{R\mu} \bullet (\nabla g\text{-}G) \, dV \ + o(\nabla g\text{-}G) \, , \tag{4.4}$$

in which the linear functional $T_{R\mu}$ acting on the deformation $\nabla g\text{-}G$ is interpreted as the *linear approximation of the change of stress due to the change in internal structure.*

Of course, Equation (4.4) can only determine the projection of $T_{R\mu}$ on the span of $\nabla g\text{-}G$. In classical continuum mechanics, the *principle of determinism for constrained materials* states that, in the presence of constraints on the deformation, the stress is determined to within a part which does no work in any deformation compatible with the constraint [11]. In the present situation, one can imagine a *principle of determinism for materials with change in internal structure,* stating that the stress due to the non-classical part of the deformation is exactly the stress which does a non-zero work in that deformation. This amounts to assume that at each point of Ω the change of stress $T_{R\mu}$ coincides with its projection on the span of $\nabla g\text{-}G$:

$$T_{R\mu} \ = \ \frac{T_{R\mu} \bullet (\nabla g\text{-}G)}{|\nabla g\text{-}G|^2} \, (\nabla g\text{-}G) \ , \tag{4.5}$$

with $T_{R\mu} = 0$ when $\nabla g\text{-}G = 0$. The explicit evaluation of $T_{R\mu}$ may be difficult; it simplifies considerably if one assumes the constitutive equation (3.5), as it will be seen later. In any case one may conclude that, at least for $\nabla g\text{-}G$ sufficienly small, the stress arising in a structured deformation is the sum of two parts, one due to the deformation of the continuum and one due to the change in the internal structure of the body:

$$T_R = T_{R\infty} + T_{R\mu} \, . \tag{4.6}$$

This decomposition is common in the theories of continua with microstructure [2][6][9]; what seems to be new here is the correlation of the stress $T_{R\mu}$ with the geometry of the microdeformation.

In view of the decomposition (4.6), the stress associated with a structured deformation is completely determined by the pair $(T_{R\infty}, T_{R\mu})$. This pair will be called a *structured stress field;* the double-field description of the stress obtained here parallels the double-field description adopted to describe the deformation of a body undergoing changes in internal structure.

5. Change of stress in a purely microscopic deformation

Given a structured deformation (κ,g,G) from Ω, consider the decomposition

$$(\kappa,g,G) = (\emptyset,i,G\nabla g^{-1}) \circ (\kappa,g,\nabla g) \tag{5.1}$$

into a simple deformation from Ω followed by a *purely microscopic deformation* [4, Sec.5]. The latter is a structured deformation from the region $\Omega_A := g(\Omega/\kappa)$, whose first two items are the empty set and the identical deformation of $g(\Omega/\kappa)$. Thus, in a purely microscopic deformation no creation of macroscopic cracks and no displacement of material points occurs; there is only a change in the internal structure, measured by the tensor G_A-I, where G_A:=$G\nabla g^{-1}$ is the deformation tensor of the continuum measured from the *actual configuration* Ω_A. By the constitutive equation (3.1), the Cauchy stress induced by the simple deformation is

$$T_o := \mathcal{K}(\nabla g). \tag{5.2}$$

Thus, to determine the stress in a structured deformation is the same as to determine the change of stress in the purely microscopic deformation $(\emptyset,i,G_A)$. The determination is easier if one takes the actual configuration as reference configuration. Indeed, with this choice, the Piola-Kirchhoff stress due to the deformation of the continuum takes the form

$$T_{A\infty} = \mathcal{H}_A(G_A), \tag{5.3}$$

where $\mathcal{H}_A$, the response function relative to the actual configuration, is related to the response function $\mathcal{H}_R$ relative to the originary reference configuration by

$$\mathcal{H}_A(G_A) = (\det G_A)\, T_o G_A^{-T} = (\det G_A G^{-1})\, \mathcal{H}_R(G)\, G^T G_A^{-T} = (\det \nabla g)^{-1} \mathcal{H}_R(G)\, \nabla g^T. \tag{5.4}$$

Note that T_o and $T_{A\infty}$ are fields over the same set Ω_A, and that their difference is the change of stress due to the change of internal structure. Note also that Equation $(5.4)_1$ implies $T_o = \mathcal{H}_A(I)$, so that the response function $\mathcal{H}_A$ restricted to the symmetric tensors admits the expansion

$$\mathcal{H}_A(U) = T_o + \boldsymbol{B}_A\,(U-I) + o\,(U-I), \tag{5.5}$$

with $\boldsymbol{B}_A$ the gradient of the restricted $\mathcal{H}_A$ at I. The requirement $(3.3)_2$ imposed by the principle of material indifference then implies

$$T_{A\infty} = \mathcal{H}_A(RU) = R\,\mathcal{H}_A(U), \tag{5.6}$$

where RU now denotes the polar decomposition of G_A. After setting R=I+W and U=I+E, it follows from Equation (5.5) that

$$T_{A\infty} = T_o + WT_o + \boldsymbol{B}_A E + o(E+W). \tag{5.7}$$

Thus, to within terms of higher order in $E+W$, the change of stress due to the deformation of the continuum is given by $WT_o + \boldsymbol{B}_A E$.

To evaluate the stress $T_{A\mu}$ due to the change in internal structure, take a determining sequence $m \mapsto (\kappa_m, f_m)$ for $(\emptyset, i, G_A)$ and suppose that the cohesive forces t_{Am} are governed by the constitutive equation (3.5). Then the assumption (4.4) implies

$$\lim_{m\to\infty} \int_{\Pi_A \cap \kappa_m} q_A(|[f_m]|)\, Q_A n_{Am} \bullet [f_m]\, dA_A = \int_{\Pi_A} T_{A\mu} \bullet (I-G_A)\, dV_A + o(I-G_A), \quad (5.8)$$

where the subscripts A denote that the actual configuration is taken as reference configuration. The fact that $m \mapsto f_m$ converges uniformly to the identity in Ω_A implies that $m \mapsto |[f_m]|$ converges uniformly to zero. Therefore, the integral on the left reduces to

$$q_A(0)\, Q_A \bullet \lim_{m\to\infty} \int_{\Pi_A \cap \kappa_m} [f_m] \otimes n_{Am}\, dA_A, \quad (5.9)$$

and the equality (4.2) written in the actual configuration tells us that

$$q_A(0)\, Q_A \bullet \int_{\Pi_A} (I-G_A)\, dV_A = \int_{\Pi_A} T_{A\mu} \bullet (I-G_A)\, dV_A + o(I-G_A). \quad (5.10)$$

Because this inequality holds for every field G_A and for every region Π_A, it follows that

$$T_{A\mu} \bullet (I-G_A) = q_A(0)\, Q_A \bullet (I-G_A) \quad (5.11)$$

at all points of Ω_A. The *principle of determinism* expressed by Equation (4.5) then allows us to conclude that

$$T_{A\mu} = q_A(0) \frac{Q_A \bullet (I-G_A)}{|I-G_A|^2} (I-G_A). \quad (5.12)$$

6. Examples

To illustrate the concepts introduced so far, I consider two two-dimensional structured deformations of the square $\Omega = (0,l)^2$. The first deformation describes a simple shear accompanied by a *slip along a dislocation line*. It is defined by the determining sequence $m \mapsto (\kappa_m, f_m)$, with

$$\kappa_m = \{ (x_1, x_2) \in (0,l)^2 \mid x_1 = \frac{pl}{m},\ p = 1,2,...m\text{-}1 \}, \quad (6.1)$$

$$f_m(x_1, x_2) = x_1 e^1 + (x_2 + \alpha\gamma x_1 + (1-\alpha)\gamma \frac{pl}{m})\, e^2 \quad for\ \ x_1 \in (\frac{p-1}{m} l, \frac{p}{m} l),\ \ p = 1,2,..m, \quad (6.2)$$

where α and γ are positive constants and $\alpha \leq 1$. This sequence determines the structured deformation (κ, g, G) with κ the empty set, g the simple shear

$$g(x_1, x_2) = x_1 e^1 + (x_2 + \gamma x_1)\, e^2, \quad (6.3)$$

and

$$G(x_1, x_2) = I + \alpha\gamma\, e^2 \otimes e^1 . \quad (6.4)$$

Note that $\nabla g - G$ is the gradient of the simple shear obtained from g after replacing γ by $(1-\alpha)\gamma$. This shows that the deformation tensor due to a change in internal structure may well be the gradient of a field, a circumstance which is excluded in those theories in which the deformation due to a change in internal structure is identified with the non-compatible part of the strain tensor, as done in [6] and in many theories of plasticity.

The tensor G_A is given by

$$G_A = (I + \alpha\gamma\, e^2 \otimes e^1)(I - \gamma\, e^2 \otimes e^1) = I - (1-\alpha)\gamma\, e^2 \otimes e^1 , \tag{6.5}$$

and Equation (5.7) tells us that the change of stress due to the deformation of the continuum is

$$T_{A\infty} - T_o = -\gamma(1-\alpha)(e^2 \otimes e^1)^W T_o + \boldsymbol{B}_A (e^2 \otimes e^1)^S + o(\gamma) , \tag{6.6}$$

where the superscripts S and W denote the symmetric and the skew-symmetric part of a tensor, respectively.

To evaluate the stress due to the change in internal structure, assume that the cohesive forces t_A are governed by the constitutive equation (3.5), and that the tensor Q_A is a rotation of $\pi/2$:

$$Q_A = e^2 \otimes e^1 - e^1 \otimes e^2 . \tag{6.7}$$

In the present example, n_R is the normal to the dislocation line e^1, and $|[f]|$ is zero; thus, Equation (3.5) associates with the tangential microslip in the direction e^1 the tangential cohesive force $t_A = q_A(0)\, e^2$. Recalling Equation (5.12) and the expression (6.5) of G_A, the tensor $T_{A\mu}$ takes the form

$$T_{A\mu} = q_A(0)\, (Q_A \bullet (e^2 \otimes e^1))\, e^2 \otimes e^1 = q_A(0)\, e^2 \otimes e^1 . \tag{6.8}$$

The second example is the *creation of voids* in a porous medium. It shows how the dilatation theories of Elasticity [7] and the theories of elastic materials with voids [8] fit into the present scheme. Consider the structured deformation defined by the determining sequence $m \mapsto (\kappa_m, f_m)$, with

$$\kappa_m = \{ (x_1,x_2) \in (0,l)^2 \mid x_1 = \frac{pl}{m}, p = 1,2,..m\text{-}1 \} \cup \\ \cup \{ (x_1,x_2) \in (0,l)^2 \mid x_2 = \frac{ql}{m}, q = 1,2,...m\text{-}1 \} , \tag{6.9}$$

$$f_m(x_1,x_2) = ((1+\alpha\gamma) x_1 + (1-\alpha)\gamma \frac{pl}{m}) e^1 + ((1+\alpha\gamma) x_2 + (1-\alpha)\gamma \frac{ql}{m}) e^2 \\ \text{for } x_1 \in (\frac{p-1}{m} l, \frac{p}{m} l), \ x_2 \in (\frac{q-1}{m} l, \frac{q}{m} l), \quad p, q = 1,2,..m . \tag{6.10}$$

This sequence determines the structured deformation (κ, g, G), with κ the empty set and

$$g(x_1, x_2) = (1+\gamma)(x_1 e^1 + x_2 e^2) , \qquad G(x_1, x_2) = (1+\alpha\gamma) I . \tag{6.11}$$

The deformation due to the continuum and that due to the change in internal structure are both homogeneous dilatations, of amount $\alpha\gamma$ and $(1-\alpha)\gamma$, respectively. The tensor G_A is

$$G_A = G \nabla g^{-1} = \frac{1+\alpha\gamma}{1+\gamma} I = (1 - (1-\alpha)\gamma + o(\gamma))\, I , \tag{6.12}$$

and therefore, by Equation (6.8), the change of stress due to the deformation of the continuum is

$$T_{A\infty} - T_o = -\gamma\,(1-\alpha)\,\mathbf{B}_A\,I + o(\gamma)\,. \tag{6.13}$$

Let us evaluate the stress due to the change in internal structure. For a porous medium, it can be assumed that $Q_A = I$, i.e., that the direction of the cohesive force is orthogonal to the surface of discontinuity. It follows then from Equation (5.12) and from the expression (6.12) of G_A that

$$T_{A\mu} = q_A(0)\,I\,. \tag{6.14}$$

Acknowledgement. This research was supported by the Italian Ministry for University and Scientific Research (MURST).

References

1. G. I. Barenblatt, *On some general concepts of the mathematical theory of brittle fracture,* PMM **28**, 630-643, 1964.

2. G. Capriz, *Continua with Microstructure,* Springer Tracts in Natural Philosophy n.35, 1989.

3. A. Carpinteri, *Linear elastic fracture mechanics as a limit case of strain-softening instability,* Meccanica **23**, 160-65, 1988.

4. G. Del Piero and D.R. Owen, *Structured deformations of continua,* Arch. Rat. Mech. Analysis **124**, 99-155, 1993.

5. G. Del Piero and D.R. Owen, *Integral-gradient formulae for structured deformations,* Arch. Rat. Mech. Analysis, forthcoming.

6. I.A. Kunin, *Internal stresses in an anisotropic elastic medium,* PMM **28**, 612-621, 1964.

7. K.Z. Markov, *Dilatation theories of elasticity,* in: *New Problems in Mechanics of Continua,* Univ. of Waterloo Press, 1983.

8. J.W. Nunziato and S.C. Cowin, *A nonlinear theory of elastic materials with voids,* Arch. Rat. Mech. Analysis **72**, 175-201, 1993.

9. T. Mura, *Continuum theory of plasticity and dislocations,* Int. J. Engng. Sciences **5**, 341-351, 1967.

10. D.R. Owen, *Disarrangements in continua and the geometry of microstructure,* in: *Recent Advances in Elasticity, Viscoelasticity, and Inelasticity,* K.R. Rajagopal ed., World Scientific, 1995.

11. C. Truesdell and W. Noll, *The non-linear field theories of mechanics,* in: *Handbuch der Physik,* **III/3**, 1965.

Continuum Models and Discrete Systems
Proceedings of 8th International Symposium, June 11-16, 1995, Varna, Bulgaria, ed. K.Z. Markov
© World Scientific Publishing Company, 1996, p. 172

MATHEMATICAL MODELLING OF STRUCTURE-SENSITIVE PROPERTIES OF MATERIALS

Ph. M. DUXBURY
Physics and Astronomy Department, Michigan State University,
East Lansing, MI 48824-1116, USA

Abstract

Almost all materials are disordered, either due to randomly occuring point defects, dislocations and grain boundaries, and/or due to a mixture of phases in their microstructure. Although elasticity, electrical conductivity and many transport properties are dependent on defects, they are usually not as sensitive to defects as are instabilities such as dielectric breakdown, electrical failure, yield and fracture. The latter properties are often dependent on rare or extreme fluctuations in the material micsrostructure and for this reason the variational and effective medium ideas which have been so successful in studies of transport and elasticity are often invalid. In this presentation, a simple lattice model (the random fuse network) will be used to illustrate the anomalous scaling properties characteristic of "extreme" properties and then simpler, mathematically tractable, parallel bar models will be described. One remarkable result found using these simpler models is that the failure probability is minimum at an optimim sample size that scales as the inverse of the applied stress. Finally extensions to include the dynamics of crack nucleation and coalescence will be presented.

Continuum Models and Discrete Systems
Proceedings of 8th International Symposium, June 11-16, 1995, Varna, Bulgaria, ed. K.Z. Markov

GEOMETRICAL ASPECTS OF UNIFORMITY IN ELASTICITY AND PLASTICITY

M. EPSTEIN
Department of Mechanical Engineering, The University of Calgary, Calgary, Alberta T2N 1N4, Canada

Abstract. An overview of the geometrical theory of inhomogeneities within the context of Continuum Mechanics is presented, including considerations on evolution laws, thermodynamics, and non-simple materials.

1. Introduction

Quite apart from any physical interpretation in terms of continuous distributions of defects, Continuum Mechanics can legitimately ask the question as to the uniformity and homogeneity of a given constitutive law. And indeed it has, thus producing one of the theoretical jewels of the discipline. The original idea is to be found in a celebrated article by Noll [1], which appeared in 1967, but reflects earlier work. The main results had already been reported in [2]. A paper by Wang [3], which appeared simultaneously with [1], admirably completes Noll's presentation, restricted as it was to local considerations, by re-casting it within the framework of the theory of principal fibre bundles. The resulting theory of inhomogeneities by Noll and Wang is a self-contained, self-explanatory, rigorous presentation within the tenets of Continuum Mechanics, namely, that the material body is a three-dimensional differentiable manifold to which a material response functional, to be adjusted according to experimental evidence, is attached. Although the treatment of Noll and Wang is restricted to simple elastic materials, it is quite obvious that their results can be extended in several directions.

By the time the articles of Noll and Wang appeared in print, there had been already an abundant literature in the theory of continuous distributions of dislocations. Without attempting to do justice to the full history of the subject, one can nevertheless mention some of the main contributions, starting with Kondo [4], and then Bilby [5] and Kröner [6]. This theory is based, at least in part, on the recognition of the discontinuous nature of matter and, particularly, on the consideration of defective crystalline lattices. The continuum results are obtained by a heuristic passage to the limit. It is both remarkable and unfortunate that both theories, the continuum-based theory of inhomogeneities and the lattice-inspired theory of dislocations, produced in some cases identical or similar geometrical constructs, thus raising expectations which are clearly beyond the terms of reference of each theory. On the other hand, such coincidences are certainly not accidental, and work remains to be done in their precise explanation in order to shed light onto the gray area of the transition between the two domains. This paper will not attempt to do so, but rather to explore a few ramifications of the continuum approach. At the expense of rigour, the presentation will be kept as elementary as possible.

2. Simple Materials

The material response of a simple material can be understood in terms of hypothetical measurements carried out on a reference "infinitesimal parallelepiped" deforming linearly into other infinitesimal counterparts. For brevity, we shall call a chosen reference infinitesimal parallelepiped the *reference crystal*. Mathematically, the meaning of material simplicity is that, given a deformation history for a finite body, the material response of the body can be characterized in terms of point-wise material responses and, moreover, the material response at a point depends only on the history of the first gradient of the deformation evaluated at that point. A further simplification is obtained in the case of elasticity, whereby it is only the present value of the deformation gradient which enters the constitutive equation, rather than its whole past history. In the case of *hyperelasticity*, to which we confine our present attention, the constitutive equation consists of a single scalar function

$$\mathcal{W} = \mathcal{W}(\mathbf{H}), \tag{2.1}$$

representing the *strain energy* per unit volume of the reference crystal as a function of its linear deformation $\mathbf{H}$. For a hyperelastic simple body in a given global reference configuration $\mathcal{R}$, the strain energy per unit volume will, accordingly, be of the form

$$\mathcal{W}_{\mathcal{R}} = \mathcal{W}_{\mathcal{R}}(\mathbf{F}, \mathbf{X}), \tag{2.2}$$

where $\mathbf{X}$ labels points in $\mathcal{R}$, and $\mathbf{F}$ is the value of the deformation gradient at $\mathbf{X}$. If all the points of the body are made of the same material as the reference crystal, the functions $\mathcal{W}$ and $\mathcal{W}_{\mathcal{R}}$ must be related, and it is not difficult to understand that this relation must be such that a suitable deformation $\mathbf{P}(\mathbf{X})$ of the reference crystal exists for each point $\mathbf{X}$ so that, for all non-singular $\mathbf{F}$,

$$\mathcal{W}_{\mathcal{R}}(\mathbf{F}, \mathbf{X}) = J_P^{-1} \mathcal{W}(\mathbf{F}\,\mathbf{P}(\mathbf{X})). \tag{2.3}$$

In other words, after a suitable transplant operation $\mathbf{P}(\mathbf{X})$, the reference crystal will behave exactly as the point $\mathbf{X}$ for all further deformations $\mathbf{F}$. The appearance of J_P, the absolute value of the determinant of $\mathbf{P}$, is designed to account for the fact that we have insisted on defining the strain energy function per unit volume (rather than per unit mass). We shall call $\mathbf{P}(\mathbf{X})$ a *uniformity field*, and we will say that the body is *materially uniform* if such a field exists. A body is *smoothly uniform* if for every point $\mathbf{X}$ there exists a neighbourhood such that in it $\mathbf{P}(\mathbf{X})$ can be chosen smoothly.

An alternative way of looking at a uniformity field is by fixing a basis in the reference crystal and considering its images by the maps $\mathbf{P}(\mathbf{X})$. The field of bases thus obtained is sometimes called a *crystallographic basis.*

Uniformity fields are, in general, not unique. Indeed, let $\mathbf{G}$ be a member of the (unimodular) *material symmetry group* $\mathcal{G}$ of the reference crystal, i.e.,

$$\mathcal{W}(\mathbf{F}\,\mathbf{G}) = \mathcal{W}(\mathbf{F})\,, \tag{2.4}$$

for all non-singular $\mathbf{F}$. Then, if $\mathbf{P}(\mathbf{X})$ is a uniformity map at $\mathbf{X}$, so is $\mathbf{P}(\mathbf{X})\mathbf{G}$, as can be verified directly. The totality $\mathcal{P}(\mathbf{X})$ of uniformity maps at $\mathbf{X}$ can be spanned in this way, viz.

$$\mathcal{P}(\mathbf{X}) = \mathbf{P}(\mathbf{X})\,\mathcal{G}\,, \tag{2.5}$$

or, equivalently,

$$\mathcal{P}(\mathbf{X}) = \mathcal{G}_{\mathcal{X}}\,\mathbf{P}(\mathbf{X})\,, \tag{2.6}$$

where $\mathcal{G}_{\mathcal{X}}$ is the symmetry group of $\mathbf{X}$ in $\mathcal{R}$.

Once the smooth uniformity of a body has been established, the question as to its possible *homogeneity* can be formulated as follows: does there exist a global change of reference configuration such that the uniformity field can be chosen as a constant, independent of $\mathbf{X}$? An answer in the affirmative will correspond to the notion of homogeneity, since in the new (*homogeneous*) reference configuration all points have, effectively, the same constitutive law. A weaker version (*local homogeneity*) would correspond to the existence, for each point $\mathbf{X}$, of a reference configuration such that in a neighbourhood of $\mathbf{X}$ the uniformity maps can be chosen as constant. The standard example for the latter situation is an initially homogeneous strip of material which is later welded so as to form a ring. Unless the ring is cut open, the body is only locally homogeneous. To answer the question of (local) homogeneity Noll noticed that a uniformity field defines a distant parallelism, whose Christoffel symbols in reference coordinates X^I are given by

$$\Gamma^I_{JK} = (P^{-1})^{\alpha}_{J,K}\,P^I_{\alpha}\,, \tag{2.7}$$

where Greek indices denote components in a reference crystal basis. These Christoffel symbols characterize a so-called *material connection.* Local homogeneity is now shown to be equivalent to the integrability conditions implied by the vanishing of the Cartan torsion (or skew-symmetric part) of a material connection, a result similar to that obtained in the lattice-based approach. It is important to realize, however, that when the material symmetry group is continuous, the material connection is not unique, so that the non-vanishing of a material torsion may very well be a reflection of having chosen the "wrong" uniformity field. It can be shown [7] that this situation can be better understood in terms of the theory of G-structures, whereby local homogeneity corresponds to the notion of integrability of the material G-structure. Roughly speaking, this G-structure is the construct that results from attaching to each point $\mathbf{X}$ its uniformity set $\mathcal{P}(\mathbf{X})$ as defined above. The object obtained is, technically, a principal fibre bundle whose typical fibre is the symmetry group $\mathcal{G}$. A *section* of this bundle is tantamount to the choice of a uniformity field, and integrability is equivalent to the existence of a local coordinate system on the reference configuration such that its natural basis is a crystallographic basis corresponding to some section of the G-structure. If that is the case, an appropriate change of reference configuration induces a (local) "straightening" of the crystallographic basis.

3. Evolving Inhomogeneities and the Eshelby Tensor

In his 1951 study of the force on an elastic singularity, Eshelby [8] introduced the idea of what he called "the Maxwell tensor of Elasticity", now commonly known as the *Eshelby tensor*. This important concept has been extended to the continuous case both using the structural [6] and the continuum [9, 10] approaches. To see how in the latter the concept emerges rather naturally, we consider the possibility of evolving inhomogeneity patterns, such as in Plasticity theory, whereby there is a stored energy function retaining the original functional form (2.3), but with $\mathbf{P}$ now also possibly depending on a time-like parameter. The variable tensor $\mathbf{P}$ functions now as an internal variable whose associated driving force is

$$-\frac{\partial \mathcal{W}_{\mathcal{R}}}{\partial \mathbf{P}} = (\mathcal{W}_{\mathcal{R}}\,\mathbf{I} - \mathbf{F}^T\,\mathbf{T})\,\mathbf{P}^{T^{-1}} = \mathbf{b}\,\mathbf{P}^{T^{-1}}, \tag{3.1}$$

where $\mathbf{b}$ is the Eshelby tensor and $\mathbf{T} = \dfrac{\partial \mathcal{W}_{\mathcal{R}}}{\partial \mathbf{F}}$ is the first Piola-Kirchhoff stress.

As originally intended by Eshelby, the tensor $\mathbf{b}$ represents a measure of the change of elastic energy involved in "moving" the inhomogeneity. Though not in the realm of the theory of uniform materials, a similar result was anticipated in [11]. With the added background of material uniformity, a direct calculation reveals that the divergence of the Eshelby tensor is given (in Cartesian reference coordinates) by

$$b^J_{I,J} = b^K_M\,\Gamma^M_{KI} \tag{3.2}$$

in the static case and in the absence of body forces. This equation results form enforcing the balance of linear momentum. Similarly, the balance of angular momentum results in the C-symmetry

$$\mathbf{bC} = \mathbf{Cb}^T, \tag{3.3}$$

where $\mathbf{C}$ is the right Cauchy-Green tensor, $\mathbf{C} = \mathbf{F}^T\mathbf{F}$.

A further important symmetry of the Eshelby tensor results when the material symmetry group $\mathcal{G}_X$ is continuous. Then it can be shown [9] that the Eshelby tensor is orthogonal to the Lie-algebra of $\mathcal{G}_X$. Thus, for instance, in the case of a fully isotropic solid, the Eshelby tensor is symmetric with respect to the Riemannian metric given by

$$\mathbf{D} = (\mathbf{PP}^T)^{-1} \tag{3.4}$$

in the reference configuration $\mathcal{R}$.

Suppose now that an evolution law of some kind is proposed for the inhomogeneity pattern. This would mean that the uniformity field evolves in time as driven, say, by the value of the Eshelby tensor. For definiteness, assume a simple law of the form

$$\Phi(\mathbf{P}, \dot{\mathbf{P}}, \mathbf{b}) = 0. \tag{3.5}$$

One immediately realizes that the function Φ must be subject to some restrictions [12]. The first such restriction is one of consistency with the material symmetry group $\mathcal{G}$. Let $\mathbf{P}(\mathbf{X}, t)$ and $\mathbf{Q}(\mathbf{X}, t)$ be two smoothly time-dependent uniformity fields, that is, two time dependent sections of the same instantaneous material G-structure. We say that they evolve *in parallel* if there exists a fixed element $\mathbf{G} \in \mathcal{G}$ such that

$$\mathbf{Q}(\mathbf{X}, t) = \mathbf{P}(\mathbf{X}, t)\,\mathbf{G} \tag{3.6}$$

for all times t within the interval of interest. The following *principle of G-covariance* can then be laid down: *A law of evolution must be form-invariant under parallel changes of sections within the structural (material symmetry) group.* To see the implications of this principle we obtain from the above equation, by differentiation with respect to time and elimination of $\mathbf{G}$,

$$\dot{\mathbf{Q}}\mathbf{Q}^{-1} = \dot{\mathbf{P}}\mathbf{P}^{-1}. \tag{3.7}$$

In other words: Sections evolving in parallel have at all times the same *inhomogeneity velocity gradient*, defined as

$$\mathbf{L}_P = \dot{\mathbf{P}}\mathbf{P}^{-1}. \tag{3.8}$$

Conversely, one can show that if two evolving uniformity fields $\mathbf{P}(\mathbf{X}, t)$ and $\mathbf{Q}(\mathbf{X}, t)$ satisfy at all times

$$\mathbf{L}_P = \mathbf{L}_Q, \tag{3.9}$$

and if at some particular time t_0

$$\mathbf{Q}(\mathbf{X}, t_0) = \mathbf{P}(\mathbf{X}, t_0)\mathbf{G}, \tag{3.10}$$

then they must evolve in parallel at all times. We conclude that the G-covariance principle is equivalent to the restriction that $\mathbf{P}$ and $\dot{\mathbf{P}}$ enter the evolution law only through the combination $\mathbf{L}_P$, i.e., the suggested law must be rephrased as

$$\phi(\mathbf{L}_P, \mathbf{b}) = 0. \tag{3.11}$$

A second kind of restriction is to be considered in cases where the material symmetry group is continuous, namely, that the proposed evolution law actually prescribe a change of uniformity field to counterparts in other G-structures. Otherwise, a putative evolution could stay within the same G-structure by taking advantage of the smooth degree of freedom afforded in the choice of uniformity field by the continuity of the group. This undesirable apparent evolution happens if, and only if, the section at time t is related to the section at a fixed time t_0 by

$$\mathbf{P}(\mathbf{X}, t) = \mathbf{G}_X(\mathbf{X}, t)\mathbf{P}(\mathbf{X}, t_0), \tag{3.12}$$

where $\mathbf{G}_X(\mathbf{X},t)$ is a time-dependent member of the symmetry group $\mathcal{G}_X$ in the reference configuration. By differentiation at time t_0, and noting that $\mathbf{G}_X(\mathbf{X},t_0)$ must be the identity map, we obtain the following form of the *principle of actual evolution: The functional form of the evolution law ϕ must guarantee that, whenever evolution is expected, $\mathbf{L}_P$ not belong to the Lie-algebra of the instantaneous symmetry group $\mathcal{G}_X$.* Consider, for instance, the case of a fully isotropic solid, already mentioned in connection with the Eshelby tensor itself. Then the principle of actual evolution requires that for every D-symmetric value of the argument $\mathbf{b}$ the resulting value of $\mathbf{L}_P$ not be D-skew-symmetric.

4. Some Thermodynamic Considerations

By including the absolute temperature θ and its referential gradient $\nabla_{\mathcal{R}}\theta$ in the list of independent constitutive variables, a formulation of thermoelastic inhomogeneities can be obtained [13, 14] with the Eshelby tensor now defined in terms of the free energy $\psi_{\mathcal{R}}$ per unit volume of $\mathcal{R}$, viz.

$$\mathbf{b} = -\frac{\partial \psi_{\mathcal{R}}}{\partial \mathbf{P}}\mathbf{P}^T = \psi_{\mathcal{R}}\mathbf{I} - \mathbf{F}^T\mathbf{T}. \tag{4.1}$$

Using standard arguments on the Clausius-Duhem inequality, the form of the residual inequality is obtained as

$$\operatorname{tr}(\mathbf{b}^T\mathbf{L}_P) + \theta^{-1}\mathbf{q}_{\mathcal{R}} \cdot \nabla_{\mathcal{R}}\theta \leq 0, \tag{4.2}$$

where $\mathbf{q}_{\mathcal{R}}$ is the referential heat-flux vector. In considering the term $tr(\mathbf{b}^T\mathbf{L}_P)$ we must bear in mind that $\mathbf{b}$ has two canonical symmetries. The first one, or C-symmetry, is a consequence of the balance of angular momentum, as we have seen. The second symmetry, of material origin, consists of being in the orthogonal complement of the Lie-algebra of $\mathcal{G}_X$. We have already established that the evolution law must produce an outcome in terms of an inhomogeneity velocity gradient which does not entirely lie in this Lie-algebra. Thermodynamics now requires more, namely, that for every C-symmetric value of the argument $\mathbf{b}$ which lies in the orthogonal complement of the Lie-algebra of $\mathcal{G}_X$, $\mathbf{L}_P$ must be restricted by the residual thermodynamic inequality.

5. Non-simple Materials

The extension of the continuum theory of inhomogeneities to non-simple materials, such as second-grade materials and Cosserat media, has been accomplished recently [15,16,17,18]. It involves, as one might expect, the appearance of further differential-geometric measures of inhomogeneity. Consider, for instance, a generalized Cosserat medium, namely, a body $\mathcal{B}$ to each of whose points an arbitrarily deformable linearly independent vector triad is attached. Let the mechanical response be characterized by a scalar function

$$\mathcal{W}_{\mathcal{R}} = \mathcal{W}_{\mathcal{R}}(\mathbf{F}, \mathbf{K}, \nabla_{\mathcal{R}}\mathbf{K}, \mathbf{X}), \tag{5.1}$$

where $\mathbf{F}$ is, as before, the deformation gradient of $\mathcal{B}$, $\mathbf{K}$ is a non-singular linear transformation representing the deformation of the triad, and $\nabla_{\mathcal{R}}$ is its gradient. The uniformity condition involves now the existence of three fields of maps $\mathbf{P}$, $\mathbf{Q}$ and $\mathbf{R}$, whose action is expressed in Cartesian reference coordinates as

$$\mathcal{W}_{\mathcal{R}}(F^i_I, K^i_I, K^i_{I,J}, X^K) = J_P^{-1}\mathcal{W}(F^i_I P^I_\alpha, K^i_I Q^I_\alpha, K^i_{I,J} P^J_\beta Q^I_\alpha + K^i_I R^I_{\alpha\beta}), \tag{5.2}$$

where $\mathcal{W}$ is the strain-energy function of a "reference crystal" of a more complicated nature than before. The form of the action of the uniformity fields in the right-hand side of this equation is a direct consequence of the law of composition of deformations for polar materials [19].

As far as the question of homogeneity is concerned, there are two possible definitions: homogeneity with respect to a given fixed reference crystal, or homogeneity with respect to some arbitrary reference crystal. In the first case, homogeneity would be equivalent to the possibility of finding a change of reference configuration such that the uniformity fields $\mathbf{P}$, $\mathbf{Q}$ and $\mathbf{R}$ become $\mathbf{1}$, $\mathbf{1}$ and $\mathbf{0}$, respectively. In the second case, it is enough to demand that these fields become arbitrary constants, since a suitable change of reference crystal will effect the transformation of these constants to $\mathbf{1}$, $\mathbf{1}$ and $\mathbf{0}$. Note that in the case of simple materials both definitions coincide, since the passage from a constant uniformity field to the identity can, in that case, be effected with a global affine deformation, without changing the reference crystal. Using either definition, there are three linear connections at play, whose Christoffel symbols are given by

$$\Gamma^I_{JK} = (P^{-1})^\alpha_{J,K} P^I_\alpha\,, \tag{5.3}$$

$$\Delta^I_{JK} = (Q^{-1})^\alpha_{J,K} Q^I_\alpha\,, \tag{5.4}$$

and

$$\Lambda^I_{JK} = -R^I_{\alpha\beta}(Q^{-1})^\alpha_J (P^{-1})^\beta_K\,. \tag{5.5}$$

The homogeneity conditions can be expressed in terms of the vanishing of the torsion of $\mathbf{\Gamma}$ and the tensor difference $\mathbf{\Sigma} = \mathbf{\Delta} - \mathbf{\Lambda}$, representing the *Cosserat inhomogeneity tensor*. The case of second-grade materials [15,18] can be seen as the particular case when $\mathbf{K} = \mathbf{F}$ and $\nabla_{\mathcal{R}}\mathbf{K}$ is symmetric. Generalizations of Eshelby's tensor in this context [20] as well as in the context of electrified materials [21] are also possible.

Acknowledgements. This work has been partially supported through a grant of the National Science and Engineering Research Council of Canada.

References

1. W. Noll, *Arch. Rat. Mech. Anal* **27** (1967) 1.
2. C. Truesdell and W. Noll, *The Non-Linear Field Theories of Mechanics*, Handbuch der Physik, **III**/3, Springer Verlag, Berlin-New York, 1965.
3. C. C. Wang, *Arch. Rational Mech. Anal.* **27** (1967) 33.

4. K. Kondo, *Geometry of Elastic Deformation and Incompatibility*, Memoirs of the Unifying Study of the Basic Problems in Engineering Sciences by means of Geometry, Tokyo Gakujutsu Benken Fukyu-Kai, **1C**, 1955.
5. B.A. Bilby, Continuous Distributions of Dislocations, In *Progress in Solid Mechanics*, **1**, North-Holland, Amsterdam, 1960, p. 329.
6. E. Kröner, *Arch. Rational Mech. Anal.* **4** (1960) 273.
7. M. Elzanowski, M. Epstein and J. Sniatycki, *J. Elasticity* **23** (1990) 167.
8. J.D. Eshelby, *Phil. Trans. Royal Soc. London* **A244** (1951) 87.
9. M. Epstein and G. A. Maugin, *Acta Mechanica* **83** (1990) 127.
10. M. Epstein and G. A. Maugin, *C. R. Acad. Sci. Paris* II **310** (1990) 675.
11. A. Golebiewska-Herrmann, *Int. J. Solids Structures* **17** (1981) 1.
12. M. Epstein and G. A. Maugin, On the Geometrical Structure of Anelasticity, *Acta Mechanica*, in press.
13. M. Epstein, In *Nonlinear Thermomechanical Processes in Continua*, eds. W. Muschik and G. A. Maugin, Heft 61, TUB-Dokumentation, Berlin, 1992, p. 147.
14. M. Epstein and G. A. Maugin, *C. R. Acad. Sci. Paris* II **320** (1995).
15. M. de León and M. Epstein, *Reports Math. Phys.* **33** (1993) 419.
16. M. de León and M. Epstein, *C. R. Acad. Sci. Paris* I **319** (1994) 617.
17. M. Epstein and M. de León, In *Proceedings Colloquium on Differential Geometry*, July 25-30, 1994, Debrecen, Hungary, 1994.
18. M. Epstein and M. de León, The Geometry of Uniformity in Second-Grade Elasticity, *Acta Mechanica*, (in press).
19. A. C. Eringen and Ch. B. Kafadar, In *Continuum Physics*, ed. A. C. Eringen, vol. IV, Academic Press, New York, 1976, p. 1.
20. M. Epstein and G. A. Maugin, In *Analysis, Manifolds and Physics, Y. Choquet-Bruhat Colloquium*, ed. R. Kerner *et al.*, Klüwer, 1993, p. 331.
21. G. A. Maugin and M. Epstein, *Proc. R. Soc. London* **A433** (1991) 299.

Continuum Models and Discrete Systems
Proceedings of 8th International Symposium, June 11-16, 1995, Varna, Bulgaria, ed. K.Z. Markov

MODELS OF THE INTRINSIC CONVECTION IN A SETTLING SUSPENSION OF SPHERES

F. FEUILLEBOIS, J. BLAWZDZIEWICZ[1]
PMMH, ESPCI, 10 rue Vauquelin, F-75231 Paris Cedex 05, France

D. BRUNEAU
LEPT-ENSAM, Esplanade des Arts et Métiers, F-33405 Talence Cedex, France

and

R. ANTHORE
UFR des Sciences, URA 808, F-76821 Mont Saint Aignan, France

Abstract. Mazur and co-workers have shown that an "intrinsic convection" should develop in a homogeneous settling suspension, so that spherical particles would fall at different velocities in the center and near the sides of the container. This phenomenon is simply modeled here by replacing the spheres by independent Stokeslets located at their centers. The driving force is provided by the steric effect, the Stokeslets being at least one radius away from the walls. The intrinsic convection velocity profile then amounts to a Poiseuille flow with a slip velocity at the wall. The exact account of sphere-wall interactions provides only a small correction to the slip velocity. But an estimate of concentration effects shows that the intrinsic convection might be largely decreased in nondilute suspensions.

1. Introduction

Geigenmueller and Mazur [1] have proved theoretically that the velocity of sedimentation of a suspension depends on the shape of the container: they showed that a general motion of the suspension, that they called an "intrinsic convection", is superimposed on the sedimentation velocity of the particles with respect to the fluid. The model introduced here is simpler than Geigenmueller and Mazur's one. Due to its simplicity, it provides a more physical insight into the phenomenon.

The model equations for a dilute suspension are presented in Section 2. These equations are solved using a boundary layer type solution. A more refined model in which sphere-wall interactions are exactly taken into account is then developed in Section 3. Finally an estimate of concentration effects is obtained in Section 4.

[1]On leave from Institute of Fundamental Technological Research, Polish Academy of Sciences, Świętokrzyska 21, PL-00 049 Warsaw, Poland.

2. The Model of Distributed Stokeslets

In Geigenmueller and Mazur model, the influence of spherical particles on the fluid is taken into account by a uniform distribution of induced forces on the surfaces of the particles. The particle-particle and wall-particle interactions are assumed to be negligible.

Similar assumptions are kept in the present model. However, the influence of the spheres on the fluid is modeled in a simpler way: each sphere is simply replaced by a single Stokeslet, or point force, at its center. Nevertheless, the condition that the spheres do not overlap with the walls of the container is taken into account: it is enforced by requiring that the Stokeslets be kept at a distance at least one sphere radius away from these walls. Apart from this condition the concentration is assumed to be constant throughout the system.

The governing equations for the motion of a fluid containing this force distribution then are [2]:

$$\nabla \cdot \mathbf{V} = 0\,, \tag{1}$$

$$\eta \nabla^2 \mathbf{V} - \nabla p + \rho \mathbf{g} = n_0 H\big[d(\mathbf{r}) - a\big]\mathbf{K}\,, \tag{2}$$

in which $\mathbf{V}$ is the fluid velocity, p is its pressure, η and ρ are the fluid viscosity and density, $\mathbf{g}$ is the body force per unit mass, $d(\mathbf{r})$ is the distance from the current point (described by a vector $\mathbf{r}$) to the nearest wall, a is a sphere radius, H is the Heaviside step function. In Eq. (2), $\mathbf{K} = -6\pi a \eta \mathbf{v}_{\mathbf{ps}}$ is the Stokes drag force on a sphere with Stokes velocity $\mathbf{v}_{\mathbf{ps}}$ and n_0 is the number of Stokeslets per unit volume in the bulk of the suspension (n_0 is a constant). The concentration n_0 can be expressed in terms of the volume fraction of particles, ϕ_0, by $n_0 = (3\phi_0)/(4\pi a^3)$.

To Eqs. (1) and (2), we add a no-slip condition for the fluid on the walls of the container:

$$\mathbf{V}\Big|_w = 0\,. \tag{3}$$

For a dilute suspension, $\mathbf{V}$ can approximately be considered as a certain average velocity of the mixture of fluid and particles. As a consequence of the incompressibility of the suspension, the flux of $\mathbf{V}$ across any horizontal plane is assumed to be zero:

$$\int_S V_z \, dS = 0\,, \tag{4}$$

V_z being the projection of $\mathbf{V}$ on the vertical axis. Actually, V_z is the only non-zero component of $\mathbf{V}$. The top and bottom walls of the container, which are assumed to be widely separated, are taken into account by this zero-flux condition.

It is essential to note that the force distribution in (2) is lower near the walls than in the bulk of the container. The fluid containing this force distribution, that is the suspension, will then fall down preferentially in the middle of the container, and rise in the regions close to the walls. This is the triggering mechanism for the intrinsic convection. It is thus clear that any intrinsic convection model must contain the condition that the spheres do not overlap with the walls. In a sense, the model presented here is the minimal one satisfying this condition.

As shown in [2], a simple solution of the model equations can be obtained by means of a boundary layer analysis, by using the fact that the sphere radius a is usually small as compared to the characteristic container dimension, say b. Let us define a small parameter

$$\epsilon = \frac{a}{b} \ll 1 \tag{5}$$

and the nondimensional quantities

$$\mathbf{U} = \frac{\mathbf{V}}{v_{\mathrm{ps}}}, \qquad P = \frac{b}{\eta v_{\mathrm{ps}}}(p - \rho g z + n_0 K z), \tag{6}$$

where v_{ps} and K are the projections of $\mathbf{v}_{\mathrm{ps}}$ and $\mathbf{K}$, respectively, on the z-axis.

Eqs. (1)–(4) can be rewritten in a nondimensional form now using these variables. In the first approximation, $\epsilon \to 0$, Eq. (2) then gives

$$\nabla^2 U_z - \frac{dP}{dz} = 0 . \tag{7}$$

The solution of Eq. (7), to be called the outer solution, is simply a Poiseuille flow plus a constant, representing the unknown slip velocity U_w:

$$U_z = U_{\mathrm{Poiseuille}} + U_w . \tag{8}$$

The zero-flux condition (4) provides the expression for the pressure gradient dP/dz in terms of U_w. The no-slip condition (3) on the walls, that is $U_z = 0$, cannot be applied to the outer solution. Thus, an inner solution has to be constructed. This solution consists of two parts, because of the Heaviside function in (2): one part in $d(\mathbf{r}) \leq a$ for which the no-slip condition applies and one part in $d(\mathbf{r}) \geq a$ which matches with the outer solution. The velocity and its derivative should also be continuous across $d(\mathbf{r}) = a$.

As a result of the matching, the slip velocity of the outer flow is found to be

$$U_w = -\frac{9}{4}\phi_0 , \tag{9}$$

in the limit $\epsilon \to 0$. The results from this boundary layer analysis are shown in Fig. 1 for two vertical parallel plane walls. They are seen to be close to Geigenmueller and Mazur model. Similar results are obtained in a vertical cylinder with circular cross section.

It is found from our analysis that the velocity in the center of the container is independent of ϵ at the lowest order. We thus recover an essential feature of intrinsic convection, that is the value of the velocity of the global motion of the suspension in the center of the container stays constant if for a given particle radius the walls recede to infinity.

It is important to note that our boundary layer analysis can be generalized to any type of container with sufficiently smooth walls, except possibly in regions of negligible extent (like some corners). In cylindrical containers for which the expression of the Poiseuille flow is readily available, the intrinsic convection velocity can thus be obtained explicitly.

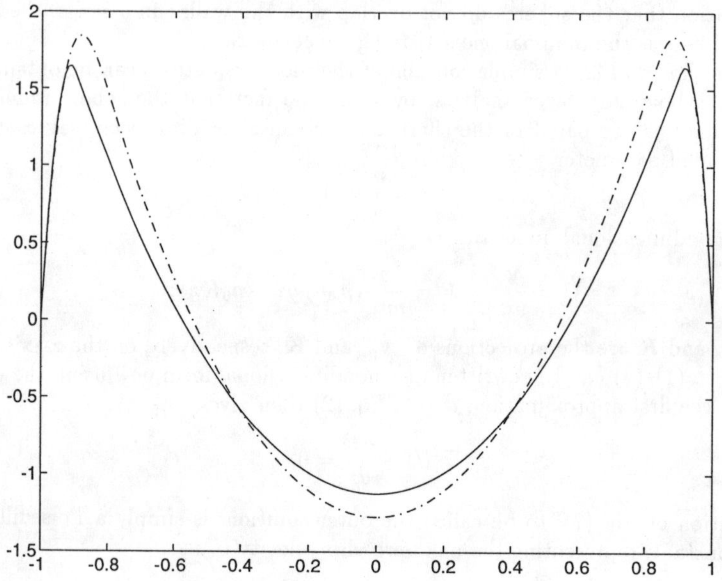

Fig. 1. Normalized velocity $V_z(X)/(\phi_0 \mid \mathbf{v}_{ps}|)$ of the global motion of a suspension settling between two vertical parallel planes vs. the nondimensional coordinate X normal to the planes. The planes are represented by $X = \pm 1$. Here, $\epsilon = 0.1$. The dash-dotted line is Geigenmueller and Mazur result, the continuous line is from the present model.

3. Exact Effect of Sphere-Wall Hydrodynamic Interactions

In the previous Section we have illustrated the physical mechanism of the intrinsic convection. Summarizing, we have shown that the intrinsic convection is driven by a natural nonuniformity of the force distribution close to the wall, resulting from the particle-wall non-overlap condition. A similar physical mechanism remains valid also for more realistic suspension models. We now present some results for a dilute hard-sphere suspension with full particle-wall hydrodynamic interactions taken into account. The inter-particle interactions are neglected since they lead to higher order density corrections to the intrinsic convection. As previously, the suspension is enclosed in a large vertical cylinder. We consider the suspension in the limit of a thin boundary layer, $\epsilon \to 0$.

In this Section, we use the method of induced forces [3]. This method is based on an observation that the velocity field of an array of solid spheres immersed in a viscous fluid can be described in terms of the Stokes equation valid in the whole space (in our case the interior of the container), with the induced forces distributed at the surface of the spheres. The induced force distribution is determined from a condition that the velocity field evaluated from the Stokes equation in the region occupied by the spheres should match the sphere rigid-body motion.

Due to the linear character of the problem, the average suspension motion can be evaluated from the Stokes equation in which the induced force density has been averaged with respect to the particle distribution. As in the previous model, the particle density is assumed to be uniform everywhere except for the particle-wall overlap configurations, where it is zero. For a sufficiently large container, the average force distribution in the bulk of the system tends to a constant value. Therefore, as previously, the solution in this region (the outer solution) is a Poiseuille flow plus a slip velocity U_w, which should be determined form matching to the inner solution.

If the curvature radius of the container walls is much larger than the particle radius, the induced force distribution in the inner region can be obtained by considering the motion of spherical particles in presence of an infinite plane wall. To determine the force distribution induced on a freely rotating sphere moving parallel to the wall we have used the solutions of O'Neill [4] and of Dean and O'Neill [5] in bipolar coordinates.

From matching the inner and outer solutions the following general expression for the slip velocity U_w has been derived:

$$U_w = \frac{9}{2}\phi_0 \int_0^\infty dX\, X[f_z(X) - 1] \,, \tag{10}$$

where $-n_0 K f_z(X)$ is the z-component of the average force density at a distance aX from the wall. By integrating the induced force calculated in bipolar coordinates we have obtained the following result:

$$U_w = -1.60\phi_0 \,. \tag{11}$$

We have verified that an approximate value of U_w obtained by retaining only the lowest order term in the expansion of f_z in the inverse powers of X (such an expansion results from from a method of reflections) is very close to our accurate value (11).

4. Estimate of Concentration Effects

An accurate evaluation of the magnitude of the intrinsic convection in a non-dilute suspension is very difficult due to the complex nature of the hydrodynamic interactions between many particles and to the unknown form of the particle distribution close to the wall in non-equilibrium suspension states. To give some estimate of the magnitude of the concentration effects we now return to the simple point-force model introduced in Section 2. In this model there are no many-particle hydrodynamic interactions and the concentration effects result solely from the concentration dependence of the particle distribution close to the wall. For illustration, we have evaluated the density dependence of the slip velocity U_w corresponding to the equilibrium distribution of hard spheres.

In our calculations we have used the equilibrium particle-wall correlation function calculated in the Percus-Yevick approximation [6]. Examples of particle concentration profiles calculated in this approximation are given in Fig. 2. One can see that even for small non-zero particle volume fractions the particle concentration exhibits characteristic oscillations in the wall region.

The intrinsic-convection slip velocity U_w has been calculated from the expression (10) with the normalized force distribution $f_z = n(x)/n_0$, resulting from the point-force model, where $n(x)$ denotes the concentration of particles at a distance x from the wall. The large oscillations of the particle concentration with the distance from the wall (and the corresponding oscillations of the force density) result in a substantial decrease of the slip velocity U_w, illustrated in Fig. 3.

Thus, although the hydrodynamic interactions between spheres have been omitted here, it can be inferred that the intrinsic convection effects might be largely decreased in nondilute suspensions.

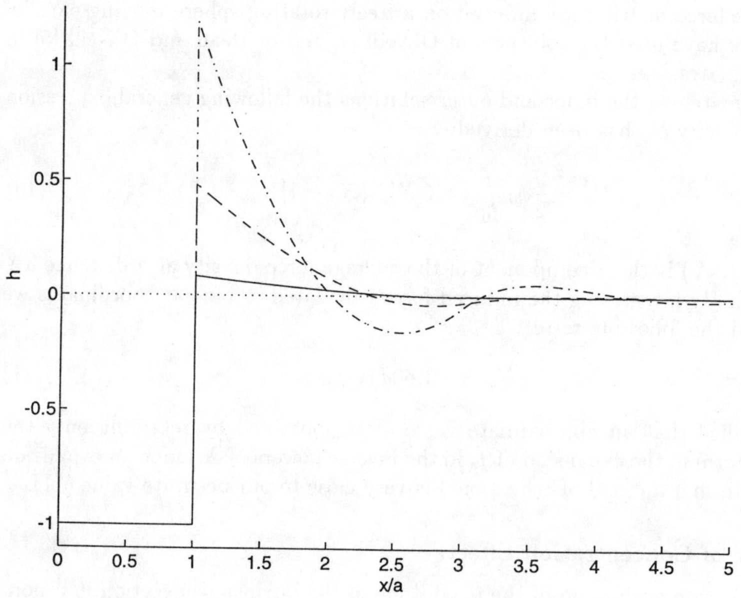

Fig. 2. Deviation of the reduced particle concentration near the wall from the bulk value, $h(x/a) = n(x/a)/n_0 - 1$, for the hard-sphere equilibrium distribution, calculated in the Percus-Yevick approximation, for various values of the volume concentration ϕ_0 in the bulk of the suspension, viz. $\phi_0 = 0.02$ (solid line), 0.10 (dashed line), 0.20 (dash-dotted line).

5. Conclusions

It has been shown that the effect of the intrinsic convection may be understood in terms of a fluid containing a particular distribution of Stokeslets. The equations have been solved using a boundary layer analysis for a small ratio of the radius of the spheres to the distance between the walls. The outer solution is then simply a Poiseuille flow with a constant slip velocity $-9/4\,\phi_0 \mathbf{v}_{\mathrm{ps}}$ at the wall (where ϕ_0 is the bulk volume concentration and $\mathbf{v}_{\mathrm{ps}}$ is the Stokes sedimentation velocity). This model has been validated against Geigenmueller and Mazur's one. When the sphere-wall interactions are exactly taken into account, the slip velocity becomes $-1.60\,\phi_0 \mathbf{v}_{\mathrm{ps}}$, a relatively small change. But an estimate of concentration effects shows that the intrinsic convection is very sensitive to the particle-wall correlation function. Although an exact calculation of the concentration effects remains to be done, this suggests that the intrinsic convection might be largely decreased in nondilute suspensions.

We tried to observe the intrinsic convection with an X-ray absorption technique [7,8]. Unfortunately, the technique was not precise enough to provide the required $O(\phi_0)$ correction terms at all points in the suspension. The experimental observation of this effect is still unresolved, to the best of our knowledge.

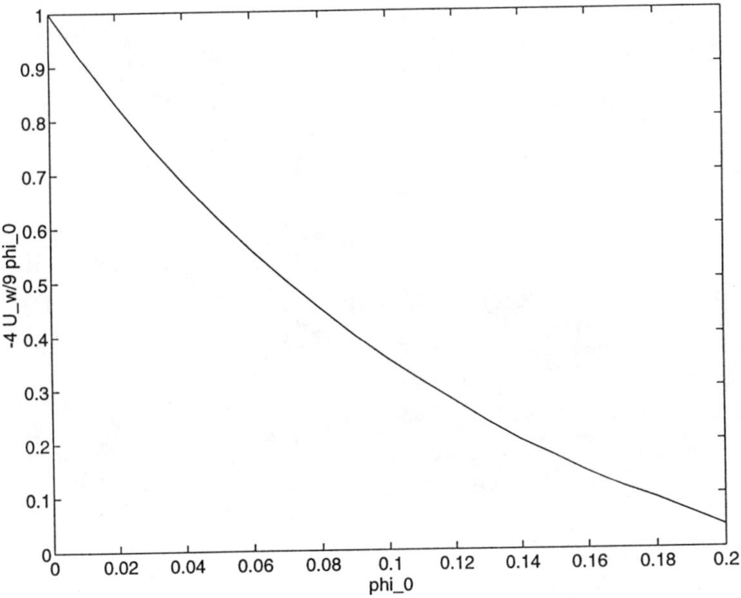

Fig. 3. Normalized slip velocity $-4U_w/9\phi_0$ at a wall for various values of the volume concentration in the bulk of the suspension.

References

1. U. Geigenmueller and P. Mazur, *J. Stat. Phys.* **53** (1988) 137.
2. D. Bruneau, F. Feuillebois, R. Anthore and E. J. Hinch, *proposed for publication* (1995).
3. P. Mazur and D. Bedeaux, *Physica* **76** (1974) 235.
4. M. E. O'Neill, *Mathematika* **11** (1964) 67.
5. W. R. Dean and M. E. O'Neill, *Mathematika* **10** (1963) 13.
6. D. Henderson, F. F. Abraham and J. Barker, *Molec. Phys.* **31** (1976) 1291.
7. D. Bruneau, *PhD thesis*, Rouen University, France, 1991.
8. D. Bruneau, R. Anthore, F. Feuillebois, X. Auvray and C. Petipas, *J. Fluid Mech.* **221** (1990) 577.

Continuum Models and Discrete Systems
Proceedings of 8th International Symposium, June 11-16, 1995, Varna, Bulgaria, ed. K.Z. Markov
© World Scientific Publishing Company, 1996, pp. 189–198

FIBRE-REINFORCED COMPOSITES WITH ENTIRELY PLASTIFIED MATRICES—A MODEL OF THERMOMECHANICAL RESPONSE

K. P. HERRMANN
Laboratory for Technical Mechanics,
University of Paderborn, Pohlweg 47-49, D-33098 Paderborn, Germany

and

I. M. MIHOVSKY
Faculty of Mathematics and Informatics,
"St. Kl. Ohridski" University of Sofia, 5 blvd J. Bourchier, BG-1126 Sofia, Bulgaria

Abstract. An advanced unified approach to the investigation of the inelastic thermomechanical response of a class of continuously reinforced fibrous composites with partially plastified matrices has been recently developed by the authors. In this study the approach is extended to include regimes of deformation of composites with entirely plastified matrix phases. Such regimes are of great interest especially in cases of thermal loadings which are, as a rule, always involved in the fabrication of these man-made materials. Matrix plasticity seems to be a reason for certain strange effects of composites behaviour in such regimes, for example, vanishing axial elongation under progressive heating. The proposed extended version of the unified approach is shown to deliver quite a natural explanation of this nontrivial effect as well as to specify necessary conditions for its occurrence.

1. Introduction

A series of recent papers of the authors has been devoted to the study of the effects of matrix plasticity on the thermomechanical response and the failure of composites with ductile matrices reinforced unidirectionally by stiff continuous fibres. Authors' attention has been focussed on the behaviour of a representative unit composite cell consisting of a circular cylindrical linear-elastic fibre with a perfectly bonded coaxial coating of elastic-perfectly-plastic matrix material obeying a standard plasticity theory—associated flow rule with a von Mises' yield condition. Two axisymmetric model problems of thermal and external mechanical loading of an initially stress free unit composite cell (matrix cooling and longitudinal cell extension, respectively) have been considered in Refs. 1 and 2. As a result of the analysis of the interactions between the principal effects of fibre reinforcement and matrix ductility a model of the matrix plastification process has been proposed. The model suggests validity of a plane cross-sections hypothesis and involves a plastic zone in the form of a circular cylindrical domain which surrounds the fibre and spreads with progressive loading into the matrix phase. It has been shown in Refs. 3 and 4 that along with the above mentioned plasticity theory the model provides detailed approximate quantitative solutions (even in analytical forms for cases of small fibre volume fractions) of both these problems. It relates the increasing plastic zone size to the current loading factor and implies, in particular, the unit cell (and overall composite) response as a function of the plastic zone size.

In Ref. 5 the model has been incorporated into a general unified approach which allows to reduce thermomechanical problems of a relatively large class, including the model problems already mentioned, to particular cases of a certain general problem of the mathematical theory of plasticity. The latter problem is in much similar in statement and features of its solutions to the classical plane-stress perfect-plasticity problem as formulated and interpreted, for example, in Refs. 6 and 7. The primary features and the potentials of the unified approach are briefly summarized if Ref. 8 and presented in considerable detail in Ref. 9. An important limitation of both the model and the unified approach lies in the fact that in their present form they become unfit to the stages of deformation upon the instant at which the entire matrix phase plastifies.

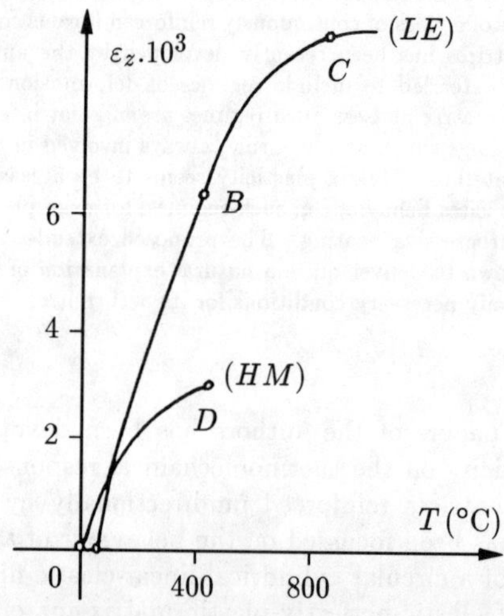

Fig. 1. Predicted (curve HM) and actual (curve LE, after Larsson [10]) inelastic response to uniform heating of a $SS/W\text{-}2\%ThO_2$ fibrous composite.

It has been briefly reported in Ref. 5 and shown in more detail in Refs. 8 and 9 that the predictions of the unified approach are in a surprisingly good agreement with typical observations on real composites and, in particular, with the observations made by Larsson [10] on heating a $SS/W-2\%ThO_2$ (SS stays for a type 304 stainless steel) fibrous composite from its as fabricated state at room temperature, i.e. from a state with considerable internal stresses induced by the process of composite fabrication. This agreement reflects the fact that the arc BC of the experimental $\varepsilon_z(T)$-curve (LE) (ε_z – axial strain, T – composite temperature) in Fig. 1 with point B corresponding to the free of axial stress state of the fibre, i.e. to the approximately stress free composite state, is simply congruent with the arc OD of the predicted (HM)-curve with points O and D corresponding to the assumed initially stress free composite state and to the predicted state of entire matrix plastification, respectively.

This congruence suggests that at point C the real composite enters a stage of deformation with entirely plastified matrix phase to which, as was pointed out above, the unified approach is unfit in its present form. As the actual (LE)-curve illustrates the rate of axial elongation abruptly decreases at this point and under further heating the composite keeps approximately constant length. This strange at first sight effect is thus certainly due to the entire plastification of the matrix phase and needs a realistic explanation.

In this paper the unified approach is extended to include regimes of deformation of composites with entirely plastified matrices. Its extended version reveals the mechanism of initiation of this strange effect and allows to predict whether the latter effect will occur under given loading (matrix cooling in the case considered below) of a given composite material.

2. The Unified Approach (Partial Matrix Plastification)

Let a unit composite cell be referred to cylindrical coordinates (r, θ, z) with axis z coinciding with the fibre axis. The fibre and matrix cross-sections occupy regions $0 \leq r \leq r_f$ and $r_f \leq r \leq r_m$, respectively. The Young's moduli E_i, Poisson's ratios ν_i, thermal expansion coefficients α_i of matrix $(i = m)$ and fibre $(i = f)$ as well as the tensile yield stress of the matrix σ_Y are temperature independent.

For each of the above mentioned model problems the normal stresses σ_i^k, $i = r, \theta, z$, in both the matrix $(k = m)$ and the fibre $(k = f)$ are principal ones. As is shown in the cited authors' works the analysis of the plastic deformation process in the expanding plastified matrix domain should necessarily take into account the current elastic component ε_z^e of the total axial strain ε_z which involves also a plastic ε_z^p and, in thermal problems, a temperature induced component $\varepsilon_z^{\rm temp}$ as well. The assumption that the limited by itself ε_z^e-strain keeps a specific for a given "composite-loading"-combination constant value ε_z^* allows to define in the stress space a specific plane by means of the Hooke's law relation

$$\sigma_z = E_m \varepsilon_z^* + \nu_m(\sigma_r + \sigma_\theta), \tag{2.1}$$

where $\sigma_i \equiv \sigma_i^m$, $i = r, \theta, z$ are now the stresses in the plastifying matrix domain and the approximate ε_z^*-value in the thermal problem is estimated as

$$\varepsilon_z^* = \sigma_Y E_c/(1 + E_c)E_m\sqrt{3}, \tag{2.2}$$

with

$$E_c = E_f r_f^2 / E_m r_m^2. \tag{2.3}$$

The projection on the $(\sigma_r, \sigma_\theta)$-plane of the curve of intersection of the plane defined by Eq. (2.1) with the von Mises' yield cylinder is the ellipse shown in Fig. 2. Its equation reads

$$(\sigma_r - \sigma_\theta)^2 + \left(\sigma_r + \sigma_\theta - \frac{2E_m\varepsilon_z^*}{1 - 2\nu_m}\right)^2 \tan^2\Phi - \frac{4}{3}\sigma_Y^2 = 0, \tag{2.4}$$

where

$$\tan\Phi = (1 - 2\nu_m)/\sqrt{3}. \tag{2.5}$$

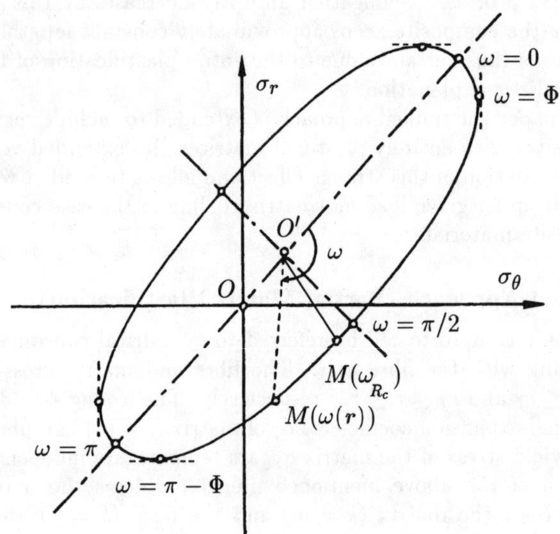

Fig. 2. Yield ellipse.

Along with the notation

$$\sin\omega = (\sigma_\theta - \sigma_r)\sqrt{3}/2\sigma_Y, \tag{2.6}$$

the stresses σ_i, $i = r, \theta$, in the form

$$\left.\begin{matrix}\sigma_r\\ \sigma_\theta\end{matrix}\right\} = \frac{E_m\varepsilon_z^*}{1 - 2\nu_m} + \frac{\sigma_Y}{\sqrt{3}\sin\Phi}\cos(\omega \pm \Phi), \tag{2.7}$$

identically satisfy the yield condition, Eq. (2.4). The angle ω, being itself a function of r, defines the positions over the yield ellipse of the points $M(\omega(r))$ with coordinates $(\sigma_\theta(r), \sigma_r(r))$ which are representative in the stress space for the states of stress in the points (r, θ) of the plastifying matrix region.

Upon integration with approximate boundary condition (cf., for example, Ref. 1) $\omega_{R_c} = \arccos[-E_m\varepsilon_z^*/\sigma_Y(1+\nu_m)]$, the equilibrium equation

$$\frac{d\sigma_r}{dr} + \frac{\sigma_r - \sigma_\theta}{r} = 0, \tag{2.8}$$

implies the function $\omega(r)$ which defines further the $R_c(\omega_{r_f})$-dependence as

$$\frac{R_c^2}{r_f^2} = \frac{\sin\omega_{r_f}}{\sin\omega_{R_c}} \exp[(\omega_{r_f} - \omega_{R_c}) \operatorname{cotan}\Phi], \tag{2.9}$$

where R_c is the current plastic zone radius, $\omega_{r_f} \equiv \omega(r_f)$, and $\omega_{R_c} \equiv \omega(R_c) > \pi/2$ is (approximately) the ω_{r_f}-value at initial matrix plastification at the interface.

The set of Eqs. (2.1)–(2.7) is easily seen to specify, irrespectively of the ε_z^*-value chosen in Eq. (2.2), a "plane-stress"-like perfect-plasticity problem. Depending upon the positions of the $M(\omega(r))$-points on the yield ellipse corresponding hyperbolic ($\Phi < \omega < \pi - \Phi$ or $\Phi < \omega - \pi < \pi - \Phi$), elliptic ($|\omega| < \Phi$ or $|\omega - \pi| < \Phi$), and parabolic ($|\omega| = \Phi$, $|\omega - \pi| = \Phi$) regimes of plastic redistribution of stresses may develop in the respective points (r, θ) of the plastifying matrix region (for details see Refs. 1–4).

The analysis of this set in the thermal problem proves that the point $M(\omega_{r_f})$ which represents the matrix stresses $\sigma_i(\omega_{r_f})$, $i = r, \theta$ at the interface $r = r_f$ moves upon the instant of initial matrix plastification along the hyperbolic arc $\Phi < \omega < \pi - \Phi$ from the point $M(\omega_{R_c})$ towards the parabolic point $M(\pi - \Phi)$ which, as is shown in Refs. 1–4, is a critical one from the point of view of eventual composite failure due to plastic instability at the interface. Thus, increasing ω_{r_f}-values imply through Eq. (2.9) corresponding increase in the plastic zone radius $R_c(\omega_{r_f})$. As Eq. (2.9) predicts the latter achieves a maximum possible value R_c^* for $\omega_{r_f} = \pi - \Phi$. Depending on the parameters of the specific "composite-loading"- combination this value may appear formally to be smaller or larger than (or equal to) r_m. In the cases $R_c^* < r_m$ entire matrix plastification is impossible. Provided $R_c^* \geq r_m$ regimes of deformation with entirely plastified matrix phase develop upon the instant defined by the relation $\omega_{r_f} = \omega_{r_f}^*$, where the angle $\omega_{r_f}^*$, being a root of the equation $R_c = r_m$, belongs in the considered $R_c^* \geq r_m$-cases to the interval $(\omega_{R_c}, \pi - \Phi)$.

3. Thermal Loading—Entirely Plastified Matrix

3.1. Preliminary Considerations

It would be instructive to precede the analysis of the unit cell behaviour in a regime of deformation with entirely plastified matrix phase under the same thermal loading (matrix cooling) by the following remark. As is generally adopted in mechanics of composites reinforced by continuous, i.e. long enough, fibres the axial fibre-matrix load transfer is localized within end unit cell portions and induces in the latters interfacial shear stress distributions. The increasing lengths of these portions remain small compared to the length of the remaining main unit cell portion to which the plane cross-sections hypothesis is actually referred. This main portion remains thus long enough for to be further considered as continuously reinforced. Its response is representative for the response of the unit cell and of the composite, respectively. As is easily observed the unified approach not only complies with this basic adoption but verifies it with the above mentioned experimental agreement of the predicted composite response.

This remark allows to conclude that in the deformation regime in question the unit cell might be viewed as involving a few specific portions. Two of them are located at the very ends of the cell and have the same length, namely the length that these load transfer portions have achieved at the instant of entire matrix plastification. There are two other portions next to them. They are again of equal but increasing length. These portions occur due to the load transfer between the fibre and the entirely plastified matrix. Finally, a main middle portion of a decreasing length exists between the latter portions. This portion is long enough compared to the other portions and its cross-sections remain plane.

It becomes evident now that the above mentioned strange effect, when referred to the here considered case of matrix cooling, under which the entire cell should be expected rather further to shorten than to keep practically constant length, might be only due to a certain mechanism of additional stretching within the decreasing in its own length middle cell portion. As it will be shown below compositions "fibre-fully plastified matrix" with appropriate mechanical properties and r_f/r_m-ratios may really activate under matrix cooling such a mechanism. Note that the analysis of more realistic cases of cooling or heating (as in Larsson's [10] experiments) of the entire composite (unit cell) corresponds, roughly speaking, to that of the model problem of matrix cooling (heating) with α_m-coefficient replaced by the $(\alpha_m - \alpha_f)$-difference, cf. Ref. 9.

3.2. Entirely Plastified Matrix Phase

Consider a subportion of an arbitrary length l of the main middle portion of the unit cell. There are no specific reasons to reject any of the assumptions upon which the associated with the unified approach analysis of cases with partially plastified matrix is based. Only the plastic incompressibility condition should be now referred to the entire matrix phase. According to this condition an infinitesimal matrix temperature drop $dT_m < 0$ should imply changes in r_f, r_m, and l which satisfy the relation

$$\frac{dr_f}{r_f} + \frac{1}{2}\frac{dl}{l} + \frac{1}{2}\frac{d(r_m^2/r_f^2)}{r_m^2/r_f^2 - 1} = \frac{3}{2}\alpha_m dT_m. \tag{3.1}$$

The volume change of the respective subportion of the linear-elastic fibre satisfies the Hooke's law relation

$$\frac{dr_f}{r_f} + \frac{1}{2}\frac{dl}{l} = \frac{1-2\nu_f}{2E_f}\left(d\sigma_z^f + 2d\sigma_r\Big|_{r_f}\right). \tag{3.2}$$

The equilibrium of axial forces requires that

$$d\sigma_z^f = -E_m\varepsilon_z^*\, d\left(\frac{r_m^2}{r_f^2}\right) + 2\nu_m d\sigma_r\Big|_{r_f}. \tag{3.3}$$

Following the scheme of deriving the continuity condition for the radial displacement u_r at the interface $r = r_f$ developed in Refs. 1 and 3 one can reduce this condition in the here considered case to the form

$$\frac{dl}{l} = -\frac{1}{1-2\nu_f+\tan\omega_{r_f}\operatorname{cotan}\Phi}$$

$$\times\left[\frac{2(1+\nu_f)(1-2\nu_f)}{E_f}d\sigma_r\Big|_{r_f} + \alpha_m dT_m(3+\tan\omega_{r_f}\operatorname{cotan}\Phi)\right]. \tag{3.4}$$

Note that Eqs. (3.1)–(3.4) involve already the condition of continuity of the radial stress at the interface. They prove, in addition, that r_f, r_m, l, and, respectively, the (r,θ,z)-coordinates should be regarded now as current values of the respective quantities which yields through the equilibrium equation, Eq. (2.8), the relation

$$d\left(\frac{r_m^2}{r_f^2}\right) = \frac{r_m^2}{r_f^2}\frac{\sin(\omega_{r_f}+\Phi)}{\sin\omega_{r_f}\sin\Phi}d\omega_{r_f}, \tag{3.5}$$

or, respectively,

$$r_i^2\sin\omega_{r_i}\exp[\omega_{r_i}\operatorname{cotan}\Phi] = F(T_m), \quad i=f,m, \tag{3.6}$$

where $\omega_{r_m} \equiv \omega_{R_c}$ and the function F is to be, in principle, further defined.

Finally, it follows from the yield condition (Eqs. (2.4) and (2.7), cf. Fig. 2) that

$$d\sigma_r\Big|_{r_f} = -\frac{\sigma_Y}{\sqrt{3}}\frac{\sin(\omega_{r_f}+\Phi)}{\sin\Phi}d\omega_{r_f}. \tag{3.7}$$

Now, Eqs. (3.1) and (3.2) imply through Eqs. (3.3) and (3.5) the relation

$$d\omega_{r_f}\frac{\sin(\omega_{r_f}+\Phi)}{\sin\Phi}\left(\frac{r_m^2/r_f^2}{r_m^2/r_f^2-1}\frac{1}{\sin\omega_{r_f}} - \kappa_\omega\right) = 3\alpha_m dT_m, \tag{3.8}$$

where the notation κ_ω reads

$$\kappa_\omega = E_m\varepsilon_z^*\frac{1-2\nu_f}{E_f}\frac{r_m^2/r_f^2}{\sin\omega_{r_f}} + 2\frac{(1+\nu_m)(1-2\nu_f)}{E_f}\frac{\sigma_Y}{\sqrt{3}}. \tag{3.9}$$

It is seen that $\kappa_\omega > 0$, $\kappa_\omega \approx \sigma_Y/E_f \ll 1$.

Eq. (3.9) with r_m^2/r_f^2-ratio defined from Eqs. (3.5) and (3.6) as

$$\frac{r_m^2}{r_f^2} = \frac{\sin \omega_{r_f}}{\sin \omega_{R_c}} \exp[(\omega_{r_f} - \omega_{R_c}) \operatorname{cotan} \Phi] . \tag{3.10}$$

implies now the conclusion that upon entire matrix plastification further matrix cooling ($dT_m < 0$) induces negative $d\omega_{r_f}$-values, i.e. a decrease in ω_{r_f}. In accordance with Eq. (3.7) (see also Fig. 2) the corresponding negative by itself $\sigma_r|_{r_f}$-stress increases ($d\sigma_r\,|_{r_f} > 0$), i.e. the shrinkage effect becomes weaker. Upon integration Eq. (3.8) provides, in principle, the specific form of the $\omega_{r_f}(T_m)$-dependence, or, which is the same, the law of motion of the point $M(\omega_{r_f})$ along the yield ellipse. The integration of Eq. (3.8) will not be considered here. It will be just pointed out that upon entire matrix plastification this point moves from the point $M(\omega_{r_f}^*)$ along the hyperbolic arc $(\Phi, \pi - \Phi)$ towards the point $M(\omega_{R_c})$ of initial matrix plastification at the interface.

Eqs. (3.4) and Eq. (3.8) imply now that

$$\frac{dl}{l} = d\omega_{r_f} \frac{\sin(\omega_{r_f} + \Phi)}{\sin \Phi} \frac{1}{1 - 2\nu_f + \tan \omega_{r_f} \operatorname{cotan} \Phi}$$

$$\times \left[\left(\frac{r_m^2/r_f^2}{r_m^2/r_f^2 - 1} \frac{1}{\sin \omega_{r_f}} - \kappa_\omega \right) \frac{3 + \tan \omega_{r_f} \operatorname{cotan} \Phi}{3} + \kappa_l \right] , \tag{3.11}$$

where

$$\kappa_l = \frac{2(1 + \nu_f)(1 - 2\nu_f)\sigma_Y}{\sqrt{3} E_f} , \tag{3.12}$$

and $\kappa_l > 0$, $\kappa_l \approx \sigma_Y/E_f \ll 1$.

It is easy to observe that $1 - 2\nu_f + \tan \omega_{r_f} \operatorname{cotan} \Phi < 0$ for $\omega_{r_f} \in [\pi/2, \pi - \Phi]$ while $3 + \tan \omega_{r_f} \operatorname{cotan} \Phi$ is negative, positive, and zero provided $\omega_{r_f} \in [\pi/2, \hat{\omega})$, $\omega_{r_f} \in (\hat{\omega}, \pi - \Phi]$, and $\omega_{r_f} = \hat{\omega}$, respectively, where $\hat{\omega} \in (\omega_{R_c}, \pi - \Phi)$ and

$$\tan \hat{\omega} = -3 \tan \Phi . \tag{3.13}$$

Note that the term in square brackets in Eq. (3.11) changes sign when ω_{r_f} runs through a value $\hat{\omega}_l = \hat{\omega} - \varepsilon_l$, where $\varepsilon_l > 0$ is a small (actually of the order of σ_Y/E_f) quantity. Thus, for values of ω_{r_f} decreasing within the interval $[\pi - \Phi, \hat{\omega}_l)$ the quantity dl/l is positive. It vanishes at $\omega_{r_f} = \hat{\omega}_l$ and is then negative for further decrease in ω_{r_f}, i.e. under further matrix cooling. In other words, provided $\omega_{r_f}^* > \hat{\omega}_l$, then just upon entire matrix plastification further matrix cooling, i.e. further ω_{r_f}-decrease, will induce first stretching ($dl/l > 0$) of the considered unit cell subportion. Upon the instant defined by the relation $\omega_{r_f} = \hat{\omega}_l$, i.e. for $\omega_{r_f} < \hat{\omega}_l$, the stretching will be followed by a regime of further shortening of the subportion under consideration and of the entire cell, respectively.

Such a stretching, when superposed with the decrease in length of the main middle unit cell portion (which decrease, as adopted above, is due to the increasing length of the short by themselves load transfer cell portions) will certainly result in a sudden abrupt decrease in the rate of shortening of the entire unit cell length. Evidently enough this rate decrease will be observed physically as a *strange* at first sight effect of approximate (to a certain extent) constancy of the unit cell and, respectively, of the composite specimen length. The latter effect is thus observable within the interval $(\hat{\omega}_l, \omega^*_{r_f})$ and its occurrence necessarily requires fulfillment of a relation $\omega^*_{r_f} > \hat{\omega}_l$.

If the combination of geometrical and mechanical parameters of a composite material provides validity of a relation $\omega^*_{r_f} \leq \hat{\omega}_l$, then the specific mechanism of stretching will not be activated under the considered thermal loading and thus the composite will not manifest such a strange effect.

It is a matter now of an appropriate shift in the scale of ω_{r_f}-measuring and of some additional computations for to prove that for the composite studied by Larsson [10] the $\omega^*_{r_f}$-value is approximately 0.81π while the corresponding $\hat{\omega}_l$-value is 0.74π. In the light of the foregoing considerations these $\omega^*_{r_f}$ and $\hat{\omega}_l$ estimates explain the strange effect observed on Larsson's [10] composite. It will be mentioned, for completeness, that the approximate ω_{R_c} and Φ values for this composite are 0.53π and 0.1π, respectively, while the value of R^*_c is $1.12r_m$.

3.3. Further Implications

Eqs. (3.1), (3.3), (3.9) and (3.11) imply an expression for dr_f/r_f which is similar to Eq. (3.11) for dl/l but is of a more complicated form. Its analysis proves that dr_f/r_f is negative when ω_{r_f} decreases within the interval $(\pi - \Phi, \hat{\omega}_r)$, where $\hat{\omega}_r = \hat{\omega} + \varepsilon_r$ and ε_r is, similar to ε_l, a small positive quantity of the order of $\sigma_Y/E_f \ll 1$. The dr_f/r_f-ratio vanishes when the decreasing ω_{r_f}-angle achieves the $\hat{\omega}_r$-value and is then positive under further matrix cooling, i.e. for further ω_{r_f}-decrease. It is evident enough that $\hat{\omega}_r > \hat{\omega}_l$.

4. Concluding Remarks

The present analysis allows to conclude that if, for a given composite material, the relation $\omega^*_{r_f} > \hat{\omega}_r$ is valid, then upon entire matrix plastification the decreasing in its own length main middle portion of the representative unit cell will first enter a regime of stretching accompanied by a lateral contraction of the fibre. The fibre will start thickening upon the instant defined by the relation $\omega_{r_f} = \hat{\omega}_r$ while the middle cell portion will further remain in a regime of stretching up to the instant corresponding to $\omega_{r_f} = \hat{\omega}_l$. At this instant the stretching will turn into shortening again (as in the stage preceding entire matrix plastification). The latter instant will be then followed by a regime of a simultaneous shortening of the middle cell portion and thickening of the fibre. Thus, the response of a given composite upon entire plastification of its matrix phase depends on the specific $\omega^*_{r_f}$, $\hat{\omega}_r$ and $\hat{\omega}_l$ values which are themselves predetermined by the combination of geometrical and mechanical parameters of the composite.

It should be emphasized that the present study is far from delivering a complete analysis of the general problem of the behaviour of fibrous composites with entirely plastified matrices. It rather indicates the complexity of the problem on the whole and, in particular, of the study of the important and nontrivial interactions between the above introduced main middle and load transfer unit cell portions. The approach proposed in this paper concerns the behaviour of the middle unit cell portion and implies a reasonable prediction of the overall composite response, including the considered strange effect. It is hoped for this reason that this approach might form a useful contribution to the complete study of the effects of entire matrix plastification on fibrous composites response.

Acknowledgements. The support to one of the authors (I.M.M.) of the Bulgarian National Research Fund under Grant No MM 416-94 is gratefully acknowledged.

References

1. K. Herrmann and I. Mihovsky, *Engn. Transactions – Rozprawy Inžynierskie* **31** (2) (1983) 165.
2. K. Herrmann and I. Mihovsky, *Mechanika Teoretyczna i Stosowana* **22** (1/2) (1984) 25.
3. K. Herrmann and I. Mihovsky, *Ann. Univ. Sofia, Fac. Math. Inf., Livre 2, Méchanique* **83** (1989) 5.
4. K. Herrmann and I. Mihovsky, *Ann. Univ. Sofia, Fac. Math. Inf., Livre 2, Méchanique* **83** (1989) 21.
5. K. Herrmann and I. Mihovsky, in *Developments in the Science and Technology of Composite Materials* (*Proc. Fourth ECCM, Stuttgart, Sept. 25-28, 1990*) eds. J. Füller *et al.*, Elsevier Applied Science, London and New York, 1990, p. 717.
6. K. Herrmann and I. Mihovsky, *ZAMM* **68**(4) (1988) T193.
7. R. Hill, *Mathematical Theory of Plasticity*, Clarendon Press, Oxford, 1950.
8. K. Herrmann and I. Mihovsky, In *Continuum Models of Discrete Systems,* (*Proc. 7th Int. Symp. CMDS, Paderborn, June 14-19, 1992*), eds. K.-H. Anthony and H.-J. Wagner, Trans Tech Publications, Switzerland, 1993, p. 331.
9. K. Herrmann and I. Mihovsky, In *Recent Advances in Mathematical Modelling of Composite Materials*, ed. K. Z. Markov, World Sci., 1994, p. 141.
10. L. O. K. Larsson, In *Proc. Int. Conf. on Composite Materials, ICCM/2* (*Toronto, April 16-20, 1978*) eds. B. Norton *et al.*, The Metallurgical Soc. of AIME, Warendale, 1978, p. 805.

Continuum Models and Discrete Systems
Proceedings of 8th International Symposium, June 11-16, 1995, Varna, Bulgaria, ed. K.Z. Markov
©World Scientific Publishing Company, 1996, pp. 199–206

MORPHOLOGICAL MODELING OF RANDOM COMPOSITES

D. JEULIN
Centre de Morphologie Mathématique, ENSMP, 35 rue St-Honoré,
F-77300 Fontainebleau, France

and

A. Le COËNT
Centre Commun de Recherche, Aérospatiale,
12 rue Pasteur, F-92152 Suresnes Cédex, France

Abstract. We review recent developments in modeling the morphological textures of random two phase composites. Probabilistic properties, including correlation functions are known for the following basic models using a primary random grain: mosaic on a random tessellation, Boolean model, Dead Leaves. These models can also be combined to produce random media with different scales. Their potential use to predict the macroscopic physical properties of random media is illustrated by the calculation of functions occurring in third-order upper and lower bounds in different situations, enabling us to compare the textures according to their overall physical properties.

1. Introduction

Modeling complex composite microstructures by means of random sets [12,14, 18,5–7] is a useful way to summarize microstructural information in a few parameters, and also to propose practical methods of simulations. For appropriate models it is possible to calculate some of their probabilistic properties such as correlation functions that can be used for the prediction of their macroscopic properties.

An estimation of the overall properties of random composites can be obtained from bounds based on a limited amount of statistical information [2,9]. The commonly used bounds for isotropic media are Hashin and Shtrikman (H-S) bounds based on the volume fractions [3]. However many different morphologies can be encountered for a given volume fraction. Tighter bounds can be derived from additional statistical information, such as the infinite set of their correlation functions [2,9,4]. In practice it is difficult to obtain results beyond the third-order correlation functions, which already provide interesting bounds that are useful for the comparison of microstructures according to their incidence on the behaviour of composite media.

After a recall of the third-order bounds, we examine their practical derivation for some generic random sets and their combination. This can be used as guidelines to select the most appropriate microstructures and to design materials for given physical properties.

2. Third-Order Bounds

We consider random composites made of two phases A_1 (with fraction p) and A_2 (with fraction $q = 1 - p$) having a scalar dielectric permittivity ε_1 and ε_2 (with $\varepsilon_2 > \varepsilon_1$) (the same approach could be followed for other physical properties such as the thermal or electrical conductivity, the permeability of porous media, etc.). The composite is modelled by a stationary and isotropic random set A (with $A = A_2$ and $A^c = A_1$). The third-order bounds (upper bound ε_+ and lower bound ε_-) were derived from a perturbation expansion of the electric field and a variational principle on the stored electrostatic energy [2]. These can be expressed in R^d, $d = 1, 2, 3$, as follows [21]

$$\varepsilon_- = \varepsilon_1 \frac{1 + ((d-1)(1+q) + \zeta_1 - 1)\,\beta_{21} + (d-1)\,(((d-1)q + \zeta_1 - 1))\,\beta_{21}^2}{1 - (q + 1 - \zeta_1 - (d-1))\,\beta_{21} + ((q - (d-1)p)\,(1 - \zeta_1) - (d-1)q)\,\beta_{21}^2}, \quad (1)$$

$$\varepsilon_+ = \varepsilon_2 \frac{1 + ((d-1)(p + \zeta_1) - 1)\,\beta_{12} + (d-1)\,(((d-1)p - q)\,\zeta_1 - p)\,\beta_{12}^2}{1 - (1 + p - (d-1)\zeta_1)\,\beta_{12} + (p - (d-1)\zeta_1)\,\beta_{12}^2}. \quad (2)$$

In Eqs. (1) and (2), β_{12} and β_{21} are given by:

$$\beta_{ij} = \frac{\varepsilon_i - \varepsilon_j}{\varepsilon_i + (d-1)\varepsilon_j}. \quad (3)$$

The function $\zeta_1(p)$ [16] is deduced from the probability $P(h_1, h_2)$ that the three points $\{x\}$, $\{x + h_1\}$, $\{x + h_2\}$ belong to A_1. Denoting this probability $P(|h_1|, |h_2|, \theta)$, with $u = \cos\theta$, θ being the angle between the vectors h_1 and h_2, we have:

$$\begin{aligned} \zeta_1(p) &= \frac{9}{4pq} \int_0^{+\infty} \frac{dx}{x} \int_0^{+\infty} \frac{dy}{y} \int_{-1}^{+1} (3u^2 - 1) P(x, y, \theta)\, du \text{ in 3D}, \\ \zeta_1(p) &= \frac{4}{\pi pq} \int_0^{+\infty} \frac{dx}{x} \int_0^{+\infty} \frac{dy}{y} \int_0^{\pi} P(x, y, \theta) \cos(2\theta)\, d\theta \text{ in 2D}. \end{aligned} \quad (4)$$

The evaluation of the integrals in Eq. (4) can be made in some cases analytically, but most often numerically. In that last case, it is convenient to replace in the integrands $P(x, y, \theta)$ by $P(x, y, \theta) - P(x)P(y)/p$, with $P(h)$ being the probability that the two points $\{x\}$, $\{x + h\}$ belong to A_1. We used the Gauss-Legendre quadrature, adapting the number of points (usually a few tens) to the convergence [11,10].

Exchanging phases A_1 and A_2 enables us to define the function $\zeta_2(q)$ with $\zeta_2(q) = 1 - \zeta_1(p)$. The function $\zeta_1(p)$ satisfies the inequalities $0 \leq \zeta_1(p) \leq 1$. For $\zeta_1(p) = 1$ or 0 (and only in these cases), the two bounds ε_+ and ε_- coincide and are equal to the H-S upper ($\zeta_1(p) = 1$) or lower ($\zeta_1(p) = 0$) bound. For given p, ε_1 and ε_2, the two bounds increase with the function $\zeta_1(p)$ (while the H-S bounds remain fixed), so that higher values of the effective properties are expected. This point can be used to compare the expected properties of materials with different morphologies, as seen later.

If the two phases A_1 and A_2 are symmetric, the case of $p = 0.5$ produces an autodual random set (the two phases having the same probabilistic properties), for which $\zeta_1(0.5) = 0.5$. This case corresponds to a third-order central correlation function equal to zero. Therefore in two and three dimensions, the third-order bounds of a symmetric medium present a fixed point at $p = 0.5$. In addition, it is known [12,13] that for an autodual random set in two dimensions the effective permittivity is equal to the geometrical average of the two permittivities.

3. Basic Random Sets Models

A broad range of random structures models are available [12,14,18,5-7]. We concentrate here on random sets (or two-phase media) models noted A. These are characterized by probability distributions defined on compact sets K:

$$T(K) = 1 - Q(K) = P\{K \cap A \neq \emptyset\} = 1 - P\{K \subset A^c\}. \tag{5}$$

For various models, the theoretical expression of $T(K)$ is available, as a function of their parameters. On the other hand, $T(K)$ can be estimated from image analysis on pictures of the material. This enables us to estimate the parameters of the model, and to test which model may be the most appropriate to describe a given material, as illustrated in [17] for the case of WC-Co composites. For the calculation of third-order bounds, we need to calculate $T(K)$ or $Q(K)$ when $K = \{x, x + h_1, x + h_2\}$. Examples are given now.

3.1. The Mosaic Model

The mosaic model [13,5-7], or "cell" model [15], is built up in two steps: starting with a random tessellation of space into cells, every cell A' is affected independently to the random set A with the probability p and to A^c with the probability q. The medium is symmetric in A and A^c, which may be exchanged (changing p into q). From a direct calculation of third-order bounds, based on the third order central correlation function, Miller defined a parameter G depending only on the random cell geometry by means of the function

$$s(|h_1|, |h_2|, \theta) = \frac{\overline{\mu}_d(A' \cap A'_{h_1} \cap A'_{h_2})}{\overline{\mu}_d(A')}, \tag{6}$$

where μ_d is the Lebesgue measure in R^d, $\overline{\mu}_d$ its average over the realizations of A', and A'_h is obtained by translation of A' by the vector h. Then

$$\begin{aligned} G &= \frac{1}{9} + \frac{1}{2} \int_0^{+\infty} \frac{dx}{x} \int_0^{+\infty} \frac{dy}{y} \int_{-1}^{+1} (3u^2 - 1) s(x, y, \theta)\, du \quad \text{in 3D}, \\ G &= \frac{1}{4} + \frac{1}{\pi} \int_0^{+\infty} \frac{dx}{x} \int_0^{+\infty} \frac{dy}{y} \int_0^{+\pi} s(x, y, \theta) \cos(2\theta)\, d\theta \quad \text{in 2D}. \end{aligned} \tag{7}$$

We have $\frac{1}{9} \leq G \leq \frac{1}{3}$ in 3D and $\frac{1}{4} \leq G \leq \frac{1}{2}$ in 2D. For the mosaic model Eq. (4) becomes [16]:

$$\zeta_1(p) = p + a(q-p) \quad \text{with} \quad a = \frac{d^2G-1}{d-1}, \qquad \zeta_2(p) = \zeta_1(p), \tag{8}$$

and therefore $\zeta_1(p)$ varies linearly with the proportion p, with a slope $1-2a$ ($0 \leq a \leq 1$). The two extreme cases are the following: $a = 0$ for $G = \frac{1}{9}$ in R^3 and for $G = \frac{1}{4}$ in R^2 (this corresponds to spherical and disk cells respectively); $a = 1$ for $G = \frac{1}{3}$ in R^3 and for $G = \frac{1}{2}$ in R^2 (this corresponds to spheroidal cells of plate and needle shapes respectively). However, no correct construction of random tessellations corresponding to these extreme values seems to exist.

A particular random mosaic can be obtained from a Poisson tessellation of space by Poisson random planes in R^3 (with the intensity λ, which is a scale parameter that does not appear in the calculation of G) and by Poisson lines in R^2 [14,18]. The cells of these tessellations are Poisson polyhedra and Poisson polygons. For the Poisson mosaic model built up from this tessellation, we have:

$$s(|h_1|,|h_2|,\theta) = \exp\left(-\lambda\left(|h_1|+|h_2|+\sqrt{|h_1|^2+|h_2|^2-2|h_1||h_2|\cos\theta}\right)\right). \tag{9}$$

From Eqs. (7), G can be calculated analytically: in R^2, $G = 1 - \ln 2$ [13] and in R^3, $G = \frac{1}{6}$, see Ref. 10. Therefore, we have $a = 3 - 4\ln 2$ in R^2 and $a = \frac{1}{4}$ in R^3.

Anisotropic cells (but distributed with a uniform orientation) are studied in [10,11]: G (and consequently $\zeta_1(p)$ for $a \leq \frac{1}{2}$) increases with the cell anisotropy.

3.2. The Boolean Model

The Boolean model [12,14] is obtained by implantation of random primary grains A' (with possible overlaps) on Poisson points x_k with the intensity λ: $A = \bigcup_{x_k} A'_{x_k}$. For this model, we have:

$$T(K) = 1 - Q(K) = 1 - \exp\left(-\lambda\overline{\mu}_d(A' \oplus \check{K})\right) = 1 - q^{\frac{\overline{\mu}_d(A' \oplus \check{K})}{\overline{\mu}_d(A')}}, \tag{10}$$

where $A' \oplus \check{K} = \bigcup_{-x \in K} A'_x$ is the result of the dilation of A' by K [18]. Particular cases of Eq. (10) give the covariance $Q(h) = P\{x \in A^c, x+h \in A^c\}$ and the three point probability $P(x,y,\theta)$ entering into Eqs. (4). Any shape (convex or non convex, and even non connected) can be used for the grain A'. Most often in the literature Boolean models of spheres are considered; in [17], a Boolean model with Poisson polyhedra was found appropriate for WC-Co composites. Contrary to the mosaic model, the Boolean model is not symmetric (and not autodual for $p = 1/2$). Therefore, different sets of bounds are obtained when exchanging the properties of A and A^c.

The known functions $\zeta_1(p)$ for the Boolean model were obtained by numerical integration of these equations. With a good approximation, linear functions of p are obtained: $\zeta_1(p) \simeq \alpha p + \beta$. The value $\zeta_1(0)$ (β for the approximation) is deduced for

$p \to 0$ (or $\lambda \to 0$ in Eq. (10)): $\zeta_1(0) = a$ with a given by Eq. (8) for the random grain A' (and consequently for any size distribution of this grain, since G is invariant by similarity of A'). Therefore, in R^3, $\zeta_1(0) = 0$ for spheres, and $\zeta_1(0) = \frac{1}{4}$ for Poisson polyhedra. In R^2, $\zeta_1(0) = 0$ for disks and $\zeta_1(0) = 3 - 4\ln 2 \simeq 0.2274$ for Poisson polygons. The function $\zeta_1(p)$ was estimated in [19,20], for spheres and in [8,10,11] for discs. A direct estimation of bounds for discs is presented in [1]. It is given in [10,11] for polygons in R^2, including Poisson polygons. These results (and $\zeta_2(p) = 1 - \zeta_1(1-p)$ obtained by exchange of the two phases) are summarized by:

$$\begin{gathered}
\zeta_1(p) \simeq 0.5615p \text{ for spheres [19,20]}, \\
\zeta_2(p) \simeq 0.5615p + 0.4385 \text{ for (spheres)}^c , \\
\zeta_1(p) \simeq \tfrac{2}{3}p \text{ for discs in } R^2 \text{ [8, 10, 11]}, \\
\zeta_2(p) \simeq \tfrac{2}{3}p + \tfrac{1}{3} \text{ for (discs)}^c \text{ in } R^2 , \\
\zeta_1(p) \simeq 0.5057p + 0.2274 \text{ for Poisson polygons in } R^2 \text{ [10, 11]}, \\
\zeta_2(p) \simeq 0.5057p + 0.2669 \text{ for (Poisson polygons)}^c \text{ in } R^2 .
\end{gathered} \tag{11}$$

For the models presented in Eqs. (11), we have $\zeta_2(p) > \zeta_1(p)$, so that the third-order bounds increase when $\varepsilon(x) = \varepsilon_2 > \varepsilon_1$ for $x \in A^c$. This is due to the fact that it is easier for the "matrix" phase A^c to percolate than for the overlapping inclusions building A. For the sphere and disks models, $\zeta_2(p)$ is larger than the $\zeta_1(p)$ of the corresponding mosaic models.

3.3. The Dead Leaves Model

The Dead leaves model [5-7] is obtained sequentially by implantation of random primary grains $A'(t)$ on a Poisson point process: in every point x is kept the last occurring value $\varepsilon(x,t)$ during the sequence. In this way, non symmetric random sets are obtained if two different families of primary grains are used for A and for A^c. When using the same family of primary grains for the two sets, a mosaic model built on a random tessellation is obtained. The shape of the resulting cell is non convex (and even non connected!), due to the overlaps occurring during the construction of the model. For the symmetric case, the calculation of G can be made from the knowledge of the function $s(x,y,\theta)$:

$$s(|h_1|, |h_2|, \theta) = \frac{\overline{\mu}_d(A' \cap A'_{h_1} \cap A'_{h_2})}{\overline{\mu}_d(A' \cup A'_{h_1} \cup A'_{h_2})} . \tag{12}$$

Calculations were made for Poisson polyhedra as primary grains [10]: $G \simeq 0.170$ in R^3 and $G \simeq 0.311$ in R^2, which is slightly larger than for the Poisson mosaic.

4. Combination of the Basic Random Sets Models

It is interesting to combine the previous basic models in order to describe more complex structures. For instance, fluctuations of morphological properties (like of the local volume fraction of one phase, p) may exist in real materials. We consider now various cases of random sets with different scales.

4.1. A Hierarchical Model

A simple hierarchical model with two separate scales can be built up in two steps [7]: we start with a primary random tessellation of space into cells; every cell is intersected by a realization of a secondary random set, admitting the function $\zeta_1(p)$, and with random parameters (the realizations in distinct cells being independent). Any random tessellation can be used in this construction. Any type of random set (mosaic model, Boolean model, dead leaves, etc.) can be used in the second step. For cells much larger than the scale of the secondary random set, implanted with a random volume fraction P (with expectation $E\{P\} = p$ and variance $D^2\{P\}$), the function $\zeta_{H1}(p)$ corresponding to the hierarchical model is approximated by:

$$pq\zeta_{H1}(p) = E\{P(1-P)\zeta_1(P)\} + pD^2\{P\} + aE\{(P-p)^3\}. \qquad (13)$$

Eq. (13) recovers $\zeta_1(p)$ when P is non random, and the mosaic result (Eq. (8)) when $P = 1$ with the probability p and $P = 0$ with the probability q. It is instructive to examine the case where $P = p_1$ with the probability p_2 and $P = 0$ with the probability $1 - p_2$. For this example, $p = p_1p_2$, a proportion p_2 of cells is filled with a mixture of A_1 (in a proportion p_1) and A_2, and a proportion $1 - p_2$ of cells is filled with the phase A_2 alone. If $\zeta_1(p) = \alpha p + \beta$ (which is the case of mosaic models, and approximately the case of some Boolean models), Eq. (13) becomes:

$$pq\left(\zeta_{H1}(p) - \zeta_1(p)\right) = p(p_1 - p)\left(p(1-3\alpha) + \alpha - \beta + (p_1 - 2p)(a - \alpha)\right). \qquad (14)$$

It is interesting to consider situations where the increment $\zeta_{H1}(p) - \zeta_1(p)$ can reach an optimal value when varying p_1 for a given p. It can admit a positive optimum, corresponding to some optimal value $\zeta_{H1\mathrm{opt}}(p)$ when $a < \alpha$, obtained for $p_1 = \dfrac{p(1-3a) + \alpha - \beta}{2(a-\alpha)}$:

$$\zeta_{H1\mathrm{opt}}(p) - \zeta_1(p) = \frac{\left(p(a + 2\alpha - 1) + \beta - \alpha\right)^2}{4q(\alpha - a)}. \qquad (15)$$

For instance, when the secondary random set is a mosaic model built up on the same type of tessellation as the primary random tessellation (but on a much smaller scale), the increment is always positive for $a < \frac{1}{3}$ (that is for $G < \frac{1}{3}$ in R^2 and $G < \frac{5}{27}$ in R^3). In these conditions, the hierarchical model provides higher bounds than the initial one scale mosaic, and we obtain $\zeta_{H1\mathrm{opt}}(p)$ for $p_1 = (p+1)/2$:

$$\zeta_{H1\mathrm{opt}}(p) = p\left(\frac{3-5a}{4}\right) + \frac{a+1}{4}. \qquad (16)$$

When the second scale is a Boolean model of spheres in R^3, the optimal values given by Eq. (15) can be reached for $p < 0.25$ for $a = 0$ or $a = \frac{1}{4}$ (Poisson tessellation).

The corresponding bounds are higher than the bounds of the mosaic for $a = 0$, and equivalent to them for $a = \frac{1}{4}$. As a result of the improvement of $\zeta_1(p)$ by the two scales model, the third-order upper bound is usually increased relatively more than the lower bound.

4.2. Bounds for Union and Intersection of Random Sets

Another way to combine random sets is to consider the union or the intersection of two independent random sets A_1 and A_2. Since we have $(A_1 \cup A_2)^c = A_1^c \cap A_2^c$, we can limit our purpose to the intersection. For this model, $p = p_1 p_2$ and $P(K) = P\{K \subset (A_1 \cap A_2)\} = P\{K \subset A_1\}P\{K \subset A_2\}$. When using the same type of random set (with $\zeta_1(p) = \alpha p + \beta$) for the two primary structures, with the scale of A_2 much smaller than the scale of A_1, we get the following approximate result:

$$pq\left(\zeta_{H1}(p) - \zeta_1(p)\right) = \frac{p^2}{p_1^2}(p_1 - p)(1 - p_1)(\alpha - \beta) \tag{17}$$

and for the two scale model $\zeta_{H1}(p) > \zeta_1(p)$ when $\alpha \geq \beta \geq 0$. This is the case when the basic model is a mosaic model with $a < \frac{1}{3}$ or when it is one of the studied Boolean models (or its complementary set). The increment of Eq. (17) can be optimized over the values of p_1 for a given p. For the intersection of two Boolean models (or of their complementary sets), when $\alpha \geq \beta \geq 0$ we get an optimum for $p_1 = \dfrac{2p}{1+p}$:

$$\zeta_{H1opt}(p) = p\frac{3\alpha + \beta}{4} + \frac{3\beta + \alpha}{4}. \tag{18}$$

For the two scale mosaic model with $a < \frac{1}{3}$, it becomes

$$\zeta_{H1opt}(p) = p\frac{3 - 5a}{4} + \frac{a + 1}{4}. \tag{19}$$

A last combination of structures is obtained for $p_1 = p_2$ and for widely separate scales. In that particular case, we have $p = p_1^n$, $\zeta_{H_1}(p) \simeq \zeta_1(p_1) = \dfrac{1+p}{1+p^{\frac{1}{2}}}\zeta_1(p^{\frac{1}{2}})$. This process can be iterated and at the order n (for n well separate scales) we have $\zeta_{H1}^{(n)}(p) \simeq \dfrac{1+p}{1+p^{1/n}}\zeta_1(p^{1/n})$. Since $p^{1/n} \geq p$, we get $\zeta_{H_1}^{(n)}(p) \geq \zeta_1(p)$ and $\zeta_{H_1}^{(n+1)}(p) \geq \zeta_{H_1}^{(n)}(p)$ if $\zeta_1(p) = \alpha p$ with $\alpha > 0$: the two bounds are improved by iteration of the intersection of random sets with different scales. A limiting case is obtained by iterations for $n \to \infty$: we get $\zeta_{H_1}^{(n)}(p) \to \dfrac{1+p}{2}\zeta_1(1)$. If the basic structure is a mosaic model with $a = 0$, or the complementary set of a Boolean model of spheres in R^3 or of discs in R^2 we obtain $\zeta_{H_1}^{(n)}(p) \to \dfrac{1+p}{2}$. This corresponds to a Boolean model of spheres with an infinite range of widely separate sizes.

5. Conclusion

Combining basic models of random structures can provide different bounds of the effective properties, even when considering the third-order bounds.

The derivations presented in this paper were limited to the scalar case. Their extension to the elasticity case can be made by similar derivations, following the developments given by Milton [16], as was made for some basic models [20,21].

The models introduced in this paper have a multiphase and even a continuous version (scalar or multivariate) [6,7], for which the calculation of third-order bounds can be made using the general derivation based on the third-order correlation function [2,15,4].

References

1. G. Babos and D. Chassapis, *J. Phys. Chem. Solids* **51** (1990) 209.
2. M. J. Beran, *Statistical Continuum Theories*, J. Wiley, New York, 1968.
3. Z. Hashin and S. Shtrikman, *J. Appl. Phys.* **33** (1962) 3125.
4. M. Hori, *J. Math. Phys.* **14** (1973) 1942.
5. D. Jeulin, In *Continuum Models of Discrete Systems*, ed. A. J. M. Spencer, Balkema, Rotterdam, 1987, p. 217.
6. D. Jeulin, *Modèles morphologiques de structures aléatoires et de changement d'échelle*, Thèse de Doctorat d'Etat, University of Caen, 1991.
7. D. Jeulin, *Morphological Models of Random Structures*, CRC press, in preparation.
8. C. Joslin and G. Stell, *J. Appl. Phys.* **60** (1986) 1607.
9. E. Kröner, *Statistical Continuum Mechanics*, Springer-Verlag, Berlin, 1971.
10. A. Le Coënt, *Observations et modélisation statistique de matériaux composites à agrégats*, PhD thesis, Paris School of Mines, March 1995.
11. A. Le Coënt and D. Jeulin, Bounds of effective physical properties for random polygons composites, in preparation.
12. G. Matheron, *Eléments pour une théorie des milieux poreux*, Paris, 1967.
13. G. Matheron, *Rev. IFP* **23** (1968) 201.
14. G. Matheron, *Random Sets and Integral Geometry*, J. Wiley, New York, 1975.
15. M. Miller, *J. Math. Phys.* **10** (1969) 1988.
16. G. Milton, *J. Mech. Phys. Solids* **30** (1982) 177.
17. J. L. Quenec'h, J. L. Chermant, M. Coster and D. Jeulin, In *Mathematical Morphology and its Applications to Image Processing*, eds. J. Serra and P. Soille, Kluwer Academic Pub., Dordrecht, 1994, p. 225.
18. J. Serra, *Image Analysis and Mathematical Morphology*, Academic Press, London, 1982.
19. S. Torquato and G. Stell *J. Chem. Phys.* **79** (1983) 1505.
20. S. Torquato and F. Lado, *Phys. Rev. B* **33** (1986) 6428.
21. S. Torquato, *Appl. Mech. Rev.* **44** (1991) 37.

Continuum Models and Discrete Systems
Proceedings of 8th International Symposium, June 11-16, 1995, Varna, Bulgaria, ed. K.Z. Markov

SOLIDS WITH CAVITIES OF VARIOUS SHAPES: EFFECTIVE MODULI AND PROPER DENSITY PARAMETERS

M. L. KACHANOV
Department of Mechanical Engineering, Tufts University, Medford, MA 02155, USA

Abstract

Effective elastic properties of solids with cavities of various shapes are derived in two approximations: the approximation of non-interacting cavities (basic building block for various effective media theories) and Mori-Tanaka's scheme (that appears appropriate if the mutual positions of cavities are random).

The structure of the elastic potential of a solid with cavities dictates the proper parameters of cavity density. Such an approach covers cracks as a special case in which no degeneracies arise. It also covers mixtures of cavities of diverse shapes.

Continuum Models and Discrete Systems
Proceedings of 8th International Symposium, June 11-16, 1995, Varna, Bulgaria, ed. K.Z. Markov

ON THE ESSENTIAL IMPORTANCE OF DISTINGUISHING BETWEEN THE EFFECTS OF DEVIATORIC AND ISOTROPIC PARTS OF STRESS TENSORS IN THE PROBLEM OF STRENGTH OF DISCRETE MATERIALS

V. KAFKA
Institute of Theoretical and Applied Mechanics of the Academy of Sciences of the Czech Republic,Prosecká 74, 190 00 Praha 9, Czech Republic

Abstract. The paper presents explanation and description of some seemingly paradoxical phenomena using the simple tools of classical mechanics. It is shown that making full use of splitting the stress tensors into their deviatoric and isotropic parts offers interesting possibilities of interpreting - on a unified theoretical basis - tearing under uniaxial compression of very different materials, their only common feature being their discrete structure.

1. Introduction

It is well known that the classical approach of elasticity working with mascroscopic stress fields does not give satisfactory results in many problems of strength. This gave rise to fracture mechanics that successfully solved a number of questions, but is very demanding as to the input information. The aim of my communication is to show that some problems of strength can be clarified and described even with the simple classical approach if the splitting of stress tensors into their isotropic and deviatoric parts is made full use of.

The discussion is limited to cases with homogeneous macroscopic stress fields and with negative (compressive) isotropic parts of the stress tensors. This is the very area where some experimental findings seem to be paradoxical.

2. Failure of brittle bodies under compressive loading

One of the apparent paradoxes in mechanics is the type of failure of brittle materials under the following stressing:

$$\sigma_{xx} = \sigma_{yy} < 0, \qquad \sigma_{zz} = 0 \tag{1}$$

Experiments of this kind were performed and described by Bridgman [1]. The scheme of such an experiment is shown in Fig. 1.

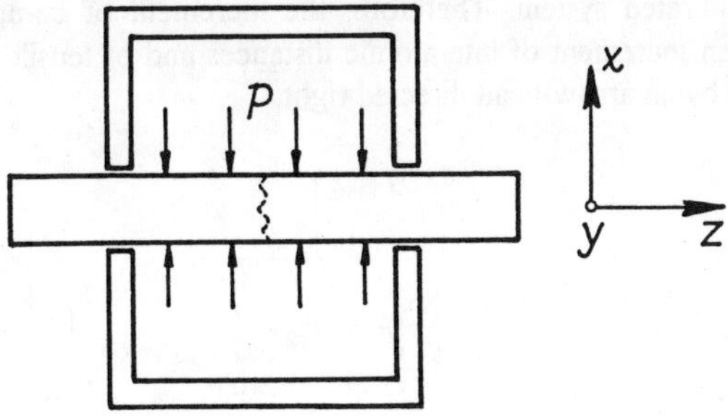

Fig. 1. Bridgman's experiment: sample of a brittle material with free ends in a chamber with high pressure of liquid p

It was realized that failure proceeds exactly in the same way as under the stressing

$$\sigma_{xx} = \sigma_{yy} = 0, \qquad \sigma_{zz} > 0, \tag{2}$$

i.e. as in the case of simple tension (even with the phenomenon that the two separated parts spring off each other). The type of failure has all the symptoms of pulling-apart, but there does not exist any tensile stress. Our common experience influences us to expect that there should exist some tensile stress acting on the plane of separation (normal to the z-axis) and as there is none, this seems to be a paradox.

If we want to grasp this mode of failure it is useful to analyse it at first on the microscale. Macroscopic stresses mean some statistical representation of interatomic forces and these forces depend on the interatomic distances in the way that is demonstrated in Fig.2a.

The process corresponding to stressing of type (1) is schematically shown in two dimensions in Fig. 2b. The initial self-equilibrated interatomic forces (without external loading) between the atomic couples β–β and β–α are illustrated in Fig. 2a. Compressive loading p brings nearer firstly the atomic couples that have the same coordinate z, which, however, is irrelevalent for our discussion. But it brings nearer also the couples β–α, which causes an increase of compressive forces between them. In Fig. 2a this is illustrated by an arrow-head directed left. The pivotal point of our problem are the interatomic distances and forces between the couples β–β. The ends of the sample (cf. Fig. 1) are free of constraint and free of any loading and therefore the forces α–β and β–β crossing any plane z = const.

must form an equilibrated system. Therefore, the increment of compressive interatomic forces α–β causes an increment of interatomic distances and of tensile forces β–β. In Fig. 2a this is illustrated by an arrow-head directed right.

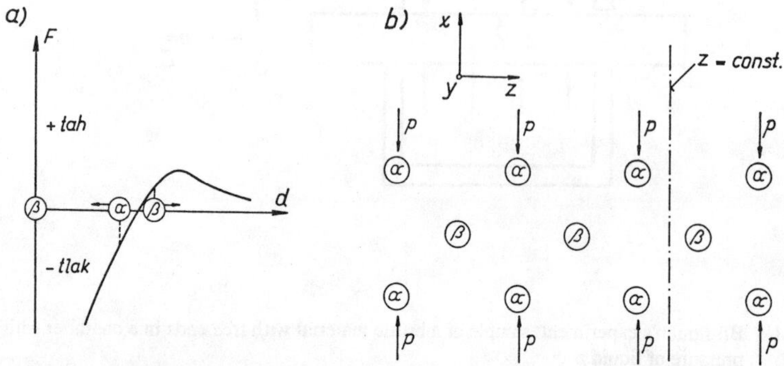

Fig. 2. a) Dependence of the interatomic force ***F*** on the interatomic distance ***d***
b) Scheme of the atomic configuration for the discussion concerning the changes of interatomic distances and forces

It is clear from Fig. 2a that the compressive forces β–α can increase practically without limits. On the other hand the tensile forces β–β can increase to some limit only and then they decrease with their distance increasing . If this limit is reached, the compressive forces β–α prevail and in the weakest spot the sample fails and its separated parts spring off. It is clear enough that our reasoning is not bound to the special demonstrative scheme used in Fig. 2b and to two dimensions only.

It is one thing to seize the problem on the microscale and it is another thing to model it on the macroscale. On the microscale there appear tensile interatomic forces (vectors), but on the macroscale there does not appear any tensile stress, the scheme is not vectorial, but tensorial and if we want to model the process, it is inevitable to split the respective tensor into its deviatoric and isotropic parts and to realize that *the main role is played by the deviatoric part.* This statement can easily be substantiated. Let us imagine that the compressive isotropic part disappears: With only the deviatoric part in action the failure will proceed exactly in the same way and more easily (with lower absolute values of the deviatoric stress components).Now let us imagine that the deviatoric part disappears: With only the compressive isotropic part in action there will arise no failure.

Our statement fully agrees with the observation that with the stressing of type (1) the failure proceeds in exactly the same way as with the stressing of type (2). In both these cases the type of the deviatoric matrix is namely the same:

$$\begin{vmatrix} -a/2 & 0 & 0 \\ 0 & -a/2 & 0 \\ 0 & 0 & a \end{vmatrix}$$

with a > 0.

The isotropic part influences only the absolute values of the deviatoric stress components.

Of course that in real materials the situation is not so simple as in the ideal scheme of Fig. 2b. There exist defects of different sorts and the theoretical strength that would correspond to perfect crystals is substantially reduced by their influence. Different approaches of fracture mechanics study the processes caused by these defects (especially of cracks), but our point is that even without going into such sophisticated analysis it is possible to explain and describe such phenomena as e.g. the Bridgman experiment, remaining on the macroscale. Of course that it is necessary to take into account the influence of the defects existing in the material in question, which means reduction of the strength parameters.

In the third part of our monograph [2] a simple strength criterion is presented that takes into account the non-local effect of the stress field and the quality of the surface layer. The main feature of the criterion is that it distinquishes between the areas from which the elastic energy of the deviatoric and the isotropic parts of stress tensors is supplied for creating failure. The criterion has meaning especially for heterogeneous stress fields and for thin bodies; for homogeneous stress fields and large bodies the criterion reduces to a quite simple form:.

$$as_{ij}s_{ij} + b\sigma^2 \leq C^2 \tag{3}$$

where s_{ij}, $\delta_{ij}\sigma$ are the deviatoric and isotropic parts of the stress tensor respectively, *b* is negative for isotropic compressive loading, *a* is positive. This agrees with the conclusions of the micromechanical analysis discussed above, with the fact that the deviatoric part must be considered separately from the isotropic part and with the phenomenon that tearing of brittle discrete materials appears without tensile stress.

3. Cracks in articular cartilage

The problem of cracks in articular cartilage - cracks that are normal to its surface - calls also for an analysis connecting the micromechanical processes with the possibilities of macroscopic modelling. Such cracks - together with splitting at the interface between normal and calcified cartilage ("tide mark") - have been observed in joints with primary osteoorthritis [3]. Whereas the splitting at the tide mark was easy to explain by a "coated elastic sphere model" [4], the reasons for the appearance of the cracks normal to the surface are not so easy to find. The "coated elastic sphere model" gave no such reasons. Hlaváček [5] discussed the problem from the point of view of interstitial fluid motion and its possible influencing upon propagation of an already existing crack. He came to the conclusion that this effect "could lead to the matrix failure at the crack edge".

Nevertheless, it seems that the pivotal point of the problem consists in the initial creation of the cracks. The situation is different from that of technical materials, where small cracks exist a priori due to solidification, working, machining and corrosion and the heart of the problem is under what conditions they propagate. None of the factors enumerated above exists in articular cartilage and the heart of the problem is how the cracks originate.

Our model that is able to explain and describe it, is close to reality considering cartilage as a three-phase medium. One phase (elastic) corresponds to collagen fibres, the second phase (elastic, but substantially more compliant than the fibres) corresponds to the matrix in which the fibres are embedded and the third phase (viscous) corresponds to the thin constituent of synovial fluid that infiltrates the cartilage. All the phases are considered incompressible, but due to the filtration of the viscous liquid out of cartilage under the action of pressure the first and the second phases together (phase "non-fibres") behave generally as quasi-compressible. However, there are two singular cases in which the phase "non-fibres" behaves as incompressible:

(i) Under a quick impact loading, where the filtration of the liquid out of cartilage does not take place because there is no time for it.

(ii) Under a heavy long-lasting static loading, where the filtration of the liquid out of cartilage does not take place any more, because all the liquid is gone.

From these two singular cases the first one is probably more dangerous as the impact loading leads to higher stresses.

Close to the surface of articular cartilage the fibres are parallel with the surface and aligned in one direction (corresponding to the direction of normal joint movement) [6]. Then the deformation of the superficial zone under compressive loading can be demonstrated by the scheme shown in Fig. 3.

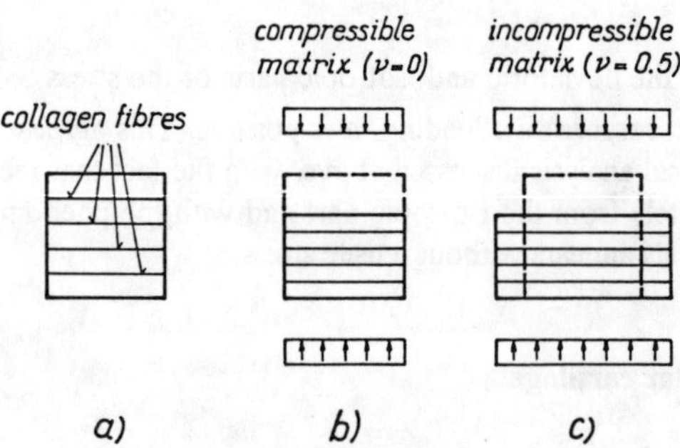

Fig. 3. Demonstrative scheme of the influence of compressibility and incompressibility upon the tensile stress in collagen fibres

The respective mathematical model based on the general author's concept [2] can easily be formulated proceeding similarly as in the case of the model of tendon [7]. However, even in the simple variant assuming the surface zone of cartilage to be transversely isotropic with the axis of symmetry parallel with the fibres, this three-phase model is rather demanding on the experimental data that are not available so far. Therefore, at present - having in mind the real scheme - we are going to limit ourselves to a simpler model that retains the main characteristics: to a two-phase elastic transversely isotropic

body. Its axis of symmetry x is parallel with the fibres, direction z is normal to the surface and parallel with
the compressive loading, direction y is parallel with the surface. One phase is formed by the fibres (f), the other (n) is the "non-fibres" phase (matrix infiltrated with the thin constituent of the synovial fluid). The volume fraction and Young's moduli are V_f, V_n, E_f, E_n respectively, Poisson's ratio ν is assumed equal for both phases. Another simplification is that deformation in the direction of fibres is homogeneous:

$$\varepsilon_{xxf} = \varepsilon_{xxn} = \bar{\varepsilon}_{xx} \tag{4}$$

and normal stress in the other two directions is homogeneous:

$$\sigma_{zzf} = \sigma_{zzn} = \bar{\sigma}_{zz} = p \tag{5}$$

$$\sigma_{yyf} = \sigma_{yyn} = \bar{\sigma}_{yy} = 0 \tag{6}$$

Then it is straightforward to deduce the expression for the resulting tension in the fibres:

$$\sigma_{xxf} = -\nu V_n \frac{E_f - E_n}{V_n E_n + V_f E_f} p = -\nu V_n \frac{\hat{e} - 1}{V_n + V_f \hat{e}} p \tag{7}$$

where

$$\hat{e} = \frac{E_f}{E_n} \tag{8}$$

This expression - in spite of the severe simplification - shows very clearly the main features of the problem:

(i) For the fictitious case $E_f = E_n$ (homogeneity) the tensile stress in the fibres disappears. The existence of this tension is bound to heterogeneity.

(ii) For the case of incompressibility (ν = 0.5) the tension reaches its maximum with regard to the possible variations of ν (cf. Fig. 3c).

(iii) For $\nu = 0$ the tension disappears (maximum compressibility - cf. Fig. 3b)

Now let us try to make a quantitative estimate of the maximum value of σ_{xxf}. Not knowing the exact values of V_f, V_n in the superficial zone of cartilage let us assume that they are near to the values found for caartilage in [8], i.e. V_f= 0.3, V_n= 0.7.

The value of $\hat{e}$ is estimated to be about 1000. This follows from the Young modulus of the matrix that is - according to [8] - $E_m \doteq 0.8$ MPa and the Young modulus of collagen $E_f \doteq 1000$ MPa - given in [10]. This relation of the moduli $(\hat{e} \doteq 1000)$ agrees with such a relation found in [7] for the human tendon of musculus flexor hallucis longus and with this value found for the rat-tail tendon in [11]. The value of E_f given in [8] is substantially lower than 1000 MPa, but the reason is probably that it corresponds to a slow quasi-static process in which straightening of fibres influences the value and makes it lower.

To estimate the maximum value of p let us state that the authors of [9] measured by a telemetric device the maximum dynamic force in hip joint in the value of 3.2 x body weight in normal walking. In the case of running or jumping it is of course substantially higher. Let its maximum value be estimated 4000 N and the resolved contact area on the femoral head (assuming homogeneous stress distribution) about 100 mm^2. With these estimates it results:

$$p = -40 \text{ MPa}$$

This value of maximum dynamic stress is substantially higher than the static ultimate compressive stress measured on samples of cartilage in mortuo (cf. e.g. [6]) but it is not contradictory; here it is instantaneous dynamic stress and the cartilage is filled with the thin constituent of synovial fluid which strengthens it.

Then for σ_{xxf} it turns out:

$$\sigma_{xxf} = +47 \text{ MPa}$$

The tensile strength of collagen fibres is - according to [10] - from 50 to 100 MPa and hence, this mechanism can really lead to tearing of the collagen fibres and in this way to superficial cracks.

But let us turn again to the question whether this process can be modelled on the macroscale. The answer is affirmative if the stress tensor is split into its deviatoric and isotropic parts. It is easy do deduce that the deviatoric part gives:

$$\left(\sigma_{xxf}\right)_d = -\frac{1}{3} \cdot \frac{\nu V_n(\hat{e}-1)+\hat{e}}{V_n+V_f\hat{e}} p$$

and the isotropic part:

$$\left(\sigma_{xxf}\right)_i = -\frac{1}{3} \cdot \frac{2\nu V_n(\hat{e}-1)-\hat{e}}{V_n+V_f\hat{e}} p$$

For our estimate of the maximum dynamic values it leads to:

$$\left(\sigma_{xxf}\right)_d \doteq 60 \text{ MPa}$$

$$\left(\sigma_{xxf}\right)_i \doteq -13 \text{ MPa}$$

This means that it is only the deviatoric part that is responsible for the tensile stress in the fibres and if it is dealt with separately from the isotropic part - with different and higher weight - the problem in question can be modelled on the macroscale omitting the micromechanical analysis.

Conclusion

The current way of dealing with the whole stress tensors is limiting and sometimes even misleading. The deviatoric and the isotropic parts of the stress tensors must be dealt with as two different physical factors with different weights and if doing so some problems can be clarified and modelled on the macroscale (without any sofisticated micromechanical analysis), problems that otherwise seem paradoxical and out of the reach of the macroscopic analysis.

References

1. P. W. Bridgman, *Collected Experimental Papers*, Harward University Press - Cambridge, Massachusetts, 1964.
2. V. Kafka, *Inelastic Mesomechanics*, World Scientific Publ. Co. - Singapore, New Jersey, London, Hong Kong, 1987.
3. G. Meachim and G. Bentley, *Arthritis and Rheumatism* **21** (1978)669.
4. A. W. Eberhard, J. L. Lewis and L. M. Keer, *J. Biomech. Engng* **113** (1991)410.
5. M.Hlaváček, In Proc. Conf. *Biomechanics of Man '94*, Benešov, 1994, p. 33.
6. N. G. Shrive and C. B. Frank, In *Biomechanics of the Musculo-skeletal System*, eds. B. M. Nigg and W. Herzog, Wiley - Chichester, New York, Brisbane, Toronto, Singapore, 1994, p.79.
7. V. Kafka, J. Jírová and V. Smetana, *Clinical Biomechanics* **10** (1995)50.
8. M.H. Schwartz, P.H. Leo and J.L. Lewis, *J. Biomechanics* **27** (1994) 865.
9. F. Greichen and G. Bergmann, *Biomed. Technik* **33** (1988)305.
10. Y. C. Fung, *Biomechanics - Mechanical Properties of Living Tissues*, Springer - Verlag, New York - Heidelberg - Berlin, 1981, p.202.
11. H.K. Ault and A.H. Hoffman, *J. Biomech. Engng* **114** (1992) 137.

Continuum Models and Discrete Systems
Proceedings of 8th International Symposium, June 11-16, 1995, Varna, Bulgaria, ed. K.Z. Markov
©World Scientific Publishing Company, 1996, pp. 216–224

SELF-CONSISTENT APPROACH IN THE PROBLEM OF WAVE PROPAGATION THROUGH INHOMOGENEOUS MEDIUM

S. K. KANAUN
División de Graduados e Investigación, ITESM, Campus Edo. de México,
Apdo. Postal 18, Atizapan, Edo. de México, 52926, México

Abstract. Method of the effective medium is applied to calculation of velocities and attenuation factors of elastic waves in polycrystalline materials and scalar waves in homogeneous media containing sets of isolated inclusions. The area of application of the method is analyzed on the base of obtained numerical results.

1. General Consideration

The problem of propagation of monochromatic waves through an inhomogeneous medium can be reduced to the solution of the integral equation of the following type

$$u(x) = u_0(x) - (K_0C_1u)(x), \tag{1}$$

see Ref. 1; Here $u(x)$ is the amplitude of the wave field, $x(x_1, x_2, x_3)$ is a point of the medium, $C_1(x)$ is fluctuation of a dynamic characteristic of the medium

$$C_1(x) = C(x) - C_0, \tag{2}$$

$C(x)$ is such a characteristic, C_0 is a characteristic of a certain homogeneous medium, K_0 is an integral operator whose kernel is defined through the Green's function of the medium with the property C_0. In Eq. (1) u_0 is an exitating field which could have existed in the medium if $C_1 = 0$ under the same conditions at infinity.

The so-called method of the effective medium is used for the construction of the first statistical moment of the solution of Eq. (1) (the mean wave field in the medium). The method can be applied if it is possible to distinguish a typical particle in the given inhomogeneous medium. It can be a grain of polycrystalline material, an inclusion in a matrix composite, etc.

The main hypothesis of the method can be formulated as following (see [1,2] and a number of references there).

Every particle in inhomogeneous medium behaves as an isolated one embedded in a homogeneous medium with the effective properties of the considered material. Conditions at infinity coincide with the originally given ones.

Thus these assumptions reduce the problem of interactions between many particles in the inhomogeneous material to the one-particle problem. Integral equation of the latter takes form similar to Eq. (1)

$$u(x) = u_*(x) - (K_* C_{*1} u)(x), \quad C_{*1}(x) = C(x) - C_* . \tag{3}$$

Here C_* is an effective characteristic of the inhomogeneous medium, the operator K_* is defined through Green's function of the medium with the properties C_*, $C_{1*}(x) = 0$, if x is outside v, where v is the region occupied by a typical inclusion.

Let us suppose that the general solution of the integral equation (3) is known and has the form

$$u(x) = (\Lambda u_*)(x), \quad \Lambda = \Lambda(C_*, C, a), \tag{4}$$

where Λ is a certain operator that depends on the property C of the inclusion, property C_* of the effective medium and on certain parameters a that specify the shape of the particle.

Let us consider now a finite volume V of the inhomogeneous medium embedded in the homogeneous medium with effective properties C_*. If we assume that inside every particle the wave field has the form of Eq. (4), we can represent the wave field similarly to Eq. (3), namely,

$$u(x) = u_*(x) - [K_* C_{*1} \Lambda(C_*, C, a) u_*](x). \tag{5}$$

The second term in the right side of this equation is a field scattered on the volume V of the inhomogeneous medium. The condition of self-consistency in the framework of the effective field method can be formulated in the following way [2].

The characteristic C_* should be as chosen so that the mean scattered field from the volume V vanishes.

After averaging both sides of Eq. (5) over the ensemble of realization of inhomogeneities in the volume V, we get

$$\langle u(x) \rangle = u_*(x) - (K_* \langle C_{*1} \Lambda(C_*, C, a) \rangle u_*)(x) = u_*(x). \tag{6}$$

Thus the equation for the characteristic C_* which is a consequence of Eq. (6) takes the form

$$\langle C_{*1}(x) \Lambda(C_*, C, a) \rangle = \langle (C(x) - C_*) \Lambda(C_*, C, a) \rangle = 0. \tag{7}$$

Having solved this equation, one can find the effective dynamic characteristics of the inhomogeneous medium and then calculate the velocities and attenuation factors of the mean wave field in it.

2. Elastic Waves in Polycrystalline Materials

Let us consider a statistically homogeneous polycrystalline material composed of a set of grains ideally bonded along the interfaces. Every grain is a homogeneous monocrystal with random orientation of crystallographic axes. The grains have quasispherical shapes of random radii a. If $C(x)$ is the tensor of elastic moduli of the polycrystal amplitude of strain tensor ε in the medium satisfies an equation similar to Eq. (1) [2]

$$\varepsilon_{ij}(x) = \varepsilon_{*ij}(x) - \int K_{*ijkl}(x-x')C_{*1klmn}(x')\varepsilon_{mn}(x')\,dx',$$

$$K_{*ijkl}(x) = -\left[\nabla_i \nabla_k G_{*jl}\right]_{(ij)(kl)}, \quad \varepsilon_{ij}(x) = \nabla_{(i}u_{j)}, \tag{8}$$

where $u_i(x)$ is the displacement vector, $G_*(x)$ is Green's tensor of the effective medium by which the given composite material is replaced.

The first step when applying the effective medium method to the considered problem is to solve the one-particle problem. It is a problem of diffraction of elastic waves on a typical grain embedded into a homogeneous medium with the effective elastic properties of the polycrystal. The equation of this problem has the form

$$\varepsilon_{ij}(x) = \varepsilon_{*ij}(x) - \int_v K_{*ijkl}(x-x')C_{*1klmn}\varepsilon_{mn}(x')\,dx'. \tag{9}$$

Here we integrate over the volume v of a typical inclusion, the tensor C_1 is constant inside v.

In order to construct an approximate solution of the one-particle problem for an anisotropic grain, let us assume that for plain excitation field $\varepsilon_*(x)$ with a wave vector q_*

$$\varepsilon_*(x) = D_* e^{-iq_*\cdot x}, \tag{10}$$

(D_* is a constant second-rank symmetric tensor), the wave field inside a typical grain (i.e., the solution of Eq. (9)) is also a plane wave with the same wave vector q_*

$$\varepsilon(x) = De^{-iq_*\cdot x}. \tag{11}$$

In order to obtain a constant tensor D, let us substitute this expression for $\varepsilon(x)$ in Eq. (9) and average both parts of the latter over the volume v of the inclusion and over all possible directions of the wave vector q_*. As a result a linear equation for the tensor D is obtained whose solution has the form

$$D = \Lambda D_*, \quad \Lambda = [I + K(q_*)C_1]^{-1},$$

$$K(q_*) = \frac{1}{4Piv}\int_\Omega \int_v \int_v K(x-x')\, e^{iq_* m\cdot(x-x')}\, dx'dxdm, \quad q_{*i} = q_* m_i. \tag{12}$$

Here I is the fourth rank unit tensor, Ω is the unit sphere, $m \in \Omega$, i.e., $|m| = 1$. The integral in Eq. (12) can be represented as

$$K_{ijkl}(q_*) = K(q_*)\delta_{ij}\delta_{kl} + 2M(q_*)\left(\delta_{i(k}\delta_{j)l} - \frac{1}{3}\delta_{ij}\delta_{kl}\right), \tag{13}$$

where $K(q_*)$ and $M(q_*)$ are scalar integrals which can be calculated explicitly. Their values will be different for longitudinal and transversal waves. Therefore the effective dynamical tensor C_* will be different for different types of wave propagating through the polycrystalline material. The dispersion equation for the effective wave vectors q_* in the polycrystal takes the form

$$C_{*ijkl}(q_*)q_{*i}q_{*k} - \rho\omega^2\delta_{jl} = 0\,. \tag{14}$$

Here ρ is the density of the polycrystal, ω is the frequency of oscillations. The real part of the vector q_* defines the phase velocity v_* of the mean wave field in the polycrystal ($v_* = \omega/\mathrm{Re}\, q_*$) and the imaginary part of q_* ($\gamma = -\mathrm{Im}\, q_*$) represents the attenuation factor of the corresponding wave.

The results of velocities and attenuation factors calculation for elastic waves in the polycrystal of a stainless steel are presented in Fig. 1.

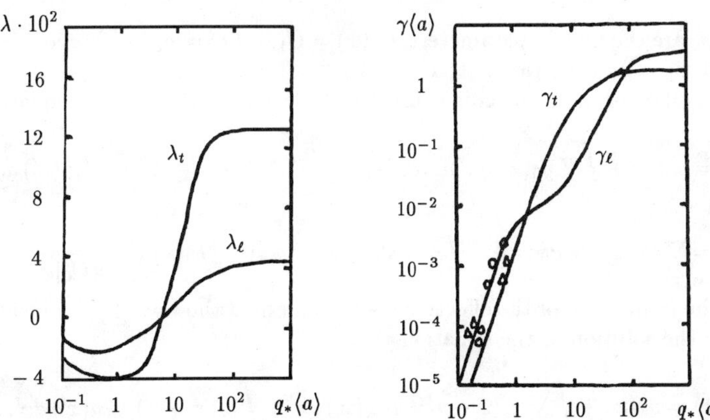

Fig. 1. The velocities and attenuation factors for elastic waves in a stainless steel polycrystal, see the text.

The parameters λ_l, λ_t in Fig. 1a are defined as

$$\lambda_l = \frac{v_{*l} - v_{0l}}{v_{0l}}\,, \quad \lambda_t = \frac{v_{*t} - v_{0t}}{v_{0t}}\,, \tag{15}$$

where v_{0l} and v_{0t} are Voigt's velocities of longitudinal and transversal waves [2].

Fig. 1b shows the dependence of nondimensional attenuation factors $\gamma_l\langle a\rangle$ and $\gamma_t\langle a\rangle$ of longitudinal and transversal waves on the real part of the corresponding nondimensional wave number $q_*\langle a\rangle$, with $\langle a\rangle$ denoting the average radius of the grains.

Circles and triangles on Fig. 1b are the experimental data of attenuation factors of longitudinal ($\circ$) and transversal ($\triangle$) waves given in [3].

3. Propagation of Scalar Waves through Composite Materials

In the solution of the wave problem for polycrystals we had to accept two types of hypotheses. The first ones were the hypotheses of the effective medium method itself. The second ones were connected with the solution of the one-particle problem. In order to analyze the error of the latter we consider here the problem of scalar wave propagation through a homogeneous medium containing an array of spherical homogeneous inclusions (matrix composite). In this case one can built up an exact solution of the one-particle problem and therefore there is a possibility to analyze the influence of the proposed approximations on the final results

In the case of matrix composites the method of effective medium needs certain modifications. The differential equation of the scalar problem for such a medium has the form

$$\nabla_i C(x)\nabla_i u(x) + \rho(x)\omega^2 u(x) = 0,$$

$$C(x) = C_0 + C_1(x), \quad \rho(x) = \rho_0 + \rho_1(x), \tag{16}$$

where C_0, ρ_0 are the matrix parameters, $C_1(x) = C_1$, $\rho_1(x) = \rho_1$ inside every inclusion and $C_1 = 0$, $\rho_1 = 0$ inside the matrix.

This differential equation is equivalent to the following integral equation [1]

$$u(x) = u_0(x) + \int \nabla_i g_0(x - x')C_1(x')\varepsilon_i(x')\,dx' + \int g_0(x - x')\rho_1(x')u(x')\,dx',$$

$$\varepsilon_i(x) = \nabla_i(x), \quad L_0 u_0 = 0, \quad L_0 = \nabla_i C_0 \nabla_i + \rho_0\omega^2, \quad g_0(x) = \frac{1}{4\pi C_0|x|}\,. \tag{17}$$

In the framework of the effective medium method the wave field u inside every inclusion is the solution of the equation

$$u(x) = u_*(x) + \int_v \nabla_i g_*(x - x')C_{*1}\varepsilon_i(x')\,dx' + \int g_*(x - x')\rho_{*1}u(x')\,dx',$$

$$C_{*1} = C_0 + C_1 - C_*, \quad \rho_{*1} = \rho_0 + \rho_1 - \rho_*\,. \tag{18}$$

Here C_*, ρ_* are the parameters of the effective medium, v is the area occupied by the inclusion and $u_*(x)$ is the wave field in the effective medium.

Let $u_0(x)$ be a plane wave. The condition of self-consistency can be formulated in this case as

$$u_*(x) = \langle u(x)\rangle = e^{-iq_*\cdot x}, \tag{19}$$

where q_* is the wave vector of the effective medium.

Due to the linearity of the problem, the field $u(x)$ inside the inclusion can be represented in the form $u(x) = \Lambda[e^{-iq_*\cdot x}]$, where Λ is a certain linear operator. That is why the field inside the inclusion with the center at a point x_0 takes form

$$u(x) = \Lambda[e^{-iq_*\cdot(x-x_0)}]e^{-iq_*\cdot x_0} = \Lambda_u(x - x_0)e^{-iq_*\cdot x} = \Lambda_u(x - x_0)\langle u(x)\rangle,$$

$$\varepsilon(x) = \Lambda_\varepsilon(x - x_0)\langle u(x)\rangle, \quad \Lambda_u(x - x_0) = \Lambda[e^{-iq_*\cdot(x-x_0}]e^{-iq_*\cdot(x-x_0)}. \tag{20}$$

Note that the functions $\Lambda_u(x)$, $\Lambda_\varepsilon(x) = \nabla\Lambda_u(x)$ can be found upon solving the problem for an inclusion centered at the point $x_0 = 0$.

Let us substitute the obtained expression for the fields inside inclusions into the right side of Eq. (17) and average the result over the ensemble of inclusions. As a result we get

$$\langle u(x)\rangle = u_0(x) + p\int \nabla_i g_0(x - x')C_1(-iq_{*i})Hc(q_*)\langle u(x')\rangle\, dx'$$

$$+p\int g_0(x - x')\rho_1 Hr(q_*)\langle u(x')\rangle\, dx',$$

$$Hr(q_*) = \frac{1}{v}\int \Lambda_u(x)\, dx, \quad -iq_* Hc(q_*) = \frac{1}{v}\int \Lambda_\varepsilon(x)\, dx, \tag{21}$$

where p is the volume concentration of inclusions.

After applying the operator L_0 on both sides of this equation it is possible to demonstrate that $\langle u(x)\rangle$ is a plain wave with wave number q_* which is the solution of the following dispersion equation

$$q_*^2 C_*(q_*) - \rho_*(q_*)\omega^2 = 0,$$

$$C_*(q_*) = C_0 + pC_1 Hc(q_*), \quad \rho_*(q_*) = \rho_0 + p\rho_1 Hr(q_*). \tag{22}$$

Using the known exact solution of the one-particle problem [4] one can obtain the expressions for Hr and Hc in the form

$$Hr(q_*) = \sum_{m=0}^{\infty} A_m(q_*)g_m(q_*), \quad Hc(q_*) = \sum_{m=0}^{\infty} A_m(q_*)g_{1m}(q_*),$$

$$A_m(q_*) = \frac{(-i)(2m+1)C_*}{q_*[C_* q_* j_m(q)Dh_m^{(2)}(q_*) - CqDj_m(q)h_m^{(2)}(q^*)]}, \quad q^2 = \omega^2\frac{\rho}{C},$$

$$\rho = \rho_0 + \rho_1, \quad C = C_0 + C_1, \quad g_m(q_*) = \frac{3}{q - q_*}[qj_{m+1}(q)j_m(q_*) - q_* j_{m+1}(q_*)j_m(q)],$$

$$g_{1m}(q_*) = g_m(q_*) + \frac{3}{q_*}j_m(q)Dj_m(q_*), \quad Df(x) = \frac{d}{dx}f(x). \tag{23}$$

Here $j_m(x)$ and $h_m^{(2)}(x)$ are respectively the spherical Bessel function and spherical Bessel function of the second rank of order m.

An approximate solution of the one-particle problem similar to that already utilized in Section 2 can be obtained after substituting the following expressions for the fields inside the inclusion

$$u(x) = Ue^{-iq_*\cdot x}, \quad \varepsilon(x) = \Sigma e^{-iq_*\cdot x}, \tag{24}$$

in Eq. (18) and in the equation for $\varepsilon(x)$ which is a consequence of Eq. (18)

$$\varepsilon_i(x) = \varepsilon_{*i}(x) + \int_v \nabla_i\nabla_j g_*(x-x')C_{*1}\varepsilon_j(x')\,dx' + \int \nabla_i g_*(x-x')\rho_{*1}u(x')\,dx'.$$

Averaging these equations over the volume v of the inclusion, we get a liner system for the unknowns U and Σ. The solution of this system gives approximate expressions for $Hr(q_*)$ and $Hc(q_*)$, namely,

$$Hr(t) = \frac{De(t) - t^2c_1Q(t)}{\Delta(t)}, \quad Hc(t) = \frac{de(t) - t^2r_1Q(t)}{\Delta(t)},$$

$$D(t) = 1 - c_1K(t), \quad de(t) = 1 - t^2r_1q(t), \quad c_1 = C_1/C_0, \quad r_1 = \rho_1/\rho_0,$$
$$\Delta(t) = De(t)de(t) + t^4c_1r_1Q^2(t). \tag{25}$$

Here $q(t)$, $Q(t)$ and $K(t)$ are the following integrals

$$q(t) = \int_0^2 e^{-itz}f(z)j_0(tz)z\,dz, \quad Q(t) = -\int_0^2 e^{-itz}[Df(z)j_1(tz) + tf(z)j_0(tz)]z\,dz,$$

$$K(t) = \int_0^2 e^{-itz}\big[(Df(z) - zf(z)t^2)j_0(tz) + (zDDf(z) - Df(z)$$
$$-2z^2t^2Df(z))\frac{j_1(tz)}{tz} - (zDDf(z) - Df(z))j_2(tz)\big]\,dz, \quad f(z) = 1 - 3z/4 + z^3/16. \tag{26}$$

Note that these integrals can be explicitly evaluated.

The dependence of the velocities and attenuation factors of the mean wave field on the effective wave number for porous medium is represented in Fig. 2 at $C_0 = 1$, $\rho_0 = 1$, $C = 0$, $\rho = 0$ and $a = 1$. The solid lines correspond to the exact solution of the one-particle problem, the dashed lines are drawn on the base of the obtained approximate solution.

The same dependences for the rigid inclusions are represented in Fig. 3 at $C_0 = 1$, $\rho_0 = 1$, $C = 10^3$, $\rho = 10$ and $a = 1$.

4. Conclusions

Let us formulate here some consequences of the obtained results.

1. Application of the effective medium method to the problem of monochromatic wave propagation through polycrystalline materials clearly shows that the method allows to describe all important features of the considered phenomena. It is known [3] that the dependence of attenuation factors γ on the frequency of elastic waves has three typical areas: Raleigh or long wavelength area where $\gamma \sim \omega^4$, stochastic area where $\gamma \sim \omega^2$ and diffusive area where γ =const. All these areas can be seen in Fig. 1b, where logarithmic scale is used along both axes and $q_* \sim \omega$. In the Raleigh area the method leads to a good coincidence with the experimental data (see also [4]).

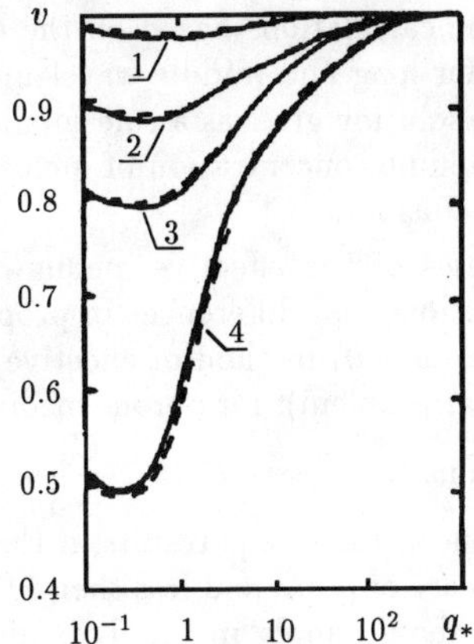

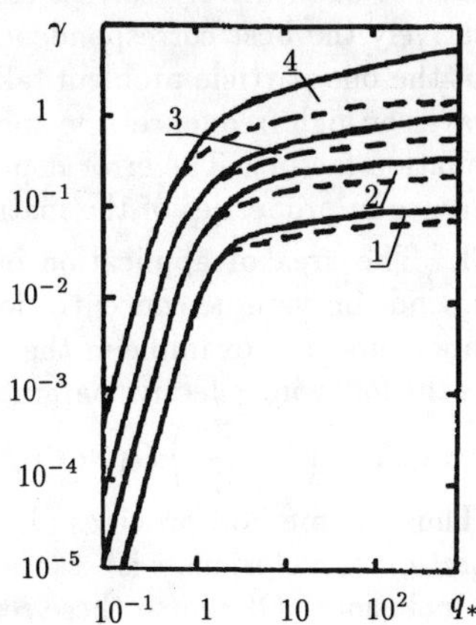

Fig. 2. Velocity v and attenuation factor γ on the effective wave number γ_* for a porous medium at $C_0 = 1$, $\rho_0 = 1$, $C = 0$, $\rho = 0$ and $a = 1$; pore fractions $p = 0.1$ (curve '1'), $p = 0.3$ (curve '2'), $p = 0.4$ (curve '3') and $p = 0.6$ (curve '4').

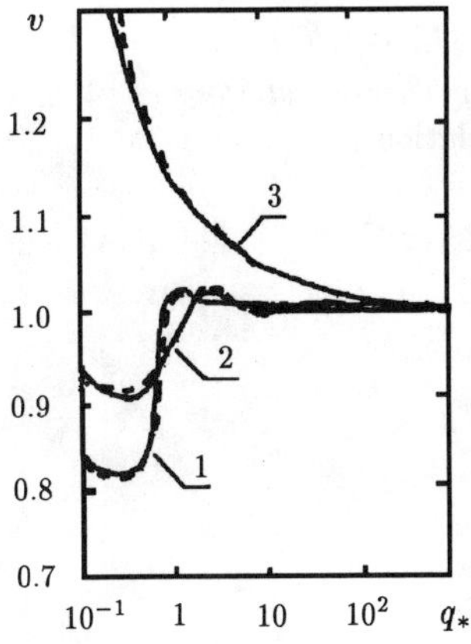

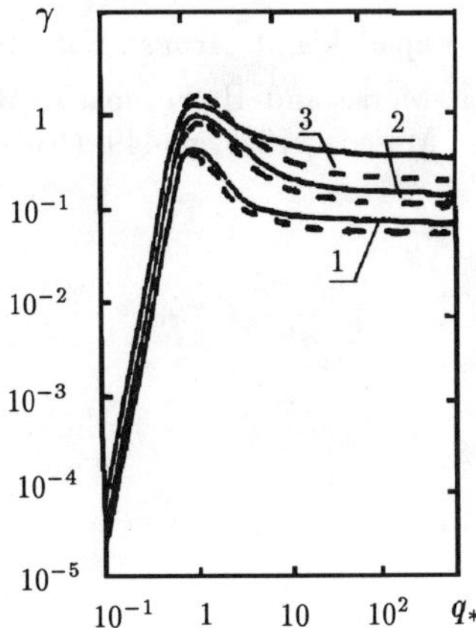

Fig. 3. The same as in Fig. 2, but for rigid inclusions, at $C_0 = 1$, $\rho_0 = 1$, $C = 10^3$, $\rho = 10$ and a =1; inclusion fractions $p = 0.1$ (curve '1'), $p = 0.2$ (curve '2') and $p = 0.3$ (curve '3').

2. An approximate solution of the one-particle problem proposed in this article provides qualitatively correct results for all possible values of frequences. But quantitatively the best correspondence with the calculations based on the exact solution of the one particle problem takes place for long and middle wavelengths. For short waves or high frequences the approximate solution gives as a rule smaller values of attenuation factors. The error depends on volume concentration of inclusions and on difference in properties of the matrix and inclusions.

3. The area of application of hypotheses of the effective medium method itself depends on volume concentration of inclusions and differences in properties of the components. For example in the static case ($\omega = 0$) method of effective medium provides the following effective parameters (scalar problem): for porous medium ($C = 0$) we have $C_* = \left(1 - \frac{3p}{2}\right) C_0$; for rigid inclusions ($C = \infty$)—$C_* = \frac{C_0}{1-3p}$.

Thus the method provides physically noncontradictious results, if the volume concentration of inclusions p is less then 2/3 in case of pores and less then 1/3 in case of rigid inclusions. Of course these restrictions should apply in the dynamic case as well.

References

1. S. K. Kanaun and V. M. Levin, *Effective Field Method in the Mechanics of Composite Materials*, Petrosavodsk University, Petrasovodsk, 1993. (in Russian.)
2. F. J. Sabina and J. R. Willis, *Eur. J. Mech., A/Solids* **12** (1993) 265.
3. E. P. Papadakis, *J. Acoust. Soc. Am.* **37** (1965) 703, 711.
4. Ph.M. Morse and H. Feshbach, *Methods of Theoretical Physics, V.2.* Inostr. Liter., Moscow, 1960, p. 449. (Russian translation.)

Continuum Models and Discrete Systems
Proceedings of 8th International Symposium, June 11-16, 1995, Varna, Bulgaria, ed. K.Z. Markov
©World Scientific Publishing Company, 1996, pp. 225–232

THE OVERALL PROPERTIES OF PIEZOACTIVE MATRIX COMPOSITE MATERIALS

V. M. LEVIN
Department of Mechanics, University of Petrosavodsk,
Lenin pr. 33, 185640 Petrosavodsk, Russia

Abstract. This paper is concerned with the composite materials of matrix type with piezoactive components. A variant of self consistent scheme (effective field method) is used for the theoretical prediction of overall elastic and electric properties of these materials, taking into account the electromechanical coupling. The explicit formulas are estimated for the effective elastic and electric constants of composites reinforced by the spherical inclusions and unidirectional continuous fibers.

1.Introduction

Composite materials made from two or more constituents find an increasingly wide application in modern technology. An important class of such materials is so-called matrix composites that comprise a homogeneous phase (matrix) containing a random array of filling particles (inclusions). One of the central problems of mechanics and physics of such materials is investigation of the microstructure influence on their overall (macroscopic) properties. The treatment of this problem deals mostly with uncoupled elastic and electric responses of composites (see, for example, the survey articles [1,2]). In the case of composite constituents exhibiting electromechanical coupling the external electric field can produce mechanical deformations which generate an internal stress field in the composite. These fields can influence the material's macroscopic response, so that the coupling effects must be taken into account when estimating the overall properties of composites with piezoactive components. The publications on this subject are comparatively rare. We cite only the book [3] and the paper [4] in which the piezoactive polycrystals were considered and one of the self-consistent schemes (the effective medium method) was used for prediction of their coupling effective electroelastic constants.

In this paper we study the effective elastic and electric properties of piezoactive matrix composites with the help of another self-consistent procedure (effective field method [2]) which allows to take into account not only the interactions between inclusions but some geometrical details of their distributions in the matrix.

2. Equilibrium of Homogeneous Electroelastic Medium with a Single Inclusion

Let us consider a homogeneous elastic piezoelectric material under isothermal conditions. The electromechanical response of such materials can be written in the form

$$\sigma_{ij} = C_{ijkl}\varepsilon_{kl} - e_{ijk}E_k\,,$$
$$\frac{1}{4\pi}D_i = e^t_{ikl}\varepsilon_{kl} + \eta_{ik}E_k\,, \tag{2.1}$$

where σ, ε are the stress and strain tensors, E, D are the electric field and electric displacement, $C = C^E$ denotes the tensor of elastic module at fixed E, $\eta = \eta^\varepsilon$ is the dielectric tensor at fixed strain and e is the third-rank piezoelectric tensor (superscript t represents the transposition of tensor). It is convenient to write the relations (2.1) in the following short form

$$J = \mathbf{L}F\,, \quad J = \begin{bmatrix} \sigma \\ \frac{1}{4\pi}D \end{bmatrix}, \quad \mathbf{L} = \begin{bmatrix} C & -e \\ e^t & \eta \end{bmatrix}, \quad F = \begin{bmatrix} \varepsilon \\ E \end{bmatrix}, \tag{2.2}$$

where the "matrix" $\mathbf{L}$ has to be considered as a linear operator which transforms the tensor-vector pair $[\sigma, D]$ into another tensor-vector pair $[\varepsilon, E]$.

Because the effective field method is based on the solution of one-particle problem [2], let us consider now the unbounded medium with the electromechanical characteristics $\mathbf{L^0}$ containing the closed region V with different elastic and electric properties $\mathbf{L}$. The system of coupled equations of electric and elastic equilibrium can be written as

$$\nabla\mathbf{L}(x)F(x) = 0, \quad \mathbf{L}(x) = \mathbf{L^0} + \mathbf{L^1}(x),$$
$$\mathbf{L^1}(x) = \mathbf{L^1}V(x), \quad \mathbf{L^1} = \mathbf{L} - \mathbf{L^0}, \quad \nabla_i = \frac{\partial}{\partial x_i}\,. \tag{2.3}$$

Here $V(x)$ is the characteristic function of the region V, occupied by the inclusion.

If the matrix and inclusion are assumed to be ideally bounded, it can be shown that the "vector" $F(x)$ in the medium with an inhomogeneity satisfies an integral equation which is equivalent to Eq. (2.3)

$$F(x) = F^0(x) + \int_V \mathbf{P}(x - x')\mathbf{L^1}F(x')\,dx', \quad x \in V\,,$$
$$\mathbf{P}(x) = \mathbf{D\Gamma}(x)\mathbf{D}, \quad \mathbf{D} = \begin{bmatrix} \mathrm{def} & 0 \\ 0 & \mathrm{grad} \end{bmatrix}. \tag{2.4}$$

Here $F^0(x)$ denotes the electric and elastic fields that would exist in the medium without the inclusion under given conditions at infinity, $\mathbf{\Gamma}(x)$ is the Green's function of the coupled electroelastic differential operator for the unbounded medium with the properties $\mathbf{L^0}$. This function is defined by the expressions

$$\mathbf{\Gamma}(x)=\frac{1}{8\pi^2}\int_{|\xi|=1}\mathbf{\Gamma}(\xi)\delta(\xi\cdot x)dS_\xi,\quad \mathbf{\Gamma}(x)=\begin{bmatrix} G_{ij}(\xi) & -\Gamma_i(\xi)\\ -\gamma_j(\xi) & g(\xi)\end{bmatrix},$$

$$G_{ij}=\left(\Lambda_{ij}-\frac{1}{\lambda}H_ih_j\right)^{-1},\quad \gamma_j=\frac{1}{\lambda}h_iG_{ij},\quad \Lambda_{ij}(\xi)=C^0_{ikjl}\xi_k\xi_l,$$

$$g=\left(\lambda-h_i\Lambda^{-1}_{ij}H_j\right)^{-1},\quad \Gamma_i=\Lambda^{-1}_{ij}H_jg,\quad H_i(\xi)=e^0_{ikl}\xi_k\xi_l,$$

$$h_j(\xi)=e^{0t}_{ljk}\xi_k\xi_l,\quad \lambda(\xi)=\eta^0_{kl}\xi_k\xi_l. \tag{2.5}$$

Let the inclusion with constant elastic and electric characteristics occupy an ellipsoidal domain V. It can be shown in the same manner as in [5], that the integral operator in (2.4) with the kernel $\mathbf{P}(x)$ has the property of polynomial conservations. In particular, a homogeneous in V external field $F^0(x)$ leads to a homogeneous field inside the domain V

$$F=\mathbf{A}F^0,\quad \mathbf{A}=\left(\mathbf{I}+\mathbf{PL^1}\right)^{-1},$$

$$\mathbf{P}=\frac{|\det a|}{4\pi}\int_{|\xi|=1}\xi\mathbf{\Gamma}(\xi)\xi\frac{dS_\xi}{\rho^3(\xi)},\quad \rho(\xi)=\sqrt{\xi_i(a^2)_{ij}\xi_j}, \tag{2.6}$$

where $a_{ij}=a_i\delta_{ij}$, a_1,a_2,a_3 are ellipsoid's semiaxes, $\mathbf{I}$ is the unit operator.

3. Piezoactive Medium With Random Fields of Inhomogeneities

3.1. Effective Electroelastic Operator

Let us now examine an unbounded elastic piezoelectric medium, containing a random set of ellipsoidal inclusions. As before we denote by $V(x)$ the characteristic function of the region V, occupied by inclusions and $x\,(x_1,x_2,x_3)$ a point of the inhomogeneous medium. Then the strain ε and electric field E in composite satisfy the following system of integral equations

$$F(x)=F^0(x)+\int\mathbf{P}(x-x')\mathbf{L^1}(x')V(x')F(x')\,dx', \tag{3.1}$$

where $\mathbf{L^1}(x)$ has the constant value $\mathbf{L^1}(\omega_k)$ when the point x is in a region V_k occupied by the k-th inclusion (ω_k are the set of parameters which characterize the principal anisotropy axes orientation in the space).

To solve the homogenization problem with the help of this equations let us use the self-consistent scheme named the effective field method [2]. This method is based on the following assumptions. Let us consider an arbitrary i-th inclusion in a fixed typical realization of random set of inhomogeneities. We denote by $F^*_{(i)}(x)$ the local external fields (ε^* and E^*) acting on this inclusion. The field $F^*_{(i)}(x)$ is composed of the external field $F^0(x)$ and disturbances of the fields due to surrounding inclusions.

Let the function $F^*(x)$ coincide with the local external field $F^*_{(i)}(x)$ when $x \in V_k$. Then from (3.1) it follows

$$F^*(x) = F^0(x) + \int \mathbf{P}(x - x')\mathbf{L}^1(x')V(x;x')F(x')\,dx', \tag{3.2}$$

where $V(x;x')$ is the characteristic function of the region V_x defined by the relation

$$V_x = \bigcup_{i \neq k} V_i \quad \text{when } x \in V_k. \tag{3.3}$$

Let us suppose that the field $F^*(x)$ has the same structure in any region occupied by the inclusions (hypothesis H_1 of effective field method [2]). Let in particular $F^*(x)$ be constant inside each region V_i (but may vary randomly from one inclusion to another). Then the connection between the field $F(x)$ $(x \in V)$ and $F^*(x)$ may be defined by the relation

$$F(x) = \mathbf{A}(x)F^*(x), \tag{3.4}$$

which has been obtained above when solving the one-particle problem. In Eq. (3.4) $\mathbf{A}(x)$ is the function that coincides with the constant operator $\mathbf{A}(\omega_k)$, defined in Eq. (2.6), when $x \in V_k$.

Substitution of the expression (3.4) to the right side of Eqs. (3.1) and (3.2) allows us to express elastic and electric fields $F(x)$ in an arbitrary point x via the local external field

$$F(x) = F^0(x) + \int \mathbf{P}(x - x')\mathbf{L}^A(x')F^*(x')dx' \tag{3.5}$$

and to obtain the self-consistent equation for $F^*(x)$

$$F^*(x) = F^0(x) + \int \mathbf{P}(x - x')\mathbf{L}^A(x,x')F^*(x')dx'. \tag{3.6}$$

In Eqs. (3.5) and (3.6)

$$\mathbf{L}^A(x') = \mathbf{L}^1(x')\mathbf{A}(x')V(x'), \quad \mathbf{L}^A(x,x') = \mathbf{L}^1(x')\mathbf{A}(x')V(x;x'). \tag{3.7}$$

Let us take the ensemble expectation of both sides of Eq. (3.5). Supposing that the random field $F^*_{(i)}(x)$ does not depend statistically on the geometrically characteristics and physical properties of i-th inclusion (hypothesis H_2 of the effective field method), we obtain

$$\langle F(x)\rangle = F^0(x) + p\int \mathbf{P}(x - x')\mathbf{L}^A\langle F^*(x')|x'\rangle dx', \quad \mathbf{L}^A = \langle \mathbf{L}^A(x)\rangle, \tag{3.8}$$

where p is the volume concentration of inclusions ($p = \langle V(x)\rangle$) and $\langle\cdot \mid x\rangle$ denotes averaging under the condition $x \in V$. The $\langle F^*(x) \mid x\rangle$ will be named in following by the effective field. Hence the average field $\langle F(x)\rangle$ in the composite material is expressed via the effective field. Eq. (3.6) is the basic for this field determination. Taking the conditional expectation of Eq. (3.6) and using the main hypotheses of effective field method [2], we obtain

$$\langle F^*(x)|x\rangle = F^0(x) + \int \mathbf{P}(x-x')\Psi(x,x')\mathbf{L}^A\langle F^*(x')|x'\rangle dx',$$

$$\Psi(x,x') = \langle V(x;x')|x\rangle. \tag{3.9}$$

Here the function $\Psi(x,x')$ defines the shape of the "correlation hole" inside which a typical inclusion is located. For statistically homogeneous composites this function depends only on the difference $x-x'$.

The problem now is to find the dependence between the fields $F^0(x)$ and $\langle F(x)\rangle$. Eliminating for this purpose the effective field from Eqs. (3.8) and (3.9) (the details of the derivation one could see in [2]), we have

$$\langle F(x)\rangle = F^0(x) + \int \mathbf{P}(x-x')\mathbf{L}^A\mathbf{D}\langle F(x')\rangle\, dx',$$

$$\mathbf{D} = (\mathbf{I}+p\Pi\mathbf{L}^A)^{-1}, \quad \Pi = \int \mathbf{P}(x)\left(1-\frac{1}{p}\Psi(x)\right)dx. \tag{3.10}$$

Let us apply the operator $\nabla\mathbf{L^0}$ to the both sides of this expression. Taking into account the obvious relations $\nabla\mathbf{L^0}F^0(x)=0$, $\nabla\mathbf{L^0}\nabla\Gamma(x) = -\mathbf{I}\delta(x)$ we obtain that the average field $\langle F(x)\rangle$ satisfies the equation

$$\nabla\mathbf{L}^*\langle F(x)\rangle = 0, \quad \mathbf{L}^* = \mathbf{L^0}+p\mathbf{L}^A\mathbf{D}, \tag{3.11}$$

in which the value $\mathbf{L}^*$ represents the operator of effective electroelastic characteristics of composite materials with piezoactive components.

Let us consider some special cases.

3.2. Medium with an Array of Spherical Inclusions

Assume that the matrix in the composite is isotropic and the inclusions are spherical of the same radius. Then the operator $\mathbf{P}$ in Eq. (6) takes the form

$$\mathbf{P} = \begin{bmatrix} P_{ijkl} & 0 \\ 0 & p_{ik} \end{bmatrix},$$

$$P_{ijkl} = \frac{1}{9k_P}\delta_{ij}\delta_{kl} + \frac{1}{2\mu_P}\left(I_{ijkl} - \frac{1}{3}\delta_{ij}\delta_{kl}\right), \quad p_{ik} = \frac{1}{3\eta_0}\delta_{ik},$$

$$k_P = k_0 + \frac{4}{3}\mu_0, \quad \mu_P = \frac{5\mu_0(3k_0+4\mu_0)}{6(k_0+2\mu_0)}, \quad I_{ijkl} = \delta_{i(k}\delta_{l)j}, \tag{3.12}$$

where k_0, μ_0 are the volume and shear elastic module and η_0 is the dielectric coefficient of the matrix.

Let the elastic and electric properties of inclusions belong to the cubic classes $\bar{4}3m$ and 23. Then these properties are characterized by the elastic moduli (bulk k and two shear moduli μ and m), dielectric coefficient η and piezoelectric constant e.

If the inclusions in the matrix are homogeneously and isotropically distributed so that the correlation hole has spherical symmetry and the operator Π in Eq. (3.10) coincides with $\mathbf{P}$ defined by Eqs. (3.12). Let us consider two limiting cases.

a) Let the orientation of the principal crystallographic axes of inclusions be random. The coupling electric and elastic effects in such material are absent and its properties are characterized by the two effective elastic module k^* and μ^* and coefficient of dielectric permeability η^*. These values can be represented in the form

$$k^* = k_0 + p\left(\frac{1}{k_A} - \frac{p}{k_P}\right)^{-1}, \quad \mu^* = \mu_0 + p\left(\frac{5}{2m_A + 3\mu_A} - \frac{p}{\mu_P}\right)^{-1},$$

$$\eta^* = \eta_0 + p\left(\frac{1}{\eta_A} - \frac{p}{3\eta_0}\right)^{-1}, \tag{3.13}$$

where

$$k_A = k_1\left(1 + \frac{k_1}{k_P}\right)^{-1}, \quad \mu_A = \bar{\mu}\left(1 + \frac{\bar{\mu}}{\mu_P}\right)^{-1}, \quad m_A = m_1\left(1 + \frac{m_1}{\mu_P}\right)^{-1},$$

$$\eta_A = \bar{\eta}\left(1 + \frac{\bar{\eta}}{3\eta_0}\right)^{-1}, \quad \bar{\mu} = \mu_1 + \frac{e^2}{\eta_1 + 3\eta_0}, \quad \bar{\eta} = \eta_1 + \frac{e^2}{\mu_1 + \mu_P},$$

$$k_1 = k - k_0, \quad \mu_1 = \mu - \mu_0, \quad m_1 = m - \mu_0, \quad \eta_1 = \eta - \eta_0. \tag{3.14}$$

b) Suppose the principal anisotropic axes of the inclusions are oriented in one direction. In this case such composite has the same symmetry as inclusions and is characterized by the following effective elastic moduli

$$k^* = k_0 + p\left(\frac{1}{k_1} - \frac{1-p}{k_P}\right)^{-1}, \quad m^* = \mu_0 + p\left(\frac{1}{m_1} - \frac{1-p}{\mu_P}\right)^{-1},$$

$$\mu^* = \mu_0 + p\left[\frac{3\eta_0 + (1-p)\eta_1}{3\eta_0\mu_1 + (1-p)(\mu_1\eta_1 + e^2)} + \frac{1-p}{\mu_P}\right]^{-1}, \tag{3.15}$$

the coefficient of dielectric permeability

$$\eta^* = \eta_0 + p\left[\frac{\mu_P + (1-p)\mu_1}{\eta_1\mu_P + (1-p)(\eta_1\mu_1 + e^2)} + \frac{1-p}{3\eta_0}\right]^{-1} \tag{3.16}$$

and the piezoelectric constant

$$e^* = 3p\eta_0\mu_P e\left[(\mu_P + (1-p)\mu_1)(3\eta_0 + (1-p)\eta_1) + (1-p)^2 e^2\right]^{-1}. \tag{3.17}$$

3.3. Medium Reinforced By Unidirectional Fibers

Let us consider a composite material with the inclusions in the form of continuous unidirectional cylindrical fibers with a circular cross-section. Both materials are assumed to be transversely isotropic and the axes of fibers coincide with the symmetry axes of the matrix and inclusions. The tensors C^0, e^0 and η^0 for such medium can be represented as

$$C^0_{ijkl} = k_0\theta_{ij}\theta_{kl} + 2m_0\left(\theta_{i(k}\theta_{l)j} - \frac{1}{2}\theta_{ij}\theta_{kl}\right) + l_0\left(\theta_{ij}m_km_l + m_im_j\theta_{kl}\right)$$

$$+4\,\mu_0\theta_{i)(k}m_{l)}m_{(j} + n_0m_im_jm_km_l, \quad \theta_{ij} = \delta_{ij} - m_im_j,$$

$$e^0_{ijk} = e^0_1\theta_{ij}m_k + 2e^0_2m_{(i}\theta_{j)k} + e^0_3m_im_jm_k, \quad \eta^0_{ij} = \eta^0_1m_im_j + \eta^0_2\theta_{ij}. \tag{3.18}$$

Here k_0, m_0, l_0, μ_0, n_0 are the five independent elastic modules, e^0_1, e^0_2, e^0_3 are the three piezoelectric constants and η^0_1, η^0_2 are the two dielectric coefficients of the transversely isotropic medium.

The same relations as (3.18) with the omission of the superscript '0' can be written for the tensors C, e and η of the inclusions.

The infinite cylindrical fibers can be considered as a limiting case of ellipsoidal inhomogeneities when one of their semi axes tends to infinity. As result the operator **P** in (2.6) takes the form

$$\mathbf{P} = \begin{bmatrix} P_{ijkl} & r_{ijk} \\ -r^t_{ikl} & p_{ik} \end{bmatrix},$$

$$P_{ijkl} = P_1\theta_{ij}\theta_{kl} + P_2\left(\theta_{i(k}\theta_{l)j} - \frac{1}{2}\theta_{ij}\theta_{kl}\right) + \frac{1}{2\overline{\mu}_0}\theta_{i)(k}m_{l)}m_{(j}, \quad r_{ijk} = \frac{\gamma}{2}m_{(i}\theta_{j)k},$$

$$p_{ij} = \frac{1}{2\overline{\eta}^0_2}\theta_{ij}, \quad P_1 = \frac{1}{4(k_0+m_0)}, \quad P_2 = \frac{k_0+2m_0}{4m_0(k_0+m_0)},$$

$$\gamma = \frac{e^0_2}{\mu_0\eta^0_2 + (e^0_2)^2}, \quad \overline{\mu}_0 = \mu_0 + \frac{(e^0_2)^2}{\eta^0_2}, \quad \overline{\eta}^0_2 = \eta^0_2 + \frac{(e^0_2)^2}{\mu_0}. \tag{3.19}$$

In the case when correlation hole has the form of a continuous cylinder parallel to the fibers the general formula (3.11) transforms to

$$\mathbf{L}^* = \mathbf{L}^0 + p\mathbf{L}^1\left[\mathbf{I} + (1-p)\mathbf{P}\mathbf{L}^1\right]^{-1}, \tag{3.20}$$

where **P** is defined by the expression (3.19).

In accordance with this formula the composite material is macroscopically transversely isotropic and characterized by the following effective elastic modules

$$k^* = k_0 + pk_1d(p), \quad m^* = m_0 + pm_1\left[1 + (1-p)\frac{m_1(k_0+2m_0)}{2m_0(k_0+m_0)}\right]^{-1},$$

$$l^* = l_0 + pl_1 d(p), \quad \mu^* = \mu_0 + \frac{p}{\Delta(p)}\left[\mu_1 + \frac{(1-p)f}{2\overline{\eta}_2}\right],$$

$$n^* = n_0 + p\left[n_1 + \frac{(1-p)l_1^2 d(p)}{k_0 + m_0}\right], \quad d(p) = \frac{k_0 + m_0}{k_0 + m_0 + (1-p)k_1},$$

$$\Delta(p) = [1 + (1-p)b]\,[1 + (1-p)B] - (1-p)^2 Qq, \quad f = \mu_1\eta_2^1 + (e_2^1)^2,$$

$$b = \frac{1}{2}\left(\frac{\eta_2^1}{\overline{\eta}_2} + \gamma e_2^1\right), \quad B = \frac{1}{2}\left(\frac{\mu_1}{\overline{\mu}_0} + \gamma e_2^1\right),$$

$$q = \frac{1}{2}\left(\frac{e_2^1}{\overline{\eta}_2} - \gamma\mu_1\right), \quad Q = \frac{1}{2}\left(\gamma\eta_2^1 - \frac{e_2^1}{\overline{\mu}_0}\right), \tag{3.21}$$

three piezoelectric constants

$$e_1^* = e_1^0 + pe_1^1 d(p), \quad e_2^* = e_2^0 + \frac{p}{\Delta(p)}\left[e_2^1 + \frac{1}{2}(1-p)\gamma f\right],$$

$$e_3^* = e_3^0 + p\left[e_3^1 - \frac{(1-p)l_1 e_1^1 d(p)}{k_0 + m_0}\right] \tag{3.22}$$

and by two coefficients of dielectric permeability

$$\eta_1^* = \eta_1^0 + p\left[\eta_1^1 - \frac{(1-p)(e_1^1)^2 d(p)}{k_0 + m_0}\right], \quad \eta_2^* = \eta_2^0 + \frac{p}{\Delta(p)}\left[\eta_2^1 + \frac{(1-p)f}{2\overline{\mu}_0}\right]. \tag{3.23}$$

As it follows from these expressions, coupling of the electric and elastic fields influences only the values of the effective longitudinal shear modulus μ^*, the piezoelectric constants e_i^*, $i = 1, 2, 3$, and the dielectric coefficients η_k^* $(k = 1, 2)$. As far as the elastic moduli k^*, m^*, l^* and n^* are concerned, they are the same [2] as in the case of deformation without electric and elastic coupling.

References

1. J. R. Willis, *Advances in Applied mechanics* **21** (1981) 1.
2. S. K. Kanaun and V. M. Levin, In *Advances in Mathematical Modeling of Composite Materials*, ed. K. Z. Markov, World Sci., 1994, p. 4.
3. L. P. Choroshun, B. P. Maslov and P. V. Leschenko, *The Prediction of Effective Properties of Piezoactive Composite Materials*, Kiev, 1989, p. 208. (in Russian.)
4. T. Olson, *Overall Properties of Granular Piezoelectrics: Bounds and Effective Medium Approximations*, Thesis, N.Y. University, 1991, p. 57.
5. R. J. Asaro and D. M. Barnet, *J. Mech. Phys. Solids* **23** (1975) 77.

Continuum Models and Discrete Systems
Proceedings of 8th International Symposium, June 11-16, 1995, Varna, Bulgaria, ed. K.Z. Markov

APPLICATION OF THE HOMOGENIZATION METHOD TO THE VISCOSITY OF A SUSPENSION OF FIBRES

T. LÉVY and F. PÉRIN
Laboratoire de Modélisation en Mécanique,
Université Pierre et Marie Curie (CNRS URA 229),
4 place Jussieu, F-75252 Paris Cédex 05, France

Abstract. The aim of the paper is to obtain approximations for the viscosity of a nondilute ordered suspension of rods. The assembly of cylindrical, infinitely long, parallely aligned fibres suspended in a viscous incompressible fluid is considered in situations where the bidimensional structure of the suspension is preserved. The cases of bulk flows in the direction of the fibres, or perpendicular to the fibres are investigated by the homogenization method.

1. Introduction

The effective viscosity of a suspension of rigid particles in a viscous incompressible fluid has been the focus of a number of studies (see for references [1] and [2]). Most of them concern the case of dilute suspensions. Less numerous theoretical works exist for nondilute suspensions, one of them investigates the problem using the homogenization process for periodic media [3]. Without any phenomenological assumption this method gives the equations governing the bulk medium and proves that generally the constitutive equation of the suspension is that of an incompressible anisotropic non-Newtonian fluid. The effective viscosity is defined by a fourth-rank tensor which relates the average stress to the average rate of strain. The components of the tensor are completely determined by the instantaneous microstructure. The non-Newtonian behaviour is the result of the dependance of the medium deformation on the bulk velocity gradient. Thus the computation of the effective viscosity tensor is a too much complicated problem.

The aim of this study is to apply the homogenization method to the bidimensional flows of a fibres suspension. Namely an assembly of infinitely long, parallel cylinders suspended in a viscous incompressible fluid is considered in situations where the bidimensional structure of the suspension is preserved. We believe the results will be relevant to the flows of ordered suspensions of rods. Section 2 deals with the formulation of the problem and the main features of the homogenization method, the specificity of its application to a bidimensional microstructure being taken into account. Section 3 is devoted to the parallel to the fibres Poiseuille-like flows and Section 4 to the bulk flows which are perpendicular to the fibres direction. The programmes of these two sections are similar: determination of the local variations of the flow (at the level of the microstructure), obtaining of the macroscopic flow equations specially the effective constitutive relation, and determination of the microstructure evolution. Next the instantaneous viscosity of the suspension is examined. When the microstructure exhibits symmetries, bounds of the viscosity coefficient may be determined and some numerical results allow to think that these bounds set up good approximations. The last section sums up the informations of Sections 3 and 4 on the viscosity coefficients and reads them as approximations of the fourth-order viscosity tensor of a long-fibres ordered suspension.

2. Formulation of the Problem

We investigate a suspension of infinitely long, parallelly aligned cylinders in a viscous incompressible fluid. The $\vec{e}_3$ axis of a rectangular Cartesian system of coordinates $x(x_1, x_2, x_3)$ is chosen to coincide with the direction of the fibres. We suppose that at the considered instant t the spatial distribution of the fibres is locally periodic and that the section of the periodicity cell by a plane $\vec{e}_1$, $\vec{e}_2$ is a small parallelogram $\varepsilon Y(t, x_1, x_2)$ (Fig. 1). The small parameter ε characterizes both the smallness of the fibres section and their mutual distance. So we deal with a nondilute suspension, the fibres concentration is $c = \mid Y_S \mid / \mid Y \mid$.

With the usual notations, the velocity $\vec{V}^\varepsilon$ and the pressure P^ε satisfy the following equations in the fluid:

$$\operatorname{div} \vec{V}^\varepsilon = 0 \,; \tag{1}$$

$$\rho_0 \left(\frac{\partial V_i^\varepsilon}{\partial t} + V_j^\varepsilon \, \frac{\partial V_i^\varepsilon}{\partial x_j} \right) = \frac{\partial \sigma_{ij}^\varepsilon}{\partial x_j} \,, \tag{2}$$

$$\sigma_{ij}^\varepsilon = -P^\varepsilon \delta_{ij} + 2\mu \; D_{ij}(\vec{V}^\varepsilon) \,; \tag{3}$$

In the fibres:

$$D_{ij}(\vec{V}^\varepsilon) = 0 \,; \tag{4}$$

On the surface of the fibres:

$$\vec{V}^\varepsilon \text{ is continuous} \tag{5}$$

and for each arbitrary portion S^ε of a fibre defined by $x_3 \in [a, b]$ with mass centre G, the dynamic law implies:

$$\int_{S^\varepsilon} \rho_S^\varepsilon \frac{d\vec{V}^\varepsilon}{dt} \, dv = -\int_{\partial S^\varepsilon} \hat{\sigma}_{ij}^\varepsilon n_j \vec{e}_i \, d\sigma \,, \tag{6}$$

$$\int_{S^\varepsilon} \rho_S^\varepsilon (\vec{x} - \vec{x}_G) \times \frac{d\vec{V}^\varepsilon}{dt} \, dv = -\int_{\partial S^\varepsilon} (\vec{x} - \vec{x}_G) \times \hat{\sigma}_{ij}^\varepsilon n_j \vec{e}_i \, d\sigma \,. \tag{7}$$

In (6), (7), $\hat{\sigma}_{ij}^{\varepsilon}$ denote the components of the stress tensor in the whole medium defined as $\hat{\sigma}_{ij}^{\varepsilon} = \sigma_{ij}$ given by (3) in the fluid and extended symmetrically in the solids, satisfying the momentum equation in the fibres, and the stress continuity condition on their surfaces.

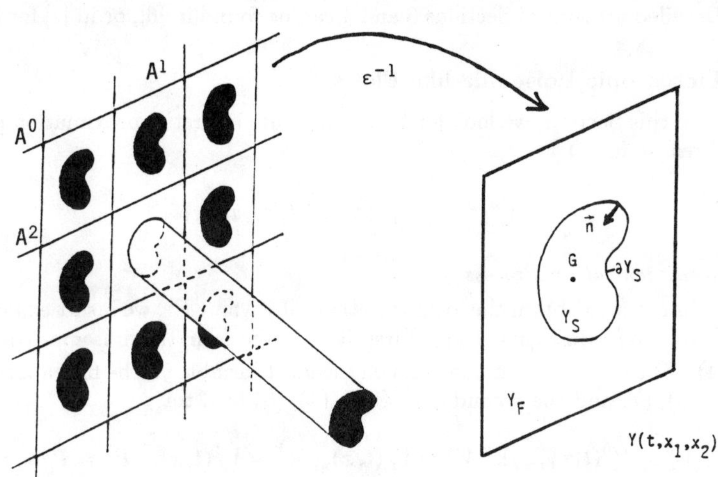

Fig. 1. Locally periodic distribution of the fibres and basic period $Y(t,\, x_1,\, x_2)$.

The homogenization method is based on a two scale procedure [4,5], with the standard variable x and the microscopic one $y = (y_1, y_2)$ which is associated with the bidimensional microstructure of the problem by: $y_\alpha = (x_\alpha - x_{G\alpha})/\varepsilon$, $\alpha = 1,2$. In the following, Latin indices $i, j, k, \ldots$ can take values $1,2,3$, wheras Greek indices $\alpha, \beta, \ldots$ can take values $1,2$. We look for $\vec{V}^\varepsilon$ and P^ε solutions of (1)–(7) in the form of double scale asymptotic expansions when ε tends to zero. As we limit our investigations to suspension flows that remain, in first approximation, bidimensional, we postulate expansions of the form:

$$\vec{V}^\varepsilon(t,x) = \vec{V}^0(t,x_1,x_2,y) + \varepsilon \vec{V}^1(\varepsilon,x,y) + \cdots\,, \qquad (8)$$
$$P^\varepsilon(t,x) = P^0(t,x,y) + \varepsilon P^1(t,x,y) + \cdots\,, \qquad (9)$$

where $\vec{V}^i$, P^i, $i = 0,1,\ldots$, according to the periodicity of the structure, are $Y-$periodic with respect to the local stretched variable y. We insert expansions (8) and (9) in Eqs. (1)–(7), taking care that, when applied to a function $f(x,y)$, $\dfrac{\partial}{\partial x_\alpha}$ becomes $\dfrac{\partial}{\partial x_\alpha} + \dfrac{1}{\varepsilon}\dfrac{\partial}{\partial y_\alpha}$. By identifying the like powers of ε, we obtain successive differential equations with respect to the variable y to be solved in the basic period $Y(t, x_1, x_2)$ with periodicity boundary conditions on ∂Y. At this stage x is a parameter. As classically in homogenization process [5], we obtain the macroscopic conservation laws by applying the flux method. The homogenized constitutive law is obtained by taking the averaged values in a periodicity cell of the first approximation of the stresses $\hat{\sigma}^{\varepsilon}_{ij}$. Lastly the deformation of the structure is determined by (8). Detailed account of Sections 3 and 4 can be found in [6], or in [7] for Section 3.

3. Macroscopic Poiseuille-like Flows

In this Section, we look for flows which are in first approximation parallel to the fibres, so in (8)

$$\vec{V}^0(t, x_1, x_2, y) = V_3^0(t, x_1, x_2, y)\vec{e}_3 \,. \tag{10}$$

3.1. Homogenization Process

In order to obtain the local variations of $\vec{V}^{\varepsilon}$ and P^{ε}, we insert expansions (8) with (10) and (9) in Eqs. (1)–(7). First, it is proved from the first approximations of (2), (4), (5) that $\vec{V}^0$ does not depend on the local variable $\vec{y}$. The first approximation of (1), (6), (7) and the second one of (2), (4), (5) lead to:

$$\begin{gathered} V_3^0 = V_3^0(t, x_1, x_2), \quad V_1^1 = V_1^1(t, x), \quad V_2^1 = V_2^1(t, x), \quad P^0 = P^0(t, x)\,, \\ V_3^1(t, x, y) = A(t, x) - \frac{\partial V_3^0}{\partial x_\alpha}(t, x_1, x_2)\psi^\alpha(t, x_1, x_2, y)\,, \end{gathered} \tag{11}$$

where ψ^α, completely determined by the microstructure geometry, is the unique Y-periodic solution of: $\Delta\psi^\alpha = 0$ in Y_F, $\psi^\alpha = y_\alpha$ in Y_S, ψ^α continuous on ∂Y_S.

In the next step of the process, we apply the momentum conservation law to the mixture contained in an arbitrary macroscopic bounded cylinder parallel to the fibres. As usual [3,5] we deduce by using (11), (12), the first order contribution of each term and then the macroscopic momentum law, with the notation $\tilde{\cdot} = \dfrac{1}{|Y|}\displaystyle\int_Y \cdot \, dy$

$$\frac{\partial \tilde{\sigma}^0_{\alpha j}}{\partial x_j} = 0\,; \quad \rho\frac{\partial V_3^0}{\partial t} = \frac{\partial \tilde{\sigma}^0_{3j}}{\partial x_j}\,. \tag{12}$$

The macroscopic stress symmetric tensor $\tilde{\sigma}^0$ is expressed, using the first approximation of (3) with (11), in the fluid and its extension in the solid part Y_S satisfying the momentum equation and the stress continuity on ∂Y_S. Thus we obtain the macroscopic constitutive law

$$\begin{gathered} \tilde{\sigma}^0_{\alpha\beta} = -P^0(t, x)\delta_{\alpha\beta} \quad , \quad \tilde{\sigma}^0_{33} = -P^0(t, x) + \sigma^*_{33}(t, x_1, x_2)\,, \\ \tilde{\sigma}^0_{\alpha 3} = a_{\alpha\beta}\frac{\partial V_3^0}{\partial x_\beta} \quad \text{with} \quad a_{\alpha\beta} = \mu(1 + c)\delta_{\alpha\beta} + \frac{\mu}{|Y|}\int_{Y_F}\frac{\partial\psi^\alpha}{\partial y_\gamma}\frac{\partial\psi^\beta}{\partial y_\gamma}\,dy\,. \end{gathered} \tag{13}$$

The undetermined σ_{33}^* does not appear in the momentum equation (12).

Finally the microstructure evolution is determined by:

$$\frac{d\,\overrightarrow{A^0 A^\alpha}}{dt}\cdot\vec{e}_\beta = V_\beta^\varepsilon(t, x_{A^\alpha}) - V_\beta^\varepsilon(t, x_{A^0}),\quad \vec{\Omega}^\varepsilon = \frac{1}{2}\mathrm{rot}\,\vec{V}^\varepsilon, \tag{14}$$

which give the Y-parallelogram deformation and the rotation rate $\vec{\Omega}^\varepsilon$ of the solid Y_S. Using the asymptotic expansion of $\vec{V}^\varepsilon$, we find $O(\varepsilon^2)$ for the first variations and $O(\varepsilon)$ for the solid rotation. So, in first approximation the microstructure does not evolve in time: $Y(t, x_1, x_2) = Y(x_1, x_2)$. Consequently, the behaviour of the suspension is that a Newtonian anisotropic inhomogeneous fluid according to (13).

3.2. *The Macroscopic Viscosity Coefficients*

The tensor $a_{\alpha\beta}$ is given by (13). It is symmetric and elliptic, completely determined by the local period $Y(x_1, x_2)$. If we assume that Y admits a symmetry axis, for example Oy_1, then $a_{12} = 0$ and a lower bound can be found for a_{22} [6,7]

$$\frac{a_{22}}{\mu} \geq m = 1 + \frac{1}{2A}\int_{y_1^-}^{y_1^+} \frac{f_1(y_1)}{B - f_1(y_1)}\,dy_1\,. \tag{15}$$

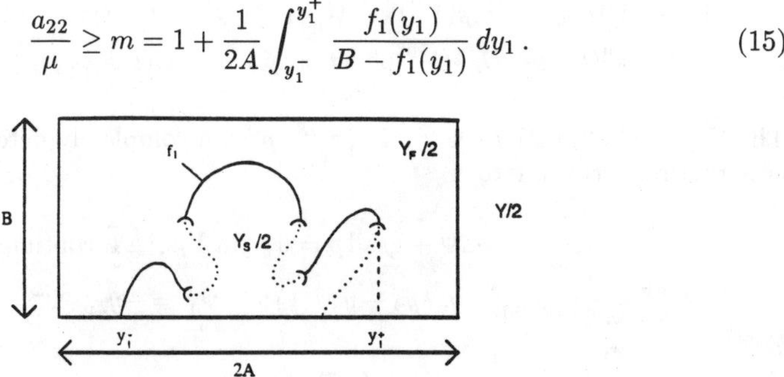

Fig. 2. Half basic period and graph of f_1.

Let us remark that this lower bound (see Fig. 2) is not influenced by the concave parts of the fibre section.

If the period Y admits two axis of symmetry and is self-invariant by a $\pi/2$ rotation, the tensor $a_{\alpha\beta}$ is isotropic: $a_{\alpha\beta} = \mu^h\delta_{\alpha\beta}$, and μ^h/μ satisfies (15). Numerical computations of μ^h/μ and its lower bound m are performed using the MODULEF code for a square shaped period including a square shaped fibre section or a circular fibre section. From these numerical results, we can conclude that when the solid concentration c is greater than 0.5 the viscosity μ^h of the suspension is highly dependent on the microstructure geometry. Furthermore it appears in these two simple cases that the lower bound μm is a rather good approximation of μ^h whatever the solid concentration is. Let us make some remarks about the viscosity of a dilute suspension. From the computations, it may be suggested that when the fibre section is convex the influence of its shape is negligible. This is not true when the fibre section involves concave parts, as it can be deduced from the lower bound (15) for a T cross-shaped section.

4. Transverse Flows

In this Section, we look for flows which remain in first approximation in a plane perpendicular to the fibres

$$\vec{V}^0(t, x_1, x_2, y) = V_\alpha^0(t, x_1, x_2, y)\vec{e}_\alpha . \tag{16}$$

4.1. Homogenization Process

We insert expansions (8) with (16) and (9) in Eqs. (1)–(7). The first approximation proves that $\vec{V}^0$ does not depend on y, and the second one determines the local variations of V_3^1, V_1^1, V_2^1 and P^0. Namely we obtain [6]

$$\begin{gathered} V_\alpha^0 = V_\alpha^0(t, x_1, x_2), \qquad V_3^1 = V_3^1(t, x_1, x_2), \\ V_\gamma^1 = A_\gamma(t, x) - D_{\alpha\beta}(\vec{V}^0)\Psi_\gamma^{\alpha\beta}(t, x_1, x_2, y) + \omega(t, x_1, x_2)\Psi_\gamma^\omega(t, x_1, x_2, y), \\ P^0 = \pi^0(t, x) - D_{\alpha\beta}(\vec{V}^0)p^{\alpha\beta}(t, x_1, x_2, y) + \omega(t, x_1, x_2)p^\omega(t, x_1, x_2, y) . \end{gathered} \tag{17}$$

The Y-periodic functions $\vec{\Psi}^{\alpha\beta}$, $\vec{\Psi}^\omega$, $p^{\alpha\beta}$, p^ω are completely determined by the microstructure according to

$$\begin{gathered} \operatorname{div} \vec{\Psi} = 0 , \qquad -\mu\Delta\vec{\Psi} + \operatorname{grad} p = 0 \quad \text{in } Y_F , \quad \vec{\Psi} \text{ continuous on } \partial Y_S , \\ \Psi_\gamma^{\alpha\beta} = \frac{1}{2}(y_\beta\delta_{\alpha\gamma} + y_\alpha\delta_{\beta\gamma} - y_\gamma\delta_{\alpha\beta}); \qquad \Psi_1^\omega = -y_2, \quad \Psi_2^\omega = y_1 \quad \text{in } Y_S . \end{gathered} \tag{18}$$

From the second approximation of (7)

$$\omega = D_{\gamma\delta}(\vec{V}^0) \frac{\int_{Y_F} D_{\alpha\beta}(\vec{\Psi}^{\gamma\delta})D_{\alpha\beta}(\vec{\Psi}^\omega)\,dy}{\int_{Y_F} D_{\alpha\beta}(\vec{\Psi}^\omega)D_{\alpha\beta}(\vec{\Psi}^\omega)\,dy} . \tag{19}$$

The mean value of the second approximation of (1) proves the incompressiblity of the suspension. And the macroscopic momentum equation is obtained by the same method as in Section 3.1 and the use of the asymptotic expansions (17):

$$\tilde{\rho}\left(\frac{\partial V_\alpha^0}{\partial t} + V_\beta^0 \frac{\partial V_\alpha^0}{\partial x_\beta}\right) - b_{\alpha\beta\gamma\delta} V_\beta^0 D_{\gamma\delta}(\vec{V}^0) = \frac{\partial \tilde{\sigma}_{\alpha j}^0}{\partial x_j}, \quad 0 = \frac{\partial \tilde{\sigma}_{3j}^0}{\partial x_j} . \tag{20}$$

Let us point out the unusual non-linear terms in the acceleration of the suspension. The macroscopic coefficients $b_{\alpha\beta\gamma\delta}$ are determined by the microstructure [6].

The macroscopic constitutive relation is obtained as in Section 3.1:

$$\tilde{\tilde{\sigma}}^0_{\alpha\beta} = -P^0_h(t,x)\delta_{\alpha\beta} + a_{\alpha\beta\gamma\delta} D_{\gamma\delta}(\vec{V}^0),$$

with $$a_{\alpha\beta\gamma\delta} = \mu(1+c)(\delta_{\alpha\gamma}\delta_{\beta\delta} + \delta_{\alpha\delta}\delta_{\beta\gamma}) + \frac{2\mu}{|Y|}\int_{Y_F} D_{\xi\eta}(\vec{\Psi}^{\alpha\beta}) D_{\xi\eta}(\vec{\Psi}^{\gamma\delta})\, dy$$

$$-\frac{2\mu}{|Y|}\frac{\int_{Y_F} D_{\xi\eta}(\vec{\Psi}^{\alpha\beta}) D_{\xi\eta}(\vec{\Psi}^{\omega}) dy \int_{Y_F} D_{\xi\eta}(\vec{\Psi}^{\gamma\delta}) D_{\xi\eta}(\vec{\Psi}^{\omega})\, dy}{\int_{Y_F} D_{\xi\eta}(\vec{\Psi}^{\omega}) D_{\xi\eta}(\vec{\Psi}^{\omega})\, dy},$$

$$\tilde{\tilde{\sigma}}^0_{\alpha 3} = 0, \quad \tilde{\tilde{\sigma}}^0_{33} = -P^0_h(t,x) + \sigma^*_{33}(t,x_1,x_2). \tag{21}$$

As in Section 3, σ^*_{33} is undetermined.

The microstructure evolution is determined by (14) which gives in this section:

$$\frac{d\,\overrightarrow{A^0A^\alpha}}{dt}\cdot\vec{e}_\beta = \frac{\partial V^0_\beta}{\partial x_\gamma}(t,x_{A^0})\left(\overrightarrow{A^0A^\alpha}\right)\cdot\vec{e}_\gamma + o(\varepsilon),$$

$$\vec{\Omega}^\varepsilon = \frac{1}{2}\operatorname{rot}\vec{V}^0 + \omega\vec{e}_3 + O(\varepsilon), \tag{22}$$

with ω given by (19). So the microstructure evolves in time by keeping the locally periodic character. Consequently the evolution of the suspension is consistent with the homogenization process performed at each instant t with the basic period $Y(t,x_1,x_2)$. The evolution of the period Y is, according to (22), determined by the microstructure itself and the mean velocity gradient.

From the dependence of the basic period on grad $\vec{V}^0$ and the determination of the viscosity coefficients $a_{\alpha\beta\gamma\delta}$ in (21) arises the non-Newtonian behaviour of the suspension. The suspension is a non-Newtonian anisotropic incompressible fluid.

4.2. The Macroscopic Viscosity Coefficients

The four tensor $a_{\alpha\beta\gamma\delta}$ satisfies the usual properties of symmetry and ellipticity. Furthermore, taking into account the form (21) of the $a_{\alpha\beta\gamma\delta}$ and the properties (18) of the functions $\vec{\Psi}^{\alpha\beta}$, it can be proved that this tensor involves just three different coefficients: a_{1111}, a_{1112} and a_{1212}. As a matter of last the following equalities and minorations occur

$$\begin{aligned} a_{1111} = a_{2222} &= 2\mu(1+c) - a_{1122} \geq 2\mu(1+c), \\ a_{1112} = -a_{1222}, &\qquad a_{1212} \geq \mu(1+c). \end{aligned} \tag{23}$$

As the period of the microstructure is distorted by the flow, the influence of period symmetries on the computations of the $a_{\alpha\beta\gamma\delta}$ (as in (3.2)) may be apply only at one instant. If at one instant Y admits Oy_1 as a symmetry axis $a_{1112} = 0$ and a better lower bound than (23), may be found for a_{1111} in a form similar to (15), which emphasizes as in section 3.2 that in some sense the influence of the concave parts of the fibre section is not important.

5. Conclusions. Viscosity of a Long-Fibres Suspension

The case of a long-fibres ordered suspension is relevant of the general three dimensional theory of suspensions. The macroscopic constitutive relation is that of a non-Newtonian anisotropic incompressible fluid [3,4]

$$\tilde{\sigma}^0_{ij} = -P^0\delta_{ij} + A_{ijk\ell}D_{k\ell}(\vec{V}^0)\,.$$

The results of sections 3 and 4 may be considered as approximations of those of the $A_{ijk\ell}$ that are implied in motions keeping the fibres parallelly aligned in the $\vec{e}_3$ direction. This can be summed up in Table 1, where "?" stands for a coefficient non-involved in Section 3 or 4.

TABLE 1

Approximation of the viscosity coefficients $A_{ijk\ell}$

	$D_{11}(\vec{V}^0)$	$D_{12}(\vec{V}^0)$	$D_{22}(\vec{V}^0)$	$D_{13}(\vec{V}^0)$	$D_{23}(\vec{V}^0)$	$D_{33}(\vec{V}^0)$
$\tilde{\sigma}^0_{11}$	a_{1111}	a_{1112}	$2\mu(1+c)-a_{1111}$	0	0	?
$\tilde{\sigma}^0_{12}$	a_{1112}	a_{1212}	$-a_{1112}$	0	0	?
$\tilde{\sigma}^0_{22}$	$2\mu(1+c)-a_{1111}$	$-a_{1112}$	a_{1111}	0	0	?
$\tilde{\sigma}^0_{13}$	0	0	0	a_{11}	a_{12}	?
$\tilde{\sigma}^0_{23}$	0	0	0	a_{12}	a_{22}	?
$\tilde{\sigma}^0_{33}$	?	?	?	?	?	?

References

1. H. Brenner, *Ann. Rev. Fluid Mech.* **2** (1970) 137.
2. H. Phan-Thien, T. Tran-Cong and A.L. Graham, *J. Fluid Mech.* **228** (1991) 275.
3. T. Lévy and E. Sanchez-Palencia, *J. Non-Newt. Fluid Mech.***13** (1983) 63.
4. E. Sanchez-Palencia, *Non-Homogeneous Media and Vibrations Theory, Lectures Notes in Physics*, **127**, Springer, Berlin, 1980.
5. T. Lévy, In *Homogenization Techniques for Composites Media, Lectures Notes in Physics* **272**, ed. E. Sanchez-Palencia and A. Zaoui, Springer, Berlin, 1987.
6. F. Périn, *Application de la méthode d'homogénéisation aux suspensions de fibres longues et aux suspensions multidisperses de particules,* Thèse de l'Université Pierre et Marie Curie, Paris, 1994.
7. F. Périn and T. Lévy, *Int. J. Engng. Sci.* **32** (1994) 1253.

Continuum Models and Discrete Systems
Proceedings of 8th International Symposium, June 11-16, 1995, Varna, Bulgaria, ed. K.Z. Markov

ON A STATISTICAL PARAMETER IN THE THEORY OF RANDOM DISPERSIONS OF SPHERES

K. Z. MARKOV
Faculty of Mathematics and Informatics,
"St. Kl. Ohridski" University of Sofia, 5 blvd J. Bourchier, BG-1126 Sofia, Bulgaria

Abstract. A two-point statistical parameter which naturally appears in variational bounds in the absorption problem for random media is studied. For random dispersions of nonoverlapping sphere an analytic formula for the parameter is first given through the radial distribution function for the spheres. Analyzing the asymptotic behaviour of the parameter, two kinds of formulae are derived: i) Simple relations between the values of the two-point correlation function and its derivatives at $r = 0$ with the values of radial distribution function and its derivatives at the "touching distance" $r = 2a$. ii) Relations between the moments of the two-point correlation on $(0, \infty)$ and the moments of the radial distribution function. As a simple application, the failure of the well-stirred approximation for sphere fractions higher than 1/8 is finally demonstrated.

1. Introduction

In the theory of random media, when evaluation of their effective macroscopic properties is the aim, the internal random constitution shows up in the final results through certain statistical parameters that incorporate, in an integral form, the multipoint correlations in the media. Presumably the first such parameter appeared in Brown's study [1] of the effective conductivity of weakly inhomogeneous two-phase media. The same Brown's parameter entered later on the well-known variational bounds of Beran [2]. The counterparts of the Beran bounds in the elasticity context and/or Hashin-Shtrikman variational principle involved other and more complicated statistical parameters, see, e.g., the surveys [3,4] for details and references.

If a context, different from conductivity or elasticity, is chosen, different kinds of statistical parameters appear. Consider, for instance, the absorption problem

$$\Delta c(x) - k^2(x)\, c(x) + K = 0,$$

where $c(x)$ is the concentration of a diffusing species absorbed with different rates k_1^2 and k_2^2 in the constituents '1' and '2' respectively of a two-phase random medium (so that $k^2(x)$ is a random field taking the values k_1^2 and k_2^2 depending on whether x lies in '1' or '2'); K is the fixed rate of creation of the species in the bulk of the specimen.

The variational bounds of Beran's type for the effective absorption coefficient of the medium involve, in addition to an integral containing the three-point correlation function in its integrand, the dimensionless two-point statistical parameter (i.e., such in which only two-pair correlation takes part):

$$i_2(p) = p^2 \int_0^\infty re^{-pr}\gamma_2(r)\,dr, \quad p \in (0,\infty), \tag{1.1}$$

where $\gamma_2(r) = \langle I_1'(0)I_1'(x)\rangle/\eta_1\eta_2$ is the usual two-point correlation function, $r = |x|$. (The medium is assumed statistically homogeneous and isotropic); the brackets $\langle\cdot\rangle$ signify ensemble averaging. Here $I_1(x)$ is the characteristic function of the region, occupied by the constituent '1' so that $\langle I_1(x)\rangle = \eta_1$, where η_1 is its volume fraction and $\eta_2 = 1-\eta_1$, $I_1'(x) = I_1(x) - \eta_1$ is the fluctuating part of the field $I_1(x)$. For details we refer the reader to the recent papers [5-7]. Our aim here is to study the parameter (1.1) for random dispersions of equal and nonoverlapping spheres and to extract from its asymptotic behaviour (at $p \to \infty$ and $p \to 0$) certain simple and useful relations and facts concerning this important class of random media.

2. The Evaluation of $i_2(p)$ for Random Dispersions

Hereafter we shall deal with a random dispersion of equal and nonoverlapping spheres. Their centers $\{x_k\}$ form a system of random points, characterized by the usual probability density functions $f_k(y_1,\ldots,y_k)$ [8]. In particular, for the two-point probability density we have $f_2(y_1,y_2) = f_2(r) = n^2 g(r)$, $r = |y_1 - y_2|$, where $g(r)$ is the radial distribution function, $f_1 = n$ is the number density of the spheres, $n = \eta_1/V_a$, $V_a = \frac{4}{3}\pi a^3$; η_1 is their volume fraction. Then $I_1(x) = \sum_k h_a(x-x_k) = \int h(x-y)\psi(y)\,dy$, where $h_a(x)$ is the characteristic function of a single sphere of radius a, located at the origin, $\psi(x) = \sum_k \delta(x-x_k)$ is the so-called random density field for the dispersion [8], $\delta(x)$ denotes the Dirac delta function and the integration is over the whole R^3. Using the fact that $\langle\psi(y)\rangle = n$, $\langle\psi(y_1)\psi(y_2)\rangle = n\delta(y_1-y_2)+f_2(y_1,y_2)$, one easily gets the two-point correlation function in the integral form

$$\gamma_2(x) = \frac{1}{\eta_1(1-\eta_1)}\iint h_a(x-y')h_a(y'')\langle\psi'(y')\psi'(y'')\rangle\,dy'\,dy''$$

$$= \frac{1}{\eta_1(1-\eta_1)}\Big\{n\int h_a(x-z)h_a(z)\,dz + n^2\iint h_a(x-y')h_a(y'')\nu_2(y'-y'')\,dy'\,dy''\Big\}, \tag{2.1}$$

where $\nu_2(z) = g(z) - 1$ is the so-called binary correlation function,

Introducing (2.1) into (1.1) allows to evaluate, after some efforts, the parameter $i_2(p)$. The details of the calculations are given, as a matter of fact in [6]. The final result reads

$$i_2(p) = \frac{A(\tau)-\eta_1 B(\tau)}{1-\eta_1}, \quad A(\tau) = 1 - 3\frac{1+\tau}{\tau^3}e^{-\tau}(\tau\cosh\tau - \sinh\tau),$$

$$B(\tau) = 1 - \frac{36(\tau\cosh\tau - \sinh\tau)^2}{\tau^4}I, \quad I = I(\tau) = \int_1^\infty se^{-2s\tau}g(s)\,ds, \tag{2.2}$$

where $s = r/2a$ and $\tau = ap$ is dimensionless and I is the statistical parameter, that appeared in Talbot and Willis' [9] bounds on the effective absorption coefficient of the dispersion.

Hence from Eqs. (1.1) and (2.2) it is clear that the Laplace transforms of the functions $r\gamma_2(r)$ and $sg(s)$ are comparatively simply connected. This fact allows us to find a number of useful relations between the two-point correlation and the radial distribution function for a dispersion.

3. Asymptotics of $i_2(p)$ as $p \to \infty$ and its Consequences

Consider first the quantity $e^{2\tau}I$

$$e^{2\tau}I = \int_1^\infty se^{-2\tau(s-1)}g(s)\,ds. \tag{3.1}$$

As $p \to \infty$, i.e., $\tau = pa \to \infty$, the function $e^{-2\tau(s-1)}$ tends pointwisely to 0, if $s-1 \geq 0$ and equals 1, if $s - 1 = 0$. Therefore only the behaviour of $g(s)$ around $s = 1$ will matter in the limit $\tau \to \infty$. Let

$$\nu_2(s) = g(s) - 1 = g_0 + g_1(s-1) + g_2(s-1)^2 + \cdots, \quad s \geq 1, \tag{3.2}$$

be the Taylor expansion of the binary correlation at the point $s = r/2a = 1$, i.e., $r = 2a$; the coefficients g_N depend in general on the sphere fraction η_1, $g_N = g_N(\eta_1)$. Obviously

$$g_N = \frac{1}{N!}(2a)^N \nu_2^{(N)}(2a), \tag{3.3}$$

so that knowledge of g_N determines immediately the derivatives of the radial distribution function $g(r)$ at the "touching" distance $r = 2a$.

Inserting (3.2) in (3.1) gives

$$e^{2\tau}I = \sum_{N=0}^\infty \frac{G_N}{(2\tau)^{N+1}}, \quad G_N = N!\,(g_{N-1} + g_N) \ \text{ at } \ N \geq 2, \tag{3.4}$$

$G_0 = 1+g_0$, $G_1 = 1+g_0+g_1$. Note that (3.4) holds only asymptotically at $\tau = ap \gg 1$, since the binary correlation $\nu_2(r)$ is not obliged in general to be analytical for all $r \geq 2a$—the series (3.2) may converge to $\nu_2(r)$ only in a vicinity of the point $s = 1$.

Note that the parameter I for the Percus-Yevick (PY) approximation is analytically known due to Wertheim [10] and hence the coefficients G_N can be easily found. In turn, using (3.3) and (3.4), one can obtain the values of the PY radial distribution function and its derivatives at $r = 2a$, in particular,

$$g_0 = g(2a) = \frac{2+\eta_1}{2(1-\eta_1)^2}, \quad g_1 = -\frac{9}{2}\frac{\eta_1(1+\eta_1)}{(1-\eta_1)^3},$$

$$g_2 = \frac{3\eta_1(1+2\eta_1)^2}{2(1-\eta_1)^4}, \quad g_3 = \frac{\eta_1(1+2\eta_1)^2}{2(1-\eta_1)^4},$$

$$g_4 = -\frac{3\eta_1^2(2+\eta_1)^3}{4(1-\eta_1)^6}\,, \quad g_5 = \frac{3\eta_1^2(8+5\eta_1+5\eta_1^2)}{20(1-\eta_1)^7}\,, \tag{3.5}$$

etc. The coefficients g_N at $N \geq 6$ can be also found analytically, using a symbolic algebra package, but their form will be more and more complicated with k increasing.

Note that the first of these values, i.e. $g(2a)$, was pointed out by Lebowitz [11].

An obvious application of the formulae (3.5) consists in an approximate evaluation of the PY function $g(r)$ in a vicinity of the point $r = 2a$. To this end, truncate the series (3.2), say, after the term $g_5(s-1)^5$, use the values of g_k at $k \leq 5$, see Eq. (3.5), and denote the result by $g_5^{\rm ap}(r)$. The function $g_5^{\rm ap}(r)$ is plotted in Fig. 1 for the values $8na^3 = \frac{\pi}{6}\eta_1 = 0.2$, 0.5 and 1, i.e., for sphere fractions $\eta_1 \approx 0.105$, $\eta_1 \approx 0.262$ and $\eta_1 \approx 0.523$; the dots correspond to the numerical solution of the PY equation, due to Throop and Bearman [12]. Obviously, the higher the sphere fraction, the smaller is the region where the approximation $g_5^{\rm ap}(r)$ is useful. Nevertheless, the latter provides a very good fit to the numerical data in the region $2a \leq r \leq 3a$, if $\eta_1 \approx 0.105$, and in the region $2a \leq r \leq 2.5a$, if $\eta_1 \approx 0.524$.

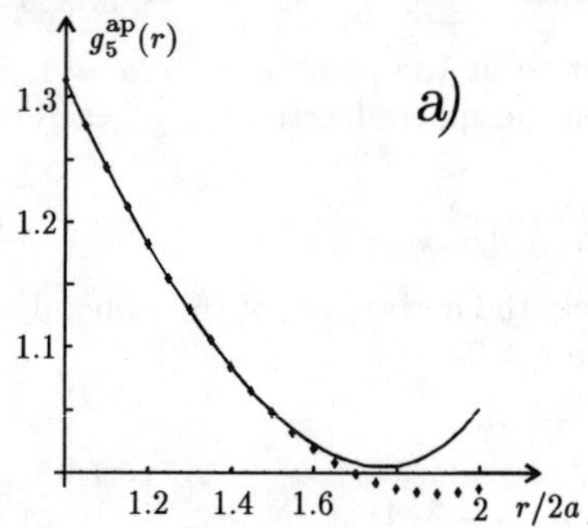

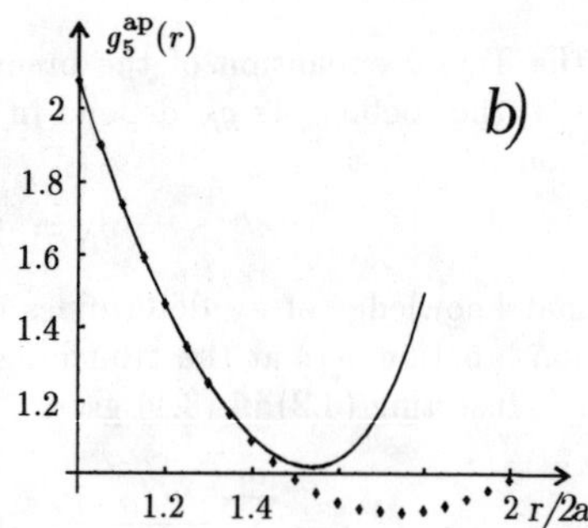

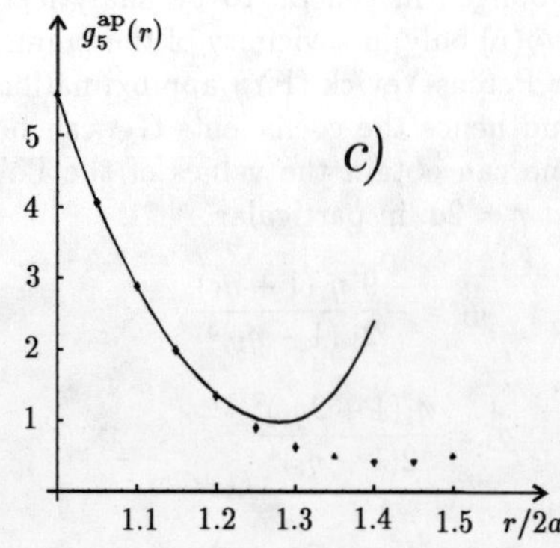

Fig. 1. Plots of the approximation g_5^{ap}. The dots correspond to the Throop and Bearman [12] numerical solution for the PY function. Sphere fractions:
a) $\eta_1 = \pi/20 \approx 0.105$;
b) $\eta_1 = \pi/12 \approx 0.262$;
c) $\eta_1 = \pi/6 \approx 0.523$.

In the well-stirred case $g(r) = 1$ at $r \geq 2a$, so that $g_N = 0$, $\forall N$, and (3.4) yields

$$I = I^{\text{ws}} = \frac{1+2\tau}{4\tau^2}\, e^{-2\tau}, \quad \tau \in (0, \infty), \tag{3.6}$$

which can be directly obtained from (3.1) by elementary integration.

Assume that $i_2(p)$ admits the expansion

$$i_2(p) = 1 + \frac{C_1}{p} + \frac{C_2}{p^2} + \cdots + \varepsilon(\tau), \quad \tau = ap \gg 1, \tag{3.7}$$

$C_0 = 1$. Then

$$\gamma_2^{(N)}(0) = \frac{C_N}{N+1} = \frac{1}{N+1} \lim_{p\to\infty} p^N \left[i_2(p) - \sum_{j=0}^{N-1} \frac{C_j}{p^j} \right], \tag{3.8}$$

$N = 0, 1, \ldots$, which easily follows from the definition (1.1) of $i_2(p)$ and the well-known properties of the Laplace transform. Hereafter $\varepsilon(\tau)$ denotes terms that decrease exponentially as $\tau \to \infty$.

To find the coefficients C_j in the expansion (3.7), note first that

$$A(\tau) = 1 - \frac{3}{2}\left(\frac{1}{\tau} - \frac{1}{\tau^3}\right) + \varepsilon(\tau), \quad \tau \gg 1, \tag{3.9}$$

and rewrite next the coefficient B, see (2.2), in the form

$$B(\tau) = 1 - F(\tau)(e^{2\tau} I),$$

$$F(\tau) = \frac{36(\tau \cosh\tau - \sinh\tau)^2}{\tau^4}\, e^{-2\tau} = 9\,\frac{(\tau-1)^2}{\tau^4} + \varepsilon(\tau), \quad \tau \gg 1. \tag{3.10}$$

The asymptotic expansion of $e^{2\tau} I$ is given in (3.4). Combining the latter with (3.9) and (3.10) and inserting the result into the formula (1.1) for the statistical parameter $i_2(p)$ gives after some algebra

$$i_2(p) = 1 - \frac{3}{2(1-\eta_1)}\frac{1}{\tau} + \frac{3(1+3\eta_1+3g_0\eta_1)}{2(1-\eta_1)}\frac{1}{\tau^3} + \sum_{N=4}^{\infty} \frac{T_N}{\tau^N} + \varepsilon(\tau), \quad \tau \gg 1, \tag{3.11}$$

with the coefficients

$$T_N = \frac{9\eta_1}{2^{N-2}(1-\eta_1)} \Big(G_{N-3} - 4G_{N-4} + 4G_{N-5} \Big), \tag{3.12}$$

$N = 4, 5, \ldots$, assuming $G_j = 0$ at $j < 0$. Using in turn the formula for G_N, see (3.4), gives

$$T_N = \frac{9\eta_1}{2^{N-2}(1-\eta_1)} \Big((N-3)!\, g_{N-3} + (N-7)(N-4)!\, g_{N-4}$$

$$-4(N-5)(N-5)!\,g_{N-5}+4(N-5)!\,g_{N-6}\Big). \tag{3.13}$$

From (3.8) and (3.11) one finds, first of all,

$$\gamma_2'(0)=-\frac{3}{4(1-\eta_1)a}, \quad \gamma_2''(0)=0. \tag{3.14}$$

The first of Eqs (3.14) is a simple consequence of Debye *et al.* formula [13], which connects the specific surface of a two-phase material with $\gamma_2'(0)$. Thus for any dispersion of *nonoverlapping* spheres $\gamma_2'(0)$ is not sensitive to the sphere statistics depending, at a fixed radius a, on the sphere fraction η_1 only. A stronger and more curious fact is embodied into the second relation of (3.14), namely, the vanishing at the origin of the second derivative of the correlation function for such dispersion whatever be the sphere statistics. The assumed spherical shape of the particles is not important here; the fact that $\gamma_2''(0)=0$ is essentially connected with the assumption of nonoverlapping. Indeed, consider a statistically isotropic dispersion of particles of fixed shape, whose location and orientation are both random but not interconnected statistically; the particles should not overlap whatever their orientations at fixed locations. An averaging with respect to orientation first (which is possible, due to the statistical independence of the latter), leads just to a dispersion of nonoverlapping spheres. Each one is obtained through rotation of the particle, centered at the same location; the rotation represents simply the averaging with respect to all possible orientations of the particle. Note that the fact that $\gamma_2''(0)=0$ for a dispersion of nonoverlapping particles was first noticed by Kirste and Porod [14] using different and more complicated geometrical arguments; they also assumed that there are no corner points on the particle's surfaces. This assumption is not necessary, as easily seen from the foregoing reasoning. The results of Kirste and Porod were rederived and extended by Frisch and Stillinger [15] who expanded directly the two-point correlation function $\gamma_2(r)$ at $r=0$ starting, as a matter of fact, with its integral representation (2.1).

According to (3.12), the statistics of the dispersion, that is, the radial distribution function, shows up only in the derivatives $\gamma_2^{(N)}$ at $N\geq 3$. Indeed, from (3.12) and (3.13) it follows

$$\gamma_2'''(0)=\frac{3(1+3\eta_1+3g_0\eta_1)}{8(1-\eta_1)a^3}, \quad \gamma_2^{(4)}(0)=\frac{9(g_1-3(1+g_0))\eta_1}{20(1-\eta_1)a^4},$$

$$\gamma_2^{(5)}(0)=\frac{3(g_2-g_1)\eta_1}{8(1-\eta_1)a^5}, \quad \gamma_2^{(6)}(0)=\frac{9(3g_3-g_2-2g_1+2(1+g_0))\eta_1}{56(1-\eta_1)a^6}, \tag{3.15}$$

and, in general,

$$\gamma_2^{(N)}(0)=\frac{T_N}{(N+1)a^N}, \quad N=7,8,\dots, \tag{3.16}$$

where T_N is expressed in (3.13) by the coefficients g_{N-3}, g_{N-4}, g_{N-5} and g_{N-6}, connected with the local behaviour of the binary correlation $\nu_2(r)$ at the "touching" distance $r=2a$. Note that the first of expressions (3.15)—the value of $\gamma_2'''(0)$—coincides with that given by Kirste and Porod [14] and Frisch and Stillinger [15].

In the well-stirred case all g_N vanish. From (3.13), (3.15) and (3.16) one finds the needed values of $\gamma_2'''(0)$, $\gamma_2^{(4)}(0)$ and $\gamma_2^{(6)}(0)$; all the rest of the derivatives $\gamma_2^{(N)}(0) = 0$ at $N = 4$ and $N \geq 7$ vanish in this case. Thus in a certain vicinity of the origin the two-point correlation function of the well-stirred dispersion is the polynomial

$$\gamma_2(r) = 1 - \frac{3}{4(1-\eta_1)}\frac{r}{a} + \frac{1+3\eta_1}{16(1-\eta_1)}\left(\frac{r}{a}\right)^3 - \frac{9\eta_1}{160(1-\eta_1)}\left(\frac{r}{a}\right)^4 + \frac{\eta_1}{2240(1-\eta_1)}\left(\frac{r}{a}\right)^6. \tag{3.17}$$

Note that the function $\gamma_2(r)$ should vanish at $r = 4a$ in the well-stirred case under study. The polynomial (3.17) does not possess this property which means that $\gamma_2(r)$ is not analytical on the whole semiaxis $(0, \infty)$ and hence (3.17) holds in a certain vicinity of the origin $r = 0$. Indeed, a direct analytical computation, details of which will be reported elsewhere, shows that (3.17) holds only at $r \leq 2a$. In the point $r = 2a$, $\gamma_2^{(4)}(r)$ is discontinuous. It is to be mentioned that the above computation allows us to claim that the general formula for $\gamma_2^{(4)}(0)$, given by Frisch and Stillinger [15], is not correct, since in the well-stirred case it does not yield the respective value in (3.17).

4. Asymptotics of $i_2(p)$ as $p \to 0$ and its Consequences

Note immediately that at small $p \ll 1$:

$$i_2(p) = \theta_1\tau^2 - \theta_2\tau^3 + \cdots = \tau^2\sum_{N=0}^{\infty}\frac{(-1)^N}{N!}\theta_{n+1}\tau^N, \tag{4.1}$$

$$\theta_N = \int_0^\infty t^N \rho_2(r/a)\,dt, \quad t = r/a, \tag{4.2}$$

so that θ_N are the moments of the correlation function $\gamma_2(r)$ on the semiaxis $(0, \infty)$. To connect these moments with the appropriate moments of the binary correlation note first that

$$I = I(\tau) = \int_1^\infty se^{-2\tau s}g(s)\,ds = \frac{1+2\tau}{4\tau^2}e^{-2\tau} + \int_1^\infty se^{-2\tau s}\nu_2(s)\,ds, \tag{4.3}$$

using the definition $\nu_2(s) = g(s) - 1$ of the binary correlation, $s = r/2a$. The first term in the right side of (4.3) is just the parameter $I = I^{\rm ws}$ in the well-stirred case, already known, see (3.6).

Expand next $I(\tau)$, as given in (4.3), around $\tau = 0$:

$$I(\tau) = \frac{1+2\tau}{4\tau^2}\left(1 - 2\tau + \frac{(2\tau)^2}{2!} - \cdots\right) + \left(m_1 - 2\tau m_2 + \frac{(2\tau)^2}{2!}m_3 - \cdots\right)$$

$$= \frac{1}{4\tau^2}\left(1+\sum_{N=2}^{\infty}\frac{(-1)^{N-1}(1-Nm_{N-1})}{N\,(N-2)!}\,(2\tau)^N\right)$$

$$= \frac{1}{4\tau^2}\left(1-2(1-2m_1)\tau^2+\frac{8}{3}(1-3m_2)\tau^3-2(1-4m_3)\tau^4+\cdots\right), \tag{4.4}$$

where

$$m_l = \int_1^\infty s^l\,\nu_2(s)\,ds, \quad s=r/2a, \tag{4.5}$$

$l=0,1,\ldots$, are the moments of the binary correlation on the semiaxis $(1,\infty)$.

Insert the series (4.4) for I into the formula (1.1) for the statistical parameter $i_2(p)$, and use that

$$A(\tau)=12\tau^2\sum_{N=0}^{\infty}\frac{(-1)^N\,(N+1)\,(N+4)}{(N+5)!}\,(2\tau)^N=\tau^2\left(\frac{2}{5}-\frac{1}{3}\tau+\frac{6}{35}\tau^2-\cdots\right),$$

$$F(\tau)e^{2\tau}=9\sum_{N=0}^{\infty}\frac{(N+1)\,(2N+5)\,2^{2(N+3)}}{(2(N+3))!}\,(2\tau)^N=4\tau^2\left(1+\frac{1}{5}\tau^2+\frac{3}{175}\tau^4+\cdots\right),$$

see (2.2) and (3.10). Then

$$i_2(p)=\tau^2\left[\frac{2/5-\eta_1(9/5-4m_1)}{5(1-\eta_1)}-\frac{1-8\eta_1(1-3m_2)}{3(1-\eta_1)}+\cdots\right],$$

which, when compared to (4.1), gives the interconnection between the moments θ_N of the two-point correlation $\gamma_2(r)$ and the moments m_l of the binary correlation $\nu_2(r)$ for a dispersion of nonoverlapping spheres. In particular,

$$\theta_1=\frac{2-\eta_1(9-20m_1)}{5(1-\eta_1)}, \quad \theta_2=\frac{1-8\eta_1(1-3m_2)}{3(1-\eta_1)}, \quad \text{etc.} \tag{4.6}$$

The formulae (4.6) are very convenient, if the binary correlation is given analytically. For instance, in the well stirred case $\nu_2(r)=0$ at $r\geq 2a$, so that all the moments m_l vanish and hence, in particular,

$$\theta_1=\frac{2-9\eta_1}{5(1-\eta_1)}, \quad \theta_2=\frac{1-8\eta_1}{3(1-\eta_1)}, \tag{4.7}$$

in this case.

Note that for any statistically homogeneous and isotropic random medium the moments θ_1 and θ_2 should be nonnegative. (As a matter of fact, this follows from the Bochner-Khinchine theorem which states that the two-point correlation function $\gamma_2(r)$ should be positive-definite for such media [16,17].) An elementary proof of this fact consists in introducing the random fields

$$\chi(x)=\int\frac{1}{4\pi|x-y|}\,I_1'(y)\,dy, \quad \phi(x)=\int I_1'(x-y)\,dy,$$

and noting that $\langle|\nabla\chi|^2\rangle \geq 0$ and $\langle\phi^2\rangle \geq 0$.

The nonnegativeness of θ_1 and θ_2 imposes, through Eq. (4.6), restrictions on the moments m_1 and m_2 of the binary correlation for any realistic dispersion of spheres, namely,

$$m_1 \geq \frac{9\eta_1 - 2}{20\eta_1}, \quad m_2 \geq \frac{8\eta_1 - 1}{24\eta_1}. \tag{4.8}$$

Hence the well-stirred approximation, for which $m_1 = m_2 = 0$, is realistic only at $\eta_1 \leq 1/8$—something conjectured by Willis [18], who noticed that a certain well-known scheme of mechanics of composites in the wave propagation context yields unrealistic predictions for this approximation, if $\eta_1 > 1/8$.

Acknowledgements. The support of the Bulgarian Ministry of Education and Science under Grant No MM 416-94 is gratefully acknowledged.

References

1. W. F. Brown, *J. Chem. Physics* **23** (1955) 1514.
2. M. Beran, *Nuovo Cimento* **38** (1965) 771.
3. K. Z. Markov and Kr. D. Zvyatkov, *Advances in Mechanics (Warsaw)* **14**(4) (1991), p. 3.
4. K. Z. Markov and Kr. D. Zvyatkov, In: *Recent Advances in Mathematical Modelling of Composite Materials*, ed. K. Z. Markov, World Sci., 1994, p. 59.
5. K. Z. Markov and M. K. Kolev, *Int. J. Engng Sci.* **32** (1994) 1859.
6. M. K. Kolev and K. Z. Markov, *Annuaire Univ. Sofia, Fac. Math. Inf.* **86/1992** (1994) 30.
7. K. Z. Markov, D. R. S. Talbot and J. R. Willis, *IMA J. Appl. Math.*, submitted.
8. R. L. Stratonovich, *Topics in Theory of Random Noises*, Volume 1, Gordon and Breach, New York, 1967.
9. D. R. S. Talbot and J. R. Willis, *Mech. Materials* **3** (1984) 171, 183.
10. M. S. Wertheim, *Phys. Rev. Lett.* **10** (1963) 321.
11. J. L. Lebowitz, *Phys. Rev.* **133** (1964) A895.
12. G. H. Throop and R. J. Bearman, *J. Chem. Phys.* **42** (1965) 2408.
13. P. Debye, H. R. Anderson, Jr. and H. Brumberger, *J. Appl. Phys.* **28** (1957) 679.
14. R. Kirste and G. Porod, *Kolloid-Z.* **184** (1962) 1.
15. H. L. Frisch and F. H. Stillinger, *J. Chem. Phys.* **38** (1963) 2200.
16. A. M. Yaglom, *An Introduction to the Theory of Stationary Random Functions*, Dover, New York, 1973.
17. E. Vanmarcke, *Random Fields: Analysis and Synthesis*, MIT Press, Cambridge, Massachusetts and London, England, 1983.
18. J. R. Willis, *J. Mech. Phys. Solids* **28** (1980) 307.

Continuum Models and Discrete Systems
Proceedings of 8th International Symposium, June 11-16, 1995, Varna, Bulgaria, ed. K.Z. Markov
©World Scientific Publishing Company, 1996, pp. 250–257

MICROBUCKLING AND DELAMINATION EFFECTS ON THE COMPRESSION BEHAVIOR OF MATERIALS WITH LAYERED MICROSTRUCTURE

O. I. MINCHEV
Department of Solid Mechanics, Institute of Mechanics,
Bulgarian Academy of Sciences, Acad. G. Bonchev str., bl. 4,
BG-1113 Sofia, Bulgaria

and

F. G. RAMMERSTORFER
Christian Doppler Laboratory "Micromechanics of Materials" at the
Institute of Lightweight Structures and Aerospace Engineering,
Vienna Technical University, Gusshausstrasse 27-29/317, A-1040 Vienna, Austria

Abstract. The present paper addresses the behavior of materials with layered microstructure subjected to compression in the layers direction. The available experimental evidence shows that microbuckling is one of the dominant compressive failure mechanisms for such materials. In some cases it can be accompanied or preceded by a delamination along the interface boundaries. The process is modeled by considering a micro-section of the material containing an imperfection in the form of a waviness. The formulation accounts for material and geometrical nonlinearities. As a particular case, two-phase titanium aluminide intermetallic alloys are considered. They exhibit, in an as-cast state, a nearly fully lamellar microstructure within the material grains. In respect with their practical application the microstructure must be changed by a quasi-static pressing from a lamellar into a globular, fine-grained one. This causes the lamellar microstructure to deform by microbuckling.

1. Introduction

The dominant compressive failure mechanism observed at materials with layered microstructure is microbuckling. In some cases it can be accompanied or preceded by a delamination along the interface boundaries. The delamination effects on the composite behavior are to be studied because, depending on the material and microgeometrical characteristics, they may dominate completely and lead to an essential reduction of the composite's compressive strength. Microbuckling is characterized by the formation of kink bands in which one or both constituents undergo material instability and/or localized failure. It is believed that the bands formation starts in a vicinity of existing imperfections in the form of misalignments or wavinesses.

In the present paper, as a particular case, two-phase titanium aluminide intermetallic alloys with a specific lamellar microstructure in an as-cast state are considered, consisting of alternating layers of γ-TiAl and α_2-Ti_3Al platelets. Such titanium aluminides show a significant potential of applications at high temperatures, but their poor ductility and brittle fracture at room temperature make them unlikely to find extensive practical application in an as-cast configuration. Nowadays, many attempts for finding ways to overcome these drawbacks are under development. The improvements are to be attributed to mechanisms related to the decomposition of the lamellar microstructure into a fine-grained one. This can be done by a compressive deformation during the primary forming process. Under significant compression the layers that are parallel to the loading axis show a locally buckled configuration and the refinement of the lamellar microstructure takes place.

However, the detailed mechanisms by which the lamellar structure affects the material properties are not clearly understood and need further investigation. For this purpose the modeling and the understanding of the deformation behaviour under compression is very important. The investigation of this process can therefore contribute for estimating the overall strength of layered composites and, as in the case of titanium aluminides, to give valuable information for further alloy design towards improved mechanical properties. The aim of the present investigation is to examine the main mechanisms that are the origin of the failure of the layered microstructure.

2. Problem Formulation

A detailed overview of several studies concerned with the compressive failure of aligned fiber and layered composites can be found, e.g., in [1-3]. As noted by Budiansky and Fleck [1], an adequate understanding of these phenomena requires more than the traditional ideas of stability and bifurcation of equilibrium paths. Motivated by the available experimental evidence, many authors proposed microbuckling as the dominant failure mechanism. The process is characterized by a concentration of deformation near certain stress raisers leading to localized bending. This, in turn, results in the local formation of microbuckles that develop into kink bands, i.e. kinking can be viewed as a next stage of that what might be called microbuckling process rather than a separate mechanism. The kink bands are of distinct widths oriented at a non-zero angle to the loading axis. The deformation within the buckled/kinked region grows to significant values and after a smooth attainment of a peak value the overall compressive stress lowers. It is considered that an overall material instability appears when this peak is reached. As an example, in Fig. 1 are shown typical compressive strain-stress curves of titanium aluminides (data from [5]). It is believed that the formation of the above mentioned zones of localized bending originates in a vicinity of imperfections existing in the material. Such an idea was first proposed by Argon (see Budiansky and Fleck [1] and the references therein) for aligned fiber composites. The experiments suggest that the imperfections are in a form of wavinesses or misalignments. From here it can be concluded that the analysis must follow the non-linear response of an imperfect microstructure and is not to be a bifurcation analysis. Semi-analytical-numerical studies of bifurcation analysis of layered materials are treated by other authors (see, e.g., [7]), and studies of the static behavior of laminated composites with periodically or locally curved microstructure can be found also in [8]. Although certain bonding exists between the layers, the process can be preceded or accompanied by a delamination along the layers interfaces followed, in some cases, by a crack opening and growth. The present paper is concerned with investigating as to how the actual imperfections govern the formation of microbuckles and/or kink bands, i.e. which initial waviness is the most likely one to be the source of them, and how the interface delamination and crack opening affects the deformation of multi-layered materials.

Experiments showing the changes of the microstructure of titanium aluminides with compression deformation were reported by Tsujimoto *et al.* [4]. By a small amount of deformation, microbuckling leading to a zig-zag formation of the lamellae occurs in the grains with lamellae that are parallel to the compression axis, and with increasing the amount of deformation they decompose (affected additionally by dynamic recrystallization) into an equiaxed two-phase structure consisting of α and γ grains. In the lamellae perpendicular to the compression axis any decomposition does not occur. Some lamellae being not parallel to the compression axis rotate to the orientation being perpendicular to it. Further lamellae decomposition by deformation in this direction cannot be expected, because only lamellae perpendicular to the compression axis are remained. Therefore, a compression from different direction is needed in order to change all the lamellae into equiaxed grains. For this reason, the case when the lamellae are parallel to the compression axis was modeled and computationally investigated in the present study.

3. Computational Analysis

Due to its local nature the process is modeled by considering a small section of the material grain consisting of a few alternating thin α_2 and thick γ layers surrounded from both sides by a relatively large area of material with smeared, i.e. effective, properties. The thicknesses ratio is assumed to be 1:10. One α_2 layer contains an initial imperfection in the form of a waviness. The model accounts for material and geometrical nonlinearities. Both composite constituents undergo finite deformations and are modeled as isotropic elastic-plastic solids with hardening by the J_2 flow theory of plasticity. The experimental data from [6] indicates that the single α-phase shows work hardening at low plastic strains (less than 2%), after which the material is nearby perfectly plastic. The single γ-phase shows a relatively higher strain hardening tendency. The uniaxial stress vs. plastic strain curves of the single phases used in the analysis are given in Fig. 2.

Micromechanical models based on the Finite Element Method allow for a representation of the expected nonlinear effects due to inelastic deformations and geometrical nonlinearities. For this purpose the computer code ABAQUS is used for the solution of the nonlinear problem described above. Eight nodded biquadratic $2D$ elements for plain strain states are used for the domain discretization. The interface interactions between the constituents are modeled by interface elements providing the possibility for the account of large relative displacements and crack opening along the interface. The sliding along the interface when the layers are in contact is assumed to be frictionless. In the present analysis it is assumed that debonding is the result of normal separation only and occurs when the traction transmitted through the interface reaches the lower yield limit of the both constituents (in the present case of the γ-phase). For comparison purposes an analysis was performed when only sliding along the interface is allowed to occur. This analysis is computationally much cheaper but, as the experiments suggest and the obtained results confirm that, in some cases may not suffice for the simulation of the process.

4. Results and Discussion

As expected, the obtained response is very sensitive to the geometry of the imperfections. A detailed parametric study has been performed for investigating the influence of their geometric characteristics (length and altitude) on the onset of failure mechanisms. It was found that, for imperfections with relatively small lengths, at first a delamination of the imperfect layer occurs followed by a crack opening along the interface. The deformation concentrates in a small vicinity of the imperfection including the debonded α_2 layer and the two neighbour γ layers. This leads to an essential bending of the α_2 layer and to high plastification of the two γ layers in the zones where the α_2 layer contacts them (see Figs.3 and 4). A small loading increase causes a rapid increase of the plastic strain mainly of the debonded α_2 layer. It is reasonable to expect that its bending strength will be easily exceeded. Due to the high plastification and bending it can be expected to develop a kind of damage process within the γ layers as well (e.g., recrystallization). The calculations indicate the formation of relatively narrow bands of concentrated plastics strain that are spread across the neighbour layers. They originate from the top of the waviness and are oriented at roughly 45° to the loading axis. It can be expected that they will develop as kink bands with deformation progress. In such a way, it can be concluded that in this case the crack opening affects essentially the microstructure behavior and must not be neglected. This is illustrated in Fig. 4 where the plastification within the constituents for the cases with and without crack opening is shown. In addition, it should noted that the mechanism leading to material instability is microbuckling, and kink bands formation can occur eventually after the failure of the layers within the buckled region.

It was found that the microstructure behavior is quite different when the initial waviness is relatively long. In the beginning of the process a very small crack opening near the imperfection occurs. This crack closes after a small plastification of the γ phase in the early stage of deformation and only a sliding along this interface takes place. This permits the transmission of normal tractions across the interface and leads to a wider spread of the plastic deformations around the stress raiser. As a consequence, in this case the deformed configuration switches from a microbuckled shape to a kinking. The kinking bands originate in the areas near the both ends of the waviness where the plastic strain concentrates. They have a well defined width of the order of the waviness length. The orientation angle with respect to the loading axis is much lower than in the previous case - less than 30°. With increasing loading a

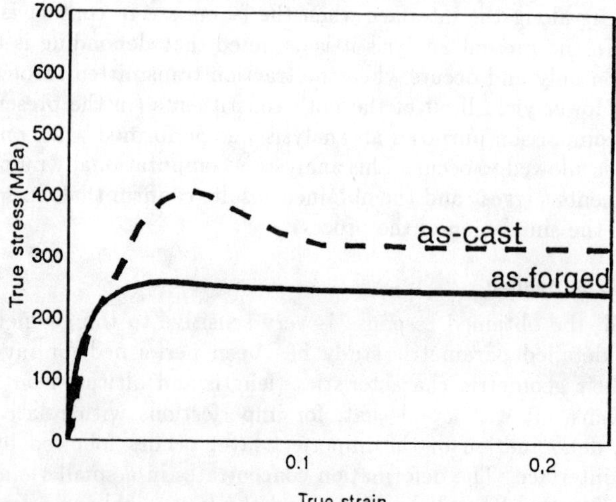

Fig.1. Compressive strain-strain curves of as-cast and as-forged $Ti48Al2Cr$ at 1200°C (data from [5]).

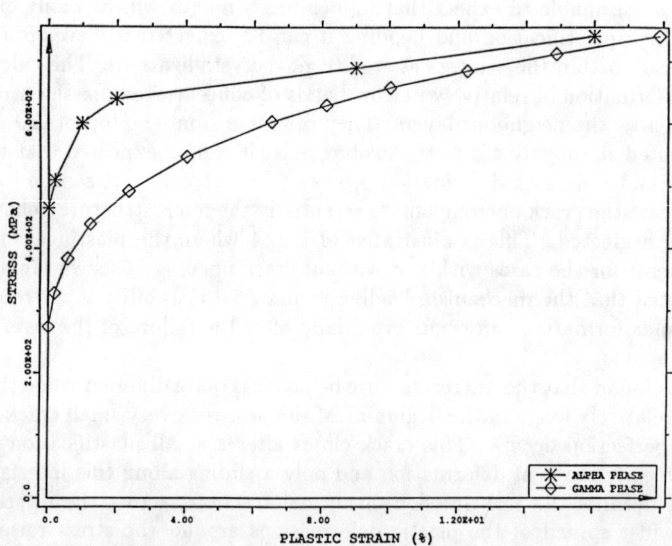

Fig.2. Uniaxial stress vs. plastic strain curves of the single phases.

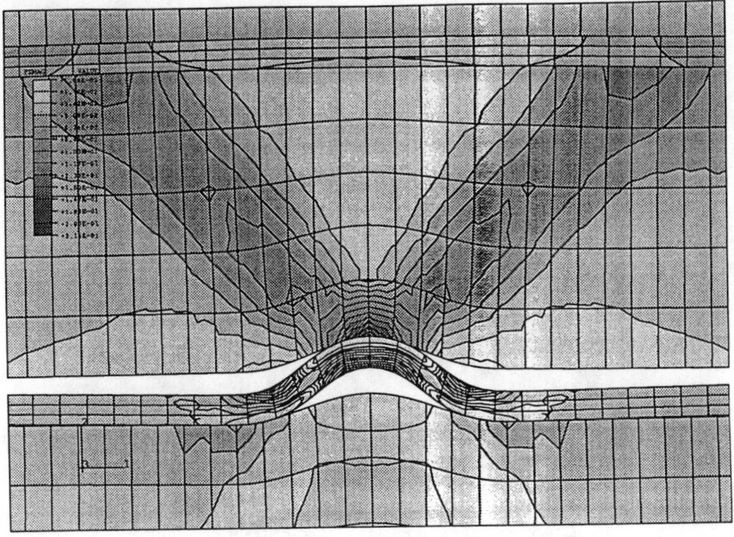

Fig.3. Crack opening and effective plastic strain distribution near initial imperfection with length 10 μm and altitude 1 μm.

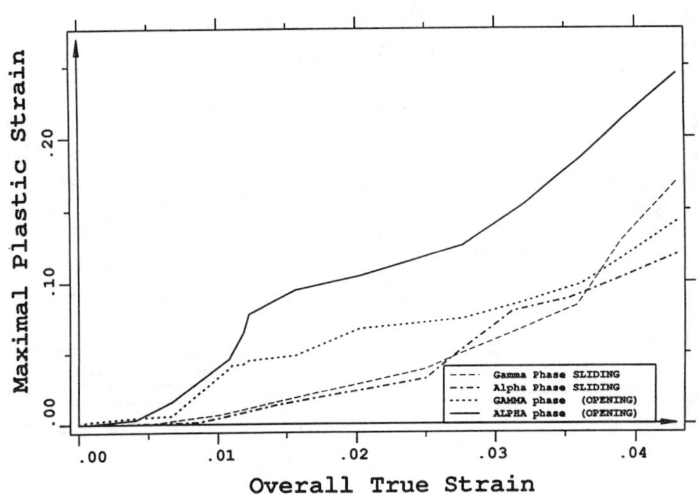

Fig.4. Maximal plastic strain vs. overall true strain of the composite constituents with and without crack opening near initial imperfection with length 10 μm and altitude 1 μm.

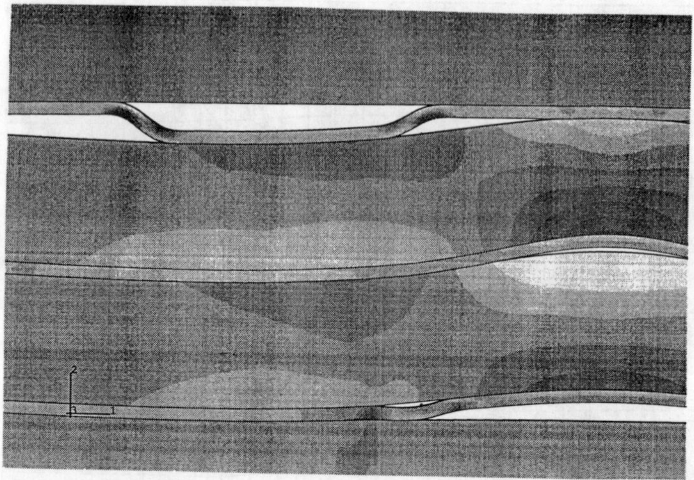

Fig.5. Crack opening and effective plastic strain distribution near initial imperfection with length 40 μm and altitude 1 μm. (the displacements are magnified by a factor of 5)

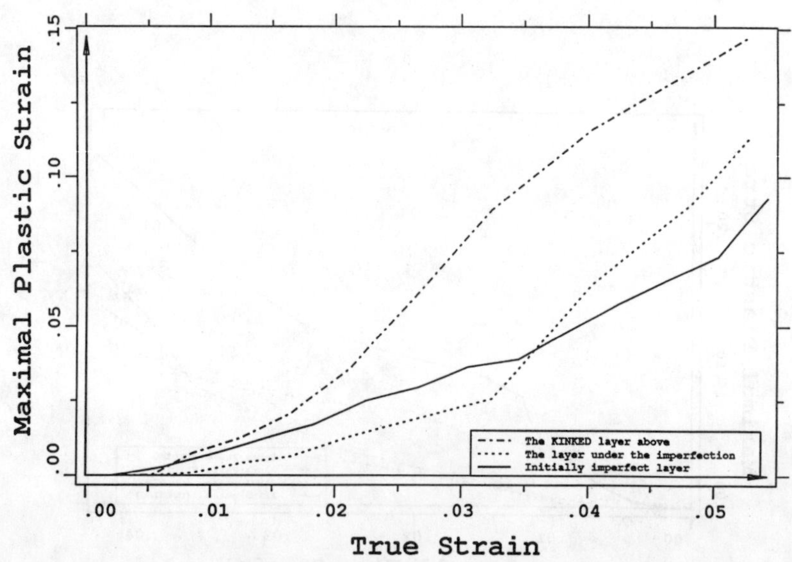

Fig.6. Maximal plastic strain vs. overall true strain of the α layers near initial imperfection with length 40 μm and altitude 1 μm.

crack opens along the neighboring α_2/γ interfaces that are within the kinking bands. This facilitates essentially the layers bending along the boundary between the kinked and unkinked region. The computations show clearly the appearance of two zones of localized bending of the thin α_2 layers near the both boundaries of the kinking band (see Fig. 5). A material instability takes place there and a breakage of the α layers can be expected to occur. The plastification of the initially imperfect α layer and the two neighbour layers is shown in Fig. 6. The γ layers suffer an essential bending and plastification, too. As in the case of short imperfections, it is realistic to expect the development of a recrystallization with further deformation.

Thus, it can be summarized that kinking is the final result of the localization process but the mechanisms that lead to it are different. The performed computational analysis shows that the imperfections of the microstructure play a decisive role on the determination of the failure mechanism.

Acknowledgments. The first author was sponsored by the the Austrian Science Foundation, FWF, as a Lise-Meitner Research Fellow, grants No. M0045-TEC and M0179-TEC. This financial support is gratefully acknowledged. This work contributes to the activities of the Christian Doppler Laboratory of Micromechanics of Materials at the Institute headed by the second author.

References

1. B. Budiansky and N. Fleck, *Appl. Mech. Rev.* **47** (1994) No. 6, part 2.
2. S. Kyriakides, R. Arseculerante, E. J. Perry and K. M. Liechti, *Int. J. Solids and Structures* **32** (1995) 689.
3. J. Lee, Z. Gürdal and O. H. Griffin, Jr., *AIAA Journal* **31** (1993) 331.
4. T. Tsujimoto, K. Hashimoto and M. Nobuki, *Materials Transactions, JIM* **33** (1992) 989.
5. K. Wurzwallner, P. Shretter, H. Clemens, In *Proceedings of the 13-th International Plansee Seminar*, eds. H. Bildstein and R. Eck, **3**, 1991, p. 537.
6. Prasad Rao and K. Tangri, *Materials Science and Engineering* **A132** (1991) 49.
7. I. A. Guz', In *Proc. ICCM/9*, ed. A. Miravete, Vol. VI, 1993, p. 377.
8. S. D. Akbarov and A.N. Guz', *Appl. Mech. Rev.* **45** (1992) 17.

Continuum Models and Discrete Systems
Proceedings of 8th International Symposium, June 11-16, 1995, Varna, Bulgaria, ed. K.Z. Markov
© World Scientific Publishing Company, 1996, pp. 258–267

ELASTIC PERCOLATION BEHAVIOR OF SOLID-LIQUID COMPOSITES

PING SHENG
Exxon Research and Engineering Co., Rt. 22 East, Clinton Twp., Annandale, New Jersey 08801, USA
and Department of Physics, Hong Kong University of Science and Technology, Clear Water Bay, Kowloon, Hong Kong

and

MINYAO ZHOU
Exxon Research and Engineering Co., Rt. 22 East, Clinton Twp., Annandale, New Jersey 08801, USA

Abstract. Elastic percolation of solid-vacuum composites involves both the bulk and the shear moduli approaching zero at the percolation threshold. For solid-liquid composites, however, only the shear modulus vanishes at the percolation threshold. Through both physical argument and finite-size scaling calculations, we show that the critical shear rigidity exponent of 2D solid-liquid composites differs significantly from the usual solid-vacuum elastic exponent. In fact, the new shear rigidity exponent is almost identical to the 2D electrical exponent, although the underlying physics are very different.

Introduction

Percolation is a universal phenomenon in two-component composites. As the volume fraction p of one of the components, let us say component 1, varies, there is a critical value p_c at which component 1 forms a barely-connected infinite network that spans the whole sample, and below which forms only disconnected clusters. While the value of p_c can vary widely from one system to the next, the manner in which the composite properties vary in the vicinity of the percolations threshold has been shown to universally follow a power law behavior as a function of $p - p_c$. For example, the electrical conductivity of a conductor-insulator composite behaves as

$$\sigma \propto (p - p_c)^t \tag{1}$$

near the threshold, where p denotes the volume fraction of the conductor, and $t \approx 1.2$ for two-dimensional samples, and 2 for three-dimensional ones. The exponent t is known as one of the critical exponents, which normally depend only on the spatial dimensionality of the sample.

About a decade ago, it was found that if instead of electrical conductivity, which is a scalar property, one was to look at elasticity, which is a tensorial property in the sense that the stress and strain are related to each other via the elasticity tensor, the critical exponent of how the shear and bulk moduli of a solid-vacuum composite would vanish at the threshold differs significantly from the conductivity exponent. In particular, for 2D samples the critical exponents for the bulk modulus and the shear modulus are demonstrated, both experimentally and through simulations, to be the same and have the value of ≈ 3.5 [1-4]. This difference in the critical behavior is attributed to the vector character of elasticity [5-8], as opposed to the scalar character of conductivity.

A simple physical argument can be presented which shows that the elastic exponent must be much larger than the conductivity exponent. However, in order to appreciate this argument it is necessary to first make a digression on the "finite size scaling" approach [9] for studying composite properties near the percolation threshold. In this approach, the aim is to see how the particular composite property under consideration varies as a function of the sample size. The scaling hypothesis states that, *very close to the percolation threshold* the particular property, let us say the conductivity, must vary as

$$\sigma = \sigma_0 \left(\frac{L}{a}\right)^{-X} f\left(\frac{L}{\xi}\right), \tag{2}$$

where a is the lattice constant for lattice systems and the minimun feature size for the continuum system, ξ is the structure correlation length for the connected infinite network, $f(x)$ is a certain universal function of the dimensionless variable L/ξ. The basic physics underlying the scaling hypothesis is that there are only two length scales in the problem, a and ξ, and the variation of the material property as a function of the sample size L can only appear in the normalized dimensionless varibles L/a and L/ξ. The power-law form $(L/a)^{-X}$ in Eq. (2) is a plausible assumption based on the fact that the geometry underlying the material properties is fractal-like *inside* a structure correlation length, and this geometric characteristics is manifested in the relevant material properties. For this assumption to be true, it is necessary for the function $f(x)$ in Eq. (2) to have the property that

$$f\left(\frac{L}{\xi}\right) \rightarrow \text{const} \quad \text{for} \quad \xi > L \rightarrow 0\,.$$

On the other hand, for $\xi > L$ it is expected that the intensive material properties, such as the conductivity, should be independent of the sampling size, since the system is homogeneous on a scale larger than ξ. This can be true only if

$$f\left(\frac{L}{\xi}\right) \rightarrow \left(\frac{L}{\xi}\right)^{X} \quad \text{for} \quad L \gg \xi\,, \tag{3}$$

so that the variable L disappears from Eq. (2) in that limit. However, precisely in that limit it becomes possible to relate the exponent with the exponent X; because we know that for infinite samples

$$\sigma = \sigma_0(p - p_c)^t = \sigma_0(a/\xi)^X \propto (p - p_c)^{X\nu}, \tag{4}$$

where we have used the known fact that the structure correlation length of the infinite connected network diverges near the percolation threshold as

$$\xi = a(p - p_c)^{-\nu}. \tag{5}$$

It follows from Eq. (4) that

$$X = \frac{t}{\nu}. \tag{6}$$

In this way, one can do simulations on finite samples (at $p \approx p_c$ to insure that $L < \xi$) and obtain the critical exponent t from the size dependence of the simulated conductances, assuming the exponent ν to be known. By using similar argument, it follows that if the bulk and shear moduli $\overline{\kappa}$ and $\overline{\mu}$ of solid-vacuum composites vary with the volume fraction as

$$\overline{\kappa},\ \overline{\mu} \propto (p - p_c)^T, \tag{7}$$

then T can be obtained from the sample size dependence L^{-Y} of the finite-size simulations near $p \approx p_c$ as

$$T = Y\nu. \tag{8}$$

On the basis of the finite-size effect, a rough estimate of the elastic critical exponent T can be obtained from the following physical argument. First, since the connected network is like a tortuous chain, either compression or shearing of this chain will be similarly translated into local shearing, or bending, of the solid. This is basically the reason why the shear response must scale similarly as the compressional response, thus insuring the two moduli must have the same exponent. Second, if we approximate the force required for the shear distortion of the network by that of a plate (in 2D) of length L (which is a very rough model that may be inaccurate), then from elasticity theory it is known that the force F required to give rise to a bending displacement b at one end of the plate, with the other end fixed, is

$$F \propto DL^{-3}b, \tag{9}$$

where D is the stiffness constant. Since the effective shear modulus is measured as

$$\overline{\mu} = \frac{FbL}{b^2} \propto L^{-2} \tag{10}$$

and the value of ν is approximately 4/3, it follows that a rough estimate of T is on the order of 3 for 2D systems, which is indeed much larger than the 2D conductivity exponent of 1.2.

The central concern of this paper, however, is on the alternate elastic percolation problem of solid-liquid composites [10], where only the shear modulus approaches zero at the percolation threshold. The basic motivation of this research is to explore the question of whether there is new critical behavior afforded by the additional degrees of freedom in elasticity problems. What we found is that although liquid has no shear modulus, its presence can nevertheless induce an entirely new shear rigidity percolation behavior. Surprisingly, the new shear rigidity critical exponent is almost identical to the scalar conductivity exponent, although the problem here is still vector in character. Below we first discuss the physics of the problem, followed by the presentation of numerical finite-size scaling results.

Physics of the Solid-Liquid Percolation Problem

Consider a 2D checkerboard in which each square can be either solid, with probability p, or liquid (or empty), with probability $1-p$. Because of the underlying square symmetry of the system, there are three effective elastic constants. However, in this work we will be concerned only with two, $\overline{C}_{11}$ and $\overline{\mu}$. At $p > p_c$, liquid cannot form a continuous infinite network in 2D. Therefore, it is inevitably segmented into isolated pockets. A liquid pocket can offer resistance to shear deformation only when its area is altered to the *same order* as the shear displacement. Thus, a pocket in the shape of a parallelogram will offer no resistance to shear, since the change in area is only to the second order of the shear displacement. However, in the case of the percolation cluster geometry, the application of a macroscopic shear strain ε to the sample would result in area changes of the pockets that are first order in ε, since the random geometry would make it very unlikely to have it otherwise. That means the liquid can exert a force on the connected solid network through the perimeters of the pockets, thus stiffening the network against shear deformation. Therefore, the new exponent τ for the shear modulus in a percolating liquid-solid composite, defined as

$$\overline{\mu} \propto (p - p_c)^{\tau},$$

must be smaller in value than T [10]. It should be noted that from the point of view of the solid network the liquid pockets introduce a nonlocal means of stress transmission. As p approaches p_c from above, the average pocket size in the percolation cluster is known to grow in a power-law manner. Since the pocket size controls the range of stress transmission through the liquid, it follows that the presence of the liquid should introduce a new critical behavior to the shear rigidity percolation problem.

Finite-Size Scaling Calculations

Instead of discrete springs with bond-bending restoring force as the basic percolating components, studied in all the previous works, here we consider continuum elastic solid and liquid. This has the advantage that there is no ambiguity about the value of the elastic percolation threshold, since the geometric, electric (if solid is assumed to be conducting and liquid or vacuum to be insulating), and elastic percolation thresholds all coincide and occur at $p_c = 0.5927$ for 2D samples. To carry out the finite-size scaling calculation, we wish to calculate the effective moduli of a strip of width L (in units of the square size) and length $N \gg L$ by using the finite difference method. Here the width of the strip is defined to be in the y-direction, and the length in the x-direction. The value of p is fixed at $p_c = 0.5927$. As the width of the strip increases, the effective moduli are expected to vary as $L^{-T/\nu}$ for the solid-vacuum composites and as $L^{-t/\nu}$ for the shear modulus of the solid-liquid composites.

The starting point of the calculation is the 2D stress-strain relations:

$$\tau_{xx} = (\kappa + \mu)\frac{\partial u_x}{\partial x} + (\kappa - \mu)\frac{\partial u_y}{\partial y}\,, \tag{11}$$

$$\tau_{yy} = (\kappa + \mu)\frac{\partial u_y}{\partial y} + (\kappa - \mu)\frac{\partial u_x}{\partial x}\,, \tag{12}$$

$$\tau_{xy} = \mu\left(\frac{\partial u_x}{\partial y} + \frac{\partial u_y}{\partial x}\right). \tag{13}$$

Here τ_{xx}, τ_{yy} and τ_{xy} ($= \tau_{yx}$) are the three components of the stress tensor, u_x, u_y are the displacements in the two directions, κ is the bulk modulus, and μ is the shear modulus. The shear modulus is set equal to zero for the liquid. For both liquid and solid, $\kappa - \mu =$ Lamé constant λ is taken to be 1 for simplicity. Given the stress tensor components, the force balance equation is simply

$$\nabla \cdot \boldsymbol{\tau} = 0\,. \tag{14}$$

In component form, Eq. (14) can be written as

$$\frac{\partial \tau_{xx}}{\partial x} + \frac{\partial \tau_{xy}}{\partial y} = 0\,, \qquad \frac{\partial \tau_{yy}}{\partial y} + \frac{\partial \tau_{xy}}{\partial x} = 0\,.$$

At the liquid-solid interface, the boundary conditions are the continuity of normal stress, the continuity of normal displacement, and $\tau_{xy} = 0$. The boundary condition at the solid-vacuum interface is simply the vanishing of all the stress components.

For the purpose of finite-difference numerical calculations, the discretization of each square is done with $\Delta x = \Delta y = 1/4$ (sixteen discretization units per square) and a staggered-grid scheme [11], where τ_{xx} and τ_{yy} are defined at the center for each discretization unit square, τ_{xy} is defined at the corners, and the displacements are defined at the midpoints of the sides of every discretization square. To evaluate the effective compressional modulus $\overline{C}_{11}$ or the shear modulus $\overline{\mu}$ of the composite strip, a uniform displacement $u_y = 0.1$ or $u_x = 0.1$ is imposed on the top edge of the strip, while the bottom edge is held fixed. The value of L is varied from 4 to 40, while the strip length N is varied from 100 to 2000. When N is large, there can be many disjoint clusters along the length-wise direction of the strip. As a result, a single calculation in that case is equivalent to averaging over many smaller-N configurations. Once the problem is solved numerically, the modulus is defined by the total elastic energy E as follows. Since

$$E = \frac{1}{2} \times (\text{applied displacement}) \times (\text{net force } F \text{ on the top edge})\,,$$

and

$$F = \frac{1}{L} \times (\text{applied displacement}) \times (\text{effective modulus } \overline{C}_{11} \text{ or } \overline{\mu}),$$

it follows that

$$\overline{C}_{11} \text{ or } \overline{\mu} = \frac{2EL}{(\text{applied displacement } u_y \text{ or } u_x)^2}.$$

The total elastic energy E is obtained from the numerical solution as one-half the sum of the product of the stress with strain over all the discretized units.

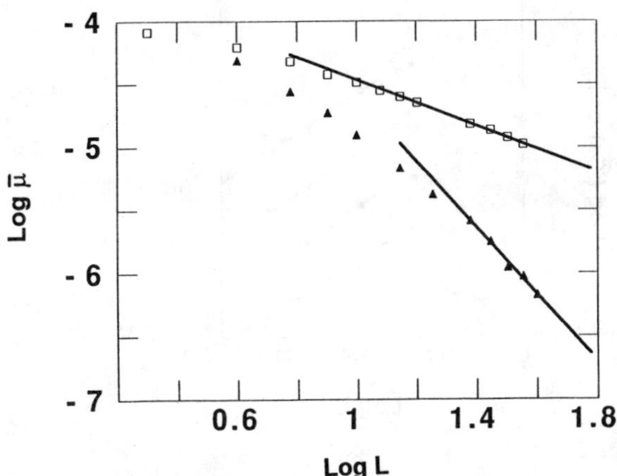

Fig. 1. Variation of the effective shear modulus $\overline{\mu}$ with L, plotted on the log-log scale. Solid triangles are for the solid-vacuum composites and open squares are for the solid-liquid composites. μ for the solid is taken to be 0.5 (in units where $\lambda = \kappa - \mu = 1$). The asymptotic slopes of the least-squares fitted lines are $-2.6\pm$ 0.3 and -0.91 ± 0.02, from which one obtains $T = 3.4\pm0.2$ and $\tau = 1.2\pm0.1$.

Results and Discussion

The main results of the finite-size calculations are shown in Fig. 1, where $\log \overline{\mu}$ is plotted as a function of $\log L$. Two cases are shown: the solid-liquid composites and the solid-vacuum composites. When L exceeds 10, $\log \overline{\mu}$ for both cases are seen to approach an asymptotic linear behavior. For the solid-vacuum composites the asymptotic regime is reached at $L > 24$. The slope obtained from the region of $L = 24-40$ gives $T = 3.4\pm0.2$, in excellent agreement with all previous lattice calculations as well as experimental measurements. For the solid-liquid composites, on the other hand, the asymptotic slope is much less than the solid-vacuum case. The value of τ obtained, 1.2 ± 0.1, is noted to coincide very closely with the electrical percolation exponent, although the underlying physics is very different. The present model also contains no stress-induced scalar elastic energy term as proposed by Alexander [12], which was one mechanism by which an elastic percolation problem can be mapped onto an electrical percolation problem. Overall, the stiffening effect of the liquid pockets is clearly dominant, and determines the critical exponent τ. However, an analytical theory demonstrating the equality of the electrical exponent and the shear rigidity exponent is still lacking. Therefore, it is not known whether the two exponents are really identical or just fortuitously close in their values.

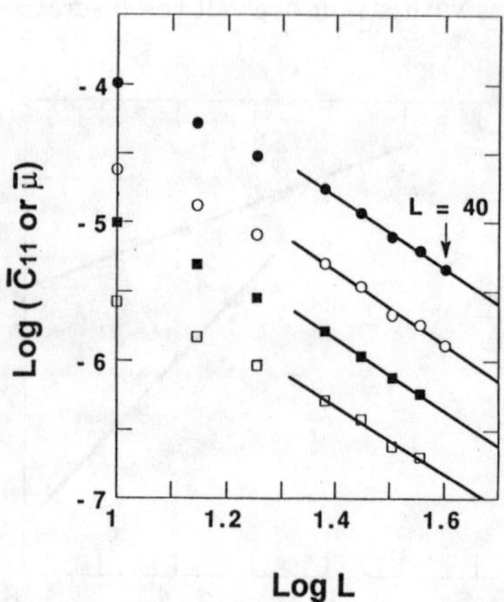

Fig. 2. Variation of $\overline{C}_{11}$ and $\overline{\mu}$ as a function of L, plotted on the log-log scale. Two sets of data are shown, one for $\mu = 1.0$ (circles) and one for $\mu = 0.1$ (squares). The solid symbols are for $\overline{C}_{11}$; the open symbols are for $\overline{\mu}$.

As a side calculation, we have also tested the accuracy of the claim that in a solid-vacuum composite, $\overline{C}_{11}$ and $\overline{\mu}$ have the same critical exponent. The test was carried out with two values of the solid shear modulus, $\mu = 0.1$ and 1. In Fig. 2 are shown the two sets of results. The critical exponents obtained by least-squares fitting of the asymptotic linear regimes are given in Table 1 below. It is seen that the values obtained are all within the numerical error bars of each other. Based on this strong evidence and the accompanying physical argument, it is clear that the critical exponents for $\overline{C}_{11}$ and $\overline{\mu}$ are indeed the same and independent of the material parameters.

Table 1. Values of the exponent T obtained for $\overline{C}_{11}$ and $\overline{\mu}$.

	$\mu = 1.0$	$\mu = 0.1$
$\overline{C}_{11}$	3.46±0.08	3.44±0.06
$\overline{\mu}$	3.52±0.15	3.3±0.23

Universal Critical Ratio Between $\overline{C}_{11}$ and $\overline{\mu}$ for Solid-Vacuum Composites?

The equality of the critical exponents means that there is a constant $\overline{C}_{11}/\overline{\mu}$ ratio at the percolation threshold of solid-vacuum composites. There have been conflicting proposals that the constant has a universal value [8] and that the constant could be dependent on the lattice type [13]. While the two proposals are demonstrated with different models, the universality claim means that if it were true, then the constant should at least be invariant with respect to the lattice type.

The $\overline{C}_{11}/\overline{\mu}$ ratio can easily be calculated by using the above-described numerical approach. The present calculation differs from the previous calculations in the use of continuum elasticity instead of lattice models. It therefore offers an independent check on the claim of universality. We have carried out an evaluation of the critical $\overline{C}_{11}/\overline{\mu}$ ratio as a function of the solid material parameters in the continuum elastic case. In order for the ratio to be independent of the material parameters, the two effective modulus must vary in the same way as a function of the solid material parameters. In Fig. 3 it is shown that indeed $\overline{C}_{11}$ and $\overline{\mu}$ are both linear functions of the solid Young's modulus. Physically, this variation can be justified by Kantor's argument [6] that in a random network, bending displacement rather than compression is the dominant local distortion. Therefore Eq. (9) applies to both types of applied strain. Since the stiffness constant $D \propto Y$, the Young's modulus, the linear relations follow. The ratio of the slopes in Fig. 3 gives $\overline{C}_{11}/\overline{\mu} = 3.4 \pm 0.2$, in excellent agreement with the prior calculation on a hexagonal lattice by Bergman [8] and the estimate by Kantor [6], based on the link to the ratio of the anisotropic radii of gyration for singly connected bonds in a percolation cluster. However, this value differs from those calculated on triangular and square lattices with nearest- and next-nearest-neighbor central forces [13]. Based on this comparison, it can be concluded that the value of the continuum elastic case is closer to the claim of a universal ratio. However, to be more certain, additional calculations using different models than the checkerboard are required. These calculations are still lacking so far.

It should be noted that the question of the $\overline{C}_{11}/\overline{\mu}$ ratio is irrelevant to the solid-liquid composites. In that case the $\overline{C}_{11}/\overline{\mu}$ ratio is always ∞ because $\overline{C}_{11}$ remains finite at the threshold.

Implications for the 3D Case

Based on the physical understanding and the evidences obtained in 2D, it is predicted that the same stiffening effect can occur in 3D composites if there is a considerable amount of isolated liquid pockets. However, 3D percolation differs from the 2D percolation in that both the solid and the liquid components can form infinite connected networks close to the threshold. Therefore, if the liquid can freely flow through the connected channels then there would be minimal effect on the sample's shear rigidity, and the behavior should be given by that of the solid-vacuum composites. But the same reasoning would allow us to argue that the critical shear rigidity behavior should be drastically different for the two cases of a liquid-filled porous medium where one has sealed surfaces and the other one has open surfaces. In the former the trapped liquid plays the same role as in the 2D case. However, it remains to be seen whether the shear rigidity exponent in that case is close to the 3D electrical exponent value of 2.

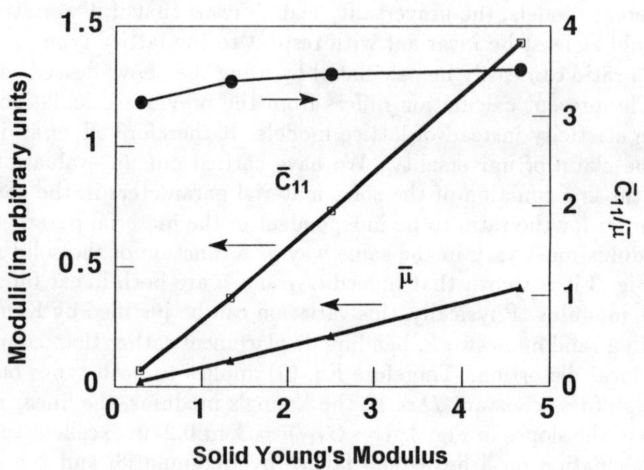

Fig. 3. Variation of $\overline{C}_{11}$ and $\overline{\mu}$ with the Young's modulus of the solid material for the solid-vacuum composite at $p=p_c$ and $L=40$. Excellent linear relations are seen. The ratio $\overline{C}_{11}/\overline{\mu}$ is plotted on top (right scale). It is almost invariant with respect to the Young's modulus and has the average value of 3.4±0.2.

References

1. C. Allain, J. C. Charmet, M. Clement and L. Limat, *Phys. Rev. B* **32** (1985) 7552.
2. L. C. Allen, B. Golding and W. H. Haemmerle, *Phys. Rev. B* **37** (1988) 3710.
3. D. Deptuck, J. P. Harrison and P. Zawadzki, *Phys. Rev. Lett.* **54** (1985) 913.
4. C. J. Lobb and M. G. Forrester, *Phys. Rev. B* **35** (1987) 1899.
5. Y. Kantor and I. Webman, *Phys. Rev. Lett.* **52** (1984) 1891.
6. Y. Kantor, *J. Phys. A* **17** (1984) L843.
7. S. Feng and M. Sahimi, *Phys. Rev. B* **31** (1985) 1671.
8. D. J. Bergman, *Phys. Rev. B* **31** (1985) 1696.

9. D. Stauffer, *Introduction to Percolation Theory*, Taylor and Francis, London, 1985.

10. M. Zhou and P. Sheng, *Phys. Rev. Lett.* **26** (1993) 4358.

11. B. Fornberg, *SIAM J. Numer. Anal.* **27** (1990) 904.

12. S. Alexander, *J. Phys. (Paris)* **45** (1984) 1939.

13. E. L. Garboczi and M. F. Thorpe, *Phys. Rev. B* **31** (1985) 7276; E. L. Garboczi and M. F. Thorpe, *Phys. Rev. B* **33** (1986) 3289.

Continuum Models and Discrete Systems
Proceedings of 8th International Symposium, June 11-16, 1995, Varna, Bulgaria, ed. K.Z. Markov
© World Scientific Publishing Company, 1996, pp. 268–279

VARIATIONAL METHODS FOR ESTIMATING THE EFFECTIVE BEHAVIOR OF NONLINEAR COMPOSITE MATERIALS

P. PONTE CASTAÑEDA
Department of Mechanical Engineering and Applied Mechanics, University of Pennsylvania, Philadelphia, Pennsylvania 19104-6315, USA

Abstract. In this paper, we review some of the methods that have been developed in recent times to estimate the effective behavior of nonlinear composite materials. Basically, there are two types of methods. The first type, proposed by Willis [21] and developed further in [17,22], consists of an extension of the Hashin-Shtrikman procedure for linear composite materials. As such, it makes use of variational principles expressed in terms of the polarization fields in the nonlinear composite relative to a linear homogeneous comparison material. The second approach, proposed by Ponte Castañeda [11] and developed further in [12-14] (see also Suquet [16]), makes use of a linear heterogeneous comparison material. Thus, this procedure allows the expression of the effective behavior of a nonlinear composite in terms of the effective behavior of a linear composite material through a variational principle for the properties of the linear comparison composite. When the nonlinear composite satisfies a certain "strong convexity" hypothesis, the first approach is found to be strictly a special case of the second. This is because the second approach may be used to obtain general types of nonlinear estimates—other than Hashin-Shtrikman—including, for example, third-order estimates of the Beran type. However, when the nonlinear composite violates the strong convexity hypothesis, it is found that a hybrid approach, proposed by Talbot and Willis [18], may be used to obtain estimates for the nonlinear composites that are stronger than those that may be obtained by the second approach. However, to be useful, this third procedure requires estimates for a "thermoelastic" comparison composite. In this paper, we introduce a fourth variational procedure, which is capable of delivering improved nonlinear estimates when the strong convexity hypothesis is violated, without the need for introducing the complication of a thermoelastic comparison composite.

1. Introduction

An important problem in classical physics is that of estimating the effective transport properties of composite materials. For materials with linear constitutive behavior, and negligible interfacial effects, there exists a well developed theory, which is reviewed, for example, in the article by Landauer [7]. However, there are numerous phenomena where nonlinear constitutive effects are very important. These include nonlinear optical phenomena in dielectrics and plasticity and creep in metals.

This paper is concerned with the theoretical prediction of the effective constitutive behavior of nonlinear composite materials. For clarity, we chose to present the results in the context of nonlinear electrostatics; however, the resulting theories will be applicable, with the necessary adjustments, to nonlinear conductivity, magnetostatics, plasticity and creep. For our purposes, a composite is a heterogeneous material with two distinct length scales: a macroscopic one which characterizes the overall dimensions of the specimen and the scale of variation of the applied loading conditions, and a microscopic one which characterizes the size of the typical heterogeneity (e.g., inclusions, fibers, cracks, etc.). By effective properties, we mean the relation between the averages of the local fields within the composite. Rigorous definitions of effective properties have been given, for example, by Bensoussan, Lions and Papanicolaou [1] and Willis [21] for composite materials with periodic and random microstructures, respectively. Below, we choose to present our results in the random context.

The nonlinear constitutive behavior of an inhomogeneous composite, occupying a region in space (of unit volume) Ω, may be characterized by means of an electric energy-density function, $w(x, E)$, depending on the position vector x and the electric field $E(x)$, such that the electric displacement field $D(x)$ is given by

$$D(x) = \frac{\partial w}{\partial E}(x, E). \tag{1}$$

We assume further that the dielectric is locally isotropic, so that

$$w(x, E) = \phi(x, |E|), \tag{2}$$

where ϕ is taken to be convex in the magnitude of the electric field $|E|$. In addition, we assume that $\phi(x, |E|) \geq 0,\ \forall\, x, |E|$ and that $\phi(x, 0) = 0,\ \forall\, x$.

It can be shown (see [6]) that the *effective* constitutive behavior of the inhomogeneous dielectric may be expressed in terms of the spatial averages of the fields, $\overline{D}$ and $\overline{E}$, via the relation

$$\overline{D} = \frac{\partial \widetilde{W}}{\partial \overline{E}}(\overline{E}), \tag{3}$$

where the effective energy-density function of the composite $\widetilde{W}$ is in turn given by the minimum energy principle

$$\widetilde{W}(\overline{E}) = \min_{E \in K} \int_{\Omega} w(x, E(x))\, dx, \tag{4}$$

where K is the set of *admissible* electric fields, specified by

$$K = \left\{ E \,|\, E = -\nabla\varphi(x) \text{ in } \Omega, \text{ and } \varphi = -\overline{E} \cdot x \text{ on } \partial\Omega \right\}. \tag{5}$$

This variational formulation of the electrostatics problem for the composite is equivalent to the standard boundary-value-problem formulation in terms of Gauss' and

Faraday's laws ($\nabla \cdot D = 0$ and $\nabla \times E = 0$, respectively), together with the uniform boundary condition

$$\varphi = -\overline{E} \cdot x \text{ on } \partial\Omega, \tag{6}$$

where φ is the electrostatic potential. The main advantage of the variational formulation is that the effective behavior of the nonlinear composite is then characterized in terms of only one scalar variable, namely, $\widetilde{W}$.

In practice, the difficulty associated with the computation of the effective energy function of the composite (4) is that the exact fields are usually difficult to determine for typical microstructures. However, numerous methods—both approximate and exact—have been devised to address this problem in the context of *linear* constitutive behavior for the composite. Next, we develop a variational principle that will allow us to make use of known estimates for linear composites to obtain corresponding estimates for nonlinear composites.

2. A Variational Principle for Nonlinear Composites

Ponte Castañeda [11] proposed a variational method to estimate the effective energy functions of *nonlinear* composites in terms of the energy functions of appropriately defined linear comparison composites. That reference was concerned with the effective behavior of nonlinearly viscous composites; the corresponding results for nonlinear dielectric composites were given in [12]. The variational principle of Ponte Castañeda (PC) centers around the change of variables $u = |E|^2$, defining a function f such that

$$f(x, u) = \phi(x, |E|) = w(x, E). \tag{7}$$

The function f has the same dependence on x as ϕ and w, satisfies the conditions $f(x,u) \geq 0 \;\forall\, x, u$ and $f(x, 0) = 0 \;\forall\, x$, but is not necessarily convex. However, if f is assumed to be convex (we then say that w is *strongly convex*), it follows that

$$w(x, E) = \sup_{\varepsilon_o \geq 0} \left\{ w_o(x, E) - v(x, \varepsilon_o) \right\}, \tag{8}$$

where

$$v(x, \varepsilon_o) = \sup_{E} \left\{ w_o(x, E) - w(x, E) \right\}, \tag{9}$$

and where $w_o(x, E) = \frac{1}{2}\varepsilon_o(x)\,|E|^2$ corresponds to the local energy-density function of a *linear comparison* composite with an arbitrary nonnegative dielectric coefficient (not constant) $\varepsilon_o(x)$. This representation is based on Legendre duality for the function f; in fact, $v(x, \varepsilon_o) = f^*(x, \frac{1}{2}\varepsilon_o)$, where f^* denotes the Legendre transform of f, given by

$$f^*(x, p) = \sup_{u} \left\{ up - f(x, u) \right\}. \tag{10}$$

When relation (8) for w is used in expression (4) for $\widetilde{W}$, and the order of the supremum over $\varepsilon_o(x)$ is interchanged with the infimum over $E(x)$, it follows that (see

[12])

$$\widetilde{W}(\overline{E}) = \sup_{\varepsilon_o(x)\geq 0} \left\{ \widetilde{W}_o(\overline{E}) - V(\varepsilon_o) \right\}, \tag{11}$$

where V is the functional generated by the function $v(x, \varepsilon_o)$, i.e.,

$$V(\varepsilon_o) = \int_\Omega v(x, \varepsilon_o(x))\, dx, \tag{12}$$

and where $\widetilde{W}_o$ denotes the effective energy function of the linear comparison composite, with local energy function w_o, such that

$$\widetilde{W}_o(\overline{E}) = \min_{E\in K} \int_\Omega w_o(x, E)\, dx\,, \tag{13}$$

cf. (4). Thus, (11), together with (12) and (13), provide an alternative variational statement for the effective energy function of the *nonlinear* composite in terms of the effective energy function of a suitably optimized *linear* composite material. We emphasize that the dielectric coefficient $\varepsilon_o(x)$ of the comparison composite in (11) is a nonnegative function of position.

The assumption of strong convexity (convex f in (7)) is essential for equality in statement (11). However, it can be shown (see [11-12]) that an analogous result exists for concave f. (This corresponds to strong convexity of the dual potential w^*). In this case, the suprema in (8) to (11) must be replaced by infima. If f is neither convex nor concave, the equality in (11) must be replaced by an inequality.

3. Relation to Other Variational Principles

3.1. Hashin-Shtrikman Variational Principle

Hashin and Shtrikman [5] introduced a variational principle for the effective energy function of a linear composite material, with local energy function $w_o(x, E) = \frac{1}{2}\varepsilon_o(x)\,|E|^2$, in terms of the polarization fields of the composite relative to a homogeneous comparison material, with energy function $w_c(E) = \frac{1}{2}\varepsilon_c\,|E|^2$. The variational principle of Hashin and Shtrikman may be expressed in the form (see [17])

$$\widetilde{W}_o(\overline{E}) = \sup_p \left\{ \min_{E\in K} \int_\Omega [w_c(E) + E\cdot p]\, dx - \int_\Omega (w_o - w_c)^*(x, p)\, dx \right\}, \tag{14}$$

where it has been assumed that $\varepsilon_c \leq \varepsilon_o(x)$, so that the function $w_o - w_c$ is convex.

Talbot and Willis [17] have shown that the above variational principle also holds for a nonlinear composite material with local and effective energy density functions w and $\widetilde{W}$ as given by relations (2) and (4), respectively, provided that ε_c can be chosen such that $(w - w_c)^*$ is bounded. This requires faster than quadratic growth (as $|E| \to \infty$) for w, which in turn implies (but does not require) strong convexity. (Analogous results have been obtained for weaker than quadratic growth; see [17].)

It has also been shown, in [12], that if w (or w^*) is strongly convex, then the Talbot-Willis variational principle for $\widetilde{W}$ may be obtained directly from the PC variational statement (11), together with the Hashin-Shtrikman expression (14) for the effective energy of the linear comparison composite $\widetilde{W}_o$. In particular, this shows that the PC variational principle includes that of Talbot and Willis, for the special case of strongly convex potentials w. However, the PC variational principle has the advantage, over the variational principle of Talbot and Willis [17], that it allows the determination of estimates other than Hashin-Shtrikman, such as Beran estimates, for example, which are not available from the Talbot-Willis procedure.

3.2. A Hybrid Procedure

Talbot and Willis [18] proposed a hybrid of the Talbot-Willis [17] and Ponte Castañeda [11] variational procedures. Such a procedure makes use of a polarization field as in [17], but instead of utilizing a homogeneous reference material, as in Hashin-Shtrikman, it makes use of linear comparison composite, as in [11]. Thus, the variational principle of Talbot and Willis [18] reads

$$\widetilde{W}(\overline{E}) = \sup_{\varepsilon_o} \sup_{p} \left\{ \min_{E \in K} \int_\Omega [w_o(x,E) + E \cdot p]\, dx - \int_\Omega (w - w_o)^*(x,p)\, dx \right\}, \quad (15)$$

where the energy function of the linear comparison composite is given by $w_o(x, E) = \frac{1}{2}\varepsilon_o(x)\,|E|^2$, and the nonlinear energy function w is assumed to be such that $(w - w_c)^*$ is bounded. (Note that w is not necessarily strongly convex). We emphasize that the above result reduces to the PC variational principle (11) when p is set equal to 0 (because $v\,(x, \varepsilon_o) = (w - w_o)^*\,(x, 0)$), and to the Talbot-Willis variational principle when $\varepsilon_o(x) = \varepsilon_c$.

Comparison of statements (11) and (15) shows that the optimal choice of p in (15) is precisely 0, if w is strongly convex, so that the variational principle of Talbot and Willis [18] reduces to that of PC [11] for strongly convex w. However, if w is not strongly convex, it was shown in [15], by means of an of an explicit example, that (15) has the potential to generate stronger results than (11). This is because the representation (15) reduces to an identity for *homogeneous* energy functions w, even if w is not strongly convex, whereas the representation (11) does not reduce to an identity for *homogeneous* w that are not strongly convex (although, of course, it does for strongly convex w; see (8)).

The main disadvantage of (15), relative to (11), is that it requires the solution of a problem for a "thermoelastic" composite (by this we mean that the mechanical counterpart of the minimum problem for E in (15) involves a linear thermoelastic composite; see [18]). Thus, the variational statement (15) has only been used thus far to obtain explicit results for two-phase nonlinear composites, by taking advantage of the result by Levin (1967) for two-phase linear thermoelastic composites. On the other hand, the variational statement (11), which makes use of a linear "elastic" composite, for which many different types of estimates are available for N-phase composites, with $N > 2$, can be used to generate corresponding estimates for N-phase nonlinear composites.

3.3. Special Variational Principles for Power-Law Materials

Let the energy function w of the nonlinear composite be of the power-law type, such that

$$\phi(x,|E|) = \frac{1}{n+1}\chi(x)\,|E|^{n+1}. \tag{16}$$

We note that the associated function f, from (7), is convex (concave) if $n > 1$ $(0 < n < 1)$. Therefore, for $n > 1$, w is strongly convex and the variational statement (11) applies exactly, with v, from (9), given by

$$v\,(x,\varepsilon_o) = \frac{(n-1)}{2\,(n+1)}\,\frac{(\varepsilon_o)^{\frac{n+1}{n-1}}}{(\chi)^{\frac{2}{n-1}}}. \tag{17}$$

Next, letting $\varepsilon_o(x) = \|\varepsilon_o\|\;\widehat{\varepsilon}_o(x)$, for some appropriate norm $\|\varepsilon_o\|$, and noting that w_o and v are homogeneous of degrees 1 and $(n+1)/(n-1)$ in $\varepsilon_o(x)$, respectively, it follows from the variational statement (11) that

$$\widetilde{W}(\overline{E}) = \sup_{\widehat{\varepsilon}_o(x)\geq 0}\;\sup_{\|\varepsilon_o\|}\left\{\|\varepsilon_o\|\;\widehat{\widetilde{W}}_o(\overline{E}) - \|\varepsilon_o\|^{\frac{n+1}{n-1}}\,V(\widehat{\varepsilon}_o)\right\}, \tag{18}$$

where $\widehat{\widetilde{W}}_o$ is the same as $\widetilde{W}_o$ in (13) with $\varepsilon_o(x)$ replaced by $\widehat{\varepsilon}_o(x)$. Then, the supremum over $\|\varepsilon_o\|$ can be easily evaluated to obtain the result that

$$\widetilde{W}(\overline{E}) = \frac{2}{n+1}\sup_{\widehat{\varepsilon}_o(x)\geq 0}\left\{\left[\widehat{\widetilde{W}}_o(\overline{E})\right]^{\frac{n+1}{2}}\left[\frac{n+1}{n-1}V(\widehat{\varepsilon}_o)\right]^{\frac{1-n}{2}}\right\}. \tag{19}$$

This is precisely the variational principle of Suquet [16], which was originally derived by means of Holder's inequality. The present derivation, directly from (11), shows that the variational principle of Suquet is strictly a special case of the PC variational principle. Like the PC variational principle, Suquet's result has the advantage over the variational principle of Talbot and Willis [17] that it may make use of any estimate for the linear comparison composites to generate corresponding estimates for the nonlinear composites, and thus it is not limited to estimates of the Hashin-Shtrikman type. However, unlike the variational principles of Talbot and Willis and Ponte Castañeda, it is restricted exclusively to power-law materials.

4. A New Variational Principle

In this section, we present a generalization of the PC variational principle that delivers improved results for nonlinear composites which are *not* strongly convex. The starting point is the inequality

$$\widetilde{W}(\overline{E}) \geq \min_{E\in K}\int_\Omega w_o(x,E)\,dx\; + \min_{E\in K}\int_\Omega (w-w_o)(x,E)\,dx, \tag{20}$$

from which it follows, by applying the classical lower bound to the second term in the right hand side of (20), that

$$\widetilde{W}(\overline{E}) \geq \sup_{\varepsilon_o(x)} \left\{ \widetilde{W}_o(\overline{E}) + \left[\int_\Omega (w - w_o)^* dx \right]^* (\overline{E}) \right\}, \tag{21}$$

where $\widetilde{W}_o(\overline{E})$ is as given by relation (13). This requires that $(w - w_o)^*$ be bounded, but not that w be strongly convex.

Noting that

$$\left[\int_\Omega (w - w_o)^* dx \right]^* (\overline{E}) \geq - \int_\Omega (w - w_o)^*(0) dx = -V(\varepsilon_o), \tag{22}$$

it follows, by comparing the right-hand sides of (11) and (21), that statement (21) will provide a better estimate for $\widetilde{W}$ than statement (11), for general energy functions w. However, we know that equality attains in (11) if w is strongly convex, and therefore equality will also attain in (21) if w is strongly convex. Therefore, statement (21) has the potential of improving the estimates of (11) when w is not strongly convex, although it reduces to (11) when w is strongly convex.

We note that the statement (21) requires faster-than-quadratic growth for w. An analogous result may be obtained for weaker-than-quadratic growth for w if the convex polars $(\cdot)^*$ in (21) are replaced by concave polars $(\cdot)_*$ (replace the sup by an inf in (10)) and the $\geq$ inequality is replaced by a $\leq$ inequality. In this case, the function $(w - w_o)_*$ is required to be bounded.

We also remark that the variational statement (21) may be interpreted as a generalization of the so-called translation method for linear composites (see [10]). Indeed if w is taken to be quadratic, corresponding to a linear composite, then w_o could be chosen quasi-convex. This suggests that it may be advantageous to incorporate a quasi-convex component in w_o to obtain improved results for nonlinear composites, but this will not be pursued here.

Finally, we note that the variational statement (15) of Talbot and Willis [18] is at least as strong as the variational statement (21). (Note that (21) may be obtained directly from (14) by breaking up the two terms in the minimum problem for E, thus introducing an additional inequality.) However, the variational statement (21) has the important advantage over (15) that it does not require the solution of a problem for a "thermoelastic" composite, but—like the variational procedure of Ponte Castañeda [11]—it only requires the solution of a problem for a linear dielectric composite. Thus, in addition to simplifying the required computations for two-phase composites, the new variational procedure would be particularly useful for N-phase composites, for which thermoelastic homogenization results are not readily available.

5. Applications to N-Phase Composites

In this Section, we consider the application of the variational principles of Sections 2 and 4 to N-phase composite dielectrics with nonlinear constitutive behavior characterized by

$$\phi(x, |E|) = \sum_{r=1}^{N} \theta^{(r)}(x)\, \phi^{(r)}(|E|), \tag{23}$$

where $\phi^{(r)}(|E|)$ and $\theta^{(r)}(x)$ $(r = 1, \ldots, N)$ are respectively the energy function and indicator function (which vanishes, unless x is in phase r, in which case it equals 1) of phase r. We assume that the corresponding volume fractions $c^{(r)}$, given by

$$c^{(r)} = \int_{\Omega} \theta^{(r)}(x)\, dx, \tag{24}$$

are fixed, and such that $\sum_{r=1}^{N} c^{(r)} = 1$.

Next we recall that, even though the properties of the nonlinear phases are homogeneous [as assumed in (23)], the solutions for the comparison dielectric coefficients $\varepsilon_o(x)$ in the variational principles (11) and (21) will not in general be constant over the individual phases, unless the actual fields happen to be constant over the phases. However, we can obtain a lower bound for $\widetilde{W}$ by restricting the class of admissible comparison dielectric coefficients to be constant within each phase, i.e., by letting $\varepsilon_o(x) = \sum_{r=1}^{N} \theta^{(r)}(x)\, \varepsilon_o^{(r)}$ (with constant $\varepsilon_o^{(r)}$). This follows from the fact that the minimum over any set is, in general, larger than the maximum over a subset of the original set. Therefore, we have, from (11) and (21), respectively, that

$$\widetilde{W}(\overline{E}) \geq \max_{\varepsilon_o^{(r)} > 0} \left\{ \widetilde{W}_o(\overline{E}) - \sum_{r=1}^{N} c^{(r)} v^{(r)}(\varepsilon_o^{(r)}) \right\}, \tag{25}$$

if w is strongly convex, and

$$\widetilde{W}(\overline{E}) \geq \max_{\varepsilon_o^{(r)} > 0} \left\{ \widetilde{W}_o(\overline{E}) + \left[\sum_{r=1}^{N} c^{(r)} (w^{(r)} - w_o^{(r)})^* \right]^* (\overline{E}) \right\}, \tag{26}$$

if w is *not* strongly convex. In the above expressions, $\widetilde{W}_o$ now corresponds to a linear comparison composite with precisely the same microstructure as the nonlinear composite (with dielectric constants $\varepsilon_o^{(r)}$ and volume fractions $c^{(r)}$).

We note that expression (25) was first given by Ponte Castañeda [11] in the context of nonlinearly viscous composites, and by Ponte Castañeda [13] for nonlinear dielectric composites. The expression (26) is new and is the most explicit and general available to date for N-phase composites with isotropic phases. It generalizes an expression given by Talbot and Willis [18] for two-phase composites with perfectly conducting inclusions (see relation (16) in that reference for the analogous case in elasticity with rigid inclusions). But the main advantage of statement (25) is that it only requires estimates for linear dielectric composites, and not for linear "thermoelastic" composites (as in the general expressions of [18]).

It is important to note that the above estimates for N-phase nonlinear composites are in the form of lower bounds for $\widetilde{W}$. Thus, lower bounds for $\widetilde{W}_o$ may be used to generate corresponding lower bounds for $\widetilde{W}$, but, on the other hand, upper bounds for $\widetilde{W}_o$ may not be used to generate upper bounds for $\widetilde{W}$. In practice, however, we are usually only interested in obtaining estimates for the effective constitutive relations (3) of a specific type of composite. Because bounds for $\widetilde{W}$ do not usually translate into bounds for the constitutive relations, we could ignore the inequalities in relations (25) and (26), and interpret them as equalities, in the sense of approximations, to obtain estimates for the effective potentials of specific types of composites. Thus, for example, Ponte Castañeda [13] (see also [14]) suggested the use of Hashin-Shtrikman upper bounds for $\widetilde{W}_o$ to generate estimates for the effective potentials $\widetilde{W}$ of *particulate* composite with the "weaker" and "stronger" phases occupying the matrix and inclusion phases, respectively. Alternatively, an effective medium estimate for $\widetilde{W}_o$ may be most appropriate to estimate $\widetilde{W}$ for composite materials with *grained* microstructures.

Because of the linearity of w_o, we may write that

$$\widetilde{W}_o(\overline{E}) = \frac{1}{2}\overline{E} \cdot (\widetilde{\varepsilon}_o \overline{E}), \tag{27}$$

where $\widetilde{\varepsilon}_o$ is the effective dielectric tensor of the linear comparison composite, which in general will be anisotropic. The optimality conditions for (25) and (26) may then be written, respectively, in the forms

$$\overline{E} \cdot \left(\frac{\partial \widetilde{\varepsilon}_o}{\partial \varepsilon_o^{(s)}} \overline{E} \right) = 2c^{(s)} \frac{\partial v^{(s)}}{\partial \varepsilon_o^{(s)}}, \quad s = 1, \ldots, N, \tag{28}$$

and

$$\overline{E} \cdot \left(\frac{\partial \widetilde{\varepsilon}_o}{\partial \varepsilon_o^{(s)}} \overline{E} \right) = 2 \frac{\partial}{\partial \varepsilon_o^{(s)}} \left[\sum_{r=1}^{N} c^{(r)} (w^{(r)} - w_o^{(r)})^* \right]^* (\overline{E}), \quad s = 1, \ldots, N. \tag{29}$$

In either case, the effective constitutive relation of the nonlinear composite can then be written in the form

$$\overline{D} = \widetilde{\varepsilon}_o \left(\widehat{\varepsilon}_o^{(1)}, \widehat{\varepsilon}_o^{(2)}, \ldots, \widehat{\varepsilon}_o^{(N)} \right) \overline{E}, \tag{30}$$

where the $\widehat{\varepsilon}_o^{(s)}$ are the optimal values of $\varepsilon_o^{(s)}$ from relations (28) and (29), respectively.

Finally, if we restrict our attention to the class of *isotropic* nonlinear composites, we are justified to consider only isotropic linear comparison composites with $\widetilde{W}_o(\overline{E}) = {}^1\!/_2 \widetilde{\varepsilon}_o |\overline{E}|^2$, where $\widetilde{\varepsilon}_o$ is now a scalar function of the corresponding constants $\varepsilon_o^{(r)}$, the volume fractions $c^{(r)}$, and of any other available information about the specific microstructure, or *class* of microstructures under consideration. There are several closely related types of bounds and estimates for linear composite materials. For the special case of a two-phase composite, these estimates may be written in the form

$$\widetilde{\varepsilon}_o = c^{(1)}\varepsilon_o^{(1)} + c^{(2)}\varepsilon_o^{(2)} - \frac{c^{(1)}c^{(2)}\left(\varepsilon_o^{(1)} - \varepsilon_o^{(2)}\right)^2}{c^{(2)}\varepsilon_o^{(1)} + c^{(1)}\varepsilon_o^{(2)} + (d-1)\varepsilon_o}, \tag{31}$$

where d stands for the dimension of the underlying space, and ε_o takes on different values for the different types of estimates, as follows. Assuming that $\varepsilon_o^{(1)} > \varepsilon_o^{(2)}$, the choices $\varepsilon_o \to 0$ and $\to \infty$ correspond, respectively, to the Wiener [20] upper and lower bounds. The choices $\varepsilon_o = \varepsilon_o^{(1)}$ and $\varepsilon_o^{(2)}$ yield the upper and lower Hashin-Shtrikman [5] bounds (also known as Maxwell-Garnet estimates for particulate microstructures with phases 1 and 2, respectively in the matrix phase). The choices $\varepsilon_o = \varsigma^{(1)}\varepsilon_o^{(1)} + \varsigma^{(2)}\varepsilon_o^{(2)}$ and $\left(\varsigma^{(1)}/\varepsilon_o^{(1)} + \varsigma^{(2)}/\varepsilon_o^{(2)}\right)^{-1}$ give the upper and lower bounds of Beran [2], in terms of the third-order parameter $\varsigma^{(1)} = 1 - \varsigma^{(2)}$ of Milton [9]. Finally, the choice $\varepsilon_o = \widetilde{\varepsilon}_o$ gives an implicit expression for $\widetilde{\varepsilon}_o$, called the effective medium approximation of Bruggeman [3].

For completeness, we present in Figs. 1 illustrative results, obtained from the expression (25), for a power-law composite with local potential w, as given by (16). For the value of n selected, it is useful to introduce reference electric fields for the two phases $E_n^{(r)} = \left(\chi^{(r)}\right)^{-1n}$ $(r = 1, 2)$, such that the energy function of the composite is also of the power-law type, with the same exponent n and with effective reference electric field $\widetilde{E}_n = \widetilde{\chi}^{-1n}$.

6. Closure

This paper was concerned exclusively with composite materials with isotropic phases, although the overall behavior of the composite could be anisotropic. However, extensions of the above procedures to composite materials with anisotropic phases is certainly possible. In fact, the original presentations of some of the above-discussed procedures includes locally anisotropic composites. Here, we only note that an extension of the original PC variational procedure for the special class of polycrystalline aggregates has been given recently by deBotton and Ponte Castañeda [4]. The extension of the new variational principle to polycrystalline materials would require the introduction of a linear comparison polycrystal akin to that of [4]. The details will be given elsewhere.

Acknowledgements. This work was supported by the National Science Foundation under its Materials Research Laboratory program at the University of Pennsylvania (Grant No. DMR-91-20668).

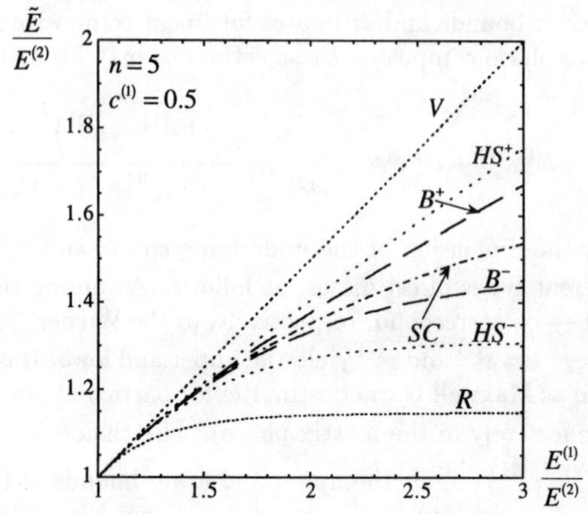

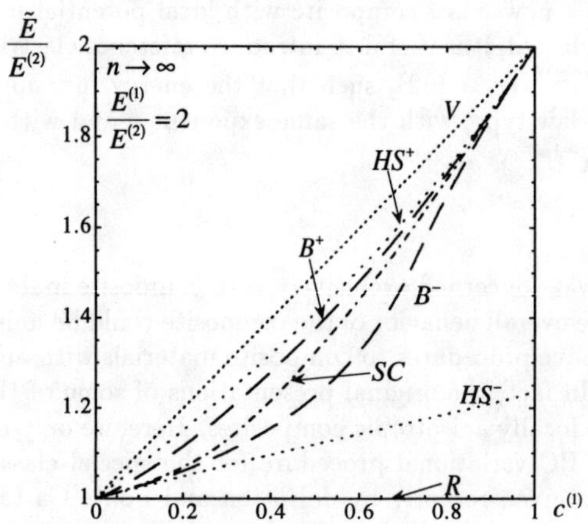

Fig. 1. Plots of $\tilde{E}_n = \tilde{\chi}^{-1/n}$ as functions of contrast $E_n^{(1)}/E_n^{(2)}$ and volume fraction $c^{(1)}$, respectively. *V* and *R* denote the rigorous classical upper and lower bounds. *HS*+ and *HS*– denote the Hashin-Shtrikman rigorous upper bound and lower estimate. *B*+ and *B*– denote the Beran upper bound and lower estimate (for $\varsigma^{(1)} = c^{(1)}$, appropriate for symmetric cell models with spherical cells). *SC* denotes the self-consistent (effective medium) estimate.

References

1. A. Bensoussan, J. L. Lions and G. Papanicolaou, *Asymptotic Analysis for Periodic Structures*, North-Holland, Amsterdam, 1978.
2. M. Beran, *Nuovo Cimento* **38** (1965) 771.
3. D. A. G. Bruggeman, *Ann. Phys.* **24** (1935) 636.
4. G. deBotton and P. Ponte Castañeda, *Proc. R. Soc. Lond.* **A 448** (1995) 121.
5. Z. Hashin and S. Shtrikman, *J. Appl. Phys.* **33** (1962) 3125.
6. R. Hill, *J. Mech. Phys. Solids* **11** (1963) 357.
7. R. Landauer, In *Electrical Transport, and Optical Properties of Inhomogeneous Media*, eds. J. C. Garland and D. B. Tanner, A. I. P., N.Y., 1978, p. 2.
8. V. M. Levin, *Mekh. Tverd. Tela.* **2** (1967) 88. (in Russian.)
9. G. W. Milton, *Phys. Rev. Lett.* **46** (1981) 542.
10. G. W. Milton, *Commun. Pure Appl. Math.* **43** (1990) 63.
11. P. Ponte Castañeda, *J. Mech. Phys. Solids* **39** (1991) 45.
12. P. Ponte Castañeda, *SIAM J. Appl. Math.* **52** (1992) 1321.
13. P. Ponte Castañeda, *Phil. Trans. R. Soc. Lond.* **A 340** (1992) 531.
14. P. Ponte Castañeda, G. deBotton and G. Li, *Phys. Rev.* **B 46** (1992) 4387.
15. P. Ponte Castañeda and J. R. Willis, In *Continuum Models of Discrete Systems (CMDS 7)*, Materials Science Forum, vols. 123-125, eds. K.-H. Anthony and H.-J. Wagner, Trans Tech, Aedermannsdorf, 1993, p. 351.
16. P. M. Suquet, *J. Mech. Phys. Solids* **41** (1993) 981.
17. D. R. S. Talbot and J. R. Willis, *IMA J. Appl. Math.* **35** (1985) 39.
18. D. R. S. Talbot and J. R. Willis, *Int. J. Solids Structures* **29** (1992) 1981.
19. J. Van Tiel, *Convex Analysis*, Wiley, New York, 1984.
20. O. Wiener, *Abhandlungen der mathematischphysischen Klasse der Koniglichen Sachsischen Gesellschaft der Wissenschaften* **32** (1912) 509.
21. J. R. Willis, *J. Appl. Mech.* **50** (1983) 1202.
22. J. R. Willis, In *Homogenization and Effective Moduli of Materials and Media*, eds. J. L. Ericksen *et al.*, Springer-Verlag, New York, 1986, p. 247.

Continuum Models and Discrete Systems
Proceedings of 8th International Symposium, June 11-16, 1995, Varna, Bulgaria, ed. K.Z. Markov
©World Scientific Publishing Company, 1996, pp. 280–289

DYNAMICS OF COMPOSITE MATERIALS

FEDERICO J. SABINA
Instituto de Investigaciones en Matemáticas Aplicadas y en Sistemas,
Universidad Nacional Autónoma de México,
Apartado Postal 20-726, 01000 México, D. F., MEXICO

Abstract. The problem of multiple scattering of waves in materials with microstructure has recently been studied for a wide variety of geometries by means of the self-consistent method. Several known static schemes used for the calculation of effective properties are now available in dynamics for the computation of wave dispersion and attenuation. The self-consistent analysis of elastic waves in composite materials composed of solid particles embedded in a solid matrix is presented. The method has been implemented for various shapes of the inclusions like spheres, ellipsoids, circular cracks and long fibres. Transversely isotropic materials can also be taken into account. The approximations made are simple enough so the final results are of interest for practical applications in nondestructive evaluation, material sciences and seismology. Comparisons made between theory and experiment produces good agreement. The results show characteristic resonance phenomena for the speeds and the attenuation of waves at finite frequencies. Besides composite materials, waves in polycrystals where there is no well-defined matrix phase have also been studied by this technique.

1. Introduction

In statics, the problem of predicting the overall elastic properties of a composite consisting of a matrix with embedded inclusions of known elastic properties has been studied and an approximate solution has been found using the self-consistent method. With this technique, the mean fields of stress and strain in the composite are related to their mean values over the inclusions. These can be estimated by embedding a single inclusion of known properties in the as yet unknown overall material of unspecified properties, and solving the resulting static problem. Budiansky [2] and Hill [8] were able to estimate the overall properties for embedded ellipsoidal or spherical inclusions via Eshelby's solution [4], which is an exact solution of the static problem. The properties are finally given by means of an implicit formulae. It is possible also to obtain these formulae by employing the Hashin-Shtrikman variational principle relative to a general uniform comparison medium. Requiring the properties of the comparison medium to coincide with those predicted by the variational scheme generates the set of self-consistent equations and also demonstrates immediately that the effective properties so obtained lie between the Hashin-Shtrikman bounds [17].

The self-consistent scheme was generalized to dynamics by Sabina and Willis [11] for the analysis of waves. In this case, also the mean fields of velocity and momentum density had to be taken into account and were expressed in terms of their mean values over the inclusions. To estimate all the mean values, an auxiliary dynamical problem had to be posed, that is, the scattering of a single inclusion embedded in a homogeneous reference medium whose properties are chosen to match those of the as yet unspecified composite. This problem has no closed-form solution so an approximate solution was sought, which was exact for ellipsoids in the static limit and sufficiently simple to be employed as a component part of a self-consistent iteration for the dynamical properties of the composite. It also produced implicit equations for the calculation of the effective elastic moduli and mass density. The Budiansky-Hill results were obtained in the limit of zero frequency and at higher frequencies the wave dispersion and attenuation show distintive features related to "resonance", even with this simple approximation, which can be applied up to frequencies such that the half-wavelength of the wave is comparable with the diameter of the inclusion.

The self-consistent method may provide a simple and important theoretical tool for the understanding of how waves of different wavelengths and inclusions (or cracks) of various sizes affect each other.

Sabina et al. [13] generalized to dynamics the static counterpart of Laws and McLaughlin [10] for a composite with aligned spheroidal inclusions randomly positioned. Smyshlyaev et al. [15] generalized to dynamics the self-consistent static scheme of Wu [18] for a random array of randomly oriented spheroidal inclusions. Two equivalent formulations for the treatment of cracks, either randomly oriented or aligned, were presented in Smyshlyaev et al. [16] which again are the dynamical generalization of the static methods of Budiansky and O'Connell [3] for randomly oriented cracks, and Hoenig [6] for aligned cracks, respectively. The two-dimensional analogue of [11] has been implemented by Bussink et al. [19] for infinitely long cylindrical fibres. Moreover, a variant of the scheme was implemented by Sabina and Willis [12] to study waves in isotropic polycrystalline media, in which each grain could be modelled as a sphere. This generalized to dynamics the static scheme of Hershey [5]. It is interesting to mention that a comparison of results obtained with this method and experimental results yields, in some cases, good to excellent match for spheres embedded in a matrix [1,11], and an isotropic polycrystal with cubic grains of a given distribution of sizes [12].

2. The Self-Consistent Method in Dynamics

2.1. Problem Formulation

A composite of $n+1$ phases is considered here. It comprises a matrix with tensor of elastic moduli L_{n+1} and mass density ρ_{n+1} and embedded inclusions having tensor of elastic moduli L_r and mass density ρ_r, where $r = 1, 2, \ldots, n$. Each phase r has the same shape. A typical inclusion r occupies the domain $x' + \Omega_r$, where Ω_r contains the origin. It is assumed that the 'center' x' of the inclusions is distributed randomly with no overlap among the inclusions.

In the absence of body forces, the response of the composite is governed by the equation of motion

$$\text{div } \sigma = \dot{p}, \tag{1}$$

understood in the weak sense, where the stress tensor σ and momentum vector density p are related to the infinitesimal strain tensor e and velocity vector $\dot{u}$ by means of the constitutive relations

$$\sigma = Le, \quad p = \rho \dot{u}, \tag{2}$$

written in a useful symbolic notation. The strain e is related to the infinitesimal displacement vector u through $e_{ij} = (u_{i,j} + u_{j,i})/2$. The comma notation for differentiation is used here, i.e., $u_{i,j} \equiv \partial u_i / \partial x_j$. In suffix notation, Eqs. (2) would read $\sigma_{ij} = L_{ijkl} e_{kl}$, $p_i = \rho \dot{u}_i$, where summation over repeated indices is understood. The tensor of elastic moduli L and mass density ρ of the composite depend on position x, taking the values L_r, ρ_r or L_{n+1}, ρ_{n+1} when x lies in the inclusion of type r or the matrix, respectively. If the exact configuration of the inclusions were known, it would be possible to find u for any particular loading, from Eqs. (1) and (2), augmented with initial and boundary conditions. However, this is not the case so, instead, the mean response of the composite is sought by an averaging procedure, say, in the mean or ensemble sense. Thus, the linear Eq. (1) becomes

$$\text{div } \langle \sigma \rangle = \langle \dot{p} \rangle. \tag{3}$$

The relations (2) can be averaged similarly. The problem of finding the mean fields $\langle \sigma \rangle$, $\langle e \rangle$, $\langle p \rangle$, $\langle \dot{u} \rangle$ and $\langle u \rangle$ would be solved, therefore, if "overall", or "effective", constitutive equations could be found, which relate $\langle \sigma \rangle$ with $\langle e \rangle$, and $\langle p \rangle$ with $\langle \dot{u} \rangle$. With this in mind, note that

$$\langle e \rangle = \sum_{r=1}^{n+1} c_r \langle e \rangle_r, \tag{4}$$

where c_r represents the probability of finding material of type r at x and $\langle e \rangle_r$, represents the expectation value of e at x, conditional upon finding material of type r at x. There is an analogous expression relating $\langle \dot{u} \rangle$ and $\langle \dot{u} \rangle_r$. The probabilities c_r are uniform, and equal to volume fractions, if the material is statistically uniform, as will be assumed in the sequel. Note also that

$$\langle \sigma \rangle = \sum_{r=1}^{n+1} c_r \langle \sigma \rangle_r = \sum_{r=1}^{n+1} c_r L_r \langle e \rangle_r \tag{5}$$

and hence, by elimination of $\langle e \rangle_{n+1}$ between Eqs. (4) and (5),

$$\langle \sigma \rangle = L_{n+1} \langle e \rangle + \sum_{r=1}^{n} c_r (L_r - L_{n+1}) \langle e \rangle_r. \tag{6}$$

A similar derivation with the other constitutive relation leads to

$$\langle p\rangle = \rho_{n+1}\langle \dot{u}\rangle + \sum_{r=1}^{n} c_r(\rho_r - \rho_{n+1})\langle \dot{u}\rangle_r. \tag{7}$$

The problem would thus be solved if the mean fields $\langle e\rangle_r$, $\langle \dot{u}\rangle_r$ could be expressed in terms of $\langle e\rangle$, $\langle \dot{u}\rangle$, respectively. Exact calculation would require complete knowledge of the composite. This is not available and the problem would in any case be too difficult. Approximate expressions will be found, instead, by solving the problem of an inclusion of properties L_r and ρ_r, centered at x', embedded in a uniform reference medium with tensor of elastic moduli L_0 and density ρ_0. The field incident on the inclusion is u_0; contact with the original problem is made by selecting L_0 and ρ_0 so that the reference medium can support the mean wave $\langle u\rangle$ and u_0 can be taken equal to $\langle u\rangle$. This is the self-consistent idea. The scattering problem having been solved, $\langle e\rangle_r$ is calculated as follows

$$\langle e\rangle_r(x,t) = \frac{1}{|\Omega_r|}\int_{U_r} e(x,t;x')dx', \tag{8}$$

where U_r is the set $\{x' : x - x' \in \Omega_r\}$ and $|\Omega_r|$ represents the volume of Ω_r. There is an analogous expression for $\langle \dot{u}\rangle_r$. These conditional mean fields depend on the fields e_0, $\dot{u}_0$, and hence on $\langle e\rangle$, $\langle \dot{u}\rangle$ once L_0 and ρ_0 are chosen self-consistently.

2.2. The Auxiliary Problem: a Single Scatterer

Estimation of $\langle e\rangle_r$ and $\langle \dot{u}\rangle_r$ is possible by means of the approximate solution of a single scatterer. The solution of this scattering problem, via integral equations, will be outlined here for one inclusion of properties L_r, ρ_r embedded in an elastic matrix of known properties whose constitutive response at radian frequency ω is

$$\sigma = L_0 e, \quad p = -i\omega\rho_0 u, \tag{9}$$

where L_0 is a fourth-order tensor and ρ_0 is a scalar; both L_0 and ρ_0 could be complex, so that viscoelastic materials can also be included in this formulation.

The stress and momentum density within the inclusion are conveniently expressed in terms of "polarizations" τ, and π

$$\sigma = L_0 e + \tau, \quad p = -i\omega\rho_0 u + \pi, \tag{10}$$

where τ and π are non-zero only over the inclusion. The substitution of Eqs. (10) into the (time-reduced) Eq. (1), shows that u could be generated in a homogeneous material with moduli L_0 and density ρ_0, as if it were driven by the body force div $\tau + i\omega\pi$. Thus, in terms of the time harmonic Green's function $G(x,\omega)$ for the homogeneous material, L_0 and ρ_0, the integral equation for the displacement is

$$u(x,\omega) = u_0 + G^{(r)} * (\text{div }\tau + i\omega\pi) = u_0 - S^{(r)} * \tau - M^{(r)} * \pi, \tag{11}$$

where the superindex (r) indicates that the operators' support is the volume of the inclusion of that type; the symbol $*$ represents the operation of convolution with respect to x; in component form, the operators are given by

$$S^{(r)}_{ikl}(x,\omega) = -G^{(r)}_{i(k,l)}(x,\omega), \quad M^{(r)}_{ij}(x,\omega) = -i\omega G^{(r)}_{ij}, \tag{12}$$

the brackets on the suffixes mean symmetrization; the field incident on the inclusion is a plane wave harmonic wave u_0 given by

$$u_0 = m \exp[i(kn \cdot x - \omega t)], \tag{13}$$

where m is the polarization vector, k the wave number and n the unit vector in the direction of incidence. This u_0 will be equated to the mean harmonic wave $\langle u\rangle$. A representation for e follows from differentiation of Eq. (11). Then, integral equations for τ and π are obtained by eliminating e and u, viz.,

$$\begin{aligned} e_0 &= (L_r - L_0)^{-1}\tau + S_x^{(r)} * \tau + M_x^{(r)} * \pi, \\ -i\omega u_0 &= (\rho_r - \rho_0)^{-1}\pi + S_t^{(r)} * \tau + M_t^{(r)} * \pi. \end{aligned} \tag{14}$$

The strain associated with u_0 is e_0, and $S_x^{(r)}$, $S_t^{(r)}$, $M_x^{(r)}$ and $M_t^{(r)}$ are operators given explicitly in [11]. Eqs. (14) are solved approximately, by a Galerkin method, taking τ, π as constant over the inclusion. It is worthwhile to recall that a constant strain is an exact explicit solution in statics, when the inclusion is ellipsoidal, for a constant load applied at infinity. In the particular case when the inclusion has a center of symmetry, it follows, since $M_x^{(r)}$ and $S_t^{(r)}$ are odd functions of x, that Eqs. (14) uncouple to yield

$$\tau = (L_r - L_0)^{-1} + \bar{S}_x^{(r)}]^{-1}\bar{e}_0, \quad \pi = -i\omega[(\rho_r - \rho_0)^{-1} + \bar{M}_t^{(r)}]^{-1}\bar{u}_0, \tag{15}$$

where $\bar{S}_x^{(r)}$ and $\bar{M}_t^{(r)}$ are mean values, obtained by integrating over the inclusion, of the functions $S_x^{(r)}\tau$ and $M_t^{(r)}\pi$, when τ and π are constant there.

Moreover, if the inclusion is centered at x', it follows that

$$\bar{e}_0 = e_0(x')h_r(kn), \quad \bar{u}_0 = u_0(x')h_r(kn), \tag{16}$$

where

$$h_r(kn) = \frac{1}{|\Omega_r|}\int_{\Omega_r} e^{ikn\cdot x}\mathrm{d}x. \tag{17}$$

This approximately solves the scattering problem. The field within the inclusion can be estimated from the relations

$$e = (L_r - L_0)^{-1}\tau, \quad -i\omega u = (\rho_r - \rho_0)^{-1}\pi. \tag{18}$$

2.3. The Self-Consistent Equations

The evaluation of $\langle e\rangle_r$ and $-i\omega\langle u\rangle_r$ now requires the calculation of the corresponding mean values $\langle\tau\rangle_r$ and $\langle\pi\rangle_r$ from Eqs. (18) and (8). It is easily found, at frequency ω, that

$$\begin{aligned} \langle e\rangle_r(x) &= h_r(kn)h_r(-kn)[I + \bar{S}_x^{(r)}(L_r - L_0)]^{-1}e_0(x), \\ \langle u\rangle_r(x) &= h_r(kn)h_r(-kn)[I + \bar{M}_t^{(r)}(\rho_r - \rho_0)]^{-1}u_0(x). \end{aligned} \tag{19}$$

Finally, through the "cancellation" of the post-multiplicative factors $\langle e\rangle$ and $\langle u\rangle$ in Eqs. (6) and (7), the self-consistent equations become

$$\begin{aligned} L_0 &= L_{n+1}+\sum_{r=1}^{n} c_r h_r(kn)h_r(-kn)(L_r-L_{n+1})[I+\bar{S}_x^{(r)}(L_r-L_0)]^{-1}, \\ \rho_0 &= \rho_{n+1}+\sum_{r=1}^{n} c_r h_r(kn)h_r(-kn)(\rho_r-\rho_{n+1})[I+\bar{M}_t^{(r)}(\rho_r-\rho_0)]^{-1}. \end{aligned} \tag{20}$$

These equations provide a convenient prescription for finding L_0 and ρ_0 uniquely for a chosen plane wave. Once they are solved, the wave number k in Eq. (13) is found by solving the associated eigenvalue problem

$$[k^2 L_0(n)-\rho_0\omega^2 I]m=0. \tag{21}$$

In the above equation, $L_0(n)$ is the acoustic tensor for propagation in the direction n; its components are $(L_0)_{ik}=(L_0)_{ijkl}n_j n_l$. In the more general situation when the eigenvalue k is complex, as may happen in a formulation for a composite, the wave speed is $\gamma=\omega/\Re(k)$, the attenuation exponent, $\Im(k)$ and the quality factor, $Q=2\Im(k)/\Re(k)$, often used in seismology. It is clear that in this formulation the waves, which propagate in a composite, are dispersive and their amplitudes attenuate with distance.

One can exploit the overall material symmetry (isotropy, cubic or transverse isotropy) to great advantage so that Eqs. (20) can be written symbolically,

$$Z=F(Z,\omega), \tag{22}$$

where Z is a complex vector. These equations are solved by iteration by a continuation method and its convergence is very rapid in most cases. For more details see [11–13, 15, 16].

3. Examples

Consider now a single population of inclusions, with label $r=1$, embedded in a matrix with label $r=2$. Let L_1 and L_2 be isotropic, with L_r, characterized by bulk modulus κ_r, and shear modulus μ_r.

3.1. Spherical Inclusions: overall isotropic medium

The inclusions are spherical of radius a. The overall properties of the composite are isotropic. The use of Hill's symbolic notation [8], $L_r=(3\kappa_r, 2\mu_r)$, and its associated algebraic properties, leads to only three equations to be solved; here Z of Eq. (22) is $Z=(\rho_0,\kappa_0,\mu_0)$, and the equations are

$$\rho_0 = \rho_2+\frac{c_1h_1(kn)h_1(-kn)(\rho_1-\rho_2)}{1+(\rho_1-\rho_0)(3-\epsilon_\alpha-2\epsilon_\beta)/(3\rho_0)}, \tag{23}$$

$$\kappa_0 = \kappa_2+\frac{c_1h_1(kn)h_1(-kn)(\kappa_1-\kappa_2)}{1+3(\kappa_1-\kappa_0)\epsilon_\alpha/(3\kappa_0+4\mu_0)}, \tag{24}$$

$$\mu_0 = \mu_2+\frac{c_1h_1(kn)h_1(-kn)(\mu_1-\mu_2)}{1+2(\mu_1-\mu_0)[2\mu_0\epsilon_\alpha+(3\kappa_0+4\mu_0)\epsilon_\beta]/[5\mu_0(3\kappa_0+4\mu_0)]}, \tag{25}$$

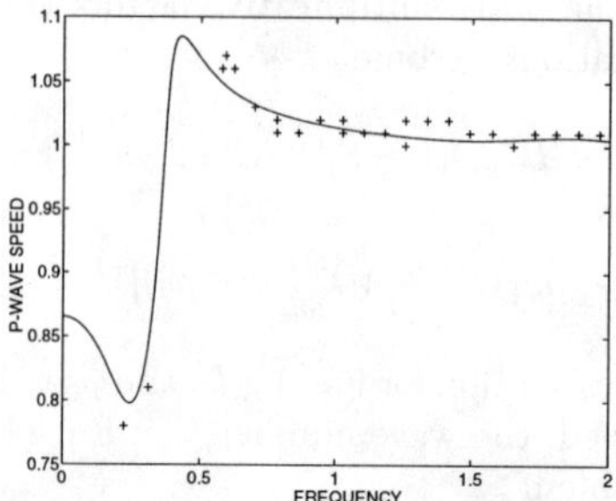

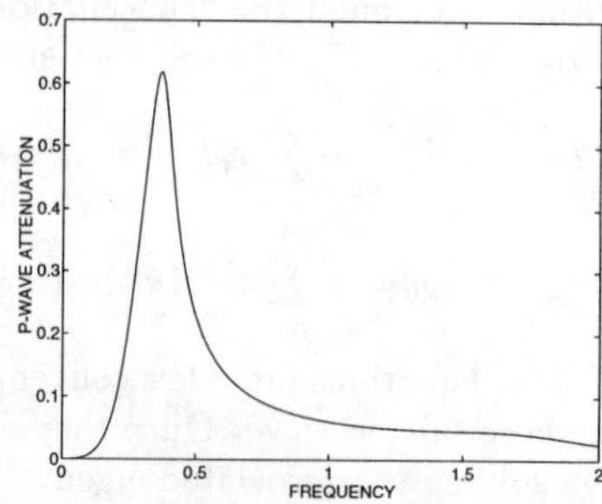

Fig 1. Plots of the variation of the dimensionless P-wave phase speed γ_1/α_2, left, and $1/Q$, right, against the normalized frequency $\omega a/\alpha_2$ for an epoxy matrix with embedded Pb spheres at 5% volume fraction. Experimental points taken from [9] are shown with plus signs.

where

$$h_1(kn) = 3[\sin(ka) - ka\cos(ka)]/(ka)^3, \; \epsilon_\gamma = (1 - ik_\gamma a)h_1(k_\gamma n)\exp ik_\gamma a, \tag{26}$$

and $\gamma = \alpha$ or β; for an incident P-wave, $k = k_\alpha = \omega[(3\kappa_0 + 4\mu_0)/3\rho_0]^{-1/2}$ or, for an incident S-wave, $k = k_\beta = \omega(\mu_0/\rho_0)^{-1/2}$. In Fig. 1 the results of the computation for a composite containing 5% of Pb spheres in Epon 828Z epoxy are displayed. Fig. 1, left, shows phase speed of P-waves, γ_1/α_2, normalized to the P-wave value of the matrix as a function of frequency, normalized as $\omega a/\alpha_2$, where α_2 is the P-wave speed of the matrix. Fig. 1, right, gives similar plot, of attenuation, presented in the form of the quality factor $1/Q$. A noteworthy feature is the presence of a "resonance" effect when the half-wavelength is about four times the diameter of the inclusion. This corresponds to the excitation of waves in the lead inclusion whose half-wavelength is of the order of the inclusion diameter. Also the experimental data of Kinra et al. [9] are plotted. The agreement between the theory and the observations is excellent.

3.2. Aligned Spheroidal Inclusions: overall transversely isotropic medium

Here the inclusions are aligned spheroids of aspect ratio δ with axis of symmetry Ox_3 as the distinguished direction. Thus, in Eqs. (20), it is consistent to take L_0 and ρ_0 to be tensors of fourth- and second-order, respectively, with transversely isotropic symmetry. That the density ρ_0 of the medium is a second-order tensor is a new characteristic, and can easily be introduced in the Green function plane wave decomposition that is used to solve the single scatterer problem. The symbolic notation is again useful, i.e., $\rho_0 = (\rho_I, \rho_{II})$, and, via Hill's notation [7], in terms of the elastic parameters of the transversely isotropic medium, $L_0 = (2k_0, l_0, l_0', n_0, 2m_0, 2p_0)$, and its associated algebra, the self-consistent Eqs. (20) can be written as Eqs. (22), where $Z = (\rho_I, \rho_{II}, k_0, l_0, l_0', n_0, m_0, p_0)$. The detailed formulae can be found in [13]. Fig. 2 shows the estimates of phase speed and attenuation for the same composite as in the

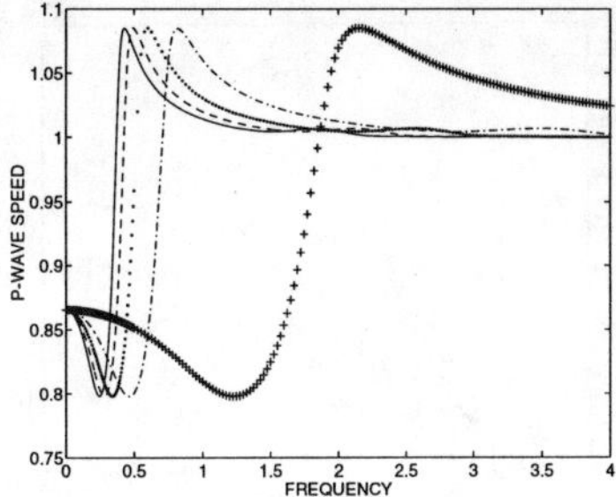

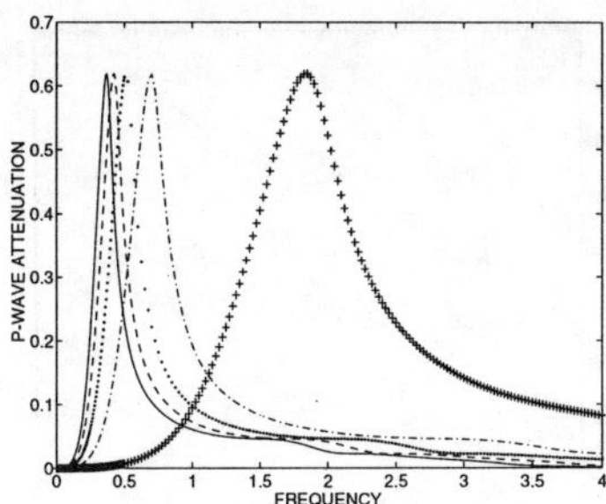

Fig 2. As in Fig. 1, except that the inclusions are ellipsoids of aspect ratio $\delta = 0.2$. Five directions of propagation are shown: $\theta = 0°$ (plus points), $\theta = 30°$ (dashed lines), $\theta = 45°$ (dotted lines), $\theta = 60°$ (dashdot lines) and $\theta = 90°$ (solid lines).

previous subsection except that the inclusions are spheroidal of aspect ratio $\delta = .2$. Fig. 2, left, shows the phase speed of quasi-P waves, normalized to the P-wave matrix value, for angles of incidence $\theta = 0°, 30°, 45°, 60°$ and $90°$. They are plotted against frequency, normalized as $\omega a/\alpha_2$, where a is the radius of the spheroids. Fig. 2, right, shows similar plots, of attenuation, presented in the form of the quality factor $1/Q$. Again, a "resonance" effect is present such that the attenuation attains a maximum value when the half-wavelength is about four times the length of the diametral chord in two directions of propagation: $\theta = 0°$ and $\theta = 90°$, like in the case of spherical inclusions. However, in the remaining directions, that value corresponds to an "effective" length, which is more than twice the size of the diametral chord, probably due to conversion into quasi-shear waves. It is also interesting to point out that the maximum and minimum values of speed and attenuation are independent of the azimuth at this concentration of 5%, and that these values are the same ones which are attained when the inclusions are spherical (Cf. Fig. 1).

3.3. Aligned Penny-Shaped Cracks

A medium permeated with penny-shaped cracks randomly positioned but aligned is now considered. The crack density parameter is $\nu = Na^3/v$, where N is the number of cracks of radius a in a volume v. In the limit when the aspect ratio δ of the ellipsoids of the previous subsection tends to zero, the relevant formulae for the cracks can be derived from Eqs. (20). The dependence on δ appears only in $c_1(\delta) = 4\pi Na^3\delta/3$, $\bar{M}_t(\delta)$ and $\bar{S}_x(\delta)$. In the asymptotic limit, when $\delta \to 0$, $\bar{M}_t(\delta) = 0$ and

$$[I + \bar{S}_x(L_1 - L_0)]^{-1} = \frac{1}{\delta}\Big(0, 0, 0, 0, 0, -\frac{1}{4\tilde{p}p_0}\Big) + O(\delta^0). \qquad (27)$$

for fluid-filled cracks with a non-viscous fluid. Only one equation has to be solved for p_0. See [15] for the definition of $\tilde{p}$ and other relevant quantities. Numerical results are obtained for water-filled cracks in granite at crack number density $\nu = 0.1$, a typical

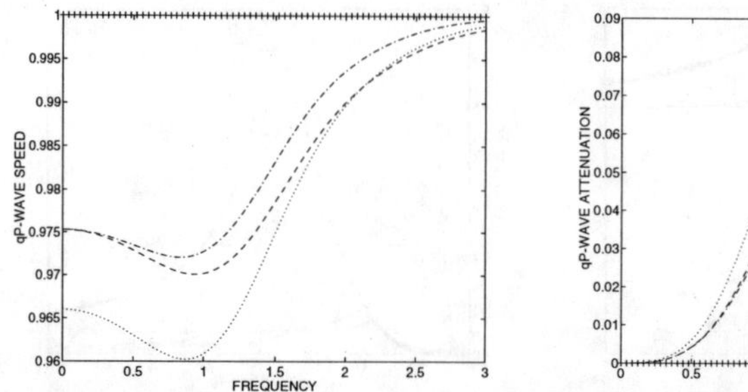

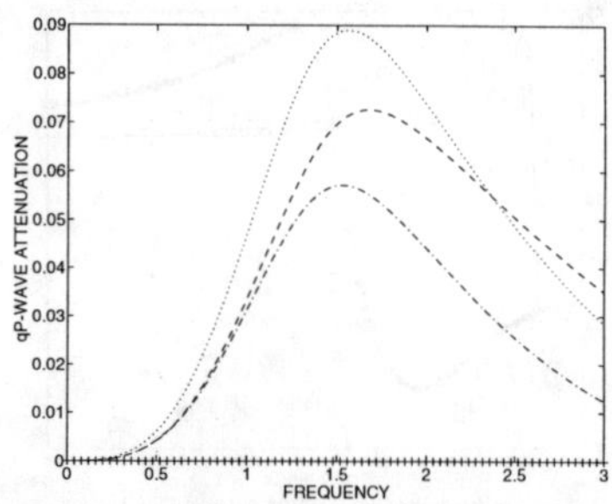

Fig 3. As in Fig. 2, for qP-waves in granite with aligned penny-shaped water-filled cracks, distributed at crack number density $\nu = 0.1$. See text discussion.

example in seismology. The parameters of the water were taken as $\rho_1 = 1$ g/cm^3; P-wave speed, $\alpha_1 = 1.5$ km/s; S-wave speed, $\beta_1 = 0$ km/s; and, correspondingly, for the granite, $\rho_2 = 2.6$ g/cm^3, $\alpha_2 = 5.8$ km/s and $\beta_2 = 3.349$ km/s. Fig. 3, left, shows the quasi-longitudinal wave speed normalized to the wave speed of the granite as a function of normalized frequency, $\omega a/\alpha_2$, where a is the radius of the crack, and right, the attenuation, $1/Q$, in the form of the quality factor. There is a resonance, which does not depend on the azimuthal angle, when half-wavelength is of the order of the diameter of the crack. On the other hand, the wave speed and the attenuation depend on the angle of incidence except for $\theta = 0°$ and $90°$ where the waves are not dispersive and not attenuated. In this case these two curves merge with the top margin, Fig. 3, left, and with the bottom one, Fig. 3, right.

4. Concluding Remarks

The self-consistent method as applied to the study of waves in polycrystals (not discussed here) [12] and composite materials (solid suspensions in solids)[11, 13–16], and the approximations made, have the property of being simple enough to be able to deal with interaction of waves and inclusions, not only in the low-frequency limit where one can compute the overall properties, but, also for finite frequencies where important resonance effects appear. This technique may provide a useful tool to deal with these interactions and provide reasonable agreement with experiments [1].

Acknowledgements. It is a pleasure to thank Prof. J. R. Willis for his always useful suggestions.

References

1. L. W. Anson and R. C. Chivers, *J. Phys. D: Appl. Phys.* **26** (1993)1566.

2. B. Budiansky, *J. Mech. Phys. Solids* **13** (1965)223.
3. B. Budiansky and R. J. O'Connell, *Int. J. Solids Struct.* **12** (1976)81.
4. J. D. Eshelby, *Proc. Roy. Soc.* **A271**(1957)376.
5. A. V. Hershey, *J. Appl. Mech.* **21** (1954)236.
6. A. Hoenig, *Int. J. Solids Struct.* **15** (1979)137.
7. R. Hill, *J. Mech. Phys. Solids* **13** (1965)189.
8. R. Hill, *J. Mech. Phys. Solids* **13** (1965)213.
9. V. K. Kinra, E. Ker and S. K. Datta, *Mech. Res. Com.*, **9** (1982)109.
10. N. Laws and R. McLaughlin *J. Mech.Phys. Solids* **27** (1979)1.
11. F. J. Sabina and J. R. Willis, *Wave Motion* **10** (1988)127.
12. F. J. Sabina and J. R. Willis, *Eur. J. Mech. A/Solids* **12** (1993)265.
13. F. J. Sabina, V. P. Smyshlyaev and J. R. Willis, *J. Mech. Phys. Solids* **10** (1993)1573.
14. F. J. Sabina, V. P. Smyshlyaev and J. R. Willis, In *Nonlinear Waves in Solids*, eds. J. L. Wegner and F. R. Norwood (Appl. Mech. Rev. Book No. 137, 1995), p. 309.
15. V. P. Smyshlyaev, J. R. Willis and F. J. Sabina, *J. Mech. Phys. Solids* **10** (1993)1589.
16. V. P. Smyshlyaev, J. R. Willis and F. J. Sabina, *J. Mech. Phys. Solids* **12** (1993)1809.
17. J. R. Willis, *J. Mech. Phys. Solids* **25** (1977)185.
18. T. T. Wu, *Int. J. Solids Struc.* **2** (1966)1.
19. P. G. J. Bussink, P. L. Iske, J. Oortwijn and G. L .M. M. Verbist, *J. Mech. Phys. Solids* in press (1995).

Continuum Models and Discrete Systems
Proceedings of 8th International Symposium, June 11-16, 1995, Varna, Bulgaria, ed. K.Z. Markov

OVERALL PROPERTIES OF NONLINEAR COMPOSITES: SECANT MODULI THEORIES AND VARIATIONAL BOUNDS

P. SUQUET
L.M.A./C.N.R.S., 31 Chemin Joseph Aiguier,
F-13402 Marseille Cedex 20, France

Abstract. This study deals with the overall properties of nonlinear composites. A modified secant moduli approach is proposed. In the classical secant moduli approach each individual nonlinear phase is replaced by a linear phase the shear modulus of which is computed by means of the average stress of the phase. The modified approach makes use of the quadratic average of the stresses over the phase. Its predictions are proved to coincide with those of Ponte Castañeda's variational procedure and, for pure power-law materials, with those of the author's variational procedure.

1. Introduction

This study is devoted to the overall behavior of nonlinear composites. Plasticity is considered in the frame of a deformation theory and the phases are assumed to be nonlinearly elastic. Their behavior is ruled by a non quadratic potential energy.

One of the most popular method to predict the overall behavior of nonlinear elastic composites is the *secant moduli approach*. In its classical version (Berveiller and Zaoui [1], Tandon and Weng [14], Bornert *et al.* [2]), this methods assigns to each individual phase a shear modulus computed by means of an effective stress of the phase under consideration. This effective stress is the von Mises stress associated with the *average stress* of the phase. The problem can alternatively be approached by means of variational principles. The overall behavior of the composite is controlled by a macroscopic potential. Several authors have recently proposed nonlinear variational procedures to bound or to estimate this overall potential using either a nonlinear extension of the Hashin-Shtrikman procedure (Willis [17]), a new variational procedure (Ponte Castañeda [7,8], or convexity inequalities to express power law energies in terms of quadratic energies (Suquet [11]).

Although the nonlinear variational procedures appeared initially to be unrelated with the secant moduli approach, it has been proved recently (Suquet [13]) that Ponte Castañeda's variational procedure can be interpreted as a modified secant moduli approach. In this modified approach the effective stress of each individual phase is the *quadratic average* of the von Mises stress over the phase. This new effective stress accounts more accurately than the classical one for the nonhomogeneous distribution of the stress field with each individual phase. It has been used in a less general context by Qiu and Weng ([10]) and Suquet ([12]). Buryachenko ([3]) also introduced this new "effective stress" independently of any consideration on bounds.

2. Secant Moduli Theories for the Overall Response of Nonlinear Elastic Composites

2.1. Secant Moduli Theories

Consider a representative volume element (r.v.e.) V of the composite material where the size of the inhomogeneities is small compared to that of V. The composite is made up of N homogeneous phases V_r, $r = 1, \ldots, N$; $\chi^{(r)}$ denotes the characteristic function of phase r, $c^{(r)}$ its volume fraction, $\langle\cdot\rangle_r$ the spatial averaging over V_r, $\langle\cdot\rangle$ the spatial averaging over V

$$\langle F \rangle_r = \frac{1}{|V_r|}\int_{V_r} F(\mathbf{x})\,d\mathbf{x}, \quad c^{(r)} = \frac{1}{|V|}\int_V \chi^{(r)}(\mathbf{x})\,d\mathbf{x},$$

$$\langle F \rangle = \frac{1}{|V|}\int_V F(\mathbf{x})\;d\mathbf{x} = \sum_{r=1}^{N} c^{(r)}\langle F \rangle_r\,.$$

The local constitutive behavior of the individual phases is assumed to be governed by a potential $\psi(\boldsymbol{\sigma})$:

$$\boldsymbol{\epsilon} = \frac{\partial \psi}{\partial \boldsymbol{\sigma}}(\boldsymbol{\sigma}), \tag{1}$$

where $\boldsymbol{\epsilon}$ and $\boldsymbol{\sigma}$ are the (infinitesimal) strain and stress fields and where the potential ψ can vary from one phase to another. For simplicity attention will be restricted to a class of isotropic materials for which ψ reads as

$$\psi(\boldsymbol{\sigma}) = \frac{1}{2k}\sigma_m^2 + f(\sigma_{\text{eq}}^2), \tag{2}$$

where

$$\sigma_m = \frac{1}{3}\text{tr}(\boldsymbol{\sigma}), \quad s_{ij} = \sigma_{ij} - \sigma_m\delta_{ij}, \quad \sigma_{\text{eq}} = \left(\frac{3}{2}s_{ij}s_{ij}\right)^{1/2}.$$

The material properties k and f depend on the position $\mathbf{x}$ inside the r.v.e. and take constant values $k^{(r)}, f^{(r)}$ in each individual phase (r):

$$k(\mathbf{x}) = \sum_{r=1}^{N} k^{(r)}\chi^{(r)}(\mathbf{x}), \quad f(\mathbf{x},\cdot) = \sum_{r=1}^{N} f^{(r)}(\cdot)\,\chi^{(r)}(\mathbf{x}).$$

With these notations, Eq. (1) can be re-written

$$\epsilon_m(\mathbf{x}) = \frac{1}{3k(\mathbf{x})}\sigma_m(\mathbf{x}), \quad e_{ij}(\mathbf{x}) = \frac{1}{2\mu_s(\mathbf{x})}s_{ij}(\mathbf{x}), \tag{3}$$

where $\epsilon_m = \frac{1}{3}\text{tr}(\epsilon)$, $\mathbf{e}$ is the strain deviator, and where the local secant moduli μ_s are given by

$$\mu_s(\mathbf{x}) = \frac{1}{6f'(\mathbf{x}, \sigma_{\text{eq}}^2(\mathbf{x}))}. \tag{4}$$

The local stress and strain fields within the r.v.e. solve a boundary value problem consisting of Eqs. (3), (4), compatibility conditions and equilibrium equations

$$\boldsymbol{\sigma} \in \mathcal{S}(\boldsymbol{\Sigma}), \tag{5}$$

where

$$\mathcal{S}(\boldsymbol{\Sigma}) = \{\boldsymbol{\tau},\ \text{div}\,(\boldsymbol{\tau}) = 0,\ \boldsymbol{\tau}\cdot\mathbf{n} = \boldsymbol{\Sigma}\cdot\mathbf{n} \text{ on } \partial V\}. \tag{6}$$

In general the boundary value problem (3), (4) and (5) cannot be solved analytically since the local secant moduli μ_s can vary from one point to another. An approximation is introduced by considering that these secant moduli are *piecewise constant* within each phase

$$\mu_s(\mathbf{x}) = \sum_{r=1}^{N} \mu_s^{(r)} \chi^{(r)}(\mathbf{x}), \quad \mu_s^{(r)} = \frac{1}{6f'(\overline{\sigma}_r^2)}, \tag{7}$$

where the "effective stress" of phase (r), $\overline{\sigma}_r$, is a constant within phase (r) and has to be specified. The (approximate) secant problem consists now of Eqs. (3)(7)(5). It is a nonlinear problem (since $\mu_s^{(r)}$ depends on $\overline{\sigma}_r$) but with piecewise constant moduli.

2.2. The Classical Theory

It remains to define the effective stress $\overline{\sigma}_r$. The classical approach consists in considering the equivalent von Mises stress associated with the average stress over the phase (r):

$$\boldsymbol{\Sigma}^r = \langle\boldsymbol{\sigma}\rangle_r, \quad \tilde{\sigma}_r = \Sigma_{\text{eq}}^r = \left(\frac{3}{2} S_{ij}^r S_{ij}^r\right)^{1/2}. \tag{8}$$

This choice has a serious limitation as pointed out by Qiu and Weng ([10]). Consider a r.v.e. comprised of an incompressible matrix with voids and subject to an hydrostatic stress. The average stress in the matrix is also hydrostatic and the effective stress evaluated with the help of (8) is always 0. Therefore the secant modulus of the matrix always coincides with its initial modulus and the secant method applied with (8) predicts a linear overall response of the porous material. However the real local stress state in the matrix is not hydrostatic (analytic calculations can be carried out on the hollow sphere model to prove this point explicitly), even if its average is hydrostatic. The occurrence of shear stresses in some regions of the r.v.e. introduces nonlinearities both in the local and overall responses of the r.v.e. which are not taken into account by (8).

2.3. The Modified Theory

Several other choices can be considered (see Thébaud *et al.* [15] or Bornert *et al.* [2] for other possible choices). The proposed theory emphasizes the role played by the second moment of the stress field

$$\overline{\sigma}_r = \left\langle \sigma_{\text{eq}}^2 \right\rangle_r^{1/2} . \tag{9}$$

This "effective stress" accounts for local shear stresses as pointed out by Qiu and Weng. The overall response of the hollow sphere predicted with this choice is nonlinear (as it should be) and close to the exact solution ([10]). The modified secant theory consists in solving Eqs. (3), (7) and (5) together with the definition (9) of the effective stress of phase (r). In practice, one has to compute $\overline{\sigma}_r$. This can be done analytically by means of an argument previously used in different contexts by several authors (see for instance Kreher ([6]). The central result is the following one.

Consider a linear *composite comprised of N phases with elastic moduli* $(3k^{(r)}, 2\mu^{(r)}$, *occupying the subdomains* V_r. *Set* $\theta^{(r)} = 1/\mu^{(r)}$, $\kappa^{(r)} = 1/k^{(r)}$, $\theta = (\theta^{(1)}, \ldots, \theta^{(N)})$ *and let* $\mathbf{M}^{\text{hom}}(\kappa, \theta)$ *be the overall compliance tensor of this* linear *composite. Then, denoting by* $\boldsymbol{\sigma}$ *the stress field in this linear composite, one has:*

$$\left\langle \sigma_{\text{eq}}^2 \right\rangle = \overline{\sigma}_r^2 = \frac{3}{c^{(r)}} \boldsymbol{\Sigma} : \frac{\partial \mathbf{M}^{\text{hom}}}{\partial \theta^{(r)}}(\kappa, \theta) : \boldsymbol{\Sigma}. \tag{10}$$

A detailed proof of this result can be found in [6] or [13] (among others). This result applies to the solution of the secant problem as well, although it is not a linear problem. Indeed, once the solution of the secant problem is known, the secant moduli $\mu_s^{(r)}$ are known. Then a *linear* problem with these specific shear moduli can be posed and its solution is precisely the solution of the secant nonlinear problem. Therefore the "effective stress" of each individual nonlinear phase can be computed from a *linear* theory.

In conclusion, the proposed modified secant moduli approach involves three steps:

1. A linear theory providing an expression for $\mathbf{M}^{\text{hom}}(\kappa, \theta)$ and its partial derivatives with respect to $\theta^{(r)}$.

2. The resolution of $2N$ nonlinear problems for the $2N$ unknowns $(\theta^{(r)}, \overline{\sigma}_r)$:

$$\theta^{(r)} = 6 f'^{(r)}(\overline{\sigma}_r^2), \quad \overline{\sigma}_r = \left(\frac{3}{c^{(r)}} \boldsymbol{\Sigma} : \frac{\partial \mathbf{M}^{\text{hom}}}{\partial \theta^{(r)}}(\kappa, \theta) : \boldsymbol{\Sigma} \right)^{1/2}, \tag{11}$$

3. Once the $2N$ nonlinear problems (11) are solved, the overall stress-strain relation is finally given by:

$$\mathbf{E} = \mathbf{M}^{\text{hom}}(\kappa, \theta) : \boldsymbol{\Sigma}. \tag{12}$$

2.4. Comments

1. The "new" effective stress (9) has been previously used in a less general context by Qiu and Weng ([10]) and Suquet ([12]) for matrices containing voids or

rigid inclusions. For general nonlinear composites Suquet has introduced in ([13]) the corresponding definition of the "effective strain" of a phase as the second moment of the strain field over this phase. But it seems that Buryachenko ([3]) had previously made use of similar definitions to predict in general the overall response of nonlinear composites and his contribution should be mentioned as one of the earliest ones. The original point in the author's derivation ([13]) is to prove that the modified theory is a "bound" for the actual overall properties of the composite and that its predictions precisely coincide with Ponte-Castañeda's variational results (see Section 3).

2. In addition to describing more accurately the "effective stress" within each individual phase, (9) has another major advantage on the classical definition (8): it can be used with *any* scheme predicting the overall linear properties of composites, without any knowledge of the local stress field. When the classical definition (8) is adopted, this is also true for a two-phase composite since the average stress over each individual phase can be deduced from the data of overall and the local stiffnesses according to Levin's result. It is also true when the predicting schemes are based on explicit expressions of these average stresses (Hashin-Shtrikman or Willis' bounds, self-consistent scheme ([16]). But it is not the case for more elaborate schemes.

2.5. Example

Consider an isotropic two-phase composite, comprised of isotropic incompressible phases. Then the compliance tensor $\mathbf{M}^{\text{hom}}$ depends on a single scalar parameter $\theta^{\text{hom}} = 1/\mu^{\text{hom}}$ and the nonlinear equations to be solved for $\overline{\sigma}_1, \overline{\sigma}_2$ reduce to

$$\overline{\sigma}_r = \left(\frac{1}{c^{(r)}} \frac{\partial \theta^{\text{hom}}}{\partial \theta^{(r)}} \right)^{1/2} \Sigma_{\text{eq}}, \quad \theta^{(r)} = 6 f'^{(r)}(\overline{\sigma}_r^2). \tag{13}$$

The coupled equations (13) are solved iteratively (starting from the arbitrary initial values $\overline{\sigma}_1 = \overline{\sigma}_2 = \Sigma_{\text{eq}}$, these relations are applied until a fixed point is reached). In the classical approach the effective strain $\tilde{\sigma}_r$ is computed by

$$\tilde{\sigma}_1 = \frac{1}{c^{(1)}} \frac{\theta^{\text{hom}} - \theta^{(2)}}{\theta^{(1)} - \theta^{(2)}}, \quad \tilde{\sigma}_2 = \frac{1}{c^{(2)}} \frac{\theta^{(1)} - \theta^{\text{hom}}}{\theta^{(1)} - \theta^{(2)}}. \tag{14}$$

To illustrate the difference between the classical and modified secant moduli approaches, an example involving pure power-law constituents has been carried out

$$\psi^{(r)}(\sigma) = \frac{\sigma_0^{(r)} \epsilon_0}{n^{(r)} + 1} \left(\frac{\sigma_{\text{eq}}}{\sigma_0^{(r)}} \right)^{n^{(r)}+1}, \tag{15}$$

with

$$\frac{\sigma_0^{(1)}}{\sigma_0^{(2)}} = 5, \quad n^{(1)} = 1, \quad n^{(2)} = 10, \quad c^{(1)} = 0.3.$$

The value $\theta^{\text{hom}} = 1/\mu^{\text{hom}}$ has been computed by means of the generalized three-phase self-consistent scheme ([4]). The results of the two predictions are compared in Fig. 1

with those of a Finite Element axisymmetric calculation on a cell model approaching a periodic array of identical spherical inclusions. The cell model consists of a spherical inclusion of phase 1 embedded in a circular cylinder of phase 2. A tensile stress is applied in the cylinder direction (see [12] for more details). The modified secant method is observed to be in a good agreement with the F.E.M. results. The classical approach yields predictions which are stiffer than the predictions of the modified approach. Gilormini ([5]), using a classical tangent approach, observed that this latter approach yields stiffer results than the modified secant approach (which will be shown to coincide with the variational bounds that he considered).

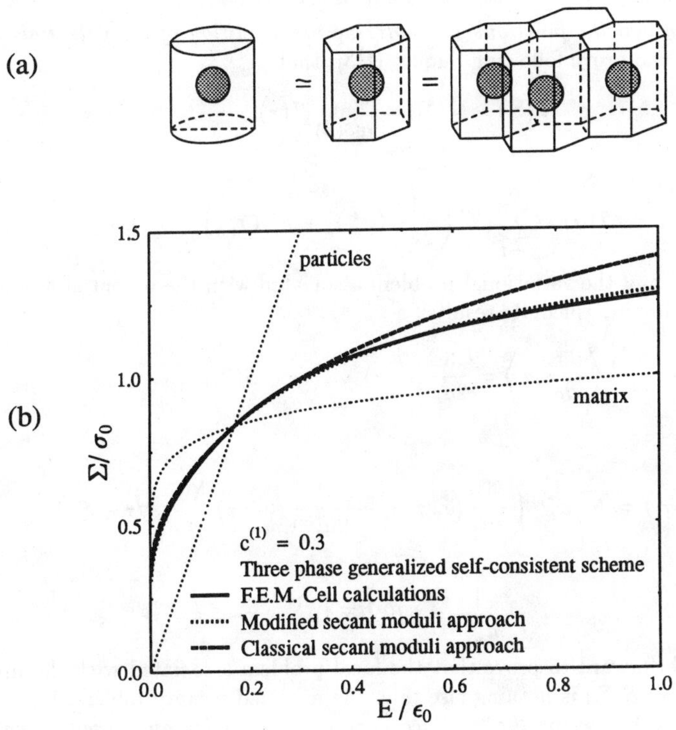

Fig. 1: (a) Axisymmetric cell model approaching a 3D periodic array. Phase 1 (grey) is the inclusion phase. (b) Overall stress strain curves: F.E.M. calculations (solid line), modified approach (broken line), classical approach (dotted line).

3. Overall Potentials and Bounds

As the local behaviour of the individual phases, the overall behaviour of the nonlinear composite can be derived from an overall potential:

$$\mathbf{E} = \frac{\partial \Psi}{\partial \boldsymbol{\Sigma}}(\boldsymbol{\Sigma}), \quad \text{where } \Psi(\boldsymbol{\Sigma}) = \inf_{\boldsymbol{\tau} \in \mathcal{S}(\boldsymbol{\Sigma})} \sum_{r=1}^{N} c^{(r)} \left\langle \psi^{(r)}(\boldsymbol{\tau}) \right\rangle_r$$

$$= \inf_{\boldsymbol{\tau} \in \mathcal{S}(\boldsymbol{\Sigma})} \sum_{r=1}^{N} c^{(r)} \left(\frac{1}{2k^{r)}} \left\langle \sigma_m^2 \right\rangle_r + \left\langle f^{(r)}(\sigma_{\text{eq}}^2) \right\rangle_r \right).$$

3.1. Bound Deduced from the Modified Secant Moduli Theory

Assuming that the functions $f^{(r)}$ entering the definition of the potentials $\psi^{(r)}$ are convex, one can prove following Suquet ([13]) that

$$\Psi(\boldsymbol{\Sigma}) \geq \underline{\Psi}(\boldsymbol{\Sigma}) = \inf_{\boldsymbol{\tau} \in \mathcal{S}(\boldsymbol{\Sigma})} \underline{\mathcal{J}}(\boldsymbol{\tau}),$$

where

$$\underline{\mathcal{J}}(\boldsymbol{\sigma}) = \sum_{r=1}^{N} c^{(r)} \left(\frac{1}{2k^{r)}} \left\langle \sigma_m^2 \right\rangle_r + f^{(r)}(\overline{\sigma}_r^2) \right).$$

The interpretation of the variational problem associated with the potential $\underline{\Psi}$ makes use of the following intermediate results:

$$\left\langle \frac{\partial \overline{\sigma}_r}{\partial \boldsymbol{\sigma}}, \boldsymbol{\tau} \right\rangle = \frac{3}{2\overline{\sigma}_r} \left\langle \mathbf{s} : \boldsymbol{\tau} \right\rangle_r, \quad \forall \boldsymbol{\tau} \in \mathcal{S}(\mathbf{0}),$$

and

$$\left\langle \frac{\partial \underline{\mathcal{J}}}{\partial \boldsymbol{\sigma}}(\boldsymbol{\sigma}), \boldsymbol{\tau} \right\rangle = \sum_{r=1}^{N} c^{(r)} \left(\frac{1}{k^{(r)}} \left\langle \sigma_m \tau_m \right\rangle_r + \frac{1}{2\mu_s{}^{(r)}} \left\langle \mathbf{s} : \boldsymbol{\tau} \right\rangle_r \right), \quad \forall \boldsymbol{\tau} \in \mathcal{S}(\mathbf{0}),$$

with

$$\frac{1}{\mu_s{}^{(r)}} = 6 f'^{(r)}(\overline{\sigma}_r{}^2).$$

It is then straightforward to prove that the local problem associated with the minimization of $\underline{\mathcal{J}}$ over $\mathcal{S}(\boldsymbol{\Sigma})$ is nothing else than the modified secant problem, Eqs. (3), (9) and (5). In other terms *the modified secant theory yields an overall potential which is a lower bound to the actual overall potential.* The secant moduli theory can be interpreted in terms of bounds on material properties. Consider for example an isotropic composite comprised of isotropic incompressible power-law materials *with the same exponent n* (see (15)), but with different flow stresses $\sigma_0^{(r)}$. Then the composite is also an isotropic incompressible power law material. Neglecting the effect of the third invariant of the overall stress, an overall flow stress σ_0^{hom} can be defined from the overall potential. Then the flow stress deduced from the modified secant theory is always an upper bound to the actual flow stress of the composite. For more general behaviours of phases, it is not straightforward to interpret inequalities on overall potentials in terms of inequalities on overall responses.

3.2. Link with Ponte-Castañeda's variational procedure

Ponte Castañeda has recently introduced a variational procedure to bound the overall potentials of nonlinear composites ([8]). His result, expressed in the present notations, reads as

$$\Psi(\mathbf{\Sigma}) \;\geq\; \inf_{\theta^{(r)}\geq 0} \left\{ W_*^{\text{hom}}(\kappa,\theta,\mathbf{\Sigma}) + \sum_{r=1}^{N} c^{(r)}\; U^{(r)}(\kappa^{(r)},\theta^{(r)}) \right\}, \tag{16}$$

where $W_*^{\text{hom}}(\kappa,\theta,\mathbf{\Sigma}) = \frac{1}{2}\mathbf{\Sigma} : \mathbf{M}^{\text{hom}}(\kappa,\theta) : \mathbf{\Sigma}$ is the complementary elastic energy of a fictitious linear composite comprised of phases with elastic moduli $(\kappa^{(r)},\mu^{(r)})$ occupying the domains V_r and where

$$U^{(r)}(\kappa^{(r)},\theta^{(r)}) = \sup_{\boldsymbol{\sigma}\,\in\,\mathbb{R}_s^{3\otimes 3}} \left\{ \psi^{(r)}(\boldsymbol{\sigma}) - \left(\frac{k^{(r)}}{2}\sigma_m^2 + \frac{\theta^{(r)}}{6}\sigma_{\text{eq}}^2 \right) \right\}.$$

To understand the link between the modified secant approach and Ponte Castañeda's variational procedure, it is useful to introduce the dual function of $f^{(r)}$. First $f^{(r)}(x)$, well defined by (2) when $x \geq 0$, is extended for negative values of x by $+\infty$. $f_*^{(r)}$ denotes the dual function of $f^{(r)}$

$$f_*^{(r)}(\theta) = \sup_x \left(x\theta - f^{(r)}(x) \right)$$

$f_*^{(r)}$ is itself a convex function. Classically:

$$f_*^{\prime(r)}(\theta) = x \Leftrightarrow \theta = f^{\prime(r)}(x). \tag{17}$$

After due account of the expression (2) of $\psi^{(r)}$, $U^{(r)}$ simplifies to

$$U^{(r)}(\kappa^{(r)},\theta^{(r)}) = \inf_{\sigma} \left(f^{(r)}(\sigma_{\text{eq}}^2) - \frac{\theta^{(r)}}{6}\sigma_{\text{eq}}^2 \right) = -f_*^{(r)}\left(\frac{\theta^{(r)}}{6}\right).$$

The optimality conditions related to the infimum problem (16) read

$$\frac{\partial W_*^{\text{hom}}}{\partial \theta^{(r)}}(\kappa,\theta,\mathbf{\Sigma}) = \frac{c^{(r)}}{6} f_*^{\prime(r)}\left(\frac{\theta^{(r)}}{6}\right) \tag{18}$$

or, according to (17) and to the expression of W_*^{hom}

$$\theta^{(r)} \;=\; 6 f^{\prime(r)}\left(\frac{3}{c^{(r)}}\mathbf{\Sigma} : \frac{\partial \mathbf{M}^{\text{hom}}}{\partial \theta^{(r)}}(\kappa,\theta) : \mathbf{\Sigma} \right). \tag{19}$$

The set of nonlinear equations (19) is identical to (11). In conclusion *the modified secant moduli approach coincides with Ponte Castañeda's variational procedure.* In other terms, Ponte Castañeda's variational procedure can be interpreted as a modified secant moduli approach.

3.3. Power Law Materials

When the individual phases are incompressible power law materials with the same exponents, Suquet ([11]) has proposed a lower bound for the overall potential Ψ of the composite. With the above notations (15), this bound reads

$$\Psi(\Sigma) \geq \frac{\epsilon_0}{n+1} \sup_{\theta^{(r)} \geq 0} \left\{ W_*^{\mathrm{hom}}(\theta, \Sigma)^{\frac{n+1}{2}} \left\langle \left(\frac{\theta}{6}\right)^{\frac{n+1}{n-1}} \sigma_0^{\frac{2n}{n-1}} \right\rangle^{\frac{1-n}{2}} \right\}. \tag{20}$$

Ponte Castañeda has proved ([9]) that these bounds can be deduced from his own bounds. Indeed a straightforward calculation shows that

$$U(\theta) = -f_*\left(\frac{\theta}{6}\right) = -\frac{n-1}{n+1}\left(\frac{2\sigma_0^n}{\epsilon_0}\right)^{\frac{2}{n-1}}\left(\frac{\theta}{6}\right)^{\frac{n+1}{n-1}},$$

and the optimization problem (16) reduces to

$$\sup_{\theta^{(r)} \geq 0} \left\{ W_*^{\mathrm{hom}}(\kappa, \theta, \Sigma) - \frac{n-1}{n+1} \left\langle \left(\frac{2\sigma_0^n}{\epsilon_0}\right)^{\frac{2}{n-1}} \left(\frac{\theta}{6}\right)^{\frac{n+1}{n-1}} \right\rangle \right\}. \tag{21}$$

With the change of variables $\theta = t\tilde{\theta}$, where $\tilde{\theta} = (\tilde{\theta}^{(1)}, \ldots, \tilde{\theta}^{(1)})$ the optimization in (21) can be performed on t first. The result of this first optimization is precisely (20).

In conclusion, for purely power law materials with the same exponent, Ponte Castañeda's and Suquet's procedures coincide. Their common predictions are also identical to those of the modified secant moduli approach.

Acknowledgments. The author is indebted to J. C. Michel for the F.E.M. cell calculations reported in Fig. 1. This study is part of the "Eurohomogenization" project supported by the SCIENCE Program of the Commission of the European Communities.

References

1. M. Berveiller and A. Zaoui, *J. Mech. Phys. Solids* **26** (1979) 325.
2. M. Bornert, E. Hervé, C. Stolz and A. Zaoui, *Appl. Mech. Rev.* **47** (1994) S66.
3. V. A. Buryachenko, The overall elastoplastic behavior of multiphase isotropic composites, *Preliminary draft,* 1994.
4. E. Hervé and A. Zaoui, *Eur. J. Mech. A/ Solids* **9** (1990) 505.

5. P. Gilormini, *C. R. Acad. Sc. Paris IIb* **320** (1995) 115.
6. W. Kreher, *J. Mech. Phys. Solids* **38** (1990) 115.
7. P. Ponte Castañeda, *J. Mech. Phys. Solids* **39** (1991) 45.
8. P. Ponte Castañeda, *J. Mech. Phys. Solids* **40** (1992) 1757.
9. P. Ponte Castañeda, *Private communication.*
10. Y. P. Qiu and G. J. Weng. *J. Appl. Mech.* **59** (1992) 261.
11. P. Suquet, *J. Mech. Phys. Solids* **41**, (1993) 981.
12. P. Suquet, In *Micromechanics of Materials*, eds J. J. Marigo and G. Rousselier, Eyrolles, 1993, p. 361.
13. P. Suquet, *C.R. Acad. Sc. Paris, IIb* **320** (1995) 563.
14. G. P. Tandon and G. J. Weng. *J. Appl. Mech.* **55** (1988) 126.
15. F. Thébaud, J. C. Michel, E. Hervé, P. Suquet and A. Zaoui. In *Proceedings of the 13th Riso International Symposium on Materials Science*, ed. S. I. Andersen, 1992, p. 467.
16. J. R. Willis. *J. Mech. Phys. Solids* **25** (1977) 185.
17. J. R. Willis. *J. Mech. Phys. Solids* **39** (1991) 73.

Continuum Models and Discrete Systems
Proceedings of 8th International Symposium, June 11-16, 1995, Varna, Bulgaria, ed. K.Z. Markov

EFFECTIVE PROPERTIES OF AN ELASTIC BODY DAMAGED BY RANDOM DISTRIBUTION OF MICROCRACKS

J.J. TELEGA and B. GAMBIN
Institute of Fundamental Technological Research,
Polish Academy of Sciences, Świętokrzyska 21, PL-00-049 Warsaw, Poland

Abstract. The aim of this contribution is to characterize the overall behaviour of an elastic solid weakened by randomly distributed microcracks. The distribution of microcracks is stochastically periodic. Unilateral behaviour of microcracks is modelled by the Signiorini's conditions without friction. By employing the method of Γ-convergence of random functionals [1] the macroscopic elastic potential is derived. The density of the complementary energy is found by the dual approach. An example is also provided.

1. Introduction

The literature on macroscopic modelling of elastic solids weakened by microfissures is already quite comprehensive. Here we shall study the case of stochastically periodic distribution of microcracks by extending mathematically rigorous approach of $\Gamma-$convergence due to Dal Maso and Modica [1].

2. Preliminaries

Prior to passing to the investigation of the title problem, let us fix notation for an elastic solid weakened by a single fissure, say $\mathbf{C}$, cf. [2,3]. An elastic solid occupies a domain $\bar{A}_0$ in $\mathbb{R}^N$ (in physical situation $N = 2, 3$); bar over a set denotes its closure. By $\boldsymbol{u}$ we denote the displacement vector, $\boldsymbol{e}(\boldsymbol{u}) = \frac{1}{2}(\nabla\boldsymbol{u} + \nabla\boldsymbol{u}^T)$ is the deformation tensor, $\boldsymbol{\sigma}$ – the stress tensor, $\mathbb{E}_s^N$ – the space of second order symmetric tensors on $\mathbb{R}^N$; moreover

$$J : \mathbb{E}_s^N \to R^+ \tag{1}$$

stands for the density of the elastic energy. The last function is assumed to be convex and satisfies the condition:

$$\lambda|\boldsymbol{E}|^2 \leq J(\boldsymbol{E}) \leq \Lambda(1 + |\boldsymbol{E}|^2), \quad 0 < \lambda \leq \Lambda < +\infty, \tag{2}$$

for each $\boldsymbol{E} \in \mathbb{E}_s^N$. Let $\mathbf{C} \subset A$ be a fissure, which is a smooth $(N-1)$–dimensional manifold. We assume that $A_0 \backslash \mathbf{C}$ is an open connected set. It means that $\mathbf{C}$ does not divide A_0 into disjoint pieces. By choosing a normal vector $\boldsymbol{n}$ to $\mathbf{C}$ we can define two sides of $\mathbf{C}$ and for any function $v \in H^1(A\backslash\mathbf{C})$ we get two traces on $\mathbf{C}$, v_1 and v_2. The jump of the function v across $\mathbf{C}$ is denoted by

$$[\![v]\!] = v_2 - v_1 \,. \tag{3}$$

For a prescribe distribution of the body forces $\boldsymbol{f} = (f_i)$, $f_i \in L^2(A_0)$ the minimum principle of the total potential energy reads:

$$\left| \quad \text{Find} \quad \min_{\mathbf{u}\in K} \left\{ \int_{A_0\backslash\mathbf{C}} J(e(\boldsymbol{u}))\, d\boldsymbol{x} - \int_{A_0\backslash\mathbf{C}} \boldsymbol{f}\cdot\boldsymbol{u}\, d\boldsymbol{x} \right\} \right. \tag{4}$$

$$\text{where} \quad K = \{\boldsymbol{u} \in [H^1(A_0\backslash\mathbf{C})]^N \mid \boldsymbol{u} = 0 \text{ on } \partial A_0 \,,\ [\![u_i n_i]\!] \geq 0 \text{ on } \mathbf{C}\} \tag{5}$$

is a convex set of kinematically admissible displacement fields. For details concerning the Euler equation of this convex minimization problem and mechanical aspects of fissured materials the reader may refer to [8], cf. also [2,4,7,9,10].

3. Periodically Stochastic Homogenization Problem for a Fissured Elastic Body

The problem of the macroscopic behaviour of an elastic material with a large number of microcracks is now solved for the case of stochastically periodic distribution of fissures. Previous results concerning applications of the homogenization methods to the determination of effective moduli of microfissured elastic solids are restricted to periodic distribution of microcracks [2–4,7,9–10].

The nonlinear stochastic homogenization approach developed in [1] is not directly applicable to finding the effective properties of elastic solids with stochastically periodic distribution of microfissures which behave according to the Signorini's condition. However, by a proper generalization, we shall perform homogenization also in this case. Toward this end we define the "geometry of fissures" as any distribution C of a countable number of smooth $(N-1)$–dimensional manifolds C_i in R^N, $i \in \mathcal{N}$, $\mathcal{N}$—the set of natural numbers. We assume that the set

$$R^N\backslash C \,, \quad \text{where } C = \bigcup_{i\in\mathcal{N}} C_i \,, \tag{6}$$

is open and connected. Following [1] we introduce the class $\mathcal{F}$ of integral functionals F in the following manner:

$$F : (L^2_{\text{loc}}(R^N))^N \times \mathcal{A}_0 \longrightarrow \mathcal{R} \cup \{+\infty\} \,, \tag{7}$$

$$F(\boldsymbol{u}, A) = \begin{cases} \displaystyle\int_A J(e(\boldsymbol{u}(\boldsymbol{x})))\, dx & \text{if} \quad \boldsymbol{u}|_A \in [H^1(A)]^N \,, \\ +\infty, \quad \text{otherwise,} \end{cases} \tag{8}$$

where $A \in \mathcal{A}^0$ and $\mathcal{A}_0$ is the family of all open bounded subsets of $\mathbb{R}^N$.

Let us define now a class $\mathcal{F}'$ of "fissured" functionals by:

$$F^C(\boldsymbol{u}, A) := F(\boldsymbol{u}, A \backslash C)\,. \tag{9}$$

To avoid technical difficulties we assume that

$$\bigcup_i C_i \cap \partial A = \emptyset\,. \tag{10}$$

In the last relation C_i are fissured contained in A. For the domain $A_0 \backslash C$ the problem:

$$\left| \quad \text{Find} \quad \min_{\boldsymbol{u} \in K} \{F(\boldsymbol{u}, A_0 \backslash C) - \int_{A_0} \boldsymbol{f} \cdot \boldsymbol{u}\, dx\}\,, \right. \tag{11}$$

describes the equilibrium state of our elastic solid weakened by fissures C_i under the action of the body forces $\boldsymbol{f}$. Similar problem can be formulated for any $A \in \mathcal{A}_0$ by a proper restriction and//or extension of $\boldsymbol{f}$ (extension by zero), depending on A_0.

The problem (11) is uniquely solvable. If $C = \emptyset$ then $F^{\emptyset} \in \mathcal{F}$. As we want to deal with a random distribution of fissures that is with measurable maps $\omega \to F(\omega)$ of a probabilistic space $(\Omega, \mathbf{S}, P)$ into $\mathcal{F}'$, we need some structure on $\mathcal{F}'$. If d is a distance on $\mathcal{F}$ defined in [1] we adopt it to measure the distance between two "fissured functionals" from $\mathcal{F}'$ as follows:

$$d(F^C, F^D) = \sum_{i,j,k=1}^{+\infty} \frac{1}{2^{i+j+k}} \left| \arctan T_{1//i} F^C(\boldsymbol{w}_j, B_k) - \arctan T_{1//i} F^D(\boldsymbol{w}_j, B_k) \right|, \tag{12}$$

where

$$T_\varepsilon F^C(\boldsymbol{u}, A) \equiv \min\left\{ F^C(\boldsymbol{v}, A) + \frac{1}{\varepsilon} \int_A |\boldsymbol{v} - \boldsymbol{u}|^2\, dx \;:\; \boldsymbol{v} \in [H^1(A \backslash C)]^N) \right\}. \tag{13}$$

Here $\boldsymbol{w}_i$ belongs to a countable dense subset of $[H^1(R^N)]^N$, while $\{B_k\}$ is a countable dense subfamily of $\mathcal{A}^0$. It is clear that

$$F^C = F^D \Longleftrightarrow C = D\,, \tag{14}$$

i.e. the distribution of microfissures is the same. Now a *random fissured functional* is defined as any measurable function

$$F^C : \Omega \to \mathcal{F}'\,, \tag{15}$$

where $\mathcal{F}'$ is endowed with the Borel $\boldsymbol{\sigma}$–field $\mathbf{S}_B$ generated by the distance d. It means that the function $\omega \to F^C(\omega)(\boldsymbol{u}, A)$ is a real extended random variable; thus it is a measurable function between $(\Omega, \mathbf{S}, P)$ and the Borel line. For any fixed $\omega \in \Omega$, the minimization problem of the functional $F^C(\omega)(\boldsymbol{u}, A_0)$ can be defined similarly as (11). Suppose now that $F^C(\omega) = F^{C(\omega)}$, which means that the randomness of our problem is defined by a random distribution of fissures. For every $z \in Z^n$ and every $\varepsilon > 0$ we define translation and homothety operators by

$$\tau_z, \rho_\varepsilon : R^N \to R^N , \quad \tau_z(\boldsymbol{x}) = \boldsymbol{x} + \boldsymbol{z} , \quad \rho_\varepsilon(\boldsymbol{x}) = \frac{\boldsymbol{x}}{\varepsilon} . \tag{16}$$

Let $(F^C_\varepsilon)(\omega)$ $(\varepsilon > 0)$ be a family of random fissured functionals defined on the same probability space Ω. We say that $(F^C_\varepsilon)(\omega)$ is a stochastic homogenization process modelled on a fixed random fissured functional $F^{C(\omega)}$ on Ω if

$$F^C_\varepsilon(\omega) \sim \rho_\varepsilon F^C(\omega) \quad \text{for every } \varepsilon > 0 , \tag{17}$$

that is $F^C_\varepsilon(\omega)$ and $\rho_\varepsilon F^C(\omega)$ have the same distribution law. We say that $F^C(\omega)$ is *stochastically periodic* with period $T > 0$ if $F^C(\omega) \sim \tau_z F^C(\omega)$ for every $\boldsymbol{z} \in TZ^N = \{\boldsymbol{x} \in R^N : \boldsymbol{x}//T \in Z^N\}$. We assume that the elastic properties of a material itself are deterministic and do not depend on ε. For the sake of simplicity we assume that $T = 1$. The main problem of the stochastic homogenization consists in passing to the limit as $\varepsilon \to 0$ in the sequence of random minimization problems

$$\inf_{K^\varepsilon(\omega)} \left\{ \int_{A_0 \backslash C^\varepsilon(\omega)} J(\boldsymbol{e}(\boldsymbol{u}(\boldsymbol{x}))) \, d\boldsymbol{x} - \int_{A_0} \boldsymbol{f}(\boldsymbol{x}) \cdot \boldsymbol{u}(\boldsymbol{x}) \, d\boldsymbol{x} \right\} , \tag{18}$$

where $C^\varepsilon(\omega)$ is a random distribution of fissures, $C^\varepsilon(\omega) = \bigcup_i C^\varepsilon_i(\omega)$, $\boldsymbol{x} \in C^\varepsilon(\omega) \Leftrightarrow \frac{\boldsymbol{x}}{\varepsilon} \in C(\omega)$ for every $\omega \in \Omega$ and $\varepsilon > 0$; moreover

$$K^\varepsilon(\omega) = \left\{ \boldsymbol{u} \in [H^1(A_0 \backslash C^\varepsilon(\omega))]^N \;\middle|\; \boldsymbol{u} = 0 \text{ on } \partial A_0 , \; [\![u_i n_i]\!] \geq 0 \text{ on } \bigcup_i C^\varepsilon_i(\omega) \right\} . \tag{19}$$

The solution of (18) for a fixed $\varepsilon > 0$ and $\omega \in \Omega$ yields the displacement field $\bar{\boldsymbol{u}}$ of the elastic solid A_0 weakened by a random distribution of the microfissures $C^\varepsilon(\omega)$.

4. Stochastic Homogenization in the Presence of Stochastically Periodic Distribution of Microfissures

Let $(F^C_\varepsilon)(\omega)$ be a stochastic homogenization process modelled on a stochastically periodic random fissured functional $F^{C(\omega)}$. Suppose that there exists $M > 0$ such that two families of random variables

$$(F^C(\cdot)(\boldsymbol{u}, A)) , \; (F^C(\cdot)(\boldsymbol{u}, B)) , \; \boldsymbol{u} \in [L^2_{loc}(R^N)]^N \tag{20}$$

are independent whenever $A, B \in \mathcal{A}^0$ with $\operatorname{dist}(A, B) \geq M$. Then $F^C_\varepsilon(\omega)$ converges in probability as $\varepsilon \to 0$ to the single functional $F_0 \in \mathcal{F}$ independent of ω (i.e. to the constant random integral functional) given by

$$F_0(\boldsymbol{u}, A_0) = \begin{cases} \displaystyle\int_{A_0} J_0(\boldsymbol{e}(\boldsymbol{u}(\boldsymbol{x}))) \, d\boldsymbol{x} , & \text{if } \; \boldsymbol{u} \in [H^1(A_0)]^N , \\ +\infty , & \text{otherwise,} \end{cases} \tag{21}$$

where

$$J_0(\boldsymbol{E}) = \lim_{\varepsilon\to 0}\int_\Omega \min_{\boldsymbol{u}}\left\{\frac{1}{|Q_{\frac{1}{\varepsilon}}|}\int_{Q_{\frac{1}{\varepsilon}}\setminus C(\omega)} J(\boldsymbol{e}(\boldsymbol{u}(\boldsymbol{x})))\,d\boldsymbol{x} \mid \boldsymbol{u}-\boldsymbol{E}\boldsymbol{x}\in[H^1(Q_{\frac{1}{\varepsilon}}\setminus C(\omega))]^N,\right.$$

$$\left.\boldsymbol{u}-\boldsymbol{E}\boldsymbol{x}=0 \quad\text{on}\quad \partial Q_{\frac{1}{\varepsilon}},\quad [\![u_i n_i]\!]\geq 0 \quad\text{on}\quad C(\omega)\right\} dP(\omega), \mathbb{E}\in\mathbb{E}^N_s \tag{22}$$

$|Q_{\frac{1}{\varepsilon}}| = \{\boldsymbol{x}\in R^N : |x_i| < \frac{1}{\varepsilon},\ i=1,\ldots,N\}$ and $|Q_{\frac{1}{\varepsilon}}| = (\frac{2}{\varepsilon})^N$. It can be shown that

$$m^\varepsilon = \min_{\boldsymbol{u}\in K^\varepsilon(\omega)} \quad Bigl\{F^C_\varepsilon(\omega)(\boldsymbol{u},A_0) - \int_{A_0}\boldsymbol{f}(\boldsymbol{x})\cdot\boldsymbol{u}(\boldsymbol{x})\,d\boldsymbol{x}\right\} \tag{23}$$

converges in probability to

$$m^0 = \min_{\boldsymbol{u}\in[H^1_0(A_0)]^N}\left\{F_0(\boldsymbol{u},A_0) - \int_{A_0}\boldsymbol{f}(\boldsymbol{x})\cdot\boldsymbol{u}(\boldsymbol{x})\,d\boldsymbol{x}\right\}. \tag{24}$$

Similar theorems can be formulated for structures like beams, plates, etc. weakened by randomly distributed microfissures, cf. [4,5]. Formula (22) determines the effective (macroscopic) elastic potential. Unfortunately, it is much more complicated then in the periodic case.

In the last case some analytical results are available for plates weakened by aligned fissures. To illustrate some differences, in Section 6 we will compare periodic and stochastic solutions.

5. Dual Homogenized Potential

5.1. Scalar Case without Fissures

We take an integral functional of the form

$$F(u,A) = \begin{cases} \displaystyle\int_A f(\boldsymbol{x}, Du(\boldsymbol{x}))\,d\boldsymbol{x}, & \text{if}\quad u|_A\in W^{1,\alpha}(A), \\ +\infty, & \text{otherwise}. \end{cases} \tag{25}$$

Then, the primal homogenized potential derived in [1] is given by

$$f_0(\boldsymbol{p}) = \lim_{\varepsilon\to 0}\int_\Omega \min_u\left\{\frac{1}{|Q_{\frac{1}{\varepsilon}}|}F(\omega)(u,Q_{\frac{1}{\varepsilon}})|u-\boldsymbol{p}\cdot\boldsymbol{x}\in W^{1,\alpha}_0(Q_{\frac{1}{\varepsilon}})\right\} dP(\omega), \tag{26}$$

$\boldsymbol{p}\in\mathbb{R}^N$, where

$$F(\omega)(u,Q_{\frac{1}{\varepsilon}}) = \begin{cases} \displaystyle\int_{Q_{\frac{1}{\varepsilon}}} f(\omega,\frac{\boldsymbol{x}}{\varepsilon},Du(\boldsymbol{x}))\,d\boldsymbol{x}, & \text{if}\quad u\in W^{1,\alpha}(Q_{\frac{1}{\varepsilon}}), \\ +\infty, & \text{otherwise}. \end{cases}$$

Let $\boldsymbol{p}^* \in \mathbb{R}^N$; $f_0^*(\boldsymbol{p}*)$ denotes the dual homogenized potential. In order to derive the function f_0^* we employ the Fenchel transformation. Thus we write

$$f_0^*(\boldsymbol{p}^*) = \sup_{\boldsymbol{p}\in\mathbb{R}^N} \{\boldsymbol{p}^* \cdot \boldsymbol{p} - f_0(\boldsymbol{p})\}, \; \boldsymbol{p}^* \in \mathbb{R}^N . \tag{27}$$

After lengthy calculations we get

$$f_0^*(\boldsymbol{p}^*) = \lim_{\varepsilon\to 0}\int_\Omega \left\{ \frac{1}{|Q_{\frac{1}{\varepsilon}}|} \inf \int_{Q_{\frac{1}{\varepsilon}}} f^*(\omega, \frac{\boldsymbol{x}}{\varepsilon}, \boldsymbol{\sigma}(\boldsymbol{x}) + \boldsymbol{p}^*)\, d\boldsymbol{x} | \boldsymbol{\sigma} \in W_\varepsilon^\perp \right\} dP(\omega), \tag{28}$$

$$\text{where} \quad f^*(\omega, \frac{\boldsymbol{x}}{\varepsilon}, \boldsymbol{p}^*) \equiv \sup_{\boldsymbol{p}\in\mathbb{R}^N} \{\boldsymbol{p}^* \cdot \boldsymbol{p} - f(\omega, \frac{\boldsymbol{x}}{\varepsilon}, \boldsymbol{p})\}, \boldsymbol{p}^* \in \mathbb{R}^N ,$$

$$W_\varepsilon^\perp = \{\boldsymbol{\sigma} \in \mathrm{L}^{\alpha'}(Q_{\frac{1}{\varepsilon}})^N \mid \operatorname{div} \boldsymbol{\sigma} = 0 \text{ in } Q_{\frac{1}{\varepsilon}}, \int_{Q_{\frac{1}{\varepsilon}}} \boldsymbol{\sigma}(\boldsymbol{x}) d\boldsymbol{x} = 0\}, \frac{1}{\alpha} + \frac{1}{\alpha'} = 1 .$$

5.2. Elastic Solid with Microfissures

The dual (complementary) elastic potential of a solid weakened by microfissures with stochastically periodic distribution is given by:

$$J_0^*(\boldsymbol{\Sigma}) = \lim_{\varepsilon\to 0}\int_\Omega \left\{ \frac{1}{|Q_{\frac{1}{\varepsilon}}|} \inf \int_{Q_{\frac{1}{\varepsilon}}\setminus C(\omega)} J^*(\boldsymbol{\sigma}(\boldsymbol{x}))\, d\boldsymbol{x} |\; \boldsymbol{\sigma} \in W_\varepsilon^C(\omega, \boldsymbol{\Sigma}) \right\} dP(\omega), \tag{29}$$

where

$$W_\varepsilon^C(\omega, \boldsymbol{\Sigma}) = \Big\{ \boldsymbol{\sigma} = (\sigma_{ij}), \sigma_{ij} = \sigma_{ji} \;\Big|\; \sigma_{ij} \in L^2(Q_{\frac{1}{\varepsilon}}\setminus C(\omega)) \operatorname{div} \boldsymbol{\sigma} = 0 \text{ in } Q_{\frac{1}{\varepsilon}}\setminus C(\omega);$$

$$\frac{1}{|Q_{\frac{1}{\varepsilon}}|}\int_{Q_{\frac{1}{\varepsilon}}\setminus C(\omega)} \boldsymbol{\sigma}(\boldsymbol{x})\, d\boldsymbol{x} = \boldsymbol{\Sigma} \Big\}, \; \boldsymbol{\Sigma} \in \mathbb{E}_s^N, \; \sigma_{ij} T_i T_j = 0 \text{ and } \sigma_{ij} N_i N_j \le 0 \text{ on } C(\omega),$$

$\boldsymbol{N}, \boldsymbol{T}$ – normal and tangent vectors to fissures $C(\omega)$ contained in $Q_{\frac{1}{\varepsilon}}$; moreover

$$J_0^*(\boldsymbol{\sigma}) = \sup_{\boldsymbol{\eta}\in\mathbb{E}_s^N} \{\boldsymbol{\sigma} \cdot \boldsymbol{\eta} - J(\boldsymbol{\eta})\} ; \boldsymbol{\sigma} \in \mathbb{E}_s^N .$$

6. Example: Kirchoff Plate Weakened by Aligned Fissures Distributed Periodically or Randomly

Let us consider an elastic, isotropic and homogeneous plate with periodic fissures characterized by the cone [9]

$$K_{YC} = \left\{ v \in V_{YC} \quad [\![v]\!] = 0\,, \quad \left[\!\!\left[\frac{\partial v}{\partial \boldsymbol{n}} \right]\!\!\right] \leq 0 \text{ on } C \right\}, \tag{30}$$

where

$$V_{YC} = \{ v \in H^2(YC),\ v \text{ and } \frac{\partial v}{\partial y_\alpha} \text{ are equal on the opposite sides of } Y,\ \alpha = 1,2 \}\,. \tag{31}$$

Here Y is the basic cell and $YC = Y \backslash C$.

The homogenized effective, elastic potential has the form:

$$W(\boldsymbol{\rho}) = \begin{cases} D_1 = D[\boldsymbol{\rho}_{11}^2 + 2\nu \boldsymbol{\rho}_{11}\boldsymbol{\rho}_{22} + \boldsymbol{\rho}_{22}^2 + (1-\nu)(\boldsymbol{\rho}_{12}^2 + \boldsymbol{\rho}_{21}^2)]/2\,, \\ \text{if } \boldsymbol{\rho}_{22} + \nu \boldsymbol{\rho}_{11} \geq 0 \text{ (the fissure is closed)}\,, \\ D_2 = D[(1-\nu^2)\boldsymbol{\rho}_{11}^2 + (1-\nu)(\boldsymbol{\rho}_{12}^2 + \boldsymbol{\rho}_{21}^2)]/2 \text{ (the fissure is open).} \end{cases} \tag{32}$$

One obtains (32) by solving the periodic local problem in the domain $Y \backslash C$. Then the local transverse displacement is a quadratic function of one variable only (perpendicular to the fissure in Y). In the case of stochastically periodic distribution of such aligned fissures the counterpart of the local problem requires solving of the following minimization problem:

$$\begin{aligned} J_0(\boldsymbol{\rho}) = \lim_{\varepsilon \to 0} \int_\Omega \min_v \Big\{ & \frac{1}{|Q_{\frac{1}{\varepsilon}}|} F^C(\omega)(v, Q_{\frac{1}{\varepsilon}}) \mid v - \tfrac{1}{2}\rho_{\alpha\beta} x_\alpha x_\beta \in H^2(Q_{\frac{1}{\varepsilon}} \backslash C(\omega))\,, \\ & v - \tfrac{1}{2}\rho_{\alpha\beta} x_\alpha x_\beta = 0 \text{ on } \partial Q_{\frac{1}{\varepsilon}}\,, \quad [\![v]\!] = 0, \left[\!\!\left[\frac{\partial v}{\partial \boldsymbol{n}} \right]\!\!\right] \leq 0 \text{ on } C(\omega) \Big\}\,, \end{aligned} \tag{33}$$

$$F^C(\omega)(v, Q_{\frac{1}{\varepsilon}}) = \begin{cases} \displaystyle\int_{Q_{\frac{1}{\varepsilon}} \backslash C(\omega)} J(D^2 v(\boldsymbol{x}))\, d\boldsymbol{x} \text{ if } v \in H^2(Q_{\frac{1}{\varepsilon}} \backslash C(\omega))\,, \\ +\infty \quad \text{otherwise}\,; \end{cases} \tag{34}$$

$$J(\boldsymbol{\kappa}) = \frac{1}{2} D_{\alpha\beta\gamma\delta} \kappa_{\alpha\beta} \kappa_{\gamma\delta} \ (\alpha, \beta = 1,2\,;\, \boldsymbol{x} = (x_1, x_2))\,,\ \boldsymbol{\kappa} = (\kappa_{\alpha\beta})\,. \tag{35}$$

For the considered isotropic material

$$D_{\alpha\beta\gamma\delta} = D_{\alpha\beta\gamma\delta}(D, \nu)\,. \tag{36}$$

Here $C(\omega)$ is a given random distribution of aligned fissures described by random variables $(X_k(\omega))$ such that

$$\begin{aligned} &P\{\omega : X_k(\omega) = 0 \text{ (if } I_k - \text{interval is without fissure)}\} = r\,,\ r \in [0,1]\,, \\ &P\{\omega : X_k(\omega) = 1 \text{ (if } I_k - \text{interval contains a fissure)}\} = 1 - r\,, \end{aligned} \tag{37}$$

where $\bigcup_{k\in Z} I_k = R^1$, $|I_k| = 1$. The independent random variables X_k can be chosen independently by the Bernoulli's law. To find the minimum in (33) for every $\varepsilon_j = \frac{1}{j}$, $j = 1, 2, \ldots$, we solve the equation of a clamped plate of dimensions $j \times a$ weakened by a number of fissures changing from 0 to j, subject to moments ρ_{ij}; here a is arbitrary.

In contrast to the periodic case, the transverse displacement is now a function of the variables x_1 and x_2. It means that though fissures are aligned yet the counterpart of the local problem remains two-dimensional. A simplification is achieved under further strong assumption that the plate is made of two materials, randomly mixed along one direction. More precisely, the first material behaves according to the elastic potential (32) whilst the other one is a virgin material with the volume fraction r. Then straightforward application of the stochastic homogenization procedure yields the following effective elastic potential

$$W(\boldsymbol{\rho}) = \begin{cases} D_1\,, & \text{if } \rho_{22} + \nu\rho_{11} \geq 0\,, \\ \dfrac{D_1 D_2}{D_2 r + D_1(1-r)}\,, & \text{if } \rho_{22} + \nu\rho_{11} \leq 0\,. \end{cases} \tag{38}$$

Hence we infer that for every $0 < r \leq 1$ the effective material of the plate is now stronger then in the case of periodically fissured plate.

Acknowledgements. The authors were supported by the Polish State Committee for Scientific Research through the grant No 3 P404 013 06.

References

1. G. Dal Maso and L. Modica, *Annali di Matematica pura ed applicata* (IV) vol. CXLIV (1986) 347.
2. E. Sanchez-Palencia, *Non-homogeneous Media and Vibration Theory*, Springer-Verlag, Berlin 1980.
3. J.J. Telega, *Mathematical Modelling and Numerical Analysis* **27** (1993) 421.
4. J.J. Telega and T. Lewiński, *Arch. Mech.* **40** (1988) 119.
5. B. Gambin, *Control and Cybernetics* **23** (1994) 671.
6. B. Gambin, *Stochastic Homogenization of First Gradient Strain Modelling of Elasticity*, 30th Polish Solid Mechanics Conference, Zakopane, 1994.
7. T. Lewiński and J.J. Telega, *Int. J. Engng Sci.* **29** (1991) 1129.
8. H. Attouch and F. Murat, *Homogenization of Fissured Elastic Materials*, Publications AVAMAC, Université de Perpignan, No 85-03, 1985.
9. T. Lewiński and J.J. Telega, *Journal of Elasticity* **19** (1988) 37.

Continuum Models and Discrete Systems
Proceedings of 8th International Symposium, June 11-16, 1995, Varna, Bulgaria, ed. K.Z. Markov
©World Scientific Publishing Company, 1996, pp. 308–315

RHEOLOGY OF DISORDERED FOAM

J. M. TURNER
Condensed Matter Theory Group, Physics Department, Imperial College,
London SW7, United Kingdom

Abstract. A simple probabilistic model of disordered foam is introduced. This model has an adjustable interaction between foam films—the simplest case is considered here. This approach is related to previous work. The results are in close agreement with accurate computer simulations, and with available experimental data, to which they are compared.

1. Introduction

1.1. Foams and Emulsions

Foams and emulsions are dispersions of fluids in fluids, and they share much in common. From now on we shall refer to foams, but both are covered. The systems we are interested are large gas fraction (ϕ) foams, with $0.9069 < \phi < 1$, see Fig. 1. Those shown in the figure are ordered and monodisperse, we consider more normal foams which are disordered both in cell size and shape.

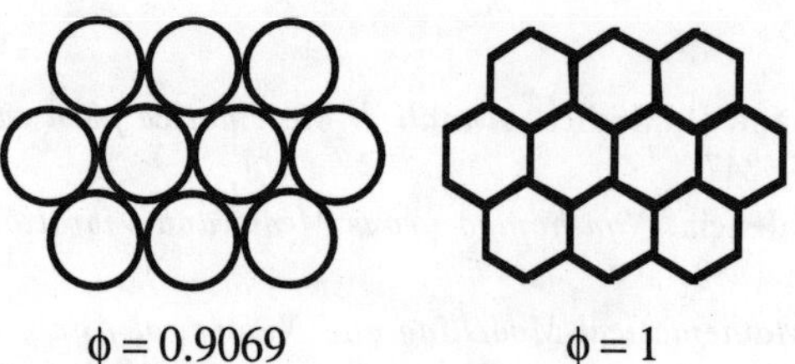

Fig. 1. Foams and emulsions.

Foams of this type are constituted of cells of gas, separated each from another by thin films, for instance of soap or some other surfactant. It is usual to assume that the films do not rupture, as such events are rare in the systems of interest. The physics of such thin films is dominated by surface tension, which therefore shapes the foam. The effect of this is that the foam constantly searches for configurations with shorter film length, and this is most marked at the vertices where films meet. They are all of the form in Fig. 2; the films meet at angles of $2\pi/3$.

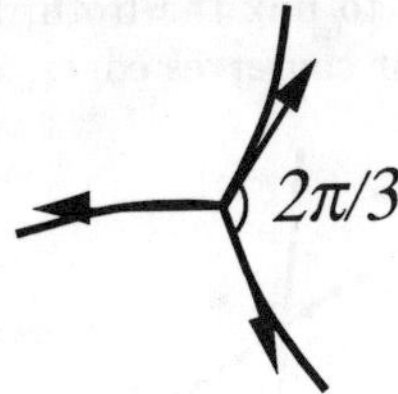

Fig. 2. Surface tension.

1.2. Previous Work

Foams have been of scientific interest for over 100 years, but over the last ten their rheology has been paid much attention. This is because these materials have entered into the engineer's armoury, and their manufacture and use often involve a knowledge of their flow properties.

The first steps were taken, analytically, by Princen and Prud'homme [14,12,13], with the shearing of ordered foam in one orientation, up to the first yield point. This work was then extended by Khan and Armstrong to arbitrary orientations, and other deformations [2]. Kraynik and Hansen then developed this formalism further to cope with the yield points, and thus large deformations [6,7], also Reinelt and Kraynik [17]. Following this, these various workers extended the formalism to deal with polydispersity [9,3,8], various viscous effects [15,4] and have begun to push forward to models of the rheology of ordered 3-dimensional foams [16]. These problems have also been tackled via numerical simulation, starting with Weaire and Kermode [20], see also [18,19]. Kawasaki *et al.* [10,1] have applied vertex models to the problem with great success. Another track has been taken by Edwards and Pithia [11], who have chosen to model the foam as a gas of edges, and have started to develop a thermodynamics of foam, using a Boltzmann equation. This work owes something to all of these foundations.

2. Edge Ensemble Model

2.1. Motivation

When looking at a foam it is evident that topology rather than geometry is its natural description. There are two possible topological events in a cell complex, bound as it is by Euler's theorem. These events are cell disappearance and cell neighbour switching, and in a foam it is normal to assume that only 3-sided cells disappear; so that we have the T1 and T2 events respectively. In this work as in others, we shall assume that the T2 disappearances occur on a timescale much longer than that considered in the model.

Thus the T1 is the most important factor at work in the foam. It relaxes stress locally in the sense that it connects two local configurations—one with a high energy, one with a lower. The T1's serve to mix the froth, keeping it in equilibrium. One notable fact about the T1 is that it conserves edges, 3 before and 3 after.

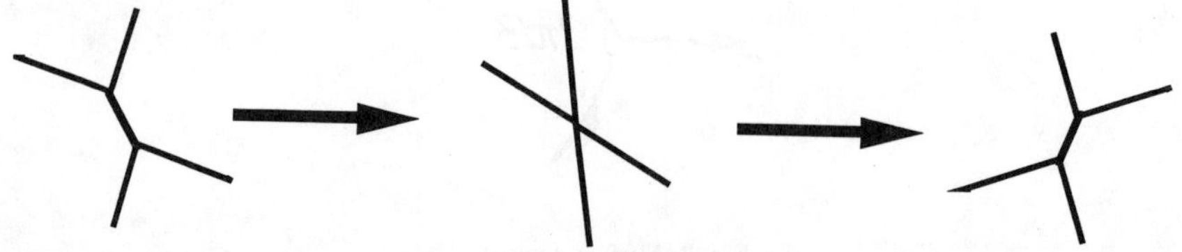

Fig. 3. T1 event.

The topological structure of foam also suggests that it is essentially an assemblage of films, where most of the important effects happen. So we might focus our attention on the films to model the rheology. Turning now to the geometry of the structure we notice two things. Firstly, glancing at Fig. 4, we imagine a foam being stretched. The same number of films (they are conserved) must now stretch a longer distance across the cell, although a shorter way upwards. This can mean two things—more films oriented lengthwise, or longer films lengthwise. In fact it means both, but there is a limit because the films must still meet at $2\pi/3$ radians, so not all films can be oriented lengthwise. Secondly we find that the films oriented perpendicular to the direction of extension shrink as the others relax, see the bottom film in Fig. 6, so that eventually they reach zero length. This means that a T1 can occur; so that the population of transverse films is maintained at some level, and the foam can creep a little.

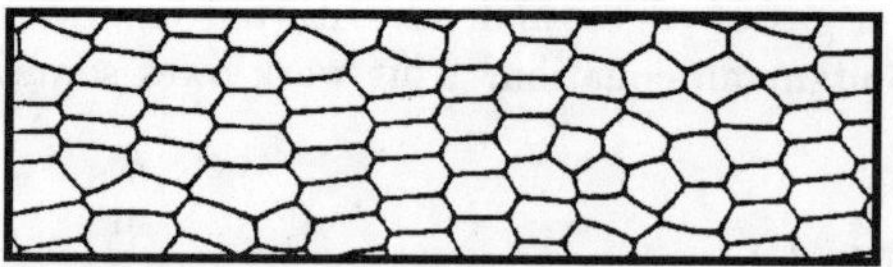

Fig. 4. Stretched foam.

All of these ideas suggest that it may be useful to model the froth as a distribution of edges, cf. [11]. This distribution is over length and orientation, where orientation is between $-\pi/2$ and $\pi/2$ as edges have a symmetry along themselves. We can either model this distribution analytically, or numerically by a representative sample, as we do here. Thus we define some ensemble of edges, and derive a numerical master equation for their evolution. This equation has three parts explained below; deformation, relaxation and T1 spotting. We express the properties of the foam as expected values over this distribution, much as in [2,6].

2.2. Algorithm

The first thing we must do is to provide an initial ensemble. Here we have used sample "foams" taken from a Voronoi-Telley simulation. We take this sample and break it up into its constituent edges, as shown in Fig. 5.

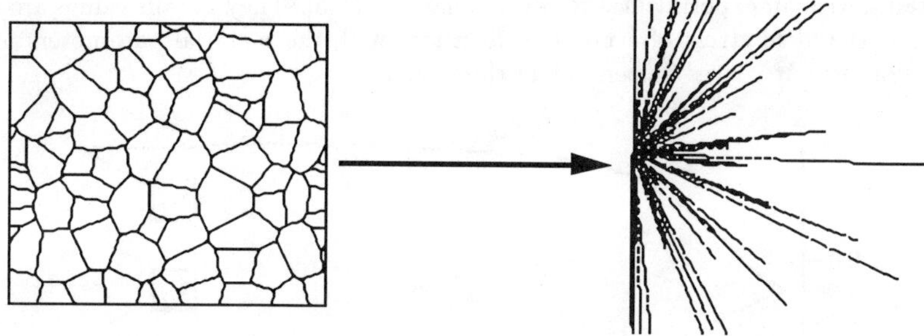

Fig. 5. Creating initial ensemble.

Now we reach the first stage of the model's cycle. We must deform the edges. At this stage we restrict ourselves to homogeneous deformations (as do most others), though this can be extended. This means that each edge experiences the same deformation, which is therefore simply applied.

Next we must relax the films. At this stage we must choose some interaction between edges; there are many possibilities, here we consider only one. We imagine that each film belongs to a triplet, which make up one vertex. They maintain this connection throughout the run; so at each stage they relax as a vertex does. Making this assumption effectively takes into Kraynik and Hansen's model [6], but averaging over a large number of orientations, sampled form a disordered foam.

Of course during this relaxation T1 events may occur, in which case we must calculate the resulting new films. Actually of course T1's require 4 films, but here we shall limit ourselves to a model where only the 3 at the vertex are affected, so that the old vertex is replaced by a new one. The whole cycle is shown in Fig. 6.

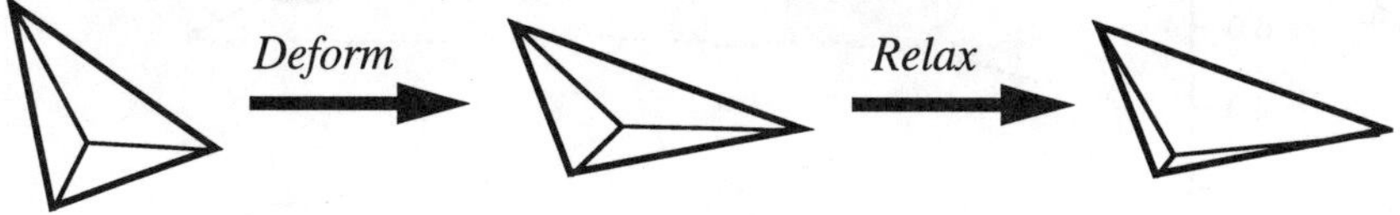

Fig. 6. Deformation and relaxation.

3. Results

In Figs. 7 and 8 are shown the components of the bulk extra stress tensors of the foams. It can be seen that all the curves have two parts, an initial piece and then a more or less constant tail. The effects of these are more apparent in Figs. 11 and 9.

Fig. 9 shows the shear stress of the sample under simple shear. Once again it has two parts, first a rapid almost linear increase, then after reaching a yield stress it flattens. This curve compares well with a much more physical vertex model [10,1], whose result is in Fig. 10. More importantly the value of the yield stress compares favourably with an experimental result of Khan *et al.* [5]. They obtain values around 17 Pa for the yield stress of a polymer foamant, with the material parameters used in our graphs; our curves flatten out in this region.

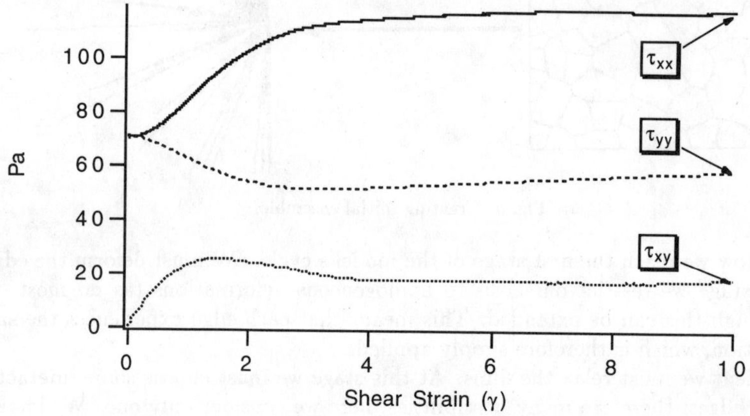

Fig. 7. Components of stress tensor for shear.

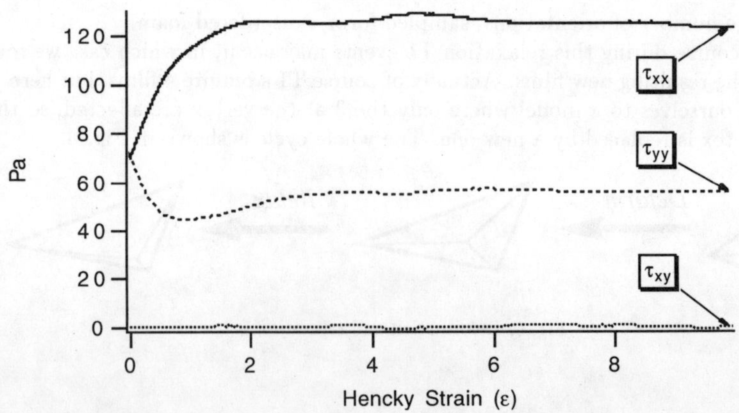

Fig. 8. Components of stress tensor for extension.

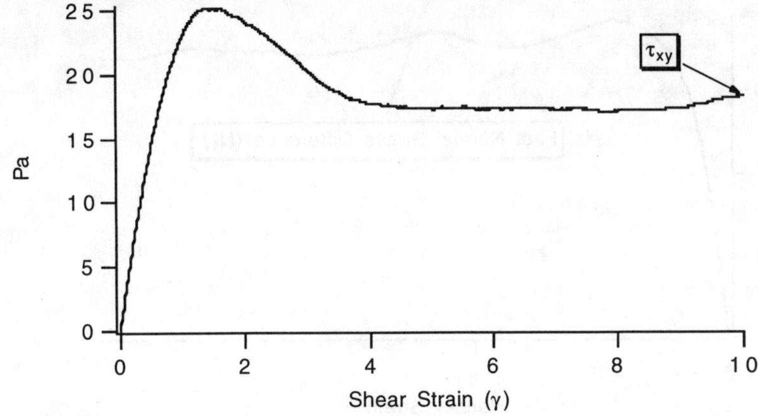

Fig. 9. Shear stress.

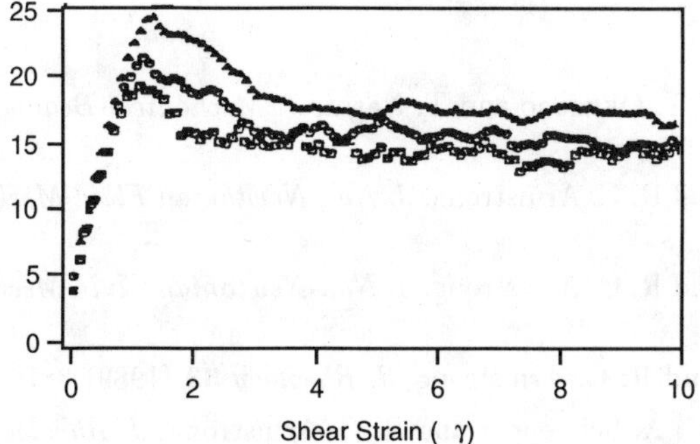

Fig. 10. Kawasaki *et al.*

The same model is used for extension, the most important variable for this geometry is the first normal stress difference which is shown in Fig. 11. This similar behaviour, but without the peak of the shear case, and with some definite oscillations in the tail. These features are under investigation.

Acknowledgements. The author would like to thank Prof. D. D. Vvedensky for his continued support and encouragement, and Dr. A. M. Kraynik for making foam rheology irresistably interesting. The support of the U.K. Engineering and Physical Sciences Research Council is gratefully acknowledged.

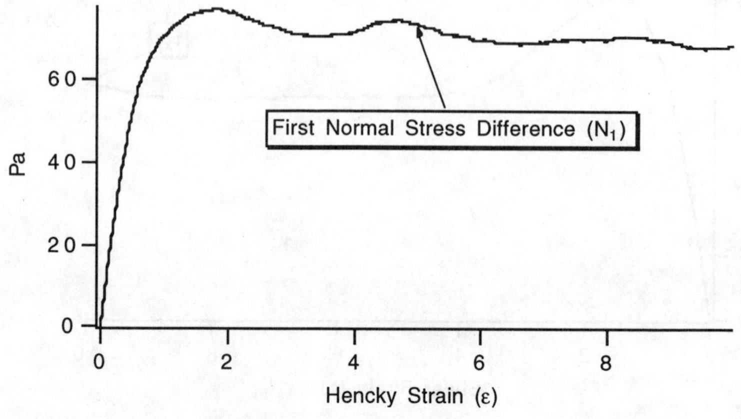

Fig. 11. First normal stress difference for extension.

References

1. K. Kawasaki, T. Okuzono and T. Nagai, *J. Mechanical Behaviour of Materials* **4** (1992) 51.
2. S. A. Khan and R. C. Armstrong, *J. Non-Newtonian Fluid Mechanics* **22** (1986) 1.
3. S. A. Khan and R. C. Armstrong, *J. Non-Newtonian Fluid Mechanics* **25** (1987) 61.
4. S. A. Khan and R. C. Armstrong, *J. Rheology* **33** (1989) 881.
5. S. A. Khan, C. A. Schnepper and R. C. Armstrong, *J. Rheology* **32** (1988) 69.
6. A. M. Kraynik and M. G. Hansen, *J. Rheology* **30** (1986) 409.
7. A. M. Kraynik and M. G. Hansen, *J. Rheology* **31** (1987) 175.
8. A. M. Kraynik and D. A. Reinelt, *Int. J. Multiphase Flow* **30** (1986) 409.
9. A. M. Kraynik, D. A. Reinelt and H. M. Princen, *J. Rheology* **35** (1991) 1235.
10. T. Okuzono and K. Kawasaki, *J. Rheology* **37** (1993) 571.
11. K. D. Pithia and S. F. Edwards, *Physica A* **205** (1994) 565.
12. H. M. Princen, *J. Coll. Interface Science* **20** (1965) 156.
13. H. M. Princen, *J. Coll. Interface Science* **71** (1979) 55.
14. H. M. Princen, *J. Coll. Interface Science* **91** (1983) 160.
15. H. M. Princen, M. P. Aronson and J. C. Moser, *J. Coll. Interface Science* **75** (1980) 246.

16. D. A. Reinelt, *J. Rheology* **37** (1993) 1117.
17. D. A. Reinelt and A. M. Kraynik, *J. Fluid Mechanics* **215** (1990) 431.
18. D. Weaire, F. Bolton, T. Herdtle and H. Aref, **66** (1992) 293
19. D. Weaire and T.-L. Fu *J. Rheology* **32** (1988) 271.
20. D. Weaire, T.-L. Fu and J. P. Kermode, **54** (1986) L39.

Continuum Models and Discrete Systems
Proceedings of 8th International Symposium, June 11-16, 1995, Varna, Bulgaria, ed. K.Z. Markov

A VARIATIONAL THEORY FOR BIAXIAL LIQUID CRYSTALS

E. VIRGA
Universita di Pisa, Istituto di Scienza delle Costruzioni,
Facolta' di Ingegneria, via Diotisalvi, 2, I-56126 Pisa, Italy

Abstract

Most liquid crystals are optically uniaxial: the existence of phases biaxial in the bulk is still disputed. On the other hand, biaxial states might be expected to occur either near defects in the bulk or near surfaces bounding the material. The lecture reviews the main outcomes of a theory intended to describe how such thin biaxial halos can arise in the equilibrium configurations on nematic liquid crystals.

Continuum Models and Discrete Systems
Proceedings of 8th International Symposium, June 11-16, 1995, Varna, Bulgaria, ed. K.Z. Markov
©World Scientific Publishing Company, 1996, pp. 317–324

POROUS MEDIA AT FINITE STRAINS

K. WILMANSKI
Institut für Mechanik, FB 10, Universität-GH Essen, D-45117 Essen, Germany

Abstract. The work contains the brief presentation of the new model describing mechanical processes in two-component immiscible mixtures (porous materials). The model is formulated in the recently proposed Lagrangian description and it is based on the assumption that the skeleton is elastic and the fluid is ideal. The additional "microscopic" field of porosity fulfills its own balance equation. The details of the construction of the model can be found in the paper [1].

1. Introduction

The growing capacity of easily available computers has changed the attitude to the highly complex models of continuum mechanics which have been avoided in the past for purely analytical reasons. This was also the case with the non-linearities appearing by the modeling of systems with large deformations and semi-microscopic interactions described by means of many components with different kinematics. Any analytical solution of boundary value problems was for such cases a hopeless task. Consequently most of such models which can be found in the literature were solely developed from the constitutive point of view (thermodynamic admissibility) without any details concerning fields fulfilling specific boundary conditions. In the best case the homogeneous processes were considered. The review of these models can be found in the extended version of the present paper [1].

It has been found that the numerical evaluation of many such models yields physically unacceptable results. One of the most popular classes of such models is based on the semi-microscopic "incompressibility" assumption. This assumption yields, for instance, erroneous relaxation properties, it does not describe waves appearing in porous materials and, in some cases, it possesses solely exceptional solutions (e.g. for the porosity $n = 1$ or $n = 0$, i.e. for the one-component systems).

In the present work a new model of the two-component immiscible mixture is presented. The details can be found in [1] and other papers quoted there.

In the next Section we present the governing equations of the model. The third Section contains a few remarks on possible simplifications of the model. The fourth Section is devoted to some basic properties of acoustic waves in porous materials described by the present model. The paper is completed with a few final remarks.

2. Governing Equations

We consider the continuous medium consisting of the skeleton and of the fluid. The motion of the skeleton is described by the function of motion

$$\mathbf{x} = \boldsymbol{\chi}^S(\mathbf{X},t), \quad \mathbf{x} \in E^3, \quad \mathbf{X} \in \mathcal{B}, \tag{2.1}$$

where $\mathcal{B}$ is a reference configuration for which the deformation gradient of the skeleton $\mathbf{F}^S$ is the identity. This gradient as well as the velocity of the skeleton are defined by the relations

$$\mathbf{F}^S(\mathbf{X},t) = \operatorname{Grad}\boldsymbol{\chi}^S(\mathbf{X},t), \quad \mathbf{x}'^S(\mathbf{X},t) = \frac{\partial \boldsymbol{\chi}^S}{\partial t}(\mathbf{X},t). \tag{2.2}$$

In the standard description the kinematics of the fluid component is given by the Eulerian velocity

$$\mathbf{v}^F = \mathbf{v}^F(\mathbf{x},t), \quad \mathbf{x} \in \boldsymbol{\chi}^S(\mathcal{B},t). \tag{2.3}$$

We use in this work the new Lagrangian description of the two-component system in which the above velocity is replaced by the following *Lagrangian velocity* of the fluid

$$\mathbf{X}'^F(\mathbf{X},t) = \mathbf{F}^{S-1}(\mathbf{x}'^F - \mathbf{x}'^S), \qquad \mathbf{x}'^F \equiv \mathbf{v}^F(\boldsymbol{\chi}^S(\mathbf{X},t),t). \tag{2.4}$$

This is the velocity of the fluid in its image on the reference configuration $\mathcal{B}$ of the skeleton.

The basic fields of the model are chosen to be

$$(\mathbf{X},t) \mapsto \{\rho^F, n, \boldsymbol{\chi}^S, \mathbf{X}'^F\} \in V^8, \quad \mathbf{X} \in \mathcal{B}, \tag{2.5}$$

where ρ^F is the fluid mass density related to the current mass density of the Eulerian description ρ_t^F by the formula

$$\rho^F = J^S \rho_t^F, \quad J^S \equiv \det \mathbf{F}^S, \tag{2.6}$$

and n denotes the *porosity*. It is understood as the volume fraction of the fluid in the microscopic control volume containing both the fluid and the skeleton. For the two-component system under considerations the volume fraction of the skeleton is $1-n$.

The mass density of the skeleton ρ^S in the reference configuration is constant and, for this reason, it does not appear as the field.

The field equations for the fields (2.5) follow from the balance equations of mass and momentum and from the balance equation of porosity. Under the assumption that the constitutive functions of the model are defined on the space of the following *constitutive variables* (elastic skeleton and ideal fluid)

$$\mathcal{C} = \{\rho^F, n, \mathbf{C}^S, \mathbf{X}'^F\}, \quad \mathbf{C}^S \equiv \mathbf{F}^{ST}\mathbf{F}^S, \tag{2.7}$$

the second law of thermodynamics and the isotropy reduce the constitutive problem to the following scalar functions

$$\{\Psi^S, \Psi^F, \Phi_0, \pi_0, \pi_1, \pi_2, \mathcal{N}\}\,, \tag{2.8}$$

which depend on the variables n, ρ^F and on the following invariants

$$\mathrm{I} = \mathbf{1}\cdot\mathbf{C}^S\,, \quad \mathrm{II} = \frac{1}{2}(\mathrm{I}^2 - \mathbf{1}\cdot\mathbf{C}^{S2})\,, \quad \mathrm{III} = \det\mathbf{C}^S \equiv J^{S2}\,,$$

$$\mathrm{IV} = \mathbf{X}'^F\cdot\mathbf{X}'^F\,, \quad \mathrm{V} = \mathbf{C}^S\cdot(\mathbf{X}'^F\otimes\mathbf{X}'^F)\,, \quad \mathrm{VI} = \mathbf{C}^{S2}\cdot(\mathbf{X}'^F\otimes\mathbf{X}'^F)\,. \tag{2.9}$$

The scalar functions Ψ^S, Ψ^F denote the Helmholtz free energies of the skeleton and of the fluid, respectively. The scalar Φ_0 describes the flux of porosity, $\mathcal{N}$— the source of the porosity and π_0, π_1, π_2 are diffusion coefficients for the non-linear diffusion processes.

These functions are restricted by the following relations

$$\rho^S\frac{\partial\Psi^S}{\partial\rho^F} + \Lambda^n\frac{\partial\Phi_0}{\partial\rho^F} = 0\,, \quad \Lambda^n \equiv \rho^S\frac{\partial\Psi^S}{\partial n} + \rho^F\frac{\partial\Psi^F}{\partial n}\,,$$

$$\rho^F\frac{\partial\Psi^F}{\partial\mathcal{A}_1} - \Lambda^n\frac{\partial\Phi_0}{\partial\mathcal{A}_1} = 0\,, \quad \mathcal{A}_1 = n, \mathrm{I}, \mathrm{II}, \mathrm{IV}, \mathrm{V}, \mathrm{VI}\,,$$

$$\rho^S\frac{\partial\Psi^S}{\partial\mathcal{A}_2} + \rho^F\frac{\partial\Psi^F}{\partial\mathcal{A}_2} = 0\,, \quad \mathcal{A}_2 = \mathrm{IV}, \mathrm{V}, \mathrm{VI}\,, \tag{2.10}$$

$$\rho^F\left(\rho^F\frac{\partial\Psi^F}{\partial\rho^F} + 2\mathrm{III}\frac{\partial\Psi^F}{\partial\mathrm{III}}\right) - \Lambda^n\sqrt{\mathrm{III}}\left[\rho^F\frac{\partial}{\partial\rho^F}\left(\frac{\Phi_0}{\sqrt{\mathrm{III}}}\right) + 2\mathrm{III}\frac{\partial}{\partial\mathrm{III}}\left(\frac{\Phi_0}{\sqrt{\mathrm{III}}}\right)\right] = 0\,.$$

These constitutive functions render explicit field equations for the fields (2.5). They have the following form

$$\frac{\partial\rho^F}{\partial t} + \mathrm{Div}\,\rho^F\mathbf{X}'^F = 0\,,$$

$$\frac{\partial n}{\partial t} + \mathrm{Div}\,\Phi_0\mathbf{X}'^F = \nu\,, \quad \rho^S\frac{\partial\mathbf{x}'^S}{\partial t} - \mathrm{Div}\,\mathbf{P}^S = \mathbf{p}^* + \rho^S\mathbf{b}^S\,, \tag{2.11}$$

$$\frac{\partial}{\partial t}(\rho^F\mathbf{x}'^F) + \mathrm{Div}\,(\rho^F\mathbf{x}'^F\otimes\mathbf{X}'^F - \mathbf{P}^F) = -\mathbf{p}^* + \rho^F\mathbf{b}^F\,,$$

where ν denotes the source of the porosity, $\mathbf{p}^*$ is the source of momentum (the diffusion force), $\rho^S\mathbf{b}^S$, $\rho^F\mathbf{b}^F$ are the partial volume forces of components. For the two former quantities we have

$$\nu = -\frac{\mathcal{N}}{\rho^F}\Lambda^n\,, \quad \mathcal{N} \geq 0\,,$$

$$\mathbf{F}^{ST}\mathbf{p}^* = (\pi_0\mathbf{1} + \pi_1\mathbf{C}^S + \pi_2\mathbf{C}^{S2})\,\mathbf{X}'^F\,, \tag{2.12}$$

$$(\pi_0\mathbf{1} + \pi_1\mathbf{C}^S + \pi_2\mathbf{C}^{S2}) \cdot (\mathbf{X}'^F \otimes \mathbf{X}'^F) \geq 0 \,.$$

The inequalities constitute the sufficient conditions for the dissipation in an arbitrary process to be non-negative.

The remaining two quantities $\mathbf{P}^S$ and $\mathbf{P}^F$ in the field equations are the Piola-Kirchoff partial stress tensors. They are connected with the usual Cauchy stresses $\mathbf{T}^S$ and $\mathbf{T}^F$ by the relations

$$\mathbf{T}^S = J^{S-1}\mathbf{P}^S\mathbf{F}^{ST} \,, \quad \mathbf{T}^F = J^{S-1}\mathbf{P}^F\mathbf{F}^{ST} \,, \tag{2.13}$$

and they are given by the constitutive relations

$$\begin{aligned} \mathbf{P}^S = 2\mathbf{F}^{S-T}\Big\{\rho^S\Big[\frac{\partial\Psi^S}{\partial \mathrm{I}}\mathbf{C}^S + \Big(\mathrm{II}\frac{\partial\Psi^S}{\partial \mathrm{II}} + \mathrm{III}\frac{\partial\Psi^S}{\partial \mathrm{III}}\Big)\mathbf{1} - \mathrm{III}\frac{\partial\Psi^S}{\partial \mathrm{II}}\mathbf{C}^{S-1}\Big] \\ + \Lambda^n\Big[\frac{\partial\Phi_0}{\partial \mathrm{I}}\mathbf{C}^S + (\mathrm{II}\frac{\partial\Phi_0}{\partial \mathrm{II}} + \mathrm{III}\frac{\partial\Phi_0}{\partial \mathrm{III}})\mathbf{1} - \mathrm{III}\frac{\partial\Phi_0}{\partial \mathrm{II}}\mathbf{C}^{S-1}\Big]\Big\}, \\ \mathbf{P}^F = -\big[\rho^F(\rho^F\frac{\partial\Psi^F}{\partial\rho^F} + \rho^S\frac{\partial\Psi^S}{\partial\rho^F}) + \Lambda^n\Phi_0\big]\mathbf{F}^{S-T} \,. \end{aligned} \tag{2.14}$$

Hence, as expected, the constitutive assumption (2.7) yields the spherical Cauchy stress tensor in the fluid.

The coupling between components is described by the function Φ_0 whose presence yields the dependence of the Helmholtz free energy of the skeleton Ψ^S on the mass density of the fluid ρ^F and on the porosity n as well as the dependence of the Helmholtz free energy of the fluid Ψ^F on the deformation of the skeleton and again on the porosity n. The additional coupling is, as usual, kinematical through the diffusion force.

It can be shown that the boundary value problem for the above set of equations can be well defined. It contains an additional constitutive element connected with the free boundary of the fluid if the fluid is flowing out through the boundary of the skeleton. We shall not consider this problem in the present work (see Ref. 1).

3. Simplifications of the Model

Let us begin with the evaluation of the identity $(2.10)_2$ for $\mathcal{A}_1 = n$. We obtain easily

$$\Lambda^n = \rho^F\frac{\partial\Psi^F}{\partial n}\left(\frac{\partial\Phi_0}{\partial n}\right)^{-1} . \tag{3.1}$$

This relation is now used in $(2.12)_1$ under the assumption of the *linearity* of the deviation from the thermodynamical equilibrium state. The latter appears for

$$\mathbf{X}'^F\Big|_E = 0\,, \quad n\Big|_E = n_0 = \mathrm{const}\,, \tag{3.2}$$

where n_0 is the initial homogeneous and constant porosity. Due to the dependence of Ψ^S, Ψ^F and $\mathbf{X}'^F$ on the velocity through the invariants (2.9), the linear approximation of these functions reduces to functions independent of the Lagrangian velocity. We limit the attention to this case.

Simultaneously due to the second law of thermodynamics in the state of the thermodynamical equilibrium the dissipation reaches its minimum. Hence

$$\Lambda^n \cong \rho^F \frac{\partial^2 \Psi^F}{\partial n^2} \left(\frac{\partial \Phi_0}{\partial n} \Big|_E \right)^{-1} (n - n_0) . \tag{3.3}$$

Substitution in $(2.12)_1$ yields

$$\nu = -\frac{n - n_0}{\tau}, \quad \tau \equiv \frac{\rho^F}{\mathcal{N}} \left(\frac{\partial \Phi_0}{\partial n} \Big|_E \right) \left(\rho^F \frac{\partial^2 \Psi^F}{\partial n^2} \Big|_E \right)^{-1} , \tag{3.4}$$

where τ is the *relaxation time* of the spontaneous evolution of the porosity.

The above considerations hold true also in the case of another approximation which is most likely justified, for instance, for some porous rocks. It has been found experimentally that for those materials the Helmholtz free energy of the skeleton is independent of ρ^F and n. It follows then from $(2.10)_2$ for $\mathcal{A}_1 = n$

$$\Phi_0 = n - n_0 + \Gamma(n_0, \mathrm{I}, \mathrm{II}, \mathrm{III}) . \tag{3.5}$$

The function Γ does not vanish if the free energy Ψ^F is dependent on the deformation of the skeleton. Namely, we have

$$\frac{\partial \Psi^F}{\partial \mathrm{I}} - \frac{n - n_0}{\tau \mathcal{N}} \frac{\partial \Gamma}{\partial \mathrm{I}} = 0 , \quad \frac{\partial \Psi^F}{\partial \mathrm{II}} - \frac{n - n_0}{\tau \mathcal{N}} \frac{\partial \Gamma}{\partial \mathrm{II}} = 0 ,$$

$$\rho^F \frac{\partial \Psi^F}{\partial \rho^F} + 2\mathrm{III} \frac{\partial \Psi^F}{\partial \mathrm{III}} - 2 \frac{n - n_0}{\tau \mathcal{N}} (\mathrm{III})^{3/2} \frac{\partial}{\partial \mathrm{III}} \left(\frac{\Gamma}{\sqrt{\mathrm{III}}} \right) = 0 . \tag{3.6}$$

In the particular case

$$\Gamma = \gamma \sqrt{\mathrm{III}} , \quad \gamma = \mathrm{const} , \tag{3.7}$$

the relations (3.6) yield

$$\Psi^F = \Psi_0^F(\rho_t^F, n_0) + \frac{1}{2\tau \mathcal{N}} (n - n_0)^2 , \tag{3.8}$$

where τ and $\mathcal{N}$ must be constant.

The Cauchy stress tensors have now the following form

$$\mathbf{T}^S = \mathcal{I}_1 \mathbf{B}^S + \mathcal{I}_0 \mathbf{1} + \mathcal{I}_{-1} \mathbf{B}^{-1} + \gamma \rho^F \frac{n - n_0}{\tau \mathcal{N}} \mathbf{1} , \quad \mathbf{B}^S = \mathbf{F}^S \mathbf{F}^{ST} ,$$

$$\mathbf{T}^F = - \left[\rho_t^{F2} \frac{\partial \Psi^F}{\partial \rho_t^F} + \gamma \rho^F \frac{n - n_0}{\tau \mathcal{N}} \right] \mathbf{1} , \tag{3.9}$$

where

$$\mathcal{I}_1 = 2\rho_t^S \frac{\partial \Psi^S}{\partial \mathrm{I}}, \quad \mathcal{I}_{-1} = -2\mathrm{III}\rho_t^S \frac{\partial \Psi^S}{\partial \mathrm{II}}, \quad \rho_t^S \equiv \rho^S J^{S-1},$$

$$\mathcal{I}_0 = 2\rho_t^S \left(\mathrm{II} \frac{\partial \Psi_0^S}{\partial \mathrm{II}} + \mathrm{III} \frac{\partial \Psi_0^S}{\partial \mathrm{III}} \right). \tag{3.10}$$

Hence the constant γ describes the only contribution of interactions with the fluid to the stress tensor of the skeleton.

Now it is easy to simplify the model for cases of small deformations of the skeleton and small changes of the mass density ρ^F and the porosity n. We obtain immediately the following set of field equations

$$\frac{\partial \rho^F}{\partial t} + \rho_0^F \mathrm{Div}\,(\mathbf{v}^F) = 0, \quad |\rho^F - \rho_0^F| \ll 1,$$

$$\frac{\partial \Delta}{\partial t} + n_0 \mathrm{Div}\,(\mathbf{v}^F) = -\frac{\Delta}{\tau}, \quad \Delta \equiv n - n_0, \quad |\Delta| \ll n_0,$$

$$\rho^S \frac{\partial^2 \mathbf{u}^S}{\partial t^2} = (\lambda^S + \mu^S)\,\mathrm{Grad\,Div}\,(\mathbf{u}^S) + \mu^S\,\mathrm{Div\,Grad}\,(\mathbf{u}^S) + \pi_3 \mathbf{w} + \rho^S \mathbf{b}^S,$$

$$\rho_0^F \frac{\partial \mathbf{v}^F}{\partial t} = -\,\mathrm{Grad}\left(K^F \rho^F + \frac{n_0 \rho_0^F}{\tau \mathcal{N}} \Delta\right) - \pi_3 \mathbf{w} + \rho^F \mathbf{b}^F,$$

$$\mathbf{w} \equiv \mathbf{v}^F - \frac{\partial \mathbf{u}^S}{\partial t}, \tag{3.11}$$

where ρ_0^F, n_0 are the initial homogeneous values of the mass density of the fluid and of the porosity, respectively. The vector $\mathbf{u}^S$ denotes the displacement of the skeleton and

$$\{\tau, \mathcal{N}, \lambda^S, \mu^S, K^F, \pi_3\}, \quad \pi_3 = \pi_0 + \pi_1 + \pi_2 = \pi_3(n_0), \tag{3.12}$$

are material constants depending on the initial porosity n_0. For the practical purposes all these constants can be found, for instance, in acoustic experiments [1]. Some numerical codes are already available for this system of equations as well.

4. Acoustic Waves

Measurements on two-component porous materials can be done either by devices penetrating into one of the components or by the bulk devices measuring the macroscopic average properties. In the case of the continuous description the procedures of the second class are certainly preferable. The most popular method of measurements is the sending and receiving of the acoustic waves. These measurements deliver the data on speeds of propagation and attenuation of various waves. It has been found, for instance, that the two-component porous material carries two direct longitudinal waves—the so-called P1- and P2-waves and the transversal shear S-wave. Their speeds depend on the bulk mechanical properties of components, on the porosity, on the saturation of pores with water or with other fluids, etc. Consequently these waves can be used for the diagnosis purposes of such complex structures.

The model discussed in the present paper admits the propagation of the above mentioned waves. The *front of the acoustic wave* is defined as the singular surface with the following properties

$$[[\chi^S]] = 0\,, \quad \left[\left[\mathbf{x}'^S\right]\right] = 0\,, \quad \left[\left[\mathbf{x}'^F\right]\right] = 0\,, \quad [[n]] = 0\,,$$
$$[[\ldots]] \equiv (\ldots)^+ - (\ldots)^-\,, \tag{4.1}$$

where the last expression describes the difference of limits from the positive and negative side of the surface.

The amplitudes of the waves are defined on this surface in the following way

$$\mathbf{a}^S U^2 = \left[\left[\frac{\partial \mathbf{x}'^S}{\partial t}\right]\right]\,, \quad \mathbf{a}^F U^2 = \left[\left[\frac{\partial \mathbf{x}'^F}{\partial t}\right]\right]\,,$$
$$rU = -\left[\left[\frac{\partial \rho^F}{\partial t}\right]\right]\,, \quad n\mathbf{N} = \left[\left[\frac{\partial n}{\partial t}\right]\right]\,, \tag{4.2}$$

where U is the *speed of propagation* of the front of the wave relative to the skeleton.

The standard evaluation of the jump of field equations across the front of the wave yields

$$r = \rho^F(\mathbf{a}^S - \mathbf{a}^F)\cdot(\mathbf{F}^{S-T}\mathbf{N})\,, \quad n = \Phi_0(\mathbf{a}^S - \mathbf{a}^F)\cdot(\mathbf{F}^{S-T}\mathbf{N})\,,$$
$$\rho^S\mathbf{a}^S U^2 = J^S(\mathbf{a}^S - \mathbf{a}^F)\cdot(\mathbf{F}^{S-T}\mathbf{N})\left\{\rho^F\frac{\partial \mathbf{T}^S}{\partial \rho^F} + \Phi_0\frac{\partial \mathbf{T}^S}{\partial n}\right\}(\mathbf{F}^{S-T}\mathbf{N}) + \mathbf{Q}^S\mathbf{a}^S\,, \tag{4.3}$$
$$\rho^F a^F U^2 = -J^S\left\{(\mathbf{a}^S - \mathbf{a}^F)\cdot(\mathbf{F}^{S-T}\mathbf{N})\left(\rho^F\frac{\partial p^F}{\partial \rho^F} + \Phi_0\frac{\partial p^F}{\partial n}\right) + 2\frac{\partial p^F}{\partial \mathbf{B}^S}\cdot(\mathbf{a}^S\otimes\mathbf{F}^S\mathbf{N})\right\}\,,$$

where

$$a^F \equiv [\mathbf{C}^{S-1}\cdot(\mathbf{N}\otimes\mathbf{N})]^{-1}\mathbf{a}^F\cdot\mathbf{F}^{S-T}\mathbf{N}\,,$$
$$\mathbf{Q}^S \equiv 2J^S\left(\frac{\partial \mathbf{T}^S}{\partial \mathbf{B}^S}\right)^{T^{23}}\cdot(\mathbf{F}^{S-T}\mathbf{N}\otimes\mathbf{F}^S\mathbf{N})\,. \tag{4.4}$$

The first two relations follow from the approximation

$$\left|\frac{X_N'^F}{U}\right| \ll 1\,, \quad X_N'^F \equiv \mathbf{X}'^F\cdot\mathbf{N}\,, \tag{4.5}$$

which seems to be quite reasonable for most practical purposes. These relations show that the changes of the mass density of the fluid and of the porosity are not carried by their own individual waves. They are propagating with the other waves caused by the discontinuity of the accelerations.

The last equation demonstrates that the amplitude of the wave in the fluid is perpendicular to the wave front in its current configuration. Consequently it is the longitudinal P2-wave.

Both the longitudinal P1-wave and the transversal (shear) S-wave are contained in the equation $(4.3)_4$. However the classical acoustic tensor $(4.4)_2$ is not alone determining the speeds of these waves. There is an additional contribution of interactions.

In order to illustrate these results let us mention that the waves described by the linear version of the model (3.11) have the following speeds

$$\begin{aligned} U_L^S &= \sqrt{\frac{\lambda^S + 2\mu^S}{\rho^S}} && \text{– longitudinal P1-wave,} \\ U_T^S &= \sqrt{\frac{\mu^S}{\rho^S}} && \text{– transversal S-wave,} \\ U_L^F &= \sqrt{K^F + \frac{n_0^2}{\tau \mathcal{N}}} && \text{– longitudinal P2-wave.} \end{aligned} \tag{4.6}$$

Solely the last wave contains the dependence on the interaction material parameter in this simple case. However all speeds depend on the porosity n_0.

The details of the theory of acoustic waves can be found in [1] and in papers quoted there.

5. Final Remarks

It has been shown that the present model of the porous material with the diffusion and the changing porosity yields reasonable results for the cases which can be at present verified experimentally. The flux term in the balance equation for the porosity as well as the source term in this equation deliver important contributions to the interactions between components. The additional constitutive laws can be obtained from the experimental data—at least for many cases of the practical bearing.

The Lagrangian description of motion of the fluid components allows quite easily to accommodate the large deformations of the skeleton. This yields as well the considerable simplifications in computer programs for typical boundary value problems. The governing set of equations can be handled numerically as the preliminary research has shown.

The present version of the work contains solely the guiding lines for the model whose full presentation is given in the paper [1] available also as a report.

References

1. K. Wilmanski, Porous Media at Finite Strains—the New Model with the Balance Equation for Porosity, *Arch. Mech.* (to appear), 1996; see also: MECH-Bericht 95/11, Universität-GH Essen, Essen, 1995.

Continuum Models and Discrete Systems
Proceedings of 8th International Symposium, June 11-16, 1995, Varna, Bulgaria, ed. K.Z. Markov

ON SOME NEW ACHIEVEMENTS IN NONLINEAR MECHANICS OF MEDIA WITH MICROSTRUCTURE

L. M. ZUBOV
Mechanics and Mathematics Department, Rostov State University,
5, Zorge str., 344104 Rostov-na-Donu, Russia

Abstract. The contributions presents the results of investigations in the field of nonlinear mechanics of deformable media possessing couple stresses and microstructure. A nonlinear theory of isolated dislocations and disclinations was developed for elastic Cosserat continuum and medium the particles of which possess microdeformation. Weingarten's theorem was extended to the case of media with microstructure, which undergoes large deformations. A series of exact solutions to problems on large deformations of elastic and inelastic Cosserat continuum was found, among them nonlinear problems on isolated dislocations and disclinations. Conditions of phase equilibrium were deduced for different models of nonlinearly elastic bodies with microstructure and the concepts of the phase transfers with microslipping and microcoherent phase transformations were clarified. New models of elastic and viscoelastic micropolar fluids have been developed on the bases of general constitutive equations of non-elastic Cosserat continuum with large deformations. These liquid media possess couple stresses and are sensitive to external volume and surface couples. The theorem of general representation of non-elastic micropolar fluid constitutive equations has been proved. This is analogous to familiar W. Noll's theorem in simple fluids mechanics.

Introduction

The classical continuum mechanics, especially the classical theory of elasticity, is based on the model of simple material [9]. The free-energy density and stresses in a given particle of a simple material are completely determined by the deformation gradient and the temperature in the particle. The Cauchy stress tensor is symmetric in simple materials. There are conditions where consideration must be given to the microinhomogeneous structure of material, even though the simple-material model describes successfully the behavior of many real media. Polycrystal grained materials, polymers, composites, suspensions, liquid crystals, geophysical structures, and others fit into these complex-structural media. To describe mathematically the physical-mechanical properties of the media mentioned above, the continuum theories dealing with the couple stresses and the rotational interaction of particles are used. The related model of continuum received the name Cosserat continuum from the brothers Cosserat, who published in 1909 the fundamental work on the theory of materials possessing couple stresses. The linear theory of couple-stress elasticity has been abundantly addressed in the literature; particular mention should be made of the works [1,4,6]. Only a small number of works is devoted to the couple-stress theory of media subject to large deformations [8,7,10,13,14,3]. In the model of Cosserat continuum, each particle in the medium has the degrees of freedom of an absolutely rigid body. In a more general model of continuum with microdeformation, each body point is endowed with the properties of deformable medium, i.e. can undergo homogeneous deformation. The linear theory of medium with microdeformation was elaborated by R. Mindlin [5], fundamental propositions of nonlinear theory were formulated by R. Toupin [8].

The report presents briefly some results of investigations of the author and collaborators in the field of nonlinear mechanics of deformable media possessing couple stresses and microstructure.

1. Nonlinear Models of a Media with Couple Stresses and Microstructure

The position of a particle of Cosserat continuum in the deformed state is specified by a radius-vector $\boldsymbol{R}$, whereas its orientation is determined by a properly orthogonal tensor $\boldsymbol{H}$ called microrotation tensor. The specific energy of elastic material (per unit volume in reference configuration) in isothermal process will be named the specific potential strain energy and symbolized by W. Using the local action and material frame-indifference principles [9], it may be proved that the specific energy W depends on deformation of a body through two second-order tensors: the strain measure, $\boldsymbol{U}$, and the tensor (more precisely pseudotensor) of bending strain, $\boldsymbol{L}$; these latter are determined from the formulae

$$\boldsymbol{U} = (\operatorname{grad} \boldsymbol{R}) \cdot \boldsymbol{H}^T, \ \boldsymbol{L} \times \boldsymbol{E} = -(\operatorname{grad} \boldsymbol{H}) \cdot \boldsymbol{H}^T, \ \boldsymbol{L} = -\frac{1}{2}\boldsymbol{r}^s \otimes \left(\frac{\partial \boldsymbol{H}}{\partial q^s} \cdot \boldsymbol{H}^T\right)_\times . \quad (1)$$

Here, $\boldsymbol{E}$ is identity tensor, grad $= \boldsymbol{r}^s \partial/\partial q^s$ is the gradient operator in the reference (undeformed) configuration, q^s are Lagrangian coordinates. The vectors $\boldsymbol{r}^s$ are deduced from the equations: $\boldsymbol{r}^s \cdot \boldsymbol{r}_k = \delta^s_k$, $\boldsymbol{r}_k = \partial \boldsymbol{r}/\partial q^k$, δ^s_k is the Kronecker symbol and $\boldsymbol{r}$ is the radius-vector of a particle in the reference configuration. The symbol $\boldsymbol{T}_\times$ denotes the vector invariant of the second-order tensor $\boldsymbol{T}$, i.e., $\boldsymbol{T}_\times = \left(T^{sk}\boldsymbol{r}_s \otimes \boldsymbol{r}_k\right)_\times = T^{sk}\boldsymbol{r}_s \times \boldsymbol{r}_k$. The virtual work principle gives the equilibrium equation and dynamic boundary conditions

$$\operatorname{div} \boldsymbol{D} + \rho_0 \boldsymbol{k} = 0, \quad \operatorname{div} \boldsymbol{G} + \left[(\operatorname{grad} \boldsymbol{R})^T \cdot \boldsymbol{D}\right]_\times + \rho_0 \boldsymbol{c} = 0\,, \quad (2)$$

$$\boldsymbol{n} \cdot \boldsymbol{D} = \boldsymbol{\varphi}_0, \quad \boldsymbol{n} \cdot \boldsymbol{G} = \boldsymbol{\psi}_0 \quad \text{on} \quad \partial v\,, \quad (3)$$

$$\boldsymbol{D} = \boldsymbol{\tau} \cdot \boldsymbol{H}, \quad \boldsymbol{G} = \boldsymbol{\mu} \cdot \boldsymbol{H}, \quad \boldsymbol{\tau} = \frac{\partial W}{\partial \boldsymbol{U}}, \quad \boldsymbol{\mu} = \frac{\partial W}{\partial \boldsymbol{L}}. \quad (4)$$

In Eqs. (2)–(4), div is the divergence operator in the reference configuration; ρ_0 is the material density in the reference configuration; $\boldsymbol{k}$ is the body force; $\boldsymbol{c}$ is the mass density of external moment; $\boldsymbol{n}$ is the unit normal to the surface ∂v of the body occupying the volume v in the reference configuration; $\boldsymbol{D}$ is the Piola-type stress tensor; $\boldsymbol{G}$ is the Piola-type couple-stress tensor; φ_0 and ψ_0 are the densities of force and moment loads distributed over the body boundary. Eqs. (2) and the boundary conditions (3) can be written with respect to the deformed configuration of the material body

$$\operatorname{Div} \boldsymbol{T} + \rho \boldsymbol{k} = 0, \quad \operatorname{Div} \boldsymbol{M} + \boldsymbol{T}_{\times} + \rho \boldsymbol{c} = 0\,, \tag{5}$$

$$\boldsymbol{N} \cdot \boldsymbol{T} = \varphi, \quad \boldsymbol{N} \cdot \boldsymbol{M} = \psi \quad \text{on} \quad \partial V\,, \tag{6}$$

$$\boldsymbol{T} = J^{-1} (\operatorname{grad} \boldsymbol{R})^T \cdot \boldsymbol{D}, \quad \boldsymbol{M} = J^{-1} (\operatorname{grad} \boldsymbol{R})^T \cdot \boldsymbol{G}, \quad J = \det (\operatorname{grad} \boldsymbol{R})\,. \tag{7}$$

Here, Div is the divergence operator in Eulerian coordinates, $\boldsymbol{T}$ and $\boldsymbol{M}$ are the Cauchy-type tensors of stresses and couple-stresses; φ and ψ are intensities of force and moment loads per unit area of the body surface ∂V in the deformed state; $\boldsymbol{N}$ is the unit normal to ∂V; ρ is the material density in the deformed state.

In the case of a medium with microdeformation, each particle of a body is characterized by its position, $\boldsymbol{R}$, and microdistorsion which is specified by a nonsingular tensor $\boldsymbol{F}$. Here, the particle microrotation is expressed in terms of microdistorsion by the formula $\boldsymbol{H} = (\boldsymbol{F} \cdot \boldsymbol{F}^T)^{-1/2} \cdot \boldsymbol{F}$. In view of the local action principle, the specific energy function for a medium with microdeformation should be taken in the form

$$W = W(\operatorname{grad} \boldsymbol{R}, \boldsymbol{F}, \operatorname{grad} \boldsymbol{F})\,. \tag{8}$$

It can be proved that the expression (8) is invariant with respect to rotations of the reference frame if and only if the specific energy depends on body motion through the three tensors

$$\boldsymbol{\Lambda} = (\operatorname{grad} \boldsymbol{R}) \cdot (\operatorname{grad} \boldsymbol{R})^T, \quad \boldsymbol{\Phi} = (\operatorname{grad} \boldsymbol{R}) \cdot \boldsymbol{F}^{-1}, \quad \boldsymbol{K} = (\operatorname{grad} \boldsymbol{F}) \cdot \boldsymbol{F}^{-1}\,, \tag{9}$$

where $\boldsymbol{\Lambda}$ is the Cauchy strain measure; $\boldsymbol{\Phi}$ is relative strain measure, while the third-order tensor $\boldsymbol{K}$ would be referred to as hyperstrain measure.

2. Nonlinear Theory of Isolated Dislocations and Disclinations

The concept of a dislocation in the classic linear theory of elasticity was introduced by Volterra in the early twentieth century. The Weingarten theorem on the nature of ambiguity of displacements in multiply connected body with unambiguous strain tensor that satisfies the compatibility equations serves as initial point for the theory of Volterra dislocations. This theorem states that the jump in the displacement vector when intersecting each cut is described by the formula of small displacement of absolutely rigid body if one draws the cuts which convert the domain into simple connected one. The papers [11,12] extended Weingarten's theorem to the case of large deformations of a simple nonlinearly elastic body. A related theorem is valid [14] also for elastic Cosserat continuum subjected to large deformations. The exact formulation of that theorem is as follows. If the strain measure $\boldsymbol{U}$ and the bending strain tensor $\boldsymbol{L}$ are continuously differentiable and single-valued within multiply connected domain v, as well as they satisfy the nonlinear compatibility equations, the values of $\boldsymbol{H}_\pm$ of the microrotation tensor on the opposite borders of each of the cuts σ_k, which convert the domain into simple connected, are related by the equation

$$\boldsymbol{H}_+ = \boldsymbol{H}_- \cdot \boldsymbol{\Omega}_k\,, \tag{10}$$

where $\boldsymbol{\Omega}_k$ is a proper orthogonal tensor having a constant value on the cut σ_k. The particle position $\boldsymbol{R}_\pm$ of the opposite borders of the cut in the deformed state are related by the equation

$$\boldsymbol{R}_+ = \boldsymbol{R}_- \cdot \boldsymbol{\Omega}_k + \boldsymbol{b}_k\,, \tag{11}$$

where $\boldsymbol{b}_k$ is a constant vector at the cut σ_k. In the case of doubly connected domain, the parameters $\boldsymbol{\Omega}$ and $\boldsymbol{b}$ are expressed in terms of the fields of strain tensors with multiplicative integral

$$\boldsymbol{\Omega} = \boldsymbol{H}^T(M_0)\cdot \widehat{\oint_{M_0}} (\boldsymbol{E} + d\boldsymbol{r}\cdot\boldsymbol{\Pi})\cdot \boldsymbol{H}(M_0)\,, \tag{12}$$

$$\boldsymbol{b} = \oint d\boldsymbol{r}'\cdot U(\boldsymbol{r}')\cdot \widehat{\int_{M_0}^{M'}} (\boldsymbol{E} + d\boldsymbol{r}\cdot\boldsymbol{\Pi})\cdot \boldsymbol{H}(M_0) + \boldsymbol{R}(M_0)\cdot(\boldsymbol{E} - \boldsymbol{\Omega}), \tag{13}$$

$$\boldsymbol{\Pi} = -\boldsymbol{r}^s \otimes \boldsymbol{E} \times (\boldsymbol{r}_s \cdot \boldsymbol{L})\,.$$

The relations (10)–(13) show that nonlinearly elastic bodies with couple-stresses may have defects in the form of Volterra dislocations, each of which combines the translational dislocation and the disclination and is described by two vector parameters: the Burgers vector $\boldsymbol{b}_k$ and Frank vector $\boldsymbol{\chi}_k$, where

$$\boldsymbol{\chi}_k = 2\left(1 + \operatorname{tr}\boldsymbol{\Omega}_k)^{-1}\right)(\boldsymbol{\Omega}_k)_\times\,. \tag{14}$$

The relations

$$\boldsymbol{F}_+ = \boldsymbol{F}_- \cdot \boldsymbol{\Omega}, \quad \boldsymbol{R}_+ = \boldsymbol{R}_- \cdot \boldsymbol{\Omega} + \boldsymbol{b} \tag{15}$$

are fulfilled at the cuts, which convert the domain v into simply connected one, for a nonlinearly elastic medium with microdeformation in circumstances where the tensor strain fields $\boldsymbol{\Lambda}$, $\boldsymbol{\Phi}$, $\boldsymbol{K}$ are single-valued and continuously differentiable.

In Eqs. (15) $\boldsymbol{\Omega}$ is a constant proper orthogonal tensor; $\boldsymbol{b}$ is a constant vector.

In the case of plane strain, it is possible to avoid the use of multiplicative integrals and to obtain the expressions for the Burgers and Frank vectors in terms of ordinary contour integrals. Furthermore, the proof of Weingarten's theorem was given for the case of plane finite strain for the Cosserat continuum, quite apart from the proof presented for three-dimensional case.

The equation system for determining the stress state of Cosserat nonlinearly elastic medium containing isolated defects with specified characteristics $\boldsymbol{b}_k$, $\boldsymbol{\chi}_k$

consists of the equilibrium equations (2) which can be transformed into equations with respect to $\boldsymbol{U}$ and $\boldsymbol{L}$, the strain compatibility equations, and the integral relations of the form of Eqs. (12), (13).

Large deformation effects as well as effects of material microstructure are essential in the domain close to the axis of an isolated defect, i.e. within the dislocation or dislocation core. To account for these effects, we can employ the solutions to the problems for screw dislocation and wedge disclination in nonlinearly elastic medium with couple stresses [14]. These solutions are constructed as follows. Let r, θ, z be the Lagrangian cylindrical coordinates, while R, Θ, Z be the Eulerian cylindrical coordinates. The unit vectors tangent to coordinate lines are denoted $\boldsymbol{e}_r$, $\boldsymbol{e}_\theta$, $\boldsymbol{e}_z$ and $\boldsymbol{e}_R$, $\boldsymbol{e}_\Theta$, $\boldsymbol{e}_Z$. Let us consider the following deformation of a Cosserat continuum

$$R = R(r), \quad \Theta = \kappa\theta + \psi z, \quad Z = \frac{b}{2\pi}\theta + \gamma z\,,$$

$$\boldsymbol{H} = \boldsymbol{e}_r \otimes \boldsymbol{e}_R + \cos\chi(r)\,(\boldsymbol{e}_\theta \otimes \boldsymbol{e}_\Theta + \boldsymbol{e}_z \otimes \boldsymbol{e}_Z) - \sin\chi(r)\,(\boldsymbol{e}_z \otimes \boldsymbol{e}_\Theta - \boldsymbol{e}_\theta \otimes \boldsymbol{e}_Z)\,, \tag{16}$$

where κ, ψ, b and γ are constants. The expressions (16) describe the formation in a circular cylinder of an isolated linear defect, for which the Burgers and Frank vectors are directed along the defect axis that coincides with the cylinder axis $\boldsymbol{e}_z$. The length of the Burgers vector equals $b\kappa^{-1}$, while the Frank vector length equals $2\tan\pi(1-\kappa^{-1})$, when $\kappa > 1$ and $2\tan\pi(1-\kappa)$, when $0 < \kappa < 1$. When a defects develops, the cylinder exhibits also twisting as well as radial and axial deformations.

If a gyrotropic Cosserat continuum with the deformation of the form of Eqs. (16) is free of body forces and mass moments, one can prove that the system of equations (2) is reduced to two ordinary differential equations with respect to the functions $R(r)$, $\chi(r)$. These equations admit exact solutions for certain particular forms of constitutive relations for a couple-stress medium. An analysis of these exact solutions show that the rigorous account of nonlinearity changes qualitatively the pattern of stressed state close to the defect axis as compared with the linear theory. In particular, an account of large deformations affects essentially the singularity order of stresses and strains near the defect axis.

3. Phase Equilibrium Conditions for Nonlinearly Elastic Media with Microstructure

Consider an elastic body involving two phases which are separated by a surface Σ and occupy volumes V_+ and V_- in the deformed configuration. The phase equilibrium conditions for a material placed in a uniform temperature field can be obtained [2] from the Gibbs variational principle by seeking for equilibrium state that provides the stationary value for the free-energy functional in an isothermal process:

$$\Pi = \int_{V_+} \rho_+\pi_+\, dV + \int_{V_-} \rho_-\pi_-\, dV + \int_\Sigma \alpha\, d\Sigma + A\,. \tag{17}$$

Here, $\rho_\pm$ are the densities of the material phases in the deformed state; $\pi_\pm$ are the mass free-energy densities; α is the surface energy density which is assumed to be constant;

A is the potential of external conservative loads. For a medium with microstructure, the strain dependence of the free energy in an isothermal process is given by the relation [2]:

$$\pi_\pm = \pi_\pm(\operatorname{grad} \boldsymbol{R}, \boldsymbol{\Gamma}, \operatorname{grad} \boldsymbol{\Gamma}),$$

where $\boldsymbol{\Gamma}$ is a microstructure parameter which characterizes inner degrees of freedom. In the case of a Cosserat continuum, the parameter $\boldsymbol{\Gamma}$ coincides with the microrotation tensor $\boldsymbol{H}$; for a medium possessing microdeformations, $\boldsymbol{\Gamma}$ is the same as microdistorsion tensor $\boldsymbol{F}$; for a nematic liquid crystal, $\boldsymbol{\Gamma} = \boldsymbol{l}$, where $\boldsymbol{l}$ is the unit particle oriented vector referred to as director. The phase-phase interface Σ is unknown beforehand and is to be varied.

The phase equilibrium conditions that follow from the stationarity of the functional, Eq. (17), include the equilibrium equations in the volumes $V_\pm$, the boundary conditions on the outer surface of the body, and the boundary conditions on the interphase boundary Σ. The latter consist of the conditions of mechanical equilibrium between two phases, which present the balance of static quantities, and the condition of thermodynamic phase equilibrium, which is necessary for determining the pre-unknown phase boundary.

The phase transition in a medium with microstructure will be said to be microcoherent if the field of microstructure parameter is continuous in the neighbourhood of the phase boundary. Otherwise we shall say about the phase transition with microslipping. Different types of phase transition with microslipping, which are distinctive in jump character of the parameter $\boldsymbol{\Gamma}$ on the interphase boundary, are possible depending upon the physical meaning of the microstructure parameter. In what follows we shall restrict ourselves to the consideration of a microslipping type such that the magnitude $[\delta\boldsymbol{\Gamma}]_-^+$ may be arbitrary. Here, the square brackets identify the jump of a corresponding quantity when intersecting the surface Σ: $[f]_-^+ = f_+ - f_-$.

In the case of a Cosserat continuum, the conditions of thermodynamic phase equilibrium for microcoherent phase transitions and phase transitions with microslipping are given by the following relations

$$\boldsymbol{N} \cdot [\boldsymbol{\beta}_1]_-^+ \cdot \boldsymbol{N} = 0, \quad \boldsymbol{N} \cdot [\boldsymbol{\beta}_2]_-^+ \cdot \boldsymbol{N} = 0,$$

$$\boldsymbol{\beta}_{1\pm} = \boldsymbol{\beta}_{2\pm} - \rho_\pm^{-1} \boldsymbol{M}_\pm \cdot \boldsymbol{H}^T \cdot \boldsymbol{L}^T \cdot (\operatorname{grad} \boldsymbol{R})^{-T}, \quad \boldsymbol{\beta}_{2\pm} = \pi_\pm \boldsymbol{E} - \rho_\pm^{-1} \boldsymbol{T}_\pm,$$

where $\boldsymbol{N}$ is the normal to the surface Σ. For a liquid crystal, the tensors $\boldsymbol{\beta}_1$, $\boldsymbol{\beta}_2$ have the form (Grad is the gradient operator in Eulerian coordinates)

$$\beta_{1\pm} = (\pi_\pm + p_\pm \rho_\pm^{-1})\boldsymbol{E}, \quad \beta_{2\pm} = \beta_{1\pm} + \frac{\partial \pi_\pm}{\partial(\operatorname{Grad} \boldsymbol{l})} \cdot (\operatorname{Grad} \boldsymbol{l})^T, \quad p_\pm = \rho_\pm^2 \frac{\partial \pi_\pm}{\partial \rho_\pm}.$$

4. The Equations for Visco-Elastic Micropolar Fluid

By considering the general constitutive relations for inelastic Cosserat continuum under large deformations, new model of elastic and visco-elastic micropolar fluid may be constructed. The general representations of stresses and couple stresses for micropolar liquid with memory have the form

$$\boldsymbol{T}(t) = Y_1\left[\rho(t), \boldsymbol{B}(t), \boldsymbol{U}_t^t(s), \boldsymbol{L}_t^t(s)\right], \quad \boldsymbol{M}(t) = Y_2\left[\rho(t), \boldsymbol{B}(t), \boldsymbol{U}_t^t(s), \boldsymbol{L}_t^t(s)\right],$$

$$\boldsymbol{B}(t) = -\frac{1}{2}\left[\operatorname{Grad} \boldsymbol{D}_k(t)\right] \times \boldsymbol{D}_k(t), \quad \boldsymbol{U}_t^t(s) = \boldsymbol{U}_t(t-s),$$

$$\boldsymbol{L}_t^t(s) = \boldsymbol{L}_t(t-s), \quad s \geq 0. \tag{18}$$

Here, $\boldsymbol{U}_t^t(s)$ is the history of the relative strain measure, which is determined taking the actual configuration as reference; $\boldsymbol{L}_t^t(s)$ is the history of the relative tensor of bending strain; $\boldsymbol{D}_k$, $k = 1, 2, 3$, are the three unit mutually orthogonal vectors that specify the orientation of a continuum particle in actual configuration; $\boldsymbol{Y}_1$, $\boldsymbol{Y}_2$ are gyrotropic operators. The representation (18) is the extension of the well-known Noll's theorem [9] on simple fluids to the case of micropolar fluids. If the fluid is at rest, Eqs. (18) become the constitutive equations for an elastic micropolar fluid. The latter model is of independent importance as a particular case of an elastic Cosserat medium with a certain material symmetry. An essential feature of elastic and visco-elastic micropolar fluids is that they can withstand couple stresses at rest.

Acknowledgements. The research described in this publication was made possible in part by Grant No MTA300 from the International Science Foundation and Russian Government.

References

1. E. L. Aero and E. V. Kuvshinsky, *Fizika Tverdogo Tela* **2** (1960) 1399. (in Russian.)
2. V. A. Eremeyev and L. M. Zubov, *Dokl. RAN* **322** (1992) 1052. (in Russian.)
3. V. A. Eremeyev and L. M. Zubov, *Izv. RAN MTT (Mechanics of Solids)* **3** (1994) 181. (in Russian.)
4. W. T. Koiter, *Proc. Neterland. Akad. Wetensh.* **B-67** (1964) 17.
5. R. D. Mindlin, *Arch. Rat. Mech. Anal.* **16** (1964) 51.
6. V. A. Pal'mov, *Prikl. Mat. Mekh.* **28** (1964) 401. (in Russian.)
7. L. I. Shkutin, *Zhurnal PMTF* **6** (1980) 111. (in Russian.)
8. R. A. Toupin, *Arch. Rat. Mech. Anal.* **17** (1964) 85.

9. C. Truesdell, *A First Course in Rational Continuum Mechanics*, Academic Press, New York, 1977.
10. P. A. Zhilin, *Trudy Leningrad. Polytekhn. Instituta.* **386** (1964) 29. (in Russian.)
11. L. M. Zubov, *Dokl. AN SSSR.* **287** (1986) 576. (in Russian.)
12. L. M. Zubov, *Izv. AN SSSR. MTT. (Mechanics of Solids)* **5** (1987) 140. (in Russian.)
13. L. M. Zubov, *Izv. AN SSSR. MTT. (Mechanics of Solids)* **6** (1990) 10. (in Russian.)
14. L. M. Zubov and M. I. Karyakin, *Zhurnal PMTF* **3** (1990) 160. (in Russian.)

Continuum Models and Discrete Systems
Proceedings of 8th International Symposium, June 11-16, 1995, Varna, Bulgaria, ed. K.Z. Markov

VARIATIONAL PRINCIPLES AND THE C^2-FORMULA FOR THE EFFECTIVE CONDUCTIVITY OF A RANDOM DISPERSION

K. D. ZVYATKOV
Faculty of Mathematics and Informatics,
"K. Preslavski" University of Shumen, BG-9700 Shumen, Bulgaria

Abstract. The problem of analytical evaluation of the effective conductivity of a dilute random dispersion of nonoverlapping spheres is revisited. The classical variational principle is employed in which a class of trial fields in the form of suitably truncated factorial series [1,2] is introduced. The class is as chosen as to contain the actual temperature field to the order c^2, where c is the sphere fraction. Minimizing the energy functional on this class we get the effective conductivity of the dispersion to the order c^2 in a form containing an absolutely convergent integral. It is shown that the so obtained formula coincides with Jeffrey's renormalized one [3]. The same form was already found in a similar, but more complicated, variational way by means of the Hashin-Shtrikman principle [4,5].

1. Introduction

Consider a statistically homogeneous dispersion of equi-sized nonoverlapping spheres of conductivity κ_f and radii a, immersed at random into a matrix of conductivity κ_m. In the heat conductivity context and absence of body sources, the temperature field, $\theta(\mathbf{x})$, in the dispersion is governed by the equations

$$\nabla \cdot (\kappa(\mathbf{x})\nabla\theta(\mathbf{x})) = 0, \quad \langle \nabla\theta(\mathbf{x})\rangle = \mathbf{G}, \tag{1.1}$$

where $\kappa(\mathbf{x})$ is the random conductivity field of the medium that takes the values κ_f or κ_m depending on whether $\mathbf{x}$ lies in a sphere or in the matrix respectively; $\mathbf{G}$ is the prescribed macroscopical value of the temperature gradient. The brackets $\langle\cdot\rangle$ denote ensemble averaging [6].

Due to the specific point-wise structure of the medium, the field $\kappa(\mathbf{x})$ allows the representation

$$\kappa(\mathbf{x}) = \langle\kappa\rangle + [\kappa]\int h(\mathbf{x}-\mathbf{y})\psi'(\mathbf{y})\,d^3\mathbf{y}, \tag{1.2}$$

where $[\kappa] = \kappa_f - \kappa_m$, $h(\mathbf{x})$ is the characteristic function of a single sphere of radius a

located at the origin, and $\psi'(\mathbf{x})$ is the fluctuating part of the random density field

$$\psi(\mathbf{x}) = \sum_j \delta(\mathbf{x} - \mathbf{x}_j),$$

generated by the random field $\{\mathbf{x}_j\}$ of sphere's centers [7]. The integrals hereafter are over the whole $\mathbf{R}^3$, if the integration domain is not explicitly indicated.

As discussed in Refs. 1,2 and 8, the solution $\theta(\mathbf{x})$ of the random problem (1.1), asymptotically valid to the order c^2, can be found in the form

$$\theta(\mathbf{x}) = \mathbf{G} \cdot \mathbf{x} + \int T_1(\mathbf{x} - \mathbf{y}) D_\psi^{(1)}(\mathbf{y})\, d^3\mathbf{y}$$

$$+ \iint T_2(\mathbf{x} - \mathbf{y}_1, \mathbf{x} - \mathbf{y}_2) D_\psi^{(2)}(\mathbf{y}_1, \mathbf{y}_2)\, d^3\mathbf{y}_1\, d^3\mathbf{y}_2, \tag{1.3}$$

with certain nonrandom kernels T_1 and T_2. In Eq. (1.3)

$$D_\psi^{(1)}(\mathbf{y}) = \psi'(\mathbf{y}), \quad D_\psi^{(2)}(\mathbf{y}_1, \mathbf{y}_2) = \psi(\mathbf{y}_1)[\psi(\mathbf{y}_2) - \delta(\mathbf{y}_1 - \mathbf{y}_2)]$$

$$-ng_0(\mathbf{y}_1 - \mathbf{y}_2)[D_\psi^{(1)}(\mathbf{y}_1) + D_\psi^{(1)}(\mathbf{y}_2)] - n^2 g_0(\mathbf{y}_1 - \mathbf{y}_2). \tag{1.4}$$

The reason is that the field $D_\psi^{(0)} = 1$, $D_\psi^{(1)}$ and $D_\psi^{(2)}$ are the first three terms in the c^2-orthogonal system formed as a result of the appropriate virial orthogonalization, see again Refs. 1, 2 and 8 for details and discussion. In Eq. (1.3) $g(\mathbf{y}) = g_0(\mathbf{y}) + o(c)$ is the well known radial distribution for the dispersion, so that g_0 is its leading part in the dilute case $n \to 0$, i.e., $g(\mathbf{y}_1 - \mathbf{y}_2) = f_2(\mathbf{y}_1, \mathbf{y}_2)/n^2$, with f_2 denoting the two-point probability density for the set of sphere centers, and $n = c/V_a$ is the number density of the spheres whose volume fraction is c, $V_a = \frac{4}{3}\pi a^3$.

The identification of the kernels T_1 and T_2 is performed in Refs. 2 and 8 by means of a procedure, proposed by Christov and Markov [9]. It consists in inserting the truncated series (1.3) into the random equation (1.1), multiplying the result by the fields $D_\psi^{(p)}$, $p = 0, 1, 2$, and averaging the results. In this way a certain system of integrodifferential equations for the needed kernels of the truncated series can be straightforwardly derived. The solution has been found analytically in Refs. 2 and hence the full statistical solution of the problem (1.1), asymptotically correct to the order c^2, was found. As a particular case, the renormalized c^2-formula of Jeffrey [3] for the effective conductivity of the dispersion was rederived, but with rigorous justification of the integration mode in the appropriate conditionally convergent integrals. Here another method for identification of the kernels T_1 and T_2 will be proposed, namely, the truncated series (1.3) will be inserted into the classical variational principle as classes of trial fields. Since these series contain the actual temperature field to the order c^2 among them, the obtained formulae for the statistical solution of (1.1) will be accurate to the same order c^2 as well. In this way the equations for the kernels T_1 and T_2, already found in Ref. 2, will be rederived, but the formula for the effective conductivity will involve only absolutely convergent integrals, without any need of "renormalizations".

2. Variational Three-point Bounds

As is well known [6], the random problem (1.1) is equivalent to the classical variational principle

$$W_A[\theta(\cdot)] = \langle \kappa(\mathbf{x}) |\nabla \theta(\mathbf{x})|^2 \rangle \longrightarrow \min, \quad \langle \nabla \theta(\mathbf{x}) \rangle = \mathbf{G}; \tag{2.1}$$

moreover $\min W_A = \kappa^* G^2$, where κ^* is the effective conductivity of the medium:

$$\langle \kappa(\mathbf{x}) \nabla \theta(\mathbf{x}) \rangle = \kappa^* \langle \nabla \theta(\mathbf{x}) \rangle = \kappa^* \mathbf{G}. \tag{2.2}$$

It is necessary to consider first of all the simpler case when the functional series (1.3) is truncated after the one-tuple integral term (see Refs. 10 and 11 for more details). This leads to the class of trial fields

$$\mathcal{K}_A^{(1)} = \left\{ \theta(\mathbf{x}) | \theta(\mathbf{x}) = \mathbf{G} \cdot \mathbf{x} + \int T_1(\mathbf{x}-\mathbf{y}) D_\psi^{(1)}(\mathbf{y})\, d^3\mathbf{y} \right\}. \tag{2.3}$$

This class was introduced and discussed in detail by Markov [6], where it was shown that maximizing $W_A[\theta(\cdot)]$ over the class $\mathcal{K}_A^{(1)}$ gives the best three-point upper bound on the effective conductivity κ^*, i.e., the most restrictive one which uses three-point statistical information for the medium.

$$\begin{aligned} W_A^{(1)}[T_1(\cdot)] &= W_A\big|_{\mathcal{K}_A^{(1)}} = \langle \kappa \rangle G^2 \\ &+ n\langle \kappa \rangle \left\{ \int |\nabla T_1(\mathbf{z})|^2\, d^3\mathbf{z} - n \int\!\!\int R_0(\mathbf{z}_1 - \mathbf{z}_2) \nabla T_1(\mathbf{z}_1) \cdot \nabla T_2(\mathbf{z}_2)\, d^3\mathbf{z}_1\, d^3\mathbf{z}_2 \right\} \\ &+ 2n[\kappa]\mathbf{G} \cdot \left\{ \int h(\mathbf{z}) \nabla T_1(\mathbf{z})\, d^3\mathbf{z} - n \int F_0(\mathbf{z}) \nabla T_1(\mathbf{z})\, d^3\mathbf{z} \right\} \\ &+ n[\kappa] \left\{ \int h(\mathbf{z}) |\nabla T_1(\mathbf{z})|^2\, d^3\mathbf{z} - n \left[\int F_0(\mathbf{z}) |\nabla T_1(\mathbf{z})|^2\, d^3\mathbf{z} \right.\right. \\ &\left.\left. + 2 \int\!\!\int h(\mathbf{z}_1) R_0(\mathbf{z}_1 - \mathbf{z}_2) \nabla T_1(\mathbf{z}_1) \cdot \nabla T_1(\mathbf{z}_2)\, d^3\mathbf{z}_1\, d^3\mathbf{z}_2 \right] \right\} + o(n^2), \end{aligned} \tag{2.4}$$

where

$$R_0(\mathbf{y}) = 1 - g_0(\mathbf{y}), \quad F_0(\mathbf{y}) = \int h(\mathbf{z}) R_0(\mathbf{y} - \mathbf{z})\, d^3\mathbf{z}.$$

The optimal kernel $T_1(\mathbf{x})$, i.e., the solution of the Euler-Lagrange equation for the functional $W_A^{(1)}$ is looked for in the virial form

$$T_1(\mathbf{x}) = T_1(\mathbf{x}; n) = T_{1,0}(\mathbf{x}) + T_{1,1}(\mathbf{x}) n + \cdots \tag{2.5}$$

The representation (2.5) induces the appropriate virial expansion of the functional (2.4):

$$W_A^{(1)}[T_1(\cdot)] = \langle \kappa \rangle G^2 + W_A^{(1,1)}[T_{1,0}(\cdot)] n + W_A^{(1,2)}[T_{1,0}(\cdot), T_{1,1}(\cdot)] n^2 + \cdots. \tag{2.6}$$

An analysis of the coefficient $W_A^{(1,1)}$ shows that

$$\delta W_A^{(1,1)}[T_{1,0}(\cdot)] = 0 \Longleftrightarrow T_{1,0}(\mathbf{x}) = T^{(1)}(\mathbf{x}), \tag{2.7}$$

where $T^{(1)}(\mathbf{x}) = 3\beta G \cdot \nabla\nabla\varphi(\mathbf{x})$ is the disturbance to the temperature field $\mathbf{G} \cdot \mathbf{x}$ in an unbounded matrix introduced by a single spherical inhomogeneity located at the origin; $\beta = [\kappa]/(\kappa_f + 2\kappa_m)$, $\varphi(\mathbf{x}) = h * \frac{1}{4\pi|\mathbf{x}|}$ is the Newtonian potential for the sphere. It turns out, however, that at $T_{1,0}(\mathbf{x}) = T^{(1)}(\mathbf{x})$ the virial coefficient $W_A^{(1,2)}$ does not depend on $T_{1,1}(\mathbf{x})$, i.e.,

$$W_A^{(1,2)}[T^{(1)}(\cdot), T_{1,1}(\cdot)] = \overline{W}_A^{(1,2)}[T^{(1)}(\cdot)] = 3\beta^2\kappa_m\left(1 + \frac{[\kappa]}{\kappa_m}m_2\right)V_a^2G^2, \tag{2.8}$$

where

$$m_2 = m_2[g_0(\cdot)] = 2\int_2^\infty \frac{\lambda^2}{(\lambda^2-1)^3}\, g_0(\lambda a)\,d\lambda, \quad \lambda = \frac{|\mathbf{y}|}{a},$$

is a statistical parameter for the dispersion. On the base of this analysis it was shown in Ref. 11 that the Beran's bounds [12] are c^2-optimal in the above explained sense. Let us recall that if

$$\frac{\kappa^*}{\kappa_m} = 1 + 3\beta c + 3\beta^2(1 + a'_{2\kappa})c^2 + \cdots \tag{2.9}$$

is the virial expansion of κ^*, the Beran bounds lead to the following estimates for $a'_{2\kappa}$ (see Refs. 10 and 11):

$$\frac{[\kappa]}{\kappa_f}m_2 \le a'_{2\kappa} \le \frac{[\kappa]}{\kappa_m}m_2\,.$$

Note that the coefficient $a'_{2\kappa}$ is just the c^2-deviation of κ^* from its value as predicted by the well known Maxwell formula.

3. Variational Derivation of the c^2-Formula for the Effective Conductivity

Consider now the series (1.3) as a class of trial fields:

$$\begin{aligned}\mathcal{K}_A^{(2)} = \Big\{\theta(\mathbf{x})\,|\,\theta(\mathbf{x}) \;&=\; \mathbf{G}\cdot\mathbf{x} + \int T_1(\mathbf{x}-\mathbf{y})D_\psi^{(1)}(\mathbf{y})\,d^3\mathbf{y} \\ &+ \iint T_2(\mathbf{x}-\mathbf{y}_1,\mathbf{x}-\mathbf{y}_2)D_\psi^{(2)}(\mathbf{y}_1,\mathbf{y}_2)\,d^3\mathbf{y}_1\,d^3\mathbf{y}_2\Big\},\end{aligned} \tag{3.1}$$

where now the kernels $T_1(\mathbf{x})$ and $T_2(\mathbf{x},\mathbf{y})$ are adjustable. Using the formulae for the moments of the fields $D_\psi^{(1)}$ and $D_\psi^{(2)}$, see [1, Eq. (4.5)], the restriction $W_A^{(2)}[T_1(\cdot), T_2(\cdot,\cdot)]$ of the functional W_A over this class becomes

$$W_A^{(2)}[T_1(\cdot), T_2(\cdot,\cdot)] = W_A\big|_{\mathcal{K}_A^{(2)}} = W_A^{(1)}[T_1(\cdot)] + \widetilde{W}_A^{(2)}[T_1(\cdot), T_2(\cdot,\cdot)],$$

where

$$\widetilde{W}_A^{(2)}\,[T_1(\cdot),T_2(\cdot,\cdot)] = n^2\kappa_m \iint g_0(\mathbf{y}_1-\mathbf{y}_2)\Big[\,|\nabla_x T_2(\mathbf{x}-\mathbf{y}_1,\mathbf{x}-\mathbf{y}_2)|^2$$

$$+\nabla_x T_2(\mathbf{x}-\mathbf{y}_1,\mathbf{x}-\mathbf{y}_2)\cdot\nabla_x T_2(\mathbf{x}-\mathbf{y}_2,\mathbf{x}-\mathbf{y}_1)\Big]\,d^3\mathbf{y}_1\,d^3\mathbf{y}_2$$

$$+2n^2[\kappa]\iint g_0(\mathbf{y}_1-\mathbf{y}_2)\Big[h(\mathbf{x}-\mathbf{y}_1)\nabla T_1(\mathbf{x}-\mathbf{y}_2)+h(\mathbf{x}-\mathbf{y}_2)\nabla T_1(\mathbf{x}-\mathbf{y}_1)\Big]$$

$$\cdot\nabla_x T_2(\mathbf{x}-\mathbf{y}_1,\mathbf{x}-\mathbf{y}_2)\,d^3\mathbf{y}_1\,d^3\mathbf{y}_2$$

$$+n^2[\kappa]\iint g_0(\mathbf{y}_1-\mathbf{y}_2)(h(\mathbf{x}-\mathbf{y}_1)+h(\mathbf{x}-\mathbf{y}_2))\,[|\nabla_x T_2(\mathbf{x}-\mathbf{y}_1,\mathbf{x}-\mathbf{y}_2)|^2$$

$$+\nabla_x T_2(\mathbf{x}-\mathbf{y}_1,\mathbf{x}-\mathbf{y}_2)\cdot\nabla_x T_2(\mathbf{x}-\mathbf{y}_2,\mathbf{x}-\mathbf{y}_1)]\,d^3\mathbf{y}_1\,d^3\mathbf{y}_2+o(n^2).$$

The optimal kernels $T_1(\mathbf{x})$ and $T_2(\mathbf{x},\mathbf{y})$ are looked for again in the virial form (2.5) for T_1 and

$$T_2(\mathbf{x},\mathbf{y}) = T_2(\mathbf{x},\mathbf{y};n) = T_{2,0}(\mathbf{x},\mathbf{y})+T_{2,1}(\mathbf{x},\mathbf{y})n+\cdots,$$

for T_2, which imply the respective virial expansion of the functional $W_A^{(2)}\,[T_1(\cdot),T_2(\cdot,\cdot)]$, namely,

$$W_A^{(2)}\,[T_1(\cdot),T_2(\cdot,\cdot)] = \langle\kappa\rangle G^2 + W_A^{(1,1)}\,[T_{1,0}(\cdot)]\,n$$
$$+W_A^{(2,2)}\,[T_{1,0}(\cdot),T_{1,1}(\cdot),T_{2,0}(\cdot,\cdot)]\,n^2+o(n^2), \tag{3.2}$$

where

$$W_A^{(2,2)}\,[T_{1,0}(\cdot),T_{1,1}(\cdot),T_{2,0}(\cdot,\cdot)] = W_A^{(1,2)}\,[T_{1,0}(\cdot),T_{1,1}(\cdot)]+\widetilde{W}_A^{(2)}\,[T_{1,0}(\cdot),T_{2,0}(\cdot,\cdot)]; \tag{3.3}$$

here $W_A^{(1,1)}$ and $W_A^{(1,2)}$ are the virial coefficients from Eq. (2.6) for which, let us recall, Eqs. (2.7) and (2.8) hold. Hence, the minimization of the functional $W_A^{(2)}$ is reduced to that of the functional

$$\widetilde{W}_A^{(2)\dagger}\,[T_{2,0}(\cdot,\cdot)] = \widetilde{W}_A^{(2)}\,[T^{(1)}(\cdot),T_{2,0}(\cdot,\cdot)]\,.$$

The Euler-Lagrange equation for the latter is

$$\kappa_m(\nabla_{\mathbf{z}_1}+\nabla_{\mathbf{z}_2})\cdot\Big\{g_0(\mathbf{z}_1-\mathbf{z}_2)(\nabla_{\mathbf{z}_1}+\nabla_{\mathbf{z}_2})\widetilde{T}_{2,0}(\mathbf{z}_1,\mathbf{z}_2)\Big\}$$

$$+[\kappa](\nabla_{\mathbf{z}_1}+\nabla_{\mathbf{z}_2})\cdot\Big\{g_0(\mathbf{z}_1-\mathbf{z}_2)(h(\mathbf{z}_1)\nabla T^{(1)}(\mathbf{z}_2)+h(\mathbf{z}_2)\nabla T^{(1)}(\mathbf{z}_1))\Big\} \tag{3.4}$$

$$+[\kappa](\nabla_{\mathbf{z}_1}+\nabla_{\mathbf{z}_2})\cdot\Big\{g_0(\mathbf{z}_1-\mathbf{z}_2)(h(\mathbf{z}_1)+h(\mathbf{z}_2))(\nabla_{\mathbf{z}_1}+\nabla_{\mathbf{z}_2})\widetilde{T}_{2,0}(\mathbf{z}_1,\mathbf{z}_2)\Big\}=0,$$

with the notation

$$\widetilde{T}_{2,0}(\mathbf{z}_1,\mathbf{z}_2) = T_{2,0}(\mathbf{z}_1,\mathbf{z}_2)+T_{2,0}(\mathbf{z}_2,\mathbf{z}_1).$$

Taking into account that $(\nabla_{\mathbf{z}_1} + \nabla_{\mathbf{z}_2})g_0(\mathbf{z}_1 - \mathbf{z}_2) = 0$, and an appropriate change of variables allows to recast Eq. (3.4) as

$$g_0(\mathbf{z})\Big\{\kappa_m \Delta_x \widetilde{T}_{2,0}(\mathbf{x}, \mathbf{x}-\mathbf{z}) + [\kappa]\nabla_x \cdot \left[h(\mathbf{x})\nabla T^{(1)}(\mathbf{x}-\mathbf{z}) + h(\mathbf{x}-\mathbf{z})\nabla T^{(1)}(\mathbf{x})\right]$$

$$+[\kappa]\nabla_x \cdot \Big[\big(h(\mathbf{x}) + h(\mathbf{x}-\mathbf{z})\big)\nabla_x \widetilde{T}_{2,0}(\mathbf{x}, \mathbf{x}-\mathbf{z})\Big]\Big\} = 0, \tag{3.5}$$

whose solution is

$$\widetilde{T}_{2,0}(\mathbf{x}, \mathbf{x}-\mathbf{z}) = T^{(2)}(\mathbf{x}; \mathbf{z}) - T^{(1)}(\mathbf{x}) - T^{(1)}(\mathbf{x}-\mathbf{z}), \tag{3.6}$$

with $T^{(2)}(\mathbf{x};\mathbf{z})$ denoting the disturbance to the temperature field $\mathbf{G}\cdot\mathbf{x}$ in an unbounded matrix of conductivity κ_m, generated by two spherical inhomogeneities of conductivity κ_f, located at the origin and at the point $\mathbf{z}$, see Ref. 2.

Making use of Eq. (3.4), the minimum value of the functional $\widetilde{W}_A^{(2)\dagger}$ can be recast now in the form in which the field $T_{2,0}(\mathbf{x}, \mathbf{y})$ enters linearly:

$$\min \widetilde{W}_A^{(2)\dagger}[T_{2,0}(\cdot,\cdot)] = n^2[\kappa] \iint g_0(\mathbf{z}_1 - \mathbf{z}_2)$$

$$\left[h(\mathbf{z}_1)\nabla T^{(1)}(\mathbf{z}_2) + h(\mathbf{z}_2)\nabla T^{(1)}(\mathbf{z}_1)\right] \cdot (\nabla_{\mathbf{z}_1} + \nabla_{\mathbf{z}_2}) T_{2,0}(\mathbf{z}_1, \mathbf{z}_2)\, d^3\mathbf{z}_1\, d^3\mathbf{z}_2 \tag{3.7}$$

$$= n^2[\kappa] \int h(\mathbf{x})\, d^3\mathbf{x} \int g_0(\mathbf{y}) \nabla T^{(1)}(\mathbf{x}-\mathbf{y}) \cdot \left[\nabla_x T^{(2)}(\mathbf{x};\mathbf{y}) - \nabla T^{(1)}(\mathbf{x}) - \nabla T^{(1)}(\mathbf{x}-\mathbf{y})\right] d^3\mathbf{y}.$$

Taking into account Eqs. (3.2), (3.3), (2.7), (2.8) and the formulae

$$\int h(\mathbf{x})\, d^3\mathbf{x} \int g_0(\mathbf{y}) \nabla T^{(1)}(\mathbf{x}-\mathbf{y}) \cdot \nabla T^{(1)}(\mathbf{x})\, d^3\mathbf{x} = 0, \tag{3.8a}$$

$$\int h(\mathbf{x})\, d^3\mathbf{x} \int g_0(\mathbf{y}) |\nabla T^{(1)}(\mathbf{x}-\mathbf{y})|^2\, d^3\mathbf{y} = 3\beta^2 V_a^2 [\kappa] m_2, \tag{3.8b}$$

for the additional term $a'_{2\kappa}$ to the Maxwell c^2-value, see (2.9), one finds

$$a'_{2\kappa} G^2 = \frac{[\kappa]}{\kappa_m} \frac{1}{V_a^2} \int h(\mathbf{x})\, d^3\mathbf{x} \int g_0(\mathbf{y}) \nabla_x T^{(1)}(\mathbf{x}-\mathbf{y}) \cdot \nabla_x T^{(2)}(\mathbf{x};\mathbf{y})\, d^3\mathbf{y}\,. \tag{3.9}$$

Let us recall that the latter result follows from the fact that the solution of the random problem (1.1), asymptotically valid to the order c^2, is one of the trial fields from the class $\mathcal{K}_A^{(2)}$, see (1.3), over which the energy functional $W_A[\theta(\cdot)]$ is minimized. Moreover, unlike Jeffrey [3], we have found here the formula for $a'_{2\kappa}$ in the form of an absolutely convergent integral. In Ref. 3 Jeffrey himself has given the formula

$$a'_{2\kappa} \mathbf{G} = \frac{[\kappa]}{\kappa_m} \frac{1}{V_a^2} \int h(\mathbf{x})\, d^3\mathbf{x} \int g_0(\mathbf{y}) \left[\nabla_x T^{(2)}(\mathbf{x};\mathbf{y}) - \nabla T^{(1)}(\mathbf{x})\right] d^3\mathbf{y}, \tag{3.10}$$

in which the integral with respect to $\mathbf{y}$ is conditionally convergent. The reason is that the function $\mathbf{U}(\mathbf{x};\mathbf{y}) = \nabla T^{(2)}(\mathbf{x}-\mathbf{y}) - \nabla T^{(1)}(\mathbf{x})$ has the asymptotics $|\mathbf{y}|^{-3}$ as $|\mathbf{y}| \gg 1$. In order to make the latter integral absolutely convergent Jeffrey made the "renormalization" extracting from $\mathbf{U}(\mathbf{x};\mathbf{y})$ the term $\beta\nabla T^{(1)}(\mathbf{x})$, which does not change the value of the integral, but makes the integrand's asymptotics $|\mathbf{y}|^{-6}$ as $|\mathbf{y}| \gg 1$ (the same as the integrand $\nabla_x T^{(2)}(\mathbf{x};\mathbf{y}) \cdot \nabla T^{(1)}(\mathbf{x})$ in our formula (3.9)).

We shall show now that the formula (3.9) can be treated as a result of a certain unobvious way of "renormalization" (if the latter is understood formally simply as a method of making the conditionally convergent integral absolutely convergent without changing its eventual value).

To this end let us replace the differential equations for the disturbances $T^{(1)}(\mathbf{x})$ and $T^{(2)}(\mathbf{x};\mathbf{y})$

$$\kappa_m \Delta T^{(1)}(\mathbf{x}) + [\kappa]\nabla \cdot \left\{h(\mathbf{x})\left(\mathbf{G} + \nabla T^{(1)}(\mathbf{x})\right)\right\} = 0,$$

$$\kappa_m \Delta_x T^{(2)}(\mathbf{x};\mathbf{y}) + [\kappa]\nabla \cdot \left\{\left(h(\mathbf{x}) + h(\mathbf{x}-\mathbf{y})\right)\left(\mathbf{G} + \nabla_x T^{(2)}(\mathbf{x};\mathbf{y})\right)\right\} = 0,$$

by the integral equations

$$T^{(1)}(\mathbf{x}) = \frac{[\kappa]}{\kappa_m}\int \nabla_x \frac{1}{4\pi|\mathbf{x}-\mathbf{w}|} \cdot \left\{h(\mathbf{w})\left(\mathbf{G} + \nabla T^{(1)}(\mathbf{w})\right)\right\} d^3\mathbf{w}, \tag{3.11}$$

$$T^{(2)}(\mathbf{x};\mathbf{y}) = \frac{[\kappa]}{\kappa_m}\int \nabla_x \frac{1}{4\pi|\mathbf{x}-\mathbf{w}|} \cdot \left\{\left(h(\mathbf{w}) + h(\mathbf{w}-\mathbf{y})\right)\left(\mathbf{G} + \nabla_w T^{(2)}(\mathbf{w};\mathbf{y})\right)\right\} d^3\mathbf{w},$$

respectively, making use of the appropriate Green function for the Laplacian. Denote the integrals in Eqs. (3.9) and (3.10) respectively as and $\mathcal{M}$ and $\mathbf{N}$, i.e.,

$$\mathcal{M} = \int h(\mathbf{x})\, d^3\mathbf{x} \int g_0(\mathbf{y}) \nabla T^{(1)}(\mathbf{x}-\mathbf{y}) \cdot \nabla_x T^{(2)}(\mathbf{x};\mathbf{y})\, d^3\mathbf{y}, \tag{3.12}$$

$$\mathbf{N} = \int h(\mathbf{x})\, d^3\mathbf{x} \int g_0(\mathbf{y}) \left[\nabla_x T^{(2)}(\mathbf{x};\mathbf{y}) - \nabla T^{(1)}(\mathbf{x})\right] d^3\mathbf{y},$$

Inserting Eq. (3.11) into Eq. (3.12), after simple algebra based on Eq. (3.8a) and the equality $h(\mathbf{x})\nabla T^{(1)}(\mathbf{x}) = -\beta \mathbf{G} h(\mathbf{x})$, we get

$$\mathbf{G}\cdot\mathbf{N} = -\frac{1}{3}\frac{[\kappa]}{\kappa_m}\mathbf{G}\cdot\mathbf{N} + \frac{1}{3\beta}\frac{[\kappa]}{\kappa_m}\mathcal{M}$$

$$-\frac{1}{3}\frac{[\kappa]}{\kappa_m}\left(1 - \frac{1}{\beta}\right)\mathbf{G}\cdot \int\!\!\int g_0(\mathbf{y}) h(\mathbf{x}) \nabla T^{(1)}(\mathbf{x}-\mathbf{y})\, d^3\mathbf{x}\, d^3\mathbf{y}. \tag{3.13}$$

The last integral in Eq. (3.13), however, vanishes which yields $\mathbf{G}\cdot\mathbf{N} = \mathcal{M}$. Hence the formulae (3.9) and (3.10) are indeed equivalent, provided the way of integration in Eq. (3.10) is fixed, see Ref. 2 for details.

Let us point out finally that the formula (3.9) for $a'_{2\kappa}$ has been derived in Ref. 4, Eq. (8.14), see also Ref. 5, Eq. (6.15), by means of the Hashin-Shtrikman variational principle, having introduced the class of trial polarization fields

$$\mathbf{p}(\mathbf{x}) = \lambda\widetilde{\kappa}(\mathbf{x})\mathbf{G} + [\kappa]\iint h(\mathbf{x}-\mathbf{y}_1)\mathbf{\Phi}(\mathbf{x}-\mathbf{y}_1,\mathbf{y}_1-\mathbf{y}_2)\,\Delta_\psi^{(2)}(\mathbf{y}_1,\mathbf{y}_2)\,d^3\mathbf{y}_1\,d^3\mathbf{y}_2, \quad (3.14)$$

$\widetilde{\kappa}(\mathbf{x}) = \kappa(\mathbf{x}) - \kappa_0$, $\kappa_0 = \kappa_m$, for the latter, with $\lambda \in \mathcal{R}$ denoting an adjustable scalar parameter. The reason is that the class (3.14) contains the actual polarization in the medium to the order c^2. (Note however that in Eq. (8.14) of Ref. 4 the validity of Eq. (3.8a) was not noticed.)

Hence we can conclude finally that the herein proposed application of the variational principles in the theory of random conductivity problems leads in passing to formulae for the effective transport properties that contain absolutely convergent integrals solely.

Acknowledgements. The support of the Bulgarian Ministry of Education and Science under Grant No MM 416-94 is gratefully acknowledged.

References

1. K. Z. Markov, *SIAM J. Appl. Math.* **51** (1991) 172.
2. K. Z. Markov, *Math. Balkanica (New Series)* **3** (1989) 399.
3. D. J. Jeffrey, *Proc. Roy. Soc. London* **A335** (1973) 355.
4. K. Z. Markov and Kr. D. Zvyatkov, In: *Recent Advances in Mathematical Modelling of Composite Materials*, ed. K. Z. Markov, World Sci., 1994, p. 59.
5. K. Z. Markov and Kr. D. Zvyatkov, *Europ. J. Appl. Math.*, in press.
6. M. Beran, *Statistical Continuum Theories*, Wiley, New York, 1968.
7. R. L. Stratonovich, *Topics in Theory of Random Noises*, Vol. 1, Gordon and Breach, New York, 1967.
8. K. Z. Markov and C. I. Christov, *Math. Models and Methods in Applied Sciences* **2** (1992) 249-269.
9. C. I. Christov and K. Z. Markov, *SIAM J. Appl. Math.* **45** (1985) 289.
10. K. Z. Markov, *SIAM J. Appl. Math.* **47** (1987) 831, 850.
11. K. Z. Markov and Kr. D. Zvyatkov, *Advances in Mechanics (Warsaw)* **14**(4) (1991), p. 3.
12. M. Beran, *Nuovo Cimento* **38** (1965) 771.
13. Z. Hashin and S. Shtrikman, *J. Appl. Phys.* **33** (1962) 3125.

PART III

CONTINUUM THEORY OF LIVING STRUCTURES

Continuum Models and Discrete Systems
Proceedings of 8th International Symposium, June 11-16, 1995, Varna, Bulgaria, ed. K.Z. Markov
© World Scientific Publishing Company, 1996, pp. 342–349

SKEW LATTICE CONTINUUM MODEL FOR CANCELLOUS BONE

T. ADACHI, Y. TOMITA
Department of Mechanical Engineering, Faculty of Engineering, Kobe University, Rokkodai, Nada, Kobe 657, Japan

and

M. TANAKA
Department of Mechanical Engineering, Faculty of Engineering Science, Osaka University, Machikaneyama, Toyonaka, Osaka 560, Japan

Abstract. An orthogonal lattice continuum model, which is proposed for cancellous bone with microstructure, is extended to a skew lattice continuum model to describe the trabecular architecture in a more general form. Based on the couple stress theory, a constitutive relation is formulated in terms of structural parameters such as lattice intervals, lattice widths and a skew angle between the lattice elements. We investigate the mechanical properties that depend on the skew angle and the orientation of the lattice. This model could be applied to mechanical bone remodeling considering not only the density change but also the change in the orientation of the trabecular architecture.

1. Introduction

Density and orientation of the trabecular architecture of cancellous bone have characteristic distributions at the remodeling equilibrium, and the architecture is believed to be closely related to the mechanical environment [1]. Based on the observation of its architecture and phenomenological hypotheses, mathematical models of mechanical bone remodeling have been proposed using a conventional continuum which describes the change of the apparent density or porosity [2-4], and the change of trabecular orientation [5]. These models use macroscopic stress and/or strain as remodeling driving forces. However, since the cancellous bone has discrete structure and the remodeling occurs on the trabecular surface [6], it is important to know the stress and strain at the trabecular level. To estimate such trabecular-level mechanical conditions from the continuum analyses, we have employed an orthogonal lattice continuum as a model of cancellous bone [7], and also proposed a model of bone remodeling as a trabecular-level stress regulation process [8]. This lattice continuum model assumes the orthogonal anisotropy of trabecular architecture *a priori* at the remodeling equilibrium. In this study, to express the change in the orientation of the trabeculae in the remodeling process, an orthogonal lattice continuum model is

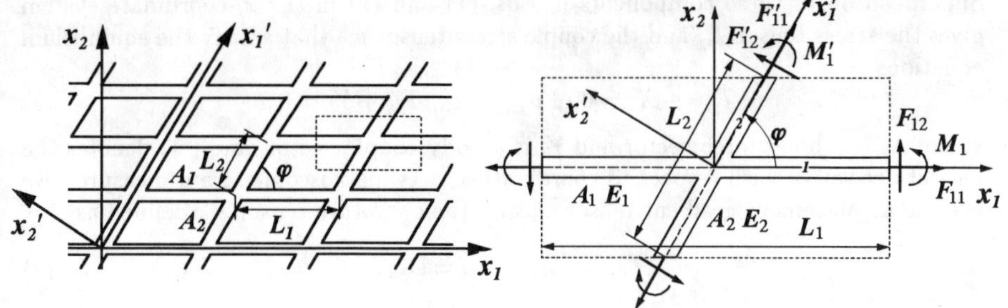

(a) Skew lattice continuum.

(b) Unit skew lattice element.

Fig. 1. Skew lattice continuum model of cancellous bone.

extended to a skew lattice continuum based on the couple stress theory [9]. We investigate the dependence of the macroscopic mechanical properties on the skew angle and the lattice orientation.

2. Skew Lattice Continuum

The skew lattice continuum is a continuous model of a discrete system of linear elastic rod or beam elements interconnected by rigid nodes forming a skew angle. Consider a two-dimensional skew lattice with unit thickness, as shown in Fig. 1a. Axis x_1 of the Cartesian orthogonal coordinate x_i is taken in the axial direction of member 1, and axis x_1' in coordinate x_i' is taken in that of member 2 forming a skew angle of φ $(0 < \varphi < 2\pi)$. Member width, lattice interval, and Young's modulus of member i are denoted as A_i, L_i, and E_i, respectively. Based on the couple stress theory [9], the constitutive relations between macroscopic stress and strain are developed for this skew lattice continuum in this Section.

2.1. Macroscopic Stress and Strain

We denote the components of the macroscopic stress tensor and the couple stress tensor in the x_i-coordinate system as T_{ij} and μ_{i3}, and those in the x_i'-coordinate system as T_{ij}' and μ_{i3}'. Axial force F_{11}, shear force F_{12} and bending moment M_1 acting on member 1, as shown in Fig. 1b, are related to the macroscopic stress and couple stress tensors as

$$T_{11} = \frac{F_{11}}{L_2 \sin\varphi}, \quad T_{12} = \frac{F_{12}}{L_2 \sin\varphi}, \quad \mu_{13} = \frac{M_1}{L_2 \sin\varphi}, \tag{1}$$

and axial force F_{11}', shear force F_{12}' and bending moment M_1' acting on member 2 are related as

$$T_{11}' = \frac{F_{11}'}{L_1 \sin\varphi}, \quad T_{12}' = \frac{F_{12}'}{L_1 \sin\varphi}, \quad \mu_{13}' = \frac{M_1'}{L_1 \sin\varphi}. \tag{2}$$

Superposition of these components in Eqs. (1) and (2) in the x_i-coordinate system gives the stress tensor T_{ij} and the couple stress tensor μ_{i3} that satisfy the equilibrium equations

$$T_{ji,j} + X_i = 0, \quad \mu_{j3,j} + e_{3jk}T_{jk} + Y_3 = 0\,, \tag{3}$$

where X_i is a body force vector and Y_3 is a body couple vector, and $(\ )_{,i}$ denotes the partial derivative with respect to coordinate x_i. As macroscopic strain measures, we use the displacement gradient tensor ε_{ij} and the curvature tensor κ_{3i} defined as

$$\varepsilon_{ij} = u_{j,i}, \quad \kappa_{3i} = \omega_{3,i}\,, \tag{4}$$

where the macrorotation vector ω_3 is related to the displacement gradient as

$$\omega_3 = \frac{1}{2}e_{3jk}u_{k,j}\,. \tag{5}$$

2.2. Deformation of Unit Element

We denote the displacement at the edge point of member i, which is marked with solid circles in Fig. 2, in the direction of the x_j axis as Δ_{ij}, and that in the direction of the x'_j axis as Δ'_{ij}. The rotation of the edge point of member i is denoted as Ω_i. Assuming that each member deforms as a rod with axial stiffness E_1A_1, E_2A_2 and as a beam with bending stiffness E_1I_1, E_2I_2, the axial and shear forces and bending moment acting on each member are related to the deformation as

$$F_{11} = \frac{2E_1A_1}{L_1}\Delta_{11}\,, \quad F_{12} = \frac{24E_1I_1}{L_1^3}\Delta_{12}\,, \quad M_1 = \frac{2E_1I_1}{L_1}\Omega_1\,, \tag{6}$$

$$F'_{11} = \frac{2E_2A_2}{L_2}\Delta'_{21}\,, \quad F'_{12} = \frac{24E_2I_2}{L_2/2}\Delta'_{22}\,, \quad M'_1 = \frac{2E_2I_2}{L_2}\Omega_2\,, \tag{7}$$

where the moment of inertia I_i is expressed as

$$I_i = \frac{A_i^3}{12}\,, \tag{8}$$

and the displacement Δ'_{2j} of member 2 is related to displacement Δ_{2j} as

$$\Delta'_{21} = \Delta_{22}\sin\varphi + \Delta_{21}\cos\varphi\,, \quad \Delta'_{22} = \Delta_{22}\cos\varphi - \Delta_{21}\sin\varphi\,. \tag{9}$$

2.3. Constitutive Model

For the individual uniform macroscopic strains of ε_{11}, ε_{12}, κ_{31}, ε_{21}, ε_{22}, and κ_{32} shown in Fig. 2, the displacements Δ_{11}, Δ_{12}, Δ_{21}, Δ_{22} and rotations Ω_1, Ω_2 of the unit skew element are determined as listed in Table 1. By substituting these displacements Δ_{ij} and rotations Ω_i into Eqs. (6), (7) and (9), the axial forces F_{11}, F'_{11}, the shear forces F_{12}, F'_{12}, and the bending moments M_1, M'_1 can be related to each uniform strain of ε_{ij} and κ_{3i}. By substituting Eqs. (6) and (7) into Eqs. (1) and (2), the stress

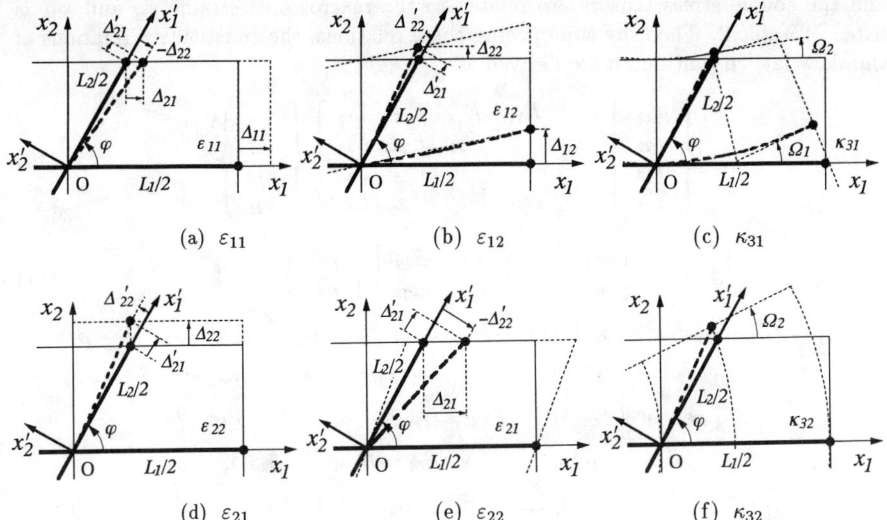

Fig. 2. Deformation of unit lattice under uniform strain.

TABLE 1.

Deformation of unit element under unit strain.

	ε_{11}	ε_{22}	ε_{12}	ε_{21}
Δ_{11}	$L_1/2$	0	0	0
Δ_{12}	0	0	$L_1/2$	0
Δ_{21}	$L_2c/2$	0	0	$L_2s/2$
Δ_{22}	0	$L_2s/2$	$L_2c/2$	0

	κ_{31}	κ_{32}
Ω_1	$L_1/2$	0
Ω_2	$L_2c/2$	$L_2s/2$

$s = \sin\varphi, \quad c = \cos\varphi$

TABLE 2.

Macroscopic stress and couple stress under unit strain.

	ε_{11}	ε_{22}	ε_{12}	ε_{21}
T_{11}	E_1A_1/L_2s	0	0	0
T_{12}	0	0	$E_1A_1^3/L_1^2L_2s$	0
T'_{11}	$E_2A_2c^2/L_1s$	E_2A_2s/L_1	E_2A_2c/L_1	E_2A_2c/L_1
T'_{12}	$-E_2A_2^3c/L_1L_2^2$	$E_2A_2^3c/L_1L_2^2$	$E_2A_2^3c^2/L_1L_2^2s$	$-E_2A_2^3s/L_1L_2^2$

	κ_{31}	κ_{32}
μ_{13}	$E_1A_1^3/12L_2s$	0
μ'_{13}	$E_2A_2^3c/12L_1s$	$E_2A_2^3/12L_1$

and the couple stress tensors are related to the macroscopic strains ε_{ij} and κ_{3i} as listed in Table 2. Then, by superposing these relations, the constitutive relations of the skew lattice continuum are derived as

$$\begin{Bmatrix} T_{11} \\ T_{22} \\ T_{12} \\ T_{21} \end{Bmatrix} = \begin{bmatrix} E_{11} & E_{12} & E_{13} & E_{14} \\ & E_{22} & E_{23} & E_{24} \\ & & E_{33} & E_{34} \\ \text{sym} & & & E_{44} \end{bmatrix} \begin{Bmatrix} \varepsilon_{11} \\ \varepsilon_{22} \\ \varepsilon_{12} \\ \varepsilon_{21} \end{Bmatrix}, \tag{10}$$

$$\begin{Bmatrix} \mu_{13} \\ \mu_{23} \end{Bmatrix} = \begin{bmatrix} \Xi_{11} & \Xi_{12} \\ \text{sym} & \Xi_{22} \end{bmatrix} \begin{Bmatrix} \kappa_{31} \\ \kappa_{32} \end{Bmatrix}, \tag{11}$$

$$E_{11} = \{\alpha_1 + \alpha_2 c^2(c^2 + \beta_2 s^2)\}/s, \quad E_{12} = \alpha_2 s c^2(1-\beta_2), \quad E_{13} = \alpha_2 c^3(1-\beta_2),$$
$$E_{14} = \alpha_2 c(c^2 + \beta_2 s^2), \quad E_{22} = \alpha_2 s(s^2 + \beta_2 c^2), \quad E_{23} = \alpha_2 c(s^2 + \beta_2 c^2),$$
$$E_{24} = \alpha_2 s^2 c(1-\beta_2), \quad E_{33} = \{\alpha_1\beta_1 + \alpha_2 c^2(s^2 + \beta_2 c^2)\}/s,$$
$$E_{34} = \alpha_2 c^2 s(1-\beta_2), \quad E_{44} = \alpha_2 s(c^2 + \beta_2 s^2),$$
$$\Xi_{11} = (\alpha_1 A_1^2 + \alpha_2 A_2^2 c^2)/12s, \quad \Xi_{12} = \alpha_2 A_2^2 c/12, \quad \Xi_{22} = \alpha_2 A_2^2 s/12,$$

where

$$\alpha_1 = E_1A_1/L_2, \quad \beta_1 = (A_1/L_1)^2, \quad s = \sin\varphi,$$
$$\alpha_2 = E_2A_2/L_1, \quad \beta_2 = (A_2/L_2)^2, \quad c = \cos\varphi.$$

In the case of skew angle $\varphi = \pi/2$, constitutive relations in Eqs. (10) and (11) reduce to those of the orthogonal lattice continuum [7].

3. Structural Dependence on Mechanical Behavior

Mechanical behavior of the skew lattice continuum model developed in the previous Section is clarified here in view of the structural dependence. For simplicity, the lattice interval, the member width, and Young's modulus are assumed as $L_1 = L_2 = L$, $A_1 = A_2 = L/4$, and $E_1 = E_2 = E = 10\,\text{GPa}$.

3.1. Dependence of Skew Angle

Dependence of the stiffness matrix E_{ij} in the constitutive relation (10) on the skew angle φ is discussed in this subsection, where E_{ij} is an influence coefficient relating unit strain ε_{ij} to stress T_{ij}.

Fig. 3a shows the dependence of E_{11} and E_{22} on the skew angle φ, that is, stress T_{11} or T_{22} under unit uniaxial strain ε_{11} or ε_{22}, respectively. When members 1 and 2 cross each other at right angle $\varphi = \pi/2$, E_{11} and E_{22} take minimum and maximum values of $E_{11} = E_{22} = EA/L = 2.5\,\text{GPa}$. Since this model assumes that the constant lattice interval L_i does not depend on the skew angle φ, E_{11} increases to infinity and E_{22} conversely decreases to zero as member 2 inclines.

Dependences of E_{14} and E_{24} on the skew angle φ are plotted as solid and dotted lines in Fig. 3b. These components E_{14} and E_{24} are the coefficients relating the unit shear strain ε_{21} to the normal stresses T_{11} or T_{22}. When the skew angle is

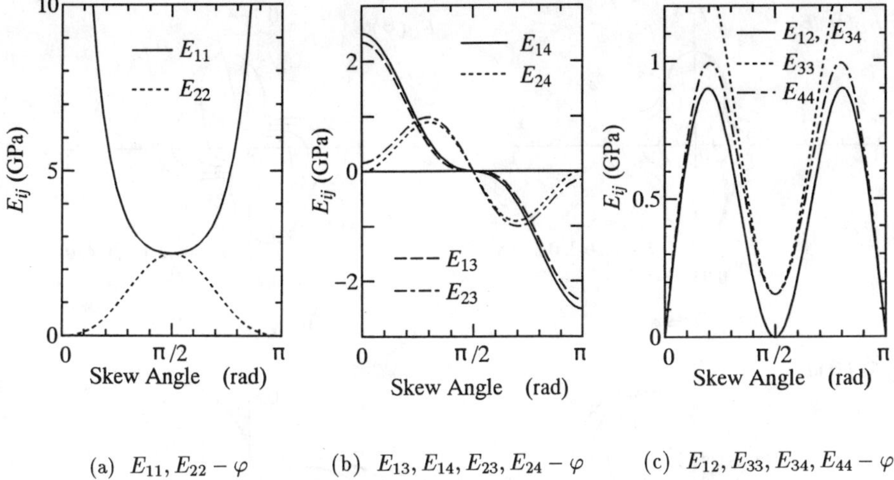

(a) $E_{11}, E_{22} - \varphi$ (b) $E_{13}, E_{14}, E_{23}, E_{24} - \varphi$ (c) $E_{12}, E_{33}, E_{34}, E_{44} - \varphi$

Fig. 3. Dependence of mechanical properties on the skew angle φ.

$\varphi = \pi/2$, coefficients E_{14} and E_{24} vanish. As member 2 inclines, E_{14} and E_{24} increase in the range of $\varphi < \pi/2$ and decrease in $\pi/2 < \varphi$. When φ approaches zero or π, E_{14} reaches ± 2.5 GPa, whereas, E_{24} takes the extremum once and becomes zero. In the same figure, E_{13} and E_{23} are plotted as broken and dash-dot lines, and they take almost the same values as E_{14} and E_{24} in the small range around $\varphi = \pi/2$.

Fig. 3c shows the dependences of E_{12}, E_{34}, E_{33} and E_{44} on the skew angle φ. In this figure, the solid line indicates the E_{12} and E_{34} that have identical values. These components are the reciprocal terms between axes x_1 and x_2, and vanish at $\varphi = 0, \pi/2$ and π. In the same figure, shear moduli E_{33} and E_{44} are plotted as broken and dash-dot lines. When the skew angle is $\varphi = \pi/2$, they take the same value of 0.156 GPa. As member 2 inclines, E_{33} tends to infinity at $\varphi = 0, 2\pi$, and E_{44} increases, reaches the maximum value once, and becomes zero at $\varphi = 0, 2\pi$.

3.2. Anisotropic Mechanical Properties

Dependence of the rotation of the lattice on the elastic modulus is investigated. Components of the elastic modulus $\hat{E}_{ij}(\theta)$ in the $\hat{x}_i$-coordinate system are derived from Eq. (10) by the coordinate transformation, where the angle between x_1 and $\hat{x}_1$ is defined as θ.

Fig. 4 shows the directional dependence of the elastic modulus $\hat{E}_{11}(\theta)$, which is normalized by Young's modulus of the lattice material E, with constant width of member 2 as $A_2/L = 0.5$ and a variety of widths of member 1 as $A_1/L = 0.2$, 0.3, 0.4, and 0.5. Figs. 4a to 4d show elastic modulus ratio $\hat{E}_{11}(\theta)/E$ with different skew

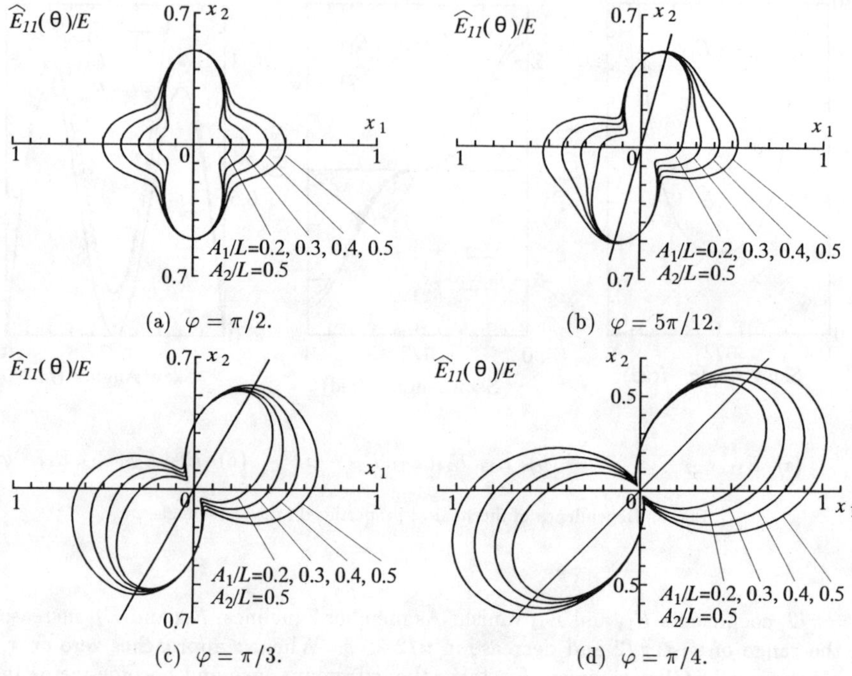

Fig. 4. Directional dependence of mechanical property $\hat{E}_{11}(\theta)$.

angles of $\varphi = \pi/2\,(90°)$, $5\pi/12\,(75°)$, $\pi/3\,(60°)$, and $\pi/4\,(45°)$.

For the orthogonal lattice shown in Fig. 4a, elastic modulus ratio $\hat{E}_{11}(\theta)/E$ takes the maximum value in the axial direction and the minimum value in the oblique direction at around $\theta = \pi/4$ and $3\pi/4$. When member 2 inclines with $\varphi = 5\pi/12$ as shown in Fig. 4b, concavity around $\theta = \pi/4$ becomes vague; conversely, it becomes clear around $\theta = 3\pi/4$. As member 2 inclines further to $\varphi = \pi/3$ and $\pi/4$, as shown in Figs. 4c and 4d, this tendency becomes clearer, and elastic modulus ratio $\hat{E}_{11}(\theta)/E$ in the range of $0 < \theta < \varphi$ becomes larger than that in the axial direction. In the case of $A_1/L = A_2/L = 0.5$ and $\varphi = \pi/3, \pi/4$, elastic modulus ratio $\hat{E}_{11}(\theta)/E$ takes maximum and minimum values at $\theta = \pi/6$ and $\theta = 2\pi/3$, respectively. As the member width ratio A_1/L decreases from 0.5 to 0.2, elastic modulus ratio $\hat{E}_{11}(\theta)/E$ decreases monotonically; however, it takes a constant value in the axial direction of x_2, since $\hat{E}_{11}(\varphi/2)/E$ is independent of A_1/L.

4. Conclusions

A skew lattice continuum model of cancellous bone is proposed as an extension of the orthogonal lattice continuum model. The constitutive relation is formulated based on the couple stress theory in terms of structural parameters such as lattice intervals, lattice widths and a skew angle between the lattice elements, in which it is assumed that this continuum is a continuous model of a discrete system of linear elastic rod or beam elements interconnected by rigid nodes making a skew angle. Structural dependence of the proposed model is investigated in view of the effect of skew angle and anisotropy. This extension to a general form enables us to permit the nonorthogonality of the trabecular architecture in the remodeling process, including its orthogonality at the remodeling equilibrium. This model can be applied to further model development of mechanical bone remodeling considering not only the density change but also the change in the orientation of the trabecular architecture.

Acknowledgements. This work was financially supported in part by Grants-in-Aid for Scientific Research on Priority Area [Biomechanics] (Nos. 04237101 and 06213225) and for Encouragement of Young Scientists (No. 06855015) from the Ministry of Education, Science and Culture, Japan.

References

1. J. Wolff, *The Law of Bone Remodeling*, Translated by P. Maquet and R. Furlong, Springer, Berlin 1986.
2. S. C. Cowin and D. H. Hegedus, *J. Elasticity* **6** (1976) 313.
3. D. R. Carter, D. P. Fyhrie and R. T. Whalen, *J. Biomech.* **20** (1987) 785.
4. R. Huiskes, H. Weinans, H. J. Grootenboer, M. Dalstra, B. Fudala and T. F. Slooff, *J. Biomech.* **20** (1987) 1135.
5. S. C. Cowin, A. M. Sadegh and G. M. Luo, *Trans. ASME. J. Biomech. Eng.* **114** (1992) 129.
6. S. C. Cowin, L. Moss-Salentijn and M. L. Moss, *Trans. ASME. J. Biomech. Eng.* **113** (1991) 191.
7. T. Adachi, Y. Tomita, and M. Tanaka, In *Proc. 37th Jpn. Cong. Mat. Res.*, 1994, p. 215.
8. T. Adachi, Y. Tomita and M. Tanaka, In *Advances in Bioengineering–1994*, ed. M. J. Askew, **BED-28**, 1994, p. 255.
9. W. T. Koiter, *Proc. K. Ned. Akad. Wet.* **B-67** (1964) 17.

Continuum Models and Discrete Systems
Proceedings of 8th International Symposium, June 11-16, 1995, Varna, Bulgaria, ed. K.Z. Markov
©World Scientific Publishing Company, 1996, pp. 350–368

WHY DOES SKIN STAY SMOOTH? THE DYNAMICS OF EPIDERMIS IN STATISTICAL EQUILIBRIUM

N. RIVIER and B. DUBERTRET
Laboratoire de Physique Théorique, Université Louis Pasteur,
3 rue de l'Université, F-67084 Strasbourg, France

Abstract. The "molecular dynamics" of cells in the epidermis of mammals is presented by generalizing a model of grain growth in polycrystals or coarsening foams due to Telley. The tissue remains in statistical equilibrium in spite of, and because of constant renewal through local elementary topological transformations, which are division (mitosis), neighbour exchange and detachment of the cells. The global evolution and response of tissues can now be described realistically, geometrically, with very few parameters and little biological input. The dynamics of cell division is modelled for the first time. This is continuous mechanics of a discrete, space-filling system (the cells).

1. Introduction

The epidermis of mammals is in a steady state throughout their life, in spite of the fact that cells constantly divide and die. This steady state can be viewed as the statistical equilibrium of an assembly of cells filling space at random. Cells explore configuration space through topological "collisions," which are here division (mitosis) and detachment from the basal layer. Death (or apopthose) occurs on the superficial, corneum layer. Statistical equilibrium can be locally upset by injury or by some proliferating skin diseases like psoriasis, and it is a property of the random structure and architecture of the epidermis that it can react efficiently and rapidly to heal the injury.

Cell division occurs on the basal layer, a corrugated two-dimensional surface. The random variables associated with each cell on the basal layer are its size A and n, the number of its neighbours. The variables n and A are shuffled when the cell or one of its neighbours divides or leaves the basal layer.

The epidermis is, in general and at the lowest level of sophistication, a fluid of cells, filling space at random and transiting from the 2D basal layer where they are born (through mitosis) to the corneum layer where they die. Each layer is in statistical equilibrium. This is the realm of statistical mechanics. The macroscopic state of the epidermis is governed by entropy, which here plays roles both as control of heat transfer (the equilibrium state is one of maximum entropy) and as a quantity of information (through prior probabilities on the basal layer, affecting, for example, the way a m-sided cell divides into two daughters with d and $m + 4 - d$ sides, or the rates in the mitotic cycle).

At this low level of sophistication, it is important that the external conditions can be modified locally, so that the statistical equilibrium is affected and tested. As with ordinary fluids or gases, it is most easily done by "molecular" dynamics, and by that we do not mean only numerical calculation of the trajectory of a molecule, but also the classical definition of pressure and temperature by Bernoulli and Maxwell.

Here, we need to define the "molecules" describing cells partitioning space at random and capable of dividing. This last requirement is crucial, and it is the subject of this paper. We will generalize only slightly a model introduced by Telley [20] to describe the coarsening of a polycrystal or a soap froth.

Like biological tissues, polycrystals and soap froths fill space at random with polyhedral cells. Their evolution is slightly different, however. They are not invariant statistically, but coarsen very slowly (a metallurgical process called Ostwald ripening). The larger grains, cells or bubbles grow at the expense of smaller ones, which ultimately disappear. The topological transformation responsible for coarsening is thus the smooth disappearance of a triangular (in 2D) or tetrahedral (in 3D) cell leaving behind a single vertex of the froth. If the molecules are the cells, the model should allow for a variable number of cells, and give a dynamical mechanism for their disappearance in polycrystals. This is what Telley [20] (Telley *et al.* [21]) succeeded in doing, by generalizing the ancient space partition of Gaultier and Laguerre. Here, we will generalize Telley's model further to include cell division.

The language used in this paper is that of continuous mechanics, not of biology. Given that cellular division is the main local, elementary topological transformation responsible for the structure of the epidermis and for its evolution (its stationary state), and that cellular division creates a pair of dislocations (the natural version of the Frank-Read source) or makes dislocations climb, why is it that the tissue remains the same, that the dislocation and its motion is neither seen nor felt? Because the tissue is random.

In an ordered structure, a local, topological defect would be immediately noticeable and the source of long-ranged stress. By contrast, a topological change ("defect" is a misnomer, even though cell division creates a pair of dislocations (Fig. 1)) does not disarrange a random structure, and its stress is screened by randomness. Tissues and froths belong to the same statistical ensemble, fully explored through topological transformations. Macroscopically, structure and distribution of cell shapes are invariant under local topological fluctuations. They are fixed points of the evolving and fluctuating tissue.

Cell division occurs because the tissue grows. "The appreciation that cells are polyhedral figures came with the very first histological report ever published" (Dormer [4]). But why should the structure of biological tissues be identical to that of a soap froth or a metallurgical aggregate, which have a large energy concentrated in the interfaces (surface tension or grain boundary energy)? What is the driving force for cellular division? A tissue, froth in statistical equilibrium, is only metastable, and requires interplay between growth, detachment and cellular division to remain in a steady state. By contrast, soap froths or metallurgical aggregates coarsen slowly (large grains swallow up smaller ones), because they are driven by topological transformations only, without growth.

We will also investigate the energetics of the physical process of cell division, given that there is a biological input (growth (stage S of the mitotic cycle) and elongation (stage G_2) of the nucleus). A general introduction can be found in CMDS7 Proceedings (Rivier [15]) or in Rivier [16]. Most of the work presented here is (to be) published (Rivier and Dubertret [18], Rivier, Schliecker and Dubertret [19]).

Thus, the epidermis can be studied as a problem in continuous mechanics. Almost all the topological activity is located on the basal layer, which is a corrugated, random, two-dimensional froth. Its elasticity and plasticity are given by the generation of dislocations and their climb, through cell division.

2. A Brief Review of the Architecture and the Evolution of Mammal Epidermis

The epidermis is divided (Montagna and Parakkal [11]) into an inner layer of viable cells (stratum Malpighi) and an outer one of anucleated dead cells (stratum corneum).

The innermost part of the stratum Malpighi, called the basal layer, is a one-cell deep structure in contact with the dermis. This layer can be regarded as a 2D topological foam. Since it is (almost) the only place where cells divide, it can be assumed that the entire dynamic evolution of the tissue is driven by the evolution of the basal layer.

The renewal of the tissue (in a steady state which depends on local external constraints) can be understood through the life of every basal cell. Three stages can be discerned: (1) mitotic cycle, (2) resting cells (G_0), (3) differentiation (detachment). The mitotic cycle is itself divided (biochemically) into 4 phases, G_1, S, G_2 and M. Only S, G_2 and M are geometrically relevant. (See Fig. 7 below.)

As seen by a physicist, basal cells are kept in contact with the basal membrane because of a downward pressure applied by the stratum corneum, and also partly because of surface adhesion properties (due to desmosomes). Division of the basal cells is controlled by the applied lateral pressure (the stronger the pressure, the less division occurs) and by their size (the smaller the size of the cell, the more it can divide (Barrandon and Green [1]). Above a certain size (which depends on the lateral pressure), the cell is pushed away from the 2D layer by its neighbours and begins its climb and differentiation in the Malpighi layer. Its ascension ends by a flattening in the stratum corneum. A cell which neither divides nor grows is called a stem cell. It is in limbo and can, later, either re-enter the mitotic cycle or differentiate. An overwhelming proportion of the basal cells are stem cells. They constitute a stock of cells, immediately available if needs be, e.g., to heal a wound.

The curvature of the basal membrane depends on the mitotic rate. Slow rates are associated with flat basal layer whereas higher mitotic rates betray a very corrugated basal layer (Montagna and Parakkal [11]). This is yet another trick used by Nature to obtain a better production per unit of epidermis' area, and to ensure that the directions of successive divisions are well randomized, to avoid long scars or appendices (see Figs. 1 and 2) (Dubertret [5], Rivier and Dubertret [18]).

3. Topological Representation of Biological Tissues. Cell Division

A 2D topological froth is a random space-filling cellular structure made of cells bounded by edges or interfaces and vertices. The numbers of cells, edges and vertices are denoted as C, E and V respectively. The number of cells meeting at a vertex is denoted by z. Unless the tissue is adjusted, we have exactly $z = 3$. The random variable is n, the number of sides of a cell. Its average $\langle n \rangle = 6$ is fixed. Also, $C = E/3 = V/2$, so that cells are the least numerous topological elements of the tissue. They are the bodies of the tissue as a many-body problem, the molecules of our statistical gas, parameterized by their size A and the number of their sides n, with a distribution p_n.

Froths are dynamical systems exploring the entire configuration space through elementary topological transformations, i.e., local topological fluctuations, which play the part of collisions in our topological gas. In principle, there are two elementary topological transformations in 2D (Weaire and Rivier [23]): neighbour switching and cell disappearance (or cell division (mitosis)).

In biological tissues only cell division occurs, because intercellular membranes, which just after division are fluid (like the interfaces in soap froths), later become rigid (Thompson [22], Errera [6]). Neighbour-switching would require fluid interfaces which can be shrunk. Apart from topological rules, the precise nature of the microscopic, biological process of mitosis is irrelevant. All that matters is that it exists, and that it shuffles the degrees of freedom locally. This is exactly like collisions in gases. The end product is a stationary distribution p_n and topological correlations (between neighbouring n- and k-cells).

The numbers of sides of the cells directly involved in the division are related,

$$m + 4 = d_1 + d_2 \,, \tag{1}$$

where m and d_i denote the numbers of sides of the mother and daughter cells, respectively. It follows that daughters should have at least 4 sides, otherwise division would increase the topological size of the cell, and go against the general trend (whereby discrete mitoses counter-balance continuous growth to leave the tissue statistically invariant). There are two other cells involved in the mitotic process, at both ends of the dividing membrane. They gain one side each. There are no triangular cells in 2D biological tissues. (Except, fleetingly, when a cell is about to detach from the basal layer: The cell sheds some of its sides by a succession of rapid neighbour-switches, as its nucleus is pushed away from the basal layer (Dubertret [5], Rivier and Dubertret

[18])). Mitosis has added one cell, three edges and two vertices to the froth, or six edges to the additional cell (since each edge separates two cells).

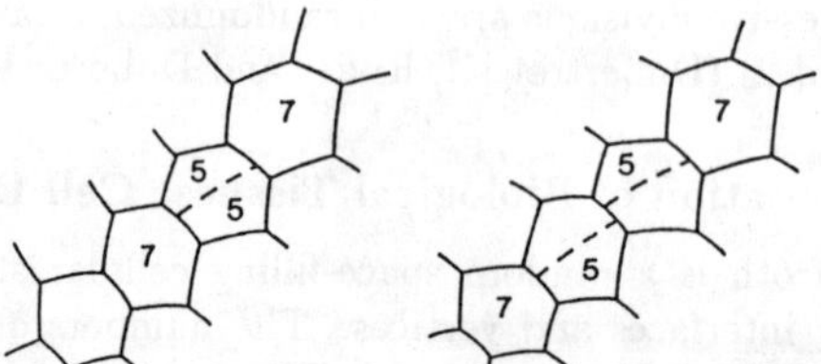

Fig. 1. Climb of dislocation by successive cellular divisions. On the left, a pair of dislocations (each consisting of a 5- and a 7- sided cell) results from the division of a hexagonal cell. On the right, a second division makes the two dislocation climb away from each other, leaving in their wake an additional layer of cells. Unlabeled cells are hexagonal.

Topological transformations change the number of sides of the cells involved. An hexagonal tissue, even random, is flat. It is in dynamical equilibrium (sides at 120°). A 6-sided cell is neutral. A 5- (7-) sided cell is a positive (negative) disclination, produced in an hexagonal lattice by cutting out (adding) a wedge and reglueing. The plane buckles into a cone (saddle), and the pentagon (heptagon) is a source of positive (negative) curvature. A dipole pentagon-heptagon is a (topological) dislocation. Accordingly, there is an elastic energy associated with local cellular configurations.

If one regards the tissue as an elastic solid, cell division is the source of two dislocations (Fig. 1) which can climb apart through further divisions. Cell division is therefore a local source of plasticity, indeed, it is responsible for the invariant growth of the tissue. Dislocations play the parts they do in elasticity, and a few more, but they do it locally, without cut or need for shoving material in from infinity. Their climb is essential in biology.

We see immediately that successive cell divisions occur randomly. If they were systematic, the tissue would show a straight scar of daughter cells (Fig. 1) or grow an appendix (Fig. 2). This randomization is the reason for the corrugation of the basal layer. Corrugation introduces points of negative curvature, where geodesics (the directions along which successive cell divisions tend to occur) are mixed. The more curved the basal layer, the more local its response can be. The resulting tissue is very random: shape correlations do not extend beyond nearest neighbour cells.

Figures 1 and 2 are *local* and internal means of adding material, mechanism by which our intestine grows (Pyshnov [13]). Gastrulation in embryology is an appendix growing inwards on a sphere. Mitoses are controlled by external forces (lateral pressure on the basal layer) such as growth in a tissue, and they can be polarized (as in Fig. 2) by positive local curvature.

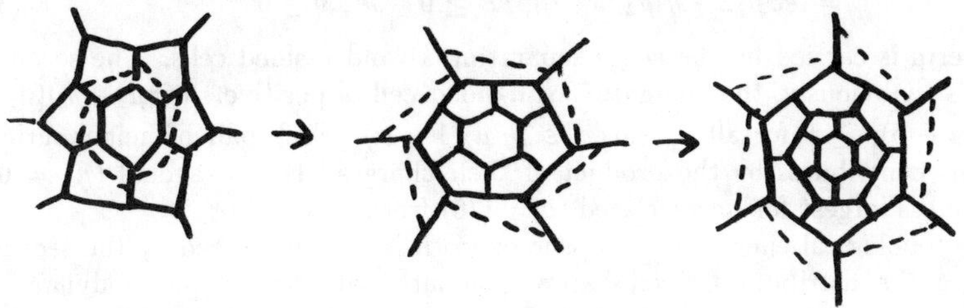

Fig. 2. Amplification of local curvature: Growth of an appendix by successive mitoses.

4. Topological Energy

In a froth, the energy is carried by the interfaces E. There are three (topological) contributions:

1) Total interfacial length.

2) Topological curvature: Disclination self-energy

3) Topological correlations: Two disclinations of opposite sign bind as a dislocation.

We also require that the topological energy is always positive and only vanishes for a hexagonal lattice (no topological defects).

A cell in 2D is described by the number n of its sides. Topological correlation is best expressed by the symmetrical coefficient $N_{kn} \geq 0$, defined through the number $C_{pk} N_{kn} p_n$ of pairs of neighbouring k- and n-sided cells. Maximum entropy (Peshkin, Strandburg and Rivier [12], Delannay, Le Caër and Khatun [3], Rivier [13])) implies that

$$N_{kn} = (k-6)\sigma(n-6) + (n+k-6) \tag{2}$$

is linear in n. Here $\mu_2 = \langle (n-6)^2 \rangle$, and σ is a structural parameter. A topological gas has $\sigma = 1/6$ and $N^0_{kn} = k(1/6)n$ factorizes (uncorrelated cell shapes). Smaller cells surrounding larger cells imposes $\sigma < 1/6$. In Nature, $\sigma < 0$.

In most froths, interfaces carry the energy (surface tension or grain boundaries). There is an elastic contribution to the energy, in that disclinations deform the tissue, and two neighbouring disclinations of opposite sign bind as a dislocation, a defect costing less energy. An n-sided cell is a disclination of "charge" $(6-n)$ in an elastic medium, it has self-energy U_{self}/E and a correlation energy U_{corr}/E (Rivier, Schliecker and Dubertret [19]), per edge,

$$U_{\text{corr}}/E + U_{\text{self}}/E = \varepsilon \sum_{n,k} p_n p_k (n-6) N_{kn} (k-6) + \eta \sum_n p_n (n-6)^2$$

$$= \sum_n p_n \left[(\varepsilon\sigma\mu_2 + \eta)(n-6)^2 + \varepsilon\mu_2(n-6) \right]$$

$$= (\varepsilon\sigma\mu_2 2 + \eta)\mu_2 = U\mathrm{top}/E \geq 0\,, \quad \varepsilon\,, \, \eta \geq 0\,. \tag{3}$$

The first term is carried by the edges separating k- and n-sided cells. The second, by the cells. Obviously, the energy of an n-sided cell is positive, $\left[(\varepsilon\sigma\mu_2 + \eta)(n - 6)^2 + \varepsilon\mu_2(n-6)\right] \geq 0$ for all n, thus $\eta/\varepsilon > \mu_2(1-\sigma)$. Each pair of neighbouring disclinations contributes by the product of their charges. For hexagonal ($\mu_2 = 0$) $U_{\mathrm{corr}} = 0$ and is largest for uncorrelated ($\sigma = 1/6$) froths.

The topological energy of the tissue or froth is thus measured by the second moment μ_2 of the distribution of cell shapes. Standard statistical (or thermodynamic) inference (maximum entropy) yields a relation between μ_2 and p_6 as structural equation of state. This relation is the equivalent of the virial expansion in real gas. It has been found to be universal in froths by Lemaitre *et al.* [9]. Its form and universality has been discussed elsewhere. (See, e.g., Rivier, Schliecker and Dubertret [19].)

The energy (3) enables us to investigate how cell division proceeds in random tissues, and to define geodesics on the structure, i.e., directions along which successive divisions occur at no additional energy cost (or gain). These geodesics are also the geodesics of the manifold supporting the froth.

Consider the change in the energy of the cells involved in the cellular division, without taking into account the details of $k - n$ correlations. The cells involved are the dividing cell (m sides), the daughters (d_1 and $d_2 = m + 4 - d_1$ sides) and the two neighbouring cells affected by the division which change from a, b sides to $a+1$, $b+1$ after division. The difference in energy (second expression of Eq. (3)) is

$$\Delta U_{\mathrm{top}}/E = 2(\varepsilon\sigma\mu_2 + \eta)[-(d_1 - 4)(m - d_1) + a + b - 9]\,. \tag{4}$$

The geodesics are defined by $\Delta U_{\mathrm{top}}/E = 0$. Obviously, they trace the path of the successive divisions described in Figs. 1 and 2: Fig. 1, straight climb of the dislocation as a long scar of softer, new tissue in an ordered, hexagonal tissue, with only the initial division costing energy. Fig. 2, growth of an appendix in a region of positive curvature (a ring of 6 pentagons), with the energy cost of the first division recovered at the sixth. These paths are indeed the geodesics of the substrate. Otherwise, small cells a, b attract the new dividing edge, and large cells repel it. Also, large cells tend to lose their energy by dividing spontaneously (and offset the steady state). Symmetric division costs the least energy. Given that $a + b \leq 12$ (otherwise the cell is likely to gain energy by dividing in the direction of 6-sided neighbours), only 6-, 7- and 8-sided cells can divide and keep the topological energy constant. This is indeed what happens in cucumber, in the absence of cell detachment (from the basal layer) which modifies these conclusions (Rivier, Schliecker and Dubertret [19]).

To avoid long scars of soft tissue, or growth of appendices, geodesics must be mixed. This is why the basal layer is corrugated. The hyperbolic points (the necessary saddle points) mix the geodesics and ensure that the tissue remains invariant, without scars or appendices, and random.

In summary, disclinations cost elastic energy, less if they bind as dislocations. Cell division creates a dipole of dislocations (a defect of very low energy, with two neighbouring dislocations of opposite sign), or makes a dislocation move (climb), at no cost of topological energy if the climb is along a geodesic of the froth substrate (as in Figs. 1 and 2). Motion in any other direction creates additional defects and is much more costly in topological energy.

If defects move at no energy cost along geodesics, it is essential that these geodesics are scrambled in the froth.

If defects can move at little energy cost along geodesics, it is essential that these geodesics are scrambled in the froth. This happens at hyperbolic points of the substrate, at the saddle-points of an egg-carton. It is only in the presence of hyperbolic points that the tissue remains random and invariant. Invariance of the tissue, and absence of scars or appendices is the probable reason for the corrugation of the basal layer and also, of its randomness. This contention is still to be proven biologically.

This is as far as one can go with topology alone. Now for the cell dynamics.

5. The Gas of Paraboloids

A tissue is a many-body problem, but what are the elementary bodies? Vertices, edges, or cells? They are connected by adjacency, incidence or boundary, so that they are not independent degrees of freedom.

Since froths and tissues have, at first sight, identical structure and similar evolution, we can look for the simplest model of a froth as a gas of *independent identical* bodies, whose "collisions" are the elementary topological transformations (ETT) needed to establish statistical equilibrium. Independence and identity is achieved simply by representing a geometrical partition of space (the radical froth (Gellatly and Finney [8], Fischer and Koch [7])) in a space of one extra dimension (Telley [20]). The independent bodies are the C grains or cells. They are the sparsest objects in the structure. Thus, in a tissue of C resting cells, there are $(D+1)C$ degrees of freedom (the position and size of the cells represented by their nuclei). Topological considerations are essential to explain the general structure and evolution of the epidermis, so that discrete topological transformations should show up dramatically in continuous dynamics of our fluid of cells.

The simplest way to generate a space-filling, random froth, is by Voronoi construction. One begins with a Poisson distribution of points, which serve as "nuclei" for the cells. A Voronoi cell contains all points in space nearest to its nucleus, and interfaces are perpendicular bisectors between nuclei. One obtains a froth (because the perpendicular bisectors of the sides of a triangle formed by three seeds are concurrent) very simply, but it has an unrealistic structure (the cells are very anisotropic, albeit convex polygons (in two dimensions)). Worse for our purpose is the fact that the number of nuclei, hence of cells, is fixed from the beginning, so there can be no coarsening or no cell division, and no evolution (only relaxation). The essential physics has been lost.

A simple generalization of the perpendicular bisector produces a froth which is both realistic and capable of evolution. Consider circles (in two dimensions) instead of points as nuclei, and define the *distance* $d(\mathbf{X}, G)$ of a point $\mathbf{X}$ to a circle $\Gamma(r, \mathbf{x}_0)$ of radius r centered at $\mathbf{X}_0$ (hypersphere if $D > 2$), as the length of the tangent to Γ through $\mathbf{X}$ (d^2 is the power of $\mathbf{X}$ with respect to the circle Γ),

$$d^2(\mathbf{X}, \Gamma) = (\mathbf{X} - \mathbf{x}_0)^2 - r^2 . \tag{5}$$

It is negative if $\mathbf{X}$ is inside Γ. The locus of points equidistant to two circles is a straight perpendicular line, the *radical axis* (hyperplane if $D > 2$). Like the perpendicular bisectors in a triangle, the three radical axes between three circles are concurrent (because the radical axis is an equivalence relation between two circles; the point of intersection of two radical axes is equidistant to the three circles and belongs therefore to the third radical axis): They are therefore the interfaces of a (radical or Laguerre) froth (Fischer and Koch [7], Gellatly and Finney [8]). The larger the circular nucleus, the larger its corresponding cell in the froth. Laguerre's froth reduces to Voronoi's if all the nucleus circles have equal radii. Cells are isotropic convex polygons, looking realistic already with small fluctuations in circle radii. Cell disappearance or division is easily accommodated, as we shall see.

5.1. The Horizon of Paraboloids. ETT

The radical axis between two circles of fixed centers and radii r_1 and r_2 depends on $r_1^2 - r_2^2$ only. This gave Telley [20] the idea of representing nuclei as identical paraboloids in one extra dimension, intersecting physical space (here basal layer) as the Laguerre circles (Fig. 3). The level of the physical (hyper)plane is arbitrary. The paraboloids are the bodies of our problem, each specified by the D+1 coordinates $\mathbf{q} = (\mathbf{x}, z)$ of their apex. $\mathbf{x}$ is the coordinate of the circle's centre in the D-dimensional physical space, and z, the height above physical space, measures the circle radius, $z \sim r^2$, or, roughly, the cell size. Laguerre's froth has $3C$ coordinates $(\mathbf{x}, z)$ in 2 dimensions, or $(D+1)C$ coordinates in D dimensions. (Note that the froth has $V = 2C$ vertices, which would have required 4C coordinates if chosen independently). Generalization to three or more dimensions is straightforward, as is the analysis of a section of the froth (a Laguerre froth itself (Rivier [14]), important in the analysis of 3D polycrystalline aggregates in metallurgy.

The set of seeds now looks like a Yosemite-type mountain profile (Fig. 4). A paraboloid below the horizon (seed too small) does not generate a cell in the froth. Conversely, a paraboloid which is too high (seed too large) obscures smaller paraboloids nearby and gobbles up their representative cell. So, local topological transformations on the basal layer can be visualized simply: A cell detaches from the basal layer (T2) when its paraboloid is pushed below the horizon; when a new paraboloid rises above the horizon, it divides the cell containing its apex. The much faster neighbours exchanges (T1) can occur when nearby nuclei move horizontally rather than vertically. ETT are therefore naturally induced by orthogonal motions of the bodies: T1 by horizontal ($\mathbf{x}$) fast motion, T2 or cell division by vertical (z) slow motion.

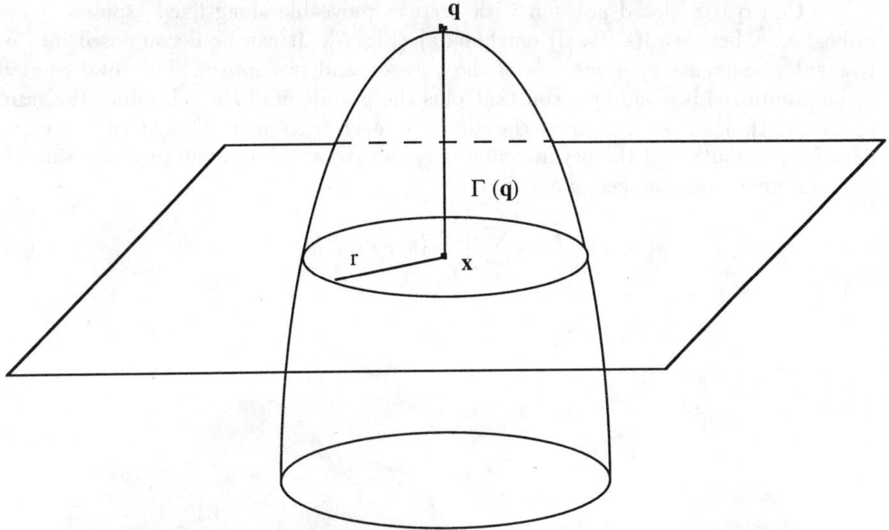

Fig. 3. The seed of a cell (a circle) as a paraboloid (Telley [20]). The physical space is horizontal. The Laguerre froth generated is independent of the level of the physical "floor."

5.2. *Energy and Slow Dynamics*

We can now compute the forces on the cell paraboloids, given that the energy U is proportional to the total interfacial length of the froth, as in soap froths (surface tension) and in metallurgical grain aggregates (grain boundaries). Let us move one seed paraboloid **q**, while keeping fixed the other paraboloids. When **q** moves, it displaces the interfaces and vertices of its own cell. Hence it changes the interfacial length and thus the energy of the froth. This defines a force

$$\mathbf{F} = -\operatorname{grad} U = \lambda_{x/z}\, \mathbf{v}\,, \tag{6}$$

on **q** and generates its dynamics (viscous, with time scales λ_x and λ_z respectively). Only interfaces between **q** and its topological neighbours are affected by the motion (and interfaces between neighbours have only their lengths affected, while their directions are unchanged). The force between paraboloids is therefore very short-ranged, and its range is only topological (it affects neighbours) rather than metric.

Paraboloid **q** can move up and down (growth or shrinkage of its grain), and sideways in physical space (mechanical equilibration of a diffusionless froth). Physically, the time constants for the two motions are very different, $\lambda_z \gg \lambda_x$. The seed relaxes rapidly to the centre of gravity of its cell in physical space **x**, whereas coarsening (motion in z) occurs at a much slower rate λ_z^{-1}. This difference in equilibration rates is familiar in thermodynamics: Two different gases separated by a piston equalize their pressure (mechanical equilibrium) much before their temperature (thermal equilibrium).

Cell $\mathbf{q}$ is a closed polygon with vertices moveable along fixed "spokes" (the radical axes between its (fixed) neighbours) (Fig. 5). It can be decomposed into n triangular segments with one side of the polygon and two spokes. The total energy to be minimized is given by a constant plus the perimeter of the cell minus the sum of the length of the spokes, or by the sum, over each triangular segment, of the outer edge l (contribution of the perimeter) minus half (to avoid overcounting) the sum of the two inner sides (spokes) a and b, say,

$$U = \sum_{n} [l - (a+b)/2] . \tag{7}$$

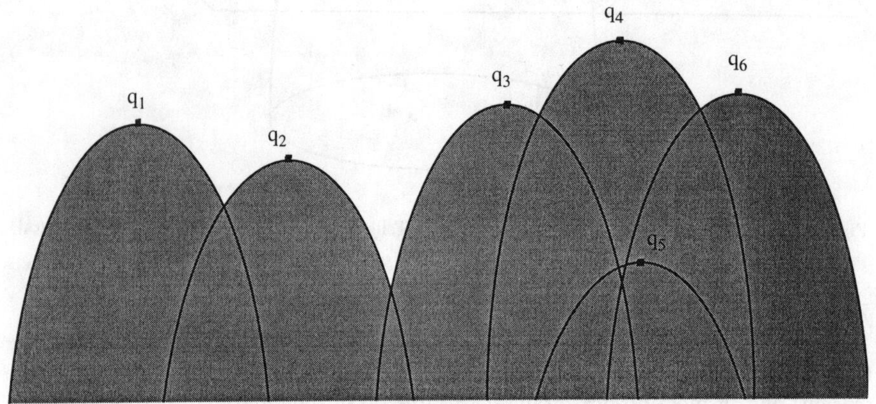

Fig. 4. The "Yosemite" horizon of paraboloids. The horizontal axis represents the physical plane. $\mathbf{q}_5$ is below the horizon and does not generate a cell in the froth. From Telley [20].

Consider equilibration (motion in $\mathbf{x}$) first. The z coordinate of the paraboloid is a suitable measure of the area of its cell. Fixing z will keep the area of its cell constant. Each term in the sum (3) is minimum at constant area when $a = b$, i.e. when the triangular segment is isosceles. This requirement is a sufficient condition for closure, which can be achieved without contradiction for all edges of a triangular cell (given that its vertices are on the radical axes of surrounding cells). (See Fig. 6 and Section 6.1). For an arbitrary cell with $n > 3$, this is not quite possible in general, without also displacing the apices of the surrounding paraboloids (local anneal). The small polygon of Fig. 5 has all its vertices very nearly on a circle, and the spokes meet at its centre. The system is equilibrated (relaxed) in physical space and the Laguerre froth is in mechanical equilibrium.

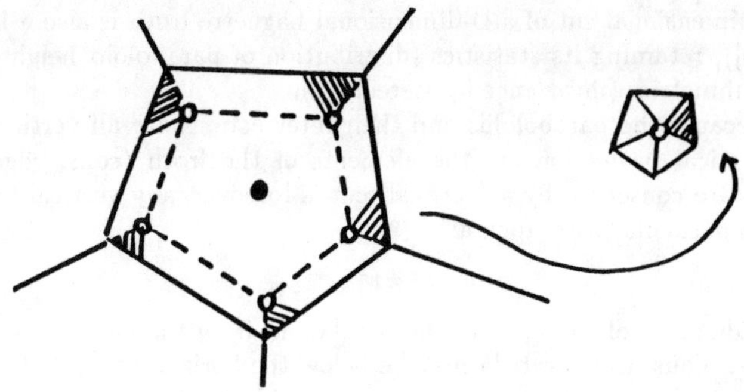

Fig. 5. Vertical motion of the central paraboloid (in mechanical equilibrium) induces parallel translation of the interfaces of its cell (left). The energy balance is the difference between the perimeter and the length of the spokes of an infinitesimal polygon (right).

Consider now vertical motion of some $\mathbf{q}_0$, all other paraboloids remaining fixed. The outer edges will be displaced parallel to themselves (since the $\mathbf{x}$ position of every paraboloid, which fixes the direction of the radical axes, remains unchanged). The energy difference dU associated with the vertical motion dz_0 of seed 0 alone, keeping $\mathbf{x}_0$ and all the other seeds $(\mathbf{x}_i, z_i)$ fixed, is proportional to the change in interfacial length when cell 0 is blown up by parallel displacement of its interfaces (Fig. 5). dU is represented geometrically by an infinitesimal polygon (with all its vertices on a circle since $\mathbf{x}_0$ is equilibrated) with n sides and n spokes, as

$$dU = \gamma\left[\text{perimeter} - \sum_n (\text{length of the spoke})\right], \tag{8}$$

so that if the perimeter $> \sum(\text{spoke})$ (as is the case if $n < 6$), the cell shrinks, whereas it grows if $n > 6$. Moreover, mechanical equilibrium implies that all spokes have the same length dR. One recovers, to a good approximation, slow coarsening through von Neumann's law $F_z = -dU/dR = \gamma(6-n)$; γ is proportional to the interfacial tension.

The same argument can be made in any dimension (except 1D, where coarsening can only be driven by coalescence). In biological tissues, cellular diffusion (Section 6.2) opposes coarsening, and the tissue is in statistical equilibrium.

5.3. Dimensional Reduction. Stereology

In our model, the dynamics of the epidermis is driven by cell growth and division in the two-dimensional basal layer: the nucleus of a cell in the latency phase, or of one about to detach from the basal layer, is pushed away by the lateral pressure. The basal layer, a cut of a Laguerre froth, is a Laguerre froth.

A (D-1)-dimensional cut of a D-dimensional Laguerre froth is also a Laguerre froth (Rivier [14]), retaining its statistics (distribution of paraboloid heights $P(z)$), geometry and symmetries (invariance by stereology).

This is because the paraboloids and their intersections are all vertical (along z), and their vertical projection are the elements of the froth (cells, edges, etc). These properties are conserved by a (vertical) cut. Moreover, any vertical (z) cut of a paraboloid is a parabola, with apex at

$$z' = z - kx_{\perp}^2 , \qquad (9)$$

where $x_{\perp}$ is the distance of the apex of the paraboloid from the cut (along $x_{\parallel}$) and k is its curvature. Thus, the parabola may be below the horizon in $x_{\parallel}$ if the apex of its original paraboloid is very far away from the cut (Mont Blanc, even though very high, does not affect the vertical gradient on the Champs-Elysées).

As we have seen, the dynamics of a Laguerre cell consists of two, independent motions with different time scales: Fast mechanical equilibration (motion in physical space $\mathbf{x}$) followed by slow evolution (motion in z). Consequently, the cell distribution $P(x, z)$ factorizes as the product of distributions of cell size $P(z)$ and of the position of their centre $\mathbf{x}$. But the latter also factorizes in distributions in $x_{\parallel}$ and $x_{\perp}$, since a random froth has no special cut direction (in a random froth all cuts are random (irrational)). Thus, $P'(z')$ also factorizes (in distributions of z and of $x_{\perp}$), hence $P(z)$ must be exponential. It also implies that $P'(z')$ is proportional to $P(z)$, and that the statistics of the froth are invariant under cut.

6. Dynamics of Cells on the Basal Layer

We now generalize the model of Telley to detaching and dividing cells. These are the slow modes, responsible for the statistical equilibrium of the epidermis. Cells are in mechanical equilibrium at all times with respect to the horizontal motion of their nuclei. We further assume that no neighbouring cells are in the dividing stage of the mitotic cycle at the same time, so that any dividing nucleus - which is no longer circular—is surrounded by "resting"—i.e., circular nuclei.

We will keep Telley's paraboloid as the "molecule" representing a cell, because the cell can and will grow, and only detaches itself from the basal layer instead of disappearing. It is now a hyperparaboloid in a 4-dimensional space, with z denoting, as before, the direction of the paraboloid's axis and measuring the size of the cell; the axes x and y are on the basal layer, and axis u cuts across the epidermis, from basal to corneum layer, and is the direction of cellular transport.

6.1. Detachment from the Basal Layer

In our model, this is essentially due to lateral pressure on the resting cell which is about to detach itself from the basal layer. This pressure is caused by the growing and dividing cells on the basal layer. The centre of the spherical "nucleus" of the detaching cell is pushed away from the cellular structure and its interaction with the basal layer, a disk, shrinks as the representative paraboloid shrinks and disappears below the horizon. The face of the cell attached to the basal membrane shrinks, it sheds sides by successive T1 transformations to its neighbours, until the remaining triangular face disappears through a T2. Note that the cell may grow while it detaches itself. This growth is along the u axis. Thus, even if its hyperparaboloid rises (z increases), its centre is pushed away from the basal layer (u increases), and the height z_b of its intersection with the basal membrane decreases as $z_b = z - ku^2$, with u, the distance from the basal layer, replacing $x_\perp$, the distance from the cut in Eq. (5). This process is controlled by the viscous dynamics of the detached cell. Coordinates x, y are always in equilibrium. So is u, but its equilibrium is a balance between:

a) Pressure due to the growing cell: edges on the basal layer of the detaching cell are pushed in, parallel to themselves (mechanical equilibrium), and the representative nucleus is pushed away (u increases) as it grows.

b) Pressure applied by the cells above, up to the corneum layer, maintained by its elastic energy (surface tension).

Note that, on the basal layer, cells with a small number of sides ($n < 6$) divide, while those with a large number of sides are in latency, lose their sides while their nuclei are pushed off the basal layer, and eventually detach themselves. The situation is opposite in botanical samples like the epithelium of cucumber (Lewis [10]), where the dividing cells have larger n (Rivier *et al.* [17]). In cucumber, there is no significant cell death.

Consider now the final stage of detachment through a triangular facet in the basal layer: We treat completely and analytically the topological transformation T2 (death of a triangular cell).

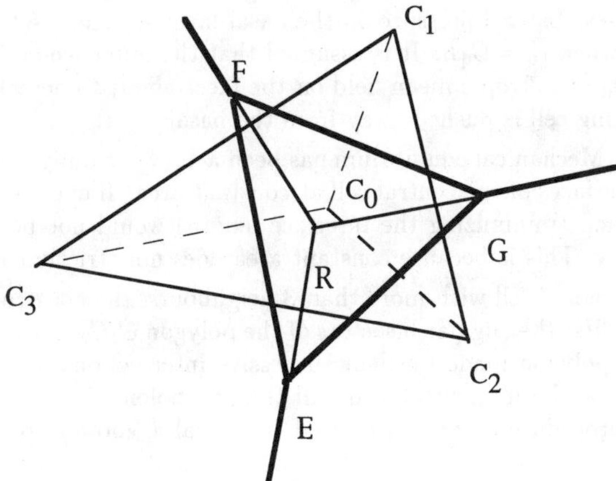

Fig. 6. Detaching cell. Mechanical equilibrium implies that triangles FRG, GRE, ERF are isoceles and that R is the centre of the circle circumscribing the triangular cell FGE; its nucleus is centered at C_0, intersection of the angular bisectors at C_1, C_2, C_3, the centers of the nuclei of neighbouring cells. Mechanical equilibrium is reached very rapidly.

The expiring triangular cell is represented by the circle of position C_0, r_0, its three neighbours by circles C_i, r_i, $i \in [1,2,3]$. It is always possible to choose the physical floor so that the three circles C_i are concurrent. They meet at the radical centre R, the confluence of the three radical axes. (The radical axis is an equivalence relation between the two circles to which it is equidistant.)

In mechanical equilibrium, the centre of the expiring cell C_0 is at the intersection of the angular bisectors of the triangle $C_1C_2C_3$, see Fig. 6. The intersection is also the centre (incentre) of the incircle to the triangle. Any other position for C_0 would fit a smaller circle inside the triangle, which is another way of understanding mechanical equilibrium.

The expiring triangular cell EFG has its vertices on the radical axes RG, RF and RE, and its edges perpendicular to C_0C_i. Mechanical equilibrium imposes that the triangles RFG, RGE and REF are isosceles. Thus $RF = RG = RE$, and the expiring cell is closed. The radical centre R is also the centre of the circumcircle to the triangle EFG, and the intersection of its perpendicular bisectors. That C_0C_1 bisects the angle C_1 if FG is in mechanical equilibrium and conversely, is now evident by construction.

As the differentiating cell leaves the basal layer (through the slow dynamic outlined above (Telley [20]), in mechanical equilibrium), its centre remains at C_0, its radius r_0 decreases, and the cell vertices E, F and G slide on the radical axes with parallel displacement of the edges. The cell disappears when its representative paraboloid, centered at C_0, dips under the horizon, that is when $r_0 = C_0R$.

The same construction can be made for detachment from the basal layer. The physical floor is lowered to keep the neighbouring cell circles meeting at R, so that r_0 effectively decreases. Lateral pressure on the basal layer shrinks the triangular facet, until it detaches when $r_0 = C_0R$. It is assumed that the influence of dividing cells is simply a pressure, an isotropic mean field on the facet about to detach. The nucleus of the differentiating cell is pushed away from the basal layer.

Remark 1. Mechanical equilibrium has been achieved simply and physically by displacing the interfaces of the central cell at constant area. If one insisted in moving C_0 only, its position (minimizing the interface length) would not be exactly at the incentre of $C_1C_2C_3$. This is because constant area does not strictly constrain z.

Remark 2. For a cell with more than 3 neighbours, the construction proceeds backwards. Draw first the angular bisectors of the polygon $C_1C_2\ldots C_n$. C_0 lies within the small convex polygon made by their successive intersections, for example at its barycenter. This can be implemented on all the paraboloids in turn, iteratively. A similar iterative procedure is frequently used to anneal a Voronoi froth (Weaire and Rivier [23]).

6.2. Cell Division

This can be discussed by generalizing the Telley dynamics to elliptic nuclei. As usual, we treat the evolution of the dividing cell surrounded by cells at rest during its mitotic cycle.

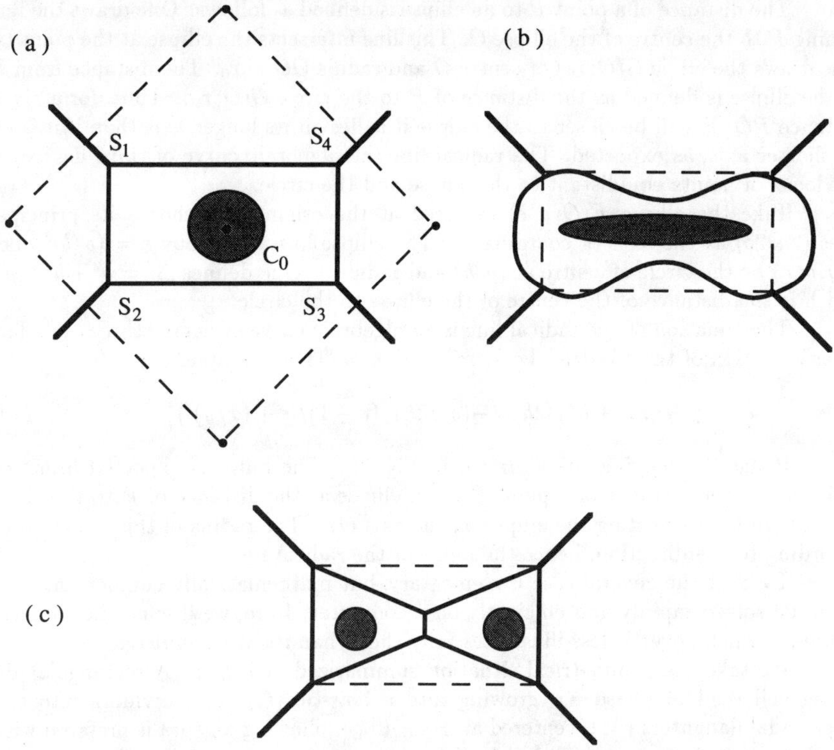

Fig. 7. Description of the 3 geometrically relevant phases of the mitotic cycle by Telley's dynamics. The nuclei of the mother and daughter cells are shaded. Full lines: radical membranes between fixed neighbouring cells. Broken lines in (b) and (c) represent the initial membranes. (a) End of phase S, the nucleus is a large circle. (b) Phase G_2, the nucleus is an ellipse. (c) After division (phase M): the two daughter cells have small circular nuclei.

We consider the three stages of the mitotic cycle which are geometrically relevant (Fig. 7). First, the circular nucleus grows by synthesizing DNA (phase S). It then elongates into an ellipse (phase G_2) through alignment and separation of the duplicated chromosomes. Finally, it divides into two daughters with circular nuclei (phase M). We will assume that from the phase S to M, the interfaces of the dividing cell are anchored at fixed positions, but they can bulge as the anisotropy of the nucleus develops (phase G_2). This bulging of the interfaces costs energy, provided by the activity in the nucleus and will be accounted for by molecular biochemistry. Our aim is to show that division can be described topologically, but accurately, by an elementary dynamics, and to see how the new membrane can lower the energy.

The distance of a point P to an ellipse is defined as follows. One draws the line joining P to the centre of the ellipse O. This line intersects the ellipse at the point Q. One draws the circle $G(O;r_0)$ of centre O and radius $OQ = r_0$. The distance from P to the ellipse is defined as the distance of P to the circle $G(O;r_0)$. Thus, for a given distance PO, P will be closer to the ellipse if it lies on its longer axis than if it is on its shorter axis, as expected. The radical line (an algebraic curve of third degree) is the locus of points equidistant to the ellipse and the circle.

Take the ellipse $E(O;a,e)$ centered at the origin, and choose its principal axes $(2a, 2b)$ as the axes of coordinates. The ellipse has anisotropy $e = (a/b)^2$. Let $\Gamma(g,h;r)$ be the circle of centre at (g,h) and radius ρ. One defines $\Delta^2 = g^2 + h^2 - \rho^2$, which is the distance of the centre of the ellipse to the circle.

The equation of the radical line is an algebraic curve of degree three, an affine transformation of the "*Witch (Versiera) of Agnesi.*" For example, if $g = 0$,

$$y - (\Delta^2 + a^2)/2h = -(a^2/2h)\{(e-1)/[e + (x/y)^2]\}\,. \tag{10}$$

It has the bulging shape drawn in Fig. 7b. The bulge is expected from the definition of the distance of a point P to an ellipse as the distance of P to the circle centered in O intersecting the ellipse on the axis OP. The radius of this circle varies according to the direction, hence the bulge of the radical line.

To treat the general case is elementary, but mathematically cumbersome and it can be solved rapidly and efficiently on a computer. Here, we discuss the simplest example, which nevertheless illustrates fully the dynamics of the mitosis.

We take the symmetrical situation summarized in Fig. 7. A rectangular dividing cell (end of phase S), growing into a bow-tie (G_2) and dividing into two trapezoidal daughters (M), centered at $\pm X_d$. (The following algebra is prefaced with the particular phases to which it refers).

The vertices S_i are fixed at $(\pm s_1, \pm s_2)$, at distance δ from the origin. Also, S_1 is at fixed Laguerre distance d from the neighbouring circles Γ_1 and Γ_2 , etc.

$[S, M]$ Since S_1 is fixed, its distance d to the nucleus gives a relation between the radii r_d of the daughter cells to the positions of their centers $(\pm X_d, 0)$.

$[S, G_2]$ Moreover, the line SC_0 cuts the ellipse at the same points αs_1, αs_2 as the circle Γ_0. The anisotropy of the ellipse e, and that of the mother cell s_1^2/s_2^2 are related

$$1 + s_1^2/s_2^2 = (r_0^2/a^2)(e + d^2)\,. \tag{11}$$

$[G_2, M]$ In the course of phase G_2, the anisotropy e increases. The elongation of the nucleus in Fig. 7b (phase G_2) costs energy supplied by biological means, but mitosis *per se* (phase M) is a lowering of interfacial length and a decrease of physical energy. This energy decrease can be optimized.

$[M]$ Let the variable w denote the depression of the dividing interface (the new vertices have ordinate $\pm(s_2 - w)$. The total interface length is minimal if $w = s_1/\sqrt{3}$. Since $w < s_2$, the anisotropy of the mother cell is bounded, $s_1^2/s_2^2 < 3$. The oblique interfaces are perpendicular to C_dC_1, where $C_1 = (0, Y_1)$ is the centre of the neighbouring circle Γ_1. This gives the positions of the centers of the daughter cells, $X_d = Y_1/\sqrt{3}$, and their radii.

The lowering of energy during mitosis proper is given by the difference in the interface lengths of Figs. 7b and 7c. Note that some energy must be supplied biologically through phase G_2: The total interface length at the end of phase S, $4(s_1 + s_2)$, is lower than after division.

7. Conclusions

The model presented here gives a direct understanding of why the epidermis is in statistical equilibrium, in spite of its constant renewal. It also explains why cells have to constantly divide in, and detach from the basal layer (as molecules of a gas must collide). This understanding is so far only theoretical and is awaiting confirmation by simulations. This model would also simplify simulations and make them more robust and conclusive, since it uses a very small number of parameters, unlike traditional simulations (Clem and Rigaut [2]) which focus on the dynamics of the nuclei only. However, the practical implementation of simulations by Telley dynamics is subtle (Telley [20]), and we have not yet been able to do it.

A good check of the model is the distribution of the numbers of sides of dividing, resting and differentiating basal cells (Dubertret [5], and to be published).

This model is essentially random space-filing by cells, in statistical equilibrium, driven by simple local topological transformations which are mitosis, neighbour exchange and detachment. As with collisions in gases, the specific microscopic physics is not directly relevant to the macroscopic dynamics. Moreover, the specific biological input is small, apart from the global geometry of the epidermis and the energy necessary to elongate the dividing nucleus in phase G (Fig. 7b), which is the *deus ex machina* in this paper.

Acknowledgements. We would like to thank L. Dubertret for his help and guidance with biological knowledge, for references and for several extremely helpful discussions. This work is supported by the EC Network, Physics of Foam, CHRXCT 94 0542, and by La Société Dior (B.D.). Part of it has been done at MIT (B.D.).

References

1. Y. Barrandon and H. Green, *Proc. Natl. Acad. Sci. USA* **82** (1985) 5390.
2. C. J. Clem and J. P. Rigaut, *Acta Biotheoretica*, 1995, to appear.
3. R. Delannay, G. Le Caër and M. Khatun, *J. Phys. A* **25** (1992) 6193.
4. K. J. Dormer, *Fundamental Tissue Geometry for Biologists*, Cambridge Univ. Press, 1980.
5. B. Dubertret, *Etude de l'influence de la mort cellulaire sur l'équilibre statistique de l'épiderme des mammifères*, Thesis, ULP, Strasbourg, 1994.

6. L. Errera, *Comptes Rendus Acad. Sci.* **103** (1886) 822.
7. W. Fischer and E. Koch, *Zeits. Kristallogr.* **150** (1979) 248.
8. B. J. Gellatly and J. L. Finney, *J. Non-cryst. Solids* **50** (1982) 313.
9. J. Lemaitre, A. Gervois, J.-P. Troadec, N. Rivier, M. Ammi, L. Oger and D. Bideau, *Phil.Mag. B* **67** (1993) 347.
10. F. T. Lewis, *Anat. Records* **38** (1928) 341; *Anat. Records* **50** (1931) 235.
11. W. Montagna et P. F. Parakkal, In *The Structure and Function of Skin*, Academic Press, 3rd edition, 1974, Chap. 2.
12. M. A. Peshkin, K. J. Strandburg and N. Rivier, *Phys. Rev. Letters* **67** (1991) 1803.
13. M. B. Pyshnov, *J. Theor. Biol.* **87** (1980) 189.
14. N. Rivier, *J. Physique, (Coll)* **51** (1990) C7-309.
15. N. Rivier, In *Continuum Models of Discrete Systems*, eds. K.-H. Anthony and H.-J. Wagner, Mater. Sci. Forum 123-125, Trans. Tech. Publ., 1993, p. 383.
16. N. Rivier, In *Solid State Phenomena 35-36 (Dislocations 93)*, eds. J. Rabier, Y. Bréchet, A. George and L. Kubin, 1994, p. 107.
17. N. Rivier, X. Arcenegui-Siemens, G. Schliecker, In *Fragmentation Physics*, eds. D. Beysens, X. Campi and E. Pefferkorn, World Scientific, 1995, p. 266.
18. N. Rivier and B. Dubertret, *Phil. Mag. B* **72** (1995) 311.
19. N. Rivier, G. Schliecker and B. Dubertret, *Acta Biotheor.*, 1995, to appear.
20. H. Telley, *Modélisation et Simulation Bidimensionnelle de la Croissance des Polycristaux*, PhD Thesis, EPFL, Lausanne, 1989.
21. H. Telley, Th. M. Liebling and A. Mocellin, *Phil. Mag.*, 1995, to appear.
22. D'A. W. Thompson, *On Growth and Form*, Cambridge Univ. Press, 2nd. ed., 1942, pp. 363, 482.
23. D. Weaire and N. Rivier, *Contemp. Physics* **25** (1984) 59.

PART IV

DEFECT DYNAMICS, SYNERGETICS, SOLITONS, COHERENT STRUCTURES

Continuum Models and Discrete Systems
Proceedings of 8th International Symposium, June 11-16, 1995, Varna, Bulgaria, ed. K.Z. Markov
©World Scientific Publishing Company, 1996, pp. 370–394

DISCRETE OUT OF CONTINUOUS: DYNAMICS OF PHASE PATTERNS IN CONTINUA

C. I. CHRISTOV[1]
Center for Nonlinear Phenomena and Complex Systems,
Université Libre de Bruxelles, Campus Plaine - CP 231,
Boulevard du Triomphe, Bruxelles 1050, Belgique

Abstract. A new paradigm is discussed in which the discrete facet of physical phenomenology is related to the existence of localized wave solutions (*solitons*) of a underlying *metacontinuum* (unified field, physical vacuum, etc.) while the action at a distance is a manifestation of the internal stresses in *metacontinuum*. The latter is considered as thin elastic 3D layer (a shell) in the 4D geometrical space. Maxwell's equations are recovered as corollaries from the equations of the motion in the middle surface. The equation for the amplitude of transverse flexural deformation appears to be a fourth order dispersive equation similar to the one derived by Schrödinger. The localized flexural (curvature) waves propagating *over* the surface of *metacontinuum* are interpreted as particles. A possible unification of the gravitation, electromagnetic phenomena and wave mechanics is demonstrated.

Introduction

The apparent duality between the point particles and field(s) underlies the foundation of modern physics. The Aristotelian tradition was connected with the notion of continuum but with the advance of the experimental physics the attention gradually shifted to the concept of point particles (corpuscles) moving freely in a void geometrical space. This new concept was heralded by Newton in his first two laws and it gave rise of a fruitful quantitative description of the motion of bodies known nowadays as *Newtonian Mechanics*. Yet it was Newton himself who admitted also an action at a distance when formulating the law of gravitation. Thus from the very beginning was planted the seed of dualism.

It is hard to imagine an action at a distance without some kind of a carrier. The nineteen-century tradition was always connected with some mechanical construct. In Cauchy's and Hamilton's vision it was a lattice whose continuum ap-

[1]On leave from National Institute of Meteorology and Hydrology, Bulgarian Academy of Sciences, Sofia, Bulgaria

proximation yielded the elastic-body model.[2] Paying tribute to the metaphysical tradition, all of the models for the presumably absolute continuous media underlying the physical world were called "aether, " although some of them (the different elastic models, for instance) were quite different from an "aetherial fluid" (where the coinage comes from). In Maxwell's imagination it was a medium with internal degrees of freedom. McCullagh and Sommerfeld quantified this idea as an elastic body with special rotational elasticity (see [1] for exhaustive review of these theories).

The downfall of the concept of aether began with the unsatisfactory results of Cauchy's "volatile aether" (elastic body with vanishing dilational modulus). After Lord Kelvin came up with the model of fluid aether and its vortex theory of matter, the coinage "aether" assumed almost exclusively fluid meaning. Then the question of entrainment of aether (aether-drift) was posed and the nil result of Michelson and Morley experiment blew down the whole edifice of aether theories. However, the notion of a material carrier of the long-distance interactions could not be dismissed so easily and the conceptual vacuum was filled by the concept of what is called nowadays "physical vacuum" which possessed all the properties of the disgraced aether (e.g., action at a distance and giving birth to particles), but was deliberately exempted from the obligation to be checked for aether-drift effect. It was then advertised as a thing in itself not connected to any "primitive mechanistism." The same conceptual load carries the coinage "field."

On a new level, the yearning for continuum description resurrected in the Hamiltonian-Schrödinger wave mechanics. From the far side de Broglie quantified the wave properties of the particles. At that moment it was not clear how the two facets of the same object—the material particle—can peacefully coexist and it led to the concept that some kind of "pilot" wave was associated with a "point particle" representing the probability to find the latter at certain spatial position. Thus the dichotomy was deeply implanted in the physical thought and is still the dominant general attitude in the modern physics. As a result there has appeared a rather intricate web of coinages (*pseudo, quasi*, etc.) when certain phenomenological fact pertaining to the one facet was being explained in the terminology of the dual facet.

In our point of view, the *field* (*physical vacuum, aether*) can only be understood from the point of view of a material continuum where the *internal stresses are the transmitter of the long-range interactions*. In order to distinguish it from the mechanical continuous media (bodies, liquids, gases, etc.), we call the continuum-mechanics model of the unified field *metacontinuum* in the sense that it is beyond (*meta*) the observable phenomena and is their progenitor. Returning to the concept of absolute continuum will be senseless without some new paradigm concerning the "point matter." It is not possible to imagine point particles pushing their way through infinitely stiff, virtually incompressible elastic body. Even if they could, the disturbances would be the predominant effect which has not been observed in any experiment. So the new paradigm has to deal with inventing the concept of a particle

[2]It is curious to mention here that Cauchy came up with the elasticity theory because of his attempts to constitute the aether.

which does not disturb the underlying continuum during its motion. The new concept we implant on the old tree of the absolute medium is that a particle is not something foreign to the field and immersed in the latter. Rather it is a localized wave of the field which *propagates over* the field (as a phase pattern) and is not moving *through* the field. Then we have to find phase patterns that are permanent enough in order to qualify for particles. Such individualized waves are well known from the soliton theory.

J. Scott Russell who was a contemporary of the golden age of the aether theories provided [2,3] the experimental evidence that a *nonlinear* field (2D surface of ideal fluid) can exhibit patterns of type of permanent waves with individualized comportment. The theoretical explanation was found by Boussinesq [4] and Lord Rayleigh [5] and it appeared that the balance between nonlinearity and dispersion is which sustains the long-living phase pattern. Zabusky and Kruskal [6] showed numerically that the solitary waves of KdV equation (close relative to Boussinesq equation) preserve their shapes and energies upon collisions and called them *solitons* to stress the particle-like behaviour ("quasi-particles").

Thus the stage for the unification was set up long time ago though it was not recognized as such. The only remaining question is which is the field (*metacontinuum*) whose solitary-wave phase patterns are the particles. A valid candidate for the luminiferous field is the elastic medium because it gives a good quantitative prediction for the shear-wave phenomena (light). As it is shown in [7] and briefly outlined here, the Maxwell equations are corollary from the linearized governing equations of the metacontinuum provided that the electric and magnetic filed are properly understood as manifestation of the "meta" internal stresses. The main difference from Cauchy's elastic aether is that we consider the opposite limiting case: an elastic continuum with infinitely large dilational modulus (virtually incompressible elastic medium).

There are other candidates for the field, e.g., the *sine*-Gordon equation considered in [8,9]. It is believed to be the meson field possessing solutions of type of "quasi-particles." Oddly enough, those localized waves are of type of fronts (kinks) which though mathematically well localized hardly fit into the common-sense picture of a particle. Yet the spatial derivatives of the solution did have the expected shape and the kinks qualify for *solitons* because the energy is conserved during the collisions. Nowadays there are many other "fields" and new are being introduced *ad hoc* for explaining one or another of the hundreds of elementary particles already known.

Here we stay firmly on the position that there is only one *metacontinuum* and the different fields are different manifestation of its internal stresses and strains. The "organization" of the matter around some terminology like "particles," "charges," "spins" is simply our way to simplify and thus to comprehend the complexity and richness of the interactions of localized phase patterns (solitons). For this reason we prefer to call the present attempt for an unified field theory "Soliton Paradigm." And the most ambitious goal is to derive the electromagnetic phenomena, the gravitation and wave mechanics from a single mechanical model of a metacontinuum.

Clearly, if one is to unify the wave function with the electromagnetic phenomena one has to consider more than three spatial dimensions. In addition, the thickness in the fourth dimension of the "material world" must be so small that it cannot be perceived by our senses. This idea was put forward by Hinton [10] but on qualitative basis only. Kalutza [11] and Klein [12] also considered additional "underdeveloped" dimensions, but they added a vector field to the Einstein equation while we add a *scalar* field to the Maxwell equations, the latter appearing here in the guise of a vector equation for shear waves in elastic continuum.

Carrying on with this idea we ask the question: "What kind of manifestation is to be expected from the fact that the material world is a thin 3D layer in the 4D geometrical space?" Naturally, the presence of "underdeveloped" dimension(s) should result into some new observables in addition to those that were sufficient to make a self-consistent picture in the Maxwellian framework. Depending on the topology of the *metacontinuum* there may be (or may not be) a number of these manifestations, connected with different spins in multi-dimensional space. The existence of at least one additional variable is inevitable, namely the amplitude of deflection of the 3D layer in direction of the fourth dimension. In Section 2 we derive an equation ("master equation of wave mechanics") governing the said amplitude for geometrically nonlinear very thin (virtually $N^{\mathcal{D}}$) elastic layers in $N+1$ dimensions. We call this kind of mechanical construct *gossamer*. Naturally, this equation is not a follow-up from the governing equations in N dimensions, just like in the real world the Schrödinger equation is not a corollary form the Maxwell equations.

The rest of the paper is devoted to arranging a cosmological picture of the solitary waves—phase patterns. In many instances they appear to be quite similar to what the long-longed unified theory is expected to bring.

1. Constituting a Metacontinuum

For small velocities the Lagrangian and Eulerian descriptions of a continuum coincide and for the displacements $\boldsymbol{u}$ of a Hookean elastic medium one has the linear vector wave equation

$$\mu_0 \frac{\partial \boldsymbol{A}}{\partial t} \equiv \mu_0 \frac{\partial^2 \boldsymbol{u}}{\partial t^2} = \eta \Delta \boldsymbol{u} + (\lambda+\eta)\nabla(\nabla\cdot\boldsymbol{u}) \equiv -\eta\nabla\times\nabla\times\boldsymbol{u} + (\lambda+2\eta)\nabla(\nabla\cdot\boldsymbol{u}), \quad (1.1)$$

where $\boldsymbol{u}$, $\boldsymbol{A}$ are the displacement and velocity vectors; η, λ are Lamé's elasticity coefficients and μ_0 is the density in material (Lagrangian) coordinates. Note that we concern ourselves for the time being only with a metacontinuum of constant elastic coefficients η, λ and density μ_0.

The full set of physical motions governed by (1.1) includes shear and compression/dilation as well. The former are controlled by the shear Lamé coefficient η, while the latter—by the dilational (second) Lamé coefficient λ, and more specifically by the sum $(\lambda + 2\eta)$. The phase speeds of propagation of the respective small disturbances are

$$c = \left(\frac{\eta}{\mu_0}\right)^{\frac{1}{2}}, \quad c_s = \left(\frac{2\eta+\lambda}{\mu_0}\right)^{\frac{1}{2}}, \quad \delta = \frac{\eta}{(2\eta+\lambda)}. \quad (1.2)$$

Here c and c_s are the speeds of shear waves (*light*) and compression waves (*sound*), respectively. In the case of very large dilational modulus, the speed of sound is much greater than the speed of light and $\delta \ll 1$. This is the opposite limiting case than the "volatile aether" of Cauchy with $\eta \gg \lambda$. Then Eq. (1.1) is recast as follows

$$\delta\left(c^{-2}\frac{\partial^2 \boldsymbol{u}}{\partial t^2} + \nabla\times\nabla\times\boldsymbol{u}\right) = \nabla(\nabla\cdot\boldsymbol{u}), \tag{1.3}$$

and displacement $\boldsymbol{u}$ can be developed into power asymptotic series with respect to δ

$$\boldsymbol{u} = \boldsymbol{u}_0 + \delta\boldsymbol{u}_1 + \cdots . \tag{1.4}$$

Introducing (1.4) into (1.3) and combining the terms with like powers we obtain for the first two terms

$$\nabla(\nabla\cdot\boldsymbol{u}_0) = 0, \tag{1.5}$$

$$c^{-2}\frac{\partial^2 \boldsymbol{u}_0}{\partial t^2} + \nabla\times\nabla\times\boldsymbol{u}_0 = \nabla(\nabla\cdot\boldsymbol{u}_1). \tag{1.6}$$

From (1.5) one can deduce

$$\nabla\cdot\boldsymbol{u}_0 = \text{const}, \quad \text{or} \quad \nabla\cdot\boldsymbol{A}_0 = 0, \tag{1.7}$$

which is also a linear approximation to incompressibility condition for a continuum. In the general model of nonlinear elasticity with finite deformations the incompressibility condition is imposed on the Jacobian of transformation from material to geometrical variables, but in the first-order approximation in δ the Eq. (1.7) holds true.

From here on we omit the index '0' without fear of confusion. We denote formally the term $(\lambda + 2\eta)\nabla\cdot\boldsymbol{u}_1$ by $(-\varphi)$ and recast (1.6) as dimensional form of linearized Cauchy balance, namely

$$\mu_0\frac{\partial \boldsymbol{A}}{\partial t} = -\nabla\varphi + \nabla\cdot\boldsymbol{\tau}, \tag{1.8}$$

where $\boldsymbol{\tau}$ is the deviator stress tensor for which the following relation is obtained from the constitutive relation (the Hooke law) for elastic body, namely

$$\boldsymbol{\tau} = \eta(\nabla\boldsymbol{u} + \nabla\boldsymbol{u}^T) - 2\eta(\nabla\cdot\boldsymbol{u})\mathbf{I}, \tag{1.9}$$

where $\mathbf{I}$ stands for the unit tensor. For the divergence of $\boldsymbol{\tau}$ one has

$$\nabla\cdot\boldsymbol{\tau} = -\eta\nabla\times(\nabla\times\boldsymbol{u}). \tag{1.10}$$

What is essential for the unification is that the linearized equations of elastic continuum admit what we call Maxwell form. The derivations here are not to be confused with McCullagh's model of pseudo–elastic continuum (see [1,13] for references and further developments) with restoring couples by means of which he tried to explain the unusual shape of Maxwell's equations apparently not fitting into the picture of continuum mechanics. Let us introduce the vector field

$$\boldsymbol{E} \stackrel{def}{=} -\nabla \cdot \boldsymbol{\tau} \equiv \eta \nabla \times (\nabla \times \boldsymbol{u}) . \tag{1.11}$$

to which the action of the purely shear part of internal stresses is reduced. It has the meaning of a point-wise distributed body force and we shall call it "electric force." In terms of $\boldsymbol{E}$, the linearized system (1.8) yields

$$\boldsymbol{E} = -\frac{\partial \boldsymbol{A}}{\partial t} - \nabla \varphi , \tag{1.12}$$

involving the well known vector and scalar potentials $\boldsymbol{A}$ and φ. In the framework of the present approach, however, these potentials are not non-physical quantities introduced merely for convenience, but rather they appear to be the most natural variables: velocity and pressure of metacontinuum. Taking the *curl* of (1.12) one obtains

$$\nabla \times \boldsymbol{E} = -\frac{\partial \boldsymbol{B}}{\partial t} \tag{1.13}$$

which is nothing else but the first of Maxwell's equations (the Faraday law) provided that a "magnetic induction" $\boldsymbol{B}$ is defined as

$$\boldsymbol{B} = \mu_0 \nabla \times \boldsymbol{A} = \mu_0 \boldsymbol{H} \quad \boldsymbol{H} \stackrel{def}{=} \nabla \times \boldsymbol{A} , \tag{1.14}$$

where $\boldsymbol{H}$ is called "magnetic field." From Eq. (1.11) one obtains

$$\frac{1}{\eta} \frac{\partial \boldsymbol{E}}{\partial t} = \nabla \times (\nabla \times \frac{\partial \boldsymbol{u}}{\partial t}) \equiv \nabla \times \boldsymbol{H} . \tag{1.15}$$

The last equation is precisely the "second Maxwell equation" provided that the shear elastic modulus of metacontinuum is interpreted as the inverse of electric permittivity $\eta = \varepsilon_0^{-1}$. This equation was postulated by Maxwell [14] as an improvement over Ampere's law incorporating the so-called displacement current $\partial \boldsymbol{E}/\partial t$ in the Biot–Savart form. For the case of a void space, however, when no charges or currents are present, the second Maxwell equation lives a life of its own and Ampere's law plays merely heuristic role for its derivation. It is broadly accepted now that the second Maxwell equation is verified by a number of experiments. Here we have shown that it is also a corollary of the elastic rheology of the metacontinuum and is responsible for the propagation of the shear stresses (action at a distance) in *metacontinuum*.

The two main equations of evolution of Maxwell's form have already been derived. The condition $\operatorname{div} \boldsymbol{H} = 0$ (the third Maxwell equation) follows directly from the very definition of magnetic field. Similarly, taking divergence of Eq. (1.11), one immediately obtains the fourth Maxwell equation $\operatorname{div} \boldsymbol{E} = 0$. Thus we have shown that Maxwell's equations follow from the linearized governing equations of a Hookean elastic medium whose dilational modulus is much larger than the shear one.

2. A Curved Metacontinuum? The Gossamer

From the point of view of material continuum a curved medium can be considered only as being embedded into a geometrical space of at least one spatial dimension larger, e.g., the Maxwell luminiferous field can be considered as 3D material hypersurface in 4D geometrical space. It is hard to imagine a material surface which has no thickness. A 3D hypersurface is a mathematical abstraction for the material construct known as thin shell (or a membrane) when the so-called middle surface of the latter is considered. If the shell is flat then the electromagnetism will be the only phenomena to be observed in three dimensions. Let us examine now the consequences of the curvature and transverse (flexural) deflections of *metacontinuum*. The existence of another spatial dimension cannot be detected in 3D shear-waves (electromagnetic) experiments.

Following the previous sections we consider an elastic shell of a 4D material whose shear Lamé coefficient is much smaller than the dilational one (their ratio given by the small parameter δ). As usual in shell/plate/membrane theories the small parameter $\varepsilon = h/L$ is the most important one, where h is the thickness of the shell and L is the length-scale of the deformations of the middle surface. The thin-layer simplification apply when $\varepsilon \ll 1$. The problem is to find a correct way to reduce the 4D continuum mechanics to an effective 3D mechanics for the middle surface of the shell.

In the technological applications an additional assumption is tacitly made, namely that L is large ("shallow shells") and *shell* is called a "reasonably" thin elastic structure whose flexural deformation (deflections) are of unit order, the strains (gradients) are $O(L^{-1})$ small and the curvatures are of second-order in smallness. Here we relieve the limitation of large L and treat the case $1 \gg L \gg h$. Then the deflections must be small, the strains (gradients)—of unit-order, and curvatures—large. The standard shell theory is not sufficient for describing such an object which is geometrically strongly nonlinear. As shown in [7], when deriving the shell equations for this limiting case one has to acknowledge more terms responsible for the geometric nonlinearity. At the same time the material nonlinearity is not so important because of the vanishing thickness h. In order to distinguish them from the classical shallow shells we call this kind of very thin elastic layers undergoing very high strains *gossamers*.

2.1. Manifestation of Underdeveloped Dimensions

We summarize here the relevant items of derivation of *gossamer*'s theory. Wherever possible we keep the derivation general enough speaking about $N^{\mathcal{D}}$ layer in $(N+1)^{\mathcal{D}}$ space, but for the purposes of the present work $N = 3$ and $N + 1 = 4$. The Cauchy form for $(N+1)^{\mathcal{D}}$ continuum reads

$$[\rho_* a^j - P^{ij}\|_i]\, \boldsymbol{g}_j = 0, \quad i, j = 1, \ldots, N\,, \tag{2.1}$$

where ρ_* is material density of the $(N+1)^{\mathcal{D}}$ continuous media filling the $N^{\mathcal{D}}$ shell.

One can call it "meta-density" in order to distinguish it from the 3D density already identified as μ_0. It is important also to note that neither ρ_* nor μ_0 have anything to do with the density of matter (number density of solitons). Here $\boldsymbol{g}_j$ are the orts of the curvilinear coordinate system; P^{ij} are the components of stress tensor; a^j are the components of the acceleration vector in the $(N+1)^{\mathcal{D}}$ space; $\|_i$ stands for the covariant derivative in $(N+1)^{\mathcal{D}}$ space. We do not consider here $(N+1)^{\mathcal{D}}$ body forces.

Upon substituting the expressions for $\|_i$ in terms of $N^{\mathcal{D}}$ covariant derivatives $|_\alpha$ (see [15,16] and the extensions in [7]), the Cauchy balance law (2.1) is recast into a system for the laminar components and a scalar equation for the $N+1$-st component. After averaging (integrating) within the surfaces of the shell one gets to the second order $O(\varepsilon^2)$ of approximation.

$$\rho_*\varphi^\alpha - \nabla_\beta \sigma^{\alpha\beta} = -2b^\alpha_\beta q^\beta - b^\beta_\beta q^\alpha , \tag{2.2}$$

$$\rho_*\varphi^{N+1} - \nabla_\beta q^\beta = b_{\beta\nu}\sigma^{\beta\nu} - c_{\beta\nu}m^{\beta\nu} + \rho_*\mathcal{F} , \tag{2.3}$$

where

$$q^\alpha = \frac{1}{h}\int P^{N+1,\alpha}\, ds \, , \quad \sigma^{\alpha\beta} = \frac{1}{h}\int P^{\alpha\beta}\, ds \, , \quad m^{\alpha\beta} = \frac{1}{h}\int sP^{\alpha\beta}\, ds,$$

$$\varphi^\alpha = \frac{1}{h}\int a^\alpha\, ds \, , \quad \psi^\alpha = \frac{1}{h}\int sa^\alpha\, ds \, , \quad \rho_*\mathcal{F} = \frac{1}{h}(P^{N+1,N+1}_{\rm up} - P^{N+1,N+1}_{\rm lo}) .$$

Here the subscripts "up" and "lo" refer to the upper and lower shell surfaces $s = h/2$ and $s = -h/2$, respectively. It is taken into account that no tractions are exerted upon the shell surfaces from the two adjacent $(N+1)^{\mathcal{D}}$ spaces. The integrals are understood as definite integrals in s between the shell surfaces while b and c are functions of the "surface" coordinates only. The notation ∇ stands for a $N^{\mathcal{D}}$ covariant derivative in which the coefficients that may possibly depend on the transverse coordinate are already averaged.

The system (2.2), (2.3) is coupled by the "momentum–of–impulses" which can be derived from (2.1) to the same asymptotic order $O(\varepsilon^2)$ after multiplying it by s and integrating across the shell, namely

$$-\rho_*\psi^\alpha + \nabla_\beta m^{\alpha\beta} = q^\alpha , \tag{2.4}$$

which allows us to exclude the quantity q^α from the governing system and to obtain

$$\rho_*\varphi^\alpha = \nabla_\beta\sigma^{\alpha\beta} - 2b^\alpha_\beta\nabla_\nu m^{\beta\nu} - b^\beta_\beta\nabla_\nu m^{\alpha\nu}, \tag{2.5}$$

$$\rho_*\varphi^{N+1} = \nabla_\beta\nabla_\nu m^{\beta\nu} + b_{\beta\nu}\sigma^{\beta\nu} - c_{\beta\nu}m^{\beta\nu} + \rho_*\mathcal{F} - \rho_*\nabla_\alpha\psi^\alpha . \tag{2.6}$$

Here the notion of the geometrization of physics becomes transparent. If the observer is confined to the $N^{\mathcal{D}}$ space of the middle surface he will appreciate the presence of the $N+1$-st dimension as additional terms in balance law (2.5), (2.6) which terms are not present in the Cauchy form for the $N^{\mathcal{D}}$ continuous media.[3] The said terms are proportional to the different curvature forms and this is the quantitative expression of Riemann–Clifford [17,18] idea that the physical laws are manifestation of deformations of the geometrical space.

2.2. Elastic Shell with Momentum Stresses

According to the Kirchhoff-Love hypothesis, the displacements u_α in the shell space are related to the $N^{\mathcal{D}}$ displacements $\tilde{u}_\alpha$ in the shell middle surface as follows

$$u_\alpha = \tilde{u}_\alpha - s\nabla_\alpha\zeta, \quad u_{N+1} = \zeta\,, \tag{2.7}$$

where ζ stands for the shape function of deformation (deflection) of the middle surface in direction of $(N+1)$-st dimension. This hypothesis is pertinent to the overall $o(\varepsilon)$ approximation since it amounts to neglecting terms proportional to s^2. Then we obtain

$$\varphi_\mu = \frac{\partial^2 \tilde{u}_\mu}{\partial t^2}, \quad \nabla^\mu \psi_\mu = -\frac{h^2}{12}\frac{\partial^2 \Delta\zeta}{\partial t^2}, \quad \varphi_{N+1} = \frac{\partial^2 \zeta}{\partial t^2}. \tag{2.8}$$

The rotational inertia $\nabla^\mu\psi_\mu$ is of second order which justifies neglecting it in comparison with the transverse inertia φ_{N+1}.

In terms of coordinates that are measured along the arcs of the middle surface (precisely the material Lagrangian coordinates), the second fundamental form assumes the following simple form

$$b_{\alpha\beta} = \nabla_\alpha\nabla_\beta\zeta\,. \tag{2.9}$$

Note that for coordinates not coinciding with the arcs, the expression of the second fundamental form involves nonlinear terms.

It is time now to couple the Cauchy equations with constitutive relations. Unlike the Cauchy form, the full nonlinear constitutive relations cannot be derived in Eulerian framework. It goes beyond the framework of the present work to derive them in full detail, especially as far as the material nonlinearity for the laminar components is concerned. We resort here to *linear* constitutive relations in the form (see [19])

$$\sigma^{\alpha\beta} = (\lambda_* + \eta_*)g^{\alpha\beta}(\nabla_\nu \tilde{u}^\nu) + \eta_*\nabla^\beta\tilde{u}^\alpha\,, \tag{2.10}$$

$$m^{\alpha\beta} = -D_*\nabla^\alpha\nabla^\beta\zeta, \quad D_* = \frac{\eta_* h^2}{12}. \tag{2.11}$$

where D_* is called stiffness of shell. Here asterisks designate the $(N+1)^{\mathcal{D}}$ material properties. If the middle surface of *gossamer* is to behave as the Maxwell luminiferous field then the following relations must hold true

$$\frac{\lambda_*}{\rho_*} = \frac{\lambda}{\mu_0}, \quad \frac{\eta_*}{\rho_*} = \frac{\eta}{\mu_0}, \quad \frac{D_*}{\rho_*} = \frac{D}{\mu_0}.$$

[3] If one was lucky enough to have already the tensor analysis developed and the notion of continuous medium (with its Cauchy balance) already established.

Upon introducing (2.10), (2.11) into Cauchy equations (2.5), (2.6) we get for the laminar components

$$\mu_0 \frac{\partial^2 \tilde{u}^\beta}{\partial t^2} = (\lambda+\eta)\nabla^\beta(\nabla_\nu \tilde{u}^\nu) + 2\eta\Delta\tilde{u}^\beta - D\,[\,2\nabla^\beta\,(\,\nabla\zeta\cdot\nabla(\Delta\zeta)\,) + (\Delta\zeta)\nabla^\beta\Delta\zeta\,]\,, \tag{2.12}$$

and for the amplitude ζ of normal deflection in direction of $N+1$–st dimension

$$\mu_0\frac{\partial^2\zeta}{\partial t^2} = \mu_0\mathcal{F} + D\left[-\Delta\Delta\zeta + (\nabla_\beta\nabla_\delta\zeta)(\nabla^\beta\nabla_\mu\zeta)(\nabla^\mu\nabla^\delta\zeta)\right] + \frac{\mu_0}{\rho_*}\sigma^{\beta\alpha}(\nabla_\beta\nabla_\alpha\zeta)\,. \tag{2.13}$$

Here we shall not consider the case of nontrivial tractions on the shell surfaces which requires additional assumptions about their asymptotic order. We keep, however, the term responsible for the normal pressure (parameter F).

If we develop with respect to the powers of the small parameter δ in the same manner as in the previous section, we get from (2.12) and (2.10) that

$$\nabla^\beta\nabla_\nu u^\nu = 0 + O(\delta) \implies \nabla_\nu u^\nu = \kappa_0 = \text{const} \implies \frac{\mu_0}{\rho_*}\sigma^{\alpha\beta} = \sigma_0 g^{\alpha\beta}\ ,\ \sigma_0 = \eta\kappa_0\ ,$$

where κ_0 is dimensionless divergence of the displacement field in the middle surface. Note that $k_0 > 0$ means uniform dilation of the middle surface, while $k_0 < 0$ reflects the case of uniform compression. Thus, in the lowest asymptotic order of δ we are faced with uniform compression/dilation in the middle surface of *gossamer* and with constant membrane stress σ acting in the middle surface. This allows to effectively decouple the laminar deformations $\tilde{u}^\alpha$ from the deflection ζ even without requiring special relation between the scales for ζ and $\tilde{u}$. In its turn the term in (2.13) containing $\sigma^{\alpha\beta}$ becomes simply $\sigma_0\Delta\zeta$. Because of the large dilational modulus, the laminar deformations appear to be orthogonal to the transverse ones to the first asymptotic order.

From here on the "tildes" denoting the laminar variables will be omitted without fear of confusion.

2.3. The Dispersive Equation of Wave Mechanics

The only term which looks unusual in (2.13) is the cubic nonlinear term. However, in dimension one the equation under consideration is exactly the cubic nonlinear Boussinesq equation. In order to benefit from the vast knowledge accumulated for equations of type of Boussinesq, we replace in a paradigmatic fashion the cubic term in (2.13) by $(\Delta\zeta)^3$. It is beyond doubt that the qualitative behaviour of the solutions will be quite similar. However, the extent to which they will be quantitatively close as well, remains to be verified. Then we arrive at

$$\mu_0\frac{\partial^2\zeta}{\partial t^2} = \mu_0\mathcal{F} + D\left[-\Delta\Delta\zeta + (\Delta\zeta)^3\right] + \sigma_0\Delta\zeta\,. \tag{2.14}$$

We render the last equation dimensionless by introducing the scales

$$\zeta = L\zeta' , \quad \boldsymbol{x} = L\boldsymbol{x}' , \quad t = \frac{L}{c_f}t' , \quad c_f = \sqrt{\frac{\sigma_0}{\mu_0}} \equiv c\sqrt{|k_0|} , \tag{2.15}$$

where c_f has dimension of velocity. Note that the scale for ζ and the length scale of the localized wave coincide (the length L), if one looks for commensurable effects (balance) of the nonlinearity and dispersion. This fits perfectly the original assumptions for *gossamer*: small deflections of order L, unit strains and large curvatures of order L^{-1}. Finally, the dimensionless form of the wave equation of *gossamer* reads

$$\frac{\partial^2 \zeta'}{\partial t^2} = \mathcal{F}' + \beta \left[-\Delta\Delta\zeta' + (\Delta\zeta')^3 \right] + \text{sign}[k_0]\Delta\zeta' , \tag{2.16}$$

where $\beta = D|\sigma_0|^{-1}L^{-2}$ is the dispersion parameter and $\mathcal{F}' = \mathcal{F}L\mu_0/|\sigma_0|$ is the dimensionless value of the normal load (hydrostatic pressure). From here on the primes denoting dimensionless variables are henceforth omitted without fear of confusion. Eq. (2.16) is our "master equation of wave mechanics."

Now β is the only intrinsic non-dimensional parameter and if it is significant, then without loosing the generality it may be set equal to one. This defines the length scale L of the particle-waves as

$$L \sim \sqrt{\frac{D}{\sigma_0}} \sim \frac{h}{\sqrt{|k_0|}} \quad \Longrightarrow \quad \varepsilon = \frac{h}{L} \simeq \sqrt{|k_0|} .$$

The last relation shows that the model is applicable only if the longitudinal compression of the *gossamer* is of order of the main small parameter, namely $\varepsilon \sim \sqrt{|k_0|} \ll 1$.

Some remarks are due concerning the "master wave equation" (2.16). Its linear part has the form originally proposed by Schrödinger (Eq. (4) from [20]). Naturally, Schrödinger himself mentioned the analogy with the plate equation and injected a remark in the cited paper. It was later on when the interpretation of the wave function as probability distribution has been introduced and the now standard form of the equation involving a complex wave function was introduced. In order to distinguish the originally derived equation from the "canonical" form we call the linear wave equation containing fourth-order dispersion Schrödinger's Schrödinger Equation (SSE). The derivation of Schrödinger was rather heuristic. Here we have arrived at qualitatively similar equation but bearing in mind a "palpable mechanical construct." The wave function now has a simple meaning: the deflection of *gossamer*'s middle surface in direction normal to it (alongside of the $(N+1)$-st dimension). These are *par excellence* curvature waves envisaged by Riemann [17] and Clifford [18].

3. The Cosmos of Localized Structures in Gossamer (Dynamics of Patterns in the Metacontinuum)

3.1. A Model for Loading the Metacontinuum

A way to achieve a homogeneous and isotropic loading of *gossamer* is to consider a large $N^{\mathcal{D}}$ bubble (hypersphere) subjected to hydrostatic pressure from the adjacent $(N+1)^{\mathcal{D}}$ spaces. We call this hypersphere *Universe*. We assume that the bubble is compressed from the outside $(N+1)^{\mathcal{D}}$ (negative membrane stress in the middle surface). A discussion on the case when the model of *Universe* is an inflated bubble can be found in [7]. It is however, a kind of artificial since to get there the familiar *sech*-solitons one has also to change the sign of the cubic term. Hence the "inflated-bubble" model appears to have only heuristic significance. The motionless (equilibrium) state is characterized by the balance between the membrane tension and the hydrostatic compression creating a negative membrane tension in the middle surface, namely

$$\sigma_0 \frac{N}{R} = -\rho_* |\mathcal{F}| \quad \Longrightarrow \quad k_0 = -\frac{|\mathcal{F}| R}{N c^2} < 0 \,.$$

Here R is the radius of the bubble made dimensionless by the scale L, and N is the dimension of the middle surface of the *gossamer* ($N = 3$ for the physical Universe).

It is clear that the dimensionless force $\mathcal{F}$ must be small enough (of order of the inverse of the dimensionless radius of Universe). We introduce the relative displacement $\bar{\zeta} = \zeta - R$. Since we are interested in an Universe whose scale is much larger than the size L of its particles then the dimensionless radius R is extremely large. Hence its derivatives are very small and can be neglected in the cubic term. For all purposes the shell can be treated as plane. Then Eq. (2.13) is reduced to the following

$$\frac{\partial^2 \bar{\zeta}}{\partial t^2} = -\Delta \bar{\zeta} + \beta [(\Delta \bar{\zeta})^3 - \Delta \Delta \bar{\zeta}] \,, \tag{3.1}$$

and $\beta = 1$ without loss of generality.

3.2. Flexural Localized Structures—The Flexons

In the stationary case the localized waves under question have spherical symmetry and for them the following dimensionless boundary value problem in infinite domain is posed

$$b - b^3 + \frac{1}{r^{N-1}} \frac{d}{dr} r^{N-1} \frac{d}{dr} b = 0, \quad b \to 0 \quad \text{for} \quad r \equiv |\mathbf{x}| \to \pm\infty \,, \tag{3.2}$$

where $b = \Delta \zeta$ is the curvature of the nontrivial transverse elevations/depressions. The linearized version of Eq. (3.2) possesses along with the trivial solution a localized non-trivial one (the *sinc* function):

$$\zeta = a r^{-1} \sin r \,, \tag{3.3}$$

where a is an arbitrary constant. This means that a linear bifurcation takes place which makes our problem different from the classical soliton problems where a hard (nonlinear) bifurcation is at hand. The *sinc*-shape solution (3.3) has been just recently interpreted as a "single event" of the Schrödinger equation [21]. In fact the present work justifies using the dispersive master equation (or Schrödinger equation) not as an equation for the probability density of a particle, but rather as a field equation (in the same fashion as *sine*-Gordon equation is used after [8]). It has to be pointed out that in our model we do not impose the shape of the potential as it is usually done in quantum mechanics.

We have solved (3.2) by means of an numerical algorithm based on the Method of Variational Imbedding (MVI) developed in [22] with application to homoclinics of Lorentz system (see also [23] for application to dissipative solitons). Because of the slow algebraic decay of the tails of the solution the interval has to be truncated at very large r. At the same time the grid has to be dense enough to allow resolving the oscillations. The results presented here employ grids with up to 40000 points and spacing 0.01. The solution turned out to be very sensitive to the mesh size and the magnitude of the "actual infinity," so some additional refinement could change the the amplitudes of solitons presented in Fig. 1 with couple of percents. Typically for nonlinear bifurcation problems, more than one nontrivial solution appear for the same values of the governing parameters. The question of the number of non-trivial solutions is of prime importance for the physical applications. Here we have found solutions with discrete set of amplitudes. This result suggests that for larger amplitudes the support of the main peak of the solution shrinks, i.e. the larger particles have shorter Compton wavelength. With the increase of the amplitude the behaviour of the solution in the origin $r = 0$ approaches r^{-1} which is a singular solution bringing a balance between the second order operator and cubic nonlinear term in (3.2). The physical significance of such kind of singular solutions is not yet understood in the framework of the proposed here paradigm and we did not go for larger amplitudes.

We identify the obtained here flexural solitary waves of the governing equations of gossamer as the *particles*. The fact that the metacontinuum is elastic does not necessarily mean elastic interaction of the particles (see the demonstration of this fact for the Boussinesq equation in [24]). As we show in what follows, the model under consideration is a conservative one, but to prove that our flexural localized waves live up to the definition of *solitons* is necessary either to find the two-soliton solution of (3.1) or to demonstrate the interactions of *flexons* numerically. We have done neither of these two things because the analytical techniques simply do not work for negative membrane tension. At the same time the 1D reduction of our model does not possess localized solutions. Numerical solution of a 2D problem with the above mentioned requirements for the grid size in each direction would have needed enormous computational resources. Until their solitonic properties are strictly established we will call the localized waves discovered here *flexons* which carries also a hint of their origin (flexural deformations or deflections).

As it will be seen later on, the amplitude of a *flexon* is directly related to the *mass* of particle. Naturally, a *flexon* with negative amplitude will be an *anti-particle*. The simplified Eq. (3.2) does not distinguish between particles and anti-particles. However, in the original model of *Universe* a slight difference between particles and anti-particles is to be expected due to to the curvature of the undisturbed bubble. Either the particles or the anti-particles will have a better chance to appear depending on which of them minimize the stored elastic energy of the shell.

One sees in Fig. 1 that the amplitudes (and hence—the masses) of flexons can be quite different. To excite a bigger *flexon*, one has to invest greater energy as initial

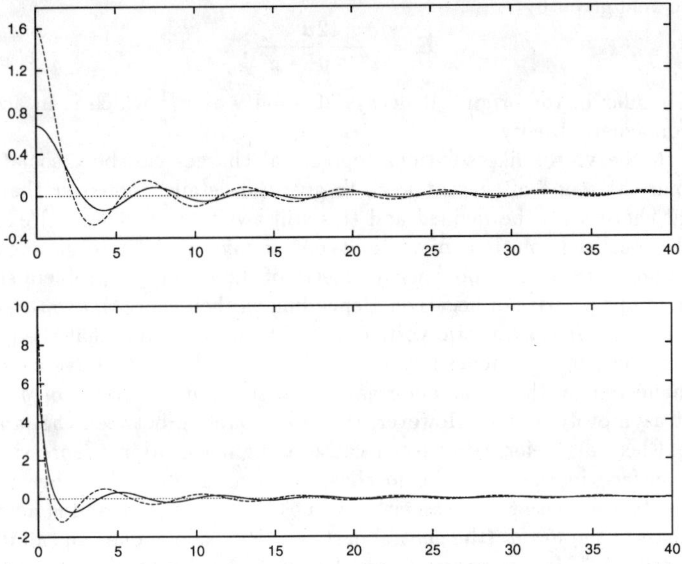

Fig. 1. Flexon solution with small amplitudes (up) and large amplitudes (down).

condition of the iterations. Respectively, in the iterative process some of the bigger particles decay to smaller emitting part of their energy. In this instance, the situation is qualitatively similar to relationship of nucleons and quarks.

3.3. Localized Shear Waves in the Middle surface: Torsion Solitons (Twistons)

Although the way we load the gossamer is the simplest one (a uniform hydrostatic pressure), it turns out that a complex "cosmological" picture appears with host of different localized waves. Bifurcation and symmetry breaking does not affect just the transverse deformation. Here we provide an example of what can happen in the middle surface. Unfortunately the material nonlinearity has not yet been incorporated in the model and the linearized equations for laminar components remain not closed due to the presence of the pressure-like term $\nabla\varphi$. For that reason the presented here solution has mainly qualitative heuristic bearing. By direct inspection one finds that the vector

$$\boldsymbol{u} = \left\{ \frac{x(z^2-y^2)}{(x^2+y^2+z^2)^{\frac{3}{2}}},\ \frac{y(x^2-z^2)}{(x^2+y^2+z^2)^{\frac{3}{2}}},\ \frac{z(y^2-x^2)}{(x^2+y^2+z^2)^{\frac{3}{2}}} \right\}$$

satisfies the "incompressibility condition" $\nabla \cdot \boldsymbol{u} = 0$. The vector lines corresponding to this solution resemble much a vortex or a bundle. This deformation field creates an electric field given by

$$E = \frac{12\boldsymbol{u}}{x^2 + y^2 + z^2},$$

which is singular in the origin. It decays at infinity as r^{-2} which is in accord with the electromagnetic theory.

With the vortex-like solutions topological charges can be associated but it goes beyond the scope of present work to give the details, moreover that different topological charges can be defined and this still awaits its mechanical explanation. We call the localized solutions of vortex type *twistons* in order to distinguish them from the fluid vortices. The obvious symmetry of the linearized problem shows that the charge can be positive or negative. Depending on their charges, two *twistons* repel or attract each other (just as two vortices do). The latter means that the presence of a *twiston* can only be experienced by another *twiston*. The neutral particles (*flexons*) remain unaffected by the shear deformations in the middle surface of *gossamer* to the lowest asymptotic order. However, the $O(\delta)$ coupling between the laminar and transverse (flexural) deformations can cause a slight elevation (depression) of the *gossamer* surface in the region of localization of a *twiston*. In other words, the *twiston* has its own mass which can either be positive or negative and is much smaller than the mass of the *flexon* (the neutral particle). By analogy one can call it "mass of electron/positron." Massive charged particles can be produced when a *twiston* "nests on" a *flexon* and then the mass is the superposition of the amplitude of the neutron (*flexon*) and the amplitude of the flexural deformation associated with the *twiston* (positron or electron).

There is no much sense going on here with the details before the material nonlinearity of *metacontinuum* is established. In addition, the singularity of vortex solutions suggests that some higher-grade elasticity is to be admitted in order to change the behaviour of the solution for $r \to 0$, e.g., in the following fashion

$$\frac{\partial^2 \boldsymbol{u}}{\partial t^2} + \Phi(\nabla \boldsymbol{u}, \Delta \boldsymbol{u}) = \Delta \boldsymbol{u} - \nabla \varphi - \chi \Delta \Delta \boldsymbol{u}, \tag{3.4}$$

where Φ stands for the yet unspecified nonlinearity and χ is the dispersion coefficient.

Here a comment is due on the limiting speed of light. Clearly, the speed of light is not a limiting celerity for a *flexon*. A truly neutral *single* particle should not be affected by the speed of shear waves (light) as being orthogonal to shear deformation. However, as far as the charged particles are concerned, the speed of light is indeed the limiting speed because the *twistons* are solutions of the nonlinear wave equations for the laminar components of displacements. Hence an atom containing charged particles cannot exceed the speed of light since above it the charges (twistons) will be disjointed from the flexural "humps" that carry them.

3.4. Density Solitons

Alongside with the flexural and torsion solitary waves, one must expect also solitary waves connected with the compressibility of *metacontinuum*. Hence a quantitative numerical solution for the density solitons has not been attempted by us. However, some qualitative conclusions can be reached on the basis of the known properties of the compression waves in solids. First of all, the speed of these waves is limited by the speed of sound of the *metacontinuum* and the steepness of a density wave increases with its celerity. This means that the swifter the movement of the wave—the smaller its spatial extent. On the other hand, the density solitons are expected to interact almost insignificantly with the matter (the *flexons* and *twistons*) and in this instance they will resemble to a great extent the behaviour of the neutrino. This qualitative description allows us to consider the all pervading compression/dilation motion (orthogonal to the matter) as quite similar to what the ancient School of Stoa called *pneuma* (see, e.g., [25]).

3.5. The Shell Membrane Tension or The Gravitation

According to the picture drawn here, the particles are localized elevations (humps) of the gossamer surface. Due to the presence of a *flexon* of shape ζ situated at certain geometrical position, the material points will experience attractive force proportional to $|k_0|\nabla\zeta$. For a single *flexon* (see Fig. 1), this force will not be monotone. The bare fact is that nobody has measured the gravitation force between two elementary particles and it is not clear whether the attractive force is monotone in the intra-atomic regions. In fact, the gravitation law is established only for bodies (ensembles of *flexons*). So one has to consider only the flexural-deformation field as averaged over the different positions of the particles-flexons. It is to be expected that the total average is positive (attraction), since the positive humps of the *flexon* amplitude are larger than the negative. Taking the *sinc*-shape for qualitative purposes one has approximately

$$\zeta = \sum_{\boldsymbol{x}_\alpha \in D_B} \frac{a_\alpha \, f|\boldsymbol{x}_\alpha - \boldsymbol{x}|}{|\boldsymbol{x}_\alpha - \boldsymbol{x}|} \approx \frac{g}{|\boldsymbol{x}|}, \quad g \sim \gamma \int f(|\boldsymbol{x}_\alpha|)\, d\boldsymbol{x}\,, \tag{3.5}$$

where D_B is the region occupied by the body. Here it is acknowledged that $|\boldsymbol{x}| \gg |\boldsymbol{x}_\alpha|$ and an assumption is made that the centers of particles are randomly distributed in the region of body with number density γ. Note that if the *flexons* were strictly *sinc*-shaped, then the constant g would have been equal to zero for randomly dispersed atoms (integral of *sine* is zero). For a *flexon* obtained here $g \neq 0$ because its amplitude in the origin is larger than of the *sinc*, so that after rescaling by $|\boldsymbol{x}|$ the *flexon* shape gives for the integral in (3.5) a positive quantity.

Thus the force that is experienced by a material point of the *gossamer* is proportional to $\nabla\zeta \sim G|\boldsymbol{x}|^{-2}$, $G = |k_0|g$ and both ingredients of G have extremely small values. The exact value of the "gravitational constant" G can be specified only after the averaging procedure is performed with the appropriate rigor. The membrane force acts to "pull" the material points of the *gossamer* towards the center of the particle system under consideration. Thus we arrive at Newton's inverse-square law of gravitation which is a manifestation of the fact that the shell is a 3D continuum.

In fact, the attraction between particles arises out of the disturbances they introduce in the uniform membrane anti-tension acting in the shell, i.e. we discover a quantitatively reversed but philosophically identical picture to the concept of Mach that the gravitation is due to the interaction with the quiescent matter at the rim of Universe. Here apply the words of Maxwell from the end of Part IV of [14] "...that the presence of dense bodies influences the medium so as to diminish this energy whatever there is a resultant attraction." Indeed, the presence of humps over the gossamer surface influences the medium so as to diminish the stored energy and there arises a resultant attractive force.

In our model there is no place for gravitational waves because the membrane tension is negative hence the deflection waves would propagate infinitely fast were the dispersion not present. Bizarrely, the particles themselves can loosely be called "gravitational waves" since part of the forces constituting the particles is the gravitation (membrane anti-tension). In other words, a particle is a localized nonlinear wave sustained from the balance between the generating effect of the membrane anti-tension on the one hand, and the restraining effect of the cubic nonlinearity and dispersion on the other.

3.6. Dispersion and "Red Shift"

It is peculiar that the Boussinesq equation (3.4) possesses *localized* solutions (see [24,26]) that propagate with the characteristic speed (photons?) and undergo some aging ("red shift") in the sense that their support increases while the amplitude decreases. Far from the source, one cannot distinguish between the red-shifting due to dispersion or to a Doppler (effect if present). This means that if a dispersion is present then the "red shift" can be alternatively explained without the help of "Big-Bang" hypothesis.

3.7. Estimating the Constants of Metacontinuum

The density of *metacontinuum* is the magnetic permeability μ_0. It is well known that the whole electrodynamics can be built without specifying the dimension of μ_0. In our paradigm it is exactly to be expected that way, since μ_0 is a *meta* quantity. Its dimension is not needed anywhere in the model and the density of matter (being the number density of solitons) has nothing to do with it.

The shear Lamé coefficient η is the inverse of the electric permittivity ε^{-1}. The thickness of *gossamer* is proportional to Plank's constant $h \sim \hbar c^{-1}\mu_0^{-1}$. If there is a dispersion coefficient in the equations for laminar displacements it can be estimated from the Hubble constant. The dilational Lamé coefficient λ can be estimated only from an "acoustic" experiment in the metacontinuum.

4. Dynamics of Patterns in Metacontinuum

4.1. The Hamiltonian Formulation

Consider Eq. (3.1) in the domain D with boundary conditions for the wave-amplitude ζ

$$\zeta = \psi_0(\boldsymbol{x}), \quad \Delta\zeta = \psi_2(\boldsymbol{x}), \quad \text{for} \quad \mathbf{x} \in \partial D. \tag{4.1}$$

The trivial boundary conditions correspond to an isolated system of wave-particles. A Hamiltonian representation is readily derived upon multiplying (3.1) by $\Delta\zeta_t$ and integrating over the infinite domain.

$$H \equiv \frac{1}{2}\int_D \left[(\nabla\zeta_t)^2 - (\Delta\zeta)^2 + \frac{1}{2}\beta(\Delta\zeta)^4 + \beta(\nabla\Delta\zeta)^2\right] d^N\boldsymbol{x}\,. \tag{4.2}$$

For boundary conditions that do not depend on time, the total energy H is conserved $dH/dt = 0$.

The energy functional (4.2) is not positive definite, but unlike the Boussinesq wave equation the solution will not blow up in finite time since the quartic nonlinearity dominates the quadratic term saturating the growth of solutions. Although the second-order term in (3.1) is of improper sign the equation is not unstable with respect to short wave lengths because of the presence of fourth spatial derivatives with the proper sign.

The wave mass and wave momentum are defined as

$$M \equiv \int_{-\infty}^{\infty} \Delta\zeta\,\mathrm{d}^N\boldsymbol{x}\,, \quad \boldsymbol{P} \equiv \int_{-\infty}^{\infty} \zeta_t \nabla\Delta\zeta\,\mathrm{d}^N\boldsymbol{x}\,. \tag{4.3}$$

The concept of *pseudomomentum* in continuum mechanics was elaborated recently [27,28] especially in connection with the interpretation of localized waves as "quasi-particles" and numerous featuring examples are presented in [29]. For this model the meaning of *mass* is as "mass of curvature" which once again is in the vain of Riemann proposal. The *mass* is conserved if $\partial\zeta_{tt}/dn = 0$ at ∂D which is a natural requirement for an isolated system. For the *pseudomomentum* we have

$$\begin{aligned} \frac{d\boldsymbol{P}}{dt} &= \oint_{\partial D} \boldsymbol{n}\left[-\frac{1}{2}(\nabla\zeta_t)^2 - \frac{1}{2}(\Delta\zeta)^2 + \frac{\beta}{4}(\Delta\zeta)^4 + \frac{\beta}{2}(\nabla\Delta\zeta)^2\right]\mathrm{d}\boldsymbol{s} \\ &\quad - \oint_{\partial D} \beta(\nabla\Delta\zeta)\frac{\partial\Delta\zeta}{\partial n}\mathrm{d}\boldsymbol{s} = \mathcal{F}_{\mathrm{psu}}\,, \end{aligned} \tag{4.4}$$

where $\mathcal{F}_{\mathrm{psu}}$ is called *pseudoforce* and $\boldsymbol{n}$ is the outer normal to the region D. The *pseudomomentum* is conserved if the *pseudoforce* is equal to zero. Note that even for an isolated system the *pseudomomentum* is not conserved if the localized patterns hit the boundary and rebound.

4.2. From Metadynamics of Underlying Continuum to Dynamics of Centers of Localized Structures

The main significance of the Hamiltonian formulation is that it provides the means to build up the dynamical model for the *discrete* phase objects (*solitons*). If the shape of localized wave is known, the Hamiltonian dynamics for the discrete system of centers of "particles" can be derived from (3.1) with a good approximation provided that they do not interact so strongly as to change appreciably their shapes. For two localized waves the wave amplitude can be decomposed as follows:

$$\zeta = F_1(\mathbf{x} - \mathbf{X}_1(t)) + F_2(\mathbf{x} - \mathbf{X}_2(t)) + F_{12}(\mathbf{x} - \mathbf{X}_1(t), \mathbf{x} - \mathbf{X}_2(t))\,. \tag{4.5}$$

Here F_i are the shape functions of the waves–particles and $\mathbf{X}_i(t)$ are the trajectories of their centers. When the particles are far from each other F_{12} is negligible. In fact F_{12} must be considered only in the cross-section of the collision of two particles and we leave the problems connected with this case for future study.

Now the time derivative of each shape function can be expressed as follows

$$\frac{dF_i}{dt} = -\nabla F_i \cdot \frac{d\mathbf{X}_i}{dt} \tag{4.6}$$

and the discrete Hamiltonian contains quadratic forms of the velocities of centers

$$\sum_i \mathcal{A}_i \cdot \dot{\mathbf{X}}\dot{\mathbf{X}}\,, \tag{4.7}$$

where the matrix $\mathcal{A}_i$ is positive definite. Then (4.7) can be interpreted as definitive relation for the kinetic energy of the center of particle. Respectively, F_2 will contribute a term which depends on the relative position of particles and plays the role of potential of interaction. In absence of interactions, the Euler–Lagrange equations for discrete system of centers of *solitons* give $\ddot{\mathbf{X}} = 0$ and a *soliton* will propagate with constant phase velocity. Thus for isolated particles the first Newton law is recovered.

4.3. FitzGerald-Lorentz Contraction and Contraction of Flexons

The FitzGerald-Lorentz contraction (FG–L, for brevity) is a standard feature with the soliton solutions of generalized wave equations, e.g., SG, Boussinesq, etc. It is especially well seen for the *sine*-Gordon equation where the "quasi-particles" are called "relativistic"[4] meaning that their contraction is exactly proportional to the Lorentz factor. Naturally, the above discussed localized shear waves are subject precisely to the same factor of contraction. This holds true also for the length scales of interaction forces (e.g., the Coulomb force).

When a *flexon* propagates with a constant celerity V its measure in the direction of propagation must be shortened by the factor (see (2.15), (3.1))

$$\left(1 + V^2/c_f^2\right)^{-\frac{1}{2}}\,, \tag{4.8}$$

where c_f is the pseudo-velocity corresponding to the negative membrane tension. Respectively the amplitudes of *flexons* (related to their particle-wave masses) are increased roughly by the same factor. Since c_f is much smaller than the speed of light, the contraction of flexons must be felt for smaller phase velocities than the Lorentz contraction. On the other hand —contrary to the FG–L contraction—there is no singularity in the expression of *flexon* contraction (4.8). However, the new type of contraction is much less important for the contraction of the bodies since it only affects the scale of gravitational interaction which is anyway the weakest and is not important in defining the intra-atomic scales. Yet, it is quite possible that some small

[4]Another specimen from the glittering panoply of coinages which is misleading in this context since SG is a model of an absolute field

effect could be felt due to the flexural contraction. It will not be a surprise if the weak but persistent deviations found in [30] from the nil effect of the Michelson-Morley experiment turn out to be the result of this additional contraction.

5. A First-Order Experiment for Detecting the Doppler Effect

To use interferometry for verification of the Doppler effect was suggested by Maxwell [31]. It was believed that discovering a Doppler effect will prove the existence of an absolute medium at rest. The experiment was implemented first by Michelson [32] and nil effect was observed. It was later on refined by Michelson and Morley [33] (MM, for brevity) and the absence of the expected type of interference was confirmed more decisively. In our opinion the nil effect of MM experiment cannot disprove the existence of absolute medium because it seeks for a second-order effect in the small parameter $d = v_e c^{-1} \approx 10^{-4}$ (v_e stands for the velocity of Earth with respect to the quiescent medium). This was specially pointed out by Maxwell [31] way before the experiment was performed. He proved that employing only a single ray with splitting and reflections inevitably renders the sought effect of second order d^2 because the light travels along a closed path. The only conclusion that can be drawn from the nil effect is that in the medium where the light is being propagated there occurs an apparent contraction of the spatial scales in the direction of motion of the source (FitzGerald-Lorentz contraction) proportional to the factor

$$1 - \frac{v_e^2}{2c^2} \approx \sqrt{1 - \frac{v_e^2}{c^2}}$$

which exactly compensates for the expected second-order effect.

In fact MM result strongly suggests the existence of an absolute medium because the FitzGerald-Lorentz contraction is a mandatory effect in a field theory based on the nonlinear generalizations of the D'Alembert wave equation (see the comments in a preceding section). Thus the soliton paradigm provides a most natural explanation of the nil effect of MM experiment. If the "material bodies" (e.g., the arms of the interferometer) are *bound states* of solitons, then they are contracted in the direction of motion because the solitons themselves and the inter-soliton distances are. The latter are defined mainly by the electromagnetic forces whose length-scales also suffer contraction. Hence the distance that must be traveled by the light through the *quiescent* metacontinuum between two soliton formations (the emitter and receptor), is indeed shorter in the direction of motion in comparison with the path traveled in transverse direction.

Note that the *metacontinuum* itself is not contracted and that is why the speed of light has a constant value in each coordinate system connected with the propagating objects—*solitons*.

The real proof of the presence of a *metacontinuum* would be an experiment for the first-order effect. Such an experiment can only be achieved if *two different* sources of light are employed with sufficiently synchronized frequencies. There are available in the market lasers with the required stabilization (e.g., the *HeNe* Model

200 of Coherent Components Group) but it goes beyond the frame of the present work to present the details on hardware. As the sought effect is of order of 10^{-4} then if the two sources are synchronized up to 10^{-6}, the accuracy would be of order of 1%. In this sense we will call such sources "identical." Here we give in Fig. 2 the principle scheme of a possible first-order experiment.

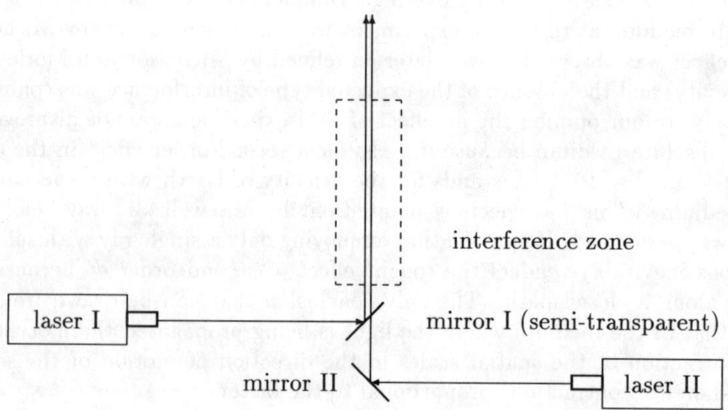

Fig. 2. Principal scheme of the interferometry experiment.

We deliberately exclude from consideration any kind of non-optical experiment and leave beyond our scope the optical experiments in dense matter (water filled columns, etc.). Consider two "identical" sources of monochromatic light which move together in the same direction with the same velocity. The first of them emits a plane wave propagating in the direction of motion of and the second one—in the opposite direction. By means of a mirror and a semi-transparent mirror the two plane waves are made co–linear. The beam of the second laser is reflected by the mirror II changing its direction on 90° and made to pass through a semi-transparent mirror whose reflecting surface serve to change the direction of the beam of first laser on 90°. Beyond the semi-transparent mirror the two beams are parallel and can produce an easily discernible interference stripped pattern. A snapshot of the region of interference would reveal strips of different intensity gradually transforming into each other and the modulation frequency can easily be measured.

It is interesting to note that Jaseda *et al.* [34] already used two lasers in interferometry experiment in order to quantitatively verify the FG–L contraction, but in their experiments the lasers beams were parallel (aiming to verify quantitatively the contraction) while in the proposed here experiment they are anti–parallel since now it is not the contraction that needs verification, but the very existence to the first-order of Doppler effect.

For the sake of self-containedness of the paper we outline here the derivation of Doppler effect (see also [35]). The harmonic waves propagating in presumably quiescent medium are given by the following formula

$$F_{\pm}(x,t) \equiv e^{i(k_r x \mp \omega_r t)}\ , \quad k_{\pm} = \frac{\omega_r}{c}\ , \quad \lambda_{\pm} = \frac{c}{\omega_r}\ , \tag{5.9}$$

where $\omega_{\pm}$ are the frequencies. The upper sign in the notations refers to the wave propagating in positive direction, while the lower sign—to the wave propagating in negative direction. These waves have to satisfy the boundary condition on the moving boundaries (the sources)

$$F_{\pm}(\pm v_e t, t) \equiv e^{i\omega_0 t}\ , \tag{5.10}$$

where v_e is the velocity of the moving frame relatively to the metacontinuum. Respectively, if the sources were at rest, then they would have produced waves with wave number $k_0 = \omega_0/c$ and wave length $\lambda_0 = k_0^{-1}$. The boundary condition (5.10) yields the following relation for the parameters of the propagating wave:

$$\omega_{\pm} = \omega_0 \left(1 \mp \frac{v_e}{c}\right)^{-1} \quad k_{\pm} = \frac{\omega_0}{c}\left(1 \mp \frac{v_e}{c}\right)^{-1} \quad \lambda_{\pm} = \lambda_0 \left(1 \mp \frac{v_e}{c}\right)\ . \tag{5.11}$$

The role of the mirrors is to change the direction of propagation of each wave without destroying its plane nature. After the reflection, the two waves are propagating as planar waves in the positive direction of z-axis (vertical in Fig. 2): $F_{\pm}(z,t) \equiv e^{i(k_{\pm} z - \omega_{\pm} t)}$. Then in the interference region one has a wave which is the superposition of two of them for a given moment of time (say, $t = 0$) so that

$$\Re\left[F_{+}(z,t) + F_{-}(z,t)\right] = 2\cos\left(\frac{k_{+} + k_{-}}{2} z\right)\cos\left(\frac{k_{+} - k_{-}}{2} z\right), \tag{5.12}$$

which is a modulated wave with a wave number of the carrier $\frac{1}{2}(k_{+} + k_{-}) = k_0 + O(d^2)$ and with a wave number for the modulation $\frac{1}{2}(k_{+} - k_{-}) = dk_0 + O(d^3)$. Respectively the expressions for wave lengths valid to the second order, are λ_0 and $\lambda_m = \lambda_0 d^{-1}$. For red-light lasers the length of the wave is $\lambda_0 \approx 6.3 \cdot 10^{-5}$ cm and then for the length of modulation wave one has $\lambda_m = 0.63$ cm and the strips produced must be easily detectable on an optical table of standard dimensions.

Alternative way around is to look for the fringes formed on the semi-transparent mirror. The length scale of a fringe would be around 6.3 cm for the red light and the laser beams should be expanded so as to cover a region of 40-50 cm when they reach the semi-transparent mirror.

6. Quasi, Pseudo, Meta: Concluding Remarks

From the point of view of soliton paradigm, the notion of discrete versus continuous (wave–particle dualism) is revisited. As a featuring example a very thin layer of Hookean elastic medium of very large dilational modulus is considered: a special kind of N-dimensional shell called *gossamer*. It is shown that the linearized equations for the laminar displacements have as a corollary the Maxwell equations for appropriately defined quantities called electric and magnetic fields. A higher-order dispersive and nonlinear Boussinesq equation is derived for the flexural deformations of *gossamer* ("master" equation of the wave mechanics). Its linear part is the linearized Schrödinger equation. Due to the conservative properties of the "master" equation, the flexural solitary waves (*flexons*) appear to be solitons. We call "Soliton Paradigm" the conceptual framework in which the solitons are identified as the particles, i.e. there is no dichotomy between particles and waves. A *particle* (or *corpuscles*) is a notion to signify our perception of the geometry and dynamics of a localized wave ("lump" of deformations, stress, energy, etc.) of *metacontinuum*. The Hamiltonian properties of wave-mechanics equation (the *metadynamics*) define the Hamiltonian (Newtonian) dynamics of the phase objects (*particles-solitons*). Thus the de-Broglie wave-particle duality is a matter of observation and perception: we are faced with a unique object—a localized wave, which is perceived either as a corpuscular object (in fact its center of wave-mass) or as a wave. Particles-Solitons are not moving *through* the metacontinuum, rather they are *phase patterns* propagating *over* it without disturbing its contiguity. Hence no aether-drift effect will be associated with the motion of a particle-soliton. A possible experimental set-up is proposed for verification of this concept.

Solutions of the "master" equation (called *flexons*) are obtained numerically by means of Method of Variational Imbedding. Localized torsional structures of integer-valued topological charge are discussed qualitatively and identified as *charges*. It is shown that the membrane tension in the *gossamer* creates attractive force (*gravitation*) between the large ensembles of localized flexural deflections-particles which is proportional to r^{1-N}.

The concept of unification based on *metacontinuum* and *soliton paradigm* gives:

- Maxwell Equations for the stress interactions in the middle surface of the *gossamer*;
- Dispersive equation (Schrödinger's Schrödinger equation) for the wave function of the transverse flexural deformations alongside the fourth spatial dimension;
- Gravitation as membrane (*anti*) tension in the middle surface;
- Charges as vortex-like solutions.

and can be summarized as follows

Metacontinuum
$\Downarrow$
Particles (Localized Waves — Solitons)
and *Fields* (Derivations of Stress- and Strain-fields of *metacontinuum*)
$\Downarrow$
Hamiltonian-Lagrangian Formalism
$\Downarrow$
Newtonian mechanics of "point particles" (centers of *solitons*)
$\Downarrow$
Mechanical continua (approximation for ensembles of particles)

The unification here is not only for the forces of interaction (the different fields), but it also fuses the Wave-particle duality into Particle-Wave *unity* subordinating the concept of a *particle* (*corpuscle*) to the one of localized phase patterns engendered by the balance between the membrane anti-tension and the nonlinearity and dispersion of the "master" equation of wave mechanics.

Acknowledgments. An European Commission Mobility Grant ERBCHBICT 940982 is gratefully acknowledged by the author.

References

1. E. Whittaker, *A History of the Theories of Aether & Electricity*, Dover, New York, 1989.
2. J. S. Russell, In *Report of 7th Meeting (1837) of British Assoc. Adv. of Sci., Liverpool*, John Murray, London, 1838, p. 417.
3. J. S. Russell, In *Report of 14th Meeting (1844) of the British Association for the Advancement of Science, York*, 1845, p. 311.
4. J. V. Boussinesq, *J. de Mathématiques Pures et Appliquées* **17**(2) (1872) 55.
5. Lord Rayleigh, *Phil. Mag.* **1** (1876) 257.
6. N. J. Zabusky and M. D. Kruskal, *Phys. Rev. Lett.* **15** (1965) 57.
7. C. I. Christov, In *Fluid Physics: Proceedings of Summer Schools*, eds. M. G. Velarde and C. I. Christov, World Scientific, Singapore, 1994, p. 33.
8. T. H. R. Skyrme, *Proc. Roy. Soc.* **A247** (1958) 260.
9. J. K. Perring and T. H. R. Skyrme, *Nuclear Physics* **31** (1962) 550.
10. *Speculations on the Fourth Dimension*, Selected Writings of C. H. Hinton, ed. R.v.B. Rucker, Dover, 1980.
11. T. Kalutza, *Sitz. Preuss. Acad. Wiss.* (1921) S966.

12. O. Klein, *Zeitsch. für Physik* **37** (1926) 895.
13. H. A. Lorentz, Aether Theories and Aether models, In *Lectures on Theoretical Physics*, vol.I, MacMillan, London, 1927.
14. J. C. Maxwell, *Trans. Roy. Soc.* **155** (1865) 469.
15. H. Neuber, *ZAMM* **29** (1949) 97; **29** (1949) 142.
16. M. Dikmen, *Theory of Thin Elastic Shells*, Pitman, Boston, 1982.
17. B. Riemann, *Gesammelte Mathematische Werke und wissenschaftlicher Nachlass*, eds. R. Dedekind and H. Weber, Teubner, Leipzig, 1892, p. 558. (See also English translation in W. Carol, *Energy Potential*).
18. W. K. Clifford, *Mathematical Papers*, ed. R. Tucker, London, 1982.
19. J. L. Ericksen and C. Trussdell, *Arch. Rat. Mech. & Anal.* **1** (1959) 295.
20. E. Schrödinger, *Annalen der Physik* **79** (1926) 743.
21. A. O. Barut, *Foundations of Physics* **20** (1990) 1233; *Phys. Lett. A* **143** (1990) 349.
22. C. I. Christov, In *Proc. 14th Conference of the Union of Bulgarian Mathematicians*, Sunny Beach, Bulgaria, 1985, p. 571.
23. C. I. Christov and M. G. Velarde, *Appl. Math. Modelling*, **17** (1993) 311.
24. C. I. Christov and M. G. Velarde, *Int. J. Bifurc. Chaos* **4** (1994) 1095.
25. S. Sambursky, *The Physical World of the Greeks*, Routledge and Keagan Paul, London, 1963.
26. C. I. Christov and G. A. Maugin, *J. Comp. Phys.* **116** (1995) 39.
27. G. A. Maugin, *J. Mech. Phys. Solids* **29** (1992) 1543.
28. G. A. Maugin, *Material Inhomogeneities in Elasticity*, Chapman and Hall, London, 1993.
29. G. A. Maugin and C. I. Christov, In *Nonlinear Wave Phenomena*, eds. D. Inman and A. Guran, World Scientific, 1996, to appear.
30. D. C. Miller, *Reviews of Modern Physics* **5** (1933) 203.
31. J. C. Maxwell, In *Enciclopedae Britanica*, Ninth Edition, 1875, p. 568.
32. A. A. Michelson, *Phil. Mag.*[5] **13** (1882) 236.
33. A. A. Michelson and E. W. Morley, *Am. J. Sci.* [3] **34** (1887) 333.
34. T. S. Jaseda, A. Javan and C. H. Townes, *Phys. Rev. Lett.* **10** (1963) 165.
35. T. P. Gill, *The Doppler Effect*, Logos Press, 1965.

Continuum Models and Discrete Systems
Proceedings of 8th International Symposium, June 11-16, 1995, Varna, Bulgaria, ed. K.Z. Markov
©World Scientific Publishing Company, 1996, pp. 395–403

NONLINEAR PROPAGATION OF WAVE PACKETS ON AN ELASTIC SUBSTRATE COATED WITH A THIN PLATE

B. COLLET and J. POUGET

Laboratoire de Modélisation en Mécanique,
Université Pierre et Marie Curie (CNRS URA 229),
4 Place Jussieu, F-75252 Paris cédex 05, France

Abstract. We study the formation of nonlinear localized states, mediated by modulational instability, on an elastic composite structure made of a nonlinear substrate coated with an elastic thin plate. In the low-amplitude limit, we show that the basic equation which governs the deflection of the elastic structure can be approximated by a two-dimensional nonlinear Schrödinger equation. The modulational instability conditions are then deduced for particular situations. They inform us about the selection mechanism of the modulus of the carrier wave vector and growth rate of the instabilities taking place in both longitudinal and transverse directions of the plate. The dynamics of the elastic structure is then investigated by means of numerical simulations which are directly performed on the basic equation. The numerics shows that, beyond the instability threshold, the initial harmonic wave is then transformed into a nonlinear localized wave which turns out to be particularly stable.

1. Introduction

Coherent excitations referred to as solitons or solitary waves require a subtle balance between nonlinearity and dispersion. These localized nonlinear waves are mainly known as: kinks, pulses and envelope modes. Most researches in this area are confined on quasi one-dimensional physical models and experiments [1]. However, some observed phenomena in plasma physics, nonlinear optics, solid state physics, hydrodynamics, etc. make the study of multidimensional models particularly exciting [2-8].

Deformable structures made of a *prestressed isotropic elastic thin plate* superposed on a *nonlinear substrate* are uniform waveguides enable to focus a high energy density such that the nonlinearities can be excited. This class of elastic composite structures is an interesting candidate for real observations of *two-dimensional wave packets with soliton-shape envelope* in elastic solids.

The purpose of the present work is to study the influences of the *geometric dispersion*, the *prestress* on the plate and *material nonlinearities* of the substrate on the modulation of flexural waves. The emphasis is placed on the localized states formation mediated by *modulational instability*. The basic equation which governs the dynamics of the structure is deduced from an energetic approach based on the Hamilton's principle. The analysis is restricted to signals which consist of slowly varying envelope in space and time modulating a harmonic carrier wave. In the limit of low amplitudes the equation is solved by means of a *reductive perturbation method*. It is shown that the complex amplitude of the envelope satisfies a *2D nonlinear Schrödinger equation* (2D-NLS) in which the dispersive and nonlinear coefficients depend on the model parameters and carrier wave vector. The modulational instability conditions are briefly examined for a simple case. Finally, we investigate by means of numerical simulations the role played by modulational instability in the evolution to *localized states* of an initial plane wave with low amplitude propagating on the elastic structure.

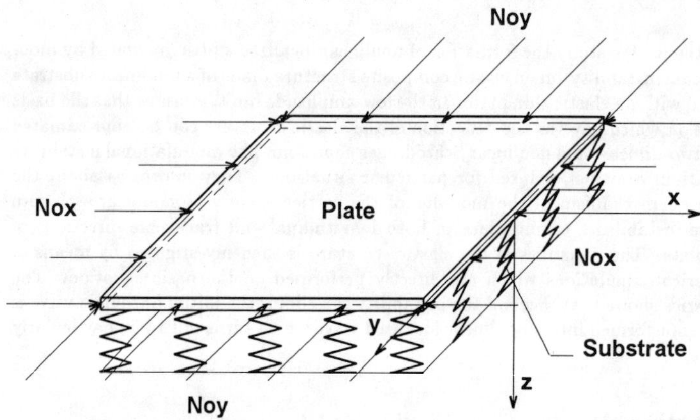

Fig. 1. Elastic structure and coordinate system.

2. The Model and Equation of Motion

Let us consider a deformable composite structure made of an infinite thin plate superposed on an elastic foundation (see Fig. 1). The attention is focussed on the flexural wave propagation. We suppose that the dynamic behavior of the plate can be described by the classical model of prestresses isotropic elastic thin plate [9]. The response of the nonlinear substrate or foundation is nonlinear and of the quartic type. In the absence of nonconservative forces and for any arbitrary part of the structure, the Hamilton principle can be written as

$$\delta \int_{t_1}^{t_2} (T-U)\,dt = 0\,, \quad U = U_B + U_P + U_F\,, \quad T = \frac{1}{2}\int_S \int_{-h/2}^{h/2} \rho\,(\bar{w}_t)^2\, dxdydz\,,$$

$$U_B = \frac{1}{2}\int_S D\left\{(\bar{w}_{xx}+\bar{w}_{yy})^2 - 2(1-\nu)\left[\bar{w}_{xx}\bar{w}_{yy} - (\bar{w}_{xy})^2\right]\right\} dxdy\,,$$

$$U_P = \frac{1}{2}\int_S \left\{N_{0x}(\bar{w}_{xx})^2 + N_{0y}(\bar{w}_{yy})^2\right\} dxdy\,, \tag{2.1}$$

$$U_F = \int_S \left\{\frac{1}{2}\bar{c}_1\bar{w}^2 + \frac{1}{3}\bar{c}_2\bar{w}^3 + \frac{1}{4}\bar{c}_3\bar{w}^4\right\} dxdy\,,$$

T, U_B, U_P and U_F denote the kinetic energy, the bending strain energy, the geometrical strain energy associated with prestresses and the potential energy of the nonlinear foundation, respectively. For a simple model of thin plate, the shear deformation and rotational inertia effects are usually neglected. In addition, $D = Eh^3/12\left(1-\nu^2\right)$ is the plate bending stiffness and E is the Young's modulus, h is the plate thickness, $\bar{w}(x,y,t)$ is the deflection of the plate in the z direction, ν is the Poisson's ratio, N_{0x} and N_{0y} are the initial prestresses (forces per unit length), ρ is the density and $\bar{c}_\alpha$ $(\alpha = 1,2,3)$ are the elastic constants of the foundation (with $\bar{c}_1 > 0$). After using the variational principle and introducing appropriate dimensionless quantities (not written here), the equation of motion takes the form

$$\nabla^4 w + 2N_x w_{xx} + 2N_y w_{yy} + w_{tt} + w + c_2 w^2 + c_3 w^3 = 0\,, \tag{2.2}$$

where ∇^4 is the bilaplacian operator in the plane of the plate.

In order to deduce important information about the linear frequency spectrum (see Fig. 2), the equation of motion (2.2) is linearized by setting $c_2 = c_3 = 0$. We look for harmonic plane wave solution to the linear equation which leads to the dispersion relation

$$\omega^2 = 1 - 2\left(N_x k_L^2 + N_y k_T^2\right) + \left(k_L^2 + k_T^2\right)^2 = 1 - 2Nk^2 + k^4\,,$$

$$N = N_x\cos^2\theta + N_y\sin^2\theta\,, \tag{2.3}$$

where $k_L = k\cos\theta$, $k_T = k\sin\theta$ and ω denote dimensionless longitudinal and transverse components of the wave vector $k = (k_L, k_T)$ and the circular frequency, respectively. The angle θ is formed by the x axis and the direction of propagation and N is an effective load parameter. The linear spectrum is given (see Fig. 2) for various values of N. We observe the presence of a cut-off frequency at the origin only for $N \le 0$ and we note a softening of the dispersion curve at a nonzero k for $0 < N < 1$. On the other hand Eq. (2.3) informs us about the existence of a particular value N_0 for a fixed k such that for $N < N_0$, the harmonic plane wave solution represents a neutral wave, while for $N > N_0$ the solution grows exponentially in time. The neutral stability curve $N = N_0(k) = \left(1+k^4\right)/2k^2$ has a minimum at N_{cr} and k_{cr} (see Fig. 3). N_{cr} and k_{cr} are the critical or buckling load and critical modulus of the wave vector [10]. In our notation we have $N_{cr} = 1$ and $k_{cr} = 1$.

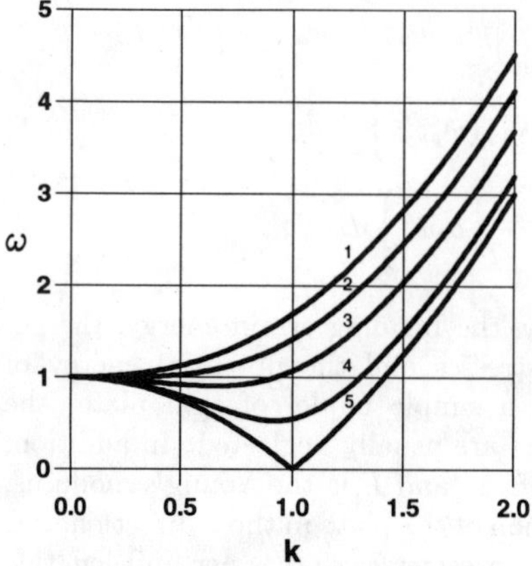

Fig. 2. Linear frequency spectrum (1: N=-0.425, 2: N=0, 3: N=0.425, 4: N=0.850, 5: N=1.0).

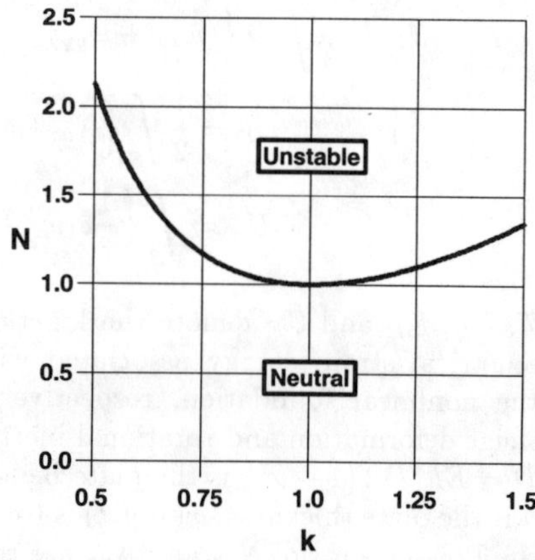

Fig. 3. Neutral stability curve (located around $k_{\rm cr}$=1).

3. Nonlinear Envelope Equation

In this section we consider the propagation of a wave packet of low amplitude centered around the dominant carrier wave vector k_c and the circular frequency ω_c. In order to obtain the slow spatio-temporal evolution of the wave packet amplitude, we use a *reductive perturbation method* [11] and we set

$$w(x,y,t) = \varepsilon A(\xi,\eta,\tau)e^{i\Phi_c} + \varepsilon^2\left(B(\xi,\eta,\tau) + C(\xi,\eta,\tau)e^{2i\Phi_c}\right) + \cdots, \tag{3.1}$$

where $\varepsilon \ll 1$ is a small dimensionless amplitude parameter; $\xi = \varepsilon\left(x - (V_{gL})_c t\right)$, $\eta = \varepsilon\left(y - (V_{gT})_c t\right)$ and $\tau = \varepsilon^2 t$ are the stretching and slow time variables of the envelope and $V_g = (V_{gL} = \omega_{k_L}, V_{gT} = \omega_{k_T})$ is the group velocity vector of the carrier wave. $\Phi_c = k_{Lc}x + k_{Tc}y - \omega t$ is the phase variable where k_{Lc}, k_{Tc} and ω_c are related through the dispersion relation (2.3). After a rather lengthy algebras we arrive finally at the equations

$$iA_\tau + P_1 A_{\xi\xi} + P_2 A_{\eta\eta} + P_3 A_{\xi\eta} + Q|A|^2 A = 0, \tag{3.2}$$

where

$$\begin{aligned}
P_1 &= \left[\left(k_c^2\left(3\cos^2\theta + \sin^2\theta\right) - N_x\right)\omega_c^2 - 2k_c^2\cos^2\theta\left(k_c^2 - N_x\right)^2\right]\omega_c^{-3},\\
P_2 &= \left[\left(k_c^2\left(3\sin^2\theta + \cos^2\theta\right) - N_y\right)\omega_c^2 - 2k_c^2\sin^2\theta\left(k_c^2 - N_y\right)^2\right]\omega_c^{-3},\\
P_3 &= \left[4k_c^2\cos\theta\sin\theta\left(\omega_c^2 - \left(k_c^2 - N_x\right)\left(k_c^2 - N_y\right)\right)\right]\omega_c^{-3},\\
Q &= \left[\left(4c_2^2 - 3c_3\right)\left(12k_c^4 - 3\right) + 2c_2^2\right]\left[2\left(12k_c^4 - 3\right)\omega_c\right]^{-1}.
\end{aligned} \tag{3.3}$$

The coefficients $P_\alpha, \alpha = 1,2,3$, and Q represent the dispersion and nonlinearity, respectively. At this point, it is important to note that, for a fixed angle θ and load parameters N_x and N_y, the sign of P_α can be modified by varying k_c. Similarly, when θ, N_x, N_y, c_2 and c_3 are chosen the sign of the nonlinear coefficient Q can be modified by varying k_c. We also note that Eq. (3.2), obtained at the third order inε, is a *two-dimensional nonlinear Schrödinger equation* (2D-NLS). The amplitudes B and C can be connected with A. We confine ourself to the particular case $\theta = \pi/4$ and $N_x = N_y = N_{iso}$, when

$$\begin{aligned} P_1 = P_2 &= \left[\left(2k_c^2 - N_{iso}\right)\omega_c^2 - k_c^2\left(k_c^2 - N_{iso}\right)^2\right]\omega_c^{-3}\,, \\ P_3 &= \left[2k_c^2\left(\omega_c^2 - \left(k_c^2 - N_{iso}\right)^2\right)\right]\omega_c^{-3}\,, \\ \omega_c &= \left(1 - 2N_{iso}k_c^2 + k_c^4\right)^{1/2}\,. \end{aligned} \tag{3.4}$$

4. Modulational Instability

Thanks to Eq. (3.2) we can investigate the stability of a plane wave propagating on the elastic structure. A linear analysis of a small disturbances of this plane wave solution yields a set of criteria of instability called *modulational or Benjamin-Feir instability* [3,7]. It is easy to show that the uniform solution $A = A_0 exp\left(iQA_0^2\tau\right)$, where A_0 is a real constant, satisfies Eq. (3.2). Now we introduce the perturbed amplitude and phase as follows

$$A = \left[A_0 + \varepsilon a(\xi,\eta,\tau)\right]\exp i\left[QA_0^2\tau + \varepsilon b(\xi,\eta,\tau)\right]\,, \tag{4.1}$$

where $\varepsilon \ll 1$ is a small parameter associated with the disturbances $a(\xi,\eta,\tau)$ and $b(\xi,\eta,\tau)$. By substituting Eq. (4.1) into Eq. (3.2) we obtain at the first order a set of two linearly coupled equations

$$\begin{aligned} a_\tau + P_1A_0b_{\xi\xi} + P_2A_0b_{\eta\eta} + P_3A_0b_{\xi\eta} &= 0\,, \\ b_\tau - P_1A_0^{-1}a_{\xi\xi} - P_2A_0^{-1}a_{\eta\eta} - P_3A_0^{-1}a_{\xi\eta} - 2QA_0a &= 0\,. \end{aligned} \tag{4.2}$$

We look for harmonic solution to the linear equations (4.2), which leads to the dispersion relation

$$\nu^2 = \left(P_1p^2 + P_2q^2 + P_3pq\right)\left(P_1p^2 + P_2q^2 + P_3pq - 2QA_0^2\right)\,, \tag{4.3}$$

where p and q are the components of the wave vector of the perturbations and ν may be complex. Modulational instability occurs at perturbation wavenumbers for which ν becomes imaginary $\nu = i\sigma$ (with $\sigma > 0$). The parameter σ is then similar to the inverse of time and it corresponds to the temporal growth rate of the perturbation or the amplification rate with a maximum given by

$$\sigma_{max} = |Q|A_0^2 \quad \text{along} \quad P_1 p^2 + P_2 q^2 + P_3 pq - QA_0^2 = 0\,. \tag{4.4}$$

The modulational instability criterion depends, of course, on the model parameters and on the carrier wave. A simple analysis of the growth rate σ allows us to distinguish, in the wave vector space (p, q), two main kinds of the instability domains: (i) unbounded domains defined by a hyperbola and its asymptotes and (ii) a bounded domain defined by an ellipse. We will not discuss in detail the instability criterion, but will illustrate only the case when $k_c = 0.8078$, $\theta = \pi/4$, $N_{iso} = 0.425$, $A_0 = 1$, $c_2 = -2.0$ and $c_3 = 1.0$. We compute $\omega_c = 0.933$, $P_1 = P_2 = 0.9$, $P_3 = 1.315$, $Q = 8.995$ and $\sigma_{max} = 8.995$. The growth rate σ as function of the perturbation wave vector has been plotted in Fig. 4, where the 2D-isoline graph shows clearly a set of ellipses defining the zone of instability.

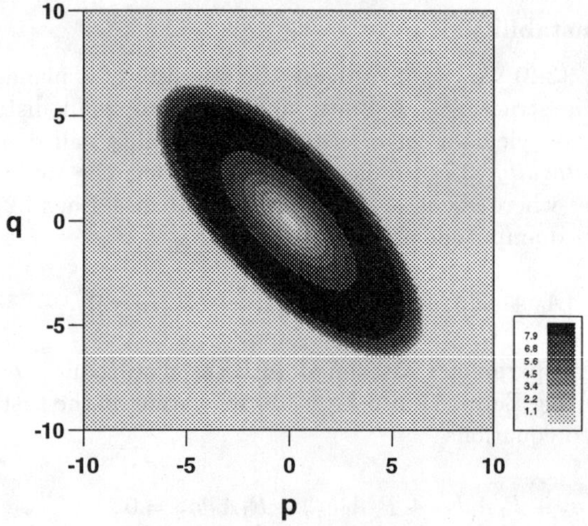

Fig. 4. Shaded-contour plot of the growth rate σ in the plane of the perturbation wave vector (p.q).

5. Numerical Simulations

The previous results show that in the low-amplitude limit, the dynamics of the elastic structure can be approximately described by a 2D-NLS equation. The latter equation allows us to predict instabilities but not the long-time evolution of the waves. Now we want to clarify the role played by the modulational instability in the response of the elastic structure to an initial homogeneous perturbation with a low amplitude. To this end, we perform numerical simulations using a finite difference scheme of the original equation of motion. We consider a square grid made of 89 × 89 points along with periodic boundary conditions on the left and right sides and on the lower and upper boundaries as well. The initial conditions are provided by a harmonic wave carrier traveling along the first diagonal direction with an amplitude w_0. The modulus of the wavenumber k_c and frequency ω_c satisfy the dispersion relation (2.3). We take $w_0 = 0.12$, $N_x = N_y = N_{iso} = 0.425$ and we have 6 periods along the diagonal which leads to $k_c = 0.8078$; the corresponding frequency is then $\omega_c = 0.933$. In order to trigger the instability, a small ($\simeq 10^{-3}$)

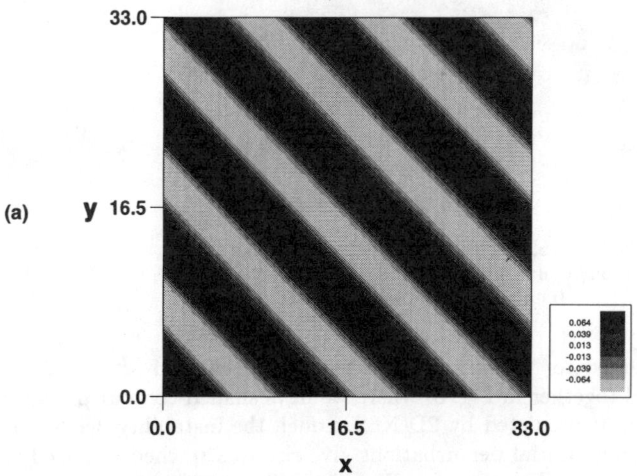

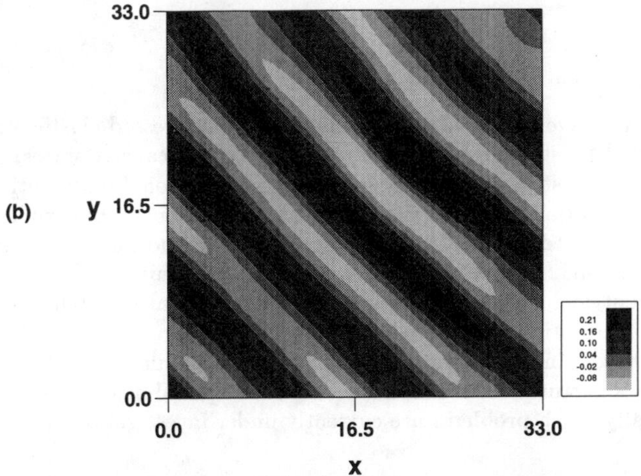

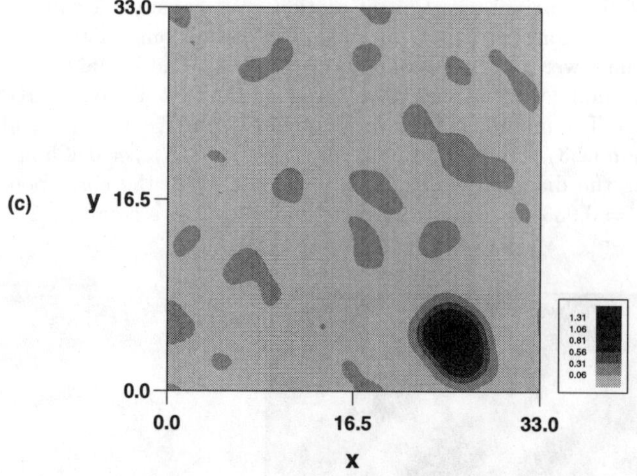

Fig. 5. Shaded-contour maps corresponding to w:
(a) initial condition; (b) the birth of localized structure at time $t = 75$;
(c) moving 2D pulse at time $t = 1000$.

random noise is added to the initial velocity w_t and removed afterwards. The results are collected together in Fig. 5, where we have shaded-contour plots for the deflection w. Then, as predicted by 2D-NLS model, the instability occurs after a lapse of time, the small initial perturbations give rise to stretched localized states along the second diagonal direction (see Fig. 5b). When time is further increased, we observe very clearly some localized objects or dynamic structures as shown in Fig. 5c. We notice that these localized structures still persist and look like stable as 2D-pulse solitons.

6. Concluding Remarks

In this work we have proposed a two-dimensional prestressed elastic structure continuum model to study the formation of nonlinear localized states. We have shown that localized states are the result of the modulational instabilities of a steady plane wave solution. The significant point, which can be underlined in the present work, is the reduction of the initial equation of motion for the elastic structure to the 2D-NLS equation in the low-amplitude limit. This standard equation is an efficient tool to predict the birth of modulational instability. Nevertheless, such an asymptotic model does not allow one to approach the long-time regime which can be examined only by means of numerical simulations. Moreover, the properties of the structures, their amplitude and their spatial distribution must be studied numerically. Such problems are currently under investigation. In short, the modulational instability is a natural vehicle for the nonlinear localized state formation. In addition, extensions of the study to specific problems such as the modulational instability near the critical point of the marginal stability curve or in the neighborhood of the zeros of the second order dispersion should be examined [12]. Finally, we note that the relative influence of moderately large amplitude displacements of the plate on the dynamic behavior of the composite elastic structure will be elucidated in a further work.

References

1. M. Remoissenet, *Waves Called Solitons—Concepts and Experiments*, Springer, Berlin, 1994.
2. E. Infeld and G. Rowlands, *Nonlinear Waves Solitons and Chaos*, Cambridge University Press, Cambridge, 1990.
3. H. C. Yuen and B. M. Lake, *Advances in Applied Mechanics* **22** (1982) 68.
4. V. Petrov, W. Rudolph and B. Wilhelm, *J. Mod. Optics* **36** (1989) 587.
5. J. V. Moloney and A. C. Newell, *Physica* **D44** (1990) 1.
6. J. E. Rothenberg, *Opt. Lett.* **17** (1992) 583.
7. J. Pouget, M. Remoissenet and J. M. Tamga, *Phys. Rev.* **B47** (1993) 14866.
8. B. Collet, *J. Phys. IV France* **4** (1994) C5-793.
9. M. Géradin and D. Rixen, *Mechanical Vibrations*, Wiley, Chichester, 1994.
10. S. Timoshenko and J. Gere, *Theory of Elastic Stability*, Mc Graw-Hill, New York, 1961.
11. A. Jeffrey and T. Kawahara, *Asymptotic Methods in Nonlinear Wave Theory*, Pitman, London, 1982.
12. C. G. Lange and A. C. Newell, *SIAM J. Appl. Math.* **21** (1971) 605.

Continuum Models and Discrete Systems
Proceedings of 8th International Symposium, June 11-16, 1995, Varna, Bulgaria, ed. K.Z. Markov

SOLITONS IN MICROPOLAR CRYSTALS

S. ERBAY
Department of Mathematics, Faculty of Science,
Istanbul Technical University, Maslak 80626, Istanbul, Turkey

Abstract. The present work considers two-dimensional wave propagation in an infinite, homogeneous, geometrically nonlinear micropolar elastic medium in the long-wave approximation. The reductive perturbation method is directly applied to a Lagrangian whose Euler-Lagrange equations give the field equations. It is shown that the behavior of nonlinear waves in the long-wave approximation is governed by two coupled modified Kadomtsev-Petviashvili equations.

1. Introduction

It is well-known that the linear waves are nondispersive in classical elasticity theory. To incorporate the dispersive wave character into the theory, various generalized continuum theories have been proposed in recent years. In this study, two-dimensional wave propagation in one of those generalized continuum theories, namely, in a micropolar elastic medium, will be considered.

The material points of a micropolar elastic medium are allowed to rotate without stretch. Hence the motion of the material points of the medium will have six degrees of freedom, three for translation and three for rotation. Because of the introduced internal structure into the model, this theory is expected to predict various phenomena such as dispersion. The dispesrive character of the model may produce solitary waves if nonlinearity is also considered.

In a previous study, one-dimensional wave propagation in a micropolar elastic medium was considered [1]. In that study one-dimensional long acoustical waves in an infinite, isotropic and homogeneous micropolar elastic solid were investigated. The reductive perturbation technique was used to describe the long-time behaviour of nonlinear field equations [2]. Expanding the dependent variables in the field equations into a power series in terms of a small parameter ε, and using the stretched coordinates, a sequence of equations to be satisfied for each order of ε is obtained. These equations lead to a single evolution equation or coupled evolution equations depending on the problem involved. It was shown that the long-time behavior of one-dimensional long acoustical waves was characterized by wo coupled nonlinear evolution equations which may be called coupled modified Korteweg-de Vries (CMKdV) equations. The Korteweg-de Vries (KdV) type equations describe the evolution of small-but-finite amplitude one-dimensional waves in the long-wave approximation. However, new evolution equations are needed to describe the long-time behavior if the waves are two-dimensional.

In this study the evolution equations describing the long-time behavior of two-dimensional wave propagation in a micropolar medium will be derived using the reductive perturbation method. Compared to the previous study [1], a different approach will be followed here. Instead of deriving the evolution equations from the field equations, they will be derived from the Lagrangian whose Euler-Lagrange equations give the original field equations [3]. In the present study it is shown that the long-time behavior of two-dimensional long acoustical waves are characterized by two coupled nonlinear evolution equations which may be called coupled modified Kadomtsev-Petviashvili (CMKP) equations. These coupled evolution equations, which are the generalized form of the KP equation for two dependent variables, include the KP equation and the CMKdV equations as special cases.

2. Preliminaries

A material point of a micropolar elastic medium has six degrees of freedom, three for translation and three for rotation. The translation and the rotation are described by

$$x_k = x_k(X_K, t), \quad \varphi_k = \varphi_k(X_K, t), \quad k, K = 1, 2, 3, \tag{1}$$

where x_k and X_K are the spatial and material Cartesian coordinates of the same material points at times t and $t = 0$, respectively, φ_k is the rotation vector. The material form of the field equations of an infinite, isotropic, homogeneous, and materially quadratic micropolar medium are given by

$$\begin{aligned} &\frac{\partial}{\partial X_K}\frac{\partial \Sigma}{\partial u_{k,K}} - \rho_0 \ddot{u}_k = 0\,, \\ &\frac{\partial}{\partial X_K}\frac{\partial \Sigma}{\partial \varphi_{k,K}} - \frac{\partial \Sigma}{\partial \varphi_k} - \rho_0 J_0 \ddot{\varphi}_k = 0\,, \end{aligned} \tag{2}$$

where $u_{k,K}$ is the displacement gradient and Σ is the strain energy. In a previous study, one-dimensional wave propagation in a micropolar elastic medium was considered [1]. In that study long-time behavior of long acoustical waves was investigated. For the study of 1D waves, the displacement vector $\boldsymbol{u}$ and the rotation vector $\boldsymbol{\varphi}$ are assumed to be functions of X and t only, i.e.,

$$u_k = u_k(X, t), \quad \varphi_k = \varphi_k(X, t), \quad k = 1, 2, 3, \tag{3}$$

where X is the coordinate along the propagation direction. In order to see the dispersive character of the model, the field equations are linearized about a constant state and harmonic wave solutions for the linearized equations give the following dispersion relation

$$\omega^4 - \omega^2(b_T^2 k^2 + c_T^2 k^2 + \omega_0^2) + c_T^2 k^2 (b_T^2 k^2 + \omega_0^2) - c_0^2 \omega_0^2 k^2/2 = 0\,, \tag{4}$$

where

$$c_T^2 = (\mu + \kappa)/\rho_0, \quad b_T^2 = \gamma/\rho_0 J_0, \quad c_0^2 = \kappa/\rho_0, \quad \omega_0^2 = 2\kappa/\rho_0 J_0\,.$$

For small wavenumbers, i.e. in the long-wave approximation, dispersion relation has the following form for the acoustical branch

$$\omega = ak + bk^3 + \mathcal{O}(k^5)\,, \tag{5}$$

where

$$a = \left(\frac{2\mu + \kappa}{2\rho_0}\right)^{\frac{1}{2}}, \quad b = -\left(\frac{\kappa J_0 + 2\gamma}{8\kappa}\right)\left(\frac{2\mu + \kappa}{2\rho_0}\right)^{\frac{1}{2}},$$

which indicates the weakly dispersive character of the waves.

Assumption of the wavenumber $k \ll 1$ suggests that $k = \mathcal{O}(\varepsilon^{\frac{1}{2}})$ where ε is a small parameter measuring the weakness of dispersion. In such a case the phase function takes the following form

$$\exp\left[i(kX - \omega t)\right] = \exp\left\{i[\bar{k}\varepsilon^{\frac{1}{2}}(X - at) + b\bar{k}^3\varepsilon^{\frac{3}{2}}t]\right\}. \tag{6}$$

This suggests the following coordinate stretching [2]

$$\xi = \varepsilon^{\frac{1}{2}}(X - at), \quad \tau = \varepsilon^{\frac{3}{2}}t\,. \tag{7}$$

To balance the nonlinearity with dispersion, the field variables are assumed to have the following series solutions

$$p_k = \varepsilon^{\frac{1}{2}} p_k^{(1)} + \varepsilon p_k^{(2)} + \mathcal{O}(\varepsilon^{\frac{3}{2}}), \quad \varphi_k = \varepsilon^{\frac{1}{2}}\varphi_k^{(1)} + \varepsilon\varphi_k^{(2)} + \mathcal{O}(\varepsilon^{\frac{3}{2}})\,, \tag{8}$$

where $p_k = \partial u_k/\partial X$. Substituting the series solutions together with the coordinate stretching into the field equations, a sequence of equations to be satisfied for each order of $\varepsilon^{\frac{1}{2}}$ is obtained.

For $\mathcal{O}(\varepsilon^{\frac{1}{2}})$, the first-order quantities $p_2^{(1)} \equiv \Phi$ and $p_3^{(1)} \equiv \Psi$ are obtained as arbitrary functions. For $\mathcal{O}(\varepsilon)$, second-order quantities are found as functions of the first-order quantities. Finally for $\mathcal{O}(\varepsilon^{\frac{3}{2}})$ the following CMKdV equations are obtained as compatibility conditions [1]:

$$\begin{aligned}
&\frac{\partial\Phi}{\partial\tau} + \Gamma\frac{\partial^3\Phi}{\partial\xi^3} + \Lambda\frac{\partial}{\partial\xi}\left[\Phi(\Phi^2 + \Psi^2)\right] = 0\,, \\
&\frac{\partial\Psi}{\partial\tau} + \Gamma\frac{\partial^3\Psi}{\partial\xi^3} + \Lambda\frac{\partial}{\partial\xi}\left[\Psi(\Phi^2 + \Psi^2)\right] = 0\,,
\end{aligned} \tag{9}$$

where Γ and Λ are constants and dependent on wavenumber and micropolar elastic constants. If a complex function, $w = \Phi + i\Psi$, is defined, then the CMKdV equations can be written as

$$\frac{\partial w}{\partial \tau} + \Gamma \frac{\partial^3 w}{\partial \xi^3} + \Lambda \frac{\partial}{\partial \xi}(|w|^2 w) = 0 \tag{10}$$

which may be called the complex MKdV equation. Therefore the evolution of small-but-finite amplitude waves can be described by KdV type equations if they are also dispersive. However, these evolution equations are strictly one-dimensional. If the variations are not only in one dimension but also in the second dimension, the evolution equation (10) cannot describe the variations in both directions. For such a case new evolution equations are needed to describe the evolution of waves in two dimensions. In the next section the derivation of the evolution equations will be given when the variations are present in both directions.

3. Two-Dimensional Long Acoustical Waves

In this section the explicit derivation of two-dimensional evolution equations will be examined. However, a different approach will be followed here [3, p.104]. Instead of deriving the evolution equations from the field equations, they will be derived from a Lagrangian whose Euler-Lagrange equations give the field equations. The Lagrangian density is defined as

$$\mathcal{L} = \Sigma - T\,, \tag{11}$$

where Σ is the strain energy function for a geometrically nonlinear solid given by

$$\Sigma = \frac{\lambda}{2} I_1^2 + \mu I_2 + (\mu + \kappa) I_4 + \frac{\alpha}{2} I_{10}^2 + \beta I_{11} + \gamma I_{13},$$

and T is the kinetic energy given by

$$T = \tfrac{1}{2}\rho_0 J_0 (\dot{\varphi}_1^2 + \dot{\varphi}_2^2 + \dot{\varphi}_3^2) + \tfrac{1}{2}\rho_0 (\dot{u}_1^2 + \dot{u}_2^2 + \dot{u}_3^2)\,.$$

Here λ, μ, κ, α, β, and γ are the material constants and I_1, $I_2, \ldots, I_{13}$ are the joint invariants of the deformation and wryness tensors [4]. The Euler-Lagrange equations corresponding to the Lagrangian (11), give the field equations of which the explicit forms will not be given here.

For the study of 2D waves, the displacement vector $\boldsymbol{u}$ and the rotation vector φ are assumed to be functions of X, Y and t, i.e.,

$$u_k = u_k(X, Y, t), \quad \varphi_k = \varphi_k(X, Y, t), \quad k = 1, 2, 3, \tag{12}$$

where X is the coordinate along the propagation direction. To find the appropriate coordinate stretching for 2D case, similarly we consider the dispersion relation (4). In the long-wave approximation, assuming the wavenumber in the X-direction is larger than the one in the Y-direction, i.e. $k_x \gg k_y$, the dispersion relation for the acoustical branch takes the following form

$$\omega = ak + bk^3 + \mathcal{O}(k^5)\,, \tag{13}$$

where $k = (k_x^2 + k_y^2)^{\frac{1}{2}}$. Therefore the wavenumbers in the $X-$ and in the Y-directions may be assumed as $k_x = \varepsilon^{1/2}\bar{k}_x$ and $k_y = \varepsilon^p \bar{k}_y$ (keeping in mind that $p > 1/2$) where ε is a small parameter measuring the weakness of the dispersion and the value of p is to be determined later by balancing nonlinearity and dispersion. If the assumptions are substituted into the phase function, the following expression is obtained

$$\exp[i(\mathbf{k}.\mathbf{r} - \omega t)] = \exp\left\{ i\left[\bar{k}_x\varepsilon^{1/2}(X - at) + \bar{k}_y\varepsilon^p Y - \left(a\frac{\bar{k}_y^2}{\bar{k}_x}\varepsilon^{2p-2} - b\bar{k}_x^3 \right)\varepsilon^{3/2} t \right]\right\}. \tag{14}$$

This suggests the following coordinate stretching with $p = 1$

$$\xi = \varepsilon^{1/2}(X - at), \quad \sigma = \varepsilon Y, \quad \tau = \varepsilon^{3/2} t \tag{15}$$

which makes each of the terms are of the same order in Eq. (14).

If the slow variables (15) together with the following power series expansions are substituted into the Lagrangian density given in (11)

$$u_k = u_k^{(1)} + \varepsilon^{1/2}u_k^{(2)} + \varepsilon u_k^{(3)} + \mathcal{O}(\varepsilon^{3/2})\,,$$
$$\varphi_k = \varepsilon^{1/2}\varphi_k^{(1)} + \varepsilon\varphi_k^{(2)} + \mathcal{O}(\varepsilon^{3/2}), \tag{16}$$

a hierarchy of Lagrangians is obtained for each power of $\varepsilon^{1/2}$

$$\mathcal{L} = \varepsilon\mathcal{L}_1 + \varepsilon^2\mathcal{L}_2 + \cdots\,, \tag{17}$$

where

$$\mathcal{L}_1 = -\frac{1}{2}\left(\lambda + \mu + \frac{\kappa}{2}\right)\left(\frac{\partial u_1^{(1)}}{\partial\xi}\right)^2 - \frac{\kappa}{2}\left[\left(\frac{\partial u_2^{(1)}}{\partial\xi}\right)^2 + \left(\frac{\partial u_3^{(1)}}{\partial\xi}\right)^2\right]$$
$$-\kappa\left[\varphi_1^{(1)2} + \varphi_2^{(1)2} + \varphi_3^{(1)2} - \varphi_3^{(1)}\frac{\partial u_2^{(1)}}{\partial\xi} + \varphi_2^{(1)}\frac{\partial u_3^{(1)}}{\partial\xi}\right],$$

and the explicit form of $\mathcal{L}_2$ is not given here due to its length. Since the Lagrangian which is the coefficient of $\varepsilon^{3/2}$ is identically zero, it is not written here.

For $\mathcal{O}(\varepsilon)$, the Euler-Lagrange equations corresponding to the first-order Lagrangian $\mathcal{L}_1$ are obtained as

$$\delta u_1^{(1)} : \frac{\partial u_1^{(1)}}{\partial \xi} = 0\,,$$

$$\delta \varphi_1^{(1)} : \varphi_1^{(1)} = 0\,,$$

$$\delta u_2^{(1)} \text{ and } \delta \varphi_3^{(1)} : \quad \frac{\partial u_2^{(1)}}{\partial \xi} = 2\varphi_3^{(1)}, \quad \frac{\partial u_2^{(1)}}{\partial \xi} \text{ arbitrary function,}$$

$$\delta u_3^{(1)} \text{ and } \delta \varphi_2^{(1)} : \quad \frac{\partial u_3^{(1)}}{\partial \xi} = -2\varphi_2^{(1)}, \quad \frac{\partial u_3^{(1)}}{\partial \xi} \text{ arbitrary function.}$$

The first two identities give that the first-order longitudinal displacement gradient and the microrotation are identically zero. The last two identities give the relations between the transverse components and arbitrariness of two transverse components.

For $\mathcal{O}(\varepsilon^2)$, the Lagrangian $\mathcal{L}_2$ depends not only on the first-order quantities but also the second-order quantities. The Euler-Lagrange equations for $\mathcal{L}_2$ give the following equations

$$\delta u_1^{(2)} : \frac{\partial^2 u_1^{(2)}}{\partial \xi^2} = -\frac{\partial^2 u_2^{(1)}}{\partial \sigma \partial \xi} - \frac{1}{4}\frac{\partial}{\partial \xi}\left[\left(\frac{\partial u_2^{(1)}}{\partial \xi}\right)^2 + \left(\frac{\partial u_3^{(1)}}{\partial \xi}\right)^2\right];$$

$$\delta \varphi_1^{(2)} : \varphi_1^{(2)} = \frac{1}{2}\frac{\partial u_3^{(1)}}{\partial \sigma};$$

$$\begin{aligned}\delta u_2^{(1)} : \ & 2\rho_0 a \frac{\partial^2 u_2^{(1)}}{\partial \xi \partial \tau} + \frac{(\rho_0 J_0 a^2 - \gamma)}{4}\frac{\partial^4 u_2^{(1)}}{\partial \xi^4} + \left(\mu + \frac{\kappa}{2}\right)\frac{\partial^2 u_2^{(1)}}{\partial \sigma^2} \\ & -\frac{1}{8}\left(\lambda + \mu + \frac{\kappa}{2}\right)\frac{\partial}{\partial \xi}\left\{\frac{\partial u_2^{(1)}}{\partial \xi}\left[\left(\frac{\partial u_2^{(1)}}{\partial \xi}\right)^2 + \left(\frac{\partial u_3^{(1)}}{\partial \xi}\right)^2\right]\right\} \\ & +\frac{1}{4}\left(\mu + \frac{\kappa}{2}\right)\left\{\frac{\partial}{\partial \xi}\left(\frac{\partial u_3^{(1)}}{\partial \xi}\frac{\partial u_3^{(1)}}{\partial \sigma}\right) - \frac{\partial}{\partial \sigma}\left(\frac{\partial u_3^{(1)}}{\partial \xi}\right)^2\right\} = 0;\end{aligned}$$

$$\begin{aligned}\delta u_3^{(1)} : \ & 2\rho_0 a \frac{\partial^2 u_3^{(1)}}{\partial \xi \partial \tau} + \frac{(\rho_0 J_0 a^2 - \gamma)}{4}\frac{\partial^4 u_3^{(1)}}{\partial \xi^4} + \left(\mu + \frac{\kappa}{2}\right)\frac{\partial^2 u_3^{(1)}}{\partial \sigma^2} \\ & -\frac{1}{8}\left(\lambda + \mu + \frac{\kappa}{2}\right)\frac{\partial}{\partial \xi}\left\{\frac{\partial u_3^{(1)}}{\partial \xi}\left[\left(\frac{\partial u_2^{(1)}}{\partial \xi}\right)^2 + \left(\frac{\partial u_3^{(1)}}{\partial \xi}\right)^2\right]\right\} \\ & +\frac{1}{4}\left(\mu + \frac{\kappa}{2}\right)\left[\frac{\partial}{\partial \xi}\left(\frac{\partial u_3^{(1)}}{\partial \sigma}\frac{\partial u_2^{(1)}}{\partial \xi} - 2\frac{\partial u_3^{(1)}}{\partial \xi}\frac{\partial u_2^{(1)}}{\partial \sigma}\right) + \frac{\partial}{\partial \sigma}\left(\frac{\partial u_3^{(1)}}{\partial \xi}\frac{\partial u_2^{(1)}}{\partial \xi}\right)\right] = 0\,.\end{aligned}$$

The first two equations give the dependence of the second-order longitudinal displacement gradient and microrotation on the first-order quantities. The last two equations appear as compatibility conditions for arbitrary functions $\partial u_2^{(1)}/\partial\xi$ and $\partial u_3^{(1)}/\partial\xi$ where, for simplicity, higher order terms are eliminated by using the previously obtained identities from the last two equations. The last two equations, then, become the two coupled partial differential equations for $u_2^{(1)}$ and $u_3^{(1)}$:

$$\begin{aligned}
&\frac{\partial^2\varphi}{\partial\xi\partial\tau}+\Gamma\frac{\partial^4\varphi}{\partial\xi^4}+\Lambda\frac{\partial}{\partial\xi}\left\{\frac{\partial\varphi}{\partial\xi}\left[\left(\frac{\partial\varphi}{\partial\xi}\right)^2+\left(\frac{\partial\psi}{\partial\xi}\right)^2\right]\right\}+\nu\frac{\partial^2\varphi}{\partial\sigma^2}\\
&\qquad+\frac{\nu}{4}\left[\frac{\partial}{\partial\xi}\left(\frac{\partial\psi}{\partial\xi}\frac{\partial\psi}{\partial\sigma}\right)-\frac{\partial}{\partial\sigma}\left(\frac{\partial\psi}{\partial\xi}\right)^2\right]=0,\\
&\frac{\partial^2\psi}{\partial\xi\partial\tau}+\Gamma\frac{\partial^4\psi}{\partial\xi^4}+\Lambda\frac{\partial}{\partial\xi}\left\{\frac{\partial\psi}{\partial\xi}\left[\left(\frac{\partial\varphi}{\partial\xi}\right)^2+\left(\frac{\partial\psi}{\partial\xi}\right)^2\right]\right\}+\nu\frac{\partial^2\psi}{\partial\sigma^2}\\
&\qquad+\frac{\nu}{4}\left[\frac{\partial}{\partial\xi}\left(\frac{\partial\psi}{\partial\sigma}\frac{\partial\varphi}{\partial\xi}-2\frac{\partial\psi}{\partial\xi}\frac{\partial\varphi}{\partial\sigma}\right)+\frac{\partial}{\partial\sigma}\left(\frac{\partial\varphi}{\partial\xi}\frac{\partial\psi}{\partial\xi}\right)\right]=0,
\end{aligned}\tag{18}$$

where $\varphi\equiv u_2^{(1)}$ and $\psi\equiv u_3^{(1)}$, and the coefficients in Eq. (18) are given as

$$\Gamma=\frac{\rho_0J_0a^2-\gamma}{8\rho_0a},\quad \Lambda=-\frac{\lambda+\mu+\kappa/2}{16\rho_0a},\quad \nu=\frac{\mu+\kappa/2}{2\rho_0a}.$$

These equations may be called *coupled modified Kadomtsev-Petviashvili equations* (CMKP). They describe two-dimensional wave propagation in the long-wave approximation when the variation in the Y-direction is smaller than the one in the X-direction.

Since the following identities hold between the first-order displacement gradient and microrotation components

$$\varphi=2\int\varphi_3^{(1)}\,d\xi\,,\quad \psi=-2\int\varphi_2^{(1)}\,d\xi\,,$$

the following coupled integro-differential equations for φ_3 and φ_2 are also valid (for convenience, the superscripts are dropped)

$$\begin{aligned}
&\frac{\partial\varphi_3}{\partial\tau}+\Gamma\frac{\partial^3\varphi_3}{\partial\xi^3}+4\Lambda\frac{\partial}{\partial\xi}[\varphi_3(\varphi_3^2+\varphi_2^2)]+\nu\frac{\partial^2}{\partial\sigma^2}\int\varphi_3\,d\xi\\
&\qquad+\frac{\nu}{2}\left[\frac{\partial}{\partial\xi}\left(\varphi_2\frac{\partial}{\partial\sigma}\int\varphi_2\,d\xi\right)-\frac{\partial}{\partial\sigma}(\varphi_2^2)\right]=0,\\
&\frac{\partial\varphi_2}{\partial\tau}+\Gamma\frac{\partial^3\varphi_2}{\partial\xi^3}+4\Lambda\frac{\partial}{\partial\xi}[\varphi_2(\varphi_3^2+\varphi_2^2)]+\nu\frac{\partial^2}{\partial\sigma^2}\int\varphi_2\,d\xi
\end{aligned}\tag{19}$$

$$+\frac{\nu}{2}\left[\frac{\partial}{\partial \xi}\left(\varphi_3 \frac{\partial}{\partial \sigma}\int \varphi_2\, d\xi - 2\varphi_2 \frac{\partial}{\partial \sigma}\int \varphi_3\, d\xi\right) + \frac{\partial}{\partial \sigma}(\varphi_2\varphi_3)\right] = 0\,.$$

The coupled evolution equations valid for transverse microrotation components, consist of the MKdV part together with some interaction terms under the integral sign. To the best knowledge of the author the evolution equations (18) and (19) are new and should be examined thoroughly.

As will be seen in the sequel, the CMKP equations include some special cases. If one of the first-order quantities, say $\psi \equiv 0$, the CMKP equations reduce to the MKP equation:

$$\frac{\partial^2 \varphi}{\partial \xi \partial \tau} + \Gamma \frac{\partial^4 \varphi}{\partial \xi^4} + \Lambda \frac{\partial}{\partial \xi}\left(\frac{\partial \varphi}{\partial \xi}\right)^3 + \nu \frac{\partial^2 \varphi}{\partial \sigma^2} = 0.$$

If Y−dependence is dropped from the equations, the CMKP equations reduce to the CMKdV equations which were already obtained for the 1D waves in a micropolar medium [1]

$$\frac{\partial \Phi}{\partial \tau} + \Gamma \frac{\partial^3 \Phi}{\partial \xi^3} + \Lambda \frac{\partial}{\partial \xi}[\Phi(\Phi^2 + \Psi^2)] = 0\,,$$

$$\frac{\partial \Psi}{\partial \tau} + \Gamma \frac{\partial^3 \Psi}{\partial \xi^3} + \Lambda \frac{\partial}{\partial \xi}[\Psi(\Phi^2 + \Psi^2)] = 0\,,$$

where

$$\Phi \equiv \frac{\partial \varphi}{\partial \xi}\,, \qquad \Psi \equiv \frac{\partial \psi}{\partial \xi}\,.$$

Acknowledgements. The author gratefully acknowledges the partial support provided by Turkish Scientific and Technical Research Council through the project TBAG-U/14-1.

References

1. S. Erbay and E. S. Şuhubi, *Int. J. Engng. Sci.* **27** (1989) 895.
2. A. Jeffrey and T. Kakutani, *SIAM Rev.* **14** (1972) 582.
3. E. Infeld and G. Rowlands, *Nonlinear Waves, Solitons and Chaos*, Cambridge University Press, Cambridge, 1990.
4. A. C. Eringen and C. B. Kafadar, in *Continuum Physics* **IV**, ed. A. C. Eringen, Academic Press, New York, 1976.

Continuum Models and Discrete Systems
Proceedings of 8th *International Symposium, June 11-16, 1995, Varna, Bulgaria, ed. K.Z. Markov*
©World Scientific Publishing Company, 1996, pp. 412–436

SYNERGETIC ORDERING OF DEFECTS IN METALS UNDER PARTICLE IRRADIATION

W. FRANK, P. HÄHNER[1)], T. KLEMM[2)], C. TÖLG, and M. ZAISER

Institut für Physik, Max-Planck-Institut für Metallforschung
Heisenbergstraße 1, D-70569 Stuttgart, Germany

and

Institut für theoretische und angewandte Physik, Universität Stuttgart
Pfaffenwaldring 57, D-70569 Stuttgart, Germany

Abstract. A spectacular phenomenon in the field of radiation damage in solids is the self-organization of ordered defect structures under non-equilibrium conditions. This article reports on defect patterns formed in metals under particle irradiation (e.g., void lattices, stacking-fault-tetrahedron lattices, or gas-filled bubble lattices), describes the underlying physical processes, and extensively makes reference to recent theoretical work in this area. Particular emphasis is laid on the rôle that is played by various mechanisms of anisotropic transport of matter in synergetic defect ordering. These break not only the rotational symmetry of the point-defect dynamics, but, for suitable values of the control parameters, also the translatory symmetry of homogeneous defect distributions. As a result, these homogeneous phases become unstable and undergo non-equilibrium phase transitions to ordered defect structures.

1. Observations and Introductory Remarks

Ordered structures of irradiation-induced defects are formed in high-dose-irradiated metals under otherwise radically different conditions. This is demonstrated by the subsequent examples, in which the irradiation conditions range from low-temperature electron irradiations to high-temperature irradiations with ions or fast neutrons.

The most thoroughly investigated irradiation-induced ordered defect structures are the (empty) void lattices [1, 2] (Fig. 1) and the (gas-filled) bubble lattices [3–7] (Fig. 2). Although some properties of the bubble lattices, which have been produced in various metals (e.g., Mo, Cu) by implantation of inert gases (e.g., He, Ne, Kr), resemble those of the void lattices, the formation mechanisms of the two structures are different. For instance, this is indicated by the fact that the

---

1) New address: Institute for Advanced Materials, Commission of the European Communities, Joint Research Centre, I-21020 Ispra, Varese, Italy

2) New address: Max-Planck-Institut für komplexe Systeme, Bayreuther Straße 40, D-01187 Dresden, Germany

bubble-lattice parameters and the bubble radii are distinctly smaller than the void-lattice parameters and the void radii, respectively. Whereas elastic bubble interactions resulting from the high gas pressure in the bubbles are expected to and, in fact, do play a rôle in the formation of bubble lattices [8], in the case of void-lattice formation such an effect can be excluded.

First, the major properties of void lattices (VLs) will be introduced. Under high-temperature irradiation (typically at about one third of the melting temperature) in body-centred cubic (bcc) metals (e.g., Mo, W, Nb, Ta, α-Fe), face-centred cubic (fcc) metals (e.g., Ni, Al), and metallic alloys (Ni–Al, Ni–Cu, stainless steel) VLs evolve from randomly distributed voids after these have passed a stage of nucleation and growth [1]. Also in ionic crystals (e.g., CaF_2) VLs are formed [9]; however, defect ordering in non-metallic solids will not be treated in this paper. In hexagonal close-packed (hcp) metals (e.g., Mg), VL formation does not occur; rather void ordering is restricted to the formation of two-dimensional void layers parallel to the basal plane [10, 11] (henceforth called "basal void layers" or "BVs").

VLs in cubic metals are three-dimensional and always possess the same crystallographic structures and orientations as their atomic host lattices. (By contrast, bubble lattices (BLs) have been found whose lattice orientations differ from those of the host lattices.) The radii R_0 of voids in VLs normally amount to about 5 nm, while the VL parameters λ are of the order of magnitude of 50 nm. In Al, void lattices with extremely large VL parameters occur ($\lambda \approx 200$ nm), which are referred to as "void hyperlattices" [12]. The ratio λ/R_0 varies from material to material between 5 and 15. It is almost independent of the conditions of irradiation, e.g., the irradiation temperature at which the VLs are produced. VLs develop easily under heavy-ion or fast-neutron irradiation, whereas electron irradiation is quite inefficient in producing VLs. The doses required for the formation of VLs range from 5 dpa (for Nb at 1075 K) to 400 dpa (for Ni at 800 K), where "dpa" is the abbreviation for "displacements per atom". They depend on temperature and on whether and to which extent gas impurities or alloying elements are present. In spite of these high threshold doses the steady-state concentrations of monovacancies during VL-inducing irradiations are quite small, in heavy-ion irradiation typically about 10^{-4}. The corresponding self-interstitial concentrations are even smaller by several orders of magnitude. This is so since at the temperatures of VL formation the thermally activated mobility of vacancies is high and still

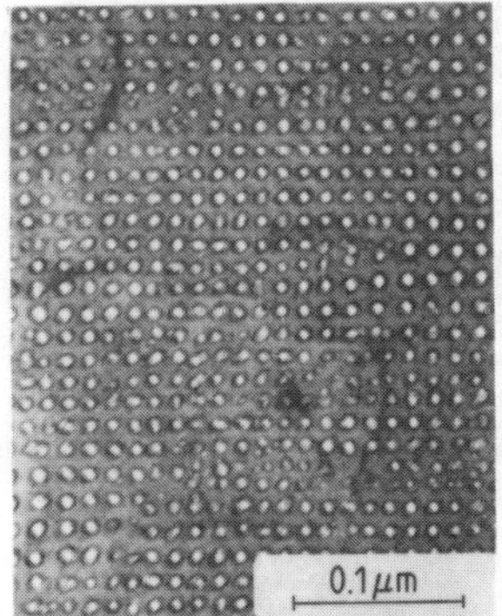

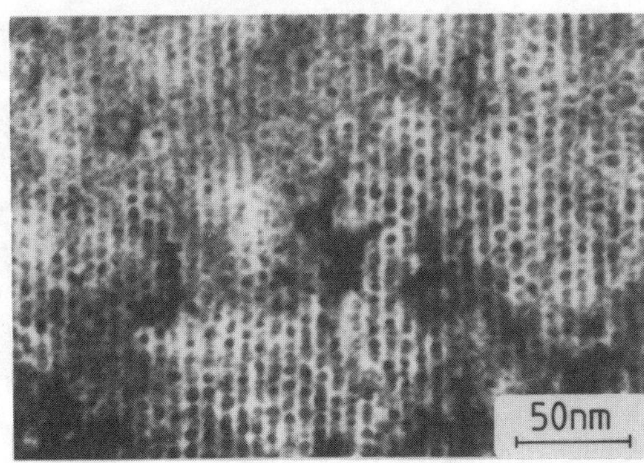

Fig. 1 (left). Void lattice in Nb–1at.%Zr after irradiation with 3.1 MeV V^+-ions at 780°C to a dose of 50 dpa. Electron micrograph after Loomis et al. [2].

Fig. 2 (right). Krypton bubble lattice in α-Zr parallel to the basal plane after irradiation with Kr^+-ions at 400°C. Electron micrograph after Evans et al. [5].

exceeded by that of self-interstitial atoms (SIAs), so that intrinsic point defects are hardly accumulated, but rapidly disappear at sinks, such as voids and dislocations.

Another spectacular kind of ordering that, like VL formation, takes place at "high" irradiation temperatures, at which both species of elementary intrinsic atomic defects generated by irradiation (isolated SIAs and monovacancies) are highly mobile, is the formation of walls of defect clusters running parallel to {100}-planes in Cu and Ni under proton irradiation [13] (Fig. 3). This phenomenon, an explanation of which has been proposed by others [14], will not be a subject of this paper. Rather, as a counterpart to VL formation at "high" temperatures, a family of ordering phenomena occurring under electron irradiation will be considered in some detail. All of them involve an alignment of stacking-fault tetrahedra (SFT) in fcc metals and alloys. They appear at "low" temperatures (no thermally activated long-range migration of point defects), "intermediate" temperatures (thermal long-range migration of SIAs only), and "elevated" temperatures (transition regime between "intermediate" and "high" temperatures), respectively, and, as may be expected, are generated via different mechanisms. (According to the preceding classification, BL formation occurs at "intermediate" temperatures.)

Ordering of SFT as result of intense electron irradiation in a high-voltage electron microscope (HVEM), i.e. with electrons whose kinetic energy is of about 1 MeV, was first seen in Pb irradiated between 50 and 165 K (formation of SFT arrays along <100>-directions) [15]. Later similar observations were made on other fcc metals, viz., Au [16, 17] and Cu [18–20]. Finally, systematic HVEM studies of Ag, Cu, Ni [21, 22], and alloys [23] brought out the important rôle of the irradiation temperature T_{irr} for SFT ordering. The results are presented in Table 1 and discussed in the following.

At very low T_{irr} (Regime A) no significant ordering of the radiation-induced defects visible by transmission electron microscopy is found, even when their number density becomes very high. In Regime B, regular SFT arrays along <100> are formed, whereas at even higher T_{irr} (Regime C) the defect density becomes very low and ordering does not take place any more.

Regime B may be divided further in two subregimes, B_1 und B_2, exhibiting different ordering features. First, we consider its low-T_{irr} part (Subregime B_1), in which SFT appear virtually *homogeneously* over the entire irradiated region and, from the very beginning, are

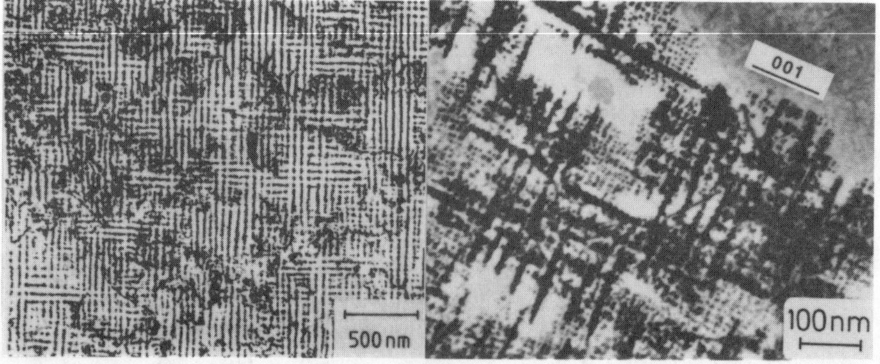

Fig. 3 (left). Periodic structure of walls of point-defect agglomerates in Cu parallel to {100}-planes after irradiation with 3.4 MeV protons at about 100°C to a dose of 0.65 dpa. Electron micrograph after Jäger et al. [13].

Fig. 4 (right). Stacking-fault-tetrahedron grid in Ag after irradiation with 6.3×10^{27} electrons/m^2 at room temperature. Electron micrograph after Jin et al. [21].

arranged in a three-dimensional primitive cubic SFT lattice (SL), whose <100>-directions are parallel to the <100>-directions of the fcc atomic host lattice and whose regularity increases during irradiation. The SFT-edge length lies between 5 and 8 nm, the centre-to-centre distance between neighbouring SFT ranges from 6 to 10 nm. The latter increases with T_{irr}, but is independent of the dose and the SFT size.

Subregime B_1 comprises the high-temperature side of the recovery stage II and the very onset of the recovery stage III of the electrical resistivity after low-dose irradiation. Hence, the high-temperature end of B_1 is related to the beginning of thermally activated migration of the Stage-III defects, whereas the onset of B_1 at low temperatures does not exhibit any features that might be associated with the migration of the SIAs taking place in the upper part of recovery stage I. This finding cannot be reconciled with the so-called one-interstitial model (OIM) [24, 25], since it is the key postulate of this model that the Stage-I interstitial configuration is mechanically stable and therefore should show up even under high-dose conditions. This makes us adopt the viewpoint of the two-interstitial model (TIM) [26–30], which assigns the free migration of *stable* SIAs to recovery stage III. According to this model the SIAs that in low-dose experiments undergo thermally activated long-range migration in Stage I are *metastable* static crowdions, which under HVEM-high-dose conditions, immediately after their production by irradiation-induced displacements of atoms, are converted to the stable dumbbell configuration (migrating in Stage III) owing to their interaction with other defects. Hence, in Subregime B_1 the surviving stable elementary point defects, vacancies and dumbbells, do not diffuse with the aid of thermal activation. Rather B_1-type ordering is terminated when the dumbbells become mobile in Stage III. Obviously, according to the classification introduced above, SL formation in Subregime B_1 falls in the category "low-temperature ordering". In marked contrast to the high-temperature conditions under which VLs are formed, during irradiation in Subregime B_1 the point defects accumulate up to their saturation concentrations (≈ 1 at.%), which are determined by athermal recombination processes. The threshold doses for SL formation amount to some dpa, i.e., they are not even so high as those for VL formation, whereas the point-defect concentrations during SL formation exceed those during VL formation by many orders of magnitude. A further difference to VL formation is that the nucleation of SL occurs, from the very beginning, almost homogeneously over the entire irradiated region of the specimens.

material	Regime A (no ordering)	Regime B (ordered SFT arrays) B_1 (homogeneous)	B_2 (inhomogen.)	Regime C (no ordering)
Cu	< 170	170–290	290–350	> 350
Ni	< 380	380–480	480–540	> 540
Ag	< 150	150–280	280–320	> 320
Ag–0.26at.%Zn	< 170	170–260	260–300	> 300
Ag–0.17at.%Cu	< 150	150–260	260–300	> 300

Table 1. Electron-microscopy observations of stacking-fault tetrahedra in fcc metals and dilute alloys during high-dose electron irradiations [21–23].

In the high T_{irr}-part of Regime B (Subregime B_2) the evolution of the SFT pattern is a *heterogeneous* process. Here the SFT nucleate preferentially in the vicinity of pre-existing defects; new SFT tend to appear next to previously formed ones along two <100>-axes, thus generating a <100> square lattice of SFT parallel to the foil plane (Fig. 4), which henceforth will be called "SFT grid" (SG). In Subregime B_2 both the SFT size (typical edge lengths ≈ 10 nm) and the distance between the SFT (exceeding the edge length by 10 to 50% only) increase with increasing dose.

Subregime B_2 comprises the temperature interval of Stage III and the temperature regime slightly above. Hence, according to the TIM (the validity of which has already been ensured in the preceding discussion of Subregime B_1) during SG formation, dumbbell interstitials are highly mobile, whereas vacancies only undergo comparatively slow temperature-assisted radiation-induced migration [15]. Thus, in the terminology introduced above, Subregime B_2 corresponds to the regime of "intermediate" temperatures.

On the high-temperature side of Subregime B_2 (regime of "elevated" temperatures), an additional ordering feature has been observed in Ni foils [23]. Under moderate magnification, HVEM studies reveal "spots" that align along <110>-directions. Upon increasing the magnification one recognizes that the spots consist of ordered planar <100>-arrays of about 10 SFT, i.e., they represent small patches of the sort of SG that is typical of Subregime B_2. In this case we thus have two interlacing ordered structures with quite different characteristic lengths, viz., patches of the planar primitive cubic <100> SG that are arranged in a planar fcc lattice lying in the foil plane and being parallel to the atomic host lattice (Fig. 5). Hereafter, this structure will be referred to as "SFT-patch grid" (PG).

Nowadays it is widely accepted that the formation of irradiation-induced ordered defect patterns in solids, as described above, cannot be understood in terms of free-energy minimization. Rather these phenomena are examples of synergetic self-organisation in open, highly dissipative, non-linear systems far from thermodynamic equilibrium [31–36]. The extent of dissipation in such systems may be realized from the fact that during VL formation the displacement probability of an atom of the host lattice amounts to 1000 to 10 000% (corresponding to 10 to 100 dpa), whereas the probability that an atom finally remains displaced is 5% only or less (corresponding to an equivalent concentration of vacancies retained in voids).

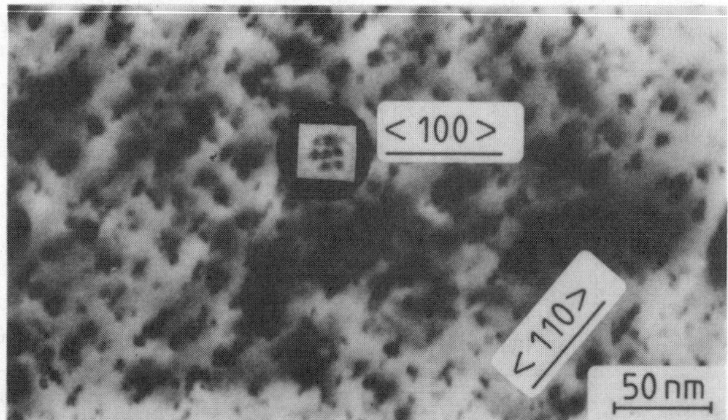

Fig. 5. Grid of patches of a stacking-fault-tetrahedron grid in Ni after irradiation with 4.4×10^{26} electrons/m^2 at 520 K. The inset shows a patch of a three times higher magnification, revealing that the patch is an ordered arrangement of stacking-fault tetrahedra. Electron micrograph after Jin et al. [23].

The occurrence of synergetic ordering of defects means that, under certain conditions, homogeneous defect distributions are unstable, i.e., defect ordering involves the breaking of symmetries. Here one has to distinguish between the breaking of the rotational symmetry, which will be considered in Sec. 2, and the breaking of the translatory symmetry, which will be postponed to Sec. 8.

2. Breaking of the Rotational Symmetry by Anisotropic Transport of Matter

In the conventional examples of self-organization the breaking of the rotational symmetry is caused by boundary conditions, stochastic fluctuations, or the propagation of the front of the ordered phase. By contrast, the finding that self-organized defect patterns in metals reflect the symmetries of the host crystals indicates that in these cases the breaking of the rotational symmetry occurs non-spontaneously and is dictated by the atomic host lattice. In the following, it will be demonstrated that there are various mechanisms of anisotropic matter transport which are inherent to crystalline metals and thus may give rise to this kind of symmetry breaking and lead to the defect structures described in Sec. 1.

The category of mechanisms of anisotropic matter transport to be discussed first is of quasi-ballistic nature: (i) *Channelling* of self-atoms may be induced by particle irradiation. However, this mechanism dies out if the energies transferred to the knocked-on atoms become smaller than about 1 keV. (ii) The transfer of energies approximately ranging from the threshold energy for atomic displacements to twice this value gives rise to *replacement collision sequences* in low-index host-lattice directions [26]. This is particularly true for the so-called *dynamic crowdions* propagating along the close-packed directions of cubic metals. The importance of this mechanism has recently been confirmed by molecular-dynamics studies done at Harwell [37, 38]. (iii) The theoretical investigations of the Harwell group just mentioned have also shown that for transferred energies above about 1 to 2 keV a considerable fraction of the dynamic crowdions ejected from the displacement cascades leave the cascades in the form "*coupled dynamic crowdions*", consisting of typically 2 to 6 crowdions. These crowdion clusters are equivalent to *small unfaulted dislocation loops*, which form from layers of SIAs inserted between close-packed planes by unfaulting with an energy gain of about 0.1 eV.

Secondly, the reader's attention is focused on anisotropic transport of matter arising from the thermally activated, anisotropic diffusion of SIAs or their agglomerates: (iv) It is a basic postulate of the TIM [26–30] and fully confirmed by the Harwell molecular-dynamics simulations that, after having dissipated their kinetic energy, dynamic crowdions may survive as metastable *static crowdions*. In a defect-poor crystalline lattice, static crowdions may undergo thermally activated one-dimensional diffusion along the close-packed directions prior to their conversion into the stable dumbbell configuration or their disappearance by other defect reactions. (v) Coupled crowdions are even more stable than isolated crowdions and thus can survive as *static coupled crowdions* and migrate one-dimensionally by gliding along close-packed directions with the aid of thermal activation [37, 38]. (vi) Jacques and Robrock [39] suggested that in Mo irradiation-induced SIAs possess the <110>-dumbbell configuration and become mobile at 40 K by migrating two-dimensionally in {110}-planes without reorientation. Although Frank and Seeger [30] pointed out that this proposal cannot account in a self-consistent way for the observations made on Mo after low-temperature irradiations, Evans [40] adopted this view by attempting to explain VL formation in terms of *two-dimensional self-interstitial migration*. Since Evans' VL theory fails to explain several outstanding features of VL formation, particularly its temperature dependence, and since the molecular-dynamics studies by Foreman and collaborators [37] show "no evidence for the planar diffusion of SIAs", two-dimensional self-interstitial migration may be safely excluded as a rotational-symmetry-breaking effect in the self-organization of defect structures in cubic metals. However, in hcp metals whose lattice-parameter ratio *c/a* is smaller than the value $(8/3)^{1/2}$ corresponding to ideal close-packing (so-called subnormal hcp metals), two-dimensional migration of SIAs in the basal plane is well established [41] and will be shown to play an important rôle in the formation of BVs.

Thirdly, drift-assisted anisotropic transport of matter arising from the interaction between defects is considered: (vii) An example is the *drift of three-dimensionally migrating vacancies and dumbbell self-interstitials in the stress fields of stacking-fault tetrahedra* in an fcc crystal. Since the orientations of the SFT and the crystal are uniquely interrelated, this drift reflects the crystal symmetry. (viii) Owing to their internal gas pressure, bubbles are surrounded by strong elastic stress fields. These also result in a drift of point defects which reflects the crystal symmetry. (ix) Dubinko et al. [42] proposed that void ordering results from the *drift of glissile interstitial-type dislocation loops along close-packed lattice directions into voids*. This mechanism can only operate if the void–dislocation-loop interaction is strong enough to overcome the glide resistance of the loops, a lower limit of which is the Peierls force. As a consequence of this requirement the diameter of the loops must be comparable to the diameter of the voids, i.e., the loops must consist of a few hundred atoms. Since it is unlikely that interstitial loops of this size are ejected from displacement cascades or formed by diffusion-controlled clustering of SIAs at the high irradiation temperatures required for VL formation, this mechanism may be ruled out.

The preceding discussion leaves us with the mechanisms (i) to (viii) of anisotropic transport of matter. Among these, one has to search for the causes of the irradiation-induced ordered defect structures reported in Sec. 1, namely SFT lattices at "low" temperatures, SFT grids and bubble lattices at "intermediate" temperatures, SFT-patch grids at "elevated" temperatures, and void lattices and basal void layers at "high" temperatures. Secs. 3 to 7 will deal with these ordering phenomena in a detailed, but mainly qualitative manner; reference to the major quantitative theoretical work will be made in the section headings. A general discussion of mechanisms that may induce instabilities of homogeneous defect distributions and, via spontaneous breaking of the translatory symmetry, result in ordered defect structures will be given in Sec. 8. The paper will be concluded by some remarks of general nature (Sec. 9).

3. Three-Dimensional SFT Lattices at "Low" Temperatures [43, 44]

The scenario prevailing in Subregime B_1, in which in fcc metals and their alloys SLs are formed during intense electron irradiation, is as follows. Since, according to the TIM, at these "low" temperatures vacancies and dumbbell interstitials do not undergo thermally activated migration, they rapidly accumulate up to concentrations at which metastable static crowdions cannot survive (Sec. 1). Thus a steady state is achieved when the production of vacancies and dumbbells is compensated by their mutual annihilation. This recombination of close vacancy–dumbbell pairs is controlled by irradiation-induced diffusion [15].

Among the mechanisms of anisotropic transport of matter listed in Sec. 2, only mechanism (ii) can operate under low-temperature electron irradiation in an electron microscope. In fact, it is the athermal propagation of dynamic crowdions along the close-packed <110>-directions which introduces non-linear and non-local couplings between the concentrations C_X of vacancies (X = V) and dumbbells (X = D) and breaks the rotational symmetry in such a way that, under additional conditions to be discussed below, SLs may self-organize. The equations of motion for the concentrations C_X [43, 44] possess spatially homogeneous steady-state solutions $C_V^0 \approx C_D^0$ $(= C^0)$ that develop from the initial condition $C_V = C_D = 0$ after irradiation to doses that are much smaller than those required for the occurrence of a self-organized spatial pattern. Therefore, C^0 may be used as an initial reference state in the description of the self-organization process. Searching for the characteristic order parameter that can become unstable we remember that vacancy–dumbbell recombination is a dominant process in the point-defect dynamics. Obviously by this reaction in regions of high dumbbell concentration a decrease of the vacancy concentration is enforced, and vice versa. Hence, only antiphase fluctuations in the concentrations of dumbbells and vacancies may become unstable, and therefore the amplitude of such fluctuations normalized to C^0,

$$\psi(\vec{r},t)=\left[C_{\mathrm{D}}(\vec{r},t)-C_{\mathrm{V}}(\vec{r},t)\right]/C^0, \tag{1}$$

may serve as order parameter ($\vec{r}$ = vector representing the space coordinates, t = time).

Fig. 6 demonstrates the breaking of the rotational symmetry by the propagation of dynamic crowdions. These are expected to be stopped by vacancies and SIAs with different probabilities. Since this crowdion stopping results in the annihilation of vacancies and the creation of SIAs, respectively, a disalignment of regions with an excess of either vacancies ($\psi < 0$) or dumbbells ($\psi > 0$) from the <110>-directions takes place. As a consequence, defect fluctuations ψ of alternating sign tend to arrange in a CsCl-type structure the orientation of which coincides with that of the fcc atomic host lattice (Fig. 7).

In order to find out the circumstances under which the steady state of a homogeneous defect distribution becomes unstable and undergoes a non-equilibrium phase transition to a CsCl-type defect structure, linear stability analysis has been applied to the evolution equation of ψ [43]. After Fourier transformation one obtains the amplification factor $\Omega(\vec{k})$, which is a function of the wavevector $\vec{k}$ in the reciprocal lattice. For $\Omega(\vec{k}) > 0$ the corresponding Fourier modes grow and thus become unstable, i.e., ordering takes place. In this way, the instability criterion

$$3\left(D_{\mathrm{V}}+D_{\mathrm{D}}\right) < \eta_{\mathrm{cd}} P\left(w_{\mathrm{D}}-w_{\mathrm{V}}\right)\left(L_{\mathrm{dc}}^{\mathrm{eff}}\right)^3 \tag{2}$$

has been derived, where P is the production rate of vacancy–dumbbell pairs, η_{cd} the fraction of

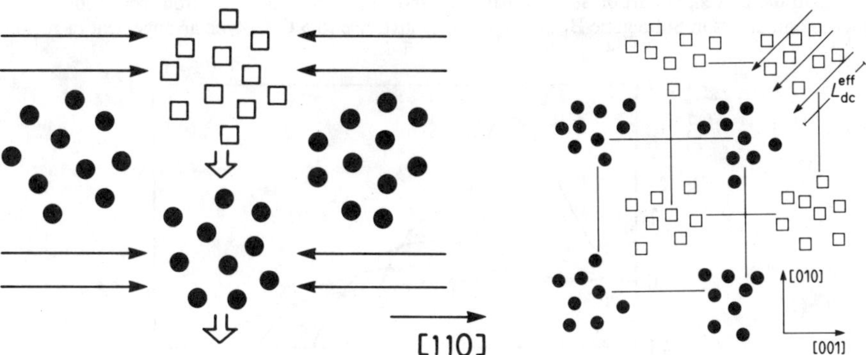

Fig. 6 (left). Destabilization of <110>-aligned accumulations of vacancies (empty squares) and self-interstitials (full circles) in an fcc host crystal by dynamic crowdions propagating along the close-packed <110>-directions.

Fig. 7 (right). Formation of a CsCl-type arrangement of zones with excesses of vacancies (empty squares) and self-interstitials (full circles) in an fcc host crystal by breaking of the translatory symmetry as a result of preferential defocusing of dynamic crowdions by self-interstitials.

dumbbells produced via dynamic crowdions, and L_{dc}^{eff} the mean effective dynamic-crowdion range. The diffusivity of vacancies (dumbbells) is denoted by D_V (D_D); w_V (w_D) is the probability with which dynamic crowdions are defocused when passing the strain field of a vacancy (dumbbell).

From (2) one recognizes that $w_D > w_V$ is a prerequisite for the occurrence of CsCl-type ordering. This is certainly fulfilled, since the strains around dumbbells are distinctly larger than around vacancies. The consequence of $w_D > w_V$ is that regions with an excess of dumbbells are less "transparent" for dynamic crowdions than vacancy-enriched regions. From Fig. 7 it is obvious that in this case alternating fluctuations ψ along <110>-directions are enhanced, i.e., the translatory symmetry is broken, and the homogeneous distributions of vacancies and dumbbells become unstable in favour of CsCl-type ordering.

In Regime A the recombination rate of vacancies and dumbbells is very low. This results in a high defect concentration. As a consequence, L_{dc}^{eff} is too small to satisfy the criterion (2), and therefore ordering does not occur in Regime A.

In the temperature regime of recovery stage II, close-pair recombination reduces the defect concentrations and thus increases the range of dynamic crowdions as the temperature increases. This effect is enhanced on the low-temperature side of recovery stage III, where the onset of thermally activated free migration of dumbbells leads to a further reduction of the defect concentrations. This regime of "long-range dynamic-crowdion propagation" strongly favours CsCl-type ordering and, according to what follows, is identified with Subregime B_1. In the centre of recovery stage III and above, D_D becomes very large. This leads to a violation of (2), and therefore on the high-temperature side of Subregime B_1 ordering is terminated.

The temperature dependence of the tendency for CsCl-type ordering just described is reflected by the Ω – $k_{<100>}$ curves shown in Fig. 8, where $k_{<100>}$ is the component of the wavevector in a <100>-direction of $\vec{k}$-space. These curves belong to various temperatures in the range $0.2 \leq T/T_{III} \leq 1.4$ (T_{III} = temperature in the centre of recovery stage III); their intersections with the abscissa shift to lower values of $k_{<100>}$ with increasing temperature. In fact, the positive values of Ω at intermediate temperatures of this regime confirm the occurrence of ordering predicted above on the basis of the criterion (2).

In the preceding part of Sec. 3 it has been shown that high-dose electron irradiations of fcc metals and alloys in Subregime B_1 are expected to give rise to a CsCl-type arrangement of regions

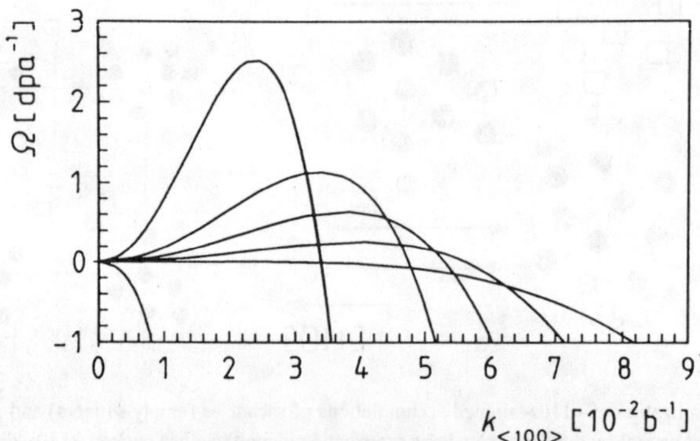

Fig. 8. Amplification factor Ω for CsCl-type ordering in an fcc host crystal as a function of the <100>-component $k_{<100>}$ of the wavevector for various temperatures in the regime $0.2 \leq T/T_{III} \leq 1.4$. The intersections of the curves with the abscissa shift to lower $k_{<100>}$ with increasing T. More details are given in [43].

with excesses of vacancies and dumbbells. The three-dimensional primitive cubic SLs observed by HVEM are interpreted as secondary ordered structures evolving from latent pre-existing CsCl-type defect arrangements by a collapse of the vacancy-rich zones to SFT.

The width of Subregime B_1 (Table 2), in which homogeneous nucleation of SLs takes place, shows an interesting correlation with the packing density ρ_{ion}/b of the host crystal (ρ_{ion} = ionic radius, b = nearest-neighbour distance of atoms). It increases with increasing packing density. This is explained in terms of a decreasing likelihood of defocusing of dynamic crowdions and thus an increase of L_{dc}^{eff} with increasing packing density [cf. criterion (2)].

4. Void Lattices and Basal Void Layers at "High" Temperatures [32–36, 44–47]

The circumstances under which VLs occur and the mechanisms of VL formation differ radically from those of SL evolution (Sec. 3).

As discussed in Sec. 3, the self-organization process leading to the CsCl-type structures that in a secondary step collapse to SLs, exclusively involves point defects, and the vacancy-rich regions constituting the precursors of SFT already nucleate on the sites of a primitive cubic lattice. By contrast, VL formation involves extended defects, namely voids, already in its pre-ordering state. It takes place in two stages. First, the voids nucleate at random; then, under continuing irradiation, order is established by preferential growth of a selection of voids located on or shifted to the sites of a three-dimensional lattice with the same structure and orientation as the host metal, and simultaneous shrinkage of the other voids.

The basically different condition, in comparison to that of SL formation, under which VLs develop is the "high" temperature of irradiation, at which all kinds of point defects undergo rapid thermally activated migration. As a result, the quasi-steady-state concentrations of the point defects are extremely low, so that mutual annihilation of vacancies and SIAs plays a negligible rôle. Rather the point-defect dynamics is sink-controlled by the absorption of point defects at voids and dislocations. Since the void dynamics is slow compared to the point-defect dynamics, the latter is "enslaved" by the former and thus may be adiabatically eliminated in the mathematical treatment [44, 46, 47].

The low point-defect concentrations established under the "high"-temperature irradiations leading to VL formation result in low internal strains. As a consequence, it is quite unlikely that metastable static crowdions are "killed" by strains; rather they may cover mean free paths L_C^{eff} of the order of magnitude of the VL parameter λ before they disappear at sinks or convert thermally to dumbbells. In the case of VL formation, the breaking of the rotational symmetry is thus due to

material	T/T_{III} of Subregime B_1	ρ_{ion}/b
Ni	0.9 – 1.2	0.28
Cu	0.65 – 1.1	0.36
Ag	0.5 – 1.05	0.42

Table 2. Correlation between crystal packing density ρ_{ion}/b and width of Subregime B_1 [43].

the one-dimensional diffusion of static crowdions along the close-packed host-lattice directions (mechanism (iv) in Sec. 2)[3]. This is confirmed by a comparison of gold, niobium, and aluminium, in which no VLs, normal VLs, and void hyperlattices have been observed, respectively (Table 3). In fact, in this order the packing density of the host lattice decreases, and hence the stability of static crowdions increases[4]. Further pieces of evidence for this interpretation are the absence of recovery stage I_E in Au and the extraordinarily low temperatures at which in Al, in comparison with other fcc metals, the recovery stages III (dumbbell migration) and IV (vacancy migration) occur [27]. According to the TIM, recovery stage I_E, which is found in most cubic metals after low-temperature irradiation, arises from the long-range migration of static crowdions. Its absence in Au thus indicates that in this metal static crowdions are unstable, in accordance with the fact that in Au VLs are not generated. In Al, in addition to the high stability of static crowdions arising from the low packing density, the relatively low stage-IV temperature, above which all kinds of elementary point defects undergo rapid thermally activated migration (a prerequisite for VL formation according to observations), favours extraordinarily large values of L_C^{eff} since thermal conversion of static crowdions becomes less important with decreasing temperature. Such large L_C^{eff} obviously account for the occurrence of void hyperlattices in Al in a straightforward manner (cf. discussion of Fig. 9).

The circumstances under which VLs in cubic metals are generated (Sec. 1) and the precedingly described implications have led Woo and Frank [32–35] to propose a theory of VL formation that is based on the TIM and does not require any assumptions going beyond this model. As already mentioned in Sec. 1, the basic idea of the TIM is that the SIAs may occupy two states. In bcc (fcc) metals the SIA ground state is the <110>-(<100>-) dumbbell configuration. In this configuration the SIAs undergo three-dimensional (i.e. isotropic) diffusion. The same is true for the vacancies. By contrast, in the excited metastable state of the SIAs, which is the <111>-(<110>-)crowdion configuration, the migration is confined to jumps along the close-packed crystallographic <111>-(<110>)-directions, i.e., it is one-dimensional. Transitions between the two SIA states can take place with the aid of thermal activation. Since a considerable fraction of the SIAs produced by irradiation are in the metastable crowdion configuration, the close-packed crystal directions are preferential directions of SIA diffusion.

material	packing density	dynamic crowdions	static crowdions	void lattices
Au	●●●	+++	–	–
Nb	●●●	++	+	+
Al	●●●	+	++	++

Table 3. Correlation between void-lattice formation and static crowdions [47].

3) In irradiations in which the transferred energies exceed 1 to 2 keV, mechanism (v) (Sec. 2) may also contribute to the transport of matter in close-packed directions [38].

4) Note that the range of dynamic crowdions decreases with decreasing packing density, i.e., dynamic-crowdion propagation may be excluded as the origin of VL formation.

On the basis of this scenario, Woo and Frank [32–35] proposed that VL formation results from the competition of voids for one-dimensionally diffusing static crowdions. This may be visualized by means of crowdion-supply cylinders (CSCs) of length L_C^{eff} possessing the void radii and extending from the void surfaces in the close-packed host crystal directions (Fig. 9). Obviously, crowdions being annihilated at a void must have been produced within one of its CSCs. Competition of voids for crowdions occurs if the CSCs of neighbouring voids overlap. Given the mean void radius R_0, CSC overlapping requires that the void number density ρ_0 exceeds a critical value ρ_{crit}. Since the onset of CSC overlapping initiates the competition of voids for crowdions, it results in a reduction of the crowdion influx per void and thus in an enhancement of the void growth rate. The minor fraction of voids whose CSCs do not overlap inspite of $\rho_0 > \rho_{crit}$ receive a net influx of SIAs and therefore shrink away. Fig. 9 shows a partial overlap of the CSCs of two neighbouring voids. In this case the crowdion fluxes in the upper part of the left void and the lower part of the right void are reduced. As a consequence, in addition to an increase of their growth rates, these voids align along the horizontal close-packed direction. We shall return to this alignment effect lateron. It is obvious that the interplay of the various effects of CSC overlapping just described leads to a kind of Darwinian selection of voids, the outcome of which is a VL with the same structure and orientation as the host lattice.

The concept of VL formation just described has originally evolved from a numerical solution of macroscopic rate equations [32]. In order to get information going beyond that obtainable from such a treatment (e.g., on the critical wavelength of void ordering) void dynamics has been treated on a mesoscopic scale defined by the mean void radius R_0 and the mean void separation $d = 2(3/4\pi\rho_0)^{1/3}$ in disordered arrangement [46, 47].

Consider a void that is exposed to fluxes $\vec{j}(\vec{r})$ of point defects which are consumed by the void just where they encounter the void surface. Thus, within an infinitesimal interval of time dt the void surface elements are shifted from their original positions $\vec{r}$ to new positions

$$\vec{r}' = \vec{r} \pm \vec{j}(\vec{r})dt. \tag{3}$$

Note that the upper (lower) sign corresponds to a flux of interstitials (vacancies) and that $\vec{j}$ is considered to be time-independent. This is a good approximation since, as already discussed, the dynamics of point defects is by orders of magnitude faster than the void dynamics. By means of

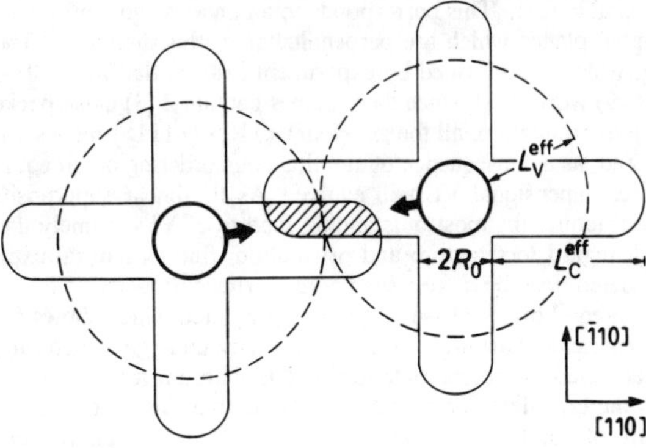

Fig. 9. Crowdion-supply-cylinder concept in the formation of void lattices, visualizing the preferential growth and alignment along <110>-directions of two neighbouring voids in an fcc host crystal.

the coordinate transformation (3), the void growth rate $\dot{R}_i$ and the drift velocity $\vec{v}_i$ of the spherical void i with radius R_i may be expressed as

$$\dot{R}_i = \pm \frac{1}{4\pi R_i^2} \int_{\text{void } i} d^3 r_i \vec{\nabla} \vec{j}(\vec{r}_i), \tag{4}$$

$$\vec{v}_i = \pm \frac{3}{4\pi R_i^3} \int_{\text{void } i} d^3 r_i \left(\vec{j}(\vec{r}_i) + \vec{r}_i \left(\vec{\nabla} \vec{j}(\vec{r}_i) \right) \right). \tag{5}$$

Here the origin has been chosen to coincide with the void centre. In Eq. (5) the second term of the integrand accounts for the alignment effect (Fig. 9). The fluxes $\vec{j}$ entering Eqs. (4) and (5) depend on the constellation of the voids in the vicinity of the void considered. In a first-order perturbation approximation, $\vec{j}$ depends in a non-local manner on the generalized sink fields

$$\rho^{(n)}(\vec{r},t) = \sum_j R_j^n(t) \delta^3(\vec{r} - \vec{r}_j(t)), \tag{6}$$

i.e., voids are modelled by monopole-type sinks. The power n of the sink strength reflects the dimensionality of diffusion (n = 2 for one-dimensional crowdion diffusion, n = 1 for three-dimensional diffusion of vacancies or dumbbells). This diffusional anisotropy difference turns out to be essential for the breaking of the translatory symmetry leading from a random void arrangement to a VL (Sec. 8).

In order to describe the non-local response of a void to the presence of another one, Green's functions of the one- and three-dimensional steady-state reaction–diffusion equations have been used, respectively. Calculation of the time derivatives of Eqs. (6) with the aid of (4) and (5) yields a hierarchical system of equations of motion for the sink fields $\rho^{(n)}(\vec{r},t)$. This hierarchy may be closed by assuming that the width of the distribution of void radii is sufficiently narrow to express R_i as $R_i = R_0 + \delta R_i$ with $\delta R_i / R_0 << 1$. Then one ends up with two coupled equations of motion for the sink fields $\rho^{(2)}(\vec{r},t)$ and $\xi(\vec{r},t) \equiv \sum \delta R_i \delta^3(\vec{r} - \vec{r}_i)$. From linear stability analysis of the spatially homogeneous, non-stationary solutions of these non-linear integro-differential equations it follows that, in bcc (fcc) metals, modes with wavevectors $\vec{k}$ in one of the <110>-(<111>)-directions become unstable first. This corresponds to an onset of void ordering on the set of close-packed crystallographic planes which are perpendicular to this direction. The deeper reason for this planar ordering, which is confirmed by experiment [48], is that in {110}-({111}-) planes the alignment effect (Fig. 9) works best, since these planes contain 2 (3) close-packed directions.

Under continuous irradiation, all the six (four) <110>-(<111>)-modes will finally satisfy the instability criterion, and, as a consequence of simultaneous ordering on all equivalent sets of close-packed planes, a three-dimensional VL will evolve. As the linear superposition of these modes leads to a bcc (fcc) structure, the most outstanding feature of VLs, namely their coincidence with the host lattices with regard to structure and orientation, finds a natural explanation. It is also noteworthy that λ is determined by d, i.e., λ is not an intrinsic property.

In Fig. 10 a non-equilibrium phase diagram is presented which shows for a bcc host crystal under which conditions random arrangements of voids may undergo structural phase transitions to VLs. The parameters used in the computation of this diagram are as follows: fraction of SIAs radiation-induced in the crowdion configuration = 50%, total dislocation density = 10^{15}m^{-2}, bias factor for the capture of dumbbells by dislocations = 0.1 (dashed curve) and 0.0 (solid curve), respectively. R_0 and d are given in units of the effective mean free path L_C^{eff} (≈ 10nm) covered by crowdions before they are annihilated at dislocations, voids, or by conversion into dumbbells. This phase diagram tells us that void ordering requires large void radii and/or small void separations, in agreement with the picture arising from the CSC concept (see above) and experimental findings. A corresponding diagram for fcc metals shows the same qualitative features.

It is noteworthy that a decrease of the dislocation bias for dumbbell capture shifts the onset of void ordering to smaller void radii (Fig. 10), since it increases the importance of the one-dimensional crowdion diffusion that promotes ordering. We shall return to this point in Sec. 8.

The dynamical stability of VLs during irradiation against a small displacement of a void from its regular lattice site has been tested. It turns out that this structural perturbation changes the point-defect fluxes in such a way that the displaced void returns to its original lattice site. Fig. 11 shows for $R_0 / L_C^{eff} = 0.0$ (uppermost curve), 0.4, and 0.6 the characteristic dose $\tilde{\Phi}$ required for the restoration of bcc VLs as a function of λ / L_C^{eff}, where here λ is the VL parameter in <100>-direction. One realizes from Fig. 11 that the range of typical characteristic doses, 10 dpa $< \tilde{\Phi} <$ 100 dpa, is the same as the dose range in which VL formation takes place.

In conclusion of the discussion of VLs in cubic metals the essential results may be summarized as follows. Like the other ordering phenomena reviewed in this paper, the irradiation-induced formation of VLs is a self-organization effect that takes place far away from thermodynamic equilibrium. It is the one-dimensional thermally activated migration of static crowdions that breaks the rotational symmetry (Sec. 2) and the difference in the diffusional anisotropy of static crowdions and vacancies/dumbbells that finally leads to a breaking of the translatory symmetry (Sec. 8) in such a way that VLs are formed. VL formation may be taken as strong evidence for the validity of the two-interstitial model of radiation damage.

The analogue to VLs in cubic metals is the layers of voids parallel to the basal plane, which have been found in subnormal hcp metals (e.g., Mg) and which in Sec. 1 have been introduced as "basal void layers" (BVs). Within these layers the voids are disordered. Klemm and Frank [49, 50] have shown that thermally activated two-dimensional migration of SIA in the basal plane (mechanism (vi) in Sec. 2) readily accounts for this kind of ordering. This is in accordance with the fact that in subnormal hcp metals the stable SIA configuration is the so-called B_0-interstitial, whose diffusion is indeed restricted to the basal plane [41].

A theoretical study of void ordering in hcp metals that goes beyond the experimental findings deals with the influence of one-dimensional diffusion of SIAs parallel to the *c*-axis [50]. This is of interest since in subnormal hcp metals the configuration C_N exists as metastable

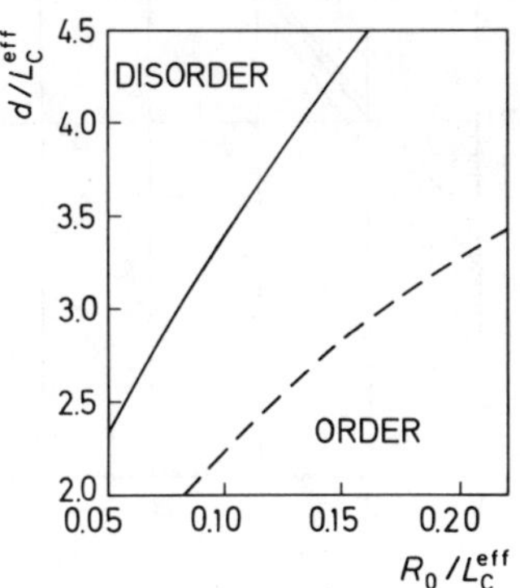

Fig. 10. Order–disorder d–R_0 phase diagram of void arrangements in a bcc host metal. The dislocation bias for dumbbell absorption is 0.0 (solid curve) and 0.1 (dashed curve), respectively. More details are found in [47].

excited SIA state and undergoes this type of anisotropic diffusion [41]. The major result of this investigation is that, if there is two-dimensional SIA diffusion in the basal plane plus one-dimensional SIA diffusion parallel to the *c*-axis, two phase transitions may occur. If the irradiation dose and thus the control parameters ρ_0 and R_0 increase, in a first transition disordered voids disappear in favour of BVs, whereas at considerably higher doses in a second transition voids arrange in columns parallel to the *c*-axis (briefly called "c-directed void columns" or "CVs"). However, a complete hexagonal void lattice is not predicted to be formed. Concerning void ordering in supernormal hcp metals with $c/a > (8/3)^{1/2}$, e.g., Cd or Zn, in which the stable SIAs are of *c*-dumbbell type and diffuse parallel to the *c*-axis [41], no BV formation, but, at very high irradiation doses, CV formation may occur.

5. SFT Grids at "Intermediate" Temperatures [44, 51]

Although, at first sight, the SGs resemble very much the SLs, the two phenomena of SFT ordering are of completely different nature. As discussed in detail in Sec. 3, the three-dimensional SLs are generated under conditions where the concentrations of vacancies and dumbbells are high and the point-defect dynamics is controlled by the irradiation-induced vacancy–dumbbell recombination; this ordering may be traced back to the athermal propagation of dynamic crowdions, which first produces CsCl-type ordering of vacancy-rich and SIA-rich regions; from

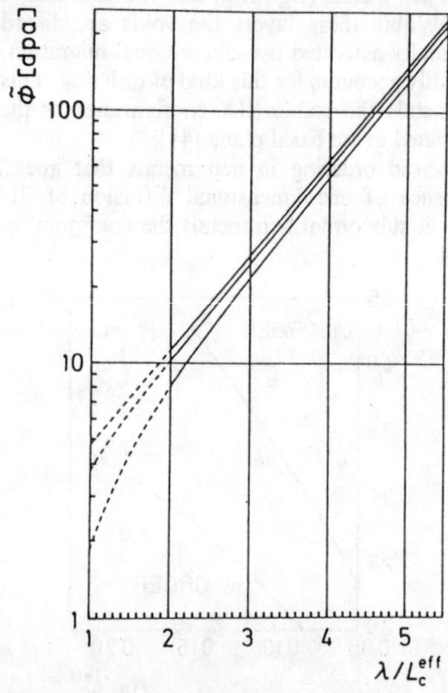

Fig. 11. Characteristic dose $\tilde{\phi}$ for the restoration of a bcc void lattice as a function of the void-lattice parameter λ. The uppermost and the lowest curve belong to $R_0 = 0$ and $R_0 = 0.6\ L_C^{eff}$, respectively. More details are given in [47].

this SLs are formed in a secondary step, namely by collapsing of the vacancy-rich regions to SFT, which is equivalent to a homogeneous SL nucleation. Subregime B_1, in which SLs are created, is terminated on its high-temperature side by the onset of thermally activated long-range migration of dumbbells in Stage III.

The formation of (planar) SGs takes place in the "intermediate"-temperature Subregime B_2, where, as a consequence of the rapid thermal long-range migration of dumbbells and, to a lesser extent, of the slower irradiation-induced diffusion of vacancies, the point-defect concentrations are low in comparison to those during SL formation. However, since for dumbbells both the mobility and the elastic attraction by the surfaces are greater than for vacancies, in the central lattice planes of the foil-shaped specimens enough vacancies are accumulated in order to give rise to the nucleation of planar random arrangements of SFT. The ordering of the SFT occurs *after* their nucleation. Therefore, the dynamics of the point defects is significantly influenced by the presence of the SFT, i.e., drift-diffusion due to the elastic SFT–point-defect interactions has to be taken into account (mechanism (vii) according to Sec. 2). Since the orientation of the SFT is uniquely related to the crystallographic directions of the host crystal and the anisotropy of the SFT strain fields is conveyed to the point-defect fluxes, these fluxes mediate an anisotropic dynamic interaction between the SFT that is, at last, dictated by the host crystal. Like in Subregime B_1, static crowdions need not be considered since, as a result of the high SFT densities, these metastable defects are athermally converted to stable dumbbell interstitials immediately after their production.

In the following a qualitative outline of SG formation is presented (Fig. 12). (α) We consider the distribution of point defects around a SFT. Due to the elastic interaction between the SFT and the point defects, which arises mainly from the strain fields of the stair-rod dislocations that form the SFT edges, this point-defect distribution is inhomogeneous. (β) In the immediate vicinity of the SFT the elastic interaction is attractive for vacancies, which tend to be absorbed by the SFT, and repulsive for dumbbell interstitials. Therefore, each SFT is surrounded by a point-defect-depleted zone, which, for simplicity, is taken to be spherical with radius R_*. (γ) In a shell $R_* < r < R_* + L_X^{\mathrm{eff}}$ outside the depleted sphere, where r is the distance from the centre of the sphere and L_X^{eff} are the effective mean free paths of vacancies (X = V) and dumbbells (X = D), drift-diffusion in the anisotropic strain field of the SFT leads to an inhomogeneous distribution of point defects, namely an accumulation of vacancies and dumbbells along <100>- and <110>-

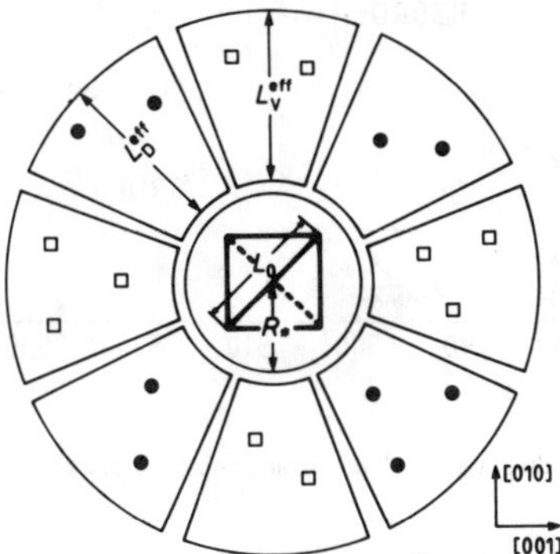

Fig. 12. Concept of the arranging of stacking-fault tetrahedra in a grid, showing the sectors with excesses of vacancies (empty squares) and dumbbells (full circles) around a stacking-fault tetrahedron in the (100)-plane of the fcc host crystal.

directions, respectively. As a consequence, the growth of new SFT is favoured in the vacancy-rich zones, i.e. in <100>-sectors around pre-existing SFT. This qualitatively explains the formation of <100>-chains and planar <100>–<010> square lattices of SFT (SGs) found by experiment. The reasons why SGs develop only in two dimensions have already been discussed above. (δ) There are two effects stabilizing SGs. If two SFT come too close to each other, their depleted spheres overlap, and thus the SFT receive a reduced influx of vacancies from the overlapping region. This results in a dynamic repulsion of the two SFT. On the other hand, two SFT aligned along a <100>-direction may be too much separated from each other in order to mutually support their growth via drift-diffusion of point defects to a significant extent. Then these SFT are attracted by regions with a higher excess of vacancies, i.e., they come closer to each other. Superposition of both effects leads to a SG with a finite lattice parameter λ. Since the strength of the interaction between SFT and point defects scales with the SFT-edge length, λ depends on the SFT size; in fact, λ increases with the SFT-edge length. (ε) Under continuing irradiation, first a patch of a SG is formed, which then spreads out in the foil plane.

In Fig. 13 the critical centre-to-centre separation d of SFT below which SGs are formed is presented as a function of temperature T for a set of material parameters typical of Ni. One realizes that d becomes smaller as T decreases. This is due to the fact that, if the temperature is lowered, owing to the higher point-defect concentrations, vacancy–dumbbell recombination leads to a decrease of the effective mean free paths L_X^{eff}. As a result, with decreasing temperature the range of the dynamic point-defect-mediated interactions between the SFT decreases and thus requires a decrease of the d at which ordering sets in. Since d must be larger than the SFT-edge length, in Ni SGs cannot form below about 490K. Concomitantly, thermally assisted drift-diffusion becomes less important with decreasing temperature, so that athermal processes, e.g., dynamic crowdion propagation, can take over. Therefore, the disappearance of SGs in Ni below 490K is accompanied with the emergence of SLs, i.e., in Ni 490K divides Regime B in the Subregimes B_1 and B_2.

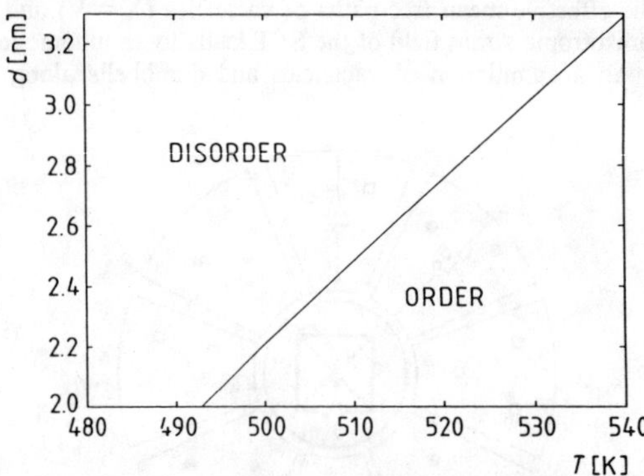

Fig. 13. Order–disorder d–T phase diagram for planar arrangements of stacking-fault tetrahedra in Ni in Subregime B_2. More details are given in [51].

6. SFT-Patch Grids at "Elevated" Temperatures [52]

During high-dose electron irradiation of thin Ni foils at 520 K, groups of about 10 SFT are formed, which are referred to as "patches" (Fig. 5). Within the patches the SFT occupy the sites of a square lattice whose sides are parallel to the two <100>-directions of the crystalline host lattice in the foil plane. This square lattice of SFT is identical with the SG, which has been the subject matter of Sec. 5. The patches themselves are arranged in a planar fcc lattice that also shares its orientation with the host lattice. In this section the theory [52] of the arranging of such SFT-grid patches in so-called SFT-patch grids (PGs) is reproduced qualitatively.

Similar to the defect-ordering phenomena reported in Secs. 3 to 5, the ordering of patches arises from a point-defect-mediated dynamic patch–patch interaction. Searching for the nature of this interaction, one must watch out for an anisotropic long-range matter-transport mechanism that can account for the structure and orientation of the PGs, their lattice parameters, and the temperature regime of their formation. Since the characteristic length of PGs ($\lambda \approx 12$ nm) differs by an order of magnitude from that of intra-patch SGs (3 nm $\leq \lambda \leq$ 4 nm), it may be anticipated that the processes leading to the formation of the two interlacing ordered structures are different and approximately independent.

PGs and VLs have in common that they possess the same lattice structure and orientation as their atomic host lattices. In addition, the lattice parameters of both structures are of the same order of magnitude. Therefore, it is not surprising that PG formation involves the same basic mechanism as VL formation, namely the thermally activated one-dimensional migration of static crowdions (mechanism (iv) in Sec. 2). The scenario of void ordering displayed in Sec. 4 may thus be translated into the language of PG formation. In particular, the CSC concept illustrating that the competition of voids in absorbing crowdions is essential in VL formation has to be adjusted to the other geometry in PG formation (Fig. 14). Since the patches are two-dimensional, run parallel to the foil plane, and lie within a thin layer in the foil centre, only crowdions migrating along the two close-packed directions in the foil plane have to be taken into account. Growth of the patches occurs by the nucleation of new SFT at the patch peripheries. If, by the analogy just mentioned, the patches are replaced by oblate spheroidal "voids", their growth corresponds to an increase of

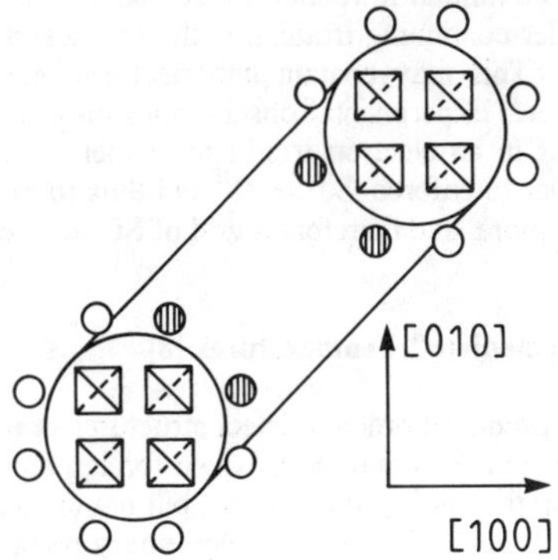

Fig. 14. Concept of ordering of stacking-fault-tetrahedron-grid patches, visualizing the preferential growth and alignment along the [110]-direction of two neighbouring patches (large circles). The small circles mark potential sites of stacking-fault tetrahedra at the peripheries of the patches. Preferential nucleation of new stacking-fault tetrahedra will occur at the locations of the shaded small circles.

the major half-axes. It is achieved by vacancies either generated inside or migrating to the patches. By contrast, the absorption of dumbbells and crowdions leads to patch shrinkage. Obviously, in PG formation the crowdion-supply volumes within which crowdions captured by a patch must have been created previously, are elliptic cylinders of length L_C^{eff} extending from the patch surface along the close-packed directions in the foil plane.

A confirmation of the importance of static crowdions in the formation of PGs is the fact that PGs have been observed in Ni, but not in Ag, Cu, or alloys. Among the materials investigated with respect to PG formation, Ni possesses the lowest packing density. This implies that in Ni the enthalpy barrier that static crowdions must surmount when thermally converting to dumbbells is particularly high and thus L_C^{eff} great enough for the operation of the PG-formation mechanism described above.

It may be readily recognized that PGs can exclusively be formed in a narrow regime of "elevated" temperatures that is squeezed in between the "intermediate" and "high" temperatures defined in Sec. 1. At these "elevated" temperatures the thermally activated mobilities of static crowdions and dumbbells are extremely high and moderately high, respectively, whereas thermal migration of vacancies just sets in. Under these circumstances, on the one hand, the large mobility difference between SIAs and vacancies still permits the nucleation of SFT in the central foil layer and thus, at least patchwise, the formation of SGs, whereas, on the other hand, at least outside the patches the defect density is low enough to ensure an L_C^{eff} that is sufficiently large for patch ordering.

VL and PG formation show not only far-reaching similarities, but also distinct differences. An obvious reason is of geometrical nature, viz., the different shapes of the extended defects (spheres versus oblate spheroids) and the different dimensionalities of the specimens (bulk samples versus foils). More important, however, are the different irradiation-temperature regimes in which VLs and PGs are formed. At the "high" temperatures at which VLs are created the effective mean free paths of all types of point defects are of the same order of magnitude, so that the void attraction mediated by crowdions and the void repulsion mediated by the combined action of vacancies and dumbbells balance each other at a finite void separation that determines the VL parameter. At the less "elevated" temperatures of PG formation, all L_X^{eff} are shorter as a result of enhanced vacancy–SIA recombination and, more important in the present context, $L_C^{eff} >> L_X^{eff}$ (X = V or D). As a result, the mutual attraction of the patches cannot be compensated by their inefficient repulsion, i.e., under continuing irradiation the patches of a PG will grow together and finally form a coherent SG. This may contain imperfections, particularly where neighbouring patches have joined each other. Experimental observations on these late stages of PGs have not yet been made. Note that, if in an electron irradiation experiment on Ni the temperature were raised far above 520K in order to enforce $L_X^{eff} \approx L_C^{eff}$ and thus to produce a "saturated" state of a PG, SGs would not form any more, and therefore a grid of SG patches could not be formed either.

7. Bubble Lattices at "Intermediate" Temperatures [8]

Among the irradiation-produced ordered defect structures in metals discussed in the present article, BLs are the only ones in the formation of which foreign atoms play a decisive rôle. (The importance of foreign atoms in the *nucleation* of voids, but not in their growth or their arranging in VLs, is comparatively minor.) As to the intrinsic elementary point defects, at the "intermediate" temperatures of BL formation vacancies are immobile whereas dumbbells undergo thermally activated three-dimensional migration. Most of the dumbbells recombine with vacancies, some of them are absorbed at dislocations, thus leading to an accumulation of vacancies. The retained vacancies make L_C^{eff} so small that static crowdions have not to be considered, and, more important for what follows, these vacancies represent the nuclei for the formation of bubbles.

The formation of a bubble begins with the capture of implanted inert gas atoms, say, He-atoms, by a vacancy (V):

$$\begin{aligned} V + He &\rightarrow VHe \\ &+ He \rightarrow VHe_2 \\ &\qquad + He \rightarrow VHe_3 \\ &\qquad\qquad \ldots \rightarrow VHe_n \,. \end{aligned} \tag{7}$$

After the vacancy has incorporated a critical number of He-atoms ($n \approx 9$ for Mo), the stresses induced in the surrounding of the VHe_n complex may be reduced by the emission of dumbbells (D), i.e., the complex VHe_n can grow though vacancies are immobile at the temperatures of bubble formation:

$$\begin{aligned} &VHe_n \rightarrow V_2He_n + D, \\ &V_2He_n \rightarrow V_3He_n + D, \\ &\text{or generally} \\ &V_mHe_n \rightarrow V_{m+1}He_n + D \text{ with } m \geq 1 \,. \end{aligned} \tag{8}$$

Then $V_{m+1}He_n$ can continue to absorb additional He-atoms etc., until a true He-bubble has developed.

The gas-pressure-induced anisotropic stress field around a bubble represents a breaking of the rotational symmetry and gives rise to an anisotropic drift-diffusion of SIAs out of the most strongly compressed regions along the close-packed directions of the host crystal (mechanism (viii) in Sec. 2). As a result, along these directions an excess of vacancies is accumulated that favours the nucleation of other bubbles. However, within a small region close to the bubble surface, emission and absorption of SIAs enforce the SIA equilibrium concentration and thus prevent the nucleation of other bubbles.

Neighbouring bubbles interact dynamically via the SIA fluxes and thus mutually influence their growth if their separation is comparable to the range of their stress fields. At a given bubble density this is the case if the control parameters "bubble radius" and "gas pressure" exceed critical values. Under these conditions bubbles possessing neighbouring bubbles in the close-packed host-lattice directions grow preferentially. As a result, bubble ordering sets in on close-packed planes. Then, under continuing He implantation, unfavourably located bubbles shrink away, and a three-dimensional BL is formed that coincides with the host lattice with regard to structure and orientation.

Fig. 15 shows the orientation dependence of the amplification factor Ω of a BL in a bcc host crystal for wavevectors in the $(1\bar{1}0)$ plane of $\vec{k}$-space. For the chosen parameter set [8] Ω has its maximum at $k_{<110>} R_0 \approx 2$, which corresponds to a BL parameter in the <110>-directions of about $3.1\, R_0$ (R_0 = bubble radius).

In bcc host crystals, BLs are dynamically stable, whereas in fcc metals in the course of irradiation BLs may domainwise undergo rotations into BL variants which are still fcc, but possess orientations that differ from that of the host lattice. A complete understanding of the occurrence of this BL domain structure in fcc metals has not yet been achieved.

A peculiar kind of bubble ordering was observed by Johnson et al. [53]. These authors reported that, as a result of 160 keV He^+-ion irradiation, in Au two interlacing BLs have been generated. On a normal BL ($R_0 \approx 2$ nm, $\lambda \approx 8$ nm) a bubble hyperlattice ($R_0 \approx 70$ nm, $\lambda \approx 150$ nm) was superimposed. Since the specimen thickness amounted to about 100 nm only, the bubble hyperlattice exclusively consisted of one {100} layer. Both BLs possessed the same fcc structure and orientation as the Au host lattice. Though it was speculated [8] that the formation of the bubble hyperlattice may involve dynamic crowdions, a final explanation of this ordering phenomenon, which resembles very much the creation of PGs in Ni under electron irradiation (Sec. 6), is not yet available.

8. Breaking of the Translatory Symmetry

The occurrence of the synergetic defect-ordering phenomena discussed in Secs. 3 to 7 requires that there are control-parameter regimes in which an amplification of fluctuations takes place that leads to instabilities of homogeneous defect distributions with regard to finite-wavelength (λ) perturbations. This obviously implies that there are mechanisms that spontaneously break the translatory symmetry by counteracting the tendency of isotropic diffusion processes to establish homogeneity.

8.1. Instabilities of Homogeneous Point-Defect Distributions

The only ordering phenomenon exclusively involving point defects (but not extended defects) which has been discussed in the present paper is the formation of CsCl-type arrangements of vacancy-enriched and SIA-enriched zones under electron irradiation (Sec. 3). (The secondary step in which the vacancy-excess zones collapse to a SL will not be considered in what follows.)

From Fig. 16 is evident that local deviations δC_X from the homogeneous distributions of vacancies (X = V) and SIA (X = I) will be amplified only if mechanisms exist that drive the point-defect fluxes $\vec{j}_X$ into regions where the concentration fluctuations δC_X are positive (x = space coordinate). According to Martin's [54] conventional proposal, such an "up-hill diffusion" may occur if a vacancy–vacancy attraction and a SIA–SIA attraction exist which exceed the vacancy–SIA attraction. In the theory of CsCl-type ordering presented in Sec. 3 this assumption has been replaced by a mechanism of *anisotropic* matter transport, namely the athermal propagation of dynamic crowdions combined with the obviously justified assumption that their defocusing probability is higher in SIA-enriched zones than in vacany-enriched zones (Fig. 7).

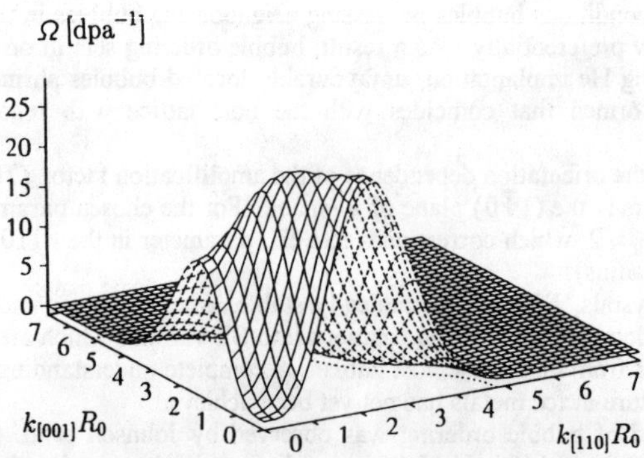

Fig. 15. Orientation dependence of the amplification factor Ω in the formation of bubble lattices in bcc host crystals for wavevectors lying in the $(1\bar{1}0)$-plane of $\vec{k}$-space. More details are given in [8].

8.2. Instabilities of Random Distributions of Extended Defects

Breaking of the translatory symmetry of a random distribution of extended defects, say, voids, requires that local fluctuations $\delta\rho > 0$ of the void density induce net influxes of vacancies, $\vec{j}_V - \vec{j}_I > 0$, into these regions and thus are amplified (Fig. 17). If both vacancies and SIAs diffuse isotropically and if, in addition, all sinks for point defects are unbiased with regard to the point-defect nature, $\vec{j}_V - \vec{j}_I = 0$, i.e., under these circumstances a disordered void arrangement cannot become unstable, and corresponding statements hold for random arrangements of other extended defects. In former OIM-based theories of VL formation [1] and other extended-defect ordering [55] this difficulty has been circumvented by ascribing to dislocations a preference for the absorption of SIAs. However, as demonstrated by Brailsford [56], the symmetry breaking enforced in this way is inefficient. If it were realized in nature, the restoration of deviations from the periodicity of a VL would require irradiation doses that exceed those actually needed for VL formation by at least an order of magnitude.

An alternative symmetry-breaking mechanism that does not suffer from this deficiency forms the basis of the TIM-based VL-formation theory reported in Sec. 4. According to this theory, it is the difference in the anisotropy of the diffusion of vacancies and dumbbells on the one hand and of static crowdions on the other hand that induces an ordering of random void arrangements. The deeper reason for this is the fact that in anisotropic diffusion of point defects the defect fluxes do not follow the steepest descents in the defect concentrations. As demonstrated by means of Fig. 11 the doses for VL restoration predicted by the novel VL theory lie in the dose range required experimentally for the creation of VLs. Finally, this theory predicts that the existence of a dislocation bias for the absorption of dumbbells is even counter-productive with regard to void ordering (Fig. 10), since it decreases the relative importance of crowdion diffusion.

In the formation of SGs, PGs, and BLs, breaking of the translatory symmetry is also achieved by anisotropic diffusion. In these cases, the diffusional anisotropies are drift-induced, and the breaking of the translatory symmetry does not require any kind of bias either.

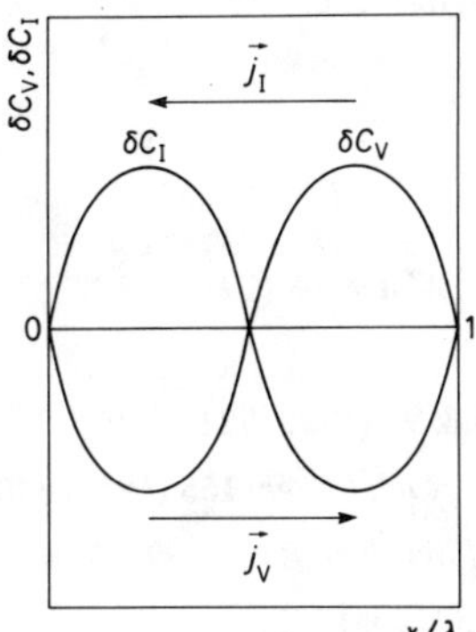

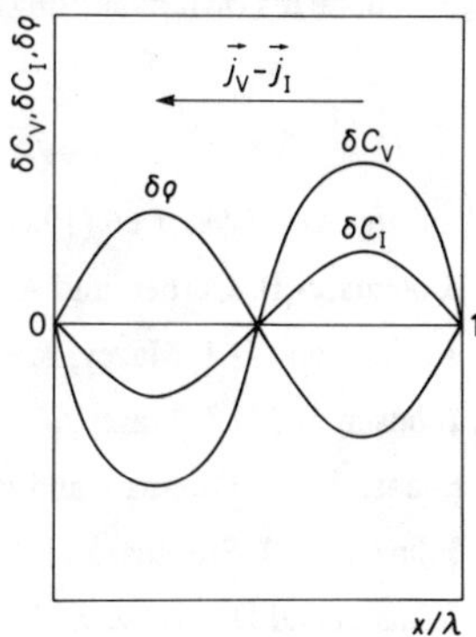

Fig. 16 (left). Breaking of the translatory symmetry of homogeneous point-defect distributions by the amplification of concentration fluctuations.

Fig. 17 (right). Breaking of the translatory symmetry of a random distribution of vacancy-type extended defects by the induction of appropriate point-defect fluxes.

9. Concluding Remarks

The formation of ordered structures takes place in nature on all size scales, reaching from spiral galaxies to the substructure of the nucleons. Most familiar is ordering in thermodynamic equilibrium, which occurs as a result of free-energy minimization if, concomitantly, the entropy is kept low. More fascinating, however, are ordered steady states, evolving far away from thermodynamic equilibrium, in open, non-linear, highly dissipative systems that are exposed to an influx of energy and/or matter provided the control parameters are chosen in such a way that the translatory symmetry is broken and internal entropy production is overcompensated by the entropy export to the environment. Prominent examples of this self-organization are the selection of optical modes in lasers and the convective patterns in liquids. The present article has reviewed a variety of ordered defect structures developing in metals under particle irradiation which belong into the same category.

Whereas lasing and pattern formation in liquids follow from the well-known basic equations of non-linear optics and hydrodynamics, respectively, corresponding equations from which defect patterning in metals might be deduced are primarily not available. Most surprisingly, this difficulty has turned out to contribute considerably to the understanding of radiation damage in metals. This is so since in each case of irradiation-induced defect patterning the underlying atomistic processes have to be "guessed" in order to be capable of establishing the corresponding equations of motion. The correctness or incorrectness of such a guess follows from checking the predictions of these equations versus the defect pattern found by experiment. In this way, we have come to the conclusion that the defect patterns produced in metals by particle irradiation may find straightforward explanations in terms of the two-interstitial model, but are incompatible with the one-interstitial model. Hopefully, as a by-product, this result may help to settle a long-lasting controversy in the field of radiation damage in metals!

Acknowledgement. We are very grateful to the director of our institute, Professor Dr. Dr. h.c. Alfred Seeger, who has stimulated and promoted research on self-organization in solids at our laboratory, leave alone his own pioneering contributions in this field.

References

1. K. Krishan, *Rad. Effects* **66** (1982) 121.
2. B.A. Loomis, S.B. Gerber, and A. Taylor, *J. Nucl. Mater.* **68** (1977) 19.
3. P.B. Johnson and D.J. Mazey, *Rad. Effects* **53** (1980) 195.
4. P.B. Johnson and D.J. Mazey, *J. Nucl. Mater.* **93 & 94** (1980) 721.
5. J.H. Evans, A.J.E. Foreman, and R.J. McElroy, *J. Nucl. Mater.* **168** (1989) 340.
6. P.B. Johnson, K.J. Stevens, and R.W. Thomson, *Phys. Rev. B* **62** (1991) 218.
7. P.B. Johnson and D.J. Mazey, *J. Nucl. Mater.* **218** (1995) 273.
8. C. Tölg, *Dr. rer. nat. Thesis*, Stuttgart University, Stuttgart, 1995.
9. E. Johnson and L.T. Chadderton, *Rad. Effects* **79** (1983) 183.
10. A. Jostsons and K. Farrell, *Rad. Effects* **15** (1972) 217.
11. A. Risbet and V. Lévy, *J. Nucl. Mater.* **50** (1974) 116.
12. A. Horsewell and B.N. Singh, *Rad. Effects* **102** (1987) 1.

13. W. Jäger, P. Ehrhart, and W. Schilling, *Solid State Phenomena* **3 & 4** (1988) 279.

14. W. Jäger and H. Trinkaus, *J. Nucl. Mater.* **205** (1993) 394.

15. K. Urban and A. Seeger, *Phil. Mag.* **30** (1974) 1395.

16. L. Kubin, A. Rocher, M.O. Ruault, and B. Jouffrey, *Phil. Mag.* **23** (1976) 293.

17. M.O. Ruault, A. Rocher, L. Kubin, and B. Jouffrey, in *Fundamental Aspects of Radiation Damage in Metals*, Vol. 2, ed. M.T. Robinson and F.W. Young, Jr., USERDA, CONF–751006–P2, Oak Ridge, TN, 1976, p. 722.

18. W. Jäger and K. Urban, In *Proc. Intern. Conf. High Voltage Electron Microscopy*, Toulouse, 1995, ed. B. Jouffrey and P. Favard, Société Française de Microscopie Electronique, Paris, 1976, p. 175.

19. N. Yoshida and M. Kiritani, *J. Phys. Soc. Japan* **38** (1975) 1220.

20. H. Fujita, T. Sakata, and H. Fukuyo, *Japanese J. Appl. Phys.* **21** (1982) L 235.

21. N.Y. Jin, F. Phillipp, and A. Seeger, *phys. stat. sol. (a)* **116** (1989) 91.

22. N.Y. Jin, F. Phillipp, and A. Seeger, In *Proc. 12th Intern. Congr. Electron Microscopy*, Vol. 4, Seattle, WA, 1990, p. 508.

23. A. Seeger, N.Y. Jin, F. Phillipp, M. Zaiser, *Ultramicroscopy* **39** (1991) 342.

24. W. Schilling, G. Burger, K. Isebeck, and H. Wenzl, In *Vacancies and Interstitials in Metals*, ed. A. Seeger, D. Schumacher, W. Schilling, and J. Diehl, North-Holland, Amsterdam, 1970, p. 255.

25. W. Schilling, P. Ehrhart, and K. Sonnenberg, In *Fundamental Aspects of Radiation Damage in Metals*, Vol. 1, ed. M.T. Robinson and F.W. Young, Jr., USERDA, CONF–751006–P1, Oak Ridge, TN, 1976, p. 470.

26. A. Seeger, In *Radiation Damage in Solids*, Vol. 1, IAEA, Vienna, 1962, p. 101.

27. A. Seeger, In *Fundamental Aspects of Radiation Damage in Metals*, Vol. 1, ed. M.T. Robinson and F.W. Young, Jr., USERDA, CONF–751006–P1, Oak Ridge, TN, 1976, p. 493.

28. W. Frank and A. Seeger, *Cryst. Latt. Defects* **5** (1974) 141.

29. W. Kienle, W. Frank, and A. Seeger, *Rad. Effects* **71** (1983) 163.

30. W. Frank and A. Seeger, *Mater. Sci. Forum* **15–18** (1987) 57.

31. H. Haken, *Synergetics*, Springer, Berlin–Heidelberg–New York, 1987.

32. C.H. Woo and W. Frank, *J. Nucl. Mater.* **137** (1985) 7.

33. C.H. Woo and W. Frank, *J. Nucl. Mater.* **140** (1986) 214.

34. C.H. Woo and W. Frank, *J. Nucl. Mater.* **148** (1987) 121.

35. C.H. Woo and W. Frank, *Mater. Sci. Forum* **15–18** (1987) 875.

36. W. Frank, *Nova acta Leopoldina NF* **67**, Nr. 281, (1992) 137.

37. A.J.E. Foreman, C.A. English, and W.J. Phythian, *Phil. Mag. A* **66** (1992) 665.

38. A.J.E. Foreman, W.J. Phythian, and C.A. English, *Phil. Mag. A* **66** (1992) 671.

39. H. Jacques and K.-H. Robrock, In *Point Defects and Defect Interactions in Metals*, ed. J. Takamura, M. Doyama, and M. Kiritani, University of Tokyo Press, Tokyo, 1982, p. 159.

40. J.H. Evans, *J. Nucl. Mater.* **132** (1985) 147.

41. W. Frank, *J. Nucl. Mater.* **159** (1988) 122.

42. V.I. Dubinko, A.V. Tur, A.A. Turkin, and V.V. Yanovskij, *J. Nucl. Mater.* **161** (1989) 57.

43. M. Zaiser, W. Frank, and A. Seeger, *Solid State Phenomena* **23 & 24** (1992) 221.

44. M. Zaiser, P. Hähner, C. Tölg, and W. Frank, *Mater Sci. Forum* **123–125** (1993) 687.

45. W. Frank, *Solid State Phenomena* **3 & 4** (1988) 315.

46. P. Hähner and W. Frank, *Mater. Sci. Forum* **62–64** (1990) 697.

47. P. Hähner and W. Frank, *Solid State Phenomena* **23 & 24** (1992) 203.

48. J.H. Evans, *Solid State Phenomena* **3 & 4** (1988) 303.

49. T. Klemm, *Diplomarbeit*, Stuttgart University, Stuttgart, 1993.

50. T. Klemm and W. Frank, *Appl. Phys. A*, to be submitted.

51. C. Tölg, M. Zaiser, P. Hähner, and W. Frank, *Appl. Phys. A* **58** (1994) 3.

52. C. Tölg, P. Hähner, M. Zaiser, and W. Frank, *Appl. Phys. A* **58** (1994) 11.

53. P.B. Johnson, R.W. Thomson, and C.J. Mazey, *Nature* **347** (1990) 265.

54. G. Martin, *Phil. Mag.* **32** (1975) 615.

55. C. Abromeit and H. Wollenberger, *J. Mater. Res.* **3** (1988) 640.

56. A.D. Brailsford, J. *Appl. Phys.* **48** (1977) 4402.

Continuum Models and Discrete Systems
Proceedings of 8th International Symposium, June 11-16, 1995, Varna, Bulgaria, ed. K.Z. Markov
© World Scientific Publishing Company, 1996, pp. 437–443

NON-RIEMANNIAN STRESS SPACE STRUCTURE OF GRAVITATION AND FERROMAGNETISM FROM DISTANT PARALLELISM

L. C. GARCIA de ANDRADE
Depto. de Física Teórica, Instituto de Física, UERJ,
Rua São Francisco Xavier, 524 – Maracanã Rio de Janeiro/RJ,
Brasil - CEP. 20.550

Abstract. Some metrics describing the non-Riemannian structure of screw dislocations are presented. The non-Riemannian stress space is shown to exhibit certain features which make it the proper arena to unify gravitation and ferromagnetism from the common language of distant parallelism and Cartan's torsion tensor. Some examples in three-dimensional gravity are given. Tod's metric of boost dislocations seems to represent a moving dislocation geometry. The metric of a twisted cylinder and its torsion components are given.

"... differential geometry of defects shows close analogies with the General Theory of Relativity."

E. Kröner

1. Introduction

Since the application of the methods of differential geometry to continuum mechanics by Elie Cartan [1], Minagawa [2], Amari [3] and Kröner [4], have applied the methods of non- Riemannian geometry to the investigation of the duality between the stress and strain space. These methods have been applied with success to the theory of defects in solids, specially to crystals by Kondo [5], Bilby and his group [6] and Kröner [7]. More recently a group of relativists and theoretical physics have applied the non-Riemannian structure of dislocations to the study of lower dimensional gravity [8]. The idea behind the application of non-Riemannian geometry to the space of stress and couple-stress is based on the equations given by Amari

$$\partial_i \sigma_{ij} = 0\,, \tag{1.1}$$

$$\partial_i \mu_{ijk} + \sigma_{[jk]} = 0\,, \tag{1.2}$$

$$\mu_{ijk} = \varepsilon^{irs} \varepsilon^{jkt} \widetilde{S}_{rst}\,, \tag{1.3}$$

where

$$\mu_{ijk} = x_j \sigma_{ik} - x_k \sigma_{ij} \,, \tag{1.4}$$

and $\widetilde{S}_{rst}$ is the dual torsion; $\sigma_{[ij]}$ is the skew-symmetric part of the stress tensor and μ_{ijk} is the couple stress. The equations $G_{ij} = \sigma_{ij} = \varepsilon^{ijk}\varepsilon^{rst}R_{rstk}$ are sometimes reffered as the Einstein's equations of continuum mechanics, where R_{ijkm} represents the Riemann curvature tensor. In this note I shall be concerned with the constructions of simple metrics describing two screw dislocations, the first associated with a long cylinder (rod) and the other with a finite cylinder of radius r. We shall observe that the case of a dislocated long cylinder (rod) yields a much simpler metric then that for a finite one, as expected. We hope that the metrics studied here will not only contribute to the development of 3-D gravity but to disclinations in space-time theories such as the Einstein-Cartan theory.

2. Dislocation Metrics and Distant Parallelism

Let us now consider the geometrical structure of the theory of elasticity given by the metric

$$g_{ij} \cong \delta_{ij} + \partial_{[i}u_{j]} \equiv \delta_{ij} + 2u_{ij} \,, \tag{2.1}$$

where u_i, $i = 1, 2, 3$, is the displacement and u_{ij} is the strain tensor. Now let us consider the stress tensor σ_{ij} and the strain vector of a rod given by

$$\sigma_{\theta z} = \frac{\mu b}{2\pi r} \,, \tag{2.2}$$

$$u_z = \frac{\mu b \theta}{\pi} \,. \tag{2.3}$$

Here b is the Burgers vector of dislocation and μ an elastic constant. Substitution of (2.3) into (2.1) yields the line element

$$ds^2 = dr^2 + r^2 d\theta^2 + 2\gamma d\theta dz + dz^2 \cong dr^2 + r^2 d\theta^2 + (dz + \gamma d\theta)^2 \,, \tag{2.4}$$

where γ is a constant parameter associated with a screw dislocation, where $\gamma = \dfrac{b}{2\pi}$. If one puts in (2.4) $dz' = dz + \gamma d\theta$, it is easily noticed that the Riemann tensor is given by $R_{rzr\theta} = -\sigma_{\theta z} = -\dfrac{\mu b}{2\pi r}$ which is singular at $r = 0$; such a behaviour is typical for disclinations. Notice that this metric is similar to Tod's [20] metric in space time representing a conical singularity in Einstein-Cartan theory of gravity. Computation of the couple-stress from expressions (2.2) and (1.4) yields

$$\mu_{\theta rz} = \widetilde{S}_{zr\theta} = \frac{\mu b}{2\pi} = \mu\gamma \,, \tag{2.5}$$

where I have used Eq. (1.3) to obtain the dual torsion.

It is clear that the dual torsion is constant and proportional to the Burgers vector. Another interesting type of metric is the one representing the screw dislocation in the case the cylinder is finite. The finite case yields the following expressions for the stress tensor and strain of the cylinder

$$\sigma_{\theta z} = \frac{\mu b}{2\pi r} - \frac{\mu b r}{\pi R^2}, \tag{2.6}$$

where

$$u_\theta = -\frac{brz}{\pi R^2} u_z = \frac{b\theta}{2\pi}. \tag{2.7}$$

A straightforward algebra yields the line element for the finite cylinder undergoing screw dislocation

$$ds^2 = dr^2 + r^2 d\theta^2 + dz^2 - \frac{2bz}{\pi R^2} dr d\theta + 2\left(\frac{b}{2\pi} - \frac{br}{\pi R^2}\right) dz d\theta. \tag{2.8}$$

Note that in both cases considered here, when $b = 0$ (no dislocation), the metric is flat. From Eq. (2.6) one obtains

$$R_{rzr\theta} = -\sigma_{\theta z} = -\frac{\mu b}{2\pi r} + \frac{\mu b r}{\pi R^2}$$

which is singular at $r = 0$ and at $\gamma = 0$, or at the surface of the cylinder the Riemann curvature is given by $R_{rzr\theta} = \frac{1}{2}\frac{\mu b}{\pi R}$ which again exhibits dislocation's type behaviour. The nice feature of metric [12] is that it admits a solution of a teleparallel of theory of gravity, where

$$\frac{1}{2}\partial_{[i} S_{jk]m} \equiv R_{[ijk]m} \equiv 0 \tag{2.9}$$

and R_{ijkm} is the curvature tensor of the Riemann-Cartan geometry. Eq. (2.9) is equivalent to

$$S_{ijm} = \partial_{[i} g_{j]m}. \tag{2.10}$$

Application of (2.10) to metric implies that torsion vanishes. Nevertheless application of (2.10) to metric (2.4) yields the only nonvanishing components of the torsion tensor

$$S_{r\theta z} = \partial_r g_{\theta z} = -\frac{b}{\pi R^2} = -S_{rz\theta} = \text{const}, \tag{2.11}$$

$$S_{z\theta r} = -\frac{2b}{\pi R^2}. \tag{2.12}$$

Let us now build up the metric of twisted long cylinder (rod) [10]. As pointed out by Bilby and Gardner [11] the term twist is sometimes used instead of the more traditional torsion of rods to avoid confusion with Cartan's geometrical torsion. Nevertheless, as I shall show here, in the case of a teleparallel theory of gravity there is no need of such a distinction since in Weitzenböck space [12,13] the metric of the twisted cylinder leads to components that are directly connected to the twist angle or torsion angle of the rod. This is easily seen since in the twisted cylinder the strain is given by

$$u_x = -\tau zy\,, \quad u_y = \tau zx \quad \text{and} \quad u_z = \tau\psi(x,y)\,, \tag{2.13}$$

where ψ is a torsion (or twist) function and τ is the twist (or torsion) angle of the elastic deformation of the rod. From Eq. (2.13) the metric in the Cartesian coordinates of the twist cylinder reads

$$ds^2_{\text{cylind}} = dx^2 + dy^2 + dz^2 + \tau\left(\frac{\partial\psi}{\partial x} - y\right)dxdz + \tau\left(\frac{\partial\psi}{\partial y} + x\right)dydz\,. \tag{2.14}$$

In turn, from the metric (2.14) it is easy to compute the torsion components using distant parallelism as

$$\begin{cases} S_{xyz} = & \tau\dfrac{\partial^2\psi}{\partial x\partial y}\,, \\ S_{yzx} = & \tau\dfrac{\partial^2\psi}{\partial y\partial x}\,, \\ S_{xyz} = & \partial_x g_{yz} - \partial_y g_{xz} = 2\tau\,, \\ S_{zyy} = & -\tau\dfrac{\partial^2\psi}{\partial y^2}\,, \\ S_{zxx} = & -\tau\dfrac{\partial^2\psi}{\partial x^2} \end{cases} \tag{2.15}$$

for the Cartan's geometrical torsion in Weitzenböck stress space. The Riemann curvature tensor for the twisted cylinder can be computed from the stress-curvature relation where the stress tensor of the cylinder is given by [10] $\sigma_{xz} = \mu\tau\left(\dfrac{\partial\psi}{\partial x} - y\right)$, $\sigma_{yz} = \mu\tau\left(\dfrac{\partial\psi}{\partial y} + x\right)$ and $R_{ijkm} = \varepsilon^{pij}\varepsilon^{qkm}\sigma_{pq}$; it reads

$$R_{zxxy} = \varepsilon^{yzx}\varepsilon^{zxy}\sigma_{yz} \equiv \sigma_{yz} = \mu\tau\left(\frac{\partial\psi}{\partial y} + x\right), \tag{2.16}$$

$$R_{zyyx} = \varepsilon^{xzy}\varepsilon^{zyx}\sigma_{xz} \equiv -\mu\tau\left(\frac{\partial\psi}{\partial x} - y\right). \tag{2.17}$$

The curvature and torsion components can be expressed in terms of the twist angle τ. When $\tau = 0$, or for a twist-free cylinder, the curvature components vanish and the space is flat. Eshelby [14] has pointed out many years ago that a twisted cylinder produces screw dislocations and vice-versa [15]. This argument agrees with the result found here for the torsion tensor of a twisted cylinder. As a final example I shall be concerned with the Katanaev-Volovich [16] line elements describing a dipole of wedge dislocations in 3-dimensional gravity far from the sources

$$ds^2 = dz^2 + \left(1 - \frac{b}{2\pi}\frac{2r\sin\theta - h}{r^2}\right)\left(dr^2 + r^2 d\theta^2\right), \tag{2.18}$$

where b is the Burgers vector and h is the separation between the dipoles, also $b = -mh$ where m is the angle of the dislocation. The only nonvanishing components of the torsion tensor for metric (2.18) if we consider distant parallelism are

$$S_{r\theta\theta} = -\frac{b\sin\theta}{\pi}, \quad S_{\theta rr} = -\frac{b}{\pi}\frac{\cos\theta}{r}. \tag{2.19}$$

Taking the Euclidean metric in cylindrical coordinates to lower and raise indices, the torsion vector is given by

$$S_r = -\frac{b\sin\theta}{\pi r^2}, \quad S_\theta = -\frac{b}{\pi}\frac{\cos\theta}{r}. \tag{2.20}$$

Expression (2.20) is similar to the magnetization potential in the case of ferromagnetism where edge the magnetic moment $\vec{\mu}$ (dipole) plays the role of the dipoles. The connection between magnetic dipoles (spin) and the torsion vector has appeared before in the work of the Sabbata and Gasperini [17] with respect to the Einstein-Cartan theory of gravity. Earlier Amari [18] has build up a geometrical model of ferromagnetism based on Finsler geometry. Nevertheless here I show that this link can be made without the recurring to Finsler geometry. Connection between spin and torsion can be established for static dislocations without the use of Finsler geometry, but simply making use of Cartan's torsion. Dipoles of disclinations have been also used here by Furtado *et al.* [19] in connection with the splitting of Landau levels. Some years ago Amari [2] gave an example of torsion imperfection with the torsion tensor given by $S_{ijk} = \varepsilon_{ijm} d_m b_k$, where d_m is the disclination line. The corresponding torsion vector is $\vec{S} = \vec{d} \times \vec{b}$ which gives support to the idea of the Burgers vector dipole-torsion relation. In four-dimensional spacetime it is possible to apply the above considerations to Tod's [20] boost dislocation metric describing conical singularities. Tod's line element is

$$ds^2 = (dt + \lambda z d\theta)^2 - (dz - \lambda t d\theta)^2 - dr^2 - r^2 d\theta^2 . \tag{2.21}$$

Computation of the torsion on a teleparallel theory of gravity leads to

$$S_{023} = -2\lambda = S_{230}, \quad S_{022} = -2\lambda^2 t, \tag{2.22}$$

for the mixed time and space components and

$$S_{122} = 2r, \quad S_{322} = -2\lambda^2 z, \tag{2.23}$$

Note that at the origin $(t = r = z = 0)$, there still exists a surviving constant torsion for the spatial components. In Amari's geometrical theory of moving dislocations the spatial components represent the screw and edge dislocation. From the case of torsion imperfection discussed in Section 2 it is possible to write down an expression for the boost dislocation and the Burgers vector b. This computation reads

$$S_{322} = \varepsilon_{321} d_1 b_2 = -\varepsilon_{123} db = -zb\,, \tag{2.24}$$

where by construction the screw dislocated cylinder axis is along the z-axis. Comparison between (2.23) and (2.24) yields

$$\lambda = \left(\frac{b}{2}\right)^{\frac{1}{2}}, \tag{2.25}$$

where boost dislocation are related to the Burgers vector though relation (2.23). Expression (2.22) and (2.23) are connected with the Amari's geometrical theory of moving dislocations [21]. Finally I should like to mention that metrics studied here can be used in the investigation of cosmic strings as cosmic dislocations [22]. For recent investigation of torsion singularities in Weitzenböck space the reader is referred to our paper [23]. There are still other interesting applications of Kondo's Non-Riemannian plasticity theory [24] to general relativity.

Acknowledgements. I would like to express my gratitude to Prof. S. Amari and Prof. E. Kröner for sending several of their reprints. Thanks are due to Prof. F.W. Hehl and Prof. P.S. Letelier for helpful discussions on the subject of this paper and CNPq for partial financial support. Special thanks go to Prof. Katanaev for his kind hospitality at the Steklov Mathematical Institute, Moscow, during a short visit.

References

1. E. Cartan, *Comptes Rendus Acad. Sci. Paris* **174** (1922) 593.
2. S. Minagawa, *Lect. Not. in Phys*, vol.249, 1986.
3. S. Amari, *Int. J. Engng. Sci.* **19** (1981) 1581.
4. E. Kröner, *Int. J. Solids* **29** (1992) 1849.
5. K. Kondo, *On the Geometrical and Physical Foundations of the Theory of Yielding*, In *Proc. 2nd. Nat. Congr. of Applied Mechanics*, Tokyo, 1952.
6. B. Bilby, R. Bulough and E. Smith, *Proc. Roy. Soc. A* **231** (1955) 263.
7. E. Kröner, *Continuum Theory of Defects*, Les Houches School on the Physics of Defects, North-Holland, 1980.
8. J. D. Brown, *Lower-dimensional Gravity*, World Scientific, 1988.
9. K. Jagannadham, M. J. Marcinkowski, *The Unified Theory of Fracture*, 1984.
10. L. Landau and I. M. Lifichitz, *Theory of Elasticity*, Course in Theoretical Physics, vol. 8, Pergamon Press, 1986.
11. B. A. Bilby and I. T. Gardner, *Proc. Roy. Soc. A* **247** (1958) 92.

12. W. Weitzenböck, *Sitzungsberichte Preuss Akad. Wiss.* (1928) 466.

13. A. Einstein, *Math Annalen* **102** (1930) 685.

14. J. D. Eshelby, Phil. Trans. A (1951) 87.

15. L. C. Garcia de Andrade and C.A. Souza Lima, Jr., *On Cartan's Differential Geometry of Torque Stress and Solid Cylinders*, IF-UERJ Preprint, 1995.

16. M. O. Katanaev and I. Volovich, *Ann. Phys.* (1992) 216.

17. V. de Sabbata and M. Gasperini, *Introduction to Gravitation*, World Scientific, 1985.

18. S. Amari, *RAAG R.N., 3rd series*, No 125, 1968.

19. C. Furtado, C. Bruno, F. Moraes, E. B. de Mello and V. B. Bezerra, *Landau Levels in the Presence of Disclinations*, preprint, IF-UFPe – Brazil, 1994.

20. K. P. Tod, *Class. and Quantum Grav.* (1994) **11** 1331.

21. S. Amari, In *RAAG Memoirs of the Unifying Study of the Basic Problems in Engineering and Physical Sciences by means of Geometry*, Vol. IV, Division D, 1968, p. 142.

22. D. V. Galtsov and P. S. Letclier, *Phys. Rev. D* **47** (1993) 4273.

23. C. M. Zhang, F. Pei and L. C. Garcia de Andrade, *Torsion Singularities in Weitzenböck Space*, *Il Nuovo Cimento B*, 1995, to appear.

24. K. Kondo, In *RAAG Memoirs*-E-XI, 1968, p. 325.

Continuum Models and Discrete Systems
Proceedings of 8th International Symposium, June 11-16, 1995, Varna, Bulgaria, ed. K.Z. Markov

LONG-WAVE VIBRATIONS LOCALIZED AT PLANAR DEFECTS AND SURFACES IN ELASTIC CRYSTALS

A. M. KOSEVICH and A. V. TUTOV
Theoretical Division,
B. Verkin Institute for Low Temperature Physics,
310164 Kharkov, Ukraine

Abstract. A consistent macroscopic description is presented for studying the dynamic properties of two-dimensional defect layer. We start from rather general assumptions and without using the microscopic models of the surface or interface layers. A total set of dynamical variables is introduced and the long-wave-length equations of motion for them are presented. These equations play the role of boundary conditions on the surface or interface in the problem for the bulk dynamic equation in contacting media. The solution of the appropriate boundary problem gives us the vibration localized at the planar defects. Such a description allows us to find the dispersion relations for the localized waves. Quasilocalized long-wave vibrations near a planar defect in an elastic isotropic medium are investigated.

1. Introduction

A number of scientific and applied problems are connected with studying physical properties of two-dimensional systems. Their dynamic and kinetic properties are subject of numerous studies. However there are interesting physical systems which are not two-dimensional, although their properties are similar to those of the two-dimensional ones, first of all, such as surfaces or interfaces of solids, as well as planar defects in crystals of stacking fault or twin boundary type.

One of the principal problems of the two-dimensional dynamics is the calculation of the spectrum of low-frequency collective excitations. However the long-wave excitations at a free surface or planar defect in crystals interact strongly with those in the bulk. A consistent macroscopic description of the planar defect dynamics consists in derivation of the boundary conditions on the surface or interface and in solution of the appropriate boundary problem for the bulk dynamic equations in contacting media. The simplest version of this programme was proposed in the papers [1,2].

Two different approaches to the derivation of the boundary conditions can be used: (1) starting from a simple discrete model of a monoatomic layer connected by elastic bonds with two semicrystals; (2) starting from the "sandwich" structure consisting of two elastic semispaces separated by an elastic layer of a finite thickness

h and performing the limit transition $h \to 0$, when the thickness of the middle layer becomes the smallest parameter of length in the problem under consideration [3].

The boundary conditions proposed allow one to find all possible vibrations localized near the plane defect and travelling along the defect plane. Those localized vibrations are similar to the Rayleigh wave at a free surface and the phase velocity of the interface wave c is smaller than the transverse sound velocity c_t. Moreover, the boundary conditions give the possibility to study the so called "pseudolocalized" vibrations. The phase velocity of the pseudolocalized wave is smaller than the longitudinal sound velocity c_l but larger than transverse sound velocity ($c_t < c < c_l$). It is shown that there is a continuum spectrum of pseudolocalized wave velocities.

2. Boundary Conditions

The main and the most difficult problem in the theory of localized vibrations is reduced to the derivation of the boundary conditions for the equations of the bulk elasticity at the interface between solids or a two-dimensional lattice defect. The planar defect is characterized by elastic properties and mass density that differ from the bulk ones. The difference between the local mass density of the planar defect material ρ and the bulk one ρ_0 can be connected either with the local stretch (or compression), or with with the impurity atoms adsorbed or segregated at the surface of the defect. In such an approximation the physical properties of the two-dimensional defect are usually described by the quantities uniform across its thickness.

2.1. Local Perturbation Theory

If the planar defect separates two identical elastic semispaces and the defect plane coincides with the coordinate plane $z = 0$ the elastic moduli and mass density of the medium under consideration in the long wave approximation can be presented in the form of the following sums:

$$\mu_0 + (\mu - \mu_0)h\delta(z), \quad \rho_0 + (\rho - \rho_0)h\delta(z), \tag{1}$$

where μ_0 and μ are the elastic moduli of the host material and defect layer material respectively.

This presentation allows one to solve the boundary problem in the infinite elastic space. Such a reduction of the problem under consideration to the vibration problem in the infinite space was proposed in the papers [1,2].

2.2. Dynamic Variables Describing the State of the Defect Layer

To describe the dynamics of the planar defect in elastic media one introduces a vector u^s giving a displacement of the center of mass of a unit area of the defect layer and a surface stress tensor σ^s_{ik} giving stresses in the defect layer. The vector u^s and two-dimensional tensor σ^s_{ik} can be defined as averaged values of the displacements and stresses in the layer respectively:

$$u^s = \frac{1}{h}\int u\, dz, \quad \sigma^s_{ik} = \frac{1}{h}\int \sigma_{ik}\, dz.$$

It is not difficult to derive the equation of motion of the interface layer which separates two contacting media 1 and 2. This equation is [4]:

$$\rho_s \frac{\partial^2 u_i^s}{\partial t^2} = (\sigma_{in}^{(2)} - \sigma_{in}^{(1)}) + \frac{\partial \sigma_{i\alpha}^s}{\partial x_\alpha}\,, \tag{2}$$

where $\rho_s = h\rho$ is the total mass of the defect layer per unit area of the interlace, $\sigma_{ik}^{(1,2)}$ are the limiting magnitudes of the bulk stress tensors in the media 1 and 2, and $\sigma_{in} = \sigma_{ik}n_k$, n_i being the normal unit vector directed from the medium 1 to the medium 2; σ_{ik}^s is the surface stress tensor. The Latin indices embrace the numbers 1,2,3 while the Greek ones embrace the numbers 1,2 and refer to the coordinate axes in the tangential to the surface plane.

Let α_s be the free energy per unit area of the planar defect. Then the total surface free energy can be written as an integral over the defect plane $E_s = \int \alpha_s\, ds$, where ds is the element of area.

The following thermodynamical identity can be derived

$$\delta\alpha_s = \sigma_{\alpha i}^s \delta u_{i\alpha}^s - \sigma_i \delta\xi_i + \langle\sigma_{in}\rangle \delta\Delta_i\,, \tag{3}$$

see Ref. 5 for details, where the following two sets of definition are used:

$$\sigma_i = \sigma_{in}^{(2)} - \sigma_{in}^{(1)}, \quad \langle\sigma_{in}\rangle = \frac{1}{2}(\sigma_{in}^{(1)} + \sigma_{in}^{(2)}), \tag{4}$$

$$u_{i\alpha} = \frac{\partial u_i^s}{\partial x_\alpha}, \quad \xi_i = u_i^s - \frac{1}{2}(u_i^{(1)} + u_i^{(2)}), \quad \Delta_i = u_i^{(2)} - u_i^{(1)}. \tag{5}$$

According to Eq. (3) the change of the free surface energy is determined by the changes of the variables (5) and it is possible to write

$$\sigma_{i\alpha}^s = \frac{\partial\alpha_s}{\partial u_{i\alpha}^s}\,, \quad \langle\sigma_{in}\rangle = \frac{\partial\alpha_s}{\partial\Delta_i}, \quad \sigma_i = -\frac{\partial\alpha_s}{\partial\xi_i}\,. \tag{6}$$

Thus the variables $u_{i\alpha}$, ξ_i and Δ_i are the independent thermodynamical variables in the problem under consideration.

2.3. Density of the Surface Free Energy

It is convenient to introduce such a description of a medium with a planar defect which admits a natural limit reduction of the problem under consideration to the one of the perfect elastic medium under the condition when the thickness h tends to zero.

Let u_{ik}^s be the distortion tensor of the elastic layer

$$u_{\alpha\beta}^s = \frac{\partial u_\alpha^s}{\partial x_\beta}, \quad u_{zz} = u_{nn} = \frac{\Delta_n}{h}\,, \tag{7}$$

$$u_{n\alpha}^s = \frac{\partial u_n^s}{\partial x_\alpha}, \quad u_{\alpha n} = \frac{\Delta_\alpha}{h}\,. \tag{8}$$

Keeping in mind the fact that the surface free energy must be invariant under the rigid rotation one can write it as a function of the variables $\epsilon_{ik} = \frac{1}{2}(u_{ik}^{(1)} + u_{ik}^{(2)})$ and ξ_i, where $i, k = 1, 2, 3$.

It is natural to consider all the variables ϵ_{ik} and ξ_i to be of the same order of magnitude. If any point of the defect layer is the point of the symmetry then the density of the surface free energy, in the quadratic with respect to the variables (8) approximation, has the form

$$\alpha_s = \frac{1}{2}\lambda_{iklm}^s \epsilon_{ik}\epsilon_{lm} + \frac{1}{2}A_{ik}\xi_i\xi_k \,. \tag{9}$$

In the isotropic approximation for the elastic properties in the defect plane the nonzero components of the tensor λ_{iklm}^s can be taken as

$$\begin{aligned} \lambda_{\alpha\beta\gamma\delta}^s &= \lambda_s\delta_{\alpha\beta}\delta_{\gamma\delta} + \mu_s(\delta_{\alpha\gamma}\delta_{\beta\delta} + \delta_{\alpha\delta}\delta_{\beta\gamma}); \quad \alpha, \beta = 1, 2, \\ \lambda_{xxzz} &= \lambda_{yyzz} = \lambda_1, \quad \lambda_{zzzz} = \lambda_2, \quad \lambda_{xzxz} = \lambda_{yzyz} = \mu_1 \,, \end{aligned} \tag{10}$$

and additionally $A_{\alpha\beta} = A_1\delta_{\alpha\beta}$, $A_{\alpha z} = 0$, $A_{zz} = A_2$. Eq. (9) is reduced to the expression

$$\alpha_s = \frac{1}{2}\lambda_s\epsilon_{\alpha\alpha}^2 + \mu_s\epsilon_{\alpha\beta}^2 + \mu_1\epsilon_{\alpha n}^2 + \frac{1}{2}A_1\xi_\alpha^2 + \frac{1}{2}A_2\xi_z^2 + \frac{1}{2}K_2\Delta_n^2 + Mh\epsilon_{\alpha\alpha}\Delta_n \,, \tag{11}$$

where $\lambda_1 = h^2 M$ and $\lambda_2 = h^2 K_2$.

To find a complete set of the boundary conditions it is necessary to combine Eqs. (2) and (6). As a result the following set of conditions appears

$$\begin{aligned} \tfrac{1}{2}(\sigma_{n\alpha}^{(1)} + \sigma_{n\alpha}^{(2)}) &= K_1\left[u_\alpha^{(2)} - u_\alpha^{(1)} + h\frac{\partial u_n^s}{\partial x_\alpha}\right], \\ \tfrac{1}{2}(\sigma_{nn}^{(1)} + \sigma_{nn}^{(2)}) &= K_2\left[u_n^{(2)} - u_n^{(1)} + \nu h\frac{\partial u_\alpha^s}{\partial x_\alpha}\right], \end{aligned} \tag{12}$$

where $\mu_1 = h^2 K_1$ and $\nu = \lambda_1/\lambda_2$, and also

$$\begin{aligned} \rho_s\frac{\partial^2 u_\alpha^s}{\partial t^2} + \sigma_{n\alpha}^{(1)} - \sigma_{n\alpha}^{(2)} - \mu_s\frac{\partial^2 u_\alpha^s}{\partial x_\beta^2} - (\lambda_s + \mu_s)\frac{\partial^2 u_\beta^s}{\partial x_\alpha \partial x_\beta} &= Mh\frac{\partial}{\partial x_\alpha}(u_n^{(2)} - u_n^{(1)}), \\ \rho_s\frac{\partial^2 u_n^s}{\partial t^2} + \sigma_{nn}^{(1)} - \sigma_{nn}^{(2)} - \mu_1\frac{\partial^2 u_n^s}{\partial x_\beta^2} &= K_1 h\frac{\partial}{\partial x_\alpha}(u_\alpha^{(2)} - u_\alpha^{(1)}), \end{aligned} \tag{13}$$

and, according to Eqs. (6) and (11), the tensor σ_i is connected with the vector ξ_i:

$$\begin{aligned} \sigma_{n\alpha}^{(1)} - \sigma_{n\alpha}^{(2)} &= A_1\left[u_\alpha^s - \tfrac{1}{2}(u_\alpha^{(1)} + u_\alpha^{(2)})\right], \\ \sigma_{nz}^{(1)} - \sigma_{nz}^{(2)} &= A_2\left[u_z^s - \tfrac{1}{2}(u_z^{(1)} + u_z^{(2)})\right]. \end{aligned} \tag{14}$$

Eqs. (12-14) give a complete set of boundary conditions permitting to find all the unknown variables $u^{(1)}$, $u^{(2)}$ and u^s.

3. Dispersion Relations for the Waves Localized near the Planar Defect in an Isotropic Medium

Two types of independent surface (interface) elastic waves can propagate in general case near the planar defect travelling along its plane, namely:

(1) pure transverse waves polarized in the plane of the defect (the so-called shear horizontal waves, SH) and

(2) the waves polarized in the saggital plane (the plane determined by the normal to the surface and the wavevector k of the surface wave), i.e., the waves with the so-called Rayleigh polarization.

3.1. The SH Interface Wave

Such a wave has the following nonzero components

$$u_x = 0, \quad u_y = u_0(z)e^{ikx-i\omega t}, \quad u_z = 0, \tag{15}$$

where

$$u_0 = \begin{cases} u_0^{(1)}e^{\kappa z}, & \text{if } z < 0, \\ u_0^s, & \text{if } z = 0, \\ u_0^{(1)}e^{-\kappa z}, & \text{if } z > 0, \end{cases} \tag{16}$$

$\kappa = \sqrt{k^2 - \omega^2/c_t^2}$ is a parameter assumed positive and c_t is the velocity of the bulk transverse waves.

It is easy to prove that only symmetric vibrations ($u_0^{(1)} = u_0^{(2)}$) can be localized near the defect and their dispersion equation is

$$(\omega^2 - c_{t_s}^2 k^2)(1 + \kappa l) = \Omega_1^2(\kappa l), \tag{17}$$

$$c_{t_s}^2 = \mu_s/\rho_s, \quad l = 2\mu_0/A_1 \tag{18}$$

and μ_0 is the bulk shear modulus.

In the long-wave limit $kl \ll 1$ the dispersion law (17) is very close to the dispersion law of the bulk shear waves:

$$\omega^2 = c_t^2 k^2 - \alpha^4 k^4; \alpha^2 = (c_t^2 - c_{t_s}^2)(c_t/l\Omega_1^2). \tag{19}$$

In the opposite limit, $kl \gg 1$, Eq. (17) takes the form:

$$\omega^2 = \Omega_1^2 + c_{t_s}^2 k^2. \tag{20}$$

Both expressions (19) and (20) are valid under the condition $c_{t_s} < c_t$.

3.2. Waves of the Rayleigh Polarization

The displacement in such a wave have the following components: $u(u_x, 0, u_z)$ and the components u_x and u_z include both the transverse u_t and longitudinal u_l parts:

$$u_i = (u_i^t e^{-\kappa_t|z|} + u_i^l e^{-\kappa_l|z|})e^{ikx-i\omega t}, \quad i = 1, 2, 3, \tag{21}$$

where $\kappa_t = \sqrt{k^2 - \omega^2/c_t^2}$ and $\kappa_t = \sqrt{k^2 - \omega^2/c_t^2}$.

If one is interested in the acoustic type of the long-wave vibrations it suffices to assume the magnitudes ξ_i to be small, put $u_i^s = \frac{1}{2}(u_i^{(1)} + u_i^{(2)})$ in Eqs. (12) and (13), and solve the following set of equations for obtaining unknown variables $u_i^{(1)}$ and $u_i^{(2)}$ [6]:

$$\begin{aligned} \tfrac{1}{2}(\sigma_{nx}^{(1)} + \sigma_{nx}^{(2)}) &= K_1 \left[u_x^{(2)} - u_x^{(1)} + \frac{1}{2} h \frac{\partial}{\partial x}(u_n^{(1)} + u_n^{(2)}) \right], \\ \tfrac{1}{2}(\sigma_{nn}^{(1)} + \sigma_{nn}^{(2)}) &= K_2 \left[u_n^{(2)} - u_n^{(1)} + \frac{\nu}{2} h \frac{\partial}{\partial x}(u_x^{(1)} + u_x^{(2)}) \right], \end{aligned} \tag{22}$$

$$\begin{aligned} \sigma_{nx}^{(1)} - \sigma_{nx}^{(2)} &= Mh \frac{\partial}{\partial x}(u_n^{(2)} - u_n^{(1)}) + \tfrac{1}{2} \left[(\lambda_s + 2\mu_s) \tfrac{\partial^2}{\partial x^2} - \rho_s \frac{\partial^2}{\partial t^2} \right] (u_x^{(1)} + u_x^{(2)}), \\ \sigma_{nn}^{(1)} - \sigma_{nn}^{(2)} &= K_1 h \frac{\partial}{\partial x}(u_x^{(2)} - u_x^{(1)}) + \tfrac{1}{2} \left[\mu_s \frac{\partial^2}{\partial x^2} - \rho_s \frac{\partial^2}{\partial t^2} \right] (u_n^{(1)} + u_n^{(2)}). \end{aligned} \tag{23}$$

After calculating the variables $u_i^{(1)}$ and $u_i^{(2)}$, the variables ξ_i are determined by Eqs. (14), namely

$$\xi_x = \frac{1}{A_1}(\sigma_{nx}^{(1)} - \sigma_{nx}^{(2)}), \quad \xi_z = \frac{1}{A_3}(\sigma_{nz}^{(1)} - \sigma_{nz}^{(2)}). \tag{24}$$

The boundary conditions (22) and (23) admit existence of solutions of two types [7], namely,

$$(1) \quad u_x^{(1)} = -u_x^{(2)}, u_z^{(1)} = u_z^{(2)}, \quad \sigma_{nx}^{(1)} = \sigma_{nx}^{(2)}, \quad \sigma_{nn}^{(1)} = -\sigma_{nn}^{(2)},$$

$$(2) \quad u_x^{(1)} = u_x^{(2)}, u_z^{(1)} = -u_z^{(2)}, \quad \sigma_{nx}^{(1)} = -\sigma_{nx}^{(2)}, \quad \sigma_{nn}^{(1)} = \sigma_{nn}^{(2)}.$$

The localized wave (21) is characterized with a dispersion relation, i.e. a dependence of the wave phase velocity $c = \omega/k$ on the wave vector. The dispersion relation in the complete form was obtained in Ref. 6. In the simplest case when $\lambda_1, \lambda_2, \mu_1 \ll h\mu_0$ and $\rho_s \ll h\rho_0$ the dispersion law for the two cases of symmetry have the following form [8,9,10]:

$$\frac{1}{2} kl_1 = \left(\frac{c}{c_t}\right)^2 D^{-1}(c) \sqrt{1 - \left(\frac{c}{c_t}\right)^2}, \quad \frac{1}{2} kl_1 = \left(\frac{c}{c_t}\right)^2 D^{-1}(c) \sqrt{1 - \left(\frac{c}{c_l}\right)^2},$$

$$D(c) = \left(2 - \left(\frac{c}{c_t}\right)^2\right)^2 - 4\sqrt{1 - \left(\frac{c}{c_t}\right)^2} \sqrt{1 - \left(\frac{c}{c_l}\right)^2},$$

where $l_1 = \mu_0/\mu_1$ and $l_2 = \mu_0/\lambda_2$.

4. Pseudolocalized Interface and Surface Waves

All the results discussed above are valid for $c < c_t < c_l$. In the case when the phase velocity is situated in the interval $c_t < c < c_l$, the stationary solutions of the dynamic equations of the theory elasticity in the half-space (+) ($z > 0$) have to be written in the form [7]:

$$\begin{aligned} u_x^+ &= [kA\exp(-\kappa_l z) + qB\cos(qz-\varphi)]\, e^{ikx-i\omega t}, \\ u_z^+ &= [i\kappa_l A\exp(-\kappa_l z) - iqB\sin(qz-\varphi)]\, e^{ikx-i\omega t}, \end{aligned} \tag{25}$$

where

$$\kappa_l^2 = k^2 - \frac{\omega^2}{c_l^2}, \quad q = \frac{\omega^2}{c_t^2} - k^2,$$

A and B are constants, and φ is an arbitrary phase.

The solutions (25) give the so-called quasilocalized vibrations near the defect layer.

In the half-space (–) ($z < 0$) it is possible to choose one of the following solutions:

$$\begin{aligned} &(1) \quad u_x^-(z) = -u_x^+(-z), \quad u_z^-(z) = u_z^+(-z) \quad \text{or} \\ &(2) \quad u_x^-(z) = u_x^+(-z), \quad u_z^-(z) = -u_z^+(-z). \end{aligned}$$

The dispersion relations for the quasilocalized waves of two types have the following form, respectively [7]:

$$\begin{aligned} kl_1 &= \left(\frac{c}{c_t}\right)^2 D^{-1}(c,\varphi)\cos\varphi\sqrt{\left(\frac{c}{c_t}\right)^2 - 1}, \\ kl_2 &= \left(\frac{c}{c_t}\right)^2 D^{-1}(c,\varphi)\sin\varphi\sqrt{1 - \left(\frac{c}{c_l}\right)^2}, \\ D(c,\varphi) &= \left(2 - \left(\frac{c}{c_t}\right)^2\right)^2 \sin\varphi - 4\sqrt{\left(\frac{c}{c_t}\right)^2 - 1}\sqrt{1 - \left(\frac{c}{c_l}\right)^2}\cos\varphi\,. \end{aligned} \tag{26}$$

Expressions (26) are reduced to a simple form in the case of the free surface. The limit transition $\mu_1, \lambda_2 \to 0$ ($l_1, l_2 \to \infty$) leads to the condition $D(c,\varphi) = 0$ and gives:

$$(k^2 - q^2)^2 = 4k^2 q\kappa_l \operatorname{cotan}\varphi\,. \tag{27}$$

The relation (27) defines a continuum spectrum of possible frequencies and phase velocities of the pseudolocalized waves at the free surface. Two velocities correspond to a fixed phase φ. It is not difficult to find the additional density of the quasilocalized stationary vibrations:

$$\delta g(\omega) = \frac{1}{2\pi}\frac{d\varphi}{d\omega} = -\frac{4\sqrt{x}(x-2)(2x^2 - x\,[3\eta - 1] + 2\,[\eta - 1]}{\eta\sqrt{x-1}\sqrt{1-x/\eta}\,[16(x-1)(1-x/\eta) + (2-x)^4]},$$

where $x = \left(\frac{c}{c_t}\right)^2$ and $\eta = \left(\frac{c}{c_l}\right)^2$.

Since the total change of the density of states $\int \delta g(\omega) d\omega = 0$, the quasilocalized vibrations do not change the number of vibrations in the interval $c_t k < \omega < c_l k$.

Acknowledgements. This research has been supported in part by the ISF Grant U2I000 and the Grant INTAS-92-1662. The support of these institutions is gratefully acknowledged.

References

1. I. M. Lifshitz and A. M. Kosevich, *Rep. Progr. in Physics* **29** (1966) 1, 217.
2. A. M. Kosevich and V. I. Khokhlov, *Fiz. Tverd. Tela* **10** (1968) 56. (in Russian.) (Engl. translation *Sov. Phys. Solid State* **10**(1968) 39.) *Fiz. Tverd. Tela* **12** (1970) 2507. (in Russian.) (Engl. translation *Sov. Phys. Solid State* **12** (1971) 2017.)
3. V. R. Velasco and Djafari-Rouhani, *Phys. Rev.* **B26** (1982) 1929.
4. A. F. Andreev and Yu. A. Kosevich, *Zh. Eksp. Teor. Fiz.* **81** (1981) 1435. (in Russian.) (Engl. translation *Sov. Phys. JEPT* **54** (1981) 761.)
5. Yu. A. Kosevich and E. S. Syrkin, *Phys. Lett.* **A122** (1987) 178.
6. A. M. Kosevich and A. V. Tutov, *Phys. Lett.* (1995) (in press).
7. A. M. Kosevich and A. V. Tutov, *Low Temp. Physics* **19** (1993) 905.
8. L. J. Pyrak-Nolte and N.G.W. Cook, *Geophys. Res. Lett.* **14** (1987) 1107.
9. Yu. A. Kosevich, *Phys. Lett.* **A155** (1991) 295.
10. L. J. Pyrak-Nolte, J. Xu and G. M. Haley, *Phys. Rev. Lett.* **68** (1992) 3650.

Continuum Models and Discrete Systems
Proceedings of 8th International Symposium, June 11-16, 1995, Varna, Bulgaria, ed. K.Z. Markov
©World Scientific Publishing Company, 1996, pp. 452–460

IDENTIFICATION OF THE UNSTABLE STATIONARY SOLUTIONS OF NAVIER–STOKES EQUATIONS AT HIGH REYNOLDS NUMBERS

R. MARINOVA
Department of Mathematics, Technical University of Varna,
BG-9010 Varna, Bulgaria

Abstract. The so-called Method of Variational Imbedding (Christov [1]) is applied to studying the viscous flow past a circular cylinder where the stationary streaming becomes unstable around $Re = 40$. For the numerical solutions of the imbedding system a difference scheme of splitting type is devised and appropriate iterative algorithm is organized. A non–uniform mesh is used in both directions. Solutions have been obtained numerically for Reynolds numbers in the range $2 \leq Re \leq 200$ and they are in very good agreement with the experimental data from the literature.

1. Introduction

It is well known that the stationary solutions of Navier–Stokes (NS) equations are stable only for sufficiently low Reynolds numbers ($Re = UL/\nu$, where U is the characteristic velocity, L the characteristic scale and ν the kinematic coefficient of viscosity). There exists a threshold for the Reynolds number beyond which the stationary solutions of Navier–Stokes equations become unstable and after undergoing different kinds of bifurcations the flow ends up eventually in a turbulent regime. The essential problem is that the stationary laminar solution ceases to be an attractor and hence it can not be obtained numerically as an initial value problem (incorrect in the sense of Hadamard).

The situation with the NS equations is much similar to that for many other dynamical systems, e.g., the Lorenz attractor, in the sense that the stationary points become unstable. Here "stationary point" stands for a point in a functional space of the solutions, i.e., a stationary solution. The fundamental question concerning the NS equations as a dynamical system is whether the stationary solution persists for higher Re or eventually disappears, in other words, whether the stationary points of the system disappear after ceasing to be attractors or do persist without attracting the solution to them. In contrast with the low–dimensional dynamical system, to answer the above posed question analytically is possible only in a very small number of situations. One of these rare examples is the Poiseuille flow where an analytical solution (the parabolic velocity profile) is known to exist for all $Re \to \infty$. In most of the cases the answers to the above question can be provided only by means of numerical solution of an incorrect problem.

On the other hand a knowledge of the pattern of the steady flow at high Reynolds numbers (even when it is not stable) is of extreme importance for gathering information about the asymptotic properties of NS equations, e.g., the asymptotics of the wake, the asymptotic sites of the points of detachment of the flow (if any), etc. One of the problems of such a type, that has attracted much attention since a long time, both theoretically and numerically, is the flow around a circular cylinder. The problem with such a flow is that it becomes experimentally (and thus numerically) unstable around $Re = 40$. There exist in the literature a number of numerical solutions to the stationary problem for $Re > 40$ and all of them explicitly or implicitly make use of smoothing techniques in order to filter out the disturbances that lead to the unsteady regime. The most successful among those are the works of Fonberg [4,5] where Raynolds numbers as high as 600 have been reached by means of an efficient smoothing technique.

The purpose of the present work is to follow a radically different way and to treat the problem of a viscous incompressible flow past a circular cylinder as an inverse one. To this end we employ a technique, introduced by C. Christov [1] and called by him Method of Variational Imbedding (MVI, for brevity). The basic idea of the method is briefly recalled in Section 3.

It should be pointed out that the flow past a circular cylinder poses some special problems when devising the difference scheme and algorithms. The worst is a result from the unboundness of the flow and the slow decay $\sim r^{-1}$ of the solution at infinity. The latter leaves an open door for different kinds of errors connected with the so-called "actual infinity". On the other hand, neither Cartesian nor polar coordinate systems are adequate enough for describing the topology of the flow when the separation takes place. These problems are aggravated with the increase of the Reynolds number. The conformal mapping as the one employed by Fornberg [4,5] do improve the topological fitness of the mesh but on the expense of introducing artificial singularities at the boundary of the body.

2. Posing the Problem

Consider the two-dimensional steady problem of a flow past a circular cylinder. In polar coordinates, the dimensionless steady Navier-Stokes and continuity equations have the form

$$\Phi \equiv u_r \frac{\partial u_\varphi}{\partial r} + \frac{u_\varphi}{r}\frac{\partial u_\varphi}{\partial \varphi} + \frac{u_\varphi u_r}{r} = -\frac{1}{r}\frac{\partial p}{\partial \varphi} + \frac{1}{Re}\left[\Delta u_\varphi - \frac{u_\varphi}{r^2} + \frac{2}{r}\frac{\partial u_r}{\partial \varphi}\right] = 0\,,$$

$$\Omega \equiv u_r \frac{\partial u_r}{\partial r} + \frac{u_\varphi}{r}\frac{\partial u_r}{\partial \varphi} - \frac{u_\varphi^2}{r} = -\frac{\partial p}{\partial r} + \frac{1}{Re}\left[\Delta u_r - \frac{u_r}{r^2} - \frac{2}{r}\frac{\partial u_\varphi}{\partial \varphi}\right] = 0\,,$$

$$\mathrm{X} \equiv \frac{\partial u_r}{\partial r} + \frac{u_r}{r} + \frac{1}{r}\frac{\partial u_\varphi}{\partial \varphi} = 0\,, \qquad (2.1)$$

where $u_r = u_r(r,\varphi)$ and $u_\varphi = u_\varphi(r,\varphi)$ are the velocity components parallel respectively to the polar axes r and φ; $p = p(r,\varphi)$ is the pressure. Respectively, the notations Φ, Ω, X for the left sides of the equations are introduced for the sake of further convenience. The Reynolds number $Re = dU_\infty/\nu$ is the governing dimensionless parameter. The cylinder diameter $d = 2a$ is the characteristic length; velocity at infinity U_∞ is the characteristic velocity; and ν is the kinematic coefficient of viscosity. In terms of dimensionless variables, the cylinder surface is represented by $r = 1$ and the velocity at infinity is taken equal to the unity. In Eq. (2.1), as usual, Δ is the Laplacian which, when written with respect to the polar coordinates (r,φ), has the form

$$\Delta = \frac{1}{r}\frac{\partial}{\partial r} r \frac{\partial}{\partial r} + \frac{1}{r^2}\frac{\partial^2}{\partial \varphi^2}.$$

The boundary conditions reflect the absence of slipping at the cylinder surface

$$u_r(1,\varphi) = u_\varphi(1,\varphi) = 0 \tag{2.2a}$$

and the asymptotic matching with the uniform outer flow at infinity

$$u_r(R_\infty,\varphi) = \cos\varphi\,, \qquad u_\varphi(R_\infty,\varphi) = -\sin\varphi\,, \tag{2.2b}$$

where R_∞ is the radial coordinate called "actual infinity".

Due to the obvious flow symmetry with respect to the line $\varphi = 0, \pi$, the computational domain may be reduced to the region $0 \le \varphi \le \pi$, $r \ge 1$ and additional boundary conditions are added on the lines $\varphi = 0$ and $\varphi = \pi$ in order to cope with the obvious symmetry of the problem, namely

$$u_\varphi(r,0) = 0\,, \quad \left.\frac{\partial u_r}{\partial \varphi}\right|_{r=0} = 0\,, \quad u_\varphi(r,\pi) = 0\,, \quad \left.\frac{\partial u_r}{\partial \varphi}\right|_{r=\pi} = 0\,. \tag{2.2c}$$

3. Variational Imbedding

For tackling inverse and incorrect problems Christov [1,3] developed the already mentioned Method of Variational Imbedding (MVI) which is a special implementation of the Least-Square-Method to ODE and PDE. The gist of MVI is to replace the direct solution of the "stiff" (unstable, incorrect, etc.) boundary or initial value problem with the problem of minimization of a quadratic functional, formed by squaring and summing the left sides of the equations that are to be solved. The necessary conditions for minimization of such a functional (called hereafter imbedding functional) yield an apparently more complicated Euler-Lagrange system, among whose solutions the solution of the original incorrect problem belongs in particular. The advantage, however, is that the imbedding system is much more tractable and its solutions are stable with respect to an iteration process. Thus the solution of the original system is "imbedded" into the solutions of some other system through a variational procedure which explains the coinage MVI.

Consider the imbedding functional of the governing equations (2.1) for our problem

$$\mathcal{J} = \int_0^\pi \int_1^\infty \left(\Phi^2 + \Omega^2 + \mathrm{X}^2\right) r \, \mathrm{d}r \, \mathrm{d}\varphi \,. \tag{3.1}$$

After some simplification the equations of Euler-Lagrange for the velocity components and pressure take the form of a conjugated system for Φ, u_φ, Ω, u_r and p:

$$\frac{1}{Re}\left(\Delta\Phi - \frac{\Phi}{r^2} + \frac{2}{r^2}\frac{\partial\Omega}{\partial\varphi}\right) + \left(u_r\frac{\partial\Phi}{\partial r} + \frac{u_\varphi}{r}\frac{\partial\Phi}{\partial\varphi} + \frac{2u_\varphi\Omega}{r} + \frac{1}{r}\frac{\partial\mathrm{X}}{\partial\varphi}\right)$$

$$+\Phi\frac{\partial u_r}{\partial r} - \frac{\Omega}{r}\frac{\partial u_r}{\partial\varphi} = 0\,, \tag{3.2}$$

$$\frac{1}{Re}\left(\Delta u_\varphi - \frac{u_\varphi}{r^2} + \frac{2}{r^2}\frac{\partial u_r}{\partial\varphi}\right) - \left(u_r\frac{\partial u_\varphi}{\partial r} + \frac{u_\varphi}{r}\frac{\partial u_\varphi}{\partial\varphi} + \frac{u_\varphi u_r}{r} + \frac{1}{r}\frac{\partial p}{\partial\varphi}\right) + \Phi = 0\,, \tag{3.3}$$

$$\frac{1}{Re}\left(\Delta\Omega - \frac{\Omega}{r^2} - \frac{2}{r^2}\frac{\partial\Phi}{\partial\varphi}\right) - \left(u_r\frac{\partial\Omega}{\partial r} + \frac{u_\varphi}{r}\frac{\partial\Omega}{\partial\varphi} - \frac{u_r\cdot\Phi}{r} + \frac{\partial\mathrm{X}}{\partial r}\right)$$

$$-\Phi\frac{\partial u_\varphi}{\partial r} - \Omega\frac{\partial u_r}{\partial r} = 0\,, \tag{3.4}$$

$$\frac{1}{Re}\left(\Delta u_r - \frac{u_r}{r^2} - \frac{2}{r^2}\frac{\partial u_\varphi}{\partial\varphi}\right) - \left(u_r\frac{\partial u_r}{\partial r} + \frac{u_\varphi}{r}\frac{\partial u_r}{\partial\varphi} - \frac{u_\varphi^2}{r} + \frac{\partial p}{\partial r}\right) + \Omega = 0\,, \tag{3.5}$$

and, finally, the Euler-Lagrange equation for the pressure p reads

$$\Delta p + \frac{2}{r}\left(\frac{\partial u_\varphi}{\partial\varphi}\frac{\partial u_r}{\partial r} - \frac{\partial u_\varphi}{\partial r}\frac{\partial u_r}{\partial\varphi} + u_\varphi\frac{\partial u_\varphi}{\partial r} + u_r\frac{\partial u_r}{\partial r}\right) = 0\,. \tag{3.6}$$

The above system of five equations is of elliptic type and second order. Therefore five boundary conditions are needed. We already posed two of them when formulating the problem, see Eqs. (2.2a) and (2.2c). The remaining three are the natural conditions for minimization of the functional (3.1) which are nothing else but $\Phi = \Omega = 0$. On the other hand, from the continuity equation we have $\partial u_r/\partial r = 0$ at the boundaries $r = 1$ and $r = r_\infty$; Respectively the symmetry conditions at the lines of symmetry $\varphi = 0, \pi$ are $\partial p/\partial\varphi = 0$, which is equivalent to the condition on the function $u_\varphi(r,\varphi)$ at the same lines, namely $-\partial^2 u_\varphi/\partial\varphi^2 = 0$.

Thus we arrive at the following boundary value problem for the system (3.2)–(3.6):

— at the cylinder boundary $r = 1$ one has

$$\Phi = 0\,, \quad \Omega = 0\,, \quad u_\varphi = 0\,, \quad u_r = 0\,, \quad \frac{\partial u_r}{\partial r} = 0\,;$$

— at infinity

$$\Phi = 0\,, \quad \Omega = 0\,, \quad u_\varphi = -\sin\varphi\,, \quad u_r = \cos\varphi\,, \quad \frac{\partial u_r}{\partial r} = 0\,;$$

— at the lines of symmetry $\varphi = 0$ and $\varphi = \pi$ one has

$$\Phi = 0\,, \quad \Omega = 0\,, \quad u_\varphi = 0\,, \quad \frac{\partial u_r}{\partial \varphi} = 0\,, \quad \frac{\partial p}{\partial \varphi} = 0\,.$$

It is clear that if we find a solution of the imbedding system for which Φ and Ω are equal to zero, then u_φ, u_r and p form the solution of the original problem.

4. Iterative Difference Solution to the Imbedding Problem

The system under consideration is nonlinear and its solution requires iterations. It can be solved by means of an iterational process in which at each stage the equations are linearized. The most consistent way is to use Newton's quasilinearization but it yields coupled systems of difference equations that require an order of magnitude more computational time than each one of the imbedding equations. For this reason we employ the simplest linearization. One can also mention additional reasons for employing exactly the procedure to be used, but the most important is, perhaps, that it preserves the vectorial (conjugated) nature of the system under study.

After introducing fictitious time, the equations for the unknown functions become parabolic difference equations. An implicit scheme for time stepping of the solution could be the following one

$$\frac{u_\varphi^{n+1} - u_\varphi^n}{\Delta\tau} = (\Lambda_{rr} + \Lambda_{\varphi\varphi})\, u_\varphi^{n+1} + \frac{1}{2}\Phi^{n+1/2} + \frac{1}{2}\Phi^n + \Lambda_{\varphi\varphi} u_\varphi^n + \Lambda_\varphi p^n + \mathcal{N}^n[u_\varphi]\,,$$

where the superscript (n) stands for the current time stage and $(n+1)$ for the "new" time stage.

The above implicit scheme in full time steps is approximated by the fractional–step scheme. For the equations for function u_φ the first half–time step is

$$\frac{u_\varphi^{n+1/2} - u_\varphi^n}{\Delta\tau} = \Lambda_{rr} u_\varphi^{n+1/2} + \frac{1}{2}\Phi^{n+1/2} + \frac{1}{2}\Phi^n + \Lambda_{\varphi\varphi} u_\varphi^n + \Lambda_\varphi p^n + \mathcal{N}^n[u_\varphi]\,.$$

The second half–time step is

$$\frac{u_\varphi^{n+1} - u_\varphi^{n+1/2}}{\Delta\tau} = \Lambda_{\varphi\varphi}(u_\varphi^{n+1} - u_\varphi^n) + \Lambda_\varphi(p^{n+1} - p^n) + \frac{1}{2}(\Phi^{n+1} - \Phi^n)\,,$$

where

$$\Lambda_{rr} u_\varphi = \frac{1}{Re}\frac{\partial}{\partial r}\left(\frac{1}{r}\frac{\partial (r u_\varphi)}{\partial r}\right), \qquad \Lambda_{\varphi\varphi} u_\varphi = \frac{1}{Re\, r^2}\frac{\partial^2 u_\varphi}{\partial \varphi^2}\,, \qquad \Lambda_\varphi p = -\frac{1}{r}\frac{\partial p}{\partial \varphi}\,.$$

The notation $\mathcal{N}[u_\varphi]$ stands for the nonlinear reminder of the equation for u_φ, namely

$$\mathcal{N}[u_\varphi] = \frac{2}{Re\, r^2}\frac{\partial u_r}{\partial \varphi} - \left(u_r \frac{\partial u_\varphi}{\partial r} + \frac{u_\varphi}{r}\frac{\partial u_\varphi}{\partial \varphi} + \frac{u_\varphi u_r}{r}\right).$$

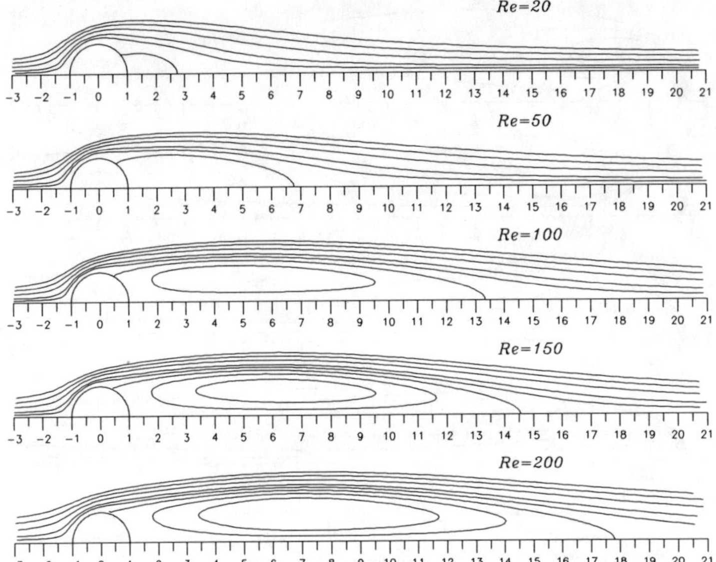

Fig. 1. Streamlines.

In a similar way we employ an implicit scheme for the equation for u_r, p, Φ and Ω. On the first half-time step (the operators with derivatives with respect to r) we solve the equations for Eqs. (3.2) and (3.3) for the "vector" $\{\Phi, u_\varphi\}$. Respectively, Eqs. (3.4), (3.5) and (3.6) are solved simultaneously for the "vector" $\{\Omega, u_r, p\}$. On the second half-time step (derivatives with respect to φ)the respective equations for the vectors $\{p, u_\varphi, \Phi\}$ and $\{u_r, \Omega\}$ are solved. The arguments for selecting the "pairs" and "triplets" of equations are obvious: Φ enters the equation for u_φ, while Ω enters the equation for u_r. The cost efficient algorithm for solving the "parabolized" imbedding boundary value problem can be built with the help of the method of coordinate splitting [6,7]. The resulting systems are either five or seven-diagonal and can be treated by the solver described in [2].

We use a non-uniform mesh in both directions. The mesh is staggered for p in direction φ. For u_r it is staggered in both directions. The number of grid lines in the two directions is N_r and N_φ, respectively. Let us denote

$$h_r = \frac{R-1}{N_r - 1}, \quad h_\varphi = \frac{\pi}{N_\varphi - 1}.$$

The coordinates of a point of the mesh are defined as follows

$$r_j = \exp\left[(j-1)h_r\right] \quad \text{and} \quad \varphi_i = \frac{1}{\pi}\left[(i-1)h_\varphi\right]^2,$$

where $1 \le j \le N_r$, $1 \le i \le N_\varphi$.

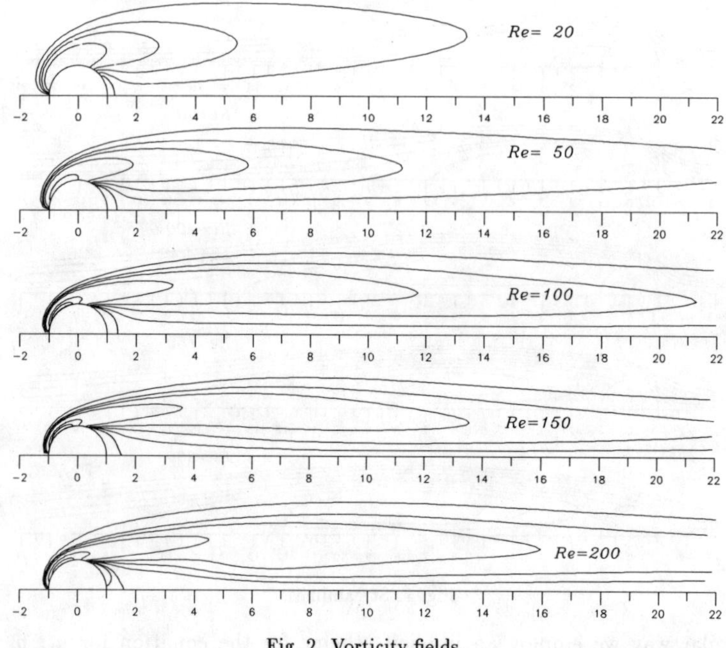

Fig. 2. Vorticity fields.

Thus the boundary conditions are evaluated with a second-order approximation.

5. Results

The problem under consideration is stationary and the stream lines coincide with the trajectories of the fluid particles. Hence the most convenient visualization of the flow are the isolines of the stream function. For this reason we calculate the stream function after the velocity components (u_r and u_φ) are obtained. The stream function Ψ is defined as

$$\frac{\partial \Psi}{\partial r} = u_\varphi \,, \qquad \frac{\partial \Psi}{\partial \varphi} = -r u_r \,,$$

and satisfies the Poisson equation

$$\Delta \Psi + \omega = 0 \,,$$

where ω is the vorticity function in polar coordinates.

Due to the symmetry of the problem, we can use in our case the Dirichlet conditions

$$\Psi = 0 \quad \text{for} \quad r = 1 \,, \quad \varphi = 0, \pi \,, \quad \text{and} \quad \Psi \to -r \sin \varphi \quad \text{for} \quad r \to \infty \,.$$

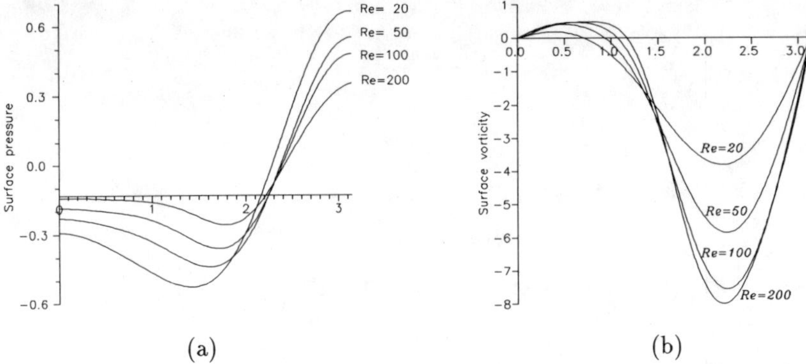

(a) (b)

Fig. 3. (a) Pressure on the body surface; (b) Vorticity on the body surface.

We employ once again the splitting method to solve the equation for the stream function.

The streamlines of the flow for Reynolds numbers $Re = 20$, 50, 100, 150 and 200 are shown in Fig. 1. Fig. 2 shows the vorticity isolines of the flow for the same values of the Reynolds number. For the stream function, the contour values, starting from the top, are $\{0.4,\ 0.3,\ 0.2,\ 0.1,\ 0.05\}$; enclosed streamlines, starting from the centre, are $\{-0.1,\ -0.05,\ 0\}$, and for the vorticity the contour values are

$$\{0.1,\ 0,\ -0.2,\ -0.4,\ -0.6,\ -1,\ -3,\ \ldots\}.$$

Fig. 3a gives the pressure distribution along the surface of the cylinder as a function of the polar coordinate angle, measured from the rare stagnation point. Fig. 3b shows the vorticity distribution on the surface.

Finally a comparative graph of the drag coefficients vs. the Reynolds number, obtained by means of our technique and by Fornberg method, is presented in Fig. 4.

The performance of the scheme is verified as well on different non-uniform meshes.

6. Conclusions

The structure of viscous steady flow past a circular cylinder at high Reynolds number forms one of the classical problems in fluid mechanics. Apart from a calculation by Fornberg, steady flow fields have been obtained numerically only up to $Re = 120$. This is natural in the light of what has been said about the instability of the stationary solution. Finding the steady solution to the problem around a circular cylinder is considered in this work as an inverse problem and we treat it by using the Method of Variational Imbedding [1]. Steady solutions have been calculated up

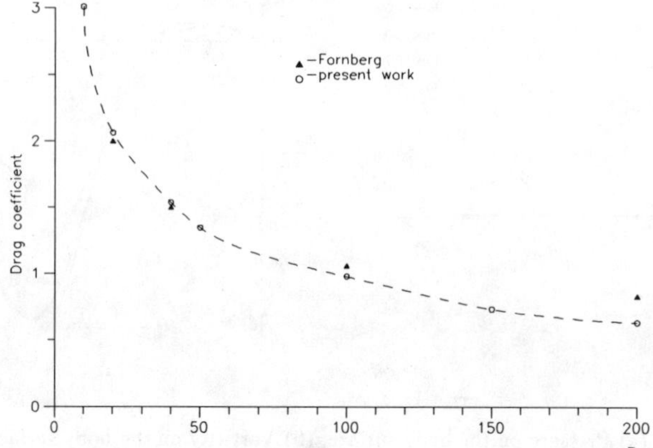

Fig. 4. Drag coefficient as function of the Reynolds number.

to $Re = 200$. The available range of Reynolds numbers of our work is wider than all of the others with the exception of Fornberg's work. But we would like to underline one more time that in our technique—the Method of Variational Imbedding, no additional procedure is applied to the solution, like filtering or smoothing.

Acknowledgments. The author cordially thanks C. I. Christov for posing the problem and many stimulating discussions.

References

1. C. I. Christov, *Comp. Rend. Acad. Bulg. Sci.* **40**(6) (1987) 5.
2. C. I. Christov, *Gaussian Elimination with Pivoting for Multidiagonal Systems*, University of Reading, Internal Report 4, 1994.
3. C. I. Christov and R. Marinova, *Bulg. J. Meteorology & Hydrology* **5** (1994) No. 3-4, to appear.
4. B. Fornberg, *J. Comput. Phys.* **61** (1985) 297.
5. B. Fornberg, *J. Fluid Mech.* **225** (1991) 655.
6. D. W. Peaceman and H. Rachford, Jr., *J. Soc. Indust. Appl. Math.* **3** (1955) 28.
7. N. N. Yanenko, *Method of Fractional Steps*, Gordon and Breach, 1971.

Continuum Models and Discrete Systems
Proceedings of 8th International Symposium, June 11-16, 1995, Varna, Bulgaria, ed. K.Z. Markov
© World Scientific Publishing Company, 1996, pp. 461–470

NONLINEAR DYNAMICS OF A TWO-DIMENSIONAL LATTICE MODEL FOR FERROELASTIC MATERIALS

J. POUGET
Laboratoire de Modélisation en Mécanique,
Université Pierre et Marie Curie (CNRS URA 229),
4 place Jussieu, F-75252 Paris Cédex 05, France

Abstract. On the basis of a two-dimensional lattice model, nonlinear dynamics and stability of elastic microstructures for ferroelastic materials are presented. The microscopic model involves nonlinear and competing interactions emerging from interatomic forces by pairs and non-central interactions. The emphasis is especially placed on the discrete and semi-discrete approaches to the model, which plays a particular role in the domain formation and phase growth. The nonlinear analysis of modulated strain structures is examined in the vicinity of the critical point of the acoustic dispersion curve. Beyond the instability process, the structure forms localized elastic domains and bands. The problem of stable states of the lattice deformation is then investigated in the framework of the microscopic model. It is shown that the problem can be written as a pseudo-spin model. All the physical conjectures are ascertained and illustrated by means of numerical simulations.

1. Introduction

The present work aims at studying the *nonlinear dynamics of microstructure patterns* involved in *phase transformations* in alloys on the basis of *lattice models*. More precisely, we are interested in the underlying microphysics which induces special kinds of strain transformations associated with nucleation of elastic domains and interface movement. The transition is to be supposed of the first order where the spontaneous strain is the order parameter [1,2], in this case we deal with proper ferroelastic materials. The main motivation of the work is to understand how spatial structure formation and related dynamics arising at the microscale are able to organize the system at the macroscale. We point out the interest of a *lattice model* because it possesses the most physical ingredients that are the background of the domain formation and movement of twin boundaries.

The starting point of the study is the construction of a two-dimensional lattice model which describes the cubic-tetragonal transformation. The model includes *nonlinear* and *competing interactions* related to the long-range particle forces. In a first step, we confine our attention to the stability of modulated-strain structures or lattice distorsion waves. We show that the instability process is due to the softening of the phonon branch at a nonzero wave number [3]. Then we take the avantage of a perturbative method to examine the situation near the instability point. We prove that from rather complicated dynamical equations for the microscopic system we can derive an amplitude equation in the semi-discrete approximation. This equation allows us to characterize the instability mechanism of a modulated-strain structure. The physical conjectures are then checked by numerical simulations exhibiting rather complex localized patterns emerging beyond the instability threshold.

In a second step, we want to tackle the nonlinear behavior of the discrete model. In this case, we suppose that the nonlinear part of the interatomic interactions can be modelled by an on-site potential with a double quadratic form with respect to the discrete strain. In addition, we keep the discrete nature of the original model. The system can be recasted into an Ising model with a long-range interactions and the deformation of the lattice is then obtained in terms of pseudo-spin. Particular situations are illustrated numerically and the obtained data exhibit fine microtextures which are usually observed on high-resolution electron micrographs [4,5].

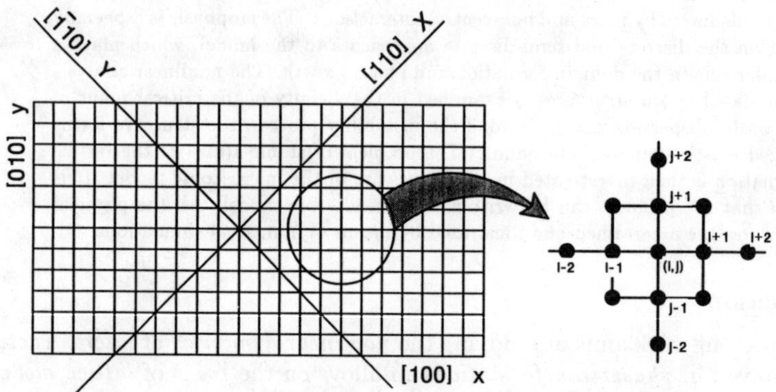

Fig. 1: Two-dimensional lattice model and interactions by pairs and noncentral forces between the first- and second-nearest neighbors.

2. Model and Equations

Let us consider an atomic plane extracted from a cubic lattice (for instance the f.c.c. symmetry of In-Tl or Fe-Pd). The geometry of the lattice, in its undeformed state, is made of squares parallel to the i and j directions (see Fig. 1). A particle of the plane is located by (i, j). After deformation of the lattice, the particles undergo displacements in the plane defined by $u(i, j)$ and $v(i, j)$, which are the displacement in the i and j directions, respectively. A further step in the simplification consists in considering transformations of the lattice involving the displacement $u(i, j)$ only. We assume that the particles interact through two types of interatomic potential: (i) interactions between first-nearest neighbors considered as nonlinear functions of *particle pairs* in the i and j directions and in the diagonal directions as well and (ii) interactions involving *noncentral forces* or three-body interactions between *first* and *second-nearest neighbors* in the i and j directions. The latter interactions provide competing interactions which are equivalent to bond bending forces due to the long-range atomic interactions [6].

Then, the equations of motion for the discrete displacement $u(i,j)$ can be deduced from the lattice and kinetic energies. These equations have the form [7]

$$\ddot{u}(i,j) = \Delta_L^+ \Sigma_L(i,j) + \Delta_T^+ \Sigma_T(i,j) - \Gamma \dot{u}(i,j) + f\,,$$

where we have defined the following discrete stresses

$$\Sigma_L(i,j) = T(i,j) - \Delta_L^- \chi(i,j) \tag{2a}$$

$$\Sigma_T(i,j) = \beta G(i,j) - \Delta_T^- \chi(i,j)\,, \tag{2b}$$

$$T(i,j) = \alpha S(i,j) - S(i,j)^2 + S(i,j)^3\,, \tag{2c}$$

$$\begin{aligned}\chi_L(i,j) = \Delta_L^+ (\delta S(i,j) + \eta(S(i+2,j) + 4S(i+1,j) + 6S(i,j) \\ + 4S(i-1,j) + S(i-2,j)))\,,\end{aligned} \tag{2d}$$

$$\begin{aligned}\chi_T(i,j) = \Delta_T^+ (\delta G(i,j) + \eta(G(i,j+2) + 4G(i,j+1) + 6G(i,j) \\ + 4G(i,j-1) + G(i,j-2))\,,\end{aligned} \tag{2e}$$

The operators Δ_L^+, Δ_T^+ and Δ_L^-, Δ_T^- are the forward and backward first-order finite differences in the i and j directions, respectively (for instance, $\Delta_L^+ f(i,j) = f(i+1,j) - f(i,j)$). Eqs. (2a) and (2b) define the macroscopic stresses while Eqs. (2d) and (2c) provide the microscopic stresses due to the noncentral interactions between the first and second-nearest neighbors characterized by the parameters δ and η. Moreover, a dissipative effect and applied force have been added to the equations of motion where Γ is the damping constant and f is the external force. The discrete deformations are defined as follows

$$S(i,j) = u(i,j) - u(i-1,j)\,, \quad G(i,j) = u(i,j) - u(i,j-1)\,. \tag{3}$$

All the dimensionless model parameters α, β, δ and η appearing in the equations depend on the interatomic potentials.

3. Instability of a Modulated Strain

By linearizing the equations of motion (1) and (2) about a uniform state of deformation S_0 corresponding to a stable minimum of the lattice potential, we search harmonic modes to the discrete linear equations. Then, we arrive at the following dispersion relation [7]

$$\omega^2 + i\Gamma\omega = 4\left[\alpha\left(S_0\right)\sin^2\left(p/2\right) + 4\delta\sin^4\left(p/2\right) + \eta\sin^4\left(p\right)\right], \tag{4}$$

where $\alpha\left(S_0\right)$ is the elastic modulus induced by the uniform strain S_0, p is the wavenumber and ω is the circular frequency. Fig. 2 shows the imaginary part of ω as a function of p. The stability condition is $\mathrm{Im}\left(\omega\right) < 0$ for all p's (curve (b)). The critical situation occurs whenever $\mathrm{Im}\left(\omega\right) = 0$ for $\alpha = \alpha_0$ at $p = p_0$ while $\mathrm{Im}\left(\omega\right)$ remains negative for all other p's (curve (c)). Then α_0 denotes the critical value of the control parameter for which a periodic state of strain with wavenumber p_0 takes place on the lattice.

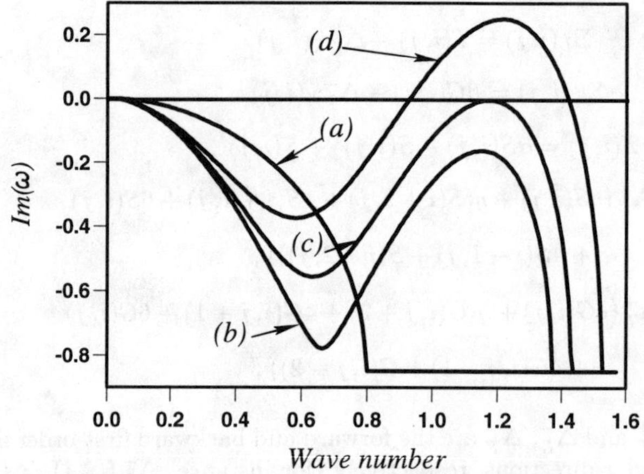

Fig. 2. Dispersion curves for $\mathrm{Im}(\omega)$: (a) linear damped model, (b) stable case $(\alpha > \alpha_0)$, (c) critical situation $(\alpha = \alpha_0)$ and (d) zone of instability around the critical wavenumber $P_0\,(\alpha < \alpha_0)$.

In the vicinity of the critical point, the linear approximation breaks down after a time of the order $1/\omega$ and the nonlinear terms must not be ignored. In order to examine the influence of the nonlinearities on the structure stability, we use a *semidiscrete approach* along with a *multiple-scale technique* [8]. We introduce thus a small parameter which accounts for the deviation of the system around the critical point

$$\alpha = \alpha_0 + \lambda\varepsilon^2 . \tag{5}$$

Then, we look for solution to the complete nonlinear equations of motion (1) as an asymptotic series of $S(n,m,t)$ in ϵ and in harmonics of the discrete phase $\theta = np_0$ and we set

$$S(n,m,t) = \varepsilon A(n,m,t)e^{i\theta} + \varepsilon^2\left(B_1(n,m,t)e^{i\theta} + B_2(n,m,t)e^{2i\theta}\right) + c.c. \tag{6}$$

The method consists of separating the fast changes of the periodic structure involving the discrete phase np_0 while the amplitudes (A, B_1 and B_2) are treated in the continuum limit since they are supposed to be slowly varying in time and space. Skipping the algebraic manipulations, we arrive at [7]

$$\Gamma A_\tau - \left(\omega^2\right)_{pp} A_{XX} - \beta A_{YY} + \bar{\lambda} A - \mu |A|^2 A = 0\,, \tag{7}$$

where we have set

$$\bar{\lambda} = 4\lambda \sin^2\left(p_0/2\right)\,, \tag{8a}$$

$$\mu = 4\sin^2\left(p_0/2\right)\left(-3 + 8\sin^2\left(p_0\right)/\omega\left(2p_0\right)\right)\,, \tag{8b}$$

here $\left(\omega^2\right)_{pp}$ represents the second derivative of ω^2 with respect to p taken at $p = p_0$. In addition, a slow time variable $\tau = \epsilon t$ and stretching spatial variables $X = \epsilon x$ and $Y = \epsilon y$ have been introduced. The amplitudes B_1 and B_2 can be connected with A. Eq. (7) is of the *Ginzburg-Landau* type and describes how the strain amplitude deviates locally from the basic steady state.

A numerical simulation using the microscopic equations (1) as the numerical scheme illustrates the time evolution of a modulated-strain structure at the birth of the instability and beyond. The initial structure made of a spatially sinusoidal strain in the x direction and homogeneous in the y direction is shown in Fig. 3a. The wavelength of the periodic structure corresponds to that of the critical wavenumber p_0, the control parameter α is slightly shifted just below the critical value α_0. After a short lapse of time, small perturbations are taking place along the transverse direction. The instability is growing further and the structure is then transformed into a strain band accompanied by disk-shaped domains (see Fig. 3b). After a rather long time, the structures merge into a large homogeneous strain band as given by Fig. 3c. Such an instability mechanism can be interpreted as a pretransformation effect.

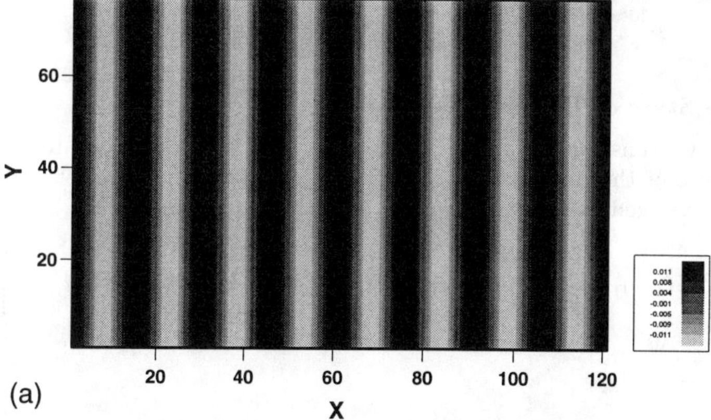

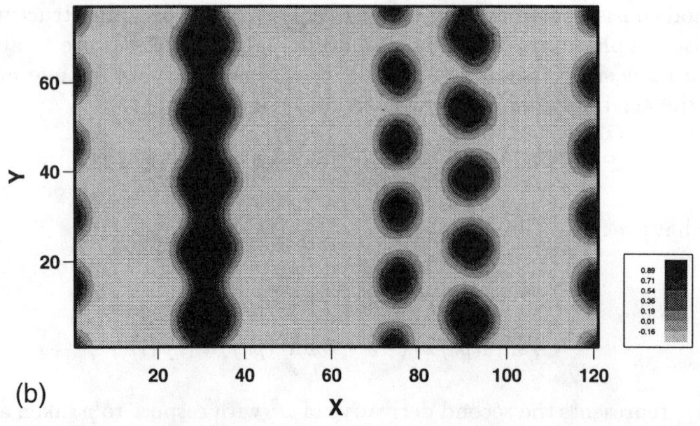

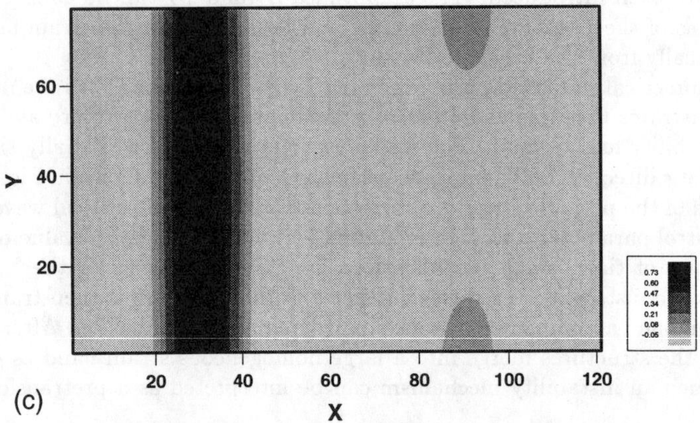

Fig. 3: Instability of a modulated-strain structure: (a) initial state, (b) birth of instability, formation of a strain band and localized structures and (c) the structure is transformed into a strain band of large amplitude.

4. Discrete State of Twinning Lattice

Here, we consider the energy of the deformed lattice. We try to keep the discrete nature of the microscopic model, consequently the long-wavelength limit will not be envisaged. Then, we adopt the following lattice energy [9]

$$\mathcal{V} = \sum_{(i,j)} \left[V(S(i,j)) + \frac{1}{2}\delta(S(i+1,j) - S(i,j))^2 + \frac{1}{2}\delta(S(i,j+1) - S(i,j))^2 \right] . \quad (9)$$

The last two terms represent the noncentral interactions in the i and j directions. The potential V depending on the strain $S(i,j)$ can be modelled by a piecewise parabolic potential as follows

$$V(S(i,j)) = \frac{1}{2}\alpha(S(i,j) - \sigma(i,j))^2 \tag{10a}$$

and

$$\sigma(i,j) = \operatorname{sign}(S(i,j)), \tag{10b}$$

where α measures the potential barrier. The potential (10a) can be deduced from a more general lattice energy by renormalisation of the deformation. The model thus considered accounts for the *frustration* caused by the competing interactions between the double-well potential and the linear noncentral interactions characterized by the parameter δ [10,11]. We notice that, for sake of simplicity, the noncentral interaction is isotropic. The static structure of the lattice model is obtained by energy minimization, which yields the following difference-differential equation

$$\begin{aligned} &V'(S(i,j)) - \delta(S(i+1,j) - 2S(i,j) + S(i-1,j)) \\ &-\delta(S(i,j+1) - 2S(i,j) + S(i,j-1)) = 0 . \end{aligned} \tag{11}$$

We introduce the discrete Fourier transforms of $S(i,j)$ and $\sigma(i,j)$

$$S(p,q) = \frac{1}{\sqrt{NM}} \sum_{(n,m)} S(n,m)e^{i(pn+qm)}, \tag{12a}$$

$$\sigma(p,q) = \frac{1}{\sqrt{NM}} \sum_{(n,m)} \sigma(n,m)e^{i(pn+qm)}, \tag{12b}$$

where N and M denote the number of particles in the i and j directions, respectively. Then Eq. (11) is transformed into

$$G(p,q)S(p,q) - \alpha\sigma(p,q) = 0, \tag{13a}$$

where

$$G(p,q) = \alpha + 2\delta(2 - \cos p - \cos q). \tag{13b}$$

If the sequence $\sigma(i,j) = \operatorname{sign}(S(i,j))$ is known, then the deformation $S(i,j)$ is obtained by (13) and the inverse of the transform (12a). We observe that the factor $G(p,q)$ is just the spectrum of the discrete linear modes, consequently the stability of the lattice is ensured if $G(p,q)$ is positive. By using the Fourier transforms (12), the energy can be rewritten in the Fourier space

$$\mathcal{V} = \sum_{(p,q)} \left\{ \frac{1}{2}\alpha\,(\sigma\sigma^* - S\sigma^* - S^*\sigma) + \frac{1}{2}GSS^* \right\}, \tag{14}$$

where the subscript * denotes the complex conjugate. All the quantities in Eq. (14) are function of p and q running on the first Brillouin segment. Thanks to Eq. (13a), the energy extremum is

$$\mathcal{V}_{extr} = \frac{1}{2}\alpha \sum_{(p,q)} (1 - \alpha/G)\sigma\sigma^* \,. \tag{15}$$

Now, from Eq. (13a), we can go back to the real deformation $S(i,j)$ by taking the inverse Fourier transform. Then we arrive at

$$S(i,j) = \sum_k \sum_l \sigma(i+k, j+l) J(k,l) \tag{16}$$

and note that the lattice deformation has the form of a double convolution product. The function $J(k,l)$ is the interaction coefficient which accounts for the long-range interactions between the pseudo-spins $\sigma(i,j)$. We suppose a very large number of particles, so that we can give an analytic form for the interaction function by using the inverse Fourier transform of $1/G(p,q)$ and we can write

$$J(k,l) = \frac{\pi}{\delta} \int_{-\infty}^{+\infty} \frac{e^{-ky}}{\sinh(y)} \, e^{-ql} \, dq \,, \tag{16a}$$

with

$$\cosh(y) = \frac{\alpha}{2\delta} + 2 - \cos(q) \,. \tag{16b}$$

In order to illustrate this part of the model, we consider a numerical simulation using the Fourier transforms of the discrete quantities for a given sequence of $\sigma(i,j)$. The latter is chosen as follows

$$\begin{cases} j = j_0, & \sigma(i,j) = \text{sign}\,(\ell - |i - i_0|) \;, \\ j \neq j_0, & \sigma(i,j) = \text{sign}\,(\cos(q_0 j)) \,, \end{cases} \tag{17}$$

where (i_0, j_0) is an arbitrary point of the lattice, ℓ is a certain characteristic lattice length and q_0 is the wavenumber of the modulation in the j direction. We first compute the Fourier transform of the sequence (17) by using Eq. (12b). Then, we deduce the Fourier transform $S(p,q)$ from Eqs. (13a) and (13b) and finally we can evaluate the real strain $S(i,j)$ by using the inverse Fourier transform. In addition, the self-consistency equation $\sigma(i,j) = \text{sign}(S(i,j))$ is checked. The computational result is given in Fig. 4. We can observe the nucleation of elastic domains around the defect point (i_0, j_0) along with a strain modulation in the j direction. This modulation is obviously given by the sequence (17) for $j \neq j_0$, but it has been

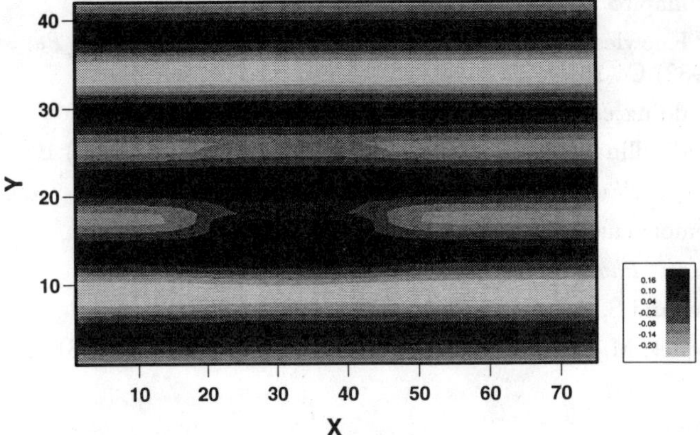

Fig. 4: Deformation state of the lattice according to the sequence $\sigma(i,j)$ given by Eq. (17).

affected by the presence of the localized structure. More details about the formation of such a structure and their physical meanings will be presented in a further work.

5. Concluding Remarks

Our main objective of the present work was merely to briefly point out some physical mechanisms of pattern formation related to phase transformations in alloys. On the basis of a two-dimensional lattice model two interesting problems associated with elastic phase twinning have been examined. The emphasis is placed especially on the importance of the microscopic description, because the instability of the modulated strain due to the softening of the acoustic branch at a nonzero wavenumber can be only explained on the basis of the discrete system. The study of the lattice deformation of the model leads to the computation of very fine microtextures which are usually observed by means of high-resolution electron microscopy in various alloys [4,5].

In further works and extensions of the model, important properties of the discrete system should be examined. In particular, we can envisage other situations using given sequences of pseudo-spins (weakly periodic or incommensurate, random sequences, for instance). It would be interesting to examine transformations involving two lattice displacements, which would lead to a more realistic physical model.

References

1. V. K. Wadhawan, *Phase Transitions* **3** (1982) 3.
2. A. L. Roitburd, In *Solid State Physics* **33**, eds. H. Ehrenreich, F. Seitz and D. Turnbull, Academic Press, New York, 1978.

3. S. M. Shapiro *et al.*, *Phys. Rev. Lett.* **57** (1986) 3199.
4. K. M. Knowles, J. M. Christian and D. A. Smith, *J. Phys. (Paris) Colloq.* **43** (1982) C4-185.
5. D. Brodding *et al.*, *Phil. Mag.* **A59** (1989) 47.
6. R. D. Mindlin and N. N. Eshel, *Int. J. Solid Struct.* **4** (1968) 109.
7. J. Pouget, *Phys. Rev.* **B46** (1992) 10554.
8. M. Remoissenet, *Phys. Rev.* **B33** (1986) 2386.
9. J. Pouget, *Phase Transitions* **34** (1991) 105.
10. F. Axel and S. Aubry, *J. Phys. C* **21** (1981) 5433.
11. E. Coquet, M. Peyrard and H. Buttner, *J. Phys. C* **21** (1988) 4895.

Continuum Models and Discrete Systems
Proceedings of 8th International Symposium, June 11-16, 1995, Varna, Bulgaria, ed. K.Z. Markov

SOLITARY–WAVE SOLUTIONS OF A GENERALIZED WAVE EQUATION WITH HIGHER–ORDER DISPERSION

Y. STEYT, C. I. CHRISTOV
Centre for Nonlinear Phenomena and Complex Systems, Université Libre de Bruxelles, Campus Plaine - CP 231, Blvd du Triomphe, 1050 Bruxelles, Belgique
and
M. G. VELARDE
Instituto Pluridisciplinar, Universidad Complutense, Paseo Juan XXIII, No 1, Madrid 28040, Spain

Abstract. A spectral technique is developed for solving the Sixth-Order Generalized Boussinesq Equation (6GBE) which arises in the well-posed models of shallow-layer inviscid flows or nonlinear chains. Stationary localized waves are considered in the frame moving to the right. Depending on the interplay between the celerity and dispersion parameters, the localized waves can be either monotone or with damped oscillatory tails. We show that the oscillatory localized waves can form bound states. The spectral technique developed here is applied also to the dynamical problem of head-on collisions of solitary waves ("quasi-particles").

1. Introduction

After Zabusky and Kruskal [10] discovered the particle-like behavior of localized solutions of KdV equation an increased attention has been paid in the literature to equations having soliton solutions. A number of these equations appear naturally in more and more diversified applied fields, such as flows in shallow layers, lattice dynamics, condensed matter, etc., and the soliton paradigm has become an important topic of the nonlinear physics furnishing the most impressive example of intrinsic connection between discrete and continuous concepts in physics.

In the present paper, we study the problem of the numerical solution of a Boussinesq equation, "improved" by adding sixth-order dispersion which is one of the options in making the model linearly stable. As argued in [5] the sixth order dispersion appears naturally both for the lattice dynamics and shallow water flows. The Sixth-Order Generalized Boussinesq Equation (6GBE) is thus mathematically sound allowing numerical investigation of permanent waves that result from the ballance between dispersion and nonlinearity. However, introducing additional dispersion term brings into existence more complicated shapes for the localized waves, e.g., with oscillatory non-monotone tails. Like some other more elaborated models, the 6GBE has not been yet shown to be fully integrable and only admits limited analytical treatment. That is why the development of adequate numerical techniques is of crucial importance. The most formidable difficulty on this way is the artificial viscosity, especially when the long time evolution of conservative systems is being investigated. In the framework of finite differences the problem of artificial viscosity was alleviated in [4] (see also [3] for generalization of the conservative scheme).

An alternative way to avoid the problem of artificial viscosity is to employ a Fourier-Galerkin spectral technique (see the discussions in [2]). To this aim is devoted the present paper. We show here that the spectral technique is an efficient tool for investigating both the stationary and transient localized solutions of 6GBE.

By means of the algorithm developed here we study the physical properties of "6GBE quasi-particles" such as collisions and formation of bound states.

2. Localized Waves in 6GBE

The simplest version of 6GBE that is paradigmatically consistent reads

$$u_{tt} = \gamma^2 u_{xx} + \alpha(u^2)_{xx} + \beta u_{x^4} + \delta u_{x^6}\,, \tag{2.1}$$

where the sixth-order dispersion coefficient is always positive $\delta > 0$. Since this coefficient can be rescaled, we will take $\delta = 1$ in the following without loss of generality.

Let us consider the stationary waves in the moving frame $\xi = x - ct$. After a double integration with respect to ξ and taking into account the localized character of the investigated solutions, one obtains the following nonlinear ODE

$$\lambda u + \alpha u^2 + \beta u_{\xi\xi} + u_{\xi^4} = 0\,, \quad \lambda = \gamma^2 - c^2\,. \tag{2.2}$$

We are looking for solutions with $u \to 0$ for $\xi \to \pm\infty$. Then $u^2 \ll u$ in the tails and the linearized version of (2.2) coincides with its linear part. The latter possesses harmonic solutions of the type $e^{k\xi}$. The corresponding dispersion relation reads

$$k^4 + \beta k^2 + \lambda = 0, \quad \text{i.e.,} \quad k^2 = \frac{1}{2}\left[-\beta \pm \sqrt{\beta^2 - 4\lambda}\right]. \tag{2.3}$$

Eq. (2.3) shows that the natural classification of stationary solutions should be based on two criteria which define together the spatial asymptotic behavior of the tails. *First*: a stationary solution can be subsonic ($\lambda > 0$) or supersonic ($\lambda < 0$); *Second*: the asymptotic tails of the localized wave can be either monotonic, purely oscillatory or damped oscillatory, depending on whether k^2 is real positive, real negative or complex.

3. Fourier-Galerkin Spectral Technique for Stationary Waves

Equation (2.2) is nonlinear and can be solved by some iterative method. We introduce an artificial time (say s) arriving at the following nonlinear equation of evolution

$$-\frac{\partial u}{\partial s} = \lambda u + \alpha u^2 + \beta u_{\xi\xi} + u_{\xi^4}\,, \tag{3.1}$$

with u vanishing at infinity together with its derivatives.

There can be used various different iterative procedures but the solitary-wave problem is a bifurcation one and the route to the stationary solution of (3.1) (or which is the same (2.2)) is of crucial importance. In fact the discretization of time-derivative in (3.1) provides some kind of relaxation procedure which prevents the algorithm from going too prematurely in the vicinity of the trivial solution. This of course introduces an artificial dissipation but it should be underlined that it has nothing to in common with the possible intrinsic artificial viscosity of the scheme for the original generalized wave equation (2.1).

After applying Fourier transform, discretizing and truncating in the Fourier space, eq.(3.1) takes the form

$$-\frac{\partial U_j}{\partial s} = (\lambda - \beta k_j^2 + k_j^4)U_j + \frac{\alpha \Delta k}{2\pi}\left\{\frac{1}{2}U_{-N}U_{j+N} + \frac{1}{2}U_N U_{j-N} + \sum_{i=-(N-1)}^{N-1} U_i U_{j-i}\right\}, \tag{3.2}$$

$$U(k) = \int_{-\infty}^{+\infty} u(\xi)e^{-ik\xi}\mathrm{d}\xi\,, \quad k_j = j\Delta k\,, \quad U_j = U(k_j)\,, \quad j = 0, \pm 1, \pm 2, \ldots, \pm N\,.$$

Here we make use of the trapezoidal rule in order to evaluate the convolution integral representing the Fourier transform of the quadratic nonlinear term. Since we seek real solutions which are also even functions of ξ, we have $U^*(k) = U(k)$.

For a bifurcation problem, nothing precludes *a priori* the solution of Eqs. (3.2) from converging to the trivial solution $U_j = 0$. We have found that the way to prevent this is to impose a constant value on one of the modes, e.g., on the zero mode U_0. Then the equation for $j = 0$ is left free and we render it as a definitive equality for one of the parameters appearing in Eq. (3.1). Since α is another scalable coefficient, we choose it as the said adaptive parameter for which we have

$$\alpha(s) = -2\pi U_0 \lambda (\Delta k)^{-1}\left\{U_0^2 + U_N^2 + 2\sum_{i=1}^{N-1} U_i^2\right\}^{-1}. \tag{3.3}$$

After convergence of the iterations, the solution obtained is renormalized to $\alpha = 1$ through multiplying by α.

4. Stationary Shapes

In the subsonic case $\lambda > 0$ for $|\beta| > 2\sqrt{\lambda}$, the square of the wavenumber k^2 given by (2.3) is real and negative for both signs of the radical and one has four purely imaginary wavenumbers. In the supersonic case $\lambda < 0$, k^2 is always real regardless to the sign of β. Eq. (2.3) gives for k^2 a positive and a negative value and we are again confronted with the problem of purely imaginary wavenumbers. In both of these special cases solutions in Fourier space include Dirac δ-functions which cannot be treated numerically without some special implementation. Furthermore, because of their persisting oscillatory nature, these solutions are obviously non-local (see [1] for illuminating discussion on this subject). For these reasons they go beyond the aim of the present paper and will make an object of a future publication. Here we consider only the subsonic localized waves for moderate values of the fourth-order dispersion parameter β when the solution can be either with monotone tails or with damped oscillatory tails. In order not to overload our discussion, we take in what follows $\gamma = 1$. Let us remind that α and δ have already been taken equal to 1.

4.1. Monotone Shapes

Such shapes appear for $\beta < -2\sqrt{\lambda}$ with λ positive. In particular, for $\beta = -1$, the condition for having monotone tails is $c > \sqrt{1 - 0.25\beta^2} \approx 0.866$. In this case an analytical solution of Eq. (2.2) is available (see, e.g., [9])

$$u_{an}(\xi) = \frac{105}{169}\frac{\beta^2}{2\alpha}\operatorname{sech}^4\left(\frac{\xi}{2}\sqrt{\frac{-\beta}{13}}\right), \quad |c| = \sqrt{\gamma^2 - \frac{36}{169}\beta^2}. \tag{4.1}$$

We have found the Fourier transform of the solution (4.1) (see [7], Sec. 3.985.2), namely

$$U_{an}(k) = \frac{105\pi}{507}\frac{k}{\theta^2}\left(\frac{k^2}{4\theta^2} + 1\right)\sinh^{-1}\left(\frac{k\pi}{2\theta}\right), \quad \theta = \frac{1}{2}\sqrt{\frac{-\beta}{13}}. \tag{4.2}$$

For the selected parameters $\gamma = 1$ and $\beta = -1$, the celerity of the analytical solution is $c \approx 0.88712$ which falls into the range $c > 0.866$ for monotone shapes. In order to verify the performance of the technique we started the numerical experiments with this particular value of the celerity. As far as the hump is concerned the solution we obtained in Fourier space virtually coincides with the Fourier transform (4.2) of the analytical solution even when only 12 (positive) modes are used. This very small number of needed modes testifies for the efficiency of our spectral technique. Nevertheless, spectral solutions are intrinsically periodic with the period $2\pi/\Delta k$ in physical space. Hence a higher number of modes may be necessary in order to obtain a good resolution of the tails in the far field. Table 1 shows the relative error in Fourier space for the six first modes as function of the number of modes N (or the Fourier spacing Δk). For the upper modes, both the analytical and approximated solutions vanish when the number of modes is between 10 and 20. It is seen that the relative error decreases exponentially with Δk. This experiment as well as all the others have been achieved with $k_{max} = 3$.

In contrast with the fourth-order Boussinesq and KdV equations, the analytical solution of Eq. (4.1) is available only for a single value of c. This is a limitation of the analytical methods rather than a new result about the spectrum of the eigenvalue problem, because there is no reason to expect such a drastic change of the latter. We were indeed able to recover numerically the monotone shapes for $0.866 < c < 1$. The

TABLE 1

Checking the algorithm for stationary shapes: relative error between analytical and approximated solutions ($c = 0.88712$).

N	10	12	14	16	18	20
Δk	0.300	0.250	0.214	0.188	0.167	0.150
$j = 0$	−0.004689	−0.000720	−0.000092	−0.000007	0.000004	0.000005
$j = 1$	0.010882	0.000988	0.000093	0.000009	0.000001	0.000001
$j = 2$	0.019644	0.001631	0.000141	0.000014	0.000002	0.000002
$j = 3$	0.029568	0.002459	0.000210	0.000210	0.000004	0.000003
$j = 4$	0.038981	0.003331	0.000296	0.000023	0.000006	0.000004
$j = 5$	0.046875	0.004691	0.000485	0.000052	0.000022	0.000000

presented results are in a very good quantitative agreement with the difference solution of [5] for monotone shapes. Similarly to the case of the fourth-order proper Boussinesq equation, subsonic humps have a larger amplitude when they are slower. In Fig. 1 the monotone solutions we obtained for different celerities are presented.

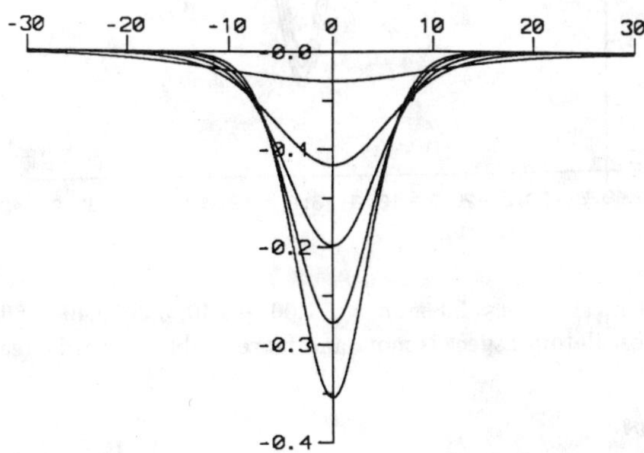

Fig. 1. Monotone shapes for different celerities $c = 0.87$, 0.90, 0.93, 0.96 and 0.99. Slower waves have larger amplitude.

4.2. Damped Oscillatory Shapes

This class of subsonic waves corresponds to $-2\sqrt{\lambda} < \beta < 2\sqrt{\lambda}$. As β is increased from the lowest limit to the upper one, the soliton develops small-amplitude damped periodic tails and the damping is lesser with the increase of β. The damped oscillatory solutions are named after Kawahara [8] who discovered them numerically in the Fifth-order KdV equation. The Kawahara solitons are intermediate case between the monotone shapes and the above mentioned non-local waves. The transfer from one extreme case to the other is made through the progressive selection of a wavenumber of the tails. Indeed, as β approaches its upper limit, the amplitude of a selected Fourier mode becomes increasingly high. Fixing $\gamma = 1$ and $c \approx 0.866$ the "β-interval" corresponding to Kawahara's solitons is $[-1, 1]$. In order to show the progressive transformation from monotone shapes to persisting oscillatory shapes, we present in Fig. 2 different Kawahara's solitons obtained with $c \approx 0.866$ and $|\beta| < 1$. Similarly to the case of monotone shapes, the following results are in full quantitative agreement with the corresponding ones obtained in [5] by means of finite differences method.

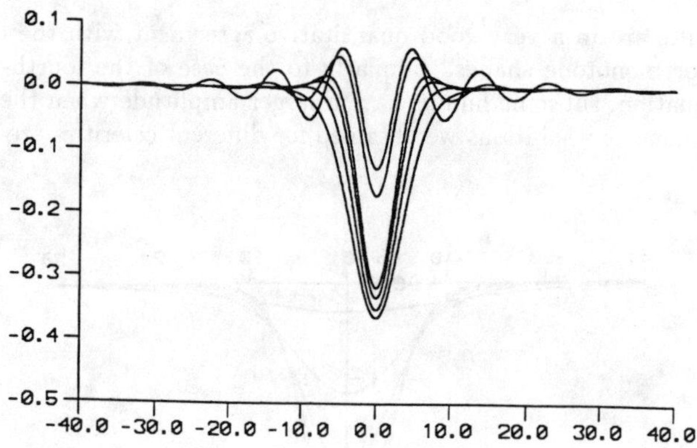

Fig. 2. Kawahara shapes $c \approx 0.866$, $\beta = -1.00, -0.40, 0.00, 0.40, 0.60, 0.80, 0.95$. The oscillatory aspect is more and more visible as β is increased.

4.3. Bound States

In the following we use the expression "bound state" to describe a stationary solution of (3.1) consisting of two main humps linked through their interacting tails and being asymptotically stable with respect to a small perturbation of their position (see [6] for discussion about the so-called multi-solitons). A bound state should be distinguished from a "pair of solitary waves" which differ from bound states by the absence of interaction between them. Such a solution is stable in the sense of Lyapunov with respect to a perturbation of the position of the humps. In order to capture bound states, we used the technique described above without any change. The only difference resides in the initial conditions. Indeed, rather than imposing arbitrary ones we took as initial conditions two identical stationary solitary shapes solutions of (3.1) and we shifted them symetrically with respect to the origin of the moving frame. Playing with the initial distance (noted d_0 in the following) between Kawahara's solitons our experiments showed that 6GBE admits two bound states (represented in bold in Fig. 3) and an infinity of pairs of solitary waves (thin lines on the same figure). This result allows us to define a finite distance of interaction (increasing function of c and β) as the minimum initial distance between humps which leads to a pair of solitary waves rather than to a bound state. While increasing d_0 from 0, the first bound state to appear is always the trivial one (e.g. initial waves collapse into the original one). After a first threshold of d_0, the waves form the second bound state, and so on until another threshold. Further increase of d_0 will give rise to the pairs of solitary waves or to the trivial bound state again if the distance of interaction is big enough. In the second case, still increasing d_0 will make the solitons forming again the second bound state and finally a pair of solitary waves.

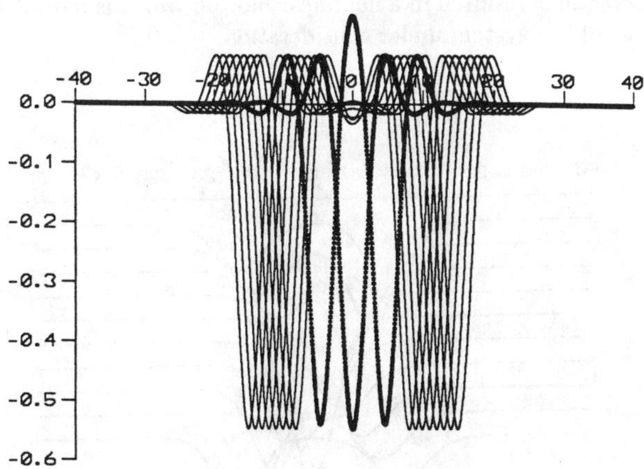

Fig. 3. Bold lines: the two bound states formed by Kawahara's solitons with $\beta = 0.80$, $c = 0.70$ and $d_0 \in [0, 2.60]$ for the trivial one and $d_0 \in [2.70, 6.70]$ for the second one. Thin lines: a series of pairs of solitary waves corresponding to $d_0 = 8.00, 9.00, \ldots, 15.00$.

5. Some Results for the Dynamics and Interactions

It is not the aim of this paper to present a detailed and fully classified description of dynamical properties of 6GBE solitons. Rather, we briefly outline here some results obtained with our spectral technique about dynamics. Now we extend the technique to the case of second-order in time equation, and we still impose a constant value to the zero mode. Eq. (2.1) is solved beginning from the initial condition consisting in the linear superposition of two solutions u_1 and u_2 differing from their celerities c_1 and c_2, respectively centered in x_1 and x_2, $|x_1 - x_2|$ being supposed to be large enough. Then in Fourier space initial conditions have the form

$$U(k,0) = e^{-ikx_1}U_1(k) + e^{-ikx_2}U_2(k), \quad (5.1)$$

$$V(k,0) = -ik(c_1 e^{-ikx_1}U_1(k) + c_2 e^{-ikx_2}U_2(k)). \quad (5.2)$$

Fig. 4 shows the propagation and head-on collision between two different monotone shapes. It is seen that the waves experience a phase shift during the collision. This phase shift is stronger for the faster wave (smaller amplitude). Similar results were obtained for Kawahara's solitons, but with a higher number of modes in order to increase the period of the spectral solution and thus to avoid unwanted interactions between adjacent solitons. It should be noted that there are two cases for which we never succeeded to achieve those collisions: between Kawahara's solitons having quite different amplitudes and between solitons from different classes. In both of these cases interactions resulted in a nonlinear blowup which is indeed allowed by the conservativeness of the system under consideration.

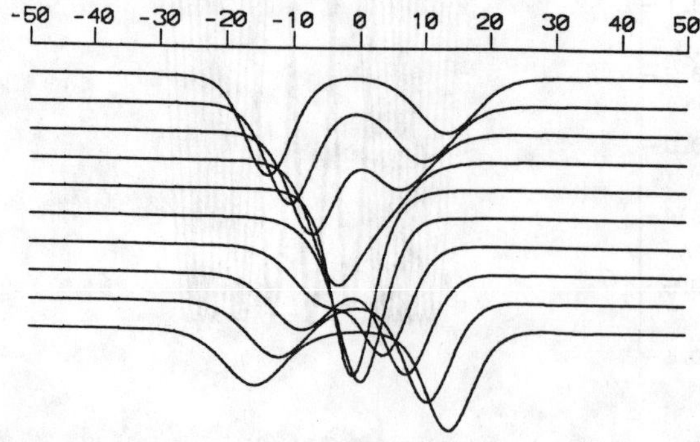

Fig. 4. Propagation and head-on collision between two monotone waves with $\beta_1 = \beta_2 = -1$ and $c_1 = 0.87$, $c_2 = 0.93$.

6. Conclusions

In the present paper we develop a spectral technique for solving a generalized wave equation (called 6GBE) containing nonlinearity and fourth- and sixth-order dispersion. Due to the bifurcative nature of the problem under consideration, special care is taken in order to prevent the algorithm from going too prematurely in the vicinity of the trivial solution. The accuracy appears to be of exponential order in the wavenumber.

We have applied the technique to both cases: stationary in the moving frame waves and dynamical interactions of the latter.

For a given subsonic celerity (c such that $\gamma^2 - c^2 = \lambda < 0$) there exist three classes of waves depending on the magnitude of the fourth-order dispersion β. We concern ourselves only with the strictly localized solutions (with monotone or damped oscillatory tails).

The so-called Kawahara's solitons constitute actually an intermediate class between monotone shapes ($\beta < -2\sqrt{\lambda}$) and purely oscillatory shapes ($\beta > 2\sqrt{\lambda}$).

We have studied also the propagation and head-on collision between 6GBE "quasi-particles" and found the presence of phase shift and nonlinear blow-up in the respective cases. We have found that besides of the one-soliton solution with oscillatory tails (Kawahara's soliton), 6GBE admits at least one quantized bound state formed by two Kawahara's solitons.

Acknowledgements. One of the authors (Y.S.) was supported by Human Capital and Mobility Grant ENP-ERBCHRXCT930107. C.I.C. was supported by E.C.C. Grant ERBCHBICT940982.

References

1. J. P. Boyd. *Physica* **D48** (1991) 129.
2. C. Canuto, M. Y. Hussaini, A. Quarteroni and T. A. Zang, *Spectral Methods in Fluid Dynamics*, Springer-Verlag, N.Y., 1987.
3. C. I. Christov. In:, *Proc. ICFD Oxford 1995*, eds. W. K. Morton and M. J. Baines, Oxford University Press, 1995, to appear.
4. C. I. Christov. In *Fluid Physics*, eds. M. G. Velarde and C. I. Christov, World Scientific, Singapore, 1994, p. 403.
5. C. I. Christov, G. A. Maugin and M. G. Velarde. *Phys. Rev. E*, 1995, submitted.
6. K. A. Gorshkov, L. A. Ostrovsky, V. V. Papko and A. S. Pikovsky. *Phys. Lett. A* **74** (1979) 177.
7. I. S. Gradshteyn and I. M. Ryzhik, *Table of Integrals, Series and Products*, 4th edition, Academic Press, N.Y., 1980.
8. T. Kawahara, *J. Phys. Soc. Japan* **33** (1972) 260.
9. Y. Yamamoto, *J. Phys. Soc. Japan* **58** (1989) 4410.
10. N. J. Zabusky and M. D. Kruskal. *Phys. Rev. Lett.* **15** (1965) 57.

Continuum Models and Discrete Systems
Proceedings of 8th International Symposium, June 11-16, 1995, Varna, Bulgaria, ed. K.Z. Markov

EQUIVALENCE TRANSFORMATIONS OF THE NAVIER-STOKES EQUATIONS AND THE INERTIAL RANGE OF TURBULENCE

G. ÜNAL
Faculty of Science and Letters,
Istanbul Technical University, Maslak 80626, Istanbul, Turkey

Abstract. The standard application of the group theory to the Navier-Stokes equations does not lead to the viscosity (ν) dependent symmetries. Hence, it finds no application in turbulence. To adopt the Lie theory to turbulence, the equivalence group G_ϵ has been found. The latter not only allowed us to investigate the group classification problem but also to study the scaling properties of inertial regime. Here we introduce the notion of Kolmogorov's invariant and determine the subgroup G_τ of the equivalence group G_ϵ such that Kolmogorov invariant is the first order differential invariant of G_τ. We have also found the new symmetries of Navier-Stokes equations by utilizing one of the operators of G_ϵ with an arbitrary function of ν. One of the these symmetries happens to be the group of projective transformations and, we discuss the physical interpretation of it. Here, we also discuss the symmetry breaking bifurcations.

1. Introduction

Richardson's experiments [1] on the small scale turbulence revealed that there exists a hierarchy of eddies of different scale sizes, for large values of the Reynolds number. According to this picture, large scale eddies appear as a result of the loss of stability of the mean flow from where they also receive their energy to survive. In this cascade mechanism, each offspring is generated due to instability of parents. Eddies of larger scale transfer energy to the smaller ones via nonlinear interactions. This energy cascade process continues until the eddies with small Reynolds number appear. Eventually, the dissipation regime dominates in which viscosity effects becomes significant on the smaller eddies [1].

Kolmogorov had a similar picture in his mind when he put forwarded his hypotheses on the inertial subrange of turbulence in an incompressible fluid. According to his picture, cascade process conserves energy and flow forgets about the nature of boundary conditions and the external forces (universality) [2]. The former implies that the dissipation of energy per unit mass (ϵ) is invariant under the scaling laws of inertial subrange [3]. And the latter suggests that the solutions will also remain invariant under the groups which are admitted by the Navier-Stokes equation. These are all based on the physical observations (visualization of the solution space). Here, we will obtain the scaling laws of inertial subrange by studying the frame **F** of the Navier-Stokes equations [4]. We will also discuss some of the new viscosity dependent transformation groups.

2. Equivalence Transformations of the Navier-Stokes Equations

2.1. Equivalence Algebra

The Navier-Stokes equations for the incompressible fluid:

$$\mathbf{u}_t + (\mathbf{u}\cdot\nabla)\mathbf{u} = -\frac{1}{\rho}\nabla p + \nu\Delta\mathbf{u}, \quad \nabla\cdot\mathbf{u} = 0 \tag{2.1}$$

where as usual $\nabla = \left[\frac{\partial}{\partial x^1}, \frac{\partial}{\partial x^2}, \frac{\partial}{\partial x^3}\right]$, $\Delta = \nabla\cdot\nabla$, $\mathbf{u} = (u^1, u^2, u^3)$, ν is the kinematic viscosity and the density $\rho > 0$ is a constant. The Navier-Stokes equations admit infinite dimensional Lie point symmetry group [5-6]. Yet, none of them comprises the viscosity ν. Thus, the scaling laws are not compatible with Kolmogorov's theory [2]. To incorporate the viscosity, one should consider it as one of the coordinates of the frame F.

The generator of the infinitesimal transformations which leaves Eq. (2.1) invariant can be written as

$$X = \xi^i(\cdot)\frac{\partial}{\partial x^i} + \eta^i(\cdot)\frac{\partial}{\partial u^i} + V(\cdot)\frac{\partial}{\partial \nu}, \quad i = 1, 2, 3, 4. \tag{2.2}$$

Here, $(\cdot) = (x^1, x^2, x^3, x^4, u^1, u^2, u^3, u^4, \nu)$ with $x^4 = t$ and $u^4 = p$. The infinitesimal invariance criterion [4] for the frame **F** given by Eq. (2.1) is

$$\underset{2}{X} F^i \Big|_{F^i} = 0, \tag{2.3}$$

where $\underset{2}{X}$ is the second prolongation of Eq. (2.2):

$$\underset{2}{X} = X + \zeta_l^k \frac{\partial}{\partial u_l^k} + \zeta_{np}^m \frac{\partial}{\partial u_{np}^m}$$

and

$$\zeta_l^k = D_l(\eta^k) - u_s^k D_l(\xi^s), \quad \zeta_{np}^m = D_p(\zeta_n^m) - u_{nr}^m D_p(\xi^r).$$

The total derivative operator **D** is defined as:

$$D_l = \frac{\partial}{\partial x^l} + u_l^\alpha \frac{\partial}{\partial u^\alpha} + u_{lt}^\beta \frac{\partial}{\partial u_t^\beta} + \cdots .$$

For the Navier-Stokes equations, infinitesimal invariance criterion, Eq. (2.3), leads to the following equations:

$$\Big\{\eta_4^i + u_4^\alpha \eta_\alpha^i - u_s^i \xi_4^s - u_s^i u_4^\alpha \xi_\alpha^s + \eta^j u_j^i + u^j \left[\eta_j^i + u_j^\alpha \eta_\alpha^i - u_s^i \xi_j^s - u_s^i u_j^\alpha \xi_\alpha^s\right]$$

$$+\frac{1}{\rho}\left[\eta_i^4 + u_i^\alpha \eta_\alpha^4 - u_s^4 \xi_i^s - u_s^4 u_i^\alpha \xi_\alpha^s\right] - \nu\left[\eta_{ll}^i + u_{ll}^\alpha(\eta_\alpha^i - u_s^i \xi_\alpha^s) - 2u_{sl}^i \xi_l^s - 2u_{sl}^i u_l^\alpha \xi_\alpha^s\right.$$

$$\left. + u_l^\alpha u_l^\beta \eta_{\alpha\beta}^i - 2u_s^i u_l^\alpha \xi_{\alpha l}^s - u_s^i u_l^\alpha u_l^\beta \xi_{\alpha\beta}^s + 2u_l^\beta \eta_{l\beta}^i - u_s^i \xi_{ll}^s\right]\Big\}\Big|_{F^i=0} = 0\,,$$

$$\left[\eta_1^1 + u_1^\alpha \eta_\alpha^1 - u_s^1 \xi_1^s - u_s^1 u_1^\alpha \xi_\alpha^s + \eta_2^2 + u_2^\alpha \eta_\alpha^2 - u_s^2 \xi_2^s - u_s^2 u_2^\alpha \xi_\alpha^s\right.$$

$$\left. + \eta_3^3 + u_3^\alpha \eta_\alpha^3 - u_s^3 \xi_3^s - u_s^3 u_3^\alpha \xi_\alpha^s\right]\Big|_{u_1^1 = -u_2^2 - u_3^3} = 0\,. \tag{2.4}$$

From Eq. (2.4) we obtain the determining equations [4] for the ξ, η and V. After having solved the determining equations and differentiating with respect to arbitrary parameters and functions, we obtain the following infinite dimensional equivalence Lie algebra L_ϵ:

$$X_1 = \frac{\partial}{\partial t}\ (Translations\ in\ time), \quad X_2 = g(t)\frac{\partial}{\partial p}\ (Pressure\ changes)\,, \tag{2.5.1}$$

$$X_3 = x^i\frac{\partial}{\partial x^i} + 2t\frac{\partial}{\partial t} - u^i\frac{\partial}{\partial u^i} - 2p\frac{\partial}{\partial p}\ (Scaling)\,, \tag{2.5.2}$$

$$X_{ij} = x^j\frac{\partial}{\partial x^i} - x^i\frac{\partial}{\partial x^j} + u^j\frac{\partial}{\partial u^i} - u^i\frac{\partial}{\partial u^j}\ (Rotations)\,, \tag{2.5.3}$$

$$X_{k+6} = h_{\underline{k}}(t)\frac{\partial}{\partial x^{\underline{k}}} + h'_{\underline{k}}(t)\frac{\partial}{\partial u^{\underline{k}}} - x^{\underline{k}}h''_{\underline{k}}(t)\frac{\partial}{\partial p} \quad k = 1,2,3,$$

$$(Transformation\ to\ an\ arbitrarily\ moving\ Coordinate\ system)\,, \tag{2.5.4}$$

$$X_{10} = t\frac{\partial}{\partial t} + x^i\frac{\partial}{\partial x^i} + \nu\frac{\partial}{\partial \nu}\ (Scaling)\,. \tag{2.5.5}$$

Here, $f(t)$, $h^1(t)$, $h^2(t)$ and $h^3(t)$ are arbitrary smooth functions of time. The summation convention does not apply to underlined indices. The operators in Eqs. (2.5.1-4) have been obtained by several authors [5-8]. The operator X_{10} has been found recently by the author [9]. The group classification aspects of the Navier-Stokes equations are discussed in [10].

2.2. Infinite Dimensional Equivalence Group

We start by proving the following lemma.

Lemma: Every system of differential equations of the form

$$\Delta_\alpha(x^1,\ldots,x^m;u^1,\ldots,u^n;\pi^1,\ldots,\pi^p;u_m^n,\ldots,u_{m_1,\ldots,m_r}^n) = 0, \tag{2.6}$$

$\alpha = 1, 2, \ldots, n$, which admits the following equivalence operator

$$X = \rho^m(x^1, \ldots, x^m; u^1, \ldots, u^n)\frac{\partial}{\partial x^m} + \theta^n(x^1, \ldots, x^m; u^1, \ldots, u^n)\frac{\partial}{\partial u^n}$$

$$+ \Pi^p(\pi^1, \ldots, \pi^p)\frac{\partial}{\partial \pi^p} \tag{2.7}$$

also admits the operator:

$$Z = f(\pi^p)X, \tag{2.8}$$

where $f(\pi^p) = f(\pi^1, \ldots, \pi^p)$ is an arbitrary function of parameters.

Proof: In order for the operator Z to leave Eq. (2.6) invariant it should fulfill the infinitesimal invariance criterion:

$$\underset{r}{Z}\Delta_\alpha \Big|_{\Delta_\alpha = 0} = 0\,. \tag{2.9}$$

Here, $\underset{r}{Z}$ is the rth prolongation of Z:

$$\underset{r}{Z} = f(\pi^p)\rho^m\frac{\partial}{\partial x^m} + f(\pi^p)\theta^n\frac{\partial}{\partial u^n} + f(\pi^p)\Pi^p\frac{\partial}{\partial \pi^p}$$

$$+\overline{\zeta}^j_k\frac{\partial}{\partial u^j_k} + \cdots + \overline{\zeta}^j_{i_1\ldots i_r}\frac{\partial}{\partial u^j_{i_1\ldots i_r}}\,. \tag{2.10}$$

In Eq. (2.10), we made use of the following quantities:

$$\overline{\zeta}^j_k = D_k(\overline{W}^j) + \overline{\xi}^s u^j_{ks}\,, \quad \overline{\zeta}^j_{i_1\ldots i_r} = D_{i_1}\ldots D_{i_r}(\overline{W}^j) + \overline{\xi}^s u^j_{si_1\ldots i_r}\,,$$

$$D^{(r)}_i = \frac{\partial}{\partial x^i} + u^n_i\frac{\partial}{\partial u^n} + \cdots + u^n_{ii_1\ldots i_{r-1}}\frac{\partial}{\partial u^n_{i_1\ldots i_{r-1}}}\,, \quad \overline{W}^j = \overline{\eta}^j - \overline{\xi}^s u^j_s\,.$$

By using the fact that $\overline{\eta}^j = f(\pi^p)\eta^j$; $\overline{\xi}^s = f(\pi^p)\xi^s$, one can now rewrite Eq. (2.10) as

$$\underset{r}{Z} = f(\pi^p)\underset{r}{X}\,, \tag{2.11}$$

where $\underset{r}{X}$ stands for the rth prolongation of the operator X, see Eq. (2.7). Since Eq. (2.6) admits the same operator X, then

$$\underset{r}{X}\Delta_\alpha \Big|_{\Delta_\alpha = 0} = 0\,.$$

Thus, upon employing Eq. (2.11) in Eq. (2.9) we obtain

$$f(\pi^p)\underset{r}{X}\Delta_\alpha \Big|_{\Delta_\alpha = 0} = 0$$

which completes the proof. This lemma guarantees that the Navier-Stokes equations possess infinite number of equivalence group of the form:

$$Z = f(\nu)X_{10} = f(\nu)t\frac{\partial}{\partial t} + f(\nu)x^i\frac{\partial}{\partial x^i} + f(\nu)\nu\frac{\partial}{\partial \nu}, \tag{2.12}$$

where $f(\nu)$ is an arbitrary function of viscosity ν.

The solutions to Lie equations allow one to obtain the equivalence transformations corresponding to Eq. (2.12). Lie equations can be written as:

$$\frac{d\overline{\nu}}{da} = \overline{\nu}f(\overline{\nu}), \quad \frac{d\overline{t}}{da} = \overline{t}f(\overline{\nu}), \quad \frac{d\overline{x}^i}{da} = \overline{x}^i f(\overline{\nu}) \tag{2.13}$$

with the initial conditions

$$\overline{\nu}(0) = \nu, \quad \overline{t}(0) = t, \quad \overline{x}^i(0) = x^i. \tag{2.14}$$

Now let us list the solutions of Eq. (2.13) for several specific types of the function $f(\nu)$ in order to obtain the new transformation groups:

$$f(\nu) = 1 \ \overline{\nu} = \nu e^a, \quad \overline{t} = te^a, \quad \overline{x}^i = x^i e^a, \tag{2.15}$$

$$f(\nu) = \nu^n, \quad \overline{\nu} = \frac{\nu}{[1 - na\nu^n]^{1/n}}, \quad \overline{t} = \frac{t}{[1 - na\nu^n]^{1/n}}, \quad \overline{x}^i = \frac{x^i}{[1 - na\nu^n]^{1/n}}, \tag{2.16}$$

$$f(\nu) = \nu^{-n}, \quad \overline{\nu} = (na + \nu^n)^{1/n} \quad \overline{t} = \frac{t}{\nu}(na + \nu^n)^{1/n}, \quad \overline{x}^i = \frac{x^i}{\nu}(na + \nu^n)^{1/n}. \tag{2.17}$$

3. Kolmogorov's Invariant and the Inertial Subrange

3.1. Kolmogorov's Invariant and Algebra L_τ

In [9], we have considered the linear combinations of operators X_3 and X_{10}. The coefficients appearing in the linear combination have been determined by resorting to Kolmogorov's invariance principle. Later, we have proven a theorem in [11] to exploit its mathematical structure. We have noticed recently that the theorem in [11] can be proven in a slightly different way to reveal an important aspect of the lemma given here.

Definition: The algebra L_τ is a subalgebra of the equivalence algebra L_ϵ with operator Eq. (2.12), such that energy dissipation rate per unit mass ϵ

$$\epsilon = \frac{\nu}{2}\sum_{r,s}(u^r_s + u^s_r) \tag{3.1}$$

is the first order differential invariant.

Theorem: The algebra L_τ is infinite dimensional and it comprises the operators in Eqs. (2.5.1), (2.5.3-4) and the operator:

$$X_\epsilon = x^i \frac{\partial}{\partial x^i} + \frac{2}{3} t \frac{\partial}{\partial t} + \frac{1}{3} u^i \frac{\partial}{\partial u^i} + \frac{2}{3} p \frac{\partial}{\partial p} + \frac{4}{3} \nu \frac{\partial}{\partial \nu} . \tag{3.2}$$

Proof: Note that Eq. (3.1) involves only the first order differentials and the first prolongations of operators X_1 and X_2 do not comprise the differential functions which appear in Eq. (3.1). Then these operators leave Eq. (3.1) invariant. Now we act with the first prolongation of the linear combination $X^* = \alpha X_3 + \beta f(\nu) X_{10}$ on ϵ:

$$\underset{1}{X^*} \epsilon = 0 ,$$

to obtain

$$X^* \epsilon = \nu(4\alpha + 3\beta f(\nu))[2(u_1^1)^2 + 2(u_2^2)^2 + 2(u_3^3)^2 + (u_2^1 + u_1^2)^2 + (u_3^1 + u_1^3)^2 + (u_3^2 + u_2^3)^2] = 0.$$

Thus we have:

$$\alpha = 1 \quad \text{and} \quad \beta = -\frac{4}{3 f(\nu)} ,$$

which yields

$$X_\epsilon = X_3 - \frac{4}{3} X_{10} .$$

Next we act with the first prolongations of the operators X_{ij} on Eq. (3.1):

$$\underset{1}{X_{ij}} \epsilon = 0$$

to get

$$\begin{gathered}
4u_1^1(u_1^j \delta_{i1} - u_i^1 \delta_{j1} + u_j^1 \delta_{i1}) + 4u_2^2(-u_2^i \delta_{j2} + u_2^j \delta_{i2} - u_i^2 \delta_{j2} + u_j^2 \delta_{i2}) \\
+4u_3^3(-u_3^i \delta_{j3} - u_i^3 \delta_{j3} + u_j^3 \delta_{i3}) + 2(u_2^1 + u_1^2)(u_2^j \delta_{i1} - u_i^1 \delta_{j2} + u_j^1 \delta_{i2} - u_1^i \delta_{j2} \\
+u_1^j \delta_{i2} - u_i^2 \delta_{j1} + u_j^2 \delta_{i1}) + 2(u_3^1 + u_1^3)(u_3^i \delta_{i1} - u_i^1 \delta_{j3} + u_j^1 \delta_{i3} - u_1^i \delta_{j3} - u_i^3 \delta_{j1} \\
+u_j^3 \delta_{i1}) + 2(u_3^2 + u_2^3)(-u_3^i \delta_{j2} + u_3^j \delta_{i2} - u_i^2 \delta_{j3} + u_j^2 \delta_{i3} - u_i^2 \delta_{j3} \\
-u_2^i \delta_{j3} - u_i^3 \delta_{j2} + u_j^3 \delta_{i2}) = 0.
\end{gathered}$$

The latter vanishes when ($i = 1$, $j = 2$), ($i = 1$, $j = 3$) and ($i = 2$, $j = 3$), due to the fact that the first prolongations of the operators X_4, X_5 and X_6 do not involve the differential functions which appear in Eq. (3.1). Therefore these operators also leave Eq. (3.1) invariant. These calculations complete the proof.

As it can been seen from the proof, regardless of the transformations in Eqs. (2.15–17) Kolmogorov's invariant yields a unique scaling law (equivalence group):

$$\overline{u}^i = e^{\frac{1}{3}a}u^i, \quad \overline{x}^i = e^a x^i, \quad \overline{t} = e^{\frac{2}{3}a}t, \quad \overline{p} = e^{\frac{2}{3}a}p, \quad \overline{\nu} = e^{\frac{4}{3}a}\nu. \tag{3.3}$$

These transformation can be easily found by constructing Eqs. (2.13) for Eq. (3.2) and, solving them with initial conditions given in Eq. (2.14).

3.2. Eddy Engine, Projective Transformations and Symmetry Breaking

The scaling law given in Eq. (3.3) allows to write

$$u^i = e^{-\frac{1}{3}a}h^i(e^a x^1, \ldots, e^a x^3, e^{\frac{2}{3}a}t, e^{\frac{4}{3}a}\nu). \tag{3.4}$$

Following Landau [12], we can write

$$u^i = e^{-\frac{1}{3}a}\sum_{p_1,p_2,p_3} A^i_{P_1\ldots P_3}(\overline{x}^1, \overline{x}^2, \overline{x}^3)\exp\left[-i\sum_{j=1}^{3} p_j(w_j e^{\frac{2}{3}a}t + \beta_j)\right]. \tag{3.5}$$

Eq. (3.5) suggests that different scales of motion can be generated by varying the group parameter (Reynolds number, *Re*) *a*. And different scales are affected differently by the viscosity. The effect of viscosity on small scale eddies are more severe than on the larger ones (the faster the fluid moves, the higher the effect of viscosity becomes).

According to the Lemma given here, we can write $Z_\epsilon = \frac{3}{4}\nu X_\epsilon$. Now forming the Lie equations for Z_ϵ and solving them with Eq. (2.14), we can write:

$$u^i = (1 - a\nu)^{\frac{1}{4}} w^i\left(\frac{x^1}{(1 - a\nu)^{1/4}}, \ldots, \frac{x^3}{(1 - a\nu)^{1/4}}, \frac{t}{(1 - a\nu)^{1/2}}, \frac{\nu}{(1 - a\nu)}\right). \tag{3.6}$$

Here, $[a] = 1/UL$ and $a\nu = 1/Re$. When $Re \gg 1$, the effect of projective transformations on the flow disappears.

Now let us consider Eq. (2.13) with (2.14). If the function $f(\overline{\nu})$ satisfies the transcritical and pitchfork bifurcation conditions [13], then we have symmetry breaking on the frame. Since these are codimension one bifurcations, they involve one control parameter. Here, the control parameter becomes Reynolds number. Thus, as the Reynolds number goes through a critical value, a bifurcation of transformation groups occur.

4. Concluding Remarks

We have investigated the equivalence transformations of the Navier-Stokes equations. It allowed us to obtain scaling group for inertial range of turbulence, and new viscosity dependent transformation groups.

Acknowledgements. This work has been supported by TUBITAK through the project TBAG-Ü/14-1. Special thanks goes to A. Kiriş.

References

1. L. F. Richardson, *Weather Prediction by Numerical Process*, Cambridge Univ. Press, New York, 1922.
2. A. N. Kolmogorov, *Dokl. Akad. Nauk S.S.S.R.* **26** (1941) 115. (in Russian.)
3. A. S. Monin and A. M. Yaglom, *Statistical Fluid Mechanics: Mechanics of Turbulence*, Nauka, Moscow, 1965.(in Russian.) (English translation published by the MIT Press-Cambridge, 1983.)
4. N. H. Ibragimov, *The Mathematical Intelligencer* **16** (1) (1994) 20.
5. V. V. Pukhnacev, *Zhurnal Prikladnoi Mekhaniki i Tekhnicheskoi Fiziki* **1** (1960) 83. (in Russian.)
6. V. V. Pukhnacev, *Dokl. Akad. Nauk S.S.S.R.* **202** (1972) 302. (in Russian.)
7. S. P. LLoyd, *Acta Mechanica* **38** (1981) 85.
8. R. E. Boisvert, W.F. Ames and U.N. Srivastava, J. Engng. Math. **17** (1983) 203.
9. G. nal, *Lie Groups and Their Applications* **1** (1) (1994) 232. In *CRC Handbook of Lie Group Analysis of Differential Equations*, ed. N. H. Ibragimov, **2**, CRC Press-Boca Raton, 1994, Ch. 11.3.3.
10. N. H. Ibragimov and G. Ünal, *Bull. Tech. Univ. İstanbul* **47** (1994) 203.
11. N. H. Ibragimov and G. Ünal, *Lie Groups and Their Applications* **1** (2) (1994) 98.
12. L. D. Landau and E. M. Lifshitz, *Fluid Mechanics*, Pergamon Press-Oxford, 1966.
13. J. Guckenheimer and P. Holmes, *Nonlinear Oscillations, Dynamical Systems, and Bifurcations of Vector Fields*, Springer-Verlag-New York, 1983.

PART V

DISLOCATIONS AND PLASTICITY

Continuum Models and Discrete Systems
Proceedings of 8th International Symposium, June 11-16, 1995, Varna, Bulgaria, ed. K.Z. Markov

DISLOCATION DYNAMICS AND PLASTICITY BY MEANS OF LAGRANGE FORMALISM

A. AZIRHI and K.–H. ANTHONY
Depart. of Theoretical Physics, University of Paderborn, Warburger Str. 100
D-33098 Paderborn, Germany

Abstract. Within the unifying framework of Lagrange Formalism (LF) we propose a phenomenological continuum theory of single crystal plasticity. In this paper our investigations are restricted to pure mechanical aspect of dislocation dynamics. Our model for plasticity is based on complex dislocation fields and on a generalized Cosserat continuum. We regard our approach as a first step towards a continuum theory of plasticity. A Lagrangian is proposed which includes the main features of dislocation dynamics.

1. General Structure of Plasticity Theory within Lagrange Formalism

1.1. Aim of the Theory

We aim in a phenomenological continuum theory of single crystal plasticity which on the macroscale takes account of dislocation dynamics. By using complex dislocation fields the theory is fitted to the microdynamics of the dislocations. Following the scheme of LF the whole information concerning the dynamics of a plastically deformed body will be included into one scalar function. Following the LF based on the concept of complex field variables $[2-5,13]$, the whole theory, i.e., the field equations, all relevant balance equations and their associated constitutive equations, can be developed.

Plastic deformation is due to dislocation dynamics giving rise to irreversible energy transfer between thermal and kinematical degrees of freedom. Thus, we have to take into account thermodynamics of irreversible processes (TIP). We have shown that TIP can be included into LF $[1-4]$. Consequently LF presents itself as an appropriate tool for an unified description of thermal and mechanical processes [5, 10, 13]. By means of LF we finally aim in a phenomenological field theory of plasticity taking account of the mechanics of plastic deformation as well as of the thermal effects. As a first step we restrict ourself to pure mechanical aspect. Thermal effects and dissipation are subject of our future investigations.

1.2. Lagrange Formalism based on Complex Field Variables

The central idea of LF lies in the fact that the whole information concerning the processes $\{\psi_K(x,t)), K = 1,\ldots,N\}$ of a physical system is included into its Lagrangian [1]:

$$\ell = \ell(\psi_K(x,t), \partial_t\psi_K(x,t), \nabla\psi_K(x,t)). \tag{1}$$

Real processes are distinguished as solutions of Hamilton's variational principle:

$$\begin{aligned} I &= \int_{t_1}^{t_2} dt \int_V \ell(\psi_K(x,t), \partial_t\psi_K(x,t), \nabla\psi_K(x,t))\, dV \\ &= \text{extremum by free variation of all fields } \psi_K(x,t). \end{aligned} \tag{2}$$

It results in

• the Euler–Lagrange equations as the fundamental field equations:

$$\partial_t \frac{\partial \ell}{\partial(\partial_t\psi_K)} + \partial_\alpha \frac{\partial \ell}{\partial(\partial_\alpha\psi_K)} - \frac{\partial \ell}{\partial\psi_K} = 0, \quad \alpha = 1,2,3, \tag{3}$$

(the summation convention is consequently applied) and in

• the balance equations due to the invariance properties of ℓ: By Noether's theorem each group parameter ε of the invariance group is associated with a balance equation:

$$\partial_t a_\varepsilon + \nabla \cdot \vec{J}_\varepsilon = 0, \tag{4}$$

where the density a_ε and flux density $\vec{J}_\varepsilon$ represent the observables associated with the group parameter ε.

One should keep in mind that the entropy concept can be incorporated in the scheme of LF as a methodological concept [4]. It is associated with the invariance of the Lagrangian with respect to a global gauge transformation of all complex fields ψ_K:

$$\psi_K \longrightarrow \psi_K e^{i\varepsilon}, \quad \varepsilon = \text{const}\,. \tag{5}$$

1.3. Complex Valued Field Variables

According to Anthony's approach each dissipative degree of freedom is associated with a complex field ψ_k. Hence heat conduction is associated with the *field of thermal excitation* $\chi(x,t)$ (thermion field [13, 14])

$$\chi(x,t) = \sqrt{T(x,t)}e^{i\varphi(x,t)}\,, \tag{6}$$

giving rise to *absolute temperature*:

$$T(x,t) = \chi^*(x,t)\chi(x,t) \geq 0. \tag{7}$$

Diffusing matter is associated with a complex matter field $\psi(x,t)$,

$$\psi(x,t) = \sqrt{\rho(x,t)}e^{i\varphi(x,t)}, \tag{8}$$

resulting in the definition of *mass density*:

$$\rho(x,t) = \psi^*(x,t)\psi(x,t). \tag{9}$$

The diffusion takes place in a deformable material carrier. This deformation is associated with a complex matter field $\Psi(x,t)$

$$\Psi(x,t) = \sqrt{R(x,t)}e^{i\Phi(x,t)}\,, \tag{10}$$

giving rise to the *mass density of the carrier*:

$$R(x,t) = \Psi^*(x,t)\Psi(x,t) \geq 0. \tag{11}$$

The phase function $\Phi(x,t)$ is part of the three Clebsch Potentials $\Phi(x,t)$, $\xi(x,t)$, $\eta(x,t)$ which define the velocity field of the carrier

$$\vec{v} = \nabla\Phi(x,t) + \xi(x,t)\nabla\eta(x,t). \tag{12}$$

The two potentials $\xi(x,t)$ and $\eta(x,t)$ may be joined together into the complex vortex field [6,7]

$$\nu(x,t) = \sqrt{\xi(x,t)}e^{i\eta(x,t)}\,. \tag{13}$$

In the case of simple diffusion in a fixed material carrier the diffusion velocity is simply defined from the phase function φ of ψ in (8)

$$\vec{v}_d = \nabla\varphi(x,t) \tag{14}$$

and due to the Fick's law the corresponding diffusion flux density is given by

$$J_d = -D\nabla\rho = \rho\vec{v}_d\,, \quad \text{D: diffusion coefficient.} \tag{15}$$

The phase function $\varphi = -D\ln\rho$ plays the role of the velocity potential. However, if diffusion takes place in a deformable carrier there is a material dragging along of the diffusing matter and we need an extension of the ansatz (14). Similar to (12) and using the Clebsch potentials φ, λ, β we get

$$\vec{v}_d = \nabla\varphi(x,t) + \lambda(x,t)\nabla\beta(x,t). \tag{16}$$

This case will get important in the subsequent problem of dislocation motion due to the driving Peach–Koehler force.

1.4. Plasticity and Complex Dislocation Fields

Basic ideas of the model: The traditional continuum theory of dislocations [8,9] is based on the concept of the dislocation density tensor $\underline{\alpha}$ which is associated with the torsion of the lattice geometry. The tensor $\underline{\alpha}$ is but a local average measure of the dislocation arrangement neglecting all correlational effects. In order to take this correlation into consideration we divide the network into a set of different classes of equal dislocations, i.e. into different slip systems (Fig. 1).

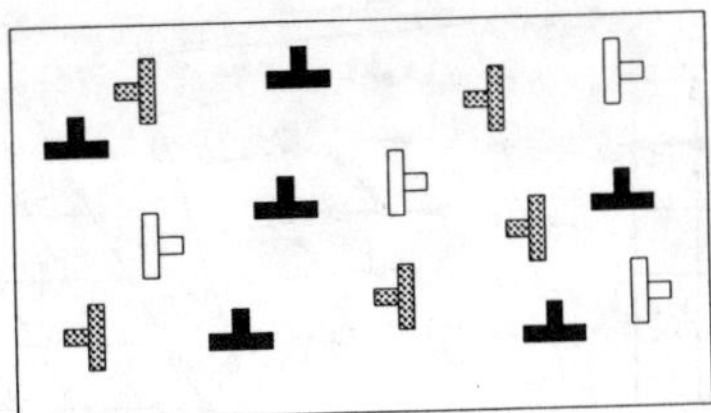

Fig. 1. Different classes of dislocations.

Each slip system is characterized by its:

{ line direction $\vec{l}$, Burgers vector $\vec{b}$, orientation $\vec{m}$ of the slip plane. }

The vectors $\vec{l}$ and $\vec{m}$ are assumed to be unit vectors with respect to the external Euclidean metric. The dynamics of the dislocations of type $\{\vec{l},\vec{b},\vec{m}\}$ is described by a *complex dislocation field* $\psi_{\{\vec{l},\vec{b},\vec{m}\}}(x,t)$ [10, 11]

$$\psi_{\{\vec{l},\vec{b},\vec{m}\}} = \sqrt{n_{\{\vec{l},\vec{b},\vec{m}\}}}\, e^{i\varphi_{\{\vec{l},\vec{b},\vec{m}\}}} \tag{17}$$

and its associated *vortex field*

$$\omega_{\{\vec{l},\vec{b},\vec{m}\}} = \sqrt{\lambda_{\{\vec{l},\vec{b},\vec{m}\}}}\, e^{i\beta_{\{\vec{l},\vec{b},\vec{m}\}}}, \tag{18}$$

giving rise to the definition of the *partial dislocation density* within the class $\{\vec{l},\vec{b},\vec{m}\}$

$$n_{\{\vec{l},\vec{b},\vec{m}\}}(x,t) = \psi^*_{\{\vec{l},\vec{b},\vec{m}\}}(x,t)\psi_{\{\vec{l},\vec{b},\vec{m}\}}(x,t) \geq 0, \tag{19}$$

and of the *dislocation flow velocity field*

$$\vec{v}_{\{\vec{l},\vec{b},\vec{m}\}} = \nabla\varphi_{\{\vec{l},\vec{b},\vec{m}\}} + \lambda_{\{\vec{l},\vec{b},\vec{m}\}}\nabla\beta_{\{\vec{l},\vec{b},\vec{m}\}}. \tag{20}$$

The (traditional) total dislocation density is a superposition of all partial dislocation densities:

$$\underline{\alpha}(x,t) = \sum_{\{\vec{l},\vec{b},\vec{m}\}} (\vec{l}\otimes\vec{b})\, n_{\{\vec{l},\vec{b},\vec{m}\}}(x,t). \tag{21}$$

The whole dynamics of dislocations is now primarily based on the dislocation fields $\psi_{\{\vec{l},\vec{b},\vec{m}\}}$ which allow for properly describing the fine–structure of the dislocation network.

2. Generalized Cosserat Fluid and Plasticity

Traditionally the plastically deformed crystal is regarded as a solid with particular flow properties. In our approach we regard the situation in a reverse order: The plastically deformed body is conceived as a "fluid" with solid properties (Fig. 2) [11].

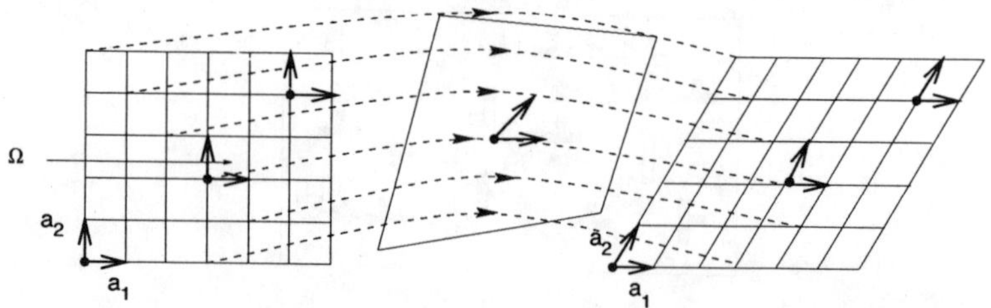

Fig. 2. Generalized Cosserat fluid for the deformation of a solid.

The aim of this procedure is to overcome the physically very dubious concept of a material manifold identified with the deformation structure of the crystal [10, 11]. The flow, i.e. the deformation of the crystal is described by means of a matter field $\Psi(x,t)$ whereas the solid properties are represented by three deformable Cosserat director fields $\{\vec{a}_\kappa(x,t), \kappa = 1,2,3\}$ which are associated with the lattice vectors of the crystal. We have to deal with a *fluid with deformable microstructure*. The deformation of Cosserat triads gives rise to local elasticity whereas their anholonomity defines the dislocation density, i.e. the torsion involved in the lattice geometry. The components of $\vec{a}_\kappa$ with respect to an external coordinate system x^k are $A^k_\kappa, k, \kappa = 1,2,3$. The matrix A^k_κ and its inverse A^κ_k satisfy the relations

$$A^k_\kappa A^\kappa_l = \delta^k_l, \qquad A^i_\kappa A^\mu_i = \delta^\mu_\kappa. \tag{22}$$

The kinematical couplings between the carrier's flow velocity $\vec{v}$, the Cosserat triads $\vec{a}_\kappa = (A^k_\kappa)$, the dislocation fields $\psi_{\{\vec{l},\vec{b},\vec{m}\}}$ and the total dislocation flux density $\vec{\mathcal{J}}$ read as follows [11]:

$$\sum_{\{\vec{l},\vec{b},\vec{m}\}} \psi^*_{\{\vec{l},\vec{b},\vec{m}\}} l^i b^j \psi_{\{\vec{l},\vec{b},\vec{m}\}} - \epsilon^{\mu\kappa\lambda} A^i_\mu A^m_\kappa \partial_m A^j_\lambda \;=\; 0\,, \tag{23}$$

$$\mathcal{J}^{kij} + \epsilon^{kim} A^\mu_m \{\partial_t A^j_\mu + v^l \partial_l A^j_\mu - A^l_\mu \partial_l v^j\} \;=\; 0\,. \tag{24}$$

The first term in (23) is the total dislocation density tensor α^{ij}. The total dislocation flux $\mathcal{J}^{kij}$ will be specified subsequently. There is another kinematical coupling

$$R\Omega(A^k_\kappa) - m_0 = 0 \tag{25}$$

which relates the mass density R of the material carrier with the volume $\Omega(A^k_\kappa)$ of the carrier: The mass density of the carrier is finally defined by the mass m_0 contained in the elementary cell of the crystal.

3. Construction Strategy for the Lagrangian of Plasticity

• Ideal Hydroelastic Fluid: The Lagrangian of an ideal hydroelastic fluid is known as

$$\ell_0 = -R(\partial_t\Phi + \Lambda\partial_t M) - \frac{1}{2}R(\nabla\Phi + \Lambda\nabla M)^2 - W(R) \tag{26}$$

with the mass density R and the Clebsch potentials Φ, Λ, M. It results in the mass balance, Bernoulli's equation and Helmholtz's law [6, 11]. $W(R)$ is the hydroelastic potential of the fluid. Following Noether's procedure, Eq. (26) results also in the traditional balance and constitutive equations for energy and momentum [11]. As an essential point in our context the flow velocity is completely defined from the mass balance without reference to a displacement field.

• Local Elasticity of the ideal Cosserat Fluid: As mentioned above the microstructure of a crystal is associated with the deformation of the Cosserat director fields $\vec{a}_\kappa$ which are embedded into the fluid material carrier. Thus there are kinematical coupling between the triads $\vec{a}_\kappa$ and the matter field Ψ. The kinematical constraints will be taken into account by means of Lagrange multipliers $L^{(n)}$. Hence the Lagrangian of our system takes the form [11]:

$$\begin{aligned} \ell_0 = & - R(\partial_t\Phi + \Lambda\partial_t M) - \frac{1}{2}R(\nabla\Phi + \Lambda\nabla M)^2 - W(A^k_\kappa) \\ & - L^{(1)}_{ij}\epsilon^{\mu\kappa\lambda}A^i_\mu A^m_\kappa \partial_m A^j_\lambda \\ & - L^{(2)}_{kij}\epsilon^{kim}A^\mu_m\{\partial_t A^j_\mu + v^m\partial_m A^k_\mu - A^m_\mu\partial_m v^j\} \\ & - L^{(3)}\{R\Omega(A^k_\kappa) - m_0\}, \end{aligned} \tag{27}$$

where $W(A^k_\kappa)$ is the density of elastic energy. The constraints associated with $L^{(1)}$ and $L^{(2)}$ refer to the vanishing densities and flux densities of dislocations. The last kinematical coupling represents the fact that the mass density is (finally) defined by the mass m_0 contained in the elementary cell of the crystal. The elastic density $W(A^k_\kappa)$ depends on the elastic strain tensor via the relations:

$$W(A^k_\kappa) = W(e_{\kappa\lambda}), \qquad e_{\kappa\lambda} = \frac{1}{2}(g_{ij}A^i_\kappa A^j_\lambda - \delta_{\kappa\lambda}). \tag{28}$$

• Elasto–plastic Interaction: Joining together the ideas developed above we are able to construct a physical model for a continuum theory of plastic deformation by means of LF. The material carrier described by the matter and vortex fields Ψ and ν is associated with the *shape deformation* of the crystal. The Cosserat director fields $\vec{a}_\kappa$ which represent locally the crystal structure are dragged along with the flowing material carrier. The dislocation fields $\psi_{\{\vec{l},\vec{b},\vec{m}\}}$ and their vortex fields $\omega_{\{\vec{l},\vec{b},\vec{m}\}}$ cause incompatibilities in the Cosserat fields resulting in plastic deformation of the crystal lattice (lattice deformation). Finally we take account of the Peach–Koehler force (PKF) as the driving force on dislocations giving rise to dislocation migration through the crystal. We look upon this force as an analogue of the thermodynamical forces which in thermodynamics of irreversible processes give rise to fluxes. The continuum version of the PKF is the third rank tensor of the specific force

$$\Gamma^k_{\cdot ij}(A) = \epsilon^{kmn}g_{mi}\sigma_{nj}(A) \tag{29}$$

based on the well-known elastic stress tensor

$$\sigma_{mn} = \frac{\partial W(A^i_\kappa)}{\partial A^m_\mu}A^l_\mu g_{ln}, \tag{30}$$

g_{ij} is the Euclidean metric tensor. The free indices k, i, j in (29) refer to the external coordinate system x^k. They are in turn physically associated with the direction of the driving force, with the line direction of the dislocations and with the Burgers vector. The density of the PKF acting on dislocation of type $\{\vec{l}, \vec{b}, \vec{m}\}$ is finally given by:

$$\Gamma^k_{\{\vec{l},\vec{b},\vec{m}\}} = \psi^*_{\{\vec{l},\vec{b},\vec{m}\}} \psi_{\{\vec{l},\vec{b},\vec{m}\}} \Gamma^k{}_{.ij} l^i b^j. \tag{31}$$

As we see, this force is by definition perpendicular to the line direction of dislocation. Following Onsager's concept of thermodynamical forces [12] the "diffusional" flux density associated with the dislocations of type $\{\vec{l}, \vec{b}, \vec{m}\}$ is assumed to be

$$j^k_{\{\vec{l},\vec{b},\vec{m}\}} = \zeta \Gamma^k_{\{\vec{l},\vec{b},\vec{m}\}} \tag{32}$$

with a material parameter ζ. By diffusional dislocation flux density we mean the flux with respect to the crystal lattice. So, the total flux density of dislocations of type $\{\vec{l}, \vec{b}, \vec{m}\}$ is given by

$$\begin{aligned} J^k_{\{\vec{l},\vec{b},\vec{m}\}} &= n_{\{\vec{l},\vec{b},\vec{m}\}} v^k + j^k_{\{\vec{l},\vec{b},\vec{m}\}} \\ &= n_{\{\vec{l},\vec{b},\vec{m}\}} \left[(\partial^k \Phi + \Lambda \partial^k M) + \zeta \Gamma^k{}_{.ij} l^i b^j \right]. \end{aligned} \tag{33}$$

We should keep in mind that the flux (33) refers to a particular picture of the dislocations. The total dislocation flux density $\mathcal{J}^{kij}$ is a superposition of all fluxes (33). However, we finally refer to a balance of the anholonomity $\underline{\alpha}$ of the Cosserat triads [11]. So we have to switch over from (33) to "weighted" flux densities $J^k_{\{\vec{l},\vec{b},\vec{m}\}} l^i b^j$. By means of (31), (32) and (33) the total dislocation flux density of all dislocation classes can be shown to have the form:

$$\begin{aligned} \mathcal{J}^{kij} &= \sum_{\{\vec{l},\vec{b},\vec{m}\}} J^k_{\{\vec{l},\vec{b},\vec{m}\}} l^i b^j \\ &= v^k \alpha^{ij} + \zeta \sum_{\{\vec{l},\vec{b},\vec{m}\}} \Gamma^k_{\{\vec{l},\vec{b},\vec{m}\}} l^i b^j. \end{aligned} \tag{34}$$

The first term $v^k \alpha^{ij}$ represents the convectional part of $\mathcal{J}^{kij}$ according to the flow of the material carrier whereas the second term of (34) represents the diffusion of dislocations (plastic flow) due to the action of Peach–Koehler force on dislocations.

Now we are prepared to establish our model of plastic deformation. Taking Eqs. (23), (24) and (34) into account the Lagrangian of plasticity has to be extended to the form:

$$\begin{aligned} \ell = &- R(\partial_t \Phi + \Lambda \partial_t M) - \frac{1}{2} R (\partial_i \Phi + \Lambda \partial_i M)^2 \\ &- W(A^i_\kappa) - \frac{1}{2} \theta^{\kappa\lambda} \delta_{ij} \dot{A}^i_\kappa \dot{A}^j_\lambda \\ &+ \sum_{\{\vec{l},\vec{b},\vec{m}\}} n_{\{\vec{l},\vec{b},\vec{m}\}} \left[-(\partial_t \varphi_{\{\vec{l},\vec{b},\vec{m}\}} + \lambda_{\{\vec{l},\vec{b},\vec{m}\}} \partial_t \beta_{\{\vec{l},\vec{b},\vec{m}\}}) - \frac{1}{2} (\partial_i \varphi_{\{\vec{l},\vec{b},\vec{m}\}} + \lambda_{\{\vec{l},\vec{b},\vec{m}\}} \partial_i \beta_{\{\vec{l},\vec{b},\vec{m}\}})^2 \right] \end{aligned}$$

$$
\begin{aligned}
&- \quad L^{(1)}_{ij}\Big\{\sum_{\vec{l},\vec{b},\vec{m}\}} n_{\{\vec{l},\vec{b},\vec{m}\}} l^i b^j - \epsilon^{\mu\kappa\lambda} A^i_\mu A^m_\kappa \partial_m A^j_\lambda\Big\} \\
&- \quad L^{(2)}_{kij}\Big\{\sum_{\{\vec{l},\vec{b},\vec{m}\}} \Big[n_{\{\vec{l},\vec{b},\vec{m}\}} l^i b^j g^{kl}(\partial_l \Phi + \Lambda \partial_l M) + \zeta \Gamma^k_{\{\vec{l},\vec{b},\vec{m}\}}(A) l^i b^j\Big] \\
&\quad + \epsilon^{kim} A^\mu_m (\partial_t A^j_\mu + v^l \partial_l A^j_\mu - A^l_\mu \partial_l v^j)\Big\} \\
&- \quad L^{(3)}\big\{R\Omega(A) - m_0\big\}. \qquad (35)
\end{aligned}
$$

The Lagrangian (35) is composed of the Lagrangian (27) of the material carrier (first row and first term of the second row), of the Lagrangian of a director continuum (second row) [1], and of the Lagrangian of the various classes of dislocations (third row). The remaining terms in (35) are due to the kinematical couplings (23)–(25); the coefficients $L^{(n)}$ are the respective Lagrange multipliers. By performing the variation one should keep in mind that the Burgers vector $\vec{b}$ and the line direction $\vec{l}$ are particular vectors of the Cosserat triad i.e. they are defined in the crystal lattice. The quantity $\theta^{\kappa\lambda}$ is the tensor of microinertia. The complete discussion of the variation of (35) is outside of this paper. Nevertheless from the ansatz (35) we see that plastic deformation is involved. Further steps of the approach are concerned with the dissipation involved by dislocation dynamics. By properly coupling of the dislocation fields $\psi_{\{\vec{l},\vec{b},\vec{m}\}}$ and the thermion field $\chi(x,t)$ in the Lagrangian the irreversible energy transfer between the kinematical and thermal degrees of freedom can be taken into account. We shall report these problems at another occasion.

References

1. K.–H. Anthony, In *Continuum Models of Discrete Systems (CMDS3)*, eds. E. Kröner and K.–H. Anthony, University of Waterloo Press, 1980.
2. K.–H. Anthony, In *Disequilibrium and Self–Organization*, ed. C. W. Kilmister, Reidel Publ. Comp., 1986, p. 75.
3. K.–H. Anthony, *Acta Physica Hungarica* **67** (1990) 321.
4. K.–H. Anthony, In *Trends in Applications of Mathematics to Mechanics*, ed. J. F. Besseling and W. Eckhaus, Springer 1988, p. 297.
5. K.–H. Anthony, *Arch. Mech. (Warsaw)* **41** (1989) 511.
6. R. L. Seliger and G. B. Whitham, *Proc. R. Soc. London* **A 305** (1968) 1
7. M. Scholle, *On Clebsch Potentials*, to be published.
8. E. Kröner, *Arch. Rat. Mech. Anal.* **4** (1959) 273.
9. E. Kröner, *Kontinuumstheorie der Versetzungen und Eigenspannungen*, Springer-Verlag, Berlin, 1958.

10. K.–H. Anthony, In *Continuum Models of Discrete Systems (CMDS 7)*, eds. K.–H. Anthony and H. J. Wagner, Materials Science Forum, **123-125**, 1993, p. 567.

11. K.–H. Anthony, A. Azirhi, *Dislocation Dynamics by Means of Lagrange Formalism of Irreversible Processes—Complex Fields and Deformation Processes* to be published in *Int. J. Engng Sci.*

12. S.R. De Groot and P. Mazur, *Non–Equilibrium Thermodynamics*, North Holland Publ. Comp., Amsterdam, 1969.

13. A. Azirhi, *Thermions: Thermal excitation fields*, to be published.

14. A. Azirhi, *Thermodynamik und Quantenfeldtheorie. Ein quantenfeldtheoretisches Modell der Wärmeleitung.* Diploma thesis (1993), University of Paderborn.

15. M. Scholle, *Hydrodynamik im Lagrange Formalismus: Untersuchungen zur Wärmeleitung in idealen Flüssigkeiten*, Diploma thesis, 1994, University of Paderborn.

Continuum Models and Discrete Systems
Proceedings of 8th International Symposium, June 11-16, 1995, Varna, Bulgaria, ed. K.Z. Markov
©World Scientific Publishing Company, 1996, pp. 499–506

ON THE NONSENSE OF INCOMPATIBILITY CONDITION IN CONTINUUM THEORY OF DISLOCATIONS

P. DŁUŻEWSKI
Institute of Fundamental Technological Research,
Polish Academy of Sciences, Świętokrzyska 21, PL-00 049 Warsaw, Poland

Abstract. In the paper it is shown that the dislocation density tensor is nothing else as only a measure of the plastic curvature of an oriented continuum. The mathematical consequence of the compatibility conditions for the continuum is discussed in relation to the elastic and plastic curvature tensors.

1. Introduction

In many papers devoted to the continuum theory of dislocations the dislocation density tensor is treated as a certain measure of incompatibilities in elastic continuum. Usually in such papers the problem of *plastic* deformations as well as the compatibility conditions between the elastic and plastic distortions are not taken into account, cf., e.g., [1,9,12].

In the next section of the present paper a multiplicative decomposition of the deformation gradient is discussed. Sections 3 and 4 are devoted to the consequences of the compatibility conditions for the elastic plastic deformation of oriented continuum. In Section 5 the relations between the dislocation density tensor and the measures of curvature of the continuum are discussed.

2. An Oriented Continuum

By an oriented continuum we mean a classical continuum on which the field of particle orientation is determined. In general, we mean the orientation field independent of the displacement field. The orientation field can be determined in the form of a mapping for angular coordinates, cf. [4,5], or in the form of the rotation tensor field, cf. [8]. The polar continua, continuum theory of dislocations and other theories, in which the elastic and/or plastic spins are predicted by means of constitutive relations are typical examples of such defined oriented continuum.

Similarly as for the elastic plastic deformation of the ordinary continuum, in the case of the oriented continuum we can assume a multiplicative decomposition of the deformation gradient **F**. In our case, it is convenient to distinguish the rotation

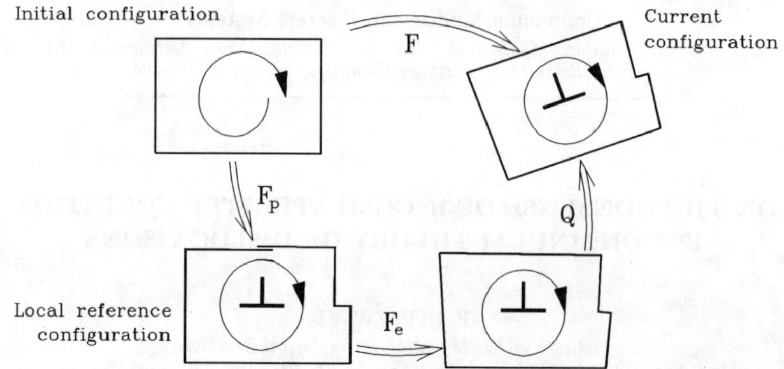

Fig. 1. Considered configurations—schematic drawing

tensor $\mathbf{Q}$ and the elastic and plastic deformation tensors $\mathbf{F}_e$ and $\mathbf{F}_p$,

$$\mathbf{F} = \mathbf{Q}\mathbf{F}_e\mathbf{F}_p\,. \tag{1}$$

This decomposition determines the respective intermediate configurations, and moreover, it assumes "ad hoc" an existence of the respective tensor fields, e.g., $\mathbf{Q}(\mathbf{x},t)$, $\mathbf{F}_e(\mathbf{x},t)$, and $\mathbf{F}_p(\mathbf{x},t)$, which leads to important consequences in the form of additional conditions that must be imposed on the gradients of the fields.

From the viewpoint of the notation used in the description of micropolar continuum the above decomposition should be rewritten with the use of the following notation

$$\mathbf{F} = \boldsymbol{\chi}\boldsymbol{C}_e\boldsymbol{C}_p, \tag{2}$$

where the micromotion tensor $\boldsymbol{\chi}$ and the elastic and plastic Cosserat (generally nonsymmetric) tensors are identified respectively with

$$\boldsymbol{\chi} \equiv \mathbf{Q}, \quad \boldsymbol{C}_e \equiv \mathbf{F}_e, \quad \boldsymbol{C}_p \equiv \mathbf{F}_p. \tag{3}$$

The continuum theory of dislocations is usually considered in the framework of symmetric elasticity, i.e. for $\boldsymbol{C}_e^T = \boldsymbol{C}_e$. However, some attempts were made to treat the dislocation theory from the viewpoint of a generally nonsymmetric continuum, i.e., not only from the viewpoint of nonsymmetric plastic deformation, but also from the viewpoint of nonsymmetric elastic deformations, cf. [10,16]. In this paper the problem of the constitutive modelling is not considered and our attention is only focused on general kinematic relations suitable for various constitutive modelling.

3. Compatibility Condition for Displacements

Let us consider the compatibility condition for the deformation gradient,

$$F^k{}_{K,L} = F^k{}_{L,K}\,. \tag{4}$$

Using the assumed multiplicative decomposition (1) the condition can be expressed with use of the respective covariant derivative of the deformation and rotation tensors. Substituting (1) in (4) gives

$$\begin{aligned} & Q^k{}_{M,L} F_{(e)}{}^M{}_N F_{(p)}{}^N{}_K + Q^k{}_O F_{(e)}{}^O{}_{P,L} F_{(p)}{}^P{}_K + Q^k{}_R F_{(e)}{}^R{}_S F_{(p)}{}^S{}_{K,L} \\ = \; & Q^k{}_{A,K} F_{(e)}{}^A{}_B F_{(p)}{}^B{}_L + Q^k{}_C F_{(e)}{}^C{}_{D,K} F_{(p)}{}^D{}_L + Q^k{}_E F_{(e)}{}^E{}_F F_{(p)}{}^F{}_{L,K} \, . \end{aligned} \tag{5}$$

Multiplying the above equation by $\overset{-1}{F}{}^K{}_m$, $\overset{-1}{F}{}^L{}_n$ and e^{mnl} we find

$$Q^k{}_{M,m} Q_n{}^M e^{mnl} = -Q^k{}_L F_{(e)}{}^L{}_{N,i} \overset{-1}{F}_{(e)}{}^N{}_O Q_j{}^O e^{ijl} - Q^k{}_P F_{(e)}{}^P{}_R F_{(p)}{}^R{}_{K,p} \overset{-1}{F}{}^K{}_r e^{prl}. \tag{6}$$

The left and right sides of the equation define the total, elastic and plastic curvature tensors, respectively,[1]

$$\boldsymbol{\alpha} \overset{df}{=} \operatorname{grad} \mathbf{Q} \overset{\cdot}{\times} \mathbf{Q}^T, \tag{7}$$

$$\boldsymbol{\alpha}_e \overset{df}{=} -\mathbf{Q} \operatorname{grad} \mathbf{F}_e \overset{\cdot}{\times} (\mathbf{Q}\mathbf{F}_e)^{-1}, \tag{8}$$

$$\boldsymbol{\alpha}_p \overset{df}{=} -\mathbf{Q}\mathbf{F}_e \operatorname{grad} \mathbf{F}_p \overset{\cdot}{\times} (\mathbf{Q}\mathbf{F}_e\mathbf{F}_p)^{-1}, \tag{9}$$

where $\overset{\cdot}{\times}$ denotes a double product: the scalar one over the first indices and the vector one over the second indices, cf. (6). So, using the introduced measures the compatibility condition (4) can be rewritten in the following form

$$\boxed{\boldsymbol{\alpha} = \boldsymbol{\alpha}_e + \boldsymbol{\alpha}_p} \tag{10}$$

The present approach does not concern directly the dislocation theory, but does concern a general, nonlinear theory of deformation of oriented continua, nevertheless, from a formal point of view we can use the Burgers relations and by analogy to the Burgers vector we can introduce following vectors for a given material surface

$$\mathbf{b} \overset{df}{=} \int_s \boldsymbol{\alpha} ds, \tag{11}$$

$$\mathbf{b}_e \overset{df}{=} \int_s \boldsymbol{\alpha}_e ds, \tag{12}$$

$$\mathbf{b}_p \overset{df}{=} \int_s \boldsymbol{\alpha}_p ds. \tag{13}$$

[1]It is easy to show that in the case of the linear theory the proposed measures correspond to

$$\left\{ \begin{aligned} \boldsymbol{\alpha} &\overset{df}{=} -\operatorname{curl} \boldsymbol{\omega} \\ \boldsymbol{\alpha}_e &\overset{df}{=} \operatorname{curl} \boldsymbol{\varepsilon}_e \\ \boldsymbol{\alpha}_p &\overset{df}{=} \operatorname{curl} \boldsymbol{\varepsilon}_p \end{aligned} \right. \quad \text{where} \quad \left\{ \begin{aligned} \boldsymbol{\omega} &\approx \mathbf{Q} - \mathbf{1} \\ \boldsymbol{\varepsilon}_e &\approx \mathbf{F}_e - \mathbf{1} \\ \boldsymbol{\varepsilon}_p &\approx \mathbf{F}_p - \mathbf{1} \end{aligned} \right. \quad \text{and} \quad \boldsymbol{\omega} + \boldsymbol{\varepsilon}_e + \boldsymbol{\varepsilon}_p = \nabla \mathbf{u},$$

where $\mathbf{u}$ is the displacement vector, $\boldsymbol{\omega}$ is an antisymmetric tensor of infinitesimal rotations, while $\boldsymbol{\varepsilon}_e$ and $\boldsymbol{\varepsilon}_p$ are the respective tensors of deformations. Note, that our meaning of the elastic deformation tensor is different than that of Kröner [13,15], where the rigid rotation of continuum is treated as a kind of elastic deformation (distortion) $\boldsymbol{\beta}_e \equiv \boldsymbol{\varepsilon}_e + \boldsymbol{\omega}$.

With respect to (10) we also find

$$\boxed{\mathbf{b} = \mathbf{b}_e + \mathbf{b}_p} \tag{14}$$

Obviously, at this moment of our considerations, we can consider only a formal analogy to dislocation theory, nevertheless, anticipating slightly the results of following considerations it is worth emphasizing now that the plastic curvature tensor $\boldsymbol{\alpha}_p$ and the vector $\mathbf{b}_p$ are just the well known dislocation density tensor and the Burgers vector, respectively, cf. (30) and (37).

Note, that the following mathematical relations hold

$$\mathbf{QF}_e \operatorname{curl}(\mathbf{QF}_e)^{-1} = \operatorname{grad}(\mathbf{QF}_e) \dot{\times} (\mathbf{QF}_e)^{-1} \tag{15}$$

$$= \operatorname{grad}\mathbf{Q} \times \mathbf{Q}^T + \mathbf{Q}\operatorname{grad}\mathbf{F}_e \times (\mathbf{QF}_e)^{-1} \tag{16}$$

$$= \boldsymbol{\alpha} - \boldsymbol{\alpha}_e. \tag{17}$$

Regarding the compatibility condition (10) we find

$$\mathbf{QF}_e \operatorname{curl}(\mathbf{QF}_e)^{-1} = \boldsymbol{\alpha}_p. \tag{18}$$

This means that the plastic curvature tensor can be defined by one of four relations

$$\boldsymbol{\alpha}_p = -\mathbf{QF}_e \operatorname{grad}\mathbf{F}_p \dot{\times} (\mathbf{QF}_e\mathbf{F}_p)^{-1} \tag{19}$$

$$= \mathbf{QF}_e\mathbf{F}_p \operatorname{grad}\mathbf{F}_p^{-1} \dot{\times} (\mathbf{QF}_e)^{-1} \tag{20}$$

$$= \operatorname{grad}(\mathbf{QF}_e) \dot{\times} (\mathbf{QF}_e)^{-1} \tag{21}$$

$$= \mathbf{QF}_e \operatorname{curl}(\mathbf{QF}_e)^{-1}. \tag{22}$$

In the analogical way we can find four relations for the elastic curvature tensor $\boldsymbol{\alpha}_e$. These relations are widely discussed in [6].

It is worth recalling that there exists a second group of the curvature measures: $\boldsymbol{\kappa}$, $\boldsymbol{\kappa}_e$, $\boldsymbol{\kappa}_p$. Due to Nye relationships [17] the kappa tensors are connected with the respective alpha tensors by linear, reversible relations:

$$\boldsymbol{\alpha}_{\dots} = -\boldsymbol{\kappa}_{\dots}^T + \mathbf{1}\operatorname{tr}\boldsymbol{\kappa}_{\dots}^T, \tag{23}$$

$$\boldsymbol{\kappa}_{\dots} = -\boldsymbol{\alpha}_{\dots}^T + \frac{\mathbf{1}}{2}\operatorname{tr}\boldsymbol{\alpha}_{\dots}^T, \tag{24}$$

where the subscript ... can be replaced here by p, e or by a space.

4. Compatibility Condition for Rotations

Compatibility for rotations corresponds to the following condition

$$\mathbf{Q}_{,KL} = \mathbf{Q}_{,LK}, \tag{25}$$

where the comma denotes the covariant derivative. With respect to the compatibility condition for displacements the compatibility for rotations can be considered in terms of second derivatives of displacement deformations, cf. Kafadar and Eringen [8]. Therefore, to hold the compatibility condition for rotation we can impose some condition on the second derivatives of deformation tensors. Namely, assuming that the plastic and elastic deformations will be predicted by means of mutually independent constitutive relations, the condition (25) can be replaced by the following two conditions

$$\mathbf{F}_{\mathrm{p},KL} = \mathbf{F}_{\mathrm{p},LK}, \tag{26}$$

$$(\mathbf{F}_{\mathrm{e}}\mathbf{F}_{\mathrm{p}})_{,KL} = (\mathbf{F}_{\mathrm{e}}\mathbf{F}_{\mathrm{p}})_{,LK}. \tag{27}$$

It is easy to show that the above conditions lead to the following relationships

$$\operatorname{div}[(\mathbf{Q}\mathbf{F}_{e}^{-1}\boldsymbol{\alpha}_{p}] = \mathbf{0}, \tag{28}$$

$$\operatorname{div}[\mathbf{Q}^{-1}(\boldsymbol{\alpha}_{e} + \boldsymbol{\alpha}_{p})] = \mathbf{0}. \tag{29}$$

These conditions can be also considered in terms of the respective measures for the intermediate configurations, this problem is widely discussed in [6].

5. Dislocation Density versus Plastic Curvature Tensor

Let us forget for a moment the measures of the elastic and plastic curvatures of the oriented continuum and focus our attention on the crystal structure and dislocation theory. According to the generally known definition the so-called true Burgers vector $\widehat{\mathbf{b}}_{d}$ is determined as an integral over a Burgers circuit c around a dislocation, cf., e.g., [9],

$$\widehat{\mathbf{b}}_{d} = \oint_{c_b} (\mathbf{Q}\mathbf{F}_{e})^{-1} d\mathbf{r}, \tag{30}$$

where according to the Kröner notation the rotation tensor $\mathbf{Q}$ and the elastic deformation tensor $\mathbf{F}_e$ are called (unfortunately) the elastic distortion tensor and denoted commonly by $\mathbf{A}$. Using the Stokes theorem we find

$$\widehat{\mathbf{b}}_{d} = \int_{s_b} \operatorname{curl}(\mathbf{Q}\mathbf{F}_{e})^{-1} d\mathbf{s}. \tag{31}$$

In the case of the continuum theory of dislocations it is assumed that in the differential form of the above equation,

$$d\widehat{\mathbf{b}}_{d} = \operatorname{curl}(\mathbf{Q}\mathbf{F}_{e})^{-1} d\mathbf{s}, \tag{32}$$

the differentials of the respective Burgers vectors in the current and reference configurations are related by

$$d\mathbf{b}_{d} = \mathbf{Q}\mathbf{F}_{e} d\widehat{\mathbf{b}}_{d}. \tag{33}$$

Substituting (33) into (32) gives the well known dependency

$$d\mathbf{b}_d = \mathbf{QF}_e \mathrm{curl}(\mathbf{QF}_e)^{-1} d\mathbf{s}. \tag{34}$$

It is worth emphasizing that in the classical, linear continuum theory of dislocations the expression placed on the left hand of $d\mathbf{s}$ is identified with the dislocation density tensor, i.e. it is assumed that

$$d\mathbf{b}_d = \boldsymbol{\alpha}_d d\mathbf{s}\,, \tag{35}$$

where

$$\boldsymbol{\alpha}_d \stackrel{df}{=} \mathbf{QF}_e \mathrm{curl}(\mathbf{QF}_e)^{-1}\,. \tag{36}$$

Obviously, at the first look, from the last equation it does not yield that $\boldsymbol{\alpha}_d$ is just the plastic curvature tensor $\boldsymbol{\alpha}_p$ defined by (9). Nevertheless, substituting the previously derived relation (22) into (36) gives finally

$$\boldsymbol{\alpha}_d \equiv \boldsymbol{\alpha}_p\,, \tag{37}$$

what leads to

$$\mathbf{b}_d \equiv \mathbf{b}_p\,. \tag{38}$$

6. Final Remarks

The considerations on the physical meaning of the dislocation density tensor are usually terminated on the derivation of (36). Obviously, using (36) as a definition it is difficult to recognize that the dislocation density tensor is a plastic curvature tensor, especially in the case, when neither the plastic deformations nor the compatibility conditions are considered. At the first look, (36) may suggest that the discussed tensor relates to the elastic curvature, however, it is easy to show, e.g. considering the elastic bending of a crystal, that this measure does not represent the elastic curvature, so, it has been generally recognized that the dislocation density tensor is a measure of an elastic incompatibility of displacements in material. For confirmation of this let us cite the remarks presented at the end of the IUTAM Symposium in Stuttgart in 1967. The Symposium was devoted in memoriam of Eugène i François Cosserat and Élie Cartan. Among participants we find A.C. Eringen, A.E. Green, W. Noll, R. Rivlin, R.A. Toupin, C.-C. Wang, P.M. Naghdi, E. Reissner and K. Kondo, B.A. Bilby, R. de Vit, K.-H. Anthony, H. Zorski. Let us recall a fragment of the closing remarks by Kröner summarizing the results of the Symposium, see [14]:

> ... *The fact that the quantities τ^{mkl} respond to a curvature and have the dimension of a moment stress has lead* Günter *[4] to his conception of a dislocated continuum as a Cosserat continuum. The idea is very tempting. Nevertheless, a complete correspondence has never been established and I am more and more convinced that the difference between* Cosserat *and dislocation theory is fundamental.*

> *The* Cosserat *continuum, as I would define it, is built up from particles which possess an inherent orientation. (...)*
>
> *In contrast to this picture, dislocations occur in crystals in which atoms need not possess an inherent orientation. Orientation enters only when the arrangement of the neighboring atoms is regarded. Hence, although one can speak of an orientation at a point both in the normal crystal and in the Cosserat continuum, the physical situation is basically different. In fact moving dislocations are not spin waves. Instead, dislocations possess the fundamental ability to produce slip.*
>
> *As a consequence of these obvious differences the geometry (and kinematics) used in the two cases should be different. ...*

So, in spite of the Nye suggestion [17], it has been stated generally that the continuum theory of dislocations and mechanics of oriented continua are to be two mutually different theories based on quite different foundations. It has to be emphasized that in many monographs devoted to the continuum theory of dislocations the problem of the compatibility condition is not stated clearly. Usually the role of the elastic deformation tensor understood as $\mathbf{A} = \mathbf{Q}\mathbf{F}_e$ is widely discussed avoiding simultaneously what role takes the plastic deformations. Note, *how important role takes the plastic deformation tensor and compatibility conditions in the continuum theory of dislocations*. Without these conditions we would not be able to show at all that the dislocation density tensor is nothing more as only a measure of the plastic curvature of continuum.

Acknowledgements. This contribution has been supported by the Bulgarian Academy of Sciences and the Polish Academy of Sciences in the framework of the Polish-Bulgarian research cooperation. These supports are gratefully acknowledged.

References

1. B. A. Bilby, In *Progress in Solid Mechanics*, eds. I. N. Sneddon and R. Hill, **1**, North-Holland, Amsterdam, 1960, p. 331.
2. J. M. Burgers, *Proc. Kon. Nederl. Akad. Wetensch.* **42** (1939) 293, 378.
3. P. H. Dłużewski, *J. Mech. Phys. Solids* **39** (1991) 651.
4. P. H. Dłużewski, *Int. J. Solids Structures* **30** (1993) 2277.
5. P. H. Dłużewski, Geometry and continuum thermodynamics of movement of structural defects, submitted to *Mechanics of Materials*, 1995.
6. P. H. Dłużewski, Habilitation thesis (in preparation, 1995).
7. P. H. Dłużewski and H. Antunez, Finite element simulations of dislocation field movement, *CAMES* **2** (1995) (in press).
8. A. C. Eringen and C. B. Kafadar, in *Continuum Physics*, ed. A. C. Eringen, vol. IV, Academic Press, New York, 1976, p. 1.

9. B. K. D. Gairola, In *Dislocations in Solids*, ed. F. R. N. Nabarro, vol. 1, North-Holland, Amsterdam, 1979, p. 223.
10. H. Günther, Zur nichtlinearen Kontinuumstheorie bewegter Versetzungen, Dissertation, Academic-Verlag, Berlin, 1967.
11. K. Kondo, In *Proc. 2nd Japan Nat. Congr. Appl. Mech.*, vol. 2, 1969, p. 41.
12. A. M. Kosevich, In *Dislocations in Solids*, ed. F.R.N. Nabarro, vol. 1, North-Holland, Amsterdam, 1979, p. 33.
13. E. Kröner, *Kontinuumstheorie der Versetzungen und Eigenspannungen*, Springer-Verlag, Berlin, 1958.
14. E. Kröner, in *Mechanics of Generalized Continua*, ed. E. Kröner, Berlin, 1968, p. 330.
15. E. Kröner, In *Physics of Defects* , eds. R. Balian, M. Kleman and J.-P. Poiries, North-Holland, Amsterdam, 1981, p. 215.
16. S. Minagawa, *Arch. Mech.* **31** (1979) 783.
17. J. F. Nye, *Acta Metall.* **1** (1953) 153.

Continuum Models and Discrete Systems
Proceedings of 8th International Symposium, June 11-16, 1995, Varna, Bulgaria, ed. K.Z. Markov
© World Scientific Publishing Company, 1996, pp. 507–513

LORENTZ SYMMETRIES IN THE THEORY OF DISLOCATIONS

H. GÜNTHER
FH Bielefeld FB Elektrotechnik, Wilhelm-Bertelsmann-Str.10, D-33602 Bielefeld

Abstract. Lorentz symmetries in solids become evident if the soliton properties of kinks on dislocations for the measurement of length and time are used. This procedure can be extended for finding out an internal space-time of the solid, which exactly is the theory of special relativity with signal velocity c_o of sine - Gordon equation. On the basis of this internal space-time the continuum theory of eigenstresses created by dislocations will be discussed. An arbitrary anisotropy of the material is included. It is the aim of this communication to point out that according to these relations the continuum theory of dislocations and eigenstresses is able to make a theoretical contribution to the theory of plasticity.

1. The Special Theory of Relativity on a crystalline Structure

We consider the crystalline solid as a *special* inertial system Σ_o. The defects of this solid supply us with natural standards for space and time measurements. To see this we consider a straight dislocation along the x-axis. Let q describe a (normalized) displacement of the dislocation line perpendicular to the x-axis, then according to Seeger [1] (cf. also Seeger and Schiller [2]) the sine-Gordon equation is valid,

$$\frac{\partial^2 q}{\partial x^2} - \frac{1}{c_o^2}\frac{\partial^2 q}{\partial t^2} = \frac{1}{\lambda_o^2}\sin q \ , \tag{1}$$

where the length L_o and the velocity c_o result from the physical constants of the lattice. With the help of this equation we find the possible static and dynamic shapes of dislocation lines, which do exist in the solid. These lines provide us with natural standards for space and time measurements in this solid.

With the help of the so–called static kink solution $q_o^I(x)$, cf. Seeger [3],

$$q_o^I = 4\arctan\exp(x/\lambda_o) \underset{\text{def}}{=} 4\arctan\exp(\pi x/L_o) \ , \quad L_o \underset{\text{def}}{=} \pi\lambda_o \ , \tag{2}$$

we define a standard of length L_o, see fig. 1 .

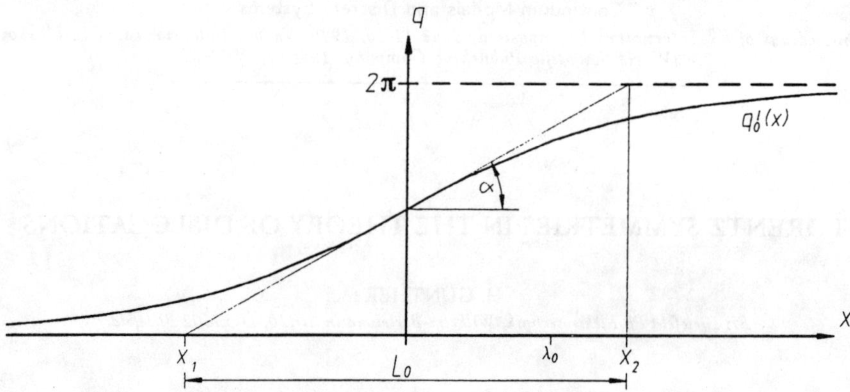

Fig. 1. The definition of a natural standard L_o with the help of kink solution (2).

For a constant velocity v we get the moving kink solution $q^I(x, t)$, see fig. 2,

$$q^I(x, t) = 4 \arctan \exp[\frac{\pi(x - vt)}{L'}] , \quad L' = \gamma L_o , \tag{3}$$

where $\gamma = \sqrt{1 - v^2/c_o^2}$. Hence the moving standard L' shows Lorentz contraction,

$$L' = \sqrt{1 - v^2/c_o^2}\, L_o . \qquad \text{Lorentz contraction} \tag{4}$$

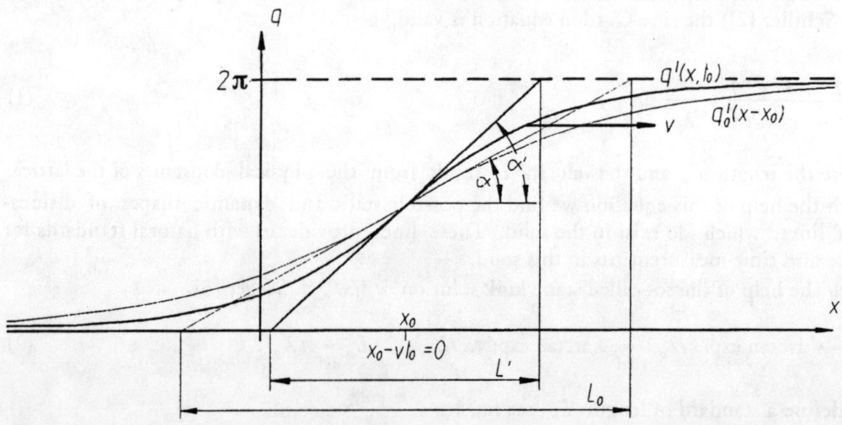

Fig. 2. The moving standard L' is related to L_o by Lorentz contraction, $L' = \gamma L_o$.

With the help of the so–called localized breather solution $q_o^{III}(x, t)$, cf. Seeger [3],

$$q_o^{III}(x, t) = 4\arctan\frac{\sin(c_o t/\lambda_o\sqrt{2})}{\cosh(x/\lambda_o\sqrt{2})} \underset{\text{def}}{=} 4\arctan\frac{\sin(2\pi t/T_o)}{\cosh(\pi x/L_o\sqrt{2})}, \quad T_o \underset{\text{def}}{=} 2\sqrt{2}\, L_o/c_o, \tag{5}$$

we define a period T_o. A standard clock which is counting these periods is measuring time t, see fig. 3 and fig. 4 .

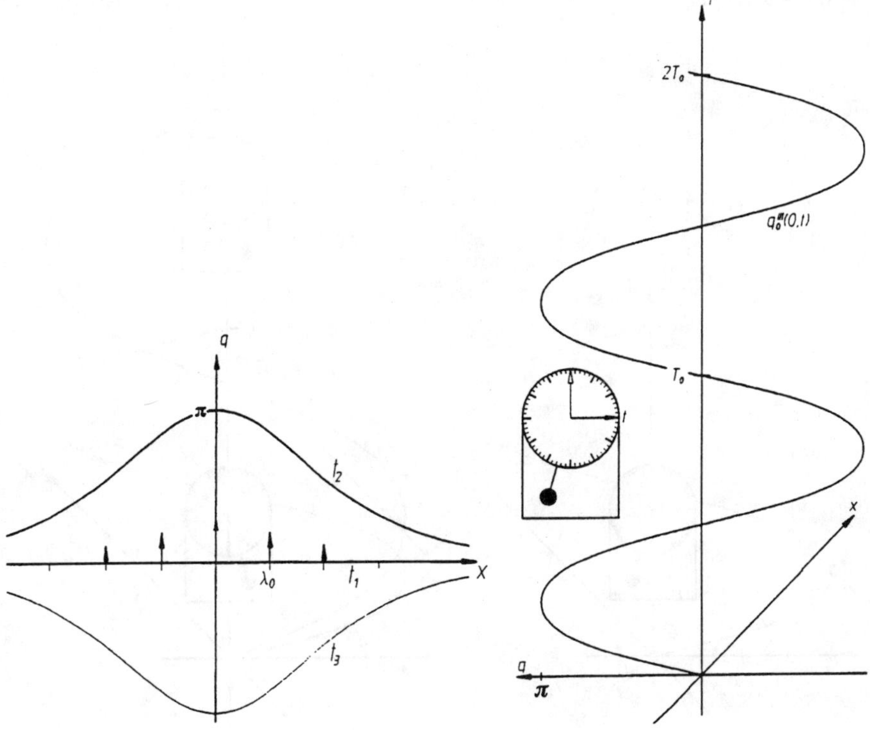

Fig. 3. The breather solution (5) for $t_1 = 0$, $t_2 = \sqrt{2}\, L_o/2c_o$, and $t_3 = 3\sqrt{2}\, L_o/2c_o$ (the arrows indicate the direction of motion for the dislocation line on the x- axis).

Fig. 4. The oscillations of the breather solution (5), here at the fixed position $x = 0$, define a standard clock that is measuring time t.

With a constant velocity v we get the moving breather solution $q^{III}(x, t)$, cf. Seeger [3],

$$q^{III}(x, t) = 4 \arctan \frac{\sin(2\pi\, t/T')}{\cosh \dfrac{\pi(x - vt)}{L'\sqrt{2}}} \;, \quad T' = T_o/\gamma \,. \tag{6}$$

Then a moving clock, measuring time by counting the extended periods T' of the moving oscillation, is measuring a time t' which is related to t by time dilatation, see fig. 5,

$$t' = \sqrt{1 - v^2/c_o^2}\; t \,. \qquad \text{Time dilatation (7)}$$

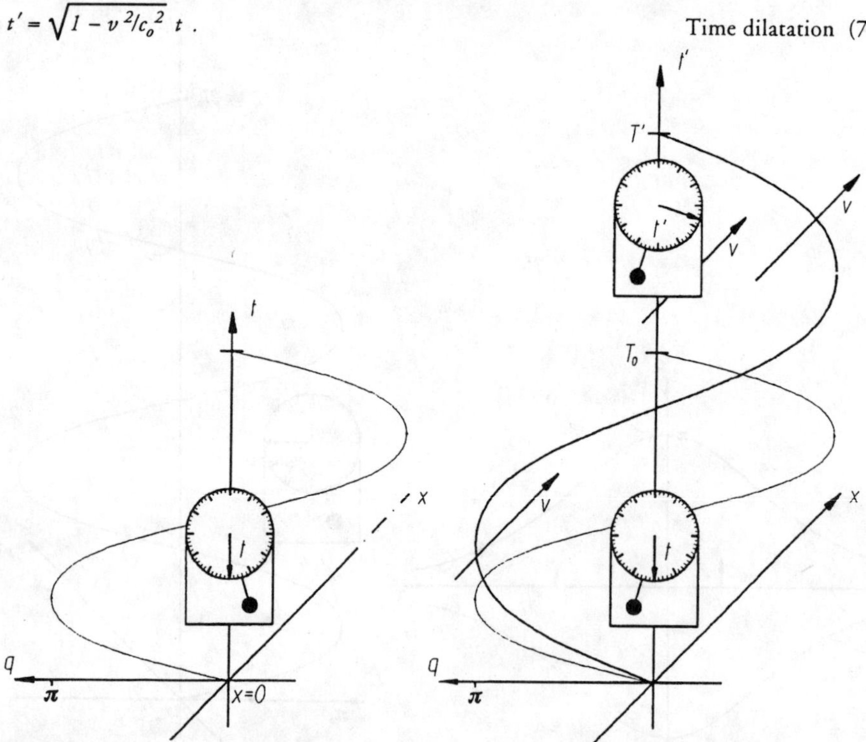

Fig. 5. The oscillations of the breather (6) are measuring a time t', which is related to t by (7).

Remember that up to this point all relations refer to the special inertial system Σ_o, which is defined by the lattice at rest. Consequently, according to common opinion, all these relations have nothing to do with special relativity. With moving standards we now consider moving systems Σ'. We refer to it as the "moving observer". Hence the moving observer of Σ' is moving relative to the lattice with the constant velocity v. However we do not demand Einsteins principle of special relativity for these systems Σ'. We only demand a rather simple and *elementary principle of relativity*:

If an observer in Σ_0 measures the velocity v for the system Σ', then the moving observer in Σ' has to synchronize his clocks in such a way that he measures the velocity $-v$ for Σ_0.
As we have shown in details, cf. Günther [4], this is a consistent procedure and at the end Einstein principle of special relativity with signal velocity c_o comes out for all the systems Σ_o and Σ'. That means: As far as the observers in the crystalline solid use the internal standards for space and time measuring, they are unable to find out any difference between the systems Σ_o and Σ'. The signal velocity c_o of sine-Gordon equation becomes an "universal constant". With the help of the internal standards the lattice no longer can be detected.

2. Hidden Lorentz symmetries

Consider an elastic continuum with symmetric stress tensor σ, elastic deformation ε, material velacity v, mass density ρ, Hooke's tensor C and volume force density f. Fundamentals of linear elasticity then read (we use $f,_r \underset{\text{def}}{=} \partial_r f = \frac{\partial}{\partial x} f$, $\partial_t f \underset{\text{def}}{=} \frac{\partial}{\partial t} f$)

$$\sigma_{ri},_r + f_i = \rho\, \partial_t v_i \ , \quad \sigma_{ik} = C_{ikrs}\, \varepsilon_{rs} \ . \tag{8}$$

Let α be the tensor of dislocation density, J the tensor of dislocation current and V the vector of dislocation velocity (which must not be confused with material velocity v). The following identities are valid (using Levi-Civita-symbol e_{ikl}),

$$\partial_r \alpha_{ri} = 0 \ , \quad J_{ik} = e_{irs} V_r \alpha_{sk} \ , \quad \partial_t \alpha_{ik} - e_{irs} J_{sk},_r = 0 \ . \tag{9}$$

Following our procedure in [5] these dislocations produce incompatible elastic strain $\varepsilon^{(i)}$ and material velocity $v^{(i)}$ according to

$$\varepsilon_{ik}^{(i)},_{rr} + \varepsilon_{rr}^{(i)},_{ik} - \varepsilon_{ir}^{(i)},_{kr} - \varepsilon_{kr}^{(i)},_{ir} = -\frac{1}{2}(e_{mni}\alpha_{nk},_m + e_{mnk}\alpha_{ni},_m + e_{mni}\alpha_{mn},_k + e_{mnk}\alpha_{mn},_i) \tag{10}$$

$$\partial_t \varepsilon_{ik}^{(i)} - \frac{1}{2}(v_i^{(i)},_k + v_k^{(i)},_i) = \frac{1}{2}(J_{ik} + J_{ki}) \ . \tag{11}$$

Left hand sides of (10), (11) are invariant with respect to the transformations

$$\varepsilon_{ik}^{(i)} \rightarrow \varepsilon_{ik} = \varepsilon_{ik}^{(i)} + \frac{1}{2}(a_i,_k + a_k,_i) \ , \quad v_i^{(i)} \rightarrow v_i = v_i^{(i)} + \partial_t a_i \ , \tag{12}$$

for an arbitrary vector field $a = a(x)$. We identify the quantity a with an elastic displacement vektor, which generates a compatible elastic strain $\varepsilon^{(c)}$ and compatible velocity $v^{(c)}$ according to

$$\varepsilon_{ik}^{(c)} = \frac{1}{2}(a_i,_k + a_k,_i) \ , \quad v_i^{(c)} = \partial_t a_i \ , \tag{13}$$

so that complete elastic strain ε as well as complete elastic velocity $\boldsymbol{v}$, for wich the fundamentals (8) are valid, will be decomposed according to (12) into elastic and inelastic components,

$$\varepsilon_{ik} = \varepsilon_{ik}^{(i)} + \varepsilon_{ik}^{(c)}, \quad v_i = v_i^{(i)} + v_i^{(c)} . \tag{14}$$

Notice that the decomposition (14) is not unique. In [5] we have explained that there is a group of transformations (12), so that the equations (8) - (13) decompose into seperate equations for incompatible elastic strain $\boldsymbol{\varepsilon}^{(i)}, \boldsymbol{v}^{(i)}$ on the one hand and equations for the elastic displacement fied $\boldsymbol{a}$ on the other hand, which no longer are coupled togehter.
Let p be the arbitrary decomposition parameter. The equations (8) - (13) are equivalent to the following two equation for incompatible strain and velocity on the one hand,

$$\varepsilon_{ik}^{(i)},_{rr} - \frac{\rho}{p}\partial_{tt}\varepsilon_{ik}^{(i)} = -\frac{1}{2}(e_{mni}\alpha_{nk,m} + e_{mnk}\alpha_{ni,m} + e_{mni}\alpha_{mv,k} + e_{mnk}\alpha_{mv,i}) + \frac{\rho}{p}\partial_t(J_{ik} + J_{ki}) , \tag{15}$$

$$v_i^{(c)},_{rr} - \frac{\rho}{p}\partial_{tt}v_i^{(c)} = -J_{ik,k} - J_{ki,k} - J_{kk,i} , \tag{16}$$

together with an other equation for the elastic displacement vector on the other hand,

$$C_{rimn}\frac{1}{2}\partial_r(a_{m,n} + a_{n,m}) - \rho\partial_{tt}a_i = -f_i - p(-2\partial_r\tilde{S}_{irab} + \tilde{S}_{rrab})C_{abmn}\varepsilon_{mn}^{(i)} . \tag{17}$$

Here the tensor $\tilde{S}_{ikab}$ is defined with the help of inverse Hooke's tensor S_{ikab} for elastic compliances (notice $\varepsilon_{ik} \underset{def}{=} S_{ikab}\sigma_{ab}$) according to

$$\tilde{S}_{ikab} = S_{ikab} - \frac{1}{4p}(\delta_{ia}\delta_{kb} + \delta_{ib}\delta_{ka} - 2\delta_{ik}\delta_{ab}) . \tag{18}$$

As is well known there are at least two different velocities for elastic wave propagation in any solid. Hence there is no Lorentz group in the domain of elastic phenomena. Equations (15) - (17) explain, what really happens. The only observable quantity is the complete elastic deformation $\boldsymbol{\varepsilon}$ (as well as velocity $\boldsymbol{v}$). Nevertheless we can define an incompatible elastic deformation $\boldsymbol{\varepsilon}^{(i)}$ (as well as velocity $\boldsymbol{v}^{(i)}$), which is generated by dislocations and dislocation currents due to the Lorentz - invariant wave equations (15) and (16). However the displacement vector $\boldsymbol{a}$ results from the equations of classical elasticity (17), hence breaking Lorentz symmetry for the observable complete elastic strain $(\boldsymbol{\varepsilon}, \boldsymbol{v})$. The numerical value of the decomposition parameter p according to (18) is not fixed. We find:

For any value of the parameter p there exists an elastic displacement field $\boldsymbol{a}$ which connects the Lorentz invariant incompatible elastic strain $(\boldsymbol{\varepsilon}^{(i)}, \boldsymbol{v}^{(i)})$ with the observable elastic strain $(\boldsymbol{\varepsilon}, \boldsymbol{v})$ according to (12).

This enables us to couple incompatible elastic strain with micropolstic deformations. To this end we choose a special value for the decomposition parameter p according to

$$p = \rho\, c_o^2 \ , \tag{19}$$

where c_o is the signal velocity of Lorentz invariant sine - Gordon - equation (1). This gives

$$\varepsilon_{ik}^{(i)}{}_{,rr} - \frac{1}{c_o^2}\partial_{tt}\varepsilon_{ik}^{(i)} = -\frac{1}{2}(e_{mni}\alpha_{nk,m} + e_{mnk}\alpha_{ni,m} + e_{mni}\alpha_{mn,k} + e_{mnk}\alpha_{mn,i}) + \\ + \frac{1}{c_o^2}\partial_t(J_{ik} + J_{ki}) \ , \tag{20}$$

$$v_i^{(c)}{}_{,rr} - \frac{1}{c_o^2}\partial_{tt}v_i^{(c)} = -J_{ik,k} - J_{ki,k} - J_{kk,i} \ . \tag{21}$$

The equations (1), (20) and (21) now obey one and the same Lorentz symmetry.

3. Outlook

The equations (1), (20) and (21) determine an interaction between microplastic deformations with incompatible elastic strain on the basis of relativistic field theory. Consider a given distribution of solutions for the sine - Gordon - equation (1), say kinks, kink pairs, breathers, etc. These quantities define a dislocation density $\boldsymbol{\alpha}^m$ and dislocation current J^m of microplastic deformations. Introduce these quantities for $\boldsymbol{\alpha}$ and $\boldsymbol{J}$ into the right hand sides of (20) and (21),then an incompatible elastic strain ($\boldsymbol{\varepsilon}^{(i)}$, $\boldsymbol{v}^{(i)}$) will be generated. The question arises if relativistic quantum field theoretical methods will work here in order to explain kink pair creation and annihilation in the same way as electron-positron pair creation and annihilation is a consequence of quantum electrodynamics. As we have seen in the first section, the crystal lattice provides us with a complete model for special relativity. In the same way it should also illustrate us, how quantum relativistic phenomena work.

REFERENCES

[1] A. Seeger, Diplomarbeit, Universität Stuttgart, 1948/49.

[2] A. Seeger and P. Schiller, in: *Physical Acoustics*, Vol.III A. Ed. W.P. Mason, Academic Press, New York/London, 1966, p. 361.

[3] A. Seeger, *Continuum Models of Discrete Systems*, University of Waterloo Press, 1980, 253.

[4] H. Günther, *phys. stat. sol.(b)* 1994, **185**, 335.

[5] H. Günther, *CMDS7, Materials science Forum*, vols. 123-125, p. 591.

Continuum Models and Discrete Systems
Proceedings of 8th International Symposium, June 11-16, 1995, Varna, Bulgaria, ed. K.Z. Markov
©World Scientific Publishing Company, 1996, pp. 514–521

STOCHASTICITY IN DISLOCATION DYNAMICS: AN EFFECTIVE MEDIUM APPROACH TO PLASTIC DEFORMATION PHENOMENA

P. HÄHNER
Institute for Advanced Materials, Joint Research Centre, European Commission, I-21020 Ispra (Va), Italy

Abstract. Current theories of spatio–temporal dislocation pattern formation during plastic flow suffer from an inadequate treatment of long-range dislocation interactions. In the present paper, an effective medium theory of plastic flow is developed that accounts for the geometrically necessary fluctuations of the local stress and plastic strain rate due to dynamical dislocation interactions. Stochastic differential equations (i.e., Langevin-type equations) are formulated for the temporal evolution of dislocation densities. From the corresponding Fokker–Planck equations, probability distributions are derived which are used to study the signature of dislocation patterning in probability space (noise-induced phase transitions). This gives qualitatively new insights into the physics of dislocation patterning (explicable as evolutionary processes) which cannot be obtained from deterministic reasoning.

1. Introduction

It has become evident that dislocation dynamics concepts are not only indispensable for understanding the basic microscopic mechanisms of crystal plasticity but also provide an important complement to continuum mechanics in describing the fundamental aspects of macroscopic materials performance and failure. In particular, one should think of the central role of dislocation dynamics in the ductile vs. brittle fracture process or in plastic strain localization and instability. The reason why a clearcut distinction between microscopic dislocation dynamics and macroscopic continuum mechanics approaches is not always useful, is related to the fact that the defect dynamics does not decouple on different scales. As illustrated in Fig. 1, a material driven far from equilibrium by plastic straining is an active medium that—by means of the coupled dynamics of various kinds of defects—behaves in a strongly nonlinear manner. This gives rise to self-organization phenomena that react upon the complex system of defects and thus influence the macroscopic materials performance.

As a common feature, complex systems consist of an enormous number of basic components (i.e., the dislocations). As it is impossible to cope analytically with their microscopic degrees of freedom, one must employ statistical methods. Previous

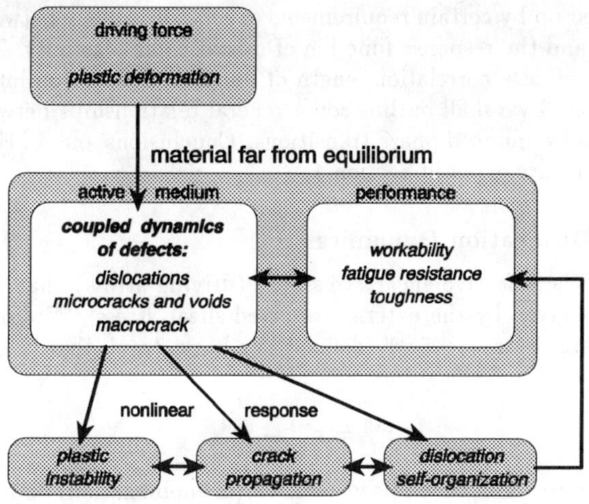

Fig. 1. A material undergoing plastic deformation constitutes a complex system far from equilibrium whose performance depends in a strongly nonlinear way on the coupled dynamics of defects.

approaches suffer from the drawback that dynamical dislocation interactions are accounted for in a very restrictive way only allowing for particular interaction events. (For instance, the formation of dislocation dipoles by mutual trapping of mobile dislocations can be treated by means of second order terms in the corresponding rate equations.) This shortcoming is problematic if dislocation patterning occurs on the same mesoscopic scale that characterizes the range of dislocation interactions. It is, therefore, safe to assume that interactions cannot be neglected for understanding how dislocation structures auto-organize.

The time scale of dislocation interactions (i.e., the average interval of time that a mobile dislocation segment interacts with the internal stress field of a certain dislocation configuration) is usually much shorter than the time scale of the macroscopic evolution of the dislocation structure (i.e., the characteristic time of strain hardening). Taking advantage of this fact, we shall consider the microscopic aspects of dislocation interactions as a background noise in terms of rapid stochastic fluctuations of the internal stress superimposed on a slowly varying external stress.

Strictly speaking, a low-dimensional stochastic dislocation dynamics should be derived from averaging the irrelevant degrees of freedom of the high-dimensional deterministic equations of motion of the dislocation ensemble. As such a rigorous projection technique is not available, we shall adopt a more phenomenological procedure. The stochastic differential equations formulated below are based upon heuristic arguments backed up by certain requirements of self-consistency between the fluctuation amplitudes and the response function of macroscopic plasticity. This leads us to a natural definition of a correlation length of dynamic dislocation interactions (Section 2). In Section 3 we shall outline some general relationships between dislocation patterning and noise-induced phase transitions. Conclusions and a brief discussion of recent applications are given in Section 4.

2. Stochastic Dislocation Dynamics

Consider the effective shear stress τ^{eff} (driving stress) that acts on a glide dislocation. It is given by the external resolved shear stress τ^{ext} diminished by the long-range internal shear stress τ^{int} (defined as a back stress) that is exerted by other crystal defects, notably dislocations:

$$\tau^{\text{eff}} = \tau^{\text{ext}} - \tau^{\text{int}} . \tag{1}$$

During the time passing between its generation (or mobilization) and its annihilation (or storage), the glide dislocation is interacting with various dislocation configurations. Consequently, its motion takes place in a spatio-temporally fluctuating effective stress field:

$$\delta\tau^{\text{eff}} = -\delta\tau^{\text{int}} . \tag{2}$$

To quantify the fluctuation amplitudes, we consider a representative mobile dislocation gliding at a velocity v (that also fluctuates with time) and calculate

$$\langle v\tau^{\text{int}}\rangle = \frac{1}{\Delta t}\int_0^{\Delta t} \mathrm{d}t\, v\, \tau^{\text{int}} = \frac{1}{\Delta x}\int_0^{\Delta x} \mathrm{d}x\, \tau^{\text{int}} = 0 . \tag{3}$$

Here, $\langle\cdot\rangle$ denotes a *temporal* average effected over a sufficiently long interval Δt of time t during the glide motion which can be either viscous or discontinuous. Note that with $\mathrm{d}x = v\,\mathrm{d}t$ and $\Delta x = \langle v\rangle\Delta t$, the temporal average is transformed into a spatial average that vanishes provided that Δt is long compared to the characteristic time that the glide dislocation spends interacting with a certain dislocation configuration (i.e. the waiting time in discontinuous glide). This is a consequence of quasi-static stress equilibrium (Albenga's rule). It is important to note, however, that the temporal average of the internal stresses does not vanish, $\langle\tau^{\text{int}}\rangle > 0$, since most of the time dislocations are held up at configurations of elevated back stress. This is schematically depicted in Fig. 2. As Eq. (3) applies to any glide dislocation, a corresponding relation holds locally for the plastic shear strain rate $\dot{\gamma}$:

$$\langle\dot{\gamma}\tau^{\text{int}}\rangle = 0 \quad \text{or} \quad \langle\dot{\gamma}\tau^{\text{eff}}\rangle = \langle\dot{\gamma}\rangle\tau^{\text{ext}} . \tag{4}$$

It is important to note that the mesoscopic continuum description in terms of $\dot{\gamma}$ requires the average dislocation spacing to be small compared to the correlation length of fluctuations (cf. Eq. (12)). Moreover, it should be stressed that Eq. (4), stating that the contribution of internal stresses to the mechanical power density is completely

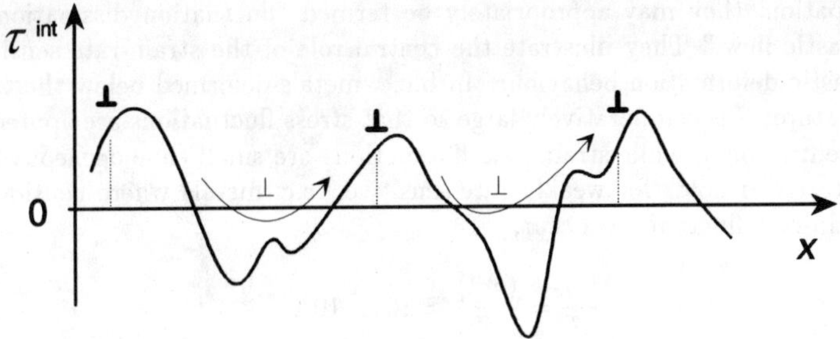

Fig. 2. Illustration of the internal stresses τ^{int} experienced by a representative glide dislocation. While the spatial average of τ^{int} vanishes, a dislocation spends most of the time waiting in positions of positive back stress: $\langle\tau^{\mathrm{int}}\rangle > 0$.

dissipative, is approximate. The reason is that dislocation reactions may intervene that impede the glide dislocations to scan a sufficient multitude of internal stress configurations (for a refined treatment accounting for the influences of dislocation multiplication, storage and cross slip, see [1]). In view of the fact, however, that the dissipation during plastic flow usually amounts to more than 90%, Eq. (4) represents a good approximation.

From Eqs. (1), (2) and (4) cross-correlations of the fluctuations of $\dot{\gamma}$ and τ^{eff} derive:

$$\langle\delta\dot{\gamma}\,\delta\tau^{\mathrm{eff}}\rangle = \langle\dot{\gamma}\tau^{\mathrm{eff}}\rangle - \langle\dot{\gamma}\rangle\langle\tau^{\mathrm{eff}}\rangle = \langle\dot{\gamma}\rangle\langle\tau^{\mathrm{int}}\rangle\,. \tag{5}$$

Here we used the fact that τ^{ext} does not fluctuate. In the focus of interest are the auto-correlations of $\dot{\gamma}$ and τ^{eff} which are readily obtained by using the definition of the strain-rate sensitivity

$$S = \frac{\partial\langle\tau^{\mathrm{eff}}\rangle}{\partial\ln\langle\dot{\gamma}\rangle}\,. \tag{6}$$

Here S represents the appropriate response function to convert strain-rate fluctuations into effective stress fluctuations or vice versa. Neglecting higher than second order correlations[1], this gives

$$\langle\delta\dot{\gamma}^2\rangle = \frac{\langle\tau^{\mathrm{int}}\rangle}{S}\langle\dot{\gamma}\rangle^2 \tag{7}$$

and

$$\langle(\delta\tau^{\mathrm{eff}})^2\rangle = S\langle\tau^{\mathrm{int}}\rangle\,. \tag{8}$$

As Eqs. (7) and (8) relate the local fluctuations of the plastic shear strain rate and of the effective shear stress to the corresponding response function and the mechanical

[1]This is consistent with assuming a Gaussian white noise, cf. Eq. (19).

power dissipation, they may appropriately be termed "fluctuation–dissipation theorems of plastic flow." They illustrate the central role of the strain-rate sensitivity S in the plastic deformation behaviour. In b.c.c. metals deformed below the transition temperature, S is comparatively large so that stress fluctuations are appreciable (stress concentrations), while strain-rate fluctuations are small (homogeneous flow). The opposite trend holds for weakly rate-sensitive f.c.c. metals where considerable relative strain-rate fluctuations occur,

$$\frac{\langle \delta\dot{\gamma}^2 \rangle}{\langle \dot{\gamma} \rangle^2} = \frac{\langle \tau^{\text{int}} \rangle}{S} \approx 10 \ldots 10^3 , \tag{9}$$

which are attributed to collective slip in groups of

$$n = (2\pi)^{1/4} \sqrt{\frac{\langle \tau^{\text{int}} \rangle}{S}} \approx 5 \ldots 50 \tag{10}$$

dislocations (coarse slip, formation of slip bands) [1].

To complete the theoretical framework of a stochastic dislocation dynamics, estimates for the spatial and temporal correlations are needed. The correlation time τ obeys the scaling law

$$\tau = \frac{b \rho_{\text{m}} L}{\langle \dot{\gamma} \rangle} , \tag{11}$$

where b is the modulus of the Burgers vector, ρ_{m} is the density of mobile dislocations and L denotes a characteristic length. If we assume that dislocations move in groups (mesoscopically collective glide) and, therefore, experience correlated deviations from the average effective stress all along their slip paths, we may identify L with the slip line length.

The physical significance of a spatial correlation length ξ is less evident, because dislocations arranged at random within an infinite medium are known to possess an infinite interaction range [2]. This is an immediate consequence of the $1/x$ type interaction law. From the dynamical point of view, however, ξ may be defined straightforwardly if one notes that the random stress fluctuations tend to reduce the dislocation pair correlations. We propose that two dislocations cannot move collectively, unless their interaction stress exceeds the stress fluctuations caused by surrounding dislocations. This is expressed by the scaling law [1]

$$\xi = \frac{\mu b}{4\pi \sqrt{\langle (\delta \tau^{\text{eff}})^2 \rangle}} = \frac{\mu b}{4\pi \sqrt{S \langle \tau^{\text{int}} \rangle}} , \tag{12}$$

where μ denotes the shear modulus.

The keynote of the present stochastic dislocation dynamics consists in considering the motion of dislocations embedded into an *effective fluctuating medium* made up by the residual dislocations. The scaling relations (7), (8), (11) and (12), though partly phenomenological in nature, are supposed appropriate to account for the major aspects of long-range interactions. They represent the starting point for formulating mesoscopic stochastic differential equations which are simple enough for analytical results to be derived. In the remainder of this paper, this procedure is elucidated by a general model of dislocation patterning. It turns out that the stochastic dislocation dynamics concept opens up new types of relationships that cannot be obtained from a deterministic line of reasoning.

3. Dislocation Patterning and Noise-Induced Transitions

In the following, we shall briefly outline the general theoretical framework of noise-induced transitions that was developed during the last two decades [3,4], and discuss some novel implications related to dislocation patterning. To keep the representation clear, let us resume a general evolution equation for some dislocation density[2] ρ,

$$\partial_t \rho = f(\rho)\,\dot{\gamma}\,, \tag{13}$$

where, for simplicity, all reaction rates are assumed to be in proportion to $\dot{\gamma}$ (which is not true for static recovery). By splitting

$$f(\rho) = f_0 + f_1(\rho) \tag{14}$$

into a part f_0 independent of ρ and the remaining $f_1(\rho)$, and writing

$$\dot{\gamma} = \langle\dot{\gamma}\rangle + \delta\dot{\gamma}\,, \tag{15}$$

a Langevin-type evolution equation is obtained,

$$\partial_t \rho = [f_0 + f_1(\rho)]\langle\dot{\gamma}\rangle + f_0\,\delta\dot{\gamma} + f_1(\rho)\,\delta\dot{\gamma}\,, \tag{16}$$

that consists of three distinct contributions. The first term on the right-hand side describes the systematic evolution as in the deterministic case. The second term represents the so-called *additive* noise (a randomly fluctuating term independent of the dynamical variable ρ), while the last term is called *multiplicative* noise, as it is ρ dependent. It is the latter contribution that may lead to qualitative changes of the solution in relation to a noise-induced transition.

The fluctuating terms of the Langevin equation (16) are the price to be payed for neglecting the microscopic details of dislocation interactions in favour of a mesoscopic effective medium description treating the interaction dynamics as an intrinsic noise. Being a stochastic differential equation, Eq. (16) cannot be integrated in a straightforward manner. It is possible, however, to analyse the probability of a certain state (characterized by ρ) and investigate the signature of dislocation patterning in probability space. To this end, Eq. (16) is transformed into a Fokker–Planck

---

[2] In general, ρ will be identified with the density of *immobile* dislocations (e.g. dislocation tangles or dipole dislocations) the average of which evolves much slower than the *mobile* dislocation density ρ_m. Therefore, we assume the latter to be eliminated adiabatically.

equation for the temporal evolution of the transition probability density. Using the Stratonovich calculus this yields [3,4]

$$\partial_t p = -\partial_\rho \left[\left(f\langle\dot{\gamma}\rangle + \frac{1}{4}\,\partial_\rho(\sigma^2 f^2) \right) p \right] + \frac{1}{2}\,\partial_\rho^2(\sigma^2 f^2 p)\,, \tag{17}$$

where $p = p(\rho, t \,|\, \rho_0, t_0)$ denotes the probability to find the system in the state ρ at time t, when it was in ρ_0 at t_0, and

$$\sigma^2 = 2\langle\delta\dot{\gamma}^2\rangle\tau \tag{18}$$

is a measure of the noise intensity which may also depend on ρ. Note that for the derivation of the Fokker–Planck equation (17), strain-rate correlations have been assigned to a Gaussian white noise process [3,4]:

$$\langle\delta\dot{\gamma}(0)\delta\dot{\gamma}(t)\rangle = \sigma^2\,\delta(t)\,. \tag{19}$$

In deterministic approaches of dislocation dynamics, the occurrence of instabilities or structural non-equilibrium phase transitions is studied by means of linear stability analyses. This often imposes conceptual problems, as bifurcations may appear under non-steady-state conditions. Here, however, it is permissible to investigate the steady-state distribution function $p_s(\rho)$, since the time scale of fluctuations (governed by the correlation time τ) is small compared to the time scale of the systematic macroscopic evolution (governed by strain hardening). This is considered an important technical advantage of the present stochastic dislocation dynamics.

Provided that the probability current vanishes at the boundaries of ρ, the steady-state solution of Eq. (17) reads:

$$p = p_s(\rho) = \frac{N}{\sigma f(\rho)} \exp\left(2\langle\dot{\gamma}\rangle \int_0^\rho \frac{\mathrm{d}\rho}{\sigma^2 f(\rho)} \right), \tag{20}$$

where N is a normalization constant. From the discussion of $p_s(\rho)$ two indications of a noise-induced transition can be derived:

1. Beyond some critical level of the multiplicative noise, $p_s(\rho)$ exhibits a non-integrable singularity. The corresponding localization in probability space can be attributed to a structural phase transition (see [5]).

2. The shape of the steady-state distribution $p_s(\rho)$ changes qualitatively at a critical multiplicative noise level. If, for example, $p_s(\rho)$ develops from a single peak to a double peak distribution, a phase separation of dislocation-poor and dislocation-rich zones occurs [6].

It is important to note that a transition cannot be induced by a purely additive noise which just causes a broadening of the probability distribution. The multiplicative noise, however, may stabilize a state that would be unstable in the absence of noise.

Conclusions

Various structure formation phenomena during plastic flow find novel explanations in terms of noise-induced transitions. Without going into the details presented elsewhere, we mention that the theory has proved useful in modelling

1. the occurrence of *dislocation cell structures* [7] during multiple slip. In this case, the stochastic formation of dislocation junctions is assumed to be the crucial dislocation reaction process [6]. The model can explain the so-called law of similitude (i.e., the inverse proportionality of the flow stress and the dislocation cell size) from a dynamical point of view. This follows immediately from the scaling law (12) if one notes that $S \sim \langle \tau^{\text{int}} \rangle \sim \tau^{\text{ext}}$ according to the Cottrell–Stokes law (see [8] for a recent discussion).

2. The formation of the so-called *matrix structure* and *persistent slip bands* (PSB) [7] as well as the transitions between these structures during single slip cyclic deformation can be related to the stochastic formation and the glide-induced decomposition of extended dipoles of primary edge dislocations [6]. Here, the salient feature of the theory is the derivation of the intrinsic plastic strain amplitude $\Delta\gamma_{\text{PSB}}$ which is accommodated by a PSB. While in former *deterministic* theories all dislocation reactions rates scale with $\Delta\gamma_{\text{PSB}}$ so that this quantity drops out in the steady state and, hence, cannot be determined, the *stochastic* dipole decomposition scales with $\sqrt{\Delta\gamma_{\text{PSB}}}$ (cf. a n-step random walk with an average diffusion path $L \sim \sqrt{n}$).

Another application of stochastic dislocation dynamics, namely, the slip localization (channel slip) in predeformed alloys, is treated in detail in the present proceedings [5]. The common aspect of all these models is the idea that dislocation patterning results from an evolutionary process driven by the intrinsic noise due to dislocation interactions.

References

1. P. Hähner, to be published in *Appl. Phys.*
2. M. Wilkens, *Acta Metall.* **17** (1969) 1155.
3. W. Horsthemke and R. Lefever, *Noise-Induced Transitions*, Springer, Berlin, 1984.
4. J. Honerkamp, *Stochastische Dynamische Systeme*, VCH, Weinheim, 1990.
5. M. Zaiser and P. Hähner, this volume.
6. P. Hähner, submitted to *Acta Metall. Mater.*.
7. L. P. Kubin, In *Treatise in Materials Science and Technology*, Vol. 6, ed. H. Mughrabi, VCH, Weinberg, 1993.
8. F.R.N. Nabarro, *Acta Metall. Mater.* **38** (1990) 161.

Continuum Models and Discrete Systems
Proceedings of 8th International Symposium, June 11-16, 1995, Varna, Bulgaria, ed. K.Z. Markov
©World Scientific Publishing Company, 1996, pp. 522–537

DISLOCATION THEORY AS A PHYSICAL FIELD THEORY

E. KRÖNER
University of Stuttgart, Institute for Theoretical and Applied Physics,
D-70550 Stuttgart, Germany

(Dedicated to Professor Gianfranco Capriz on the occasion of his 70*th* birthday)

Abstract. Dislocations are the elementary carriers in many situations of plastic flow. Since they can be seen, counted and typified, e.g. in the electron microscope, and since their presence changes the state of the (elastoplastic) medium, the dislocations have the status of a physical state quantity. Inspite of this a continuum theory of elastoplasticity can be built up which does not use the concept of dislocation. However, disregarding the dislocations implies approximation, and this approximation is not always good, inspite of the smallness of the dislocation diameter (order atomic distance). It is discussed how a field theory of dislocations can be developed which takes full account of the dislocational degrees of freedom.

1. Introduction

In recent time the field of materials with microstructure has more and more gained a central interest of material and engineering scientists as well as of mechanicists and mathematicians (see e.g. the monograph of Kunin "Theory of elastic media with microstructure I and II" [1] and the monograph of Capriz "Continua with microstructure" [2]. The latter contains also a most valuable classification of the various types of microstructure, according to the form of the order parameter, or of similar quantities, which are needed to describe the microstructure considered). In fact, real materials are not, in general, structureless continua but possess ordered or disordered structures on all possible scales from atomic scale (e.g. crystals) up to macroscopic scale (e.g. concrete). These structures have a fundamental influence on the material properties and therefore must be studied and included into general theories of material behaviour. It is tempting to try to develop a very general theory which includes all the various microstructures. However, such a theory, if it could be set up at all, would be extremely complex and of no much value for solving real physical or practical problems. Simplifications of such a theory always lead to theories whose range of validity is rather limited.

An example for such a theory of limited reach (because only one out of many microstructures is involved) is the theory of crystal dislocations, which is closely connected with the theory of elastoplasticity. The basic scale of microstructure is here the atomic (or molecular) scale. There are other important scales in crystals which develop during elastoplastic deformation. Such scales are for instance determined by the mean distance between dislocations, by the diameter of cells or grains and by the minimum distance two dislocations of opposite sign can have without collapsing. These scales are less fundamental insofar, as they imply certain deformations of the material, whereas the crystal lattice structure is inherently connected with the crystal.

In this paper we deal with the, as said limited, field of dislocation theory, and we shall make clear the limitations. We hope that the derived equations can be used for testing also more general theories of which appear more and more in recent time. Of the many serious works claiming either to contain dislocations theory as a special case or having dislocation theory itself as their aim I mention works by Maugin [3,4], Naghdi and Srinivasa [5,6], Epstein and de León [7,8] and Le and Stumpf [9]. I quote *these* works because I have studied them more than others. There are many further authors whose work would likewise deserve quotation.

Dislocation theory, as it is understood today, has the status of a physical field theory. Of course, it also has to fit within the general frame of thermodynamics, in particular of irreversible thermodynamics (internal variables, dissipation, etc.). This side of dislocation theory will not be emphasized here. We shall rather, in Section 2, discuss the nontrivial question of how well at all can a discrete structure like the crystal structure be described with the methods developed for continuous media, and we find that there are indeed problems. We then discuss, in Sections 3 and 4, how elastic and plastic deformations can be composed to give a total deformation and how, based on this idea, a simple continuum theory of elastoplastic deformation which does not need dislocations can be developed. Of course, this theory, often applied to practical problems, misses an important feature of plasticity (the dislocations). It therefore must be considered as an approximation whose accuracy is often difficult to estimate.

In Section 5 we extend this theory by introducing dislocations as dynamical variables which enter the stored elastic energy expressions as new independent variables. This gives the extended theory more motional degrees of freedom. As a consequence there appear new response quantities (stresses) in the stored energy expression. These stresses enter the equilibrium conditions for forces and moments, which are derived from the principle of virtual work in Section 6. These conditions plus the geometric constraints are not enough to solve actual problems. Rather constitutive equations are needed which connect the geometrical and statical quantities. A number of problems in statics are sketched (Section 7) which now can be solved.

To solve quasistatical and dynamical problems one must be willing to include thermodynamics because a large part of the applied external work goes into heat (the other into densification of dislocations). This is so because even very slow plastic deformations occur via dislocations whose motion is always dissipative. Incidentally, this is one of the reasons why there are difficulties when developing gauge theories of dislocations by means of the Yang-Mills principle of minimal coupling.

The dynamical theory is not yet in a satisfactory state. We say something on this point and also on the gauge theory of dislocations, which finds increasing interest, in recent time in Section 8. Section 9 concludes this article.

2. Continuum Model of Discrete Systems

In this Section we discuss how well the theory of elastoplasticity can be described in the frame of continuum mechanics. We shall in particular consider such plastic deformations which originate from the motion of dislocations. Additional mechanisms can occur under certain circumstances and then lead to more complex theories. In continuum mechanics (see Truesdell and Noll [10] and Kunin [11]) a body $\mathcal{B}$ is identified with a threedimensional differentiable manifold, and motions are families of time-dependent diffeomorphisms χ of $\mathcal{B}$ into the Euclidean space E_3. The content of this statement is close to the so-called axiom of continuity formulated by Truesdell and Noll. It is the term *differentiable* which leads over to the term *continuity*.

As a consequence of these statements one has

$$\boldsymbol{x} = \chi(\boldsymbol{X}, t), \tag{1}$$

where $\boldsymbol{x}$ denotes the position in E_3 of the particle labelled by $\boldsymbol{X}$ in some reference configuration which we may choose as stress-free. The existence of the relation (1) with differentiable χ can almost be considered as a definition of continuum mechanics in the ordinary sense.

The relation (1) implies that the body is built up from particles specified by $\boldsymbol{X}$. It is tacitly or expressis verbis assumed that these particles persist during the whole deformation motion. However this assumption needs closer inspection in the case of dislocated bodies.

Certainly the persistence is true as long as the material particles are identified with the atoms (or molecules) forming the body. It is our choice to do so. However, to a large extent plastic deformations occur by the motion of groups of dislocations along glide planes. By this motion neighbor atoms are separated by may be up to a hundred atomic distances which means that the motion of atoms, the assumed particles of our body, is extremely discontinuous. In this picture, the body is not a differentiable material manifold as emphasized by Kunin [11] in his pioneering contribution to our subject.

If, in a more continuous picture, the particles of the body are considered as material volume elements or neighborhoods, then one might hope to preserve the continuum picture. However, the groups of dislocations mentioned above, will now pass through the newly defined material particles thereby separating these into smaller elements in such a way that parts of the original particles are no longer connected. Clearly this contradicts the concept of undestructible particles which under the same neighborhood conditions form the body at all times. As a consequence the relation (1) breaks down and there is no continuum theory of elastoplasticity in the strict sense. The question then arises whether at least some of continuum mechanics can be saved for elastoplasticity.

In spite of the wild discontinuities which arise over the whole body during elastoplastic deformation there is some hope indeed, which is founded on the fact that the discontinuities, though large on the atomic scale, are small on the phenomenological scale. In fact, hundred atomic distances are usually very small on the scale of observing macroplasticity. There the discontinuities appear as fluctuations which can be neglected under certain conditions to be investigated further. If we do this then (1) acquires a new meaning as an average relation. Of course it is very tempting to postulate (1) with this interpretation because then we can apply all the powerful methods of continuum mechanics. We should be aware, however, that the accuracy of physical predictions of this average theory might be reduced, depending on the size of the fluctuations. In a sense we have a mean field theory. Here we also refer to Kunin [11] who comes to the conclusion that $\boldsymbol{\chi}$ does not describe the paths of particles. In his arguments he concludes that $\boldsymbol{\chi}$ has a meaning only in the sense of "coarse graining", thus in an average sense as said above: Coarse graining is a well-known concept of statistical mechanics.

Having a relation like (1) now with the average interpretation, the basic kinematic fields velocity $\boldsymbol{V}(\boldsymbol{X},t)$ and deformation gradient $\boldsymbol{F}(\boldsymbol{X},t)$ are defined in terms of components as

$$V^i = \frac{\partial}{\partial t}\chi^i(\boldsymbol{X},t)\,, \quad F^i_k = \frac{\partial \chi^i(\boldsymbol{X},t)}{\partial X^k}\,, \tag{2}$$

in what is called the Lagrangean description.

3. Elastic and Plastic Deformation

Accepting the described average view and, as the consequence, the average interpretation of Eqs. (1) and (2), we feel justified to speak, at least in a model understanding, of a continuum theory of elastoplasticity. Compared to other continuum theories this theory is characterized in that its deformation contains elastic and plastic contributions where only the elastic part gives rise to a specific static response. It is postulated that a decomposition formula exists which presents the deformation gradient $\boldsymbol{F}$, called *total* deformation gradient, as a product of an elastic and plastic deformation gradient ($\boldsymbol{F}^e$ and $\boldsymbol{F}^p$), so that

$$\boldsymbol{F} = \boldsymbol{F}^e\boldsymbol{F}^p \tag{3}$$

(see Bilby *et al.* [12], Kröner [13] and Lee [14]).

Obviously (3) is not the only possibility to decompose the total deformation gradient into an elastic and plastic part. For instance to have powers of $\boldsymbol{F}^p$ and $\boldsymbol{F}^e$ in (3) would also do it. However, (3) is suggested when in a thought experiment first a pure plastic (incompatible) deformation (e.g. through dislocation motion) is performed so that noncoherent volume elements are produced and then the now nonfitting elements are forced together by the inverse deformation which however goes elastically (in our thought experiment). This elastic deformation is, of course, incompatible, too. To produce it, surface forces acting on the elements are needed. After gluing together the now fitting volume elements we remove those forces so that the body as a whole can relax in a lowest energy configuration. This last (elastic) deformation is compatible because it goes from one compact state to another compact state of the body.

Incidentally, the just described thought experiment is basic for explaining incompatibility.

It is well-known that whereas $\boldsymbol{F}$ is an integrable deformation gradient, $\boldsymbol{F}^e$ and $\boldsymbol{F}^p$ are, in general, not integrable. Hence, if the deformation gradients are defined by differential forms as

$$dx^k = F^k_K dX^\kappa, \quad dx^k = F^{ek}_\kappa dx^\kappa, \quad dx^\kappa = F^{p\kappa}_K dX^K, \tag{4}$$

then the last two forms are Pfaffian (anholonomic) forms. In (4) X^K are the material coordinates, x^k the space coordinates and dx^κ are the anholonomic coordinate differentials of a fictive intermediate state which would arise if as discussed above a pure plastic deformation would be applied to the initial state. Obviously

$$F^k_K = F^{ek}_\kappa F^{p\kappa}_K \tag{5}$$

which is another form of (3).

An alternative way is to replace the decomposition (3) by

$$\boldsymbol{F} = \acute{\boldsymbol{F}}^p \acute{\boldsymbol{F}}^e \quad (F^k_K = \acute{F}^{pk}_\kappa \acute{F}^{e\kappa}_K), \tag{6}$$

i.e. to interchange the role of plastic and elastic deformation. The choice (6) is convenient if one wants to work in the Lagrangean picture. In fact, the elastic deformation gradient $\acute{F}^{e\kappa}_K$ relates the reference state K to the intermediate state κ, whereas F^{ek}_κ in (5) connects intermediate and current state. It is convenient to take the reference state as stress-free, namely as the ideal crystal state. In the following we work in the Lagrangean picture, i.e. with Eq. (6). To describe completely the (elastoplastic) deformation *process* we have to give $\boldsymbol{\chi}$, $\acute{\boldsymbol{F}}^e$, and the response quantities, all as functions of $\boldsymbol{X}$ and t. From these quantities then also $\boldsymbol{V}$, $\boldsymbol{F}$ and $\acute{\boldsymbol{F}}^p$ can be derived.

4. Stress Tensor

The question of the response quantities is now fundamental. This question is closely connected with the field energy stored in the body. Taking the temperature constant over the specimen we look for the stored (or elastic) energy per unit reference volume (in the Lagrangean picture). It is suggestive to assume that the energy density W depends continuously on the elastic deformation gradient, so that we have

$$W = W(\acute{\boldsymbol{F}}^e, \boldsymbol{X}) \tag{7}$$

if we also allow for a material inhomogeneity by introducing $\boldsymbol{X}$ in W. To simplify our presentation we shall not go into details of the $\boldsymbol{X}$-dependence, i.e. later omit $\boldsymbol{X}$ as an argument of W (not as an argument of $\acute{\boldsymbol{F}}^e$).

Eq. (7) is equally true in cases where the stress results not from plastic deformation, but from external loads. In accord with the continuity postulate we assume that $\acute{\boldsymbol{F}}^{e}$ is continuous in $\mathcal{B}$. Note that in the case of elastoplasticity $\acute{\boldsymbol{F}}^{e}$ must not be replaced by $\boldsymbol{F}$ in (7) because $\boldsymbol{F}$ is not a state quantity. In fact, in (6) $\acute{\boldsymbol{F}}^{p}$ may be arbitrary, hence also integrable (compatible). Then no elastic deformation is needed to keep the volume elements together so that $\acute{\boldsymbol{F}}^{e} = \boldsymbol{I}$ and a change of $\boldsymbol{F}$ does not change the energy density. This is the case of perfect plasticity.

The response quantity pertaining to the energy density (7) is

$$\overset{\circ}{T}{}_{\kappa}^{K} = \partial W / \partial \acute{F}^{e\kappa}_{K} . \tag{8}$$

It is analogous to the *first Piola-Kirchhoff stress tensor* (a two-point tensor). In cases of *compatible* strain fields where this tensor is defined usually, one uses $\boldsymbol{F}$ instead of $\acute{\boldsymbol{F}}^{e}$, but then $\boldsymbol{F} = \acute{\boldsymbol{F}}^{e}$.

The equilibrium equations follow by using the W of (7) in the principle of virtual work. On this occasion one can introduce further physical realities such as yield condition, incompressibility, etc., in the form of constraints. All this is well-known so that it suffices perhaps to conclude these considerations with the following remark: There does exist, indeed, a continuum theory of elastoplasticity which uses the well-known laws of continuum mechanics. This theory does not need the concept of dislocation, but renouncing it leads to a loss in the predictive power, as we shall see.

5. Dislocations and Lattice Curvature

5.1. Basic Definitions

In the foregoing Sections we have discussed a continuum theory of elastoplasticity which did not need the concept of dislocations. Since elastoplastic bodies are, in general, crystalline and therefore are filled with dislocations the question of how much we loose by disregarding them as explicit quantities of the theory arises. Recalling Frank's definition of the crystal dislocations (circuit in dislocated and undislocated crystal) we recognize the dislocation as characteristic for the *state* of the crystal. In fact, such a dislocation can be observed e.g. in the high-resolution electron microscope. Of course, also *many* dislocations can simultaneously be seen in the electron microscope. If we describe these dislocations as a density, still based on Frank's definition, then this density is a state quantity. We introduce it (in the Lagrangean picture) as a tensor valued (of rank 3) quantity $\alpha_{ML}{}^{K}$ by

$$db^{K} := \alpha_{ML}{}^{K} dS^{ML} , \tag{9}$$

where db^{K} is the Burgers vector resulting from all dislocations intersecting the (usually internal) area element dS^{ML} of the reference state. It is known for quite a long time (see Kondo [15] and Bilby *et al.* [16]) that the tensor $\alpha_{ML}{}^{K}$ can also be interpreted as the (Cartan) *torsion* of a differential-geometric space with linear connexion ${\Gamma_{ML}}^K$ (not a tensor!) whose in M, L antisymmetric part (symbol [ ]) equals ${\alpha_{ML}}^K$:

$$ {\alpha_{ML}}^K := {\Gamma_{[ML]}}^K \quad \text{or} \quad \alpha_{MLK} = \Gamma_{[ML]K}, \quad \Gamma_{MLK} \equiv g_{KH} {\Gamma_{ML}}^H . \tag{10} $$

Here g_{KL} is the (covariant) metric tensor of the crystal space in the intermediate configuration. It is used to raise and lower indices and is defined by

$$ ds^2 = g_{KL} dX^K dX^L ; \tag{11} $$

g_{KL} has the inverse g^{KL}, ds^2 is the squared distance in the intermediate state between any two neighbor points whose relative position in the reference state is given by dX^K. Note that (11) can also be written as

$$ ds^2 = g_{\kappa\lambda} dx^\kappa dx^\lambda, \quad g_{KL} = \acute{F}^{e\kappa}_{K} \acute{F}^{e\lambda}_{L} g_{k\lambda} , \tag{12} $$

where $g_{\kappa\lambda} = \boldsymbol{e}_\kappa \cdot \boldsymbol{e}_\lambda$ and the $\boldsymbol{e}$'s are the base vectors of the intermediate configuration. κ and λ assume the values 1, 2 and 3.

All this, in particular the fact that the dislocation density is a state quantity, is true if the body is a crystal as are most of the solids in practical use. ${\Gamma_{ML}}^K$ is then used to introduce the notion of parallelity into the crystal by defining it in such a way that in all three sets $\{\boldsymbol{e}_\kappa\}$ of primitive lattice vectors, i.e. in $\{\boldsymbol{e}_1\}$ $\{\boldsymbol{e}_2\}$ and $\{\boldsymbol{e}_3\}$, the pertaining vectors are parallel to each other. Of course, this parallelity is not euclidean in a deformed or dislocated crystal.

The introduction of the parallelity determines in which way we want to describe the crystal. But it has also an axiomatic feature insofar, as this parallelity is possible only in a space with teleparallelism (see, e.g., Bilby *et al.* [16]). This means that the curvature tensor ${R_{NML}}^K$ formed from ${\Gamma_{ML}}^K$ by

$$ {R_{NML}}^K = 2\left(\partial_N {\Gamma_{ML}}^K - {\Gamma_{MP}}^K {\Gamma_{NL}}^P\right)_{[NM]} \tag{13} $$

has to vanish. This is a constraint on $\boldsymbol{\Gamma}$, and therefore also on $\boldsymbol{\alpha}$ which at some place has to enter the theory.

A basic quantity of differential geometry is also the so-called *contortion*

$$ \kappa_{MLK} := \alpha_{M[LK]}, \quad \alpha_{MLK} := \kappa_{[ML]K} . \tag{14} $$

Obviously, $\boldsymbol{\alpha}$ and $\boldsymbol{\kappa}$ contain the same amount of information.

In our application (crystal with dislocations) $\boldsymbol{\Gamma}$ has the form

$$ \Gamma_{MLK} = g_{MLK} + \kappa_{MLK} \tag{15} $$

with g_{MLK} the Christoffel symbol of the 2nd kind, formed with g_{KL}. Using (15) with $(10)_3$ in the law of parallel transport, namely in

$$ dv^K = -{\Gamma_{ML}}^K v^L dX^M , \tag{16} $$

one finds that coming from the κ-part of (15) the parallel-displaced vector v^K has suffered a rotation over the distance dX^M. This rotation vanishes with the dislocation density. Hence the contortion $\boldsymbol{\kappa}$ has the meaning of a special kind of lattice curvature sometimes also called Nye's curvature, stemming from the dislocations. Obviously the formal reason of the lattice curvature is the antisymmetry in the last two indices of the contortion tensor.

For the theory of elastoplasticity we now have the full set of independent (extensive) variables, namely $\acute{\boldsymbol{F}}^e$ and $\boldsymbol{\alpha}$ (or $\boldsymbol{\kappa}$). Hence the elastic energy per unit volume of the reference state can be written as

$$W = W(\acute{\boldsymbol{F}}^e, \boldsymbol{\alpha}) \tag{17}$$

with the specific response quantities

$$\overset{\circ}{\boldsymbol{T}} = \partial W / \partial \acute{\boldsymbol{F}}^e, \quad \boldsymbol{\lambda} = \partial W / \partial \boldsymbol{\alpha}, \tag{18}$$

or

$$\overset{\circ}{T}{}_{\kappa}{}^{K} = \partial W / \partial \acute{F}^{e\kappa}_{K}, \quad \lambda^{ML}{}_{K} = \partial W / \partial \alpha_{ML}{}^{K}, \tag{19}$$

where the tensor indices κ relate to the intermediate state. Of course, one could also use the Eulerian picture and then have the space coordinates x^k instead of the material coordinates X^K. In (19) $\overset{\circ}{T}{}_{\kappa}{}^{K}$, as before, is the analogon of the first Piola-Kirchhoff tensor and $\lambda^{ML}{}_K$ the specific response tensor showing the presence of dislocations. $\boldsymbol{\lambda}$ has the dimension of a moment stress, which is easily understood from the meaning of $\boldsymbol{\kappa}$.

The next step is to derive the equilibrium equations for forces and moments by application of the principle of virtual work. We have done this under the following constraints:

(i) incompressibility of the lattice; (ii) teleparallelism of the lattice space.

The two constraints deserve their own discussion because they have a special physical content.

5.2. First Constraint: Incompressibility of the Lattice

It is well-known that dislocations can move in two modes, namely glide and climb. The glide motion can also be called *plastic glide*; it is that motion which usually is understood by the term plasticity. In particular, the volume is not changed by the gliding of dislocations. On the other hand, the climb motion, typical for instance in creep, can occur only when at the same time point defects such as vacancies and (self)-interstitials are formed or annihilated. These processes are possible only by the creation or annihilation of free volume, which implies that the lattice must locally be dilated or compressed thereby changing the local volume. This physical fact seems to be in contradiction to the statement that there exists a physical situation in which the whole inelastic motion is done only by dislocations. Such a situation, however, is possible under the restriction that climb is forbidden. This is physically possible when the temperature is low enough so that thermal activation for the creation or annihilation of point defects is not available.

In order to avoid the much more complex situation with climb we have introduced the constraint of incompressibility, namely $g \equiv \det\{g_{KL}\} = 1$, and find, when going through the calculation, a much simpler situation with this constraint. Without incompressibility point defects come in and have to be described by their own dynamical variables. This implies new degrees of freedom: The theory is no longer a dislocation theory but must include point defects. This occurs at higher temperatures (energies). Recall that in a similar manner the electrodynamics had to be extended, because the degrees of freedom of the weak interaction enter at higher energies.

5.3. Second Constraint: Teleparallelism of the Lattice Space

Above we had introduced the notion of parallelity into the crystal lattice. This in general noneuclidean parallelity is supposed to apply also to deformed and dislocated crystals. Given the linear connexion ${\Gamma_{ML}}^K$ we can by repeated parallel-displacements of the form (16) construct the pertaining configuration of the crystal. This, however, is possible only when the crystal space possesses teleparallelism (see Bilby *et al.* [16]) i.e. its curvature tensor $\boldsymbol{R}$ vanishes. In fact, if the result of the parallel-displacement ¿from atom A to atom B would be different along two different paths, then a unique crystal configuration could not be constructed. Sofar the teleparallelism ($\boldsymbol{R} = 0$) was not built into the principle of virtual work. Now it is a well-known result of differential geometry that $\boldsymbol{R} = 0$ is identically satisfied if $\boldsymbol{\Gamma}$ has the form

$$\Gamma^K_{ML} = A^K_\kappa \partial_M A^\kappa_L . \tag{20}$$

Here A^κ_L is any 2-point tensor function (with A^K_κ its inverse) based in the reference state and in some state κ. Due to the arbitrariness of A^κ_L we can identify it with $\acute{F}^{e\kappa}_L$. We then obtain the following constraint for the dislocation density $\boldsymbol{\alpha}$:

$$ {\alpha_{ML}}^K - \acute{F}^{eK}_\kappa \partial_{[M} \acute{F}^{e\kappa}_{L]} = 0. \tag{21}$$

But also the in M, L symmetric part of (20) must be satisfied. This part concerns only the Christoffel symbol in (5). The calculation shows that this condition is fulfilled provided

$$g_{KH} = \acute{F}^{e\kappa}_K \acute{F}^{e\lambda}_L g_{\kappa\lambda}. \tag{22}$$

Of course, the elastic energy depends on g_{KL} as it does on g. Since, however, the deformed state is completely determined by $\acute{\boldsymbol{F}}^e$ and $\boldsymbol{\alpha}$, we do not use g_{KL} (and g) as argument in W. As a consequence we do not introduce (22) and $g = 1$ as an extra constraint in the virtual work expression, but just use it, where it occurs.

6. Principle of Virtual Work

Since our interest is in dislocations, we omit the external part of the elastic energy. To simplify the presentation, also surface effects are neglected. The intergrations then go over the infinite space. Again we choose the Lagrangian picture. With

W according to (17) the internal virtual work becomes

$$\int_V \left[\frac{\partial W}{\partial \acute{F}^{e\kappa}_K} \delta \acute{F}^{e\kappa}_K + \frac{\partial W}{\partial \alpha_{ML}{}^K} \delta \alpha_{ML}{}^K + \boldsymbol{\lambda}^{ML}{}_K \delta (F^{eK}_\kappa \partial_{[M} \acute{F}^{e\kappa}_{L]} - \alpha_{ML}{}^K) \right] dV. \qquad (23)$$

Going through the procedure of the principle of virtual work, thereby using also (19), we obtain [17]

$$\overset{*}{\nabla}_L T^{LK} = -\alpha_{LM}{}^P \nabla_P \lambda^{MLK} \quad \text{(force equilibrium)} \qquad (24)$$

and

$$T^{[LK]} = \overset{*}{\nabla}_M \lambda^{M[LK]} \equiv \overset{*}{\nabla}_M \mu^{MLK} \quad \text{(moment equilibrium).} \qquad (25)$$

Here $\overset{*}{\nabla}_M \equiv \nabla_M + 2\alpha_{RM}{}^R$ is the symbol for the divergence in a space with torsion (dislocations), ∇_M denotes ordinary covariant differentiation in the crystal space. T^{LK} is the analogon of the *second* Piola-Kirchhoff stress tensor derived from the *first* one (Eq. (19)) as

$$T^{KL} = \acute{F}^{e\kappa}_M g^{ML} \overset{\circ}{T}{}^K_\kappa\,. \qquad (26)$$

In general, this tensor is not symmetric! $\boldsymbol{\lambda}$ (Eq. (19)) denotes the specific response to the presence of dislocations and $\mu^{MLK} \equiv \lambda^{M[LK]}$ is the specific response to the lattice curvature κ_{MLK}:

$$\mu^{MLK} = \partial W / \kappa_{MLK}. \qquad (27)$$

For vanishing dislocation density the equilibrium equations recover the form of the Cosserat equilibrium equations under zero external forces and moments. Since the dislocations bring the anholonomity into the picture, Eqs. (24) and (25) could perhaps be called anholonomous Cosserat equations. It would then be interesting to compare these with the more mathematical anholonomous Cosserat theory discussed recently by Epstein and de León [7,8]. Although this theory resembles our theory in many points, it is not the same theory and, if my interpretation is right, not *intended* to be the same theory.

Before we speak about the transition to dynamics let us deal with some problems in statics and quasistatics.

7. Examples in Statics and Quasistatics

It is suggestive to understand dislocation theory as a field theory. In this sense we have classified the theory by firstly introducing the motional degrees of freedom, represented by $\acute{F}^{e\kappa}_K$ and $\alpha_{ML}{}^K$. Secondly we have given the free energy per unit volume (of the reference state) a certain form. Any other form would lead us to another theory. The assumed form allowed us to derive the static equilibrium equations (24) and (25) under the geometrical constraints $\det\{g_{KL}\} = 1$ (incompressibility) and $R_{NMLK} = 0$ (teleparallelism). In general, this is not enough to solve real physical problems because almost nothing is said about the particular material of the problem.

The missing information is usually given in the form of constitutive equations and, perhaps, evolution equations which tell us, how the dislocation distribution changes with proceeding deformation. Let us give a few examples.

One group of problems consists in the calculation of the stress T^{KL}, given the distribution ${\alpha_{ML}}^K$ of the dislocations. This case occurs often when some model is developed. For instance the model could state that the critical shear stress for plastic deformation is related to the applied stress needed to make two dislocations of opposite sign on parallel glide planes pass each other. This problem can be solved by calculating how the interaction energy of the dislocations varies with their relative position.

Another problem is to find out dislocation arrangements of low energy, because these will occur more frequently than those of higher energy. Knowledge about this can be helpful in predicting the development of certain dislocation structures, such as wall structures, cell structures, polygonization and others (see e.g. work by Walgraef and Aifantis [18], Estrin *et al.* [19] and by Essmann and Differt [20]).

To solve problems of this kind one needs beside the mentioned geometric and static field equations also constitutive equations which connect the strain quantities $\acute{\boldsymbol{F}}^e$ and $\boldsymbol{\alpha}$ with the stresses $\boldsymbol{T}$ and $\boldsymbol{\lambda}$ (or $\boldsymbol{\mu}$). In most applications the specific contribution to the energy of $\boldsymbol{\alpha}$ is neglected with the apology that it is small, cf. Section 4. There are arguments which support such a proceeding, in particular the argument that a small length appears in the comparison of the two sorts of elastic moduli. This length is supposed related to the mean distance between dislocations, for instance of order 1 μm. However, in low energy arrangements the dislocations develop special patterns in which the long range stress contributions mutually cancel so that the elastic moduli connected with $\acute{\boldsymbol{F}}^e$ do not enter strongly. In such cases the moduli connected with $\boldsymbol{\alpha}$ become important. Incidentally, these moduli are also related to the cores of the dislocations.

In most applications the linearized form of the above equations has been used. The equations, however, are correct also in their nonlinear parts, and, in fact, also nonlinear problems were solved.

Problems which go beyond statics are those classified as quasistatic. Hereto belongs very slow plastic deformation. Unlike many other slow processes in physics quasiplastic deformation is irreversible. This again has to do with the interaction of different types of dislocations which can react thereby going into low energy configurations (e.g. Lomer-Cottrell dislocations). These are then permanent and form strong obstacles, so-called dislocation locks, to the further motion of the dislocations. As a result a hardening of the material is observed.

8. Dislocation Dynamics, Gauge Theories

Dislocation dynamics is not yet in a satisfactory state. Whereas in the continuum mechanics of simple materials a law of motion can easily be written down in the Cauchy form

$$\rho\frac{d\boldsymbol{v}}{dt} + \operatorname{div}\boldsymbol{\sigma} = \boldsymbol{f} \tag{28}$$

such a law is not valid in the theory of crystals with dislocations, if these are introduced as dynamical variables. First of all, as any motion, so also dislocation motion is endowed with inertia. Unlike Newtonian particles however dislocations do not have their own mass. That means that inertia (and other forces) acting on dislocations are not the common Newtonian forces, but so-called configurational forces of the kind considered by Eshelby [21]. One could think to introduce some effective dislocation mass. But this mass and the connected inertia will depend on the dislocation arrangement itself, a fact which makes the inertia problem rather complex. Furthermore, there are a number of dissipative mechanisms whose influence on a possible macroscopic dissipation function is also complex.

There are certain proposals by the quoted authors like e.g. that of the balance of pseudo-momentum and material forces in nonlinear elasticity by Maugin and Trimarco [22] or that of an inertia tensor introduced by Naghdi and Srinivasa [5,6]. One may hope that at the end of such developments will be a dynamical theory of dislocations that is more final than the present ones.

The dynamics of dislocations (and disclinations) also plays an important role in the so-called gauge theories of defects which in recent times gain an increased interest. We mention in particular the monographs of Kadić and Edelen [23], of Edelen and Lagoudas [24] as well as work by Kunin and Kunin [25].

The gauge concept has become very popular in physics since it helped to prove that electrodynamics is not a closed theory, but has to be combined with the theory of weak interaction to describe electromagnetic phenomena at very high energies. Whereas there a field theory of elementary particles was the aim, it is here a field theory of elementary defects (elementary as opposed to composed). As far as I know the first step in this direction was done by Turski [26]. Later Golebiewska [27] and others resumed the topic which then was of continuous interest up to present time (see, e.g., work of Popov [28]). In the language of gauge theories the dislocation gauge theory is a T(3)-gauge theory, where T(3) denotes the 3-dimensional translation group. This has the following physical background: The elastic energy of the crystal does not change when dislocations pass through the crystal. This passing however implies a translation *through one atomic distance* of all atoms involved and the elastic energy is invariant against such local translations. Local invariances of this kind are at the root of all gauge theories.

As far as statics is concerned the gauge theory of dislocations leads to the same results as does the field theory of dislocations which we discussed in this paper. In the dynamics the gauge theory has the same problems as has the field theory so that the usefulness of the gauge theory of defects seems to lie more on the theoretical than on the applied side.

9. Conclusion

Our world is full of solid matter which, when inspected closer, shows a microstructure in its constitution. In this work we have selected one particular microstructure namely that defining the crystal. The characteristic length of this microstructure is the atomic distance. In a way one can say that the crystal structure is, so-to-speak the body itself, whereas in other types of microstructured materials like e.g. that of Toupin [29] there is an underlying body to whose points (or particles) are affixed certain structural elements like vector triads (such a body is called Cosserat continuum by Epstein and de León [7,8]) or directors in the case of liquid crystals.

The difference between these two types of structures is fundamental, and the pertaining theories should take this in account. We had declared the presented theory as limited. The term "limited" is not meant as any limitation of the mathematical accuracy. Beside treating just crystalline materials there is another limitation in our presentation, namely insofar as we excluded the presence of point defects as dynamical variables. This implies a restriction to low temperature plasticity, also called *athermal plasticity*, because there thermal activation is disregarded. It may be mentioned that there does exist also a field theory of point defects. The elementary defects of this kind are the vacancy, the (self-)interstitial and the point stacking fault, the latter detected only recently [30]. It was proposed by Bilby *et al.* [31] to describe point defects, also called extra matter, by a nonmetric part of the linear connexion thereby keeping the prescription of teleparallelism.

Nonmetricity means that length measurements are disturbed. It is easy to see that just this occurs in the presence of point defects. In fact, when counting atomic steps along crystallographic lines to measure distances between two atoms, one feels disturbed when suddenly a vacancy or an interstitial emerges instead of another atom. This idea has been made quantitative also in the author's recent work [30]. There exists another development which deserves more attention than it has found sofar. Originally Schaefer [32] established an analogy between the static part of elasticity theory and (linearized) Riemannian geometry. This analogy was later extended by Stojanović [33] and myself [34] to apply also after introduction of moment stresses, as they occur in dislocation theory. Instead of the Riemannian curvature tensor one then considers the (linearized) Riemann-Cartan curvature tensor which is formed with the linear connexion containing torsion.

Later Ben-Abraham [35] realized that the analogy goes much farther, namely it is valid also when point defects are included. We can describe this as follows. Consider a differential-geometric space with the most general linear connexion $\Gamma_{ML}{}^{K}$. The pertaining curvature tensor is still defined by (13). That this 4-indexed tensor can be derived from the 3-indexed quantity $\Gamma_{ML}{}^{K}$ shows that $\boldsymbol{R}$, defined by (13), is not the most general curvature tensor one can think of. This means that $\boldsymbol{R}$ satisfies some extra conditions known as the first and second Bianchi identities (the latter is called by Schouten [36] the 2nd identity of the curvature tensor.)

Ben Abraham's statement is now that the statical equilibrium conditions of the extended theory (which contains dislocations *and* point defects) have exactly the form of the Bianchi identities in which now so-called (three-dimensional) stress functions play the role of the metric tensor. This result implies that the whole well-developed formalism of differential geometry can be applied also to the statics of defect theory. One realizes that geometry and statics of the defect theory have the status of mutually *dual* formulations, the term *dual* in the general sense of Hamiltonian mechanics. A last proof that all this is valid for arbitrarily large stress and strain was given by the present author [37], see also Kleinert [38] who has used the duality for the development of his so-called double gauge theory. Kleinert also proposes to take care of the discreteness of real dislocations by some quantization procedure, certainly an attractive thought. In this work we have not utilized the duality just described because an understandable introduction this topic requires the time of an extra lecture.

A last remark: Even when fully developed the discussed field equations will not suffice to solve physical and practical problems of materials behaviour, since relevant information about the materials (constitutive laws) is missing. However, the field equations form a frame within which all phenomena with dislocations, or more general, with defects have to fit just as Maxwell's equations form a frame for all electromagnetic phenomena. Research on constitutive laws goes on since a long time. Doing such research one needs models, and to test these, the field equations are of great importance. Then applications go far beyond the few examples shown in Section 7.

References

1. I. A. Kunin, *Elastic Media with Microstructure I and II*, Springer series in Solid-State Sciences **26** and **44**, Springer, Heidelberg 1982.
2. G. Capriz, *Continua with Microstructure*, Springer Tracts in Natural Philosophys **35**, Springer, Heidelberg, 1989.
3. G. A. Maugin, *Material Inhomogeneities in Elasticity*, Chapman and Hill, London, 1993.
4. M. Epstein and G. A. Maugin, *Acta Mechanica* (1995), in press.
5. P. M. Naghdi and A. R. Srinivasa, *Phil. Trans. Roy. Soc. London* **A345** (1993) 425 and 459.
6. P. M. Naghdi and A. R. Srinivasa, *Int. J. Engng. Sci.* **32** (1994) 1157.
7. M. Epstein and M. de León, *On the homogeneity of non-holonomic Cosserat media: a naive approach*, in press.
8. M. Epstein and M. de León, In *Proc. Int. Conf. on Differential Geometry*, Debrecen, Hungary, 1994, in press

9. K. C. Le and H. Stumpf, *Finite elastoplasticity with microstructure*, In *Mitteilungen Nr. 92 aus dem Institut für Mechanik*, Ruhr-Universität Bochum, Germany, 1994.
10. C. Truesdell and W. Noll, *The Non-linear Field Theories of Mechanics*, in: Encyclopedia of Physics III/3, Springer Heidelberg 1965.
11. I. A. Kunin, *Int. J. Theor. Physics* **29** (1990) 1167.
12. B. A. Bilby, L.R.T. Gardner and A. N. Stroh, *Continuous distributions of dislocations and the theory of plasticity*, In *Extrait des actes du IXe congrès international de mécanique appliquée*, Bruxelles, 1957, pp. 35-44.
13. E. Kröner, *Arch.Rat.Mech.Anal.* **4** (1960) 273.
14. E.H. Lee, *J. Appl. Mech.* **36** (1969) 1.
15. K. Kondo, In *Proc. 2nd Japan Nat. Congr. Appl. Mech.*, Tokyo, 1952, p. 41.
16. B. A. Bilby, R. Bullough and E. Smith, *Proc. Roy. Soc. London* **A231** (1955) 263.
17. E. Kröner, In *Proc. 2nd Int. Conf. Nonlin. Mechanics*, ed. Chien Wei-zang, Peking University Press, Beijing 1993, p. 59
18. D. Walgraef and E. C. Aifantis, *J. Appl. Phys.* **58** (1985) 688.
19. Y. Estrin, L. P. Kubin and E. C. Aifantis, *Scripta Met. Mater.* **29** (1993) 1147.
20. H. Differt and U. Essmann, *Mater. Sci. and Engng.* **A164** (1993) 295, U. Essmann and H. Differt, ibid, submitted.
21. J. D. Eshelby, *Phil.Trans.Roy.Soc.London* **A 244** (1951) 87.
22. G. A. Maugin and C. Trimarco, *Acta Mech.* **94** (1992) 1.
23. A. Kadić and D.G.B. Edelen, *Lecture Notes in Physics 174*, Springer, Heidelberg 1983.
24. D.G.B. Edelen and D.C. Lagoudas, *Gauge Theory and Defects in Solids*, North-Holland, Amsterdam 1988.
25. I. A. Kunin and B. I. Kunin, In *Trends in Applications of Pure Mathematics to Mechanics*, eds. E. Kröner and K. Kirchgässner, Springer, Berlin 1986, p. 1167.
26. L. Turski, *Bull. Pol. Acad. Sci., Techn. Sci.* **14** (1966) 289.
27. A.A. Golebiewska-Lasota, *Int. J. Engng. Sci.* **17** (1979) 329.
28. V. L. Popov, *Int. J. Engng. Sci.* **30** (1992) 329.
29. R. A. Toupin, *Arch. Rat. Mech. Anal.* **17** (1964) 85.
30. E. Kröner, *Int. J. Theor. Physics* **29** (1990) 1219.
31. B. A. Bilby, L.R.T. Gardner, A. Grinberg and M. Zorawski, *Proc. Roy. Soc. London* **A292** (1966) 105.
32. H. Schaefer, *Z. Angew. Math. Mech.* **33** (1953) 356.
33. P. Stojanović, *Int. J. Engng. Sci.* **19** (1963) 323.

34. E. Kröner, *Ann. Phys.* (Leipzig), 7 Ser., **11** (1963) 13.

35. S. Ben-Abraham, In *Fundam. Aspects of Dislocation Theory*, Nat. Bur. Stand. (U.S.) Spec. Publ. **317**, II (1970), p. 943.

36. J. A. Schouten, *Ricci-Calculus*, Springer, Berlin 1954.

37. E. Kröner, *Phys. Stat. Sol. (b)* **144** (1987) 39.

38. H. Kleinert, *Gauge Fields in Condensed Matter*, Vol. II, *Stresses and Defects*, World Scientific, Singapore 1989.

Continuum Models and Discrete Systems
Proceedings of 8th International Symposium, June 11-16, 1995, Varna, Bulgaria, ed. K.Z. Markov
©World Scientific Publishing Company, 1996, pp. 538–545

THEORY OF DISLOCATIONS BASED ON THE RESOLUTION $\mathbf{F} = \mathbf{F}^e\mathbf{F}^p$

LE KHANH CHAU and H. STUMPF
Lehrstuhl für Allgemeine Mechanik, Ruhr-Universität Bochum,
D-44780 Bochum, Germany

Abstract. The kinematics of elastoplastic bodies at finite strain based on the multiplicative decomposition of the deformation gradient is developed taking into account the motion of continuously distributed dislocations. In comparison with the macro-theories of finite elastoplasticity additional degrees of freedom are introduced through Cartan's torsion having the physical meaning of the dislocation density. The set of balance equations has been enlarged to account also for the internal motion of dislocations. Constitutive equations for such bodies are proposed in consistency with the entropy production inequality.

1. Introduction

It is well known that dislocations, as bearers of the crystal defects, are responsible for the slip and the plastic deformation of solids. On the level of continuum modelling, the account of distributed dislocations brings two notions into the theory: the *couple stresses* due to the dislocation density [4,5] and the *driving stresses* acting on dislocations and causing their motion [1]. These notions should be involved in any set of governing and constitutive equations of a continuum theory modelling elastoplastic bodies with dislocations.

In the present paper we start with the kinematical model of elastoplasticity based on the multiplicative decomposition of the deformation gradient [3]. Concerning the elastic part of this decomposition, we interpret its inverse as the map that maps tangent vectors defined on the current configuration to those of the crystal reference. The Burgers vector of dislocations in the crystal can then be defined, within a continuum limit, through this deformation field using the Cartan-Frank circuit [2,4,5]. This leads naturally to the definition of the dislocation density and the account of microstructure at the macroscopic level. It turns out that the elastic strain and the dislocation density determine the inverse elastic deformation uniquely up to a rigid body rotation [15]. This fact can be used to qualify them as state variables.

We lay down the statics of an elastoplastic body with dislocations based on a principle of virtual work and a functional form of the stored energy per unit crystal volume, which is supposed to depend on the elastic deformation as well as on the dislocation density bringing a new characteristic length scale into the theory. This ansatz satisfies the principle of frame indifference and also the principle of initial scaling indifference [13]. Since the stored energy enjoys the additional invariance with respect to the rescaling group, new balance equations of micromomentum and moment of micromomentum can be derived. The microstresses should be balanced by the internal driving stresses acting on the dislocations. The internal driving stresses are shown to be the sum of Eshelby's tensor (the stresses due to the surrounding elastic field) and an additional term characterizing the interaction between dislocations.

We formulate the dynamics of a body with continuously distributed dislocations by taking into account the inertial terms associated with the macro- and the dislocation motion. The balance of energy and the entropy production inequality are included in this set of equations (see also [6,11,12]). By decomposing the macro- and microstresses into the reversible and the irreversible constituents we reduce the entropy production inequality to the dissipation inequality. The latter can be used to substantiate new constitutive equations involving the strain and dislocation drift rates. It can be shown that the model proposed recently by Naghdi and Srinivasa [11,12] corresponds to a special choice of the dissipation function in our theory.

2. Kinematics

Fig. 1 depicts the resolution of the deformation gradient $\mathbf{F} = \partial\phi_t/\partial\mathbf{X}$ into its elastic and plastic parts

$$\mathbf{F} = \mathbf{F}^e\mathbf{F}^p. \tag{1}$$

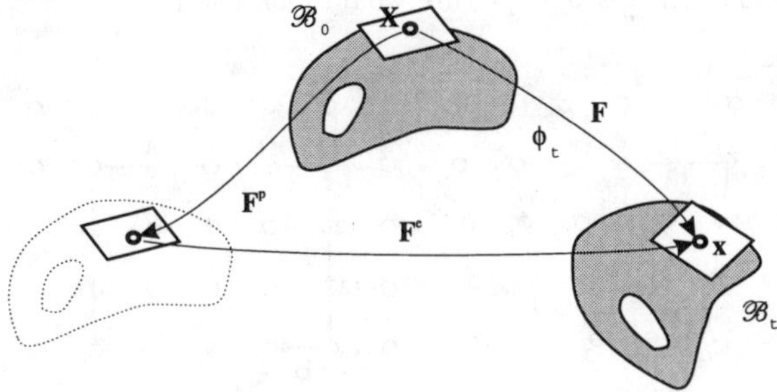

Fig. 1. Resolution of the deformation gradient.

In general, the elastic and plastic deformation fields $\mathbf{F}^e$ and $\mathbf{F}^p$ cannot be gradients of global maps (they are therefore called incompatible). We can only suppose that they are orientation preserving so that the determinants J^e and J^p of $\mathbf{F}^e$ and $\mathbf{F}^p$ are positive. This means $\mathbf{F}^p$ and $\mathbf{F}^e$ have inverse deformation fields, denoted by $\mathbf{F}^{p-1}$ and $\mathbf{F}^{e-1}$, respectively. The resolution (1) was first introduced by Bilby *et al.*

[3] as a basic assumption to develop the kinematics of elastoplastic bodies with continuously distributed dislocations. In that paper $\mathbf{F}$, $\mathbf{F}^e$, and $\mathbf{F}^p$ are called the shape deformation, the lattice deformation, and the dislocation deformation, respectively. The deformation $\mathbf{F}^{e-1}$ ($\mathbf{F}^p$) maps tangent vectors on the current (initial) configuration back (forward) to vectors of the crystal reference, where macro- and microstresses in the body elements are supposed to vanish. A viable configuration of the crystal reference does not exist.

In order to measure the change of lengths and angles we introduce the total $\mathbf{C}$, $\mathbf{c}$, elastic $\bar{\mathbf{c}}^e, \mathbf{c}^e$ and plastic $\mathbf{C}^p, \bar{\mathbf{c}}^p$ strain tensor fields, respectively, as it is usual in finite elastoplasticity [10,13]. The tensors $\mathbf{C}$ and $\mathbf{c}$ correspond to the Green and Finger deformation tensor, respectively. The knowledge of the tensor field $\mathbf{C}$ ($\mathbf{c}$) is sufficient to determine the deformation gradient $\mathbf{F}$ and subsequently the displacement field uniquely up to a rigid body motion, provided $\mathbf{C}$ ($\mathbf{c}$) satisfies the Riemann compatibility condition. In contrary, the knowledge of $\mathbf{C}^p$ ($\mathbf{c}^e$) is insufficient to determine $\mathbf{F}^p$ ($\mathbf{F}^e$). The missing variables can be introduced by appealing to the continuum theory of dislocations.

Let us take an arbitrary circuit $c(s)$ in the current configuration and consider the integral

$$(\bar{\mathbf{b}})^\alpha = -\oint_c (\mathbf{F}^{e-1})^\alpha_b \, dx^b. \tag{2}$$

If the integral (6) would vanish for all circuits, it would mean that the elastic deformation is compatible. But in the general case of plastic deformation caused by dislocation motion it is not so, and this integral measures the degree of incompatibility of the elastic deformation. One can show that (6) coincides with the Burgers vector in the limiting case of the continuum model (if we let the lattice constant approach zero). The microscopic picture would look like Fig. 2.

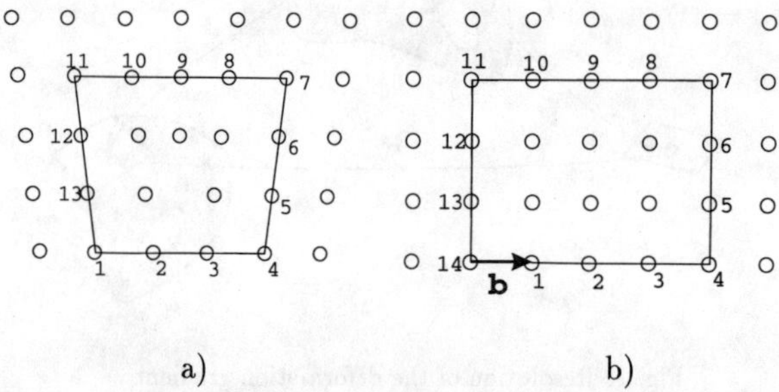

Fig. 2. Definition of the Burgers vector. a) current configuration; b) crystal reference.

Now let us apply Stokes' theorem to the contour integral (6)

$$(\bar{\mathbf{b}})^\alpha = \frac{1}{2}\int_A [(\mathbf{F}^{e-1})^\alpha_{c,b} - (\mathbf{F}^{e-1})^\alpha_{b,c}] \, dx^b \wedge dx^c, \tag{3}$$

where A denotes a surface with boundary c and $dx^b \wedge dx^c$ the wedge product of 1-forms. For infinitesimal circuits we get from (7)

$$(\bar{\mathbf{b}})^\alpha = \frac{1}{2}[(\mathbf{F}^{e-1})^\alpha_{c,b} - (\mathbf{F}^{e-1})^\alpha_{b,c}]\, dx^b \wedge dx^c. \tag{4}$$

Pulling (8) back to the initial configuration leads to the definition of the dislocation density tensor referred to the initial configuration

$$(\mathbf{T})^A_{BC} = (\mathbf{F}^{p-1})^A_\alpha[(\mathbf{F}^p)^\alpha_{C,B} - (\mathbf{F}^p)^\alpha_{B,C}]. \tag{5}$$

Here the comma preceding indices is used to denote the partial derivative with respect to the corresponding co-ordinates. We refer upper case indices to coordinates in the initial configuration, lower case indices to coordinates in the current configuration, and Greek indices to the anholonomic basis associated with the crystal reference. One can also introduce the dislocation density tensor $\bar{\mathbf{t}}$ relative to the crystal reference as follows

$$\begin{aligned}(\bar{\mathbf{t}})^\alpha_{\beta\gamma} &= (\mathbf{F}^p)^\alpha_A[(\mathbf{F}^{p-1})^A_{\beta,\gamma} - (\mathbf{F}^{p-1})^A_{\gamma,\beta}] \\ &= (\mathbf{F}^{e-1})^\alpha_a[(\mathbf{F}^e)^a_{\beta,\gamma} - (\mathbf{F}^e)^a_{\gamma,\beta}].\end{aligned} \tag{6}$$

The comma preceding Greek indices denotes the so-called relative derivative with respect to the anholonomic basis of the crystal reference defined in the following way

$$(.)_{,\alpha} = (.)_{,a}(\mathbf{F}^e)^a_\alpha = (.)_{,A}(\mathbf{F}^{p-1})^A_\alpha. \tag{7}$$

We do not present here the differential-geometric interpretation of the dislocation density tensor in terms of the crystal connection and distance parallelism, and we refer the interested reader to [13].

It turns out that the knowledge of the fields $\mathbf{C}^p$ and $\mathbf{T}$ is sufficient to determine $\mathbf{F}^p$ uniquely up to a rigid body rotation, provided some integrability condition is fulfilled. Le and Stumpf [15] have proved that this requires the curvature tensor depending on $\mathbf{C}^p$ and $\mathbf{T}$ to vanish.

For the theory developed below, the rate formulation plays an important role. The strain rates are defined as usually in finite elastoplasticity. For the dislocation density tensor $\mathbf{T}$ defined on the initial configuration we calculate its rate using the material time derivative (the partial time derivative with X kept constant). Thus,

$$\begin{aligned}2(\mathbf{Z})^A_{BC} &= (\dot{\mathbf{T}})^A_{BC} \\ &= -(\mathbf{F}^{p-1})^A_\alpha(\dot{\mathbf{F}}^p)^\alpha_D(\mathbf{T})^D_{BC} + (\mathbf{F}^{p-1})^A_\alpha[(\dot{\mathbf{F}}^p)^\alpha_{C,B} - (\dot{\mathbf{F}}^p)^\alpha_{B,C}].\end{aligned} \tag{8}$$

We call $\mathbf{Z}$ the dislocation drift rate.

3. Statics

Let us consider first an elastoplastic body with continuously distributed dislocations under the condition of constant temperature. We postulate the existence of a free energy per unit crystal volume of such a body. Relative to the crystal reference, which is assumed at the moment to be given, this free energy is supposed to depend on the elastic deformation $\mathbf{F}^e$ as well as on its relative derivative $\mathbf{F}^e_{,\alpha}$,

$$\mathfrak{w} = \bar{\mathfrak{w}}(X, \mathbf{F}^e, \mathbf{F}^e_{,\alpha}). \tag{9}$$

Since the derivative of $\mathbf{F}^e$ is involved, (17) means that the theory includes a new characteristic length associated with the distribution of dislocations.

Le and Stumpf [13] have shown that if the free energy density (17) is frame indifferent, it can depend only on $X, \bar{\mathbf{c}}^e$ and $\bar{\mathbf{t}}$

$$\mathfrak{w} = \hat{\mathfrak{w}}(X, \bar{\mathbf{c}}^e, \bar{\mathbf{t}}), \tag{10}$$

where $\bar{\mathbf{c}}^e$ is elastic strain and $\bar{\mathbf{t}}$ the dislocation density relative to the crystal reference given by (12). This result is in agreement with Kröner's requirement stating that the free energy density can depend only on the elastic strain and on the dislocation density [9].

Since $\bar{\mathbf{c}}^e$ and $\bar{\mathbf{t}}$ are the push-forward of $\mathbf{C}$ and $\mathbf{T}$ by $\mathbf{F}^p$, we can regard them as functions of the point values of $\mathbf{F}^p$, $\mathbf{C}$, and $\mathbf{T}$. Therefore, $\bar{\mathfrak{w}}$ becomes a function of X, $\mathbf{F}^p$, $\mathbf{C}$ and $\mathbf{T}$. We can say that when referred to the initial (Lagrangian) description the free energy per unit volume of the initial configuration takes the form

$$W = J^p \hat{\mathfrak{w}}(X, \bar{\mathbf{c}}^e(\mathbf{F}^p, \mathbf{C}), \bar{\mathbf{t}}(\mathbf{F}^p, \mathbf{T})) = \hat{W}(X, \mathbf{F}^p, \mathbf{C}, \mathbf{T}). \tag{11}$$

The formula (17) satisfies still another important criterion. The free energy of an elastoplastic body per unit crystal volume defined in this way depends only on the current configuration and the crystal reference, and it is insensitive to the change of the initial configuration. This means that if we superpose an arbitrary initial deformation on the initial configuration and keep the crystal reference and the current configuration fixed, the energy $\hat{\mathfrak{w}}$ will remain unchanged. We shall call this property the principle of initial scaling indifference [13].

A system of equations governing the statics of the body with dislocations can be derived by postulating a principle of virtual work [14]. Let ρ_0 be the mass density, $\mathbf{B}$ and $\mathbf{B}^d$ the external body macro- and microforce, respectively. Then except the well-known equilibrium and constitutive equations for the Piola-Kirchhoff stress tensor, we can derive also the following additional equations

$$\operatorname{Div} \mathbf{P}^d + \rho_0 \mathbf{B}^d - \mathbf{J} = \mathbf{0}, \quad \mathbf{P}^d = \frac{\partial W}{\partial D\mathbf{F}^p}, \tag{12}$$

$$\mathbf{P}^d = -\mathbf{F}^{p-T}\mathbf{S}^d, \quad \mathbf{S}^d = 2\frac{\partial \hat{W}}{\partial \mathbf{T}}, \tag{13}$$

$$\begin{aligned} (\mathbf{J})_\alpha^{\cdot A} &= \frac{\partial W}{\partial (\mathbf{F}^p)^\alpha_A} \\ &= (\mathbf{F}^{p-1})^C_\alpha \left[-(\mathbf{C})_{CD}(\mathbf{S})^{DA} + \hat{W}\delta^A_C + (\mathbf{T})^B_{CD}(\mathbf{S}^d)^{DA}_B \right]. \end{aligned} \tag{14}$$

Eq. (12) expresses the equilibrium between the microstress tensor $\mathbf{P}^d$ (contravariant of rank 2 and covariant of rank 1, with components $(\mathbf{P}^d)^{AB}_\alpha$), the external microforce tensor $\mathbf{B}^d$, and the (internal) driving stress tensor $\mathbf{J}$. Kröner has shown that the microstresses are needed to hinder the dislocation motion and keep them all in equilibrium [5, p.281]. According to (28), the driving stress tensor is the pull-back to the crystal reference of the tensor in the square brackets. This tensor is the sum of the Eshelby tensor $\mathbf{CS} - \hat{W}\mathbf{1}$, with $\mathbf{S}$ being the second Piola-Kirchhoff stress tensor, and the tensor with components $(\mathbf{T})^B_{CD}(\mathbf{S}^d)^{DA}_B$. While the physical meaning of the former is to be responsible for the action on dislocations by the surrounding elastic field [1,8], the latter can be interpreted as being caused by the interaction between dislocations.

4. Dynamics

The dynamics of the body with dislocations can be formulated either by postulating the action principle and deriving the consequences from the invariance requirements, or by laying down the balance equations at the onset. We choose the second way, because it enables us to include the heat flux and the entropy production inequality in a natural manner. Let ρ_0 be the mass density, $\mathbf{M}^d$ the micromomentum, $\mathbf{P}$ the first Piola-Kirchhoff stress tensor, $\mathbf{J}$ the driving stress tensor, $\mathbf{P}^d$ the microstress tensor, E the internal energy density per unit mass, $\mathbf{Q}$ the heat flux vector, Θ the temperature, N the entropy density per unit mass. We lay down the classical conservation of mass, balances of macromomentum and of moment of macromomentum, and also the following balance equations (see also [13,14])

balance of micromomentum

$$\rho_0 \dot{\mathbf{M}}^d = \operatorname{Div} \mathbf{P}^d + \rho_0 \mathbf{B}^d - \mathbf{J}, \tag{15}$$

balance of moment of micromomentum

$$(\mathbf{P}^d)^{BC}_\alpha = -(\mathbf{P}^d)^{CB}_\alpha, \tag{16}$$

balance of energy

$$\rho_0 \dot{E} + \operatorname{Div} \mathbf{Q} = \left\langle \mathbf{gP}, \dot{\mathbf{F}} \right\rangle + \left\langle \mathbf{J}, \dot{\mathbf{F}}^p \right\rangle + \left\langle \mathbf{P}^d, D\dot{\mathbf{F}}^p \right\rangle, \tag{17}$$

entropy production inequality

$$\rho_0(N\dot{\Theta} + \dot{\Psi}) - \left\langle \mathbf{gP}, \dot{\mathbf{F}} \right\rangle - \left\langle \mathbf{J}, \dot{\mathbf{F}}^p \right\rangle - \left\langle \mathbf{P}^d, D\dot{\mathbf{F}}^p \right\rangle + \left\langle \frac{\operatorname{Grad} \Theta}{\Theta}, \mathbf{Q} \right\rangle \le 0, \tag{18}$$

definition of the free energy

$$\Psi = E - N\Theta. \tag{19}$$

Eq. (18) is regarded as the entropy production inequality, that can and should restrict the possible functional form of the constitutive equations.

5. Constitutive equations

In [14] it has been shown that if $\mathbf{P}, \mathbf{P}^d$, $\mathbf{J}$ and N do not depend on $\dot{\mathbf{F}}, \dot{\mathbf{F}}^p, D\dot{\mathbf{F}}^p$ and $\dot{\Theta}$, then the constitutive equations for them, which are consistent with Eq. (45), must be derived from the free energy density. If this is not the case, then there is no general "recipe" of how constitutive equations could be derived from the entropy production inequality. Let us assume that the macrostress tensor $\mathbf{P}$, the microstress tensor $\mathbf{P}^d$ and the driving stress tensor $\mathbf{J}$ can be additively decomposed into two parts where the constituents with the subscript 1 as well as the entropy N are expressed through the free energy density and are therefore rate-independent. Thus, only the constituents with the subscript 2 depend on the rates of the total and plastic deformations and the dislocation drift rate according to Eq. (15). The work done by the stresses with the subscript 1 is said to be *recoverable*. In contrary, the work done by the stresses with the subscript 2 may cause an increase of entropy and is called *dissipative*. A consequence of this assumption is that

$$\sigma = \left\langle \mathbf{g}\mathbf{P}_2, \dot{\mathbf{F}} \right\rangle + \left\langle \mathbf{J}_2, \dot{\mathbf{F}}^p + \mathbf{P}_2^d, D\dot{\mathbf{F}}^p \right\rangle + \left\langle -\frac{\operatorname{Grad}\Theta}{\Theta}, \mathbf{Q} \right\rangle \geq 0. \tag{20}$$

We call σ the dissipation function. With the account of Eq. (61) the energy balance equation (44) reduces to

$$\rho_0 \dot{N} + \operatorname{Div}\left(\frac{\mathbf{Q}}{\Theta}\right) = \frac{\sigma}{\Theta}, \tag{21}$$

which can be viewed as the entropy balance equation, with $\mathbf{Q}/\Theta$ being the entropy flux and σ/Θ the entropy production [7].

Various assumptions can be made towards satisfying Eq. (61). For example, one can assume that the dissipation function σ is positive definite and depends homogeneously on the "thermodynamic fluxes" $\dot{\mathbf{F}}, \dot{\mathbf{F}}^p, D\dot{\mathbf{F}}^p$ and $\operatorname{Grad}\Theta$,

$$\sigma = \hat{\sigma}(\mathbf{F}, \mathbf{F}^p, D\mathbf{F}^p, \Theta; \dot{\mathbf{F}}, \dot{\mathbf{F}}^p, D\dot{\mathbf{F}}^p, \operatorname{Grad}\Theta). \tag{22}$$

Then the "thermodynamic forces" $\mathbf{g}\mathbf{P}_2, \mathbf{J}_2, \mathbf{P}_2^d$ and $-\mathbf{Q}/\Theta$ can be specified by the partial derivatives of $\hat{\sigma}$ with respect to the corresponding "fluxes" multiplied by some factors which should be determined from the rank of homogeneity of $\hat{\sigma}$. The derived constitutive equations can be used to describe the viscosity due to the macromotion (just as for Navier-Stokes fluids or nonlinear viscous fluids), the plastic yielding, the viscosity due to the dislocation motion, and finally, the nonlinear heat conduction in elastoplastic bodies.

References

1. J. D. Eshelby, *Phil. Trans. R. Soc. London* **A244** (1951) 87.
2. B. A. Bilby, R. Bullough and E. Smith, *Proc. Roy. Soc. (London)* **A231** (1955) 263.
3. B. A. Bilby, L. R. T. Gardner and A. N. Stroh, In *Extrait des actes du IXe congrès international de mécanique appliquée*, Brüssel, 1957.
4. E. Kröner, *Kontinuumstheorie der Versetzungen und Eigenspannungen*, Springer, Berlin, 1958.
5. E. Kröner, *Arch. Rat. Mech. Anal.* **4** (1960) 273.
6. L. I. Sedov and V. L. Berdichevsky, In *Mechanics of Generalized Continua*, ed. E. Kröner, Springer, Berlin, 1967.
7. I. Müller, *Thermodynamik*, Bertelsmann, Düsseldorf, 1973.
8. M. Epstein and G. A. Maugin, *Acta mechanica* **83** (1990) 127.
9. E. Kröner, *GAMM-Mitteilungen* **15** (1992) 104.
10. K. C. Le and H. Stumpf, *Acta Mechanica* **100** (1993) 155.
11. P. N. Naghdi and A. R. Srinivasa, *Phil. Trans. R. Soc. Lond.* **A345** (1993) 425.
12. P. M. Naghdi and A. R. Srinivasa, *Int. J. Engng Sci.* **32** (1994) 1157.
13. K. C. Le and H. Stumpf, *Finite Elastoplasticity with Microstructure*, Ruhr-Universität Bochum, 1994.
14. K. C. Le and H. Stumpf, *Int. J. Engng Sci.*, to appear.
15. K. C. Le and H. Stumpf, *Proc. R. Soc. London*, to appear.

Continuum Models and Discrete Systems
Proceedings of 8th International Symposium, June 11-16, 1995, Varna, Bulgaria, ed. K.Z. Markov
© World Scientific Publishing Company, 1996, pp. 546–554

NONLOCAL PLASTICITY OF METALS

X. LEMOINE, D. MULLER, M. BERVEILLER
Laboratoire de Physique et Mécanique des Matériaux,
ENIM - CNRS, University of Metz, Ile du Saulcy , 57045 Metz , France

Abstract. The paper presents a nonlocal theory describing dislocation cell structure induced by plastic straining and represented by a two-phase composite. The theory is based on micromechanics and thermodynamics with internal variables .

1. Introduction

Dislocation motion on crystalline lattices appears to be the main deformation mechanism responsible for plastic behavior of polycrystalline metals at low temperatures. Such a plastic activity is strongly influenced by the microstructure of the considered heterogeneous medium : at the mesoscopic scale (grain level) both crystallographic and morphological textures are essential parameters ; on the other hand induced dislocation patterns are often observed at the intragranular level [1] correspond generally to a strong intragranular heterogeneization : an organized cell structure results from high dislocation density gradients between cell interiors (low density) and walls (high density).

The usual micromechanical approaches to plasticity of polycrystalline metals (Taylor, Sachs or self-consistent models) are based on the restrictive hypothesis of strain uniformity within the grains, neglecting this way the above mentionned self-organization of dislocations and its mechanical consequences (high level of internal stresses in the walls. Furthermore the induced global plastic anisotropy results simultaneously from the so-called texture of the microstructure (organization of dislocation substructures at the polycrystalline level) as well as from the most classical morphological and crystallographical textures [2].

The present theory aims to represent one basic mechanism of intragranular heterogeneization, the so-called nonlocal hardening (resulting from a coupling between dislocation motion in cell interiors and their storage in neighboring walls), by accounting for the morphological evolution of the cells.

A nonlocal hardening formalism was introduced by the authors at CMDS7 symposium [3] in order to account for interactions between soft and hard parts of the grains. In particular a nonlocal hardening matrix is introduced, whichs relates the glide rate $\dot{\gamma}^h$ on a slip system h at point r to the critical shear stress rate $\dot{\tau}_c^g$ on system g at point r', through the nonlocal relation

$$\dot{\tau}_c^g(r)=\int_V H^{gh}(r-r')\dot{\gamma}^h(r')dV' \tag{1.1}$$

To solve the complete micromechanical problem requires the establishement of the complicated kernel H. Here, a simplified description is proposed through a two-phase approximation by considering the high gradients of dislocation density present in the dislocation cell structure. This allows to keep the main aspects of nonlocal hardening.

As a result a thermodynamical description based on the Helmholtz free energy is constructed from a finite and reduced number of internal variables describing simultaneously the usual crystalline plasticity and the microstructure evolution.

Numerical results in the polycrystalline case, where intergranular interactions are taken into account using a classical self-consistent scheme, show significant effects when dealing with an evolving dislocation cell structure.

2. Thermodynamical framework

In the following the classical thermodynamical formulation with internal variables is used in order to describe the intragranular heterogeneization and is thus restricted to the grain level. This allows to derive the elastoplastic behavior for the grain. The further scale transition to the polycrystal level is achieved in a classical manner using a self-consistent scheme.

2.1. Basic field equations

Let us consider a heterogeneous nonlocal medium subjected to a uniform displacement $u_i^d = E_{ij}x_j$ imposed at its boundary .

Classical field equations are given by :

- compatibility equation : in the small pertubation hypothesis, the total strain tensor is given by :

$$\varepsilon_{ij}=\frac{1}{2}\left(u_{i,j}+u_{j,i}\right) \tag{2.1}$$

- equilibrium : in the quasi-static situation, the Cauchy stress tensor is obtained through :

$$\sigma_{ij,j}=0 \tag{2.2}$$

- additive decomposition of ε into elastic and plastic parts ε^e and ε^p :

$$\varepsilon_{ij}(r)=\varepsilon_{ij}^e(r)+\varepsilon_{ij}^p(r) \tag{2.3}$$

- elastic constitutive relationship :

$$\sigma_{ij}=C_{ijkl}\varepsilon_{kl}^e=C_{ijkl}\left(\varepsilon_{kl}-\varepsilon_{kl}^p\right) \tag{2.4}$$

Furthermore, in case of uniform elasticity and application of Hill's lemma, the following averagings can be operated on the volume V:

$$E^p = \frac{1}{V}\int_V \varepsilon^p(r')dV' \tag{2.5}$$

$$\Sigma = \frac{1}{V}\int_V \sigma(r')dV' \tag{2.6}$$

The plastic part of the behavior is deduced from thermodynamical considerations.

2.2. General thermodynamical considerations

The evolution of a thermodynamical system derives generally from the maximal-dissipation principe [4,5] which states that the dissipation per unit volume D is such as :

$$D = P_{ext} - \phi \geq 0 \tag{2.7}$$

where P_{ext} is the power developed by external forces, per unit volume :

$$P_{ext} = \Sigma:\dot{E} \tag{2.8}$$

and ϕ is the Helmholtz free energy per unit volume.

In case of uniform elasticity and a heterogeneous plastic strain field, ϕ is expressed by :

$$\phi\left(E,\left\{\varepsilon^p(r)\right\}\right) = \frac{1}{2}\left(E - E^p\right):C:\left(E - E^p\right) - \frac{1}{2V}\int_V \sigma'(r'):\varepsilon^p(r')dV' \tag{2.9}$$

where $\sigma'(r) = \sigma(r) - \Sigma$ are the internal stresses resulting only from the field ε^p within V which verify equilibrium ($\operatorname{div}\sigma' = 0$) and the mean value ($\overline{\sigma'} = 0$) properties.

Two contributions can be distinguished in (2.9) ; the first one corresponds to the imposed conditions at the boundaries, the second one is the stored energy due to the heterogeneous plastic strain field.

2.3. Application to the two-phase situation

During plastic strain of a metal, dislocation cells are induced leading to high dislocation density gradients which allow to describe the cell structure as a two-phase composite, the cell interiors (resp. the walls) being represented by a soft (resp.hard) phase denoted by S (resp. H) [6].

In that case, the heterogeneous plastic strain field is given by :

$$\varepsilon^p(r) = \begin{cases} \varepsilon^{p_S} & \text{if } r \in V_S \\ \varepsilon^{p_H} & \text{if } r \in V_H \end{cases} \tag{2.10}$$

where f is the dislocation cell volume fraction, so that

$$E^p = f\,\varepsilon^{p_S} + (1-f)\varepsilon^{p_H} \tag{2.11}$$

Consequently the stored energy appearing in (2.9) becomes :

$$\frac{1}{2V}\int_V \sigma'(r'):\varepsilon^p(r')dV' = \frac{1}{2}f\,\overline{\sigma'}^S:\left(\varepsilon^{p_S} - \varepsilon^{p_H}\right) \tag{2.12}$$

Experimental TEM results show that the dislocation cell shape may be approximated by parallel and similar ellipsoids [7,1]. On the other hand, the small lattice misorientations from cell to cell may be neglected in each grain. This allows to consider in a micromechanical way only an averaged description of the cell structure where the cell shape is an ellipsoïd with half-axes a_1, a_2, a_3 and Euler angles α_1, α_2, α_3 describing the orientation of the ellipsoïd principal reference frame versus the crystalline lattice (Fig.1).

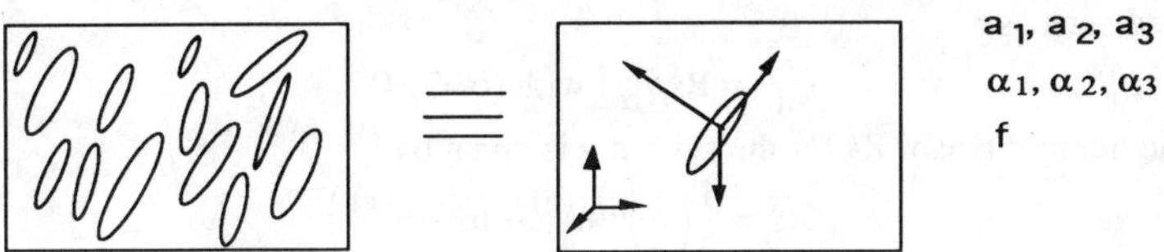

Fig 1. Averaged representation of the cell structure

The dislocation cell volume fraction f is given by $f = \dfrac{4\pi a_1 a_2 a_3 N}{3V}$ (2.13)

N is number of cells. (2.13) allows to define six independent morphologic variables $r_1 = \dfrac{a_1}{a_2}$, $r_2 = \dfrac{a_1}{a_3}$, α_1,.α_2, α_3 and f.

Under such assumptions the averaged internal stress $\overline{\sigma}^S$ inside soft phase are obtained through the Eshelby-Kröner formulation [8]:

$$\overline{\sigma}^S = (1-f)C:\left(I - S^1\right):\left(\varepsilon^{p_H} - \varepsilon^{p_S}\right) \qquad (2.14)$$

where S^1 is the Eshelby tensor accounting for shape and orientation of ellipsoid. $\overline{\sigma}^H$ is deduced from the mean values properties concerning internal stresses by :

$$(1-f)\overline{\sigma}^H = -f\,\overline{\sigma}^S \qquad (2.15)$$

ϕ becomes :

$$\phi\left(E, \varepsilon^{p_S}, \varepsilon^{p_H}, f, A\right) = \frac{1}{2}\left(E - E^p\right):C:\left(E - E^p\right) + \frac{1}{2} f(1-f)\Delta\varepsilon^p:C:\left(I - S^1(A)\right):\Delta\varepsilon^p \qquad (2.16)$$

where $\Delta\varepsilon^p = \varepsilon^{p_S} - \varepsilon^{p_H}$ and $A(r_i, \alpha_i)$ is a tensor describing the ellipsoid.

Finally the dissipation is expressed in the two-phase situation by :

$$D = f\left(\Sigma_{ij} + \overline{\sigma}_{ij}^S\right)\dot{\varepsilon}_{ij}^{p_S} + (1-f)\left(\Sigma_{ij} + \overline{\sigma}_{ij}^H\right)\dot{\varepsilon}_{ij}^{p_H}$$

$$+\left(\Sigma_{ij}\Delta\varepsilon_{ij}^p - \frac{1}{2}(1-2f)\Delta\varepsilon_{ij}^p C_{ijkl}\left(I_{klmn} - S_{klmn}^1\right)\Delta\varepsilon_{mn}^p\right)\dot{f} \qquad (2.17)$$

$$+\frac{1}{2} f(1-f)\Delta\varepsilon_{ij}^p C_{ijkl} S_{klmnpq}^2 \Delta\varepsilon_{mn}^p \dot{A}_{pq}$$

where $S^2_{ijklmn} = \frac{\partial S^1_{ijkl}}{\partial A_{mn}}$.

In the case of crystalline metals, dislocations are restricted to move on the crystallographic slip systems. As a first approximation the walls are considered as a crystallographic part of the grain, in spite of their high defect density. Furthermore the slip planes are assumed to remain continuous through the walls. Thus plastic glide γ is introduced such as :

$$\dot{\varepsilon}^{p\alpha}_{ij} = R^g_{ij}\dot{\gamma}^g_\alpha \quad \text{with } (\alpha{=}S,H) \tag{2.18}$$

where the Schmid tensor R^g for the system g is given by :

$$R^g_{ij} = \frac{1}{2}\left(m_i^{(g)} n_j^{(g)} + m_j^{(g)} n_i^{(g)} \right) \tag{2.19}$$

n_i is the unit normal to the slip plane and m_i is the slip direction,

In the same way :

$$\Delta\varepsilon^p_{ij} = R^g_{ij}\left(\gamma^g_S - \gamma^g_H\right) \tag{2.20}$$

The dissipation is generally expressed in the following manner :

$$D = F_i \dot{X}_i \geq 0 \tag{2.21}$$

where X_i are the internal variables of the thermodynamical system and F_i are the thermodynamical forces acting on X_i.

Here the internal variables of the crystal with cell structure appear to be:

- the amount of slip (γ)on the slip systems (soft and hard parts),
- the volume fraction (f) of the interior part of the cells.
- the morphological parameters (r_i, α_i) describing the shape and orientation of the cells

The evolution of the internal variable are deduced after definition of complentary relations concerning the dissipation.

3. Complementary relations for the dissipation

a) The actual irreversible evolution of the thermodynamical system, i. e. the evolution of the X_i, depends on the behavior class (elastoviscoplasticity, viscoelasticity, visco-plasticity, elastoplasticity). In the elastoplastic case, the generalized forces F_i appearing in (2.21) are restrained inside a convex domain, the so-called elasticity domain which contains the origin O due to the existence of a threshold F_i^c.

If $F_i(\Sigma, X_j) \leq F_i^c(X_j)$ then X_i=0, the response is purely elastic.

If F_i reaches F_i^c then parameters X_i may evolve following the consistency rule

$$\dot{F}_i(\Sigma, X_j) = \dot{F}_i^c(X_j) \tag{3.1}$$

This allows to obtain the following system with unknowns $\dot{X}_i$, characterizing the evolution of the system :

$$\frac{\partial F_i}{\partial \Sigma_{kl}}\dot{\Sigma}_{kl}=\left(\frac{\partial F_i^c}{\partial X_j}-\frac{\partial F_i}{\partial X_j}\right)\dot{X}_j \quad (3.2).$$

The elastoplastic constitutive relationship at the grain scale is obtained from (3.2) and the elastic behavior :

$$\dot{\Sigma}=C:\left(\dot{E}-\dot{E}^p\right) \quad (3.3)$$

where $\dot{E}^p=f\dot{\varepsilon}^{p_S}+(1-f)\dot{\varepsilon}^{p_H}+\dot{f}\left(\varepsilon^{p_S}-\varepsilon^{p_H}\right)$, and is writen

$$\dot{\Sigma}=L:\dot{E} \quad (3.4).$$

This overall behavior of the grain at the mesoscale constitutes the micromechanical basis for the further scale transition to the macroscale, i.e. the polycrystal.

b) On the other hand, the nonlocal hardening described in (1.1) through the integral formulation appears in this description in $F_i^c\left(X_j\right)$, or more explicity, :

$$\dot{\tau}^g_{c(\alpha)}=H^{gh}_{(\alpha)S}\dot{\gamma}^h_S+H^{gh}_{(\alpha)H}\dot{\gamma}^h_H \quad (3.5)$$

where $H_{\alpha\beta}$ (α,β)=(H,S) are expressed from physical parameters (Burgers vector modulus, mean free path of dislocation , ...) [9,1].

4. Applications

In the particular case of the dislocation cell structure, two main additional hypothesis are assumed in the following :

i) After a few percent of plastic strain the volume fraction f does not evolve so that f is taken constant and equal to 0,8 [7]. f is then a material parameter and no more an internal variable.

ii) The morphology adapts itself instantaneously to the loading in order to reduce the level of internal stresses. Micromechanically speaking the free energy is minimized relatively to variables A_{ij}. This assumption corresponds to the well accepted concept of Low Energy Dislocation Structures (LEDS) popular in physical metallurgy. We assume

$$F^c_{A_{ij}}=0 \text{ and } \dot{F}^c_{A_{ij}}=0 . \quad (4.1)$$

The calculation are performed for a 100 spherical grains BCC polycrystal without initial textures. Isostropic elasticity (μ=80000 MPa, ν=0.3) and slip systems of type {110}<111> are assumed. The initial critical shear stress is supposed identical for all the systems of each phase. For uniaxial tension test, Fig. 2. shows simultaneously the third order stresses, the resulting second order stresses (both described by the different clouds of points for a frozen macroscopic state of stresses after unloading) and the macroscopic response. The curves of Fig. 2. describe the means stresses of phases S and H as a function of macroscopic plastic strains. A strong heterogeneization induced by plastic strainning is observed. In agreement with experimental measurements from Mughrabi [1], the cloud of phase H is more extented than cloud of phase S .

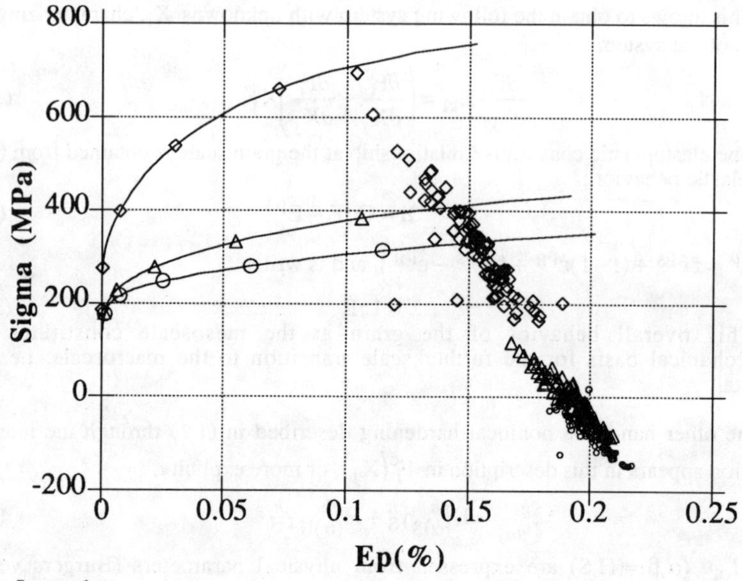

Legend :

Δ Second order stresses ◇ Stresses in thewalls o Stresses in the cells

Fig. 2. Distribution of internal stresses (2nd and 3rd order) after a tensile test

The corresponding evolution of the cell morphology in each grain is given by Fig. 3(a) for one grain and 3(b) for all the grains. Fig. 3(a) shows for one grain the evolution of the two half-axes ratio of the ellipsoid ($r_1 = \frac{a_1}{a_2}$, $r_2 = \frac{a_1}{a_3}$) during the plastification. After a few percents of plastic strain, a saturation of the cell shape evolution appears. We can explain this saturation by the stabilisation of plastic activity (cf Fig. 3(a)). In Fig 3(b), for all the grains, the initial cell shape is an oblate (2-2-1) ellipsoid. After a high evolution, the average of the higest half-axes ratio grows slowly during the plastic deformation. The cloud of points represents the higest half-axes ratio in each grain at E^p=20%. 70% of grains obtains the finally cell shape for a tensil test (4-4-1) in this case. These remarks (finally shape, and saturation) are noted experimentally by Schmitt [7].

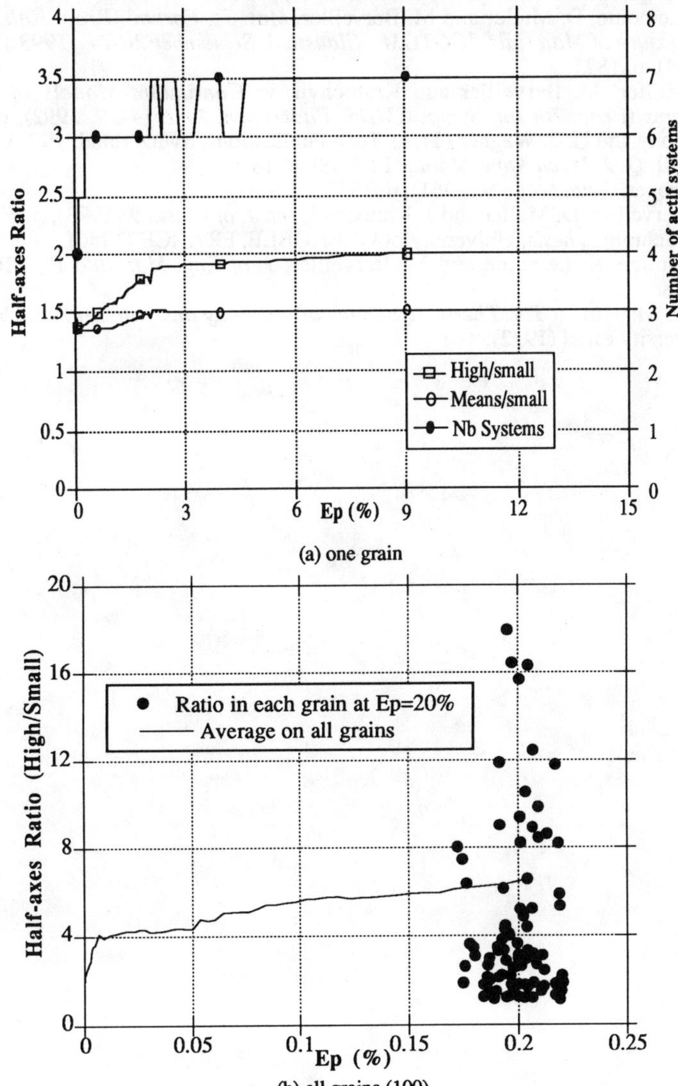

(b) all grains (100)

Fig. 3. Evolution of cell half-axes ratios during a tensile test

References

1. H. Mughrabi, *Revue Phys. Appl.*, **23** (1988), p. 367.
2. X. Lemoine, D. Muller and M. Berveiller, *Mat. Sc. Forum* (*Proc. 10th Int. conf. on Texture of Materials ICOTOM , Clausthal, September 20-24 , 1993*), **157-162** (1994), p. 1821.
3. D. Muller, M. Berveiller and Kratochvil, In *Continuum Models of Discrete Systems* (*Proc. 7th Int. Symp. CMDS, Paderborn, June 14-19, 1992*), eds K.-H. Anthony and H.-J. Wagner (Trans Tech Publications, Switzerland, 1993), p 195.
4. R. Hill, *Q. J. Mech Appl. Math.,* **1** (1948), p. 18.
5. E. Kröner, *Acta. Met.,* **9** (1961), p. 155.
6. M. Berveiller, D. Muller and J. Kratochvil, *Int. J. of Plast.*, **9** (1993), p. 633.
7. J.H. Schmitt, *Thesis*, University of GRENOBLE, FRANCE (1986).
8. D. Muller, X. Lemoine and M. Berveiller, *J. of Eng. Mat. and T.*, **116** (1994), p. 378.
9. G. A. Maugin In *The Thermodynamics of plasticity and fracture*, ed. Cambridge University Press (1992).

Continuum Models and Discrete Systems
Proceedings of 8th International Symposium, June 11-16, 1995, Varna, Bulgaria, ed. K.Z. Markov

SOME GAUGE IDEAS IN DISLOCATION THEORY

J. L. MARQUÉS
Department of Theoretical Physics, University of Paderborn,
Warburger Straße 100, D-33098 Paderborn, Germany

Abstract. A model to describe the transition in a solid body from elastic to plastic behaviour is studied. The definition of an order parameter coupled to a "gauge field," which describes the dislocation density, is used to understand the different phases by means of a partition function (statistical mechanics). It is shown that on the ground of a particular transformation there is a close analogy between the transition from elastic to plastic phase on the one hand and from hardening to plastic phase on the other. Moreover, this transformation associates hardening with a model of the elementary particle physics, the quark confinement.

1. Introduction

A solid body under an external load exhibits a behaviour, which is comprised in the stress-strain diagram in Fig. 1.

In the elastic zone the number of dislocations is low. But at the point A, the number of dislocations increases dramatically giving rise to a much greater response of the body to an external load: a small increase in the stress corresponds to a great change of strain. This is explained by the motion of a great number of dislocations through the crystal, a motion which results in a softening of the material with respect to the external loading. At the point B the density of dislocations is high enough to build up dense dislocations groups: the moving dislocations cannot go through these groups and the plastic behaviour comes to an end. In the hardening zone the behaviour of the material is similar to the elastic behaviour although the response to an external stress is even harder than in the elastic zone. But both behaviours are characterized by completely different parameters: the elastic zone is ordered (low number of dislocations, fairly perfect crystal) and the hardening zone is very disordered (macroscopic density of dislocations).

The subject of this paper is to build up a statistic model, which allows us to describe the graphic in Fig. 1. In order to achieve this, we will make use of two known models from the solid state physics and the particle physics: the superconductivity and the quark confinement. Let us start by describing a superconductor: in the high temperature phase we have a normal conductor (the electromagnetic field, the photon, is massless). At the critical temperature charged bosons, the Cooper pairs, are formed, and under this critical temperature the electric field is given a mass: the field's range is finite and the photon cannot disrupt the boson ordering (Bose condensation). This superconducting phase is an ordered phase. And in this phase the external magnetic field penetrates the material only inside certain topological configurations, which are called vortices. The coupling between two electrical charges is weaker because the photon has a finite range [1]. In the quark confinement model on the other hand, a condensation of certain field configurations (called instantons), which are topologically stable, leads to a vacuum's behaviour, which confines electrical charges.

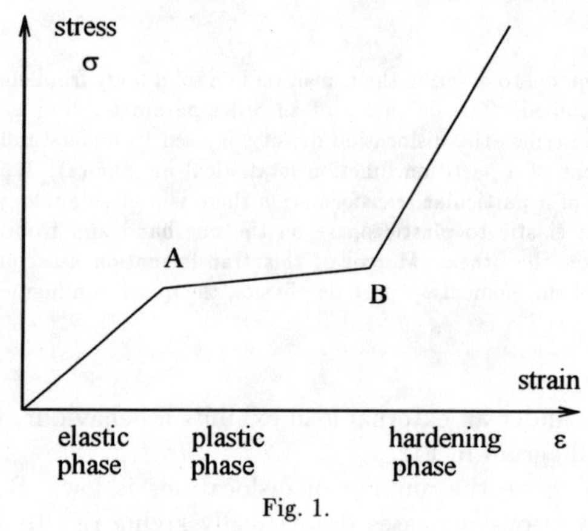

Fig. 1.

2. The Equations of Dislocation Dynamics

Now we shall write the equations of dislocation dynamics in a form analogous to Maxwell's equations. Let us begin with the static case:

$$\operatorname{div}\alpha = 0\,, \quad {\alpha_i}^a = -\epsilon_{ijk}\partial_j {A_k}^a\,,$$

where α is the dislocation density. The dislocation density α and the Burgers vector are identified with the magnetic field and the magnetic flux. The distortion tensor ${A_k}^a$ plays the part of the magnetic potential. The source for this "magnetic field" are the Frank-Read sources: let us look at a Frank-Read source and let us integrate on the area in the plane, where the source stands, see Fig. 2:

$$\oint_C \vec{\alpha}^a \cdot d\vec{l}\,.$$

Similarly to electromagnetism (Biot-Savart's law), we define an "electric current" as:

$$\oint_C \vec{\alpha}^a \cdot d\vec{l} = \vec{\mathcal{J}}_{\mathrm{el}} \cdot \hat{n} S\,. \tag{1}$$

This "electric current" is proportional to the external stress: the Frank-Read source's area becomes wider when the external stress is greater:

$$(\mathcal{J}_{\text{el}})_i^a \sim (\sigma_{\text{ext}})_i^a \,. \tag{2}$$

Note that the external stress comes into the theory as "electric current."

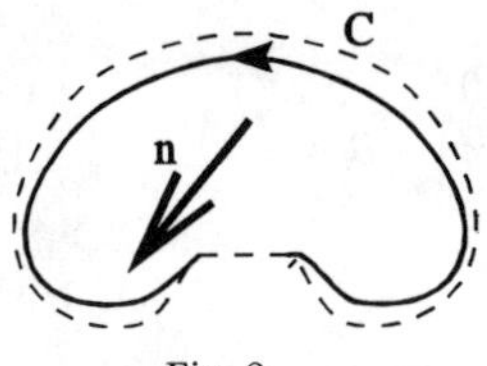

Fig. 2.

The generalization to the dynamical case is easy:

$$\text{rot}\, j_{\text{disl}} + \partial_t \alpha = 0\,,$$

$$(j_{\text{disl}})_i^a = -\partial_i v^a + \partial_t A_i^a\,,$$

where $(j_{\text{disl}})_i^a$ is the dislocation flux and v^a the dislocation velocity. The dislocation flux is similar to the electric field times the velocity of light c. The magnetic potential is here minus the distortion tensor, and c times the electric potential plays the part of the dislocation velocity. But one of the equations of the dislocation theory plays here the part of the gauge fixing equation, namely, the equilibrium condition:

$$\partial_i \sigma_i^a = \rho \partial_t v^a\,, \qquad \text{where } \sigma_i^a = \mu A_i^a$$

$$-\partial_i A_i^a + \frac{\rho}{\mu} c_s \partial_t \phi^a = 0 \qquad \text{with } \rho = \text{mass density and } \mu = \text{elastic constant}\,,$$

which is analog to the Lorentz gauge:

$$\text{div}\, A + \frac{1}{c}\partial_t \phi = 0\,,$$

where the velocity of light is here the velocity of sound:

$$\frac{\rho}{\mu} c_s = \frac{1}{c_s}\,, \quad \text{i.e.,} \quad c_s = \sqrt{\frac{\mu}{\rho}}\,.$$

This is important: there is not, in our model, the freedom to choose the gauge, since the gauge is already chosen by one of the equations of the dislocation theory. Here the electromagnetic potentials are observables (the strain and the velocity) and therefore we cannot gauge away one of these potentials. We only make use of the concept "gauge field" because the equations of the dislocations can be written in a form similar to the Maxwell equations. Let us write all these analogies:

electromagnetism	dislocation dynamics
$\operatorname{div} B = 0$	$\operatorname{div} \alpha^a = 0$
$B = \operatorname{curl} A$	$\alpha_i^a = -\epsilon_{ijk}\partial_j A_k^a$
$\operatorname{curl} cE + \partial_t B = 0$	$\operatorname{curl} j_{\text{disl}}^a + \partial_t \alpha^a = 0$
$cE = -\operatorname{grad}(c\phi) - \partial_t A$	$(j_{\text{disl}})_i^a = -\partial_i v^a + \partial_t A_i^a$
$\operatorname{curl} \frac{B}{\mu_0} = \mathcal{J}_{\text{el}}$	$\operatorname{curl}(\mu\alpha^a) = \mathcal{J}_{\text{el}}^a$, with $\mathcal{J}_{\text{el}} \sim \sigma_{\text{ext}}^a$

3. Construction of Our Model

Let us suppose we have a model, where the external stress comes into the equations as electric charge. In the elastic phase the interaction between two external stresses is Coulombian (the propagator is the Green's function of the Laplacian operator, i.e., in two dimensions lnr and $1/r$ in three dimensions). In the plastic phase the body is softer and we can describe it by means of a force between two external stresses, which is short ranged. In the hardening phase the force is even stronger than in the elastic phase and this can be explained through a confinement of external stresses, which play the part of electric charges, as we have seen.

Let us define in a deformable crystal lattice an order parameter, which we call director field, as:

$$\phi^a = |\phi^a| e^{i\varphi^a},$$

$$|\phi^a| = \sqrt{\frac{\mu b^2}{4\pi^2}}, \quad \varphi^a = \frac{2\pi}{b} u^a,$$

with $a = 1, 2, 3$, u^a is the displacement vector and b the atomic distance. The information on the shift of the crystal cell center is carried in the field's phase φ^a. The radial degree of freedom $|\phi^a|$ is frozen to a constant, which is proportional to the square root of the elastic constant μ. With this director field a distortion and a dislocation can be described as shown in Fig. 3.

Now we need an energy for this field. The elastic energy in the linear approximation is:

$$\int dv \frac{1}{2}\mu \sum_{k,a=1,2,3} (\partial_k u^a)^2$$

and we propose as energy for the director field the following formula, which is defined on a discrete lattice (for this reason we have replaced the integral by a sum):

$$E = |\phi|^2 \sum_{\vec{r}} \sum_{k,a=1,2,3} \left[1 - \cos(\varphi^a_{\vec{r}+\hat{k}} - \varphi^a_{\vec{r}})\right] . \tag{3}$$

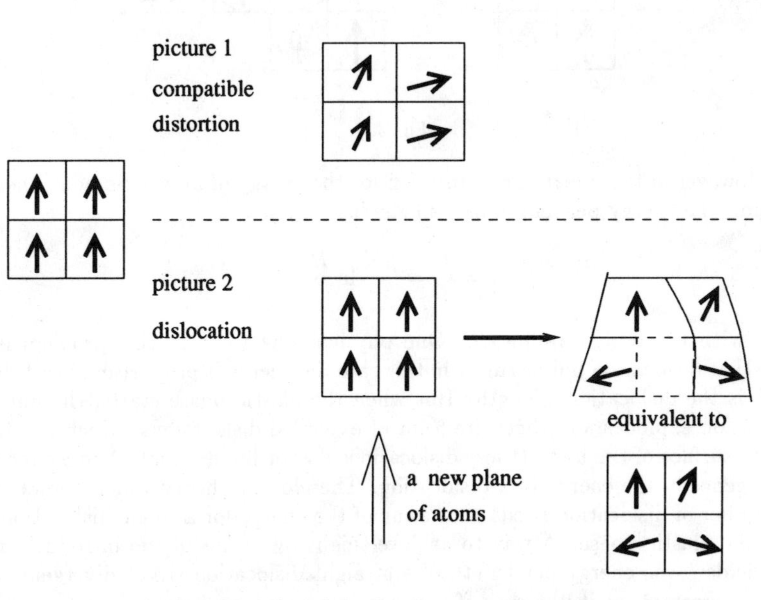

Fig. 3.

Why this form? Firstly because, when the distortion is small, we can approximate the cosine by:

$$\varphi^a_{\vec{r}+\hat{k}} - \varphi^a_{\vec{r}} = \frac{2\pi}{b}(u^a_{\vec{r}+\hat{k}} - u^a_{\vec{r}}) \approx 0 \,,$$

hence

$$|\phi|^2\{1 - \cos(\varphi^a_{\vec{r}+\hat{k}} - \varphi^a_{\vec{r}})\} \approx |\phi|^2 \frac{1}{2}(\frac{2\pi}{b})^2(u^a_{\vec{r}+\hat{k}} - u^a_{\vec{r}})^2$$

and we recover the elastic energy in the linear approximation. But Eq. (3) has another advantage: once a dislocation has crossed the crystal, the body's energy is the same as at the beginning, and we can also describe this with the energy (3), see Fig. 4.

After the dislocation has crossed the crystal, the directors above the slip plane have suffered a complete rotation.

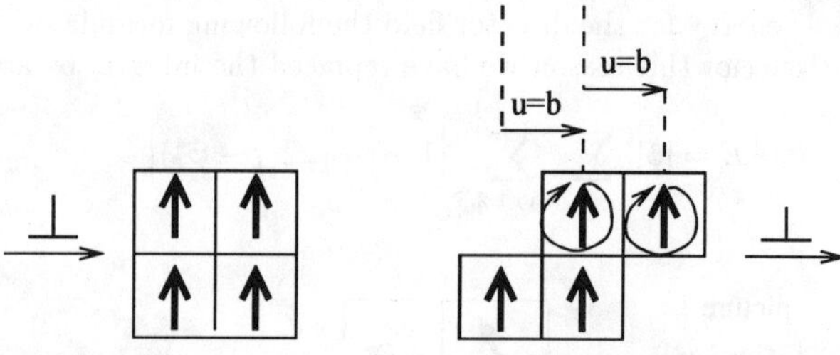

Fig. 4.

However in the linear approximation for the energy of an elastic body there is a problem: the energy per length of a single dislocation

$$\epsilon = \frac{E}{L} = \frac{\mu b^2}{4\pi} \ln \frac{R}{r_0}$$

diverges with the body's radius R. One physical way to solve this problem is to suppose that dislocation builds rings: in this case the energy is proportional to $L \cdot \ln L$, where L is the dislocation's length. But when the plastic phase starts, the number of dislocation explodes and there are a lot of extended dislocations, which run from surface to surface of the body (these dislocations do not build rings). And we recover the divergence in the energy of a dislocation. Therefore, a theory which allows for a great number of dislocation needs a lowering of the energy for a single dislocation, at least in the plastic phase. A way to achieve this is by means of the introduction of "gauge fields:" the energy per length of a straight dislocation is not divergent with $\ln R$ but a constant, as in the case of a vortex in a superconductor:

energy per length of a vortex in a superfluid:

$$\epsilon = \frac{\rho_s \kappa^2}{4\pi} \ln \frac{R}{r_0} \qquad \text{with } \kappa = \text{circulation and } \rho_s = \text{superfluid density}$$

energy per length of a vortex in a superconductor:

$$\epsilon = \frac{\lambda^{-1}\Phi_0^2}{4\pi} \ln \frac{\lambda}{\delta} ,$$

where λ is the penetration length of the magnetic field, δ the correlation length of the boson field and Φ_0 the magnetic flux, with $\lambda/\delta = \text{constant}$.

The difference between a superfluid and a superconductor is that the order parameter is coupled to a gauge field in a superconductor. In our case the gauge fields are not real gauge fields, since the phase of our director field is an observable and cannot be gauged away. This is related with the theory in Section 2, where the gauge was already chosen by one of the equations of the dislocation dynamic. What does our "gauge field" mean? The "gauge" (magnetic and electric) fields are the dislocation density and flux, as we have already seen in the Section 2. This has the advantage that the Burgers vector (here the magnetic flux) is topological quantized, as in the superconductor [2].

The energy of the body for the static case is then:

$$E = \sum_{\vec{r}} \sum_{k,a=1,2,3} \left\{ \frac{1}{2e^2} (\alpha^a_{\vec{r}})^2 + [1 - \cos(\varphi^a_{\vec{r}+\hat{k}} - \varphi^a_{\vec{r}})] \right\}, \tag{4}$$

where $\alpha^a_{\vec{r}}$ is the dislocation density and e is the coupling constant between the director field and the "gauge" field. This coupling constant must be proportional to the elastic constant μ, since the greater the elastic modulus, the stronger the elastic body and the more difficult the creation of a dislocation.

For the dynamic case the generalization is:

$$E = \sum_{\vec{r}} \sum_{k=0,1,2,3} \sum_{a=1,2,3} \left\{ \frac{1}{2e^2} [(\alpha^a_{\vec{r}})^2 + (j^a_{\mathrm{disl}\vec{r}})^2] + [1 - \cos(\varphi^a_{\vec{r}+\hat{k}} - \varphi^a_{\vec{r}})] \right\}. \tag{5}$$

But we will look at the phase transition from elastic to plastic behaviour and we do not need to deal with the dynamic case, since this transition is originated by the creation of dislocations at the Frank-Read sources and not by the flux of external dislocations. With the energy for the static case we can build a statistic by means of a partition function:

$$Z = \sum_{\text{all configurations}} \exp\left(-\frac{E_{\mathrm{conf}}}{\xi}\right), \tag{6}$$

where E_{conf} is the energy of a configuration, calculated by means of formula (4), and the parameter ξ is the energy scale (for example, the temperature): a certain configuration is probable when the configuration's energy is not much higher than this scale ξ. We will look for different behaviours in the partition function in order to find the critical points of the different transitions.

The phase transition from elastic to plastic state can be seen in the following way: the elastic phase is characterized by a long-range order of the director field. The correlation length δ of the director field (= order parameter) is infinite. In the plastic phase there is an explosion of large vortex loops, and dislocations from side to side, and the correlation length is finite. The transition takes place when the "mass" of the director field, defined as *(correlation length* $\delta)^{-1}$, becomes non zero.

We must describe the hardening phase too. In this phase a condensation of dislocations in macroscopic groups takes place and we can speak of a disorder parameter related to this dislocation condensation. This disorder parameter has a non zero average in the hardening phase. We have a similar situation in the elastic phase, where there was a condensation of an *order* parameter (the director field), but in this case the condensation is of a *disorder* parameter.

There is a mathematical transformation of the partition function, named dual transformation, which allows us to represent it in two different ways. The first one, by using an order parameter and "gauge" fields for the interaction between these order fields; and a second way by using a disorder parameter and the corresponding "gauge" fields. We can see how this transformation operates in the case of a superconductor.

In a superconductor there are electric charged bosons (the Cooper pairs) and the gauge fields (the electromagnetic fields). In the high temperature region we have a normal conductor, where the interaction between electric charges is Coulombian. In the low temperature phase a condensation of the charged bosons takes place and the magnetic field can only exist inside the vortex. Let us suppose that two magnetic monopoles (one positive monopole and the second negative) are in the superconductor: the magnetic field of the magnetic monopoles is squeezed inside a vortex. But since the vortex energy is proportional to the length, the energy of this configuration increases when the distance between the monopoles grows. The magnetic monopoles are confined.

A magnetic monopole cannot be described in an easy way, because the magnetic field is obtained by means of a curl:

$$B = \operatorname{curl} A$$

and the only way to introduce a magnetic monopole in this description:

$$\operatorname{div} B = \rho_{\mathbf{mag}}$$

is to allow a topological defect in the definition of the magnetic potential A.

Now let us suppose we have a magnetic superconductor, i.e., the bosons have magnetic charge. In the low temperature phase of this system, which we call dual-temperature, these bosons condense and the electric field can only exist inside a magnetic vortex. Therefore two electric charges (a positive charge and the second negative) in this body are confined. But a condensation of magnetic charged fields corresponds to a condensation of topological defects in a normal (=electric) superconductor. And since the magnetic charges are described as defects in a normal superconductor, a condensation of magnetic charges means a disordered system in a normal superconductor. Disorder takes place in the high temperature phase. In this way we have achieved a transformation, which associates a high temperature phase (disordered phase) in a normal superconductor with a low dual-temperature phase (condensation of topological defects) in a magnetic superconductor. This transformation has the advantage that we can describe the disordered phase with a (dis)order parameter, which condenses in the disordered phase. And then the description of the high temperature phase is formally similar to the description of the low temperature phase.

Now let us come back to our model. The elastic phase and the hardening phase are formally similar: in the first phase there is a condensation of an order parameter (the director field) and in the second one there is a condensation of topological defects (the dislocations). But we know that in the phase where the condensation of topological defects takes place, the electric charges are confined. In our model the electric currents are proportional to the external stress and, just as we would like, in the dislocation condensation phase we have the confinement of external stress, i.e., hardening: in this phase a greater force to stretch the solid body is necessary.

Let us summarize the dual transformation in our model:

our system	dual system
electric charge: external stress topological charge: dislocation	electric charge: dislocation topological charge: external stress
Low ξ phase: condensation of electric charged bosons (director field). Confinement of topological charges (dislocations build small rings). Ordered phase: elastic phase.	High dual-ξ phase: no macroscopic condensation of dislocations. Ordered phase: plastic and elastic phases.
High ξ phase: correlation length of the director field not infinite. Disordered phase: plastic and hardening phases.	Low dual-ξ phase: condensation of the dual electric charged fields (dislocations). Confinement of dual topological charges (external stresses). Disordered phase: hardening phase.

4. Conclusions

We have built up a model, which can describe the complete stress-strain diagram with only one partition function. The external stress plays the role of the elèctric charge and the dislocation density the role of the magnetic field. The elastic phase is identified with the superconducting phase (condensation of the order parameter, the director field). The plastic and hardening phases are identified with the normal conducting phase, since there is no more order. In order to differentiate the plastic of the hardening phase, we make use of the dual transformation which carries the hardening phase to a condensation phase of topological defects (the dislocations), and produces a confinement of our electric charges, the external stress.

Acknowledgments. I am very grateful to Prof. Dr. Karl-Heinz Anthony for numerous discussions and his patience.

References

1. M. E. Peskin, *Ann. Phys.* **113** (1978) 122.
2. H. B. Nielsen and P. Olesen, *Nucl. Phys. B* **61** (1973) 45.

Continuum Models and Discrete Systems
Proceedings of 8th International Symposium, June 11-16, 1995, Varna, Bulgaria, ed. K.Z. Markov
©World Scientific Publishing Company, 1996, pp. 564–572

MULTIAXIAL VISCOPLASTIC EXPERIMENTS VERSUS TENSOR FUNCTION REPRESENTATION

M. V. MIĆUNOVIĆ
Faculty of Mechanical Engineering of Kragujevac University,
Yu-34000 Kragujevac, Yugoslavia

Abstract. The paper deals with two models of viscoplastic behaviour of steels. One model is based on kinematic hardening and "universal" flow curve. The other model modifies Rice's evolution equation to include logarithmic plastic strain as representative of pattern of internal rearrangement (PIR) and applies tensor function representation. Experimental evidence from JRC-Ispra, Italy, concerning tension as well as shear tests, was confronted with these models. It is shown that the first model is unable to describe both tests simultaneously whereas the second model gives reasonably good description even with four material constants.

1. Introduction

Theoretical consideration of viscoplasticity has become an important item for finite element codes which pretend to perform calculations of complex structures with a high precision. In majority of them evolution equation for plastic strain rate is of associate type, i.e., it is perpendicular to yield surface in stress space. It should be noted that usually yield function is detected from tension tests and then applied to calculation during arbitrary stress-strain histories appearing in real structures. Such a procedure could produce significant errors destroying geometrical accuracy which FEM-codes offer.

The goal of this paper is to analyze two constitutive models and to compare plastic stretching tensors, calculated from their evolution equations, with experimentally found plastic stretchings. This would clarify admissibility of evolution equations in FEM-codes. Experiments were performed in Dynamic testing laboratory of JRC-Ispra, Italy, with specimen made of austenitic stainless steel AISI 316 in the range of small strain rates. The first model is the above explained whereas the second model is based on [1].

Before stating theories and comparing them with experiments clear stress and strain measures are necessary. This is the subject of the next Section. In the remaining Sections an analysis, results and some conclusions are given.

2. Preliminaries

Consider a crystalline body, $\mathcal{B}$, in a real configuration (ψ_t) with defects (such as dislocations) and an inhomogeneous temperature field $\theta(X,t)$ (where t stands for time and X for a particle of $\mathcal{B}$) subject to surface tractions. Corresponding to (ψ_t) there exists an observable reference configuration (κ_0) with defects differently distributed and a homogeneous temperature θ_0 without surface tractions. Due to the defects (κ_0) is not stressfree but contains an equilibrated residual stress (i.e., "back-stress").

In concordance with papers dealing with continuum representation of dislocation distributions (ψ_t) is imagined to be cut into natural state elements (ν_t) at (θ_0).

Whereas linear mapping function $\mathbf{F}(\cdot,t) : (\kappa_0) \to (\psi_t)$ is compatible (*the total deformation gradient tensor*), its constituents $\mathbf{F}_E(\cdot,t) : (\nu_t) \to (\psi_t)$ and $\mathbf{F}_P(\cdot,t) : (\kappa_0) \to (\nu_t)$ are not, and for this reason they are called *thermoelastic* and *plastic* distortion tensor, respectively. Obviously, Kröner's decomposition rule holds

$$\mathbf{F}_P(\cdot,t) := \mathbf{F}_E(\cdot,t)^{-1}\mathbf{F}(\cdot,t)\,. \tag{1}$$

According to the polar decomposition $\mathbf{F}_P = \mathbf{R}_P\mathbf{U}_P = \mathbf{V}_P\mathbf{R}_P$ (with $\mathbf{R}_P^{-1} = \mathbf{R}_P^T$), as tensor of plastic strain we choose the following logarithmic measure

$$\boldsymbol{\varepsilon}_P = \log \mathbf{V}_P = \frac{1}{2}\log(\mathbf{F}_P\mathbf{F}_P^T)\,, \tag{2}$$

whose main advantage is that it is traceless, i.e.,

$$\pi_1 = \operatorname{tr}\boldsymbol{\varepsilon}_P = 0, \quad \pi_2 = \operatorname{tr}\boldsymbol{\varepsilon}_P^2 \neq 0, \quad \pi_3 = \operatorname{tr}\boldsymbol{\varepsilon}_P^3 \neq 0\,, \tag{3}$$

provided that plastic volume change is negligible.

The *plastic stretching* is the symmetric part of the plastic "velocity gradient" tensor

$$\mathbf{D}_P = (\mathbf{L}_P)_s, \quad \mathbf{L}_P := (D\mathbf{F}_P)\mathbf{F}_P^{-1}\,, \tag{4}$$

where D stands for material derivative.

If the thermal part of $\mathbf{F}_E$ is neglected, Hooke's law reads:

$$\mathbf{S} = \mathcal{D} : \mathbf{E}_E, \quad 2\mathbf{E}_E := \mathbf{F}_E^T\mathbf{F}_E - \mathbf{1}\,, \tag{5}$$

where $\mathbf{E}_E$ is Lagrangean elastic strain, $\mathbf{S}$ the second Piola-Kirchhoff stress tensor related to (ν_t) whereas $\mathcal{D}$ is the fourth-rank tensor of elastic constants. Employing subscript d to denote deviatoric part of a second rank tensor the following set of invariants will be used in the subsequent Sections of this paper

$$\gamma := \left\{s_2, s_3, \pi_2, \pi_3, \mu_1, \mu_2, \mu_3, \mu_4\right\}\,, \tag{6}$$

where

$$s_2 = \operatorname{tr}\mathbf{S}_d^2, \quad s_3 = \operatorname{tr}\mathbf{S}_d^3, \quad \mu_1 = \operatorname{tr}\{\mathbf{S}_d\boldsymbol{\varepsilon}_P\}\,,$$

$$\mu_2 = \mathrm{tr}\,\{\mathbf{S}_d\boldsymbol{\varepsilon}_P^2\}\,, \quad \mu_3 = \mathrm{tr}\,\{\mathbf{S}_d^2\boldsymbol{\varepsilon}_P\}\,, \quad \mu_4 = \mathrm{tr}\,\{\mathbf{S}_d^2\boldsymbol{\varepsilon}_P^2\}\,. \tag{7}$$

In the language of experimentalists the following invariants

$$\overline{\sigma} \equiv \left(\frac{3}{2}s_2\right)^{1/2}, \quad \overline{\varepsilon} \equiv \left(\frac{2}{3}\pi_2\right)^{1/2}, \quad D\overline{\varepsilon}_P \equiv \left(\frac{2}{3}\mathrm{tr}\,\{\mathbf{D}_P^2\}\right)^{1/2}, \tag{8}$$

are usually called equivalent stress, equivalent plastic strain and equivalent plastic strain respectively. All of them coincide with corresponding quantities appearing during uniaxial tension.

3. Rice's Model with Tensor Representation

According to [1] the increment of plastic strain tensor is perpendicular to a loading surface Ω =const, where Ω depends on stress, temperature and *pattern of internal rearrangement* (PIR). Translating to the language of the previous Section, an evolution equation for plastic stretching should hold in the following form [1]

$$\mathbf{D}_P = \partial_{\mathbf{S}}\Omega(\mathbf{S}, \theta, \mathrm{PIR})\,. \tag{9}$$

Here PIR is described by anholonomic internal variables representing crystal slips over active slip systems.

The plastic distortion tensor $\mathbf{F}_P$ is incompatible, represents also slips and may reflect transformation of anholonomic coordinates. Thus it is assumed here that the above equation may be written in the following way

$$\mathbf{D}_P = \left[\partial_{\mathbf{S}}\Omega(\mathbf{S}, \boldsymbol{\varepsilon}_P, \theta)\right]_d\,, \tag{10}$$

where Rice's loading function depends on temperature and invariants (6), i.e.

$$\Omega = \Omega(\gamma, \theta) \equiv \Omega(s_2, s_3, \pi_2, \pi_3, \mu_1, \mu_2, \mu_3, \mu_4, \theta)\,. \tag{11}$$

Let us approximate Ω by a fourth-order polynomial with respect to $\mathbf{S}$ and second order in $\boldsymbol{\varepsilon}_P$. With such an approximation we have:

$$\begin{aligned}\Omega \approx\; & a_1\mu_1 + a_2\mu_2 + b_1 s_2 + \frac{1}{2}b_2\mu_1^2 + +b_3\mu_3 + b_4 s_2\pi_2 + b_5\mu_4 \\ & + c_1 s_2\mu_1 + c_2 s_2\mu_2 + c_3 s_3 + c_4\mu_1\mu_3 + c_5 s_3\pi_2 \\ & + d_1 s_2^2 + d_2 s_2\mu_1^2 + d_3 s_2\mu_3 + d_4 s_2\mu_4 + d_5 s_3\mu_1 + d_6 s_3\mu_2 + \frac{1}{2}d_7\mu_3^2\,.\end{aligned} \tag{12}$$

In the above equation $a_1 = 0$, $a_2 = 0$, if we accept that plastic stretching disappears at $\mathbf{S} = \mathbf{0}$, i.e.,

$$\mathbf{D}_{P|\mathbf{S}=0} = 0\,. \tag{13}$$

Therefore, tensor representation of (10) (according to [2-4]) allows to write down the following equation:

$$\mathbf{D}_P = \sum_\alpha \Gamma_\alpha(\gamma)\mathbf{H}_\alpha, \quad \alpha \in \{1,\ldots,6\}, \tag{14}$$

with the tensor generators

$$\mathbf{H}_1 = \mathbf{S} - \frac{1}{3}\mathbf{1}\operatorname{tr}\mathbf{S} \equiv \mathbf{S}_d\,, \quad \mathbf{H}_2 = (\mathbf{S}_d^2)_d\,, \quad \mathbf{H}_3 = \boldsymbol{\varepsilon}_P\,, \quad \mathbf{H}_4 = (\boldsymbol{\varepsilon}_P^2)_d\,,$$
$$\mathbf{H}_5 = (\mathbf{S}_d\boldsymbol{\varepsilon}_P + \boldsymbol{\varepsilon}_P\mathbf{S}_d)_d\,, \quad \mathbf{H}_6 = (\mathbf{S}_d\boldsymbol{\varepsilon}_P^2 + \boldsymbol{\varepsilon}_P^2\mathbf{S}_d)_d\,, \tag{15}$$

and with scalar coefficients depending on 17 material constants $b_1,\ldots,b_5$, $c_1,\ldots,c_5$, $d_1,\ldots,d_7$ as follows:

$$\begin{aligned}
\Gamma_1/2 &= b_1 + b_4\pi_2 + c_1\mu_1 + c_2\mu_2 + 2d_1s_2 + d_2\mu_1^2 + d_3\mu_3 + d_4\mu_4\,,\\
\Gamma_2/3 &= c_3 + c_5\pi_2 + d_5\mu_1 + d_6\mu_2\,,\\
\Gamma_3 &= b_2\mu_1 + c_1s_2 + c_4\mu_3 + 2d_2s_2\mu_1 + d_5s_3\,,\\
\Gamma_4 &= c_2s_2 + d_6s_3\,,\\
\Gamma_5 &= b_3 + c_4\mu_1 + d_3s_2 + d_7\mu_3\,,\\
\Gamma_6 &= b_5 + d_4s_2\,.
\end{aligned} \tag{16}$$

In the special case when $\mathbf{D}_P$ is of the second order in $\mathbf{S}$, the constants $d_1,\ldots,d_7$ should be neglected and hence only 10 material constants are needed.

Finally, if $\mathbf{D}_P$ is of the second order in $\mathbf{S}$ and linear in $\boldsymbol{\varepsilon}_P$, just four material constants are necessary, i.e., b_1 , c_1 , b_3 and c_3, whereas all the others should be rejected.

4. Kinematic Hardening and Universal Flow Curve

Almost all viscoplastic evolution equations used in finite element codes are based on associate flow rule and uniaxial tension tests. Suppose that viscoplastic region is defined by the following "dynamic" yield function:

$$f = \overline{\sigma} - F(\overline{\varepsilon}_P, D\overline{\varepsilon}_P) - B_3\mu_1 = 0\,. \tag{17}$$

Then the widely accepted associativity of flow rule gives the following evolution equation

$$\mathbf{D}_P = \Lambda\partial_{\mathbf{S}}f = \Lambda\left(\frac{3}{2\overline{\sigma}}\mathbf{S}_d - B_3\boldsymbol{\varepsilon}_P\right) \equiv \Lambda\left(\frac{3}{2\overline{\sigma}}\mathbf{H}_1 - B_3\mathbf{H}_3\right), \tag{18}$$

where Λ may be expressed by means of equivalent stress, equivalent plastic strain and equivalent strain rate as follows:

$$\Lambda = D\overline{\varepsilon}_P(1 - 2B_3\mu_1/\overline{\sigma} + B_3^2\overline{\varepsilon}_P^2)^{-1/2}. \tag{19}$$

Inserting (19) into (18) we notice that direction of $\mathbf{D}_P$ is not influenced by the shape of $F(\bar{\varepsilon}_P, D\bar{\varepsilon}_P)$.

For a complete specification of $\mathbf{D}_P$, magnitude of equivalent plastic strain rate should be given explicitly as well. Experiments in tension suggest the following scalar evolution equation:

$$D\bar{\varepsilon}_P = m_4 \exp\left\{\left[\frac{\bar{\sigma} - B_3\mu_1}{B_1(m_1 + \bar{\varepsilon}_P)^{m_2}} - \mathbf{1}\right]\frac{\exp(m_3\bar{\varepsilon}_P)}{B_2}\right\}, \tag{20}$$

where m_4 is a minimal ("control") plastic strain rate corresponding to static stress in "rate–independent" case.

The above evolution equation (18) describes so called kinematic hardening with back-stress proportional to plastic strain. If $B_3 = 0$, then isotropic hardening takes place.

In their paper [5] Zerilli and Armstrong derived scalar evolution equations for *fcc*, *bcc* and *hcp* crystals based on thermal activation analysis and dislocation theory. For example, for *fcc* crystals like ARMCO iron as well as copper, they proposed the equation

$$\bar{\sigma} = c_1\bar{\varepsilon}_P^n + c_2\bar{\varepsilon}_P^{1/2}\exp(-c_3\theta + c_4\theta\log(D\bar{\varepsilon}_P)) + \beta(\bar{\varepsilon}_P)\ell^{-1/2}\,. \tag{21}$$

The first part describes strain hardening and is equivalent to term $B_1(m_1 + \bar{\varepsilon}_P)^{m_2}$ in (20). The last term accounts for polycrystal slip band-stress concentration, ℓ being the grain diameter. The term in the middle comes from thermal activation. If we solve (21) for $D\bar{\varepsilon}_P$ and neglect temperature terms, then matching of (20) and (21) shows how micromechanical theory describes behavior of tension specimens during tests.

5. Experimental Evidence

Theories presented in the previous two Sections are confronted with experimental data achieved in Dynamic testing laboratory of JRC CEC, Ispra [6-7]. Specimen made of AISI 316 (with chemical composition 16.7% Cr, 12.4% Ni, 1.65% Mn, 2.45% Mo, 0.006% Nb, etc.) were tested at $D\bar{\varepsilon}_P \sim 10^{-3}\,\mathrm{s}^{-1}$ induced by stress rate $D\bar{\sigma} \approx 15\,\mathrm{MPa/s}$. Large plastic strain in the range $\bar{\varepsilon}_P \in [0, 0.5]$ with $\bar{\sigma} \in [Y_0, 4Y_0]$, where Y_0 is initial yield stress, are analyzed here. The specimen tested are shown at Fig. 1.

The tensile test was performed with a Hounsfield tensometer in a standard way. The shear test was accomplished by making use of a special specimen named BICCHIERINO consisting of two rigid cylinders connected by the gauge part—a thin circular crown whose shear angle is determined by the ratio x/h.

Let us denote a set of materials constants by $\mathcal{A}$. Then for Rice's model with $\mathbf{D}_P$ of M-th order in $\mathbf{S}$ and N-th order in $\boldsymbol{\varepsilon}_P$ we have

$$\mathcal{A} = \{b_1, \ldots, b_5, c_1, \ldots, c_5, d_1, \ldots, d_7\}\,, \quad M = 3,\ N = 2,$$

$$\mathcal{A} = \{b_1, \ldots, b_5, c_1, \ldots, c_5\}, \quad M = 2,\ N = 2\,, \tag{22}$$

$$\mathcal{A} = \{b_1, b_3, c_1, c_3\}, \quad M = 2,\ N = 1,$$

while the universal flow rule with kinematic hardening requires, according to (20), the following set of material constants

$$\mathcal{A} = \{B_1, \ldots, B_3, m_1, \ldots, m_4\}. \tag{23}$$

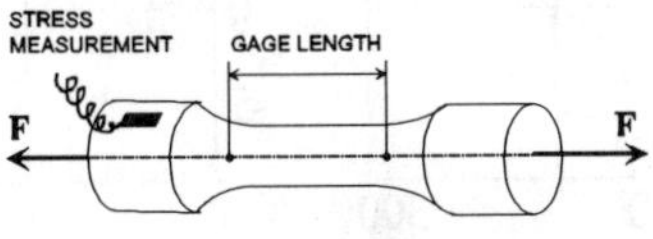

STANDARD TENSION SPECIMEN

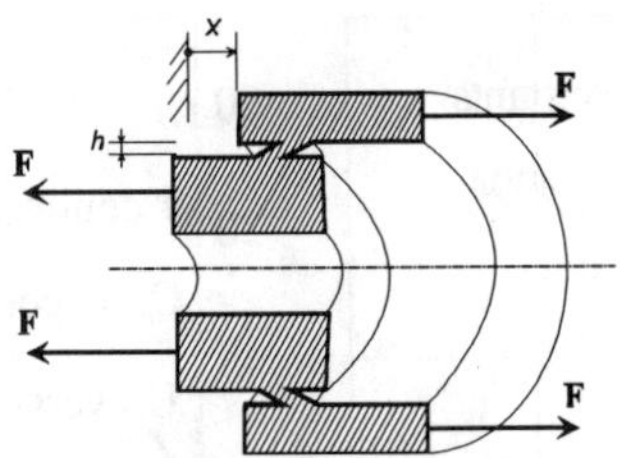

"BICCHIERINO" - SPECIMEN FOR SHEAR TEST

Fig. 1.

Now, material constants are determined by standard best-fit procedure, i.e., by minimization of χ^2-functional in $\mathcal{A}$-space. In other words, if measured plastic stretching is denoted by $\mathbf{D}_P^{exp}$, then such a minimization reads:

$$\chi^2 = \|\mathbf{D}_P^{exp} - \sum_\alpha \Gamma_\alpha(\gamma, \mathcal{A})\mathbf{H}_\alpha\| \to \min. \tag{24}$$

Firstly, the universal flow rule was confronted only with tension. Since for both tests plastic stretching has the following non-zero components

$$\mathbf{D}_P = \left\{ \begin{array}{ccc} D_{P11} & D_{P12} & 0 \\ & D_{P22} & 0 \\ \text{sym} & & D_{P33} \end{array} \right\}, \tag{25}$$

results of fitting are presented on the left-hand part of Fig. 2 (the upper part shows experimental data whereas the bottom one is depicted for calculated $\mathcal{A}$ and measured ε_P and $\mathbf{S}$). The agreement is very good judged by the commonly defined correlation coefficient $\eta = X \cdot Y/(|X||Y|)$ for two sets of random variables (which in our case correspond to $\mathbf{D}_P^{exp}$ and $\sum_\alpha \Gamma_\alpha \mathbf{H}_\alpha$ at discretized time instants $t_1, \ldots, t_n$ of the test considered).

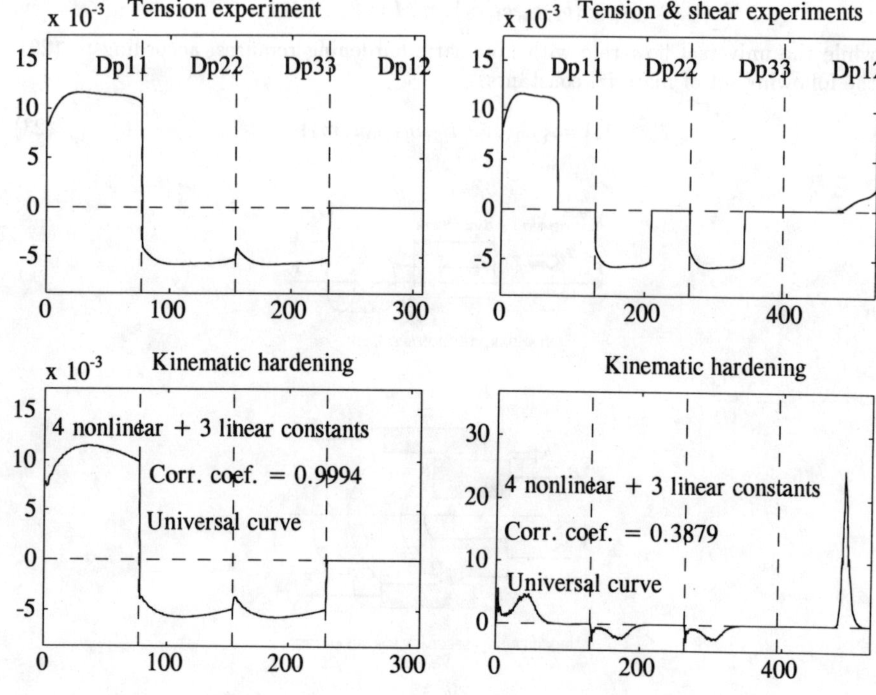

Fig. 2.

Secondly, data of both tests are shown on upper part of right side of this figure. The bottom part, again, shows fitted plastic stretching components based on (20). For each component first region represents tension while the second corresponds to shear (cf. D_{P12} for instance). The obtained results are very bad with $\eta = 0.3879$. It should be noted that even for $\eta \sim 0.8$ the correlation is not satisfactory and such a model should be rejected.

Furthermore, data of tension and shear are modelled by Rice's evolution equation for all three degrees of approximation. The results are presented in Fig. 3, using the same method of presentation as for the right-hand part of the previous figure.

6. Concluding Remarks

Although temperature distribution and its evolution were not recorded during tests described in [6] (due to experimentation difficulties) some important conclusions may be drawn:

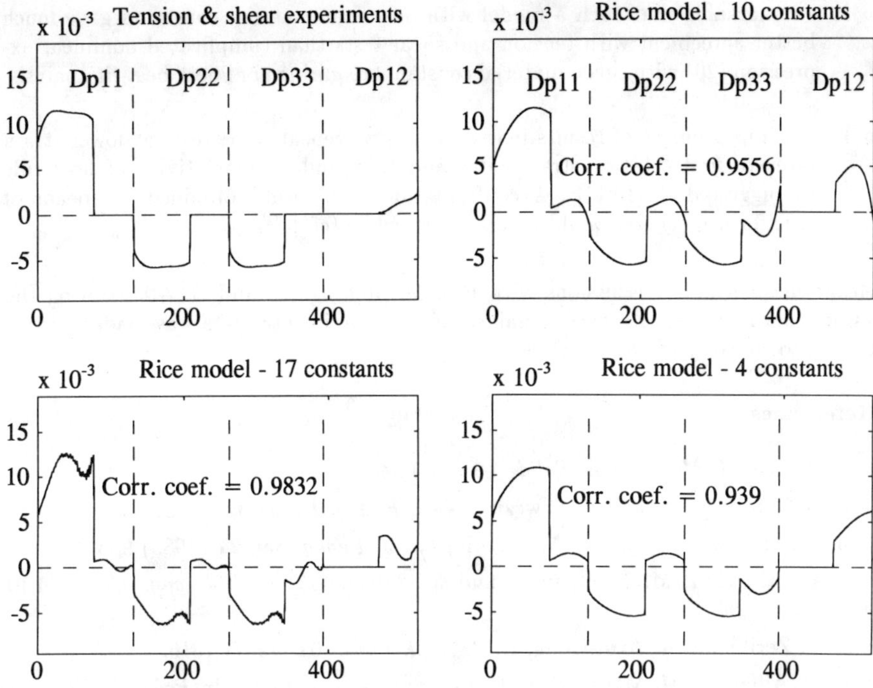

Fig. 3.

- ◇ Universal flow curve $\phi(\overline{\sigma}, \overline{\varepsilon}, D\overline{\varepsilon}) = 0$, even when corrected by kinematic hardening, is not capable to cover multiaxial stress-strain distributions. Seemingly tensorial evolution equation (18) is intrinsically scalar since it can describe successfully only tension test. The situation is even worse when cycling stresses appear like in nuclear reactors. Some authors like Bodner [8], Chaboche [9], etc., looked for a remedy in complicating Λ in (18) to account for an equivalent strain history, obtaining in this way 15 and even more material constants.
- ◇ Additional stiffness of (18) is that as flow potential the yield function must be used and this function is fixed by onset of yielding detected at fast unloading-loading sequences. Loading function Ω is much more flexible to account for real inelastic behaviour.
- ◇ Representation of anholonomic internal variables (i.e. PIR) by logarithmic plastic strain and using tensor function representation for evolution equation for plastic stretching proposed in this paper gave good agreement with multiaxial stress-strain distributions.

◇ It is remarkable that such a model with only four material constants gave much better agreement with tension and shear tests than complicated nonlinear expression (20) with seven material constants requiring nonlinear best-fit analysis.

◇ For an improvement of results it is necessary to repeat same tests at lower strain rates to find "static" strain rates and to abandon associativity of flow rule as suggested in [10,11]. A confrontation with results obtained by means of cruciform specimen would be also worthwhile (cf. [4]).

Acknowledgements. Discussions with Drs. M. Montagnani and C. Albertini on the subject are highly appreciated. Analysis of the experimental data was made possible thanks to courtesy of Dr. C. Albertini.

References

1. J. Rice, *J. Mech. Phys. Solids* **19** (1971) 433.
2. S. Murakami, S. and A. Sawczuk, *Nucl. Engng. Design* **65** (1981) 33.
3. M. Maruszewski and M. Mićunović, *Int. J. Engng. Sci.* **27** (1989) 955.
4. C. Albertini, M. Montagnani and M. Mićunović, *Nucl. Engng. Design* **130** (1991) 205.
5. F. Zerilli and R. Armstrong, R. *J. Appl. Phys.* **61** (1987) 1816.
6. C. Albertini, M. Montagnani, E. Pizzinato and A. Rodis, In *Proc. SMIRT–11*, Tokyo, 1991.
7. C. Albertini, *Private communication.*
8. S. R. Bodner, In *Unified Constitutive Equations for Plastic Deformation and Creep of Engineering Alloys*, ed. A. K. Miller, Elsevier, 1987, p. 273.
9. J. L. Chaboche and G. Rousellier, *ASME J. Press. Vessel Technol.* **105** (1983) 159.
10. M. Mićunović, *J. Mech. Behaviour Materials* **4** (1993) 331.
11. M. Mićunović, In *Continuum Models and Discrete Systems, (Proc. of CMDS7)*, ed. K.-H. Anthony and H.-J. Wagner, Materials Science Forum **123–125**, Trans Tech Publ., 1992, p. 609.
12. J. Bataille and J. Kestin, *J. Non-Equilib. Thermodyn.* **4** (1979) 229.

Continuum Models and Discrete Systems
Proceedings of 8th International Symposium, June 11-16, 1995, Varna, Bulgaria, ed. K.Z. Markov

THE CONNECTION BETWEEN MICRO- AND MACRO-PROPERTIES AT THE ELASTIC-PLASTIC DEFORMATION OF METALLIC POLYCRYSTALS

I. B. SEVOSTYANOV[1]
Max-Planck Arbeitsgruppe "Mechanik heterogener Festkörper," Hallwachstr. 3, D-01069 Dresden, Germany

Abstract. The self-consistent scheme is specified for polycrystals plasticity. The thermodynamic time scale is used for the single crystal constitutive law construction. This allows to take into account the hardening due to an action of stresses of the third kind. The generalization of the Eshelby's theorem and the effective field method allows to reduce the procedure of averaging to solving four linear integral equations.

1. Introduction.

The topic of this paper is the time independent ("athermal") plasticity of polycrystals. One of the more subtle and important problems here consists in the choice of the constitutive law for a single crystal. It is well known that the characteristic mechanism for athermal plasticity is the interaction of dislocations and their movement in glide systems. Without going into the plastic deformation physics, the single crystal will be considered here as an elastic-plastic media with a given symmetry, e.g., cubic or hexagonal.

Several length scales are important for plasticity of polycrystals. We distinguish here between macro, meso, and micro scales. The typical lengths of these scales are the dimensions of the body which in any case are large compared to the mean grain diameter, then this diameter itself, and finally the mean dislocation distance.

The length scales just discussed also lead to a subdivision of stresses into three species, usually called stresses of the first, second and third kind (see e.g. [1-3]). These are stresses that vary on the macro, meso, and micro scales, respectively. In particular, the stresses of the third kind are closely connected to the dislocation structure. The point is that gliding dislocations encounter various obstacles (e.g., Peierls barriers, forest dislocations, Lomer-Cottrell barriers, etc.). Therefore a rebuff reaction of the

[1]Permanent address: Laboratory of the Mechanics of Composites, St. Petersburg State University, Bibliotechnaya pl. 2, 198904 St. Petersburg, Russia.

obstacle will be superimposed on the external forces that press dislocation to this obstacle. This reaction is a main source of the third kind stresses. On the other hand, stresses of the second kind are often considered as the so-called back stresses which arise in a grain if its plastic strain differs from that of the neighbours. The necessity to preserve coherency through interfaces with these neighbours causes a constraint which is responsible for the back stresses.

Now, if the flow and evolution laws of the monocrystals are already known, we shall then deal with the statistical theory of grains and their interaction. There are many methods to define mechanical properties of polycrystals through those of monocrystals (see, e.g., a survey in [4]). However one cannot determine the polycrystal stress-strain state without essential simplifications. One of the most known one consists in the representation of the real polyhedron shape of the grains by ellipsoids (or even by balls) with a random spatial distribution of anisotropy axes and subsequent application of a self-consistent scheme [1,5-7].

The use of the thermodynamic time scale to obtain the constitutive law for a single crystal is described in Section 2. In Section 3 the problem of an elastic-plastic ellipsoid in an elastic medium is discussed. The solution of this problem will be used in Section 4, where the averaging scheme is considered. An example of application of the proposed approach is presented in Section 5.

2. Single Crystal Plasticity

Let θ be the specific power of dissipation, t is the laboratory time, and

$$\lambda(t) = \int_0^t \theta(\tau)\, d\tau \tag{1}$$

is the "thermodynamic" time introduced by Vakulenko [2,8]. According to the second law of thermodynamics, $\theta \geq 0$ and $\theta = 0$ if and only if the process is reversible. Therefore λ will increase always in the irreversible processes. This makes it possible to mark the sequence of the states in each of such processes. Furthermore it will permit to use the Boltzman superposition principle not only for a viscoelastic medium but for a medium with arbitrary (but admitted by the second law) rheology, even viscoelastoplastic one.

The plastic component $\varepsilon^P = \varepsilon - \mathbf{E}^{-1} : \sigma$ of the infinitesimal deformation tensor ε (where $\mathbf{E}^{-1}$ is the monocrystal compliance tensor reciprocal to the elastic tensor, σ is the stress tensor, and the colon stands for the contraction of tensors by two indexes) will be defined according to the superposition principle by the relation

$$\varepsilon^P = \int_0^\lambda \mathbf{H}(\lambda - \lambda') : \sigma(\lambda')\, d\lambda', \tag{2}$$

where $\mathbf{H}$ is a fourth rank tensor with corresponding symmetry. To satisfy the fading memory principle [9], the tensor kernel $\mathbf{H}(\lambda)$ must be a decreasing function of λ.

Upon differentiating (2) on the "laboratory" time t by taking into account that $\frac{d}{dt} = \theta \frac{d}{d\lambda}$ one obtains

$$\dot{\varepsilon}^P = \theta\, \mathbf{H}_0 : (\sigma - \mu),$$

$$\mu = \int_0^\lambda \mathbf{\Phi}(\lambda - \lambda') : \sigma(\lambda')\, d\lambda', \tag{3}$$

where $\mathbf{H}_0 = \mathbf{H}(0), \mathbf{\Phi}(\lambda) = -\mathbf{H}_0^{-1} : \frac{\partial \mathbf{H}}{\partial \lambda}$. The correction tensor μ characterizes the monocrystal hardening due to action of the third kind stresses. This tensor μ represents the mean stress of the third kind in the single crystal.

It follows from (3), that $\theta > 0 \Leftrightarrow \dot{\varepsilon}^P \neq 0$. By using the definition of dissipation power for athermal plastic deformation $\theta = (\sigma - \mu) : \dot{\varepsilon}^P$ and Eq. (3), one can obtain for $\dot{\varepsilon}^P \neq 0$:

$$1 = (\sigma - \mu) : \mathbf{H}_0 : (\sigma - \mu).$$

Since in the general case there exist crystal states for which

$$1 > (\sigma - \mu) : \mathbf{H}_0 : (\sigma - \mu)$$

(e.g. in a reversible process: unloading and subsequent loading, when $\dot{\varepsilon}^P = 0$), the plasticity condition must be represented in the more general form

$$1 \geq (\sigma - \mu) : \mathbf{H}_0 : (\sigma - \mu). \tag{4}$$

3. The Single Inclusion Problem

It was shown in [10] that if the stress field σ in the pure elastic problem inside the single inclusion of the volume V is uniform then it will be always uniform under arbitrary physical nonlinearity of the inclusion. It is only necessary that the matrix and the inclusion are homogeneous bodies separately. Let the stress field in the current moment of time t be defined by the deformation history

$$\sigma(t) = \Im(\{\varepsilon(\tau) | 0 \leq \tau \leq t\}) \equiv \mathop{\Im}_{\tau=0}^{t}(\varepsilon(\tau)). \tag{5}$$

Then by using of the Eshelby's tensor [11]

$$\mathbf{S} = \mathrm{def}\left(\int_{(V)} \nabla U(x - x')\, dV(x')\right),$$

where U denotes the Green's tensor for the linear elastic matrix, one can obtain an equation for the displacement vector u similar to that in the pure elastic case:

$$u(t) = u_0(x) + x \cdot \mathbf{S} : \left(\mathbf{E} : \varepsilon - \mathop{\Im}_{\tau=0}^{t}(\varepsilon(\tau))\right), \tag{6}$$

where $u_0(x)$ denotes the displacement field in the homogeneous body with the properties of the matrix under equivalent loading.

Let us consider the equilibrium problem for a linearly elastic media with a single elastic-plastic inclusion loaded by a homogeneous force field at infinity. Let the inclusion occupy an ellipsoidal domain V and elastic moduli of the matrix and the inclusion coincide. The ordinary conditions of continuity of the displacement vector and normal component of the stress tensor are satisfied at the interface. The relation between the full deformation and the stress tensors in the inclusion can be found by substitution $\varepsilon^P = \varepsilon - \mathbf{E}^{-1} : \sigma$ into (2). As a result the following Volterra integral equation of the second kind for the stress tensor will be obtained

$$\sigma = \mathbf{E} : \varepsilon - \mathbf{E} : \int_0^\lambda \mathbf{H}(\lambda - \lambda') : \sigma(\lambda')\, d\lambda'.$$

One can write the solution of this equation as follows:

$$\sigma = \mathbf{E} : \varepsilon - \mathbf{E}^2 : \int_0^\lambda \mathbf{\Psi}(\lambda - \lambda') : \varepsilon(\lambda')\, d\lambda', \tag{7}$$

where $\mathbf{\Psi}$ is a fourth-rank tensor kernel, resolvent for $\mathbf{H}$, and possessing analogues properties; $\mathbf{E}^2 = \mathbf{E} : \mathbf{E}$. If both of phases are homogeneous, then, for the strain tensor, generated by the displacement field (6), one can write the equation:

$$\varepsilon(t) = \varepsilon_0(x) + \mathbf{S} : \mathbf{E}^2 : \int_0^\lambda \mathbf{\Psi}(\lambda - \lambda') : \varepsilon(\lambda')d\lambda' \tag{8}$$

and its solution

$$\varepsilon(t) = \varepsilon_0(x) - \mathbf{S} : \mathbf{E}^2 : \int_0^\lambda \mathbf{H}_1(\lambda - \lambda') : \varepsilon_0(\lambda')\, d\lambda', \tag{9}$$

where the resolvent kernel $\mathbf{H}_1$ is specified by $\mathbf{\Psi}$ and therefore by $\mathbf{H}$ in a well-known manner. Thus we have obtained the Volterra's integral equation series to define mechanical fields inside a single elastic-plastic inclusion surrounded by a linear elastic material. This allows us to apply the self-consistent scheme to the problem of determining the average polycrystal properties.

4. The Effective Field Method Application

Let the polycrystal be modelled by a composite material with close packed ellipsoidal inclusions. Then we shall obtain in the limit case a problem of determination of average elastic-plastic properties of a polycrystal through those of a monocrystal. For the averaging we shall use the effective field method [12], because it does not require the knowledge of the fundamental solution for the effective elastic-plastic media. Additionally it has shown good applicability for elastic matrix composites in wide ranges of the inclusion concentration. Let us recall the main hypotheses of that method:

1. every inhomogeneity is considered as a single ellipsoid or ball, "sealed" into the main medium;

2. the strain field ε_*, in which such a single ellipsoid is immersed, consists of an external homogeneous field ε_0 and a field caused by surrounding inclusions (neighbour grains in our case);

3. ε_* is assumed spatially uniform.

By taking into account the last hypothesis, the deformation field inside a grain can be described by an equation similar to (9):

$$\varepsilon(t) = \varepsilon_*(x) - \mathbf{S} : \mathbf{E}^2 : \int_0^\lambda \mathbf{H}_1(\lambda - \lambda') : \varepsilon_*(\lambda')\, d\lambda'. \tag{10}$$

One can write the equation for ε_* by inserting (10) into (8), using the symmetrized gradient operator, and averaging the obtained result.

Consider an isotropic grain distribution (i.e. the case, when all crystallographic axes are equiprobable). Then, denoting by $\langle * \rangle$ the average value, one can write finally

$$\varepsilon_0(t) = \varepsilon_*(x) - \mathbf{E} : \mathbf{I} - \mathbf{E} : \mathbf{S} : \int_0^\lambda \langle \mathbf{H}_1(\lambda - \lambda')\delta(V) \rangle : \varepsilon_*(\lambda')\, d\lambda', \tag{11}$$

where $\mathbf{I}$ is the unit fourth rank tensor and $\delta(V)$ is the characteristic function of the domain V ($\delta(V) = 1$, if $x \in V$ and $\delta(V) = 0$ otherwise). Since the external stress field σ coincides with the averaged one throughout the whole material, one can write the main equation for the polycrystal:

$$\varepsilon_*(t) = \mathbf{E}^{-1} : \sigma(t) + \mathbf{I} - \mathbf{E} : \mathbf{S} : \int_0^\lambda \langle \mathbf{\Psi}_1(\lambda - \lambda')\delta(V) \rangle : \sigma(\lambda')\, d\lambda' \tag{12}$$

and the equality $\mathbf{\Psi}_1(0) = \mathbf{H}_1(0) = \mathbf{H}(0)$ is satisfied for the kernel $\mathbf{\Psi}_1$, resolvent to $\mathbf{H}_1$. Then by using the operations from Section 2 one can write

$$1 \geq (\sigma - \langle \mu \rangle) : (\mathbf{I} - \mathbf{E} : \mathbf{S} : \langle \mathbf{H}(0)\delta(V) \rangle : (\sigma - \langle \mu \rangle)\,, \tag{13}$$

$$\langle \mu \rangle = \int_0^\lambda \mathbf{\Psi}_2(\lambda - \lambda') : \sigma(\lambda')\, d\lambda'; \quad \mathbf{\Psi}_2 = -\langle -\mathbf{H}(0)\delta(V) \rangle^{-1} : \left\langle \frac{\partial \mathbf{\Psi}_1}{\partial \lambda} \delta(V) \right\rangle .$$

5. An Example of Calculation

As an example let us consider the case when all grains are simulated by balls and let the plastic deformation of every crystallite be connected with the shear only in one slip plane. It is a variant of the well-known Bathdorf-Budyansky sliding theory. This is typical for f.c.c. and h.c.p. lattices of the high purity monocrystals which are oriented for unit slip [13]. Then the specific power of the dissipation for a crystallite and "thermodynamic" time coincide with those for shear in the slip system. Therefore one can use the superposition principle for shear strain γ^P and tangential stress τ, which act in the slip system

$$\gamma^P = \int_0^\lambda \psi(\lambda - \lambda') : \tau(\lambda')\, d\lambda' . \tag{14}$$

The spatial orientation of the slip system is defined by a couple of unit vectors n and b—the normal to the slip plane and the slip direction vector, respectively—which depend on the Euler angles. One can define the tangential stress as follows: $\tau = P : \sigma$, where $P = \frac{1}{2}(n \otimes b + b \otimes n)$ and $\mathrm{sp}\,(P) = 0$. Then, since $D_P = \gamma^P P$ is the plastic deformation deviator, from (7) and (14), it follows

$$D_\sigma = 2G\left(D_\varepsilon - 2G\int_0^\lambda \mathbf{\Psi}(\lambda - \lambda') : D_\varepsilon(\lambda')\, d\lambda'\right), \tag{15}$$

where D_σ is the stress deviator, D_ε is the full deformation deviator, G is the shear modulus, and the fourth rank tensor $\mathbf{\Psi} = \psi P \otimes P$. Denoting by $\mathbf{C}_2$ and $\mathbf{C}_3$ the well-known isomers of the unit fourth rank tensor $\mathbf{I}$ one can write for the matrix:

$$D_\sigma = 2G\left(\frac{1}{2}(\mathbf{C}_2 + \mathbf{C}_3) - \frac{1}{3}\mathbf{I}\right) : D_\varepsilon \equiv 2G\mathbf{C}_D : D_\varepsilon .$$

The Eshelby's tensor S can be represented in our case in the form

$$\mathbf{S} = \frac{1}{20G(1-\nu)}\left(-\frac{2}{3}\mathbf{I} + \mathbf{C}_2 + \frac{-17 + 20\nu}{3}\mathbf{C}_3\right),$$

where ν is the Poisson ratio. By averaging over all possible orientations of crystallographic axes, we get

$$\left\langle \int_0^\lambda \psi_1(\lambda - \lambda')PP : D_\sigma(\lambda')\delta(V)d\lambda' \right\rangle = C_{\mathrm{in}}\mathbf{P} : \int_0^\lambda \psi_1(\lambda - \lambda')D_\sigma(\lambda')\, d\lambda', \tag{16}$$

where $\mathbf{P}$ is the isotropic fourth rank tensor $\mathbf{P} = -\frac{1}{30}\mathbf{I} + \frac{1}{20}(\mathbf{C}_2 + \mathbf{C}_3)$ and C_{in} is the volume part of the crystallites. Then the eventual plasticity condition for a polycrystal becomes

$$\frac{1}{2}\frac{22 - 25\nu}{75(1-\nu)} C_{\mathrm{in}}\mathcal{J}_2(D_\sigma - \langle D_\mu \rangle) = \tau_c^2 , \tag{17}$$

$\mathcal{J}_2$ is the second invariant of the tensor in parenthesis and $\tau_c^2 = [\psi(0)]^{-1/2}$ is the critical resolved shear stress (Schmid's stress). Comparison of the calculations by (17) with the known experimental data [14,15] is shown in Fig. 1.

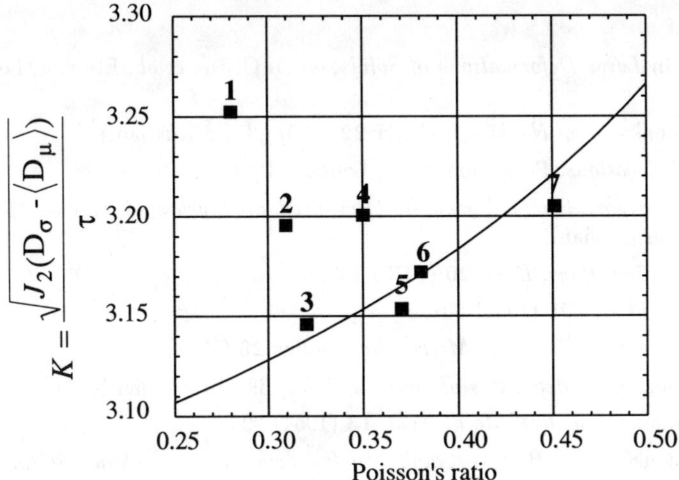

Fig. 1. Intensity of the shear stresses over the Schmid's stress vs. Poisson ratio for different metals:

1. Fe^1 (b.c.c.) 2. Cd^2 (h.c.p.) 3. Al^3 (f.c.c.) 4. Co^2 (h.c.p.)
5. Ag^3 (f.c.c.) 6. Cu^3 (f.c.c.) 7. Pb^3 (f.c.c.)

^{1} slip plane (110), slip direction [111]
^{2} slip plane (0001), slip direction $[11\tilde{2}0]$
^{3} slip plane (111), slip direction [110]

6. Conclusions

The proposed approach allows to reduce the problem of determination of average polycrystalline properties to consistent solving of linear integral equations. The influence of the third kind stresses are taken into account. In spite of its simplicity, the considered example shows a good conformity of the model to experimental data in the plasticity condition determination. Probably, to make the predictions more precise, one should replace the elastic properties of the "matrix" by the viscoelastic ones [16], using more realistic symmetry types for a single crystal, etc. Due to the application of effective field method it is possible to use a certain kind of successive approximations with taking into consideration the multiparticle interaction.

Acknowledgements. Author cordially thanks A. A. Vakulenko for stimulating and helpful discussion and comments and A. A. Gorbunov for technical assistance.

References

1. E. Kröner, In *Large Deformations of Solids*, ed. J. Gittus *et al.*, Elsevier, London, 1986.
2. A. A. Vakulenko, *Itogi Nauki i Tekhniki* **22** (1991) 3. (in Russian.)
3. J. Fridel, *Dislocations*, Pergamon Press, London, 1964.
4. T. D. Schermergor, *Teorija Uprugosti Microneodnorodnykh sred*, Nauka, Moscow, 1974. (in Russian.)
5. V. M. Levin, *Sov. Phys. Dokl.* **20** (1975) 147.
6. E. Kröner, *Z.Phys.* **151** (1958) 504.
7. M. Berveiller and A. Zaoui, *J. Mech. Phys. Solids* **26** (1979) 325.
8. A. A. Vakulenko, *J. Mechanics of Solids* 4 (1970) 69. (in Russian.)
9. B. D. Coleman, *Arch. Rat. Mech. Anal.* **13** (1964) 223.
10. A. A. Vakulenko and I. B. Sevostyanov, In *Issledovanija po mekhanike stroitelnykh materialov i konstrukcij*, ed. V. P. Ilyin, LISI, Leningrad, 1991. (in Russian.)
11. T. Mura, *Micromechanics of Defects in Solids*, 2nd edition, Martinis Nijhoff, the Haague, 1988.
12. S. K. Kanaun and V. M. Levin, In *Advances in Mathematical Modelling of Composite Materials*, ed. K. Z. Markov, World Scientific, Singapore, 1994, p. 1.
13. R. W. K. Honeykombe, *The Plastic Deformation of Metals* 2nd edition, Edward Arnold Publ., London, 1984.
14. G. E. Dieter, *Mechanical Metallurgy*, 3rd edition, McGraw Hill, New York, 1986.
15. I. S. Grigoryev and E. Z. Mejlikhov (eds), *Phisicheskie Velichiny*, Energoatomizdat, Moscow, 1988. (in Russian.)
16. J. Čadek, *Creep in Metallic Materials*, Elsevier, Amsterdam, 1988.

Continuum Models and Discrete Systems
Proceedings of 8th International Symposium, June 11-16, 1995, Varna, Bulgaria, ed. K.Z. Markov

GRAVITY FLOW OF GRANULAR MATERIALS IN CONVERGING WEDGES AND CONES

A J M SPENCER

and

N J BRADLEY

Department of Theoretical Mechanics, University of Nottingham, Nottingham NG7 2RD, UK

Abstract. We re-examine the problem of radial flow under gravity of an ideal granular material in converging wedges and cones with friction boundary conditions at the walls. The material obeys the cohesionless Coulomb-Mohr yield condition. Two types of flow rule are compared; (a) the coaxial condition, in which the principal axes of stress and strain-rate are coincident, and (b) the 'double-shearing' condition. It is shown that the double-shearing theory gives more plausible results.

1. Introduction

Many industrial processes involve the flow of granular materials such as sand, grain or powdered coal. In many cases, for example in flows down chutes or through hoppers, the flow is driven by gravity, and because the mechanism is frictional and sensitive to pressure, the gravitational force is very significant. Consequently any exact solutions that include gravitational effects are of interest. There are few such solutions in the literature. Sokolovsky [1] gives a wide range of analytical and numerical solutions to plane strain problems for granular media, but considers only stress fields and does not provide any non-trivial solutions for the flow behaviour. The earliest and most important solutions which describe gravity flows of granular materials are by Jenike [2,3,4] and Johanson [5] and relate to radial flows in converging wedges and cones. Other problems in which flow under gravity has been analysed concern rectilinear flows (Spencer and O'Mahony [6], O'Mahony and Spencer [7,8]), and flow in compression in vertical channels and cylinders and through tapering channels (Spencer and Bradley [9]).

This paper is part of a re-examination of Jenike's radial flow solutions (Bradley [10]). This study has confirmed Jenike's calculations for the stress fields in radial flows in wedges and cones and extended them somewhat to analyse flows in asymmetrical wedges and between coaxial cones. Also, consideration has been given to the flow fields. The correct choice of equations to relate the stress to the deformation rate in granular materials is still controversial, and various theories have been proposed (for an outline, see Spencer [11]). Jenike adopted the coaxiality assumption that the principal axes of stress and strain-rate are coincident. For radial flows this leads to some results which we feel are anomalous and physically unrealistic. An alternative theory, termed the 'double-shearing theory' was formulated for plane strain by Spencer [12] and in greater generality in [11]. The theory is based on the assumption that flow takes place by simultaneous shears on the two families of surfaces on which the critical shear traction is mobilised.

The main purpose of this paper is to compare results arising from the coaxiality and double-shearing theories, in the context of Jenike's radial flow problems. We find that the unsatisfactory features which arise in the coaxiality theory are not present when the double-shearing theory is used, and that the latter theory gives rise to solutions that are plausible and qualitatively in line with observed behaviour.

2. Governing Equations

2.1 *Stress Field*

We consider ideal granular materials that conform to the Coulomb-Mohr yield condition. The flows considered are either plane flows or flows with axial symmetry.

For plane flows we employ cylindrical polar coordinates (r,θ,z), with the z-axis vertical. The in-plane stress components in these coordinates are denoted $\sigma_{rr},\sigma_{r\theta},\ldots$, etc. We consider quasi-static flows, so that the equations of equilibrium apply, thus

$$\frac{\partial\sigma_{rr}}{\partial r}+\frac{1}{r}\frac{\partial\sigma_{r\theta}}{\partial\theta}+\frac{\sigma_{rr}-\sigma_{\theta\theta}}{r}-\rho g\cos\theta = 0\ ,$$

$$\frac{\partial\sigma_{r\theta}}{\partial r}+\frac{1}{r}\frac{\partial\sigma_{\theta\theta}}{\partial\theta}+\frac{2\sigma_{r\theta}}{r}+\rho g\sin\theta = 0, \qquad (1)$$

where ρ is the density and g the acceleration due to gravity.

The stress components can be expressed in the form

$$\sigma_{rr}=-p+q\cos 2\psi,\quad \sigma_{\theta\theta}=-p-q\cos 2\psi,\quad \sigma_{r\theta}=q\sin 2\psi \qquad (2)$$

Then

$$p=-\tfrac{1}{2}(\sigma_1+\sigma_3)=-\tfrac{1}{2}(\sigma_{rr}+\sigma_{\theta\theta}),\quad q=\tfrac{1}{2}(\sigma_1-\sigma_3)=\left\{\tfrac{1}{4}(\sigma_{rr}-\sigma_{\theta\theta})^2+\sigma_{r\theta}^2\right\}^{\frac{1}{2}}, \qquad (3)$$

where $\sigma_1\geq\sigma_2=\sigma_{zz}\geq\sigma_3$ are the principal stress components, and

$$\tan 2\psi=\frac{2\sigma_{r\theta}}{\sigma_{rr}-\sigma_{\theta\theta}}, \qquad (4)$$

where ψ is the angle between the principal stress axis associated with σ_1 and the radial direction, in the direction of increasing θ.

The Coulomb-Mohr condition, for a cohesionless material, has the form $q\leq p\sin\phi$, with

$$q=p\sin\phi \qquad (5)$$

when the material is deforming. Here ϕ is a material constant termed the angle of internal friction.

For axially symmetric flows we use spherical polar coordinates R,Θ,Φ, with associated stress components $\sigma_{RR},\sigma_{R\Phi}$,etc. In these coordinates, with $\Theta = 0$ the upward vertical direction, the equation of equilibrium are

$$\frac{\partial \sigma_{RR}}{\partial R}+\frac{1}{R}\frac{\partial \sigma_{R\Theta}}{\partial \Theta}+\frac{1}{R}(2\sigma_{RR}-\sigma_{\Theta\Theta}-\sigma_{\Phi\Phi}+\sigma_{R\Theta}\cot\Theta)-\rho g\cos\Theta = 0,$$

$$\frac{\partial \sigma_{R\Theta}}{\partial R}+\frac{1}{R}\frac{\partial \sigma_{\Theta\Theta}}{\partial \Theta}+\frac{1}{R}(\sigma_{\Theta\Theta}-\sigma_{\Phi\Phi})\cot\Theta+\frac{3}{R}\sigma_{R\Theta}+\rho g\sin\Theta = 0. \tag{6}$$

We again order the principal stress components so that $\sigma_1 \geq \sigma_2 \geq \sigma_3$, and in axial symmetry $\sigma_{\Phi\Phi}$ is a principal stress. We consider stress states corresponding to one of the 'Haar-von Karman' regimes, in which $\sigma_1 > \sigma_2 = \sigma_{\Phi\Phi} = \sigma_3$ and the stress can be represented as

$$\sigma_{RR} = -p+q\cos 2\Psi, \quad \sigma_{\Theta\Theta} = -p-q\cos 2\Psi, \quad \sigma_{R\Theta} = q\sin 2\Psi, \quad \sigma_{\Phi\Phi} = -p-q, \tag{7}$$

where now

$$p = -\tfrac{1}{2}(\sigma_1+\sigma_3) = -\tfrac{1}{2}(\sigma_{RR}+\sigma_{\Theta\Theta}), \quad q = \tfrac{1}{2}(\sigma_1-\sigma_3) = \left\{\tfrac{1}{4}(\sigma_{RR}-\sigma_{\Theta\Theta})^2+\sigma_{R\Theta}^2\right\}^{\frac{1}{2}}, \tag{8}$$

and Ψ is the angle between the direction of the principal axis of stress associated with σ_1 and the radial R-direction, measured in the direction of increasing Θ. The Coulomb-Mohr condition (5) again applies during flow.

2.2 *Velocity Field*

While there is quite general agreement on the equations that govern the stress field in granular materials, there are differing views on the appropriate way to represent the velocity field. The main purpose of this paper is to compare the consequences of two different theories, in the context of radial flows.

We denote the components of velocity by, v_r, v_θ and $v_z = 0$ in cylindrical polar coordinates with plane strain, and by $v_R, v_\Theta, v_\Phi = 0$ in spherical polar coordinates with axial symmetry. In fully developed steady flows of granular materials, volume changes are small, so we assume the material to be incompressible. This means, in cylindrical coordinates with $v_z = 0$

$$\frac{\partial v_r}{\partial r}+\frac{v_r}{r}+\frac{1}{r}\frac{\partial v_\theta}{\partial \theta} = 0, \tag{9}$$

and in spherical coordinates, with $v_\Phi = 0$,

$$\frac{\partial v_R}{\partial R}+\frac{1}{R}\frac{\partial v_\Theta}{\partial \Theta}+\frac{2v_R}{R}+\cot\Theta\frac{v_\Theta}{R} = 0. \tag{10}$$

To complete the system a further equation is required.

Coaxial Theory. Many authors, for example, Hill [13] and Jenike [3], have assumed that in flow of granular material the principal axes of stress and deformation rate coincide. This is sometimes, incorrectly, called a condition for material isotropy. Its mathematical expression is, for plane strain in cylindrical coordinates

$$\tan 2\psi = \frac{2d_{r\theta}}{d_{rr} - d_{\theta\theta}}, \tag{11}$$

where $d_{rr}, d_{\theta\theta}$ and $d_{r\theta}$ are rate-of-deformation components. In terms of the velocity components, (11) becomes

$$\left(\frac{1}{r}\frac{\partial v_r}{\partial \theta} + \frac{\partial v_\theta}{\partial r} - \frac{v_\theta}{r}\right)\cos 2\psi - \left(\frac{\partial v_r}{\partial r} - \frac{v_r}{r} - \frac{1}{r}\frac{\partial v_\theta}{\partial \theta}\right)\sin 2\psi = 0. \tag{12}$$

In spherical coordinates, with axial symmetry and rate-of-deformation components denoted, $d_{RR}, d_{\Theta\Theta}, d_{R\Theta}$, the coaxiality condition is

$$\tan 2\Psi = \frac{2d_{R\Theta}}{d_{RR} - d_{\Theta\Theta}}, \tag{13}$$

which in term of the velocity components gives

$$\left(\frac{1}{R}\frac{\partial v_R}{\partial \Theta} + \frac{\partial v_\Theta}{\partial R} - \frac{v_\Theta}{R}\right)\cos 2\Psi - \left(\frac{\partial v_R}{\partial R} - \frac{v_R}{R} - \frac{1}{R}\frac{\partial v_\Theta}{\partial \Theta}\right)\sin 2\Psi = 0. \tag{14}$$

Double-Shearing Theory. An alternative description of the velocity field is the double-shearing theory formulated in Spencer [11]. This is based on a physical assumption that deformation occurs by simultaneous shears on the surfaces on which the critical shear stress is mobilised. The consequence is that, for plane strain, (11) is replaced by the double-shearing condition

$$(d_{rr} - d_{\theta\theta})\cos 2\psi - 2d_{r\theta}\sin 2\psi + 2(\dot{\psi} - \omega_{r\theta})\sin\phi = 0, \tag{15}$$

where $\omega_{r\theta}$ is the local spin and $\dot{\psi}$ is the convected time derivative of ψ. For steady flow, this is equivalent to

$$\left(\frac{1}{r}\frac{\partial v_r}{\partial \theta} + \frac{\partial v_\theta}{\partial r} - \frac{v_\theta}{r}\right)\cos 2\psi - \left(\frac{\partial v_r}{\partial r} - \frac{v_r}{r} - \frac{1}{r}\frac{\partial v_\theta}{\partial \theta}\right)\sin 2\psi$$

$$+\sin\phi\left(\frac{1}{r}\frac{\partial v_r}{\partial \theta} - \frac{\partial v_\theta}{\partial r} + \frac{v_\theta}{r} + 2v_r\frac{\partial \psi}{\partial r} + 2\frac{v_\theta}{r}\frac{\partial \psi}{\partial \theta}\right) = 0. \tag{16}$$

Similarly, in the case of axial symmetry, (13) is replaced by

$$(d_{RR} - d_{\Theta\Theta})\cos 2\Psi - 2d_{R\Theta}\sin 2\Psi + 2(\dot{\Psi} - \omega_{R\Theta})\sin\phi = 0,$$

which, for steady flow, is

$$\left(\frac{1}{R}\frac{\partial v_R}{\partial \Theta}+\frac{\partial v_\Theta}{\partial R}-\frac{v_\Theta}{R}\right)\cos 2\Psi-\left(\frac{\partial v_R}{\partial R}-\frac{v_R}{R}-\frac{1}{R}\frac{\partial v_\Theta}{\partial \Theta}\right)\sin 2\Psi$$

$$+\sin\phi\left(\frac{1}{R}\frac{\partial v_R}{\partial \Theta}-\frac{\partial v_\Theta}{\partial R}+\frac{v_\Theta}{R}+2v_R\frac{\partial \Psi}{\partial R}+2v_\Theta\frac{1}{R}\frac{\partial \Psi}{\partial \Theta}\right)=0. \tag{17}$$

3. Steady Converging Radial Flow in Wedge-Shaped Channels

We consider the wedge-shaped channel illustrated in Figure 1. Material flows towards the apex of the wedge between the planes $\theta = \pm\alpha$, under the influence of gravity acting towards the apex in the direction of the channel axis.

Jenike [3] showed that there exist solutions for the stress field of the form

$$\psi = \psi(\theta), \quad q = \rho g r F(\theta). \tag{18}$$

When ψ and q are determined, the complete stress field is then given by (2) and (5). Substituting (18), (2) and (5) into (1) gives, after rearrangement, the equations

$$\frac{dF}{d\theta}=\frac{F\sin 2\psi+\sin\phi\sin(2\psi+\theta)}{\cos 2\psi+\sin\phi},$$

$$\frac{d\psi}{d\theta}=\frac{F\{\operatorname{cosec}\phi-3\sin\phi-2\cos 2\psi\}+\cos\theta+\sin\phi\cos(2\psi+\theta)}{2F(\cos 2\psi+\sin\phi)}. \tag{19}$$

We suppose the stress field is symmetrical about $\theta = 0$, and thus need to consider only the region $0 \le \theta \le \alpha$. The symmetry condition is

$$\psi = 0 \quad \text{at} \quad \theta = 0. \tag{20}$$

At the wall $\theta = \alpha$ we assume a Coulomb friction condition, so that

$$\sigma_{r\theta} = -\sigma_{\theta\theta}\tan\mu, \quad \text{at} \quad \theta = \alpha,$$

where μ is the angle of wall friction. From (2) and (5) this gives

$$\sin(2\psi-\mu)=\frac{\sin\mu}{\sin\phi} \quad \text{at} \quad \theta=\alpha, \tag{21}$$

provided $\mu \le \phi$. If $\mu \ge \phi$ then the wall is 'perfectly rough' and the material slips on itself at the wall. Then

$$\psi=\tfrac{1}{2}\phi+\tfrac{1}{4}\pi \quad \text{at} \quad \theta=\alpha. \tag{22}$$

Jenike [3] gave numerical solutions of (19) subject to the conditions (20) and (21) or (22). Since boundary conditions are specified at both $\theta = 0$ and $\theta = \alpha$, a 'shooting' technique has to be employed. Essentially it is necessary to determine the value $F(0)$ which leads to the required value of $\psi(\alpha)$. This is readily done by iteration.

Bradley [10] confirmed Jenike's calculations and extended them to include asymmetric wedges.

Jenike [3] also considered the associated radial flow field, using the *coaxial theory*. He sought solutions of (9) and (12) of the form

$$v_r = \frac{1}{r} v(\theta), \quad v_\theta = 0 , \tag{23}$$

for which (9) is trivially satisfied and (12) becomes

$$\frac{dv}{d\theta} \cos 2\psi + 2v \sin 2\psi = 0,$$

which has the solution

$$v = v_0 \exp\left\{-2\int_0^\theta \tan 2\psi \, d\theta\right\}. \tag{24}$$

Hence v can be calculated using the values of ψ obtained in the stress solution.

The expression (24) has the property that if the stress solution includes a plane θ = constant on which $\psi = \frac{1}{4}\pi$, then v is zero on that plane. This leads to implausible velocity profiles, such as that shown in Figure 2, and also implies negative plastic work in the region in which $\psi > \frac{1}{4}\pi$. This situation will always occur if $\psi(\alpha) > \frac{1}{4}\pi$, or equivalently $\sin\mu > \sin\phi \cos 2\mu$. It may or may not occur if $\psi(\alpha) < \frac{1}{4}\pi$.

If the *double-shearing* theory is employed, then by inserting the velocity field (23) into (16) rather than into (12), we obtain

$$\frac{dv}{d\theta}(\cos 2\psi + \sin\phi) + 2v \sin 2\psi = 0, \tag{25}$$

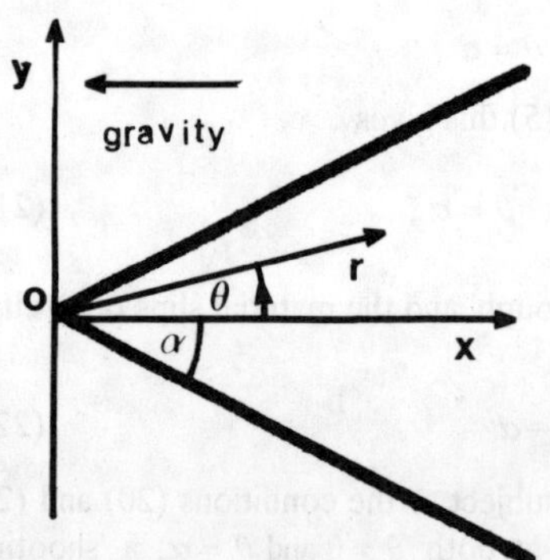

Figure 1.
Flow in a wedge-shaped channel

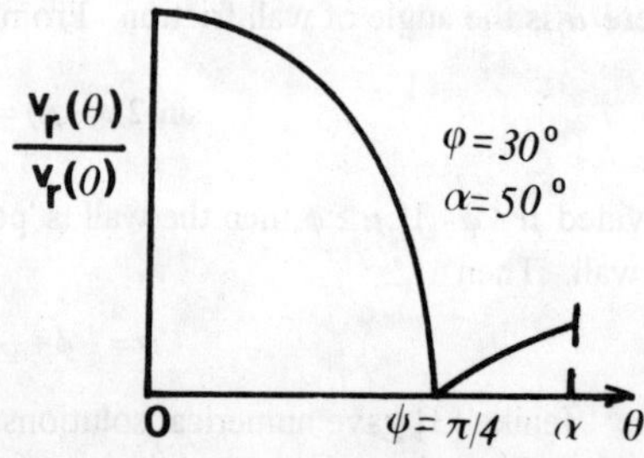

Figure 2.
Typical velocity profile in a wedge using coaxial theory

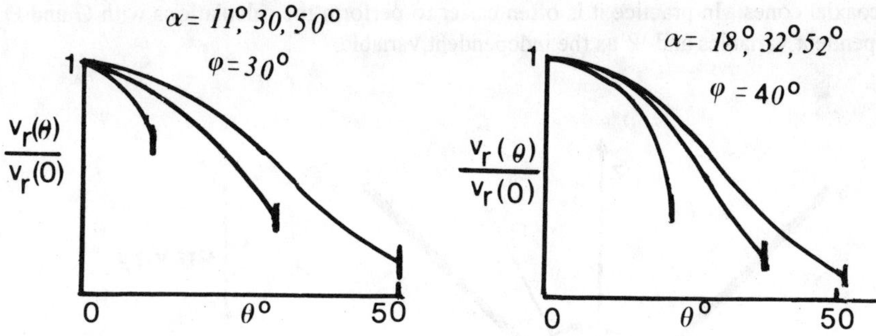

Figure 3.
Typical velocity profiles in wedges using double-shearing theory. Perfectly rough walls.

which has the solution

$$v = v_0 \exp\left\{-2\int_0^{\theta} \frac{\sin 2\psi}{\cos 2\psi + \sin\phi}\, d\,\theta\right\}. \tag{26}$$

In this case v has no singularities in the range $0 \le \psi < \frac{1}{4}\pi + \frac{1}{2}\phi$, which includes all cases of practical interest. The velocity profiles are smooth throughout the range and appear plausible. Some examples are shown in Figure 3.

4. Steady Converging Radial Flow in Conical Channels

The case of flow in a conical channel (see Figure 4) was treated by Jenike [3] in a similar manner. In this case we look for stress solution of the form

$$\Psi = \Psi(\Theta), \quad q = \rho g\, R\, G(\Theta). \tag{27}$$

Inserting these, with (5) and (7), into the equilibrium equations (1) leads to the pair of ordinary differential equations

$$\frac{dG}{d\Theta} = \frac{\sin\phi \sin(2\Psi + \Theta) + 2\sin\Psi\{\cos\Psi - \sin\phi \operatorname{cosec}\Theta \sin(\Theta + \Psi)\}G}{\cos 2\Psi + \sin\phi},$$

$$\frac{d\Psi}{d\Theta} = \frac{\cos\Theta + \sin\phi\cos(2\Psi + \Theta) + \{\operatorname{cosec}\phi - 4\sin\phi - 1 - 2\cos 2\Psi - (1 + \sin\phi)\operatorname{cosec}\Theta \sin(2\Psi + \Theta)\}G}{2(\cos 2\Psi + \sin\phi)\, G}. \tag{28}$$

Equations equivalent to these were also solved numerically by Jenike [3] and the results confirmed by Bradley [10]. The wall boundary conditions are (21) or (22) with (θ, ψ) replaced by (Θ, Ψ). Bradley [10] also extended the calculations to cases of flow between two coaxial cones. In practice it is often easier to perform the calculations with G and Θ as dependent variables and Ψ as the independent variable.

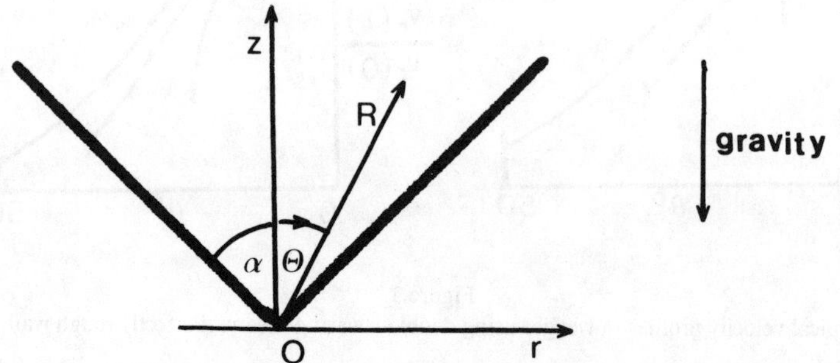

Figure 4. Flow in a converging conical channel

There is a qualitative difference, which seems not to have been noted by Jenike, between the stress solutions for a wedge and for a cone. In the case of a wedge, the calculations indicate that solutions exist for all wedge half-angles $\alpha < \frac{1}{2}\pi$ and all relevant values of the parameters ϕ and μ. The same applies in the case of a cone provided that $\sin\phi < \frac{1}{3}$. However, when $\sin\phi \geq \frac{1}{3}$, equations (28) have the singular solution

$$\Psi = -\Theta + \tfrac{1}{2}\pi - \gamma\,, \qquad G = \frac{\sin\Theta}{2\sin 2\gamma} \tag{29}$$

where $$\cos 2\gamma = \frac{1-\sin\phi}{2\sin\phi} \qquad 0 < \gamma < \tfrac{1}{4}\pi. \tag{30}$$

The calculations show that (29) is an envelope of the solution curves of (28) in the (Θ,Ψ), plane and real solutions are restricted to the region $\Psi + \Theta - \frac{1}{2}\pi + \gamma < 0$. This imposes limitations on the range of cone angles for which solutions exist; specifically, we require

$$\alpha < \tfrac{1}{2}\pi - \gamma - \Psi(\alpha)\,, \tag{31}$$

where α is the cone half-angle and $\Psi(\alpha)$ is given by the wall boundary condition. Presumably if α exceeds the limit (31), a region of stationary material bounded by conical surfaces $\theta = \frac{1}{2}\pi - \gamma - \Psi(\alpha)$ and $\theta = \alpha$ will form, and the material slides on itself on the inner of these surfaces.

Jenike [3] also calculated velocity profiles for radial flow using the *coaxial theory*. For radial flow in a cone the velocity field has the form

$$v_R = \frac{V(\Theta)}{R^2}\,, \quad v_\theta = 0\,. \tag{32}$$

Then incompressibility (10) is satisfied and the coaxiality condition (14) gives

$$\frac{dV}{d\Theta} + 3V \tan 2\Psi = 0, \tag{33}$$

with solution

$$V = V_0 \exp\left\{-3\int_0^{\Theta} \tan 2\Psi \, d\theta\right\}. \tag{34}$$

As for a wedge, V can be computed using the previously calculated values of Ψ, and again we find the integral in (34) is singular if $\Psi = \frac{1}{4}\pi$ on any conical surface within the cone. This will necessarily occur if $\Psi(\alpha) > \frac{1}{4}\pi$, and may occur even if $\Psi(\alpha) < \frac{1}{4}\pi$.

In contrast, if the *double shearing equation* (17) is adopted, we have in place of (33)

$$\left(\cos 2\Psi + \sin\phi\right)\frac{dV}{d\Theta} + 3V \sin 2\Psi = 0,$$

and the solution is now

$$V = V_0 \exp\left\{-3\int_0^{\Theta} \frac{\sin 2\Psi}{\cos 2\Psi + \sin\phi} \, d\theta\right\}. \tag{35}$$

The integral has no singularities within the cone, and smooth velocity profiles are obtained. Some examples are shown in Figure 5. A notable feature is a tendency towards plug flow, with most of the flow occurring near the axis of the cone especially for large cone angles and high internal friction. This is qualitatively in agreement with observations of hopper flows.

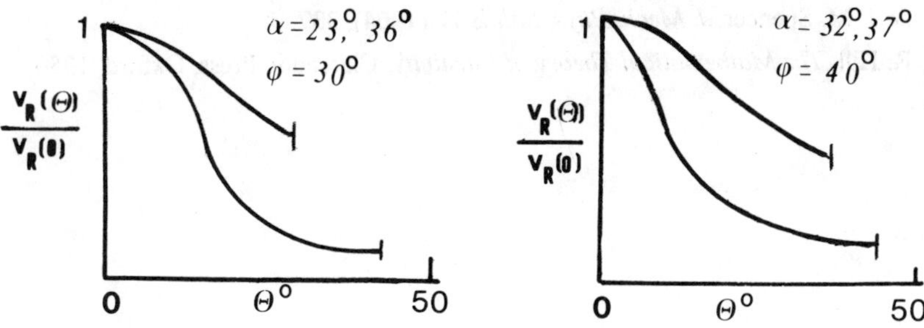

Figure 5. Typical velocity profiles in cones using double-shearing theory. $\mu = 15^0$

We conclude that for the class of flows considered, the double-shearing theory apparently provides a more realistic description of the flow than the coaxial theory.

Acknowledgement. The work of NJB was supported by an SERC CASE research studentship with British Gas plc Midlands Research Station as cooperating body. The support of SERC and British Gas is gratefully acknowledged. We also thank Mr A R Tait of British Gas for helpful discussions. Part of the contribution of AJMS was made during the tenure of an Erskine Fellowship at the University of Canterbury.

References

1. V. V. Sokolovskii, *Statics of Granular Media*, Pergamon Press, Oxford, 1965.
2. A. W. Jenike, *Trans. Inst. Chem. Engrs.* **40** (1962) 264.
3. A. W. Jenike, *J. Appl. Mech.* **31** (1964) 5.
4. A. W. Jenike, *J. Appl. Mech.* **32** (1965) 205.
5. J R Johanson, *J. Appl. Mech.* **31** (1964) 499.
6. A.J.M. Spencer and T. C. O'Mahony, In *Proc. IUTAM Symposium on Deformation and Failure of Granular Materials,* eds. P. A. Vermeer and H. J. Luger, 579, Balkema, Rotterdam, 1982.
7. T. C. O'Mahony and A.J.M. Spencer, *Int. J. Engng. Sci.* **23** (1985) 139.
8. T. C. O'Mahony and A.J.M. Spencer, *Int. J. Num. Anal. Meth. Geomechanics* **9** (1985) 225.
9. A.J.M. Spencer and N. J. Bradley, *Q. J. Mech. Appl Math.* **45** (1992) 734.
10. N. J. Bradley, *Gravity flows of granular materials,* Ph.D. Thesis, University of Nottingham, 1991.
11. A.J.M.Spencer, In *Mechanics of Solids: the Rodney Hill Anniversary Volume,* eds. H. G. Hopkins and M. J. Sewell, 607, Pergamon Press, Oxford, 1982.
12. A. J. M. Spencer, *J. Mech. Phys. Solids* **12** (1964), 337.
13. R. Hill, *The Mathematical Theory of Plasticity,* Clarendon Press, Oxford, 1950.

Continuum Models and Discrete Systems
Proceedings of 8th International Symposium, June 11-16, 1995, Varna, Bulgaria, ed. K.Z. Markov
©World Scientific Publishing Company, 1996, pp. 591–598

A THEORY OF MICROSTRUCTURAL INSTABILITIES DURING PLASTIC DEFORMATION OF LOW-TEMPERATURE PREDEFORMED B.C.C. METALS

M. ZAISER
Max-Planck-Institut für Metallforschung, Institut für Physik,
P. O. Box 800 665, D-70506 Stuttgart, Germany

and

P. HÄHNER
European Commission, Institute for Advanced Materials, Joint Research Centre
I-21020 Ispra (Va), Italy

Abstract. The paper presents a study of the microstructural instabilities observed in body-centered cubic metals that are plastically deformed after previous work-hardening at low temperatures. Under certain conditions restraining is accompanied by a pronounced localization of strain in narrow slip bands. These bands correspond, on the microstructural level, to slip channels where the dislocations that have been brought in by predeformation have vanished almost completely. It is the aim of the present investigation to clarify theoretically the processes which make slip channels emerge from an initially homogeneous dislocation microstructure. To this end, the interplay between collective dislocation motion and strain softening due to dislocation annihilation is treated in terms of a stochastic dislocation dynamics. The development of slip channels is proposed to correspond to a fluctuation-induced phase transition. The critical conditions are determined, and the results are compared to experimental findings.

1. Introduction

If body-centered cubic (b.c.c.) metals are prestrained at low temperature and then deformed at elevated temperature, plastic deformation may be accompanied by strain softening and various types of plastic instabilities. While, on a macroscopic scale, such phenomena include macroscopic strain localization and the formation of propagating deformation bands, the present investigation focuses on the *microstructural* aspects of strain softening. This may lead to pronounced localization of strain in narrow slip bands corresponding to slip channels where the dislocation microstructure produced by the prestrain is destroyed almost completely. For an understanding of the particular deformation conditions under which such instabilities may occur, it is necessary to recall some peculiarities of plastic deformation of b.c.c. metals.

1.1. *Plastic Deformation Properties of B.C.C. Metals*

The plastic behaviour of b.c.c. metals depends strongly on the deformation temperature T (for an overview see [1,2,3]) and changes qualitatively at the so-called transition temperature T_0, which is about one tenth of the melting temperature.

For $T < T_0$ the flow stress increases strongly with decreasing temperature. This is due to the specific stress and temperature dependences of the mobility of the screw dislocations with Burgers vector of type $(a/2)\langle 111\rangle$ that carry most of the slip activity. Owing to their complicated core configurations these dislocations experience a high Peierls stress, so that their glide velocity is determined by the rate of thermally assisted formation of kink pairs [4, 5]. The evolution of the dislocation microstructure is governed by the fact that edge dislocation segments are much more mobile than screw dislocations, such that their motion produces long pairs of screw dislocations of opposite sign. Since secondary dislocation sources are activated before the motion of primary screw dislocations sets in, a microstructure is established that consists of evenly distributed screw dislocations of all $(a/2)\langle 111\rangle$ Burgers vectors without excess of dislocations of one kind (Fig. 1).

For $T > T_0$, the deformation behaviour is quite different and closely resembles that of f.c.c. metals. Screw and edge dislocations now possess comparable mobilities, and their velocities are controlled by the cutting of "forest" dislocations of other slip systems. This gives rise to a weakly temperature-dependent flow stress and the formation of dislocation cell structures that consist of dislocation-rich cell walls separating depleted cell interiors. Within a cell wall, there may be surplus dislocations of one sign that give rise to a misorientation of neighbouring cells (Fig. 2).

1.2. *Behaviour During Redeformation at Higher Temperature*

The plastic behaviour of b.c.c. metals after a change of the deformation temperature has been studied extensively by Luft [6]. If b.c.c. metals are prestrained at a temperature T_1 and then restrained at a temperature $T_2 > T_1$, only dislocations of the

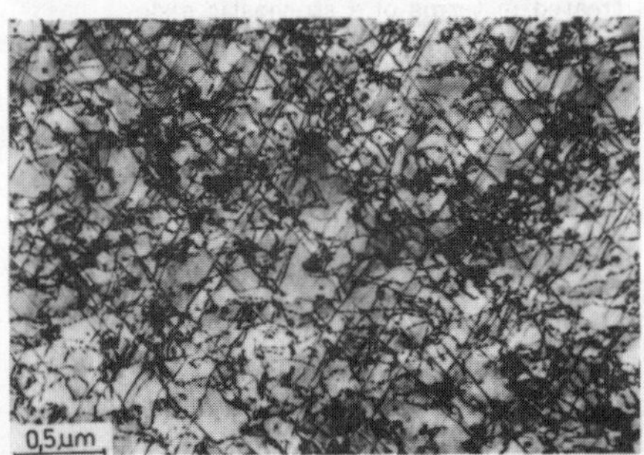

Fig. 1. Dislocation structure in a b.c.c. metal after deformation at $T < T_0$: Iron, deformed at 77 K. After Güth [6].

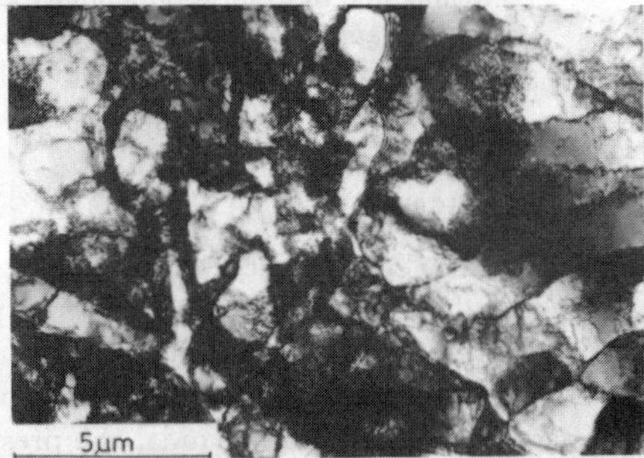

Fig. 2. Dislocation structure in a b.c.c. metal deformed at $T > T_0$: Molybdenum, deformed at 573 K. After Mühlhaus [6].

slip system(s) with maximum resolved shear stress are moving over long distances, whereas other slip systems remain almost inactive. Redeformation then leads to a transformation of the dislocation microstructure that is accompanied by a decrease of the dislocation density and by strain softening. This transformation is strongly dependent on both the temperatures of the pre- and redeformation. Three different regimes may be distinguished:

(i) If both $T_1 < T_0$ and $T_2 < T_0$, the density of dislocations decreases gradually during restraining, while their arrangement remains homogeneous. Microscopic slip localization does not occur. (ii) If both $T_1 > T_0$ and $T_2 > T_0$, the situation is quite similar. Re-organization of the cell walls is accompanied by a decrease of the dislocation density and by strain softening. However, the microstructure does not change qualitatively and, apart from some initial slip localization, slip is quite homogeneous in the course of straining.

Fig. 3. Slip channels in Molybdenum (T_1 = 173 K, T_2 = 378 K). After Luft [6].

(iii) Only for $T_1 < T_0$ and $T_2 > T_0$ qualitative changes of the microstructure occur during strain softening. Provided that the initial dislocation density is sufficiently high, a two-phase microstructure consisting of dislocation-free slip channels within a matrix of high dislocation density forms (see Fig. 3). Deformation is then localized within narrow (some μm wide) bands corresponding to the slip channels, the formation of which is associated with dislocations moving collectively in groups (dynamic pile-ups).

2. Theoretical Modelling

In order to describe theoretically the strain-softening processes during redeformation in the different temperature regimes and, in particular, the developement of dislocation-free slip channels, we start out from the following assumptions:

(i) Long-range dislocation motion occurs in one active slip system only and is, hence, characterized by the scalar plastic shear strain rate $\dot{\gamma}$.

(ii) The internal stresses τ_{int} acting against dislocation motion are caused by dislocations of the active system, τ_{a}, and by dislocations of other systems (forest dislocations), τ_{f}. The effective stress τ_{eff} available for dislocation motion is given by the difference of the external stress τ_{ext} and the internal stresses,

$$\tau_{\text{eff}} = \tau_{\text{ext}} - \tau_{\text{int}} = \tau_{\text{ext}} - \tau_{\text{a}} - \tau_{\text{f}} \,. \tag{1}$$

(iii) The mean internal stress contribution of the dislocations of the active system, $\langle\tau_{\text{a}}\rangle$, increases due to dislocation multiplication and dipole formation, while the mobile dislocation density ρ_m remains approximately constant. This corresponds to a linear hardening with plastic strain γ, which is characterized by the strain-hardening coefficient θ [7],

$$\langle\tau_a\rangle = \langle\tau_a(0)\rangle + \theta\gamma\,, \quad \rho_m = \text{const}\,. \tag{2}$$

The mean internal-stress contribution of the forest dislocations, $\langle\tau_f\rangle$, is related to their density ρ_f according to

$$\langle\tau_f\rangle = \alpha G b \sqrt{\rho_f}\,, \tag{3}$$

where $\alpha \approx 0.5$ is a numerical factor, G the shear modulus of the material, and b the modulus of the Burgers vector of the forest dislocations. Further it is assumed that the external stress τ_{ext} increases with time at a constant rate $\dot{\tau}_{ext}$ (constant-stress-rate testing). The initially applied stress is chosen such that an initial deformation rate $\dot{\gamma}(0) = \dot{\tau}_{ext}/\theta$ is reached. (iv) The shear strain rate $\dot{\gamma}$ depends on the effective stress and temperature according to an Arrhenius law,

$$\dot{\gamma} = \rho_m b v_0 \exp\left[-\frac{H(\tau_{eff})}{k_B T}\right]\,, \tag{4}$$

where v_0 is of the order magnitude of the sound velocity. The stress dependence of the activation enthalpy $H(\tau_{eff})$ is different at different deformation temperatures. At temperatures below T_0, H is given by the activation enthalpy H_{kp} for kink-pair formation, which depends on stress in a complicated way [5]. At temperatures above T_0, H is the enthalpy for cutting forest dislocations that decreases approximately linearly with stress, $H = H_0 - \tau_{eff} V_a$. Here the activation volume $V_a = C/\sqrt{\rho_f}$ is proportional to the forest dislocation spacing. (v) Although long-range dislocation motion and dislocation multiplication do not occur in the "inactive" slip systems, local rearrangements are induced by stress concentrations caused by dislocations moving in the active system. Following a suggestion of Luft [6], we assume that such rearrangements lead to the mutual annihilation of dislocations of opposite sign. This is taken into account in terms of a rate equation for the temporal evolution of the forest dislocation density:

$$\partial_t \rho = f(\rho)\dot{\gamma}\,, \quad f(\rho) = \begin{cases} -[\beta_1 G/\tau_{ext}]\,\rho\,, & T_1 < T_0\,; \\ -[\beta_2 G/\tau_{ext}]\,\rho^2/2\,, & T_1 > T_0\,. \end{cases} \tag{5}$$

Here $\rho = \rho_f/\rho_f(0)$ is the forest dislocation density scaled by its value at the start of redeformation. The annihilation kinetics depend on the dislocation structure and thus on the temperature of redeformation. For $T_1 < T_0$, dislocations of opposite sign are arranged in pairs which are stabilized by (external and internal) stress. If this is compensated by a passing mobile dislocation, the pair may annihilate. This leads to first-order kinetics with an annihilation rate that is inversely proportional to the external stress. For $T_1 > T_0$, on the other hand, the forest dislocations do not have correlated annihilation partners nearby and their annihilation is governed by second-order-kinetics. Also in this case the annihilation rate is found to be inversely proportional to the external stress [8]. The numerical factors β_1 and β_2 depend on α

as well as on orientation factors of the slip systems. Typical values are $\beta_1 \approx 3 \times 10^{-2}$ and $\beta_2 \approx 2 \times 10^{-3}$, i.e., annihilation is much more efficient in materials that have been predeformed below T_1.

2.1. Deterministic Considerations

For the treatment of the strain-softening dynamics, in a first step we neglect any fluctuations of the quantities involved and consider only their mean values. By making use of the substitution $d\gamma = \dot{\gamma} dt$ we obtain the equations of evolution

$$\partial_\gamma \rho = f(\rho)\,, \tag{6}$$

$$\partial_\gamma \tau_{\text{ext}} = \dot{\tau}_{\text{ext}} / \dot{\gamma}(\langle \tau_{\text{eff}} \rangle)\,, \tag{7}$$

where $\langle \tau_{\text{eff}} \rangle = \tau_{\text{ext}} - \langle \tau_{\text{a}}(0) \rangle - \theta\gamma - \left[\alpha G b \sqrt{\rho_{\text{f}}(0)}\right] \sqrt{\rho}$. Integration of (6) gives γ as a function of ρ. Combinination with Eq. (7) yields

$$\partial_\rho \tau_{\text{ext}} = \frac{\dot{\tau}_{\text{ext}}}{f(\rho)\dot{\gamma}(\rho, \tau_{\text{ext}})}\,, \tag{8}$$

which, in turn, may be used to determine the external stress as a function of ρ. In this way we obtain an expression for $\dot{\gamma}$ as a function of ρ only. For small ρ, $\dot{\gamma}$ approaches an asymptotic value of $\dot{\gamma}_\infty = \dot{\tau}_{\text{ext}}/\theta$ that equals the initially imposed strain rate.

According to Eq. (5), ρ approaches zero in a monotonic way. This gives rise to a *transient* strain softening if the condition

$$\left.\frac{\partial \tau_{\text{ext}}}{\partial \gamma}\right|_{\dot{\gamma}} < 0 \quad \text{or} \quad \left.\frac{\partial \dot{\gamma}}{\partial \gamma}\right|_{\tau_{\text{ext}}} > 0 \tag{9}$$

is fulfilled. Under the assumptions made above, this is true for

$$\frac{2\theta}{G}\tau_{\text{ext}}(0) < \begin{cases} \beta_1 \tau_{\text{f}}(0)\,; & T_1 < T_0, T_2 < T_0\,, \\ \beta_1[\tau_{\text{f}}(0) + \tau_{\text{eff}}(0)] \approx \beta_1 \tau_{\text{f}}(0)\,; & T_1 < T_0, T_2 > T_0\,, \\ \beta_2[\tau_{\text{f}}(0) + \tau_{\text{eff}}(0)] \approx \beta_2 \tau_{\text{f}}(0)\,; & T_1 > T_0, T_2 > T_0\,. \end{cases} \tag{10}$$

These considerations allow us to describe the evolution of the system in terms of a single variable ρ. Moreover, conditions for strain softening can be derived. However, additional information is needed for describing microstructural slip localization, since, from the experimental observations reported in Sections 1.2, it is seen that strain softening alone is *not* a sufficient condition for slip localization on the microstructural level. In order to obtain a more adequate description, it is necessary to account for the interplay between strain softening and collective dislocation motion during slip-channel formation. This is possible within the framework of stochastic dislocation dynamics [9, 10].

2.2. Stochastic Considerations

On a mesoscopic scale, dislocation interactions and correlations may be approximated by random fluctuations of the effective stress and the shear strain rate

$\dot{\gamma}$. For this purpose, Eq. (5) is re-interpreted as a stochastic differential equation by putting

$$\partial_t \rho = f(\rho)\left\{\langle\dot{\gamma}(\rho)\rangle + \delta\dot{\gamma}\right\} . \tag{11}$$

Here the mean strain rate $\langle\dot{\gamma}(\rho)\rangle$ corresponds to the deterministic case treated above. The magnitude of the fluctuations $\delta\dot{\gamma}$ is related to the strain-rate sensitivity defined by $S = \dot{\gamma}[\partial\dot{\gamma}/\partial\tau_{\rm eff}]^{-1} = k_{\rm B}T/V_{\rm a}$ and to the mean value of the internal stress [9, 10]:

$$\langle\delta\dot{\gamma}\rangle = 0\,, \quad \langle\delta\dot{\gamma}^2\rangle = \langle\dot{\gamma}\rangle^2\frac{\langle\tau_{\rm int}\rangle}{S} = \langle\dot{\gamma}\rangle^2\frac{\langle\tau_{\rm int}(0)\rangle\vartheta(\rho)}{S(0)\chi(\rho)} . \tag{12}$$

The functions χ and ϑ contain the ρ-dependence of the strain-rate sensitivity and of the mean internal stress. Their behaviour is governed by the mechanisms of dislocation motion and of dislocation annihilation, respectively. It therefore depends on the deformation temperatures T_1 and T_2. For small ρ, the leading-order terms of χ and ϑ are given by

$$\chi(\rho) \to \begin{cases} \text{const.}\,, & T_2 < T_0\,, \\ K\sqrt{\rho}\,, & T_2 > T_0\,, \end{cases} \qquad \vartheta(\rho) \to \begin{cases} \tilde{K}\ln(1/\rho)\,, & T_1 < T_0\,, \\ \bar{K}/\rho \quad , & T_1 > T_0\,. \end{cases} \tag{13}$$

For $T_1 < T_0$, $T_2 < T_0$, when dislocation motion is controlled by kink-pair formation, the magnitude of fluctuations depends only weakly on the forest density ρ. Conversely, for $T_2 > T_0$ and T_1 either above or below T_0, dislocation motion is controlled by forest-dislocation cutting. In these cases while the forest density becomes small, an increasing activation volume leads to a decrease of S and strain-rate fluctuations become appreciable. Such fluctuations can be interpreted in terms of collective slip phenomena, i.e., dislocations moving in dynamic pile-ups [9,10]. For this mode of dislocation motion the correlation time of the strain-rate fluctuations is of the order of the lifetime of mobile dislocations, $t_{\rm corr} \approx \rho_{\rm m} bL/\langle\dot{\gamma}\rangle$ (L = slip-line length) [9,10]

Idealizing $\delta\dot{\gamma}$ by a Gaussian stochastic process gives (after scaling) a Langevin-type equation for the forest density ρ,

$$d\rho = f(\rho)\langle\dot{\gamma}(\rho)\rangle dt + \sigma_0 f(\rho)\sqrt{\frac{\langle\dot{\gamma}(\rho)\rangle\vartheta(\rho)}{\chi(\rho)}}\,dw\,, \tag{14}$$

where $\sigma_0^2 = \rho_{\rm m} bL\langle\tau_{\rm int}(0)\rangle/S(0)$ and dw is the increment of a Wiener process. Stochastic integration gives the Fokker-Planck equation (Stratonovich calculus) for the probability density p to find the system in the state ρ at time t when it was in $\rho = 1$ at $t = 0$:

$$\partial_t p = -\partial_\rho\left[\left(f\langle\dot{\gamma}\rangle + \frac{\sigma_0^2}{4}\partial_\rho\left[\frac{\langle\dot{\gamma}\rangle\vartheta}{\chi}f^2\right]\right)p\right] + \frac{\sigma_0^2}{2}\partial_\rho^2\left[\frac{\langle\dot{\gamma}\rangle\vartheta}{\chi}f^2 p\right] . \tag{15}$$

This equation is to be solved on the interval $\rho \in [0,1]$ with the initial condition $p(\rho, t = 0) = \delta(\rho - 1)$. In order to assess the qualitative features of the dynamics described by Eq. (15), we study its steady-state solution

$$p_{\rm s}(\rho) = \mathcal{N}\frac{1}{f}\sqrt{\frac{\chi}{\vartheta\langle\dot{\gamma}\rangle}}\exp\left[\frac{2}{\sigma_0^2}\int\frac{\chi}{\vartheta f}d\rho'\right] , \tag{16}$$

where $\mathcal{N}$ is a normalization constant fixed such that $\int_0^1 p_s d\rho = 1$. Let us first address the question of integrability. To this end we note that p_s possesses a singularity at $\rho = 0$, the behaviour of which is compiled in the following table:

Deformation regime	Regime 1 $T_1 < T_0$, $T_2 < T_0$	Regime 2 $T_1 > T_0$, $T_2 > T_0$	Regime 3 $T_1 < T_0$, $T_2 > T_0$
Behaviour for $\rho \to 0$	$\chi \to$ const $\vartheta \sim \ln(\rho^{-1})$ $\langle\dot{\gamma}\rangle \to$ const $f \sim -\rho$	$\chi \sim \rho^{1/2}$ $\vartheta \sim \rho^{-1}$ $\langle\dot{\gamma}\rangle \to$ const $f \sim -\rho^2$	$\chi \sim \rho^{1/2}$ $\vartheta \sim \ln(\rho^{-1})$ $\langle\dot{\gamma}\rangle \to$ const $f \sim -\rho$
Integrability of p_s	not integrable	not integrable	integrable

In Regimes 1 and 2, the singularity is non-integrable. Thus the point $\rho = 0$ is *absorbing*, i.e, the probability to find the system at $\rho = 0$ becomes equal to unity for large times. This resembles the deterministic case, since the system evolves towards a single state in probability space. Another situation occurs in Regime 3, where the asymptotic probability distribution may exhibit two different shapes (Fig. 4). If the strain-softening condition (10) is not fulfilled, the asymptotic distribution exhibits a single peak at the position of the deterministic attractor $\rho = 0$, i.e., the stochastic noise leads to a "smearing out" of the deterministic dynamics, but not to a qualitatively different behaviour (Fig. 4, left). If the strain-softening condition is fulfilled, however, a noise-induced phase transition shows up: An additional peak of p_s emerges at $\rho = 1$. For large times, the system may either have a high or a low dislocation density, whereas the probability to find intermediate dislocation densities is small (Fig. 4, right). This corresponds to a two-phase separation in probability space. Obviously the behaviour of the system can then no longer be understood in terms of the deterministic dynamics.

3. Results and Interpretation

The discussion revealed that in Regime 3 strain-rate fluctuations may provoke a two-phase separation in probability space. In order to "translate" this result into physical space, we note that the fluctuations possess a finite correlation length that is typically of the order of one micrometer [9,10]. Thus different mesoscopic volume elements of the crystal behave in a statistically independent way, and the phase separation in probability space directly corresponds to a phase separation in physical space: the microstructure decomposes into a dislocation rich phase ("matrix") and a dislocation-poor phase ("slip channels"). A similar argument holds for the local strains which are large in the slip channels but small in the matrix.

The results indicate that there are three prerequisites to be fulfilled for slip-channel formation in restrained b.c.c. metals: (i) The annihilation of forest dislocations must be sufficently efficient (first-order reaction), which requires the predeformation to be done below the transition temperature. (ii) The mode of dislocation

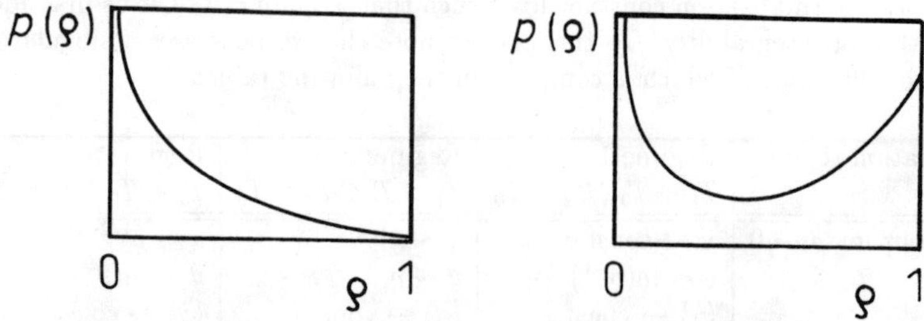

Fig. 4. Shapes of the asymptotic probability distribution in deformation regime 3. Left: As-deterministic behaviour. Right: Two-phase separation.

motion must favour the emergence of collective slip processes within regions of reduced forest density (embryonic slip channels). This is the case if forest dislocation cutting is the velocity-controlling process, i.e., for redeformation above the transition temperature. (iii) The forest density and/or the annihilation rate must be high, so that forest dislocation annihilation leads to a local strain softening. The critical values may be obtained from Eqs. (3) and (10).

Comparison with the experimental findings reported in Section 1.2 shows that the model accounts well for the observed qualitative features of slip-channel formation in low-temperature predeformed alloys. In particular, the stochastic treatment of the dislocation dynamics gives an adequate description of the interplay between microstructural softening and the emergence of collective slip phenomena. It thus seems a promising task to apply the present approach to other mesoscopic strain-localization phenomena, e.g., in precipitation- or irradiation-hardened materials.

References

1. B. Sestak and A. Seeger, *Phys. Stat. Sol. (b)* **43** (1971) 433.
2. B. Sestak and A. Seeger, In *Proc 3rd Int. Conf. Strength of Metals and Alloys,* Vol. 1, The Institute of Metals, Cambridge, 1973, p. 563.
3. B. Sestak and A. Seeger, *Z. Metallkde.* **69** (1978) 195, 355, 425.
4. A. Seeger, *Z. Metallkde.* **72** (1981) 369.
5. A. Seeger, In *Dislocations 1984*, eds. P. Veyssiere, L. Kubin and J. Castaing, C.N.R.S., Paris 1984, p. 141.
6. A. Luft, *Progr. Mat. Sci.* **35** (1991) 97.
7. P. Hähner, *Appl. Phys. A* **58** (1994) 49.
8. M. Zaiser and P. Hähner, to be published in *Scripta Metallurgica*.
9. P. Hähner, this volume.
10. P. Hähner, to be published in *Applied Physics*.

PART VI

FUNDAMENTALS OF FRACTURE MECHANICS

Continuum Models and Discrete Systems
Proceedings of 8th International Symposium, June 11-16, 1995, Varna, Bulgaria, ed. K.Z. Markov

DAMAGE ON MESO-LEVEL IN ORIENTED REINFORCED COMPOSITES

A. I. BALTOV
Institute of Mechanics, Bulg. Academy of Sciences,
Acad. G. Bontchev Str., bl. 4, BG-1113 Sofia, Bulgaria

Abstract. Oriented reinforced composites demonstrate anisotropic mechanical behaviour and loading type sensitivity. During damage this sensitivity becomes more significant, due to microcrack opening and closure. A mechanical-mathematical model of meso-damage in the considered composites is proposed on the basis of an analysis of the meso-damage mechanism. The model takes into account material mechanical anisotropy and loading type sensitivity during elastic and inelastic deformations with damage. Constant strain intensity surfaces are constructed in the stress space. Some of these surface can be experimentally obtained in different stress subspaces. This enables us to use such surfaces for parameter identification of the proposed model.

1. Introduction

We consider composite materials with an oriented meso-structure, i.e. unidirectionally or multidirectionally reinforced composites, composites with a textile structure, etc. [1]. These materials can have either a metal or a polymer matrix and can be reinforced by means of metal or glass fibers. Such materials, however, are extensively used in industry and the interest towards their mechanical property is essential [2]. Considering a homogenizing substituting continuum, the orientation of the material structure on meso-level (i.e., on "matrix-fiber" level) produces anisotropy and sensitivity to the loading type on macro-level [3]. However, meso-damage initiates and develops in such a composite over a certain stress level, see [4,5,6] *et al.*, and is characterized by

- shear and opening cracks in the matrix and fibers;
- pores in the matrix;
- cracks which spoil the cohesion between fibers and matrix;
- debonding in multilayered composites.

These defects yield on macro-level nonlinear relations between the measures of the stress and strain state and cause the occurrence of residual stresses.

2. Mechano-Mathematical Model

Consider isothermal static deformation of the composite under small strains and displacements (i.e. a linear geometric deformation theory). The material mechanical behaviour on macro level can be modelled on the following basis:

(i) Introduce measures of stress and strain as follows:

— stress vector

$$\{\Sigma_\alpha\} = \{\Sigma_1 = \sigma_{11},\ \Sigma_2 = \sigma_{22},\ \Sigma_3 = \sqrt{2}\sigma_{12},\ \Sigma_4 = \sigma_{33},\ \Sigma_5 = \sqrt{2}\sigma_{13},\ \Sigma_6 = \sqrt{2}\sigma_{23}\}^T,$$

$\alpha = 1, 2, \ldots, 6$, which corresponds to the Cauchy stress tensor $\{\sigma_{ij}\}$, $i, j = 1, 2, 3$. The set of all possible stress vectors defines a 6-dimensional stress vector space Σ. The intensity of the stress state at a body point is estimated by the quantity $\Sigma_0 = \sqrt{\Sigma_\alpha \Sigma_\alpha}$. Here summation is made over the repeating indices.

— strain vector

$$\{E_\alpha\} = \{E_1 = \epsilon_{11},\ E_2 = \epsilon_{22},\ E_3 = \sqrt{2}\epsilon_{12},\ E_4 = \epsilon_{33},\ E_5 = \sqrt{2}\epsilon_{13},\ E_6 = \sqrt{2}\epsilon_{23}\}^T,$$

$\alpha = 1, 2, \ldots, 6$, which corresponds to the small strain tensor $\{\epsilon_{ij}\}$, $i, j = 1, 2, 3$. The intensity of the strain at a body point is estimated by the quantity $E_0 = \sqrt{E_\alpha E_\alpha}$.

(ii) Since the material is sensitive to the loading type, we can distinguish a number of 2^α subspaces $(\chi_v) \subset \Sigma$. $v = 1, 2, \ldots, 2^\alpha$, where the loading type is one and the same.

(iii) We consider a fixed body point $(x_k) \in \Omega_0$, where $\Omega_0 \subset R^3$ is the region occupied by the body at the initial moment $t = t_0$ and R^3 is the three-dimensional Euclidean space. For each point (x_k) the stress vector $\{\Sigma_\alpha(t)\}$, $t \in [t_0, t_F]$, varies and t_F marks the end of the process. The image of this variation in Σ determines the loading path. Perform a time discretization of the process by step Δt corresponding to N moments of time $t(I)$, $I = 1, 2, \ldots, N$, where $t(1) = t_0$ and $t(N) = t_F$. The increments $\Delta\Sigma_\alpha(I) = \Sigma_\alpha(I+1) - \Sigma_\alpha(I)$, $\Delta E_\alpha(I) = E_\alpha(I+1) - E_\alpha(I)$, $\alpha = 1, 2, \ldots, 6$, are determined within each time interval $[t(I), t(I+1)]$. Moreover, a unit vector $\sigma_\alpha(I) = \dfrac{\Delta\Sigma_\alpha(I)}{\Delta\Sigma_0(I)}$ corresponds to each $\Delta\Sigma_\alpha(I)$ and this vector determines the subspace where the process takes place. If $\{\sigma_\alpha(I)\} = \text{const}$ for each $I = 1, 2, \ldots, N$, then loading is simple ($\{\sigma_\alpha\} \in \chi_v$, v is fixed).

(iv) The relation between the stress and strain increments is taken to have the form

$$\begin{aligned} \Delta\Sigma_\alpha(I) &= \Xi^{(v)}_{\alpha\beta}(I)\Delta E_\beta(I), \quad \text{or} \\ \Delta E_\alpha(I) &= H^{(v)}_{\alpha\beta}(I)\Delta\Sigma_\beta(I), \quad I = 1, 2, \ldots, N; \quad \alpha, \beta = 1, 2, \ldots 6. \end{aligned} \tag{1}$$

Here $\{\sigma_\alpha(I)\} \in \chi_v$ and $v = v_f$ is fixed, while the matrices $\{\Xi^{(v)}_{\alpha\beta}\} = \{H^{(v)}_{\alpha\beta}\}^{-1}$ have constant components in the interval $[t(I), t(I+1)]$ and these components can be determined experimentally [5].

(v) We assume that a relation $\Sigma_0 \leftrightarrow E_0$ exists and that it corresponds to each direction $\{\sigma_\alpha\}$ =const, $\{\sigma_\alpha\} \in \chi_v \subset \Sigma$, with fixed v, see Fig. 1. There exists a limit $\Sigma_0 = \Sigma_p^0$ over which plastic deformation and meso-damage develop in the material. This yields nonlinear relations between Σ_0 and E_0 and residual strains E_0^p occur. The relation $\Sigma_0 \leftrightarrow E_0^p$ can be built on such a basis, too. The meso-damage nucleation weakens the material and the elastic modulus of the latter decreases from its initial value $\Xi_0 = \dfrac{\Sigma_0}{E_0} =$ const for $\Sigma_0 \in [0, \Sigma_p^0]$, $\Xi_0 = \tan\alpha_0$ to $\Xi^* = \dfrac{\Sigma_0}{E_0} =$ const for $\Sigma_0 \in [\Sigma_{0m}, 0]$ and $d\Sigma_0 < 0$, $\Xi^* = \tan\alpha^*$, see Fig. 1.

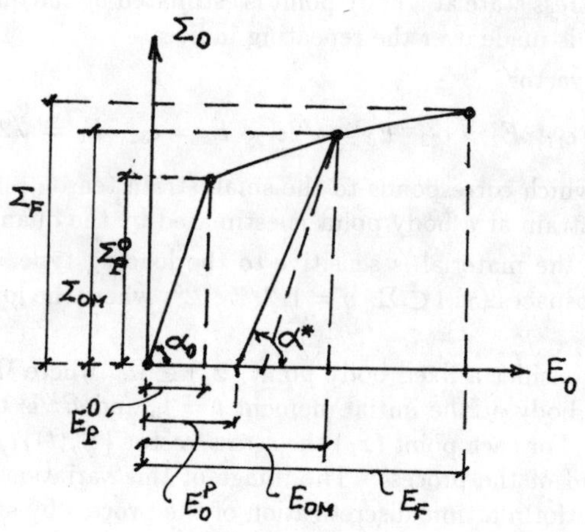

Fig. 1. Typical relation $\Sigma_0 \leftrightarrow E_0$.

Here Σ_{om} is the maximum attained value during the primary process for a loading increment $d\Sigma_0 > 0$. We introduce a damage measure $D = 1 - \dfrac{\Xi(E_{0m}^p)}{\Xi_0}$, where E_{0m}^p is the residual strain, corresponding to a stress Σ_{0m}. The variation has the form

$$dD = \Psi[E_{om}^p]dE_0^p, \quad \Psi = \frac{1}{\Xi_0}\left[-\frac{\partial \Xi^*[E_{om}^p]}{\partial E_0^p}\right]. \tag{2}$$

Upon process discretization, $D = D(J)$, $J = M, M+1, \ldots, N$, $\Sigma_0(M) = \Sigma_p^0$, $\Delta\Sigma_0(J) > 0$ and $\Delta D(J) = \Psi(J)\Delta E_0^p(J)$.

(vi) The infinite number of relations $\Sigma_0 \leftrightarrow E_0$ ($\Sigma_0 - E_0^p$, respectively) corresponding to an infinite number of directions $\sigma_\alpha =$ const in Σ are not independent. There are certain relations, however, which verify the existence of the following quadratic surfaces in Σ:

$$R^{(v)}_{\alpha\beta}(I)\Sigma_\alpha\Sigma_\beta - [E_0(I)]^2 = 0, \quad E_0(I) = \text{const},$$
$$\{\sigma_\alpha\} \in \chi_v, \quad I = 1,2,\ldots,N, \quad v = 1,2,\ldots, \tag{3}$$

and

$$\begin{aligned} F^{(v)}_{\alpha\beta}(I)\Sigma_\alpha\Sigma_\beta - [E^p_0(I)]^2 &= 0, \\ (E^p_0(I) = \text{const}, \quad I &= M, M+1,\ldots,N, \end{aligned} \tag{4}$$

where the matrices $\{R^{(v)}_{\alpha\beta}(I)\}$ and $\{F^{(v)}_{\alpha\beta}(I)\}$ have constant components for fixed (I) and (v). The latter are determined experimentally in [5].

If $E^p_0(M) \simeq 0{,}2\,\%$, then Eq. (4) for $I = M$ gives the surface of the initiation of nonelastic deformation corresponding to this initial residual strain. If $E^p_0(N) = \Delta_F$ is the residual strain at fracture, then Eq. (4) for $I = N$ gives the fracture surface of the considered material. The assumption of quadratic surfaces yields a quadratic yield condition in Σ, discontinuous with respect to the subspaces (χ_v), and a quadratic fracture criterion. They both are sensitive to the loading type.

(vii) We assume for complex loading that corrected relations $\Sigma_0 \leftrightarrow E_0$ ($\Sigma_0 \leftrightarrow E^p_0$, respectively) exist, corresponding to a specific loading path in Σ. E_0 and E^p_0 are obtained as discrete quantities with respect to the time steps Δt, when employing the following correcting denominator

$$\Phi(I) = 1 + \frac{1}{(I-1)}\sum_{J=1}^{I-1}\Big|[\Delta E^p_0[\sigma_\alpha(I)] - \Delta E^p_0[\sigma_\alpha(J)]/\Delta E^p_0[\sigma_\alpha(I)]\Big|, \tag{5}$$

where $\Delta E^p_0[\sigma_\alpha(J)]$ denotes the increment of the residual strain in the interval $[t(J), t(J+1)]$, corresponding to a relation $\Sigma_0 \leftrightarrow E^p_0$ for a $\{\sigma_\alpha(J)\}$ in Σ. Then

$$\begin{aligned} \Delta\bar{E}_0(I) &= \Phi(I)\Delta\bar{E}_0(I), \quad \bar{E}_0(I+1) = \bar{E}_0(I) + \Delta\bar{E}_0(I), \\ \Delta\bar{E}^p_0(I) &= \Phi(I)\Delta\bar{E}^p_0(I), \quad \bar{E}^p_0(I+1) = \bar{E}^p_0(I) + \Delta\bar{E}^p_0(I). \end{aligned} \tag{6}$$

(viii) The orientation of meso-damage for simple loading is estimated by using the damage vector $D_\alpha(I)$, $I = M, M+1,\ldots,N$. It is defined as follows

$$\begin{aligned} D_\alpha(I+1) &= D_\alpha(I) + \Delta D_\alpha(I), \quad \Delta D_\alpha(I) = \Psi(I)\Delta E^p_\alpha(I), \\ \Delta E^p_\alpha(I) &= \Delta E_\alpha(I) - \Delta E^e_\alpha(I), \quad \Delta E^e_\alpha(I)) = H^{*(x)}_{\alpha\beta}(I)\Delta\Sigma_\beta(I), \end{aligned} \tag{7}$$

where the matrix $H^{*(x)}_{\alpha\beta}(I)$ corresponds to the new elastic properties after the damage.

Since the loading is simple, the vectors $\Delta D_\alpha(I)$ and $\Delta E^p_\alpha(I)$ are taken to be co-linear in the time interval $t \in [t(I), t(I+1)]$.

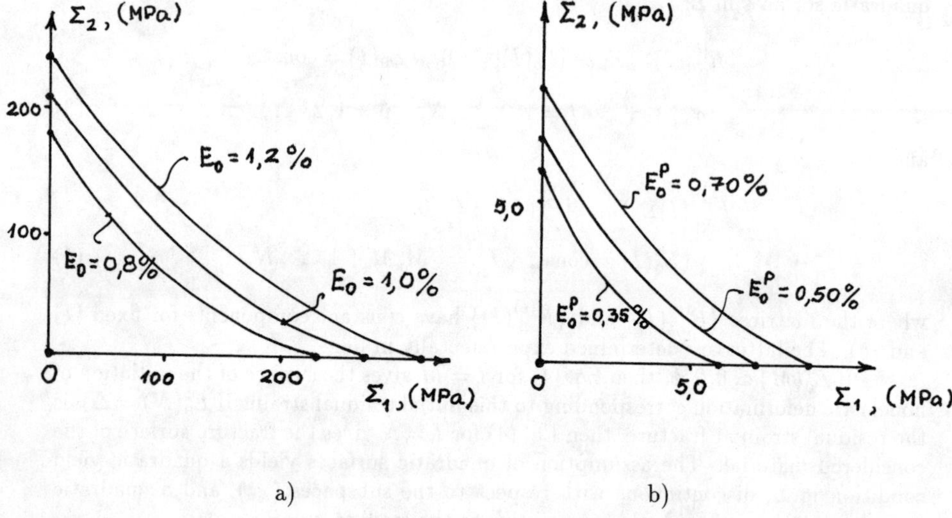

Fig. 2. a) Surfaces E_0= const for a glass/polymer composite. b) Surfaces E_0= const for a *SiC/SiC* composite.

(ix) In the case of complex loading, the oriented damage is measured by the correcting vector $\bar{D}_\alpha(I)$, $I = M, M+1, \ldots, N$. It is obtained in the following way

$$\begin{aligned} \bar{D}_\alpha(I+1) &= \bar{D}_\alpha(I) + \Delta\bar{D}_\alpha[\sigma_\beta(I)], \\ \Delta\bar{D}_\alpha[\sigma_\beta(I)] &= \Psi(I)\Delta\bar{E}^p_\alpha[\sigma_\beta(I)], \quad \alpha,\beta = 1,2,\ldots,6, \end{aligned} \tag{8}$$

for $\sigma_\beta(I)$ corresponding to the interval $[t(I), t(I+1)]$ of the path of complex loading. The loading process is simple for this time interval. Hence, the vectors $\Delta\bar{D}_\alpha(I)$ and $\Delta\bar{E}^p_\alpha(I)$ are co-linear.

3. Plane Stress

For plane stress (Σ_1, Σ_2), $v = 1,2,3,4$; $v = 1$ for $\Sigma_1 \geq 0$ and $\Sigma_2 \geq 0$; $v = 2$ for $\Sigma_1 \geq 0$, $\Sigma_2 \leq 0$; $v = 3$ for $\Sigma_1 \leq 0$, $\Sigma_2 \leq 0$; $v = 4$ for $\Sigma_1 \leq 0$, $\Sigma_2 \geq 0$, and hence the constitutive relations take the form

$$\begin{aligned} \Delta E_1(I) &= H^{(v)}_{11}(I)\Delta\Sigma_1(I) + H^{(v)}_{12}(I)\Delta\Sigma_2(I), \\ \Delta E_2(I) &= H^{(v)}_{21}(I)\Delta\Sigma_1(I) + H^{(v)}_{22}\Delta\Sigma_2(I), \end{aligned} \tag{9}$$

for $I = 1, 2, \ldots, N$, $v = 1, 2, 3, 4$. The components $H_{11}^{(v)}, \ldots$ are determined following the method given in [5]. The quadratic surfaces (3) and (4) take the from

$$\begin{aligned} R_{11}^{(v)}(I)\Sigma_1^2 + 2R_{12}^{(v)}(I)\Sigma_1\Sigma_2 + R_{22}^{(v)}(I)\Sigma_2^2 - (E_0(I)^2 &= 0, \\ F_{11}^{(v)}(I)\Sigma_1^2 + 2F_{12}^{(v)}(I)\Sigma_1\Sigma_2 + F_{22}^{(v)}(I)\Sigma_2^2 - (E_0^p(I)^2 &= 0. \end{aligned} \tag{10}$$

The experimental results (see [5] *et al.*) allow to assume the symmetries $H_{12}^{(v)} = H_{21}^{(v)}$; $R_{12}^{(v)} = R_{21}^{(v)}$; $F_{12}^{(v)} = F_{21}^{(v)}$ for fixed (I) and (v). For linear elastic behaviour these surfaces are given in [7] for a glass/polymer composite. The projections of the surfaces (10) are shown in Fig. 2 for such a composite. The meso-damaging and nonlinear behaviour of the composite are taken into account.

Acknowledgements. The author would like to thank Prof. T. Vhin, ISMCM-Saint–Ouen, for his helpful comments. This work is partially supported by The Bulgarian Science Foundation under Contract TH–409/94.

References

1. F. Lene, *Contribition a l'etude materiaux composites et de leur endommagement*, These, Univ. Paris-VI, 1984.
2. O. Allix, D. Egrand, P. Ladeveze and L. Perret, *Une novelle approche des composites par la mécanique de l'endommagement*, Cahan, 1993.
3. P. Ladeveze, In *Studies in Applied Mechanics*, No 34, Elsevier, London, 199.
4. T. Fuju, S. Amujima and F. Lin, *J. Comp. Math.* **26** (1992) 2493.
5. A.Baltov, *Modelling of Oriented Inelastic Deformation and Damage in Advanced Composite Materials*, Report No ERB-CIPA-CT-92-2132, Saint-Ouen, 1993.
6. P. Pluvinage, E. El. Bonazzaoni, B. Aurtoin and S. Baste, *Proc. X-ème Congres Fr. de Mécanique*, Paris, 1991, p. 137.
7. A. Baltov, In *Proc.7-th Int. Conf. Mech. Techn. Comp. Mater.*, Sofia, 1994, p. 28.

Continuum Models and Discrete Systems
Proceedings of 8th International Symposium, June 11-16, 1995, Varna, Bulgaria, ed. K.Z. Markov

CRACK LAYER THEORY

A. CHUDNOVSKY
The University of Illinois at Chicago,
842 W. Taylor St., 2095 ERF, Chicago, IL 60607, USA

and

R. S. LI
Motorola Inc., Automotive and Industrial Electronics Group,
4000 Commercial Ave., Northbrook, IL 60062, USA

Abstract

Recent developments in the formulation of constitutive equations for crack initiation and growth under fatigue and creep conditions are presented. A process zone (PZ) formed in the vicinity of a crack tip essentially controls fracture propagation. A system of strongly coupled crack and PZ is referred to a "Crack Layer" (CL). CL growth rates are related to the corresponding driving forces following the framework of irreversible thermodynamics. Evaluation of the CL driving forces involves solution of the crack-damage interaction. The experimental examination of the proposed constitutive equations is presented. Application of CL theory to evaluation of lifetime reliability is then discussed.

Continuum Models and Discrete Systems
Proceedings of 8th International Symposium, June 11-16, 1995, Varna, Bulgaria, ed. K.Z. Markov

ON THE ELECTROELASTIC FRACTURE THEORY

C. DASCALU
Institute of Mathematics, Romanian Academy,
P. O. Box 1-764, RO-70700 Bucharest, Romania

Abstract. The formulation of an energy-type fracture criterion for electroelastic materials is the object of this paper. Three possible forms of such a criterion are obtained from different energy balances for the fractured body. The already existing crack propagation laws are recovered and reformulated in a more general context by this technique. Then we prove the equivalence of two of them and we give an example of a particular crack problem which the third approach is unable to describe. The solution given for the fracture theory actually represents a choice between two formulations of the electroelasticity equations.

1. Introduction

Due to the practical importance of the electromechanical materials, the theory of electroelastic fracture has received a special attention in recent years. The aim of this contribution is to bring some clarifications on the formulation of an energy-type fracture criterion for electroelastic bodies. In the purely elastic theory such a criterion was sugested in the pioneering work of A. Griffith and was further developed by many authors (see [6]). The essential quantity involved in this criterion is the rate of energy released during the fracture process. When it attempts a critical value, which is a material property, the crack propagation is allowed. Different extensions of this concept to static electroelasticity have been proposed in the literature ([9,11,16]). In this paper they are obtained in quasielectrostatics (the accelerations are not neglected in the equations of motion) as the result of different global energy balances for the fractured body. Three electroelastic crack propagation laws are derived by this method. After showing the equivalence of two of them we give arguments for the inconsistency of the third approach.

We consider a linear electroelastic body containing a straight crack with traction free and electrically conducting faces. The study of this fracture problem allows us to present in a simple and unified manner some ideas contained in our previous papers [1–3].

2. The Linear Electroelastic Crack Problem

The governing equations for a linear electroelastic body, in the quasielectrostatic approximation, are the motion and no free charge equations ([5,8,17]):

$$\frac{\partial \sigma_{ij}}{\partial x_j} = \rho \frac{\partial^2 u_i}{\partial t^2}, \quad \frac{\partial D_i}{\partial x_i} = 0, \tag{1}$$

where σ_{ij}, $i,j = 1,2,3$, is the stress tensor, ρ is the mass density, u_i, $i = 1,2,3$, is the elastic displacement vector, and D_i, $i = 1,2,3$, is the electric displacement vector. The summation convention over repeated indices is used and the fields are referred to an absolute coordinate system x_1, x_2, x_3. Eqs. (1) are completed by the linear electromechanical constitutive relations:

$$\sigma_{ij} = C_{ijkl}\frac{\partial u_k}{\partial x_l} + e_{kij}\frac{\partial \varphi}{\partial x_k}, \tag{2}$$

$$D_i = e_{ikl}\frac{\partial u_k}{\partial x_l} - \epsilon_{ij}\frac{\partial \varphi}{\partial x_j}. \tag{3}$$

Here C_{ijkl} is the fourth-order tensor of elasticity coefficients, e_{ikl} is the third-order tensor of piezoelectricity coefficients and ϵ_{ij} is the tensor of dielectric constants. They are supposed to satisfy the symmetry relations:

$$C_{ijkl} = C_{klij} = C_{klji}, \quad e_{ikl} = e_{ilk}, \quad \epsilon_{ij} = \epsilon_{ji}. \tag{4}$$

In the quasielectrostatic approximation the electric field vector E_i is derived from the electric potential φ by:

$$E_i = -\frac{\partial \varphi}{\partial x_i}. \tag{5}$$

Consider the generalized plane problem, that is, the fields depend only on x_1, x_2-coordinates but having three-dimensional components. The fracture process is described by a family of reference configurations $(B(t))_{t\in[0,T)}$ (see Fig. 1), such that the open subsets $B(t)$ of R^2 differ only by the length of the crack $C(t)$, and $B(0)$ is the initial configuration of the body. The curve $C(t)$ represents the intersection of the crack surface, which front is parallel with the x_3-axis at each moment of time, with the x_1x_2-plane. Substitution of Eqs. (2) and (3) in Eq. (1) yields

$$C_{ijkl}\frac{\partial^2 u_k}{\partial x_l \partial x_j} + e_{kij}\frac{\partial^2 \varphi}{\partial x_k \partial x_j} = \rho\frac{\partial^2 u_i}{\partial t^2}, \tag{6}$$

$$e_{ikl}\frac{\partial^2 u_k}{\partial x_l \partial x_i} - \epsilon_{ij}\frac{\partial^2 \varphi}{\partial x_i \partial x_j} = 0, \tag{7}$$

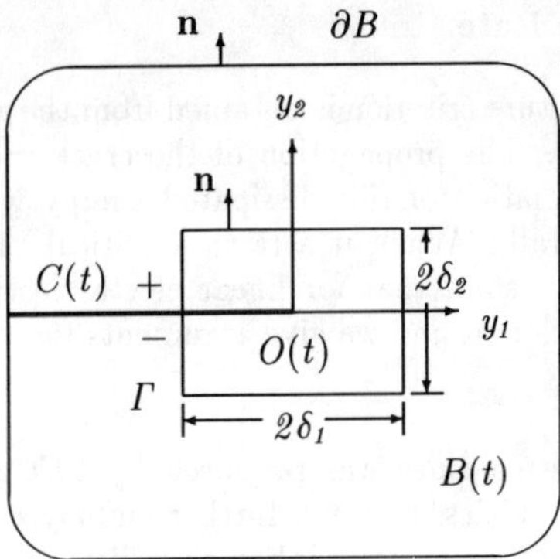

Fig. 1. Fractured electroelastic body

for $(x_1, x_2, t) \in D$ with $D = \{(x_1, x_2, t) \in R^3 \mid (x_1, x_2) \in B(t),\ t \in (0, T)\}$. At $t = 0$ initial elastic displacements $u_i(x_1, x_2, 0)$ and initial velocities $\dfrac{\partial u_i}{\partial t}(x_1, x_2, 0)$ are given for $(x_1, x_2) \in B(0)$. On the exterior boundary ∂B one can impose applied tractions $\sigma_{ij} n_j$ or elastic displacements u_i and, respectively, applied surface charge density $D_i n_i$ or an electrical potential φ. Here n_i, $i = 1, 2$, are the components of the outer unit normal to ∂B (Fig. 1). The boundary of $B(t)$ also consists of the two sides ($\pm$ in Fig. 1) of $C(t)$. The crack faces are supposed to be traction free and electrically conducting ([3,10,15]):

$$\sigma_{ij}^{\pm} n_j = 0\,, \quad \varphi^{\pm} = 0 \quad \text{on} \quad C(t)\,, \tag{8}$$

with n_i being the components of the unit normal to $C(t)$, oriented in the positive sense. We notice that another model for cracks in electroelastic materials is obtained by replacing the second condition in (8) with $D_i^{\pm} n_i = 0$. This corresponds to the electric impermeability property of the crack faces and such a model was studied in [2,12-14,16].

Near-tip asymptotic expressions for the solution of the static problem corresponding to that for Eqs. (6) and (7), were obtained by Suo [15]. A similar analysis can be done in quasielectrostatics (see [2] for electrically impermeable cracks) to deduce that in the vicinity of the crack tip u_k and φ are of $\mathcal{O}(\sqrt{r})$ and the stress and the electric displacement become singular of $\mathcal{O}\left(\dfrac{1}{\sqrt{r}}\right)$, with r being the radius from the tip $O(t)$.

The boundary value problem for Eqs. (6) and (7) must be completed with a fracture criterion describing the way in which the crack propagates through the material. Such evolution laws will be obtained in the next section from the energy analysis of the fracture process.

3. The Energy Release Rate

An energy-type fracture criterion is obtained from the global balance of energy for the electroelastic body. The propagation of the crack produces a dissipation of energy and the rate of variation of this dissipated energy with the crack length is called the energy release rate. When it attains a critical value the propagation is allowed. In this section we show that for linear electroelastic bodies such fracture parameters can be very different and we give arguments for a choice between them.

3.1. First Theory

The first theory we consider was proposed by Pak and Herrmann [11] for electroelastic materials, in statics, and was further employed for linear piezoelectric bodies in [12,16]. It extends the Eshelby's inhomogeneity analysis for elastic materials by taking into account the electric field effect. We make an energy-type analysis to obtain their result in the more general case of quasielectrostatics. For the simplicity of the exposition we work in the linear case (see [1] for the nonlinear problem). In the second part of this subsection a different form of the energy equation is used to obtain another rate of energy released during fracture and the equivalence of the two expressions is then proved.

The field equations (6) and (7) yield the following local form of the energy balance:

$$\frac{\partial}{\partial t}\left(W+\frac{\rho}{2}\frac{\partial u_k}{\partial t}\frac{\partial u_k}{\partial t}\right)=\frac{\partial}{\partial x_j}\left(\frac{\partial u_k}{\partial t}\sigma_{kj}+\frac{\partial \varphi}{\partial t}D_j\right), \tag{9}$$

where W is the electric enthalpy function

$$W = \frac{1}{2}C_{ijkl}\frac{\partial u_i}{\partial x_j}\frac{\partial u_k}{\partial x_l}+e_{ikl}\frac{\partial \varphi}{\partial x_i}\frac{\partial u_k}{\partial x_l}-\frac{1}{2}\epsilon_{ij}\frac{\partial \varphi}{\partial x_i}\frac{\partial \varphi}{\partial x_j}. \tag{10}$$

When the energy balance is written in global form for the fractured body B in Fig. 1 a standard procedure (see [7]) leads to:

$$\frac{d}{dt}\int_B\left(W+\frac{\rho}{2}\frac{\partial u_k}{\partial t}\frac{\partial u_k}{\partial t}\right)da\,+\,c\,\mathcal{G}=\int_{\partial B}\left(\frac{\partial u_k}{\partial t}\sigma_{kj}n_j+\frac{\partial \varphi}{\partial t}D_jn_j\right)ds \tag{11}$$

with

$$c\,\mathcal{G}=\lim_{\Gamma\to 0}\int_\Gamma\left(\left(W+\frac{\rho}{2}\frac{\partial u_k}{\partial t}\frac{\partial u_k}{\partial t}\right)cn_1+\frac{\partial u_k}{\partial t}\sigma_{kj}n_j+\frac{\partial \varphi}{\partial t}D_jn_j\right)ds, \tag{12}$$

where the path Γ surrounds the crack tip (in Fig. 1 Γ is a rectangle) and $c=c(t)$ is the crack speed. In order to deduce Eqs. (11), (12), we used the boundary conditions (8) on the crack faces. The quantity $\mathcal{G}$ is the energy release rate given by the energy balance (11) and the propagation criterion takes the form $\mathcal{G} = \mathcal{G}_{cr}$, where $\mathcal{G}_{cr}$ is a constant expressing the resistance to fracture of the electroelastic material. As it was shown in [2] for the antiplane problem, the fracture criterion provides a differential equation for the crack length as function of time, this equation being coupled to the system (6), (7).

The previous analysis is essentially based on the form (9) of the energy equation. However, a different form of the energy equation can be deduced from Eqs. (6) and (7) (see [5,8,17]):

$$\frac{\partial}{\partial t}\left(U + \frac{\rho}{2}\frac{\partial u_k}{\partial t}\frac{\partial u_k}{\partial t}\right) = \frac{\partial}{\partial x_j}\left(\frac{\partial u_k}{\partial t}\sigma_{kj} - \varphi\frac{\partial D_j}{\partial t}\right), \tag{13}$$

where U is the internal energy, related to the electric enthalpy W by

$$W = U - D_j E_j\,. \tag{14}$$

In global form, on B, Eq. (13) becomes

$$\frac{d}{dt}\int_B \left(U + \frac{\rho}{2}\frac{\partial u_k}{\partial t}\frac{\partial u_k}{\partial t}\right) da \;+\; c\,\overline{\mathcal{G}} \;=\; \int_{\partial B}\left(\frac{\partial u_k}{\partial t}\sigma_{kj}n_j - \varphi\frac{\partial D_j}{\partial t}n_j\right) ds\,. \tag{15}$$

wherein

$$c\,\overline{\mathcal{G}} \;=\; \lim_{\Gamma\to O}\int_\Gamma \left(\left(U + \frac{\rho}{2}\frac{\partial u_k}{\partial t}\frac{\partial u_k}{\partial t}\right) cn_1 + \frac{\partial u_k}{\partial t}\sigma_{kj}n_j - \varphi\frac{\partial D_j}{\partial t}n_j\right) ds\,, \tag{16}$$

The quantity $\overline{\mathcal{G}}$ may be a candidate for the fracture criterion, hence it is important to establish its relation to $\mathcal{G}$. With Γ being the rectangle in Fig. 1 we evaluate $\mathcal{G} - \overline{\mathcal{G}}$ when the limit $\Gamma \to O$ is taked by first letting $\delta_2 \to 0$ and then $\delta_1 \to 0$. We consider a new coordinate system y_1, y_2, centered in the crack tip and moving with it. Since there is no contribution to $\overline{\mathcal{G}}$ or $\mathcal{G}$ from the segments parallel to y_2-axis we get

$$\mathcal{G} - \overline{\mathcal{G}} \;=\; \lim_{\delta_1\to 0}\left(\lim_{\delta_2\to 0}\int_{-\delta_1}^{\delta_1}\left[\frac{\partial}{\partial t}(\varphi D_2)(y_1,\delta_2) - \frac{\partial}{\partial t}(\varphi D_2)(y_1,-\delta_2)\right] dy_1\right). \tag{17}$$

Only the higher order asymptotic terms have a contribution to (17) and for these terms the time derivative can be calculated with the formula $\dfrac{\partial}{\partial t} = -c\dfrac{\partial}{\partial y_1}$. Integrating on $(-\delta_1, \delta_1)$ and taking the limit $\delta_2 \to 0$ we obtain

$$\overline{\mathcal{G}} - \mathcal{G} \;=\; \lim_{\delta_1\to 0}\left(\left[\varphi^+(y_1,0)D_2^+(y_1,0) - \varphi^-(y_1,0)D_2^-(y_1,0)\right]\Big|_{-\delta_1}^{\delta_1}\right). \tag{18}$$

Because of the boundary conditions (8) and the continuity of φ and D_2 ahead the crack tip we deduce from Eq. (18) that $\overline{\mathcal{G}} = \mathcal{G}$. This shows that $\overline{\mathcal{G}}$ is an equivalent expression for the electroelastic energy release rate $\mathcal{G}$.

3.2. Second Theory

A new fracture parameter was proposed more recently by Maugin and Epstein [9] in electroelastostatics. They observed that the inhomogeneity method of Pak and

Herrmann [11] can be employed without taking into account the free electric field contribution. As it was shown in [1] this new rate of released energy is different from that proposed by Pak and Herrmann. In what follows we obtain their result in quasielectrostatics by the same method as previously and then we give a particular example of a fracture problem which allows us to make a selection between the two energy release rates.

From the field equations (6) and (7) one can deduce the local balance of energy:

$$\frac{\partial}{\partial t}\left(W+\frac{1}{2}E_kE_k+\frac{\rho}{2}\frac{\partial u_k}{\partial t}\frac{\partial u_k}{\partial t}\right) = \frac{\partial}{\partial x_j}\left(\frac{\partial u_k}{\partial t}\sigma_{kj}+\frac{\partial\varphi}{\partial t}P_j\right) + \frac{\partial\varphi}{\partial t}\frac{\partial E_j}{\partial x_j}, \tag{19}$$

where $P_j = D_j - E_j$ is the electric polarization vector. In global form the balance of energy becomes:

$$\begin{aligned}\frac{d}{dt}\int_B\left(W+\frac{1}{2}E_kE_k+\frac{\rho}{2}\frac{\partial u_k}{\partial t}\frac{\partial u_k}{\partial t}\right)da \;&+\; c\,\mathcal{G}^* \\ = \int_{\partial B}\left(\frac{\partial u_k}{\partial t}\sigma_{kj}n_j+\frac{\partial\varphi}{\partial t}P_jn_j\right)ds \;&+\; \int_B\frac{\partial\varphi}{\partial t}\frac{\partial E_j}{\partial x_j}\,da\end{aligned} \tag{20}$$

with

$$c\,\mathcal{G}^* = \lim_{\Gamma\to O}\int_\Gamma\left(\left(W+\frac{1}{2}E_kE_k+\frac{\rho}{2}\frac{\partial u_k}{\partial t}\frac{\partial u_k}{\partial t}\right)cn_1+\frac{\partial u_k}{\partial t}\sigma_{kj}n_j+\frac{\partial\varphi}{\partial t}P_jn_j\right)ds\,. \tag{21}$$

The existence of this balance of energy for the fractured body is the consequence of a different formulation of the electroelasticity equations. The last integral term in Eq. (20) is a volume source of energy and it is the result of the consideration of the free electric field action on matter as a volume force. Although this force does not appear in the linear equations of motion (6), the (nonlinear) energy equation (19) still contains its contribution. For a comparison with the nonlinear theory the reader is referred to [1,3].

In [1] it was showed that $\mathcal{G}$ differs from $\mathcal{G}^*$ so that we have to make a choice between them. To do that we consider a plane crack problem for a linear anisotropic dielectric with no piezoelectrical coupling, for which the second constitutive equation (3) reduces to

$$D_i = \epsilon_i E_i \tag{22}$$

(no summation over i), where ϵ_i, $i = 1, 2$, are the electric permeabilities in the directions of the y_1, y_2 coordinate axis and with $\epsilon_1 \neq \epsilon_2$. Constitutive relations of this type describe the behaviour of some important crystal classes, like those belonging to the thetragonal or hexagonal systems (see [5], pp. 248-252). In this case the problem for the electric field is decoupled from the mechanical one, and the asymptotic electric field in the vicinity of the crack tip can be explicitely obtained (see [3]).

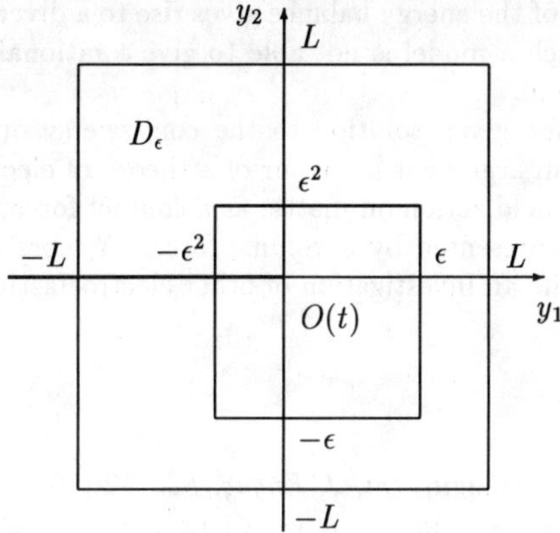

Fig. 2. Special domain D_ϵ

Consider now the energy balance (20) for this problem and let us investigate the integrability of the last source term. The singularity of the solution in the vicinity of $O(t)$ shows that it may be too singular to be integrable. When we choose an integration domain D_ϵ like that situatued between the two squares in Fig. 2, the calculation made in [3] leads to

$$\lim_{\epsilon \to 0} \int_{D_\epsilon} \frac{\partial \varphi}{\partial t} \frac{\partial E_j}{\partial x_j} \, da \; = \; \infty \, . \tag{23}$$

This shows that the considered crack problem does not allow a global energy equation of the form (20). The only reasonable energy balance in this case is (11) leading to the energy release rate $\mathcal{G}$. It is the fracture parameter to be used in the propagation criterion.

Finally we remark that the interpretation of the integral in Eq. (23) in a generalized sense, like principal value, is not physically acceptable. Indeed, it is reasonable to assume that for *each* family of subbodies B_ϵ converging to the entire body B, the sequence of the energy integrals on B_ϵ must converge to the same energy integral evaluated on B.

4. Concluding Remarks

An analysis of different energy approaches to fracture of electroelastic materials has been made in this contribution. First we showed that starting with two different forms of the energy equation, the same rate of energy released during the propagation of the crack is obtained. Then, another energy balance law, providing a different energy release rate, was compared with the first one on a concrete problem. We considered a particular case of an anisotropic dielectric with a straight crack and we proved that the last form of the energy balance gives rise to a divergent integral term. The conclusion is that such a model is not able to give a rational description of the considered fracture problem.

We remark that the given solution to the controversy on the electroelastic fracture criterion is also an argument in favour of a theory of electroelasticity, which considers the free electric field action on matter as a contact force, in opposition with an at-a-distance action represented by a volume force. A more detailed discussion on this is given in [3], while an investigation of other electroelasticity theories in this context is made in [4].

References

1. C. Dascalu and G.A. Maugin, *Int. J. Engng. Sci.* **32** (1994) 755.
2. C. Dascalu and G. A. Maugin, *Q. J. Mech. Appl. Math.* **48** (1995) 237.
3. C. Dascalu and G. A. Maugin, *J. Appl. Math. Phys. (ZAMP)*, 1995, to appear.
4. C. Dascalu, *Electroelasticity Equations and Energy Approaches to Fracture*, submitted.
5. A. C. Eringen and G. A. Maugin, *Electrodynamics of Continua I*, Springer-Verlag, New York, 1990.
6. L. B. Freund, *Dynamic Fracture Mechanics*, University Press, Cambridge, 1990.
7. M. E. Gurtin and C. Yatomi, *Arch. Rational Mech. Anal.* **74** (1980) 231.
8. G. A. Maugin, *Continuum Mechanics of Electromagnetic Solids*, Amsterdam, North-Holland, 1988.
9. G. A. Maugin and M. Epstein, *Proc. R. Soc. Lond.* A**433** (1991) 299.
10. R. M. McMeeking, *Int. J. Engng. Sci.* **28** (1990) 608.
11. Y. E. Pak and G. Herrmann, *Int. J. Engng. Sci.* **24** (1986) 1365.
12. Y. E. Pak, *J. Appl. Mech.* **57** (1990) 647.
13. H. Sosa, *Int. J. Solids Structures* **29** (1992) 2613.
14. H. Sosa and Y. E. Pak *Int. J. Solids Structures* **26** (1990) 1.
15. Z. Suo, *J. Mech. Phys. Solids* **41** (1993) 1155.
16. Z. Suo, C-M. Kuo, D. M. Barnett and J. R. Willis, *J. Mech. Phys. Solids* **40** (1992) 739.
17. H.F. Tiersten, *Linear Piezoelectric Plate Vibrations*, Plenum Press, New York, 1969.

Continuum Models and Discrete Systems
Proceedings of 8th International Symposium, June 11-16, 1995, Varna, Bulgaria, ed. K.Z. Markov

ON A MODEL OF DAMAGE IN POLYMER MATERIALS

G. DRAGANOVA
Institute of Chemical Technology and Biotechnology,
P. O. Box 110, BG-7200 Razgrad, Bulgaria

and

K. Z. MARKOV
Faculty of Mathematics and Informatics,
"St. Kl. Ohridski" University of Sofia, 5 blvd J. Bourchier, BG-1126 Sofia, Bulgaria

Abstract. The aim of the paper is to extend the well known reasoning and conclusions of the continuum damage mechanics from metals to polymer materials, when Norton's creep law is replaced by a hereditary type constitutive equation. Treating damage as volume fraction of microvoids appearing and evolving during straining, some simple models of mechanics of composites are employed. A system of coupled differential equation for the longitudinal strain and damage as functions of time is derived as a result. The brittle, ductile and mixed brittle-ductile type of failure are identified in a simplified model example.

1. Introduction

The aim of the paper is to extend the well known reasoning and conclusions of the continuum damage mechanics from metals to polymer materials, when Norton's creep law is replaced by a hereditary type constitutive equation. In the former case the assumption of incompressibility essentially simplifies the analysis. For polymer creep this assumption is inapplicable, moreover, though the constitutive relations are linear, their integral structure makes the damage analysis much more complicated in general.

Here we shall outline an approach which despite its approximate nature allows, hopefully, an adequate study of the damage processes in polymer material in interconnection with straining history. The approach follows the traditions of the continuum damage mechanics, initiated by Kachanov [1,2], invoking in particular a kinetic equation for damage evolution of the same type as in Kachanov's theory. A difference lies in a bit more "physical" interpretation of damage as volume fraction of microvoids that appear and evolve during deformation. This allows us to employ some of the simple models of mechanics of composites, treating the damaged solid as porous. In the classical problem for a deteriorating rod under a fixed tensile force, a system of coupled differential equation for the longitudinal strain and damage as functions of time is derived as a result. The brittle, ductile and mixed brittle-ductile type of failure are identified in a simplified model example.

2. Basic Equations of the Model

Consider a linear viscoelastic material governed by usual hereditary type constitutive equations:

$$\varepsilon(t) = \frac{1}{E}\sigma(t) + \int_0^t \Gamma(t-s)\,\sigma(s)\,ds\,; \tag{2.1}$$

only $1D$ case will be treated hereafter, in the context of a uniaxial tension of circular cylindrical rod, whose current length is l and its radius is r; the initial values (at $t = 0$) of these quantities will be denoted by l_0 and r_0 respectively. In Eq. (2.1) $\varepsilon = dl/l$ is the longitudinal strain, $\sigma = F/A$, $A = \pi r^2$, is the stress, Γ is the creep kernel and E—the Young modulus.

Imagine that due to loading damage appears and evolves in the rod, in the form of spherical microvoids (porosity), homogeneously and isotropically distributed throughout it. The volume concentration of the voids will be denoted by $c = c(t)$. Hence the rod becomes a microporous material whose effective macroscopic properties, say, the creep kernel, the Young modulus, etc., are strongly affected by the magnitude of c. Moreover, though the loading force F remains constant, the tensile stress σ increases since the radius r of the cross-section shrinks. To simplify the calculations hereafter, keeping at the same time the basic features of the problem under study, we shall replace Eq. (2.1) with the simplified relation

$$\varepsilon(t) = K(t,c)\,\sigma(t)\,, \tag{2.2}$$

where, at fixed c, $K(t,c)$ is just the creep curve of the material with porosity c, corresponding to unit stress. The functional dependence of $K(t,c)$ upon c is supplied by the theory of two-phase viscoelastic composite materials in the particular case when one of the constituents represents voids. The appropriate results and approximate analytical relations for the function $K(t,c)$ are discussed in Section 3 below, so that we shall treat this function as known for the moment.

In the case under study

$$\sigma(t) = \frac{F}{\pi r^2} = \frac{\sigma_0}{\rho^2}\,, \quad \rho = \frac{r}{r_0}\,, \tag{2.3}$$

where $\sigma_0 = F/(\pi r^2)$ is the initial value of the stress and $\rho = \rho(t) \le 1$ is the dimensionless radius of the cross-section.

Differentiating (2.2) with respect to time, we get

$$\dot{\varepsilon}(t) = \frac{dl/dt}{l} = \frac{\dot{\lambda}}{\lambda} = \frac{d}{dt}\Big[K(t,c)\frac{\sigma_0}{\rho^2}\Big]$$

$$= \frac{\sigma_0}{\rho^3}\left[\rho\frac{\partial K(t,c)}{\partial t} + \rho\dot{c}\frac{\partial K(t,c)}{\partial c} - 2K(t,c)\dot{\rho}\right], \quad \lambda = \frac{l}{l_0}, \tag{2.4}$$

so that $\lambda = \lambda(t) \geq 1$ is the dimensionless length of the rod.

In turn, the increments of λ and ρ are interconnected through the Poisson ratio $\nu = \nu(c)$:

$$\frac{d\lambda}{\lambda} = -\nu(c)\frac{d\rho}{\rho}, \quad \text{i.e.} \quad \frac{\dot{\lambda}}{\lambda} = -\nu(c)\frac{\dot{\rho}}{\rho}, \tag{2.5}$$

since $d\lambda/\lambda$ is the longitudinal and $d\rho/\rho$ is the transverse strains. We again underline the fact that $\nu = \nu(c)$ depends on the void ratio c; the dependence $\nu(c)$ will also be discussed in Section 3 below.

Finally, an evolution law for the "damage" variable $c = c(t)$ is required. In the best tradition of the continuum damage mechanics we postulate that its rate $\dot{c}$ is determined by the magnitude of the current stress:

$$\dot{c} = B\sigma^n, \tag{2.6}$$

i.e., the same as in Kachanov's damage model [1,2], with B and n denoting material parameters. The difference is only in the fact that while in the latter $\sigma = \sigma_0/(1-\omega)$, where ω is Kachanov's damage, here $\sigma = \sigma_0/\rho^2$ and ρ is a much more complicated function of c that depends implicitly on the creep curve $K(t,c)$, see (2.4) and (2.6).

Note that Eq. (2.6) is tantamount to the equation of the kinetic theory of fracture as developed in detail in [4]. In the latter, one considers the number N of broken bonds in a polymer, assuming that their rupture results from thermal fluctuations induced by the actual applied stress σ which is increased due to the very appearance of the broken bonds. An analysis of the frequency and magnitude of such fluctuations yields that the rate dN/dt is proportional to $\exp(\sigma)$. But broken bonds generate obviously microvoids so that their number is closely connected with the microporosity, i.e., damage in our interpretation, in the polymer. Hence formally we can replace (2.6) by the law

$$\dot{c} = A_1 e^{A_2\sigma}, \tag{2.7}$$

with material parameters A_1, A_2. From the formal point of view, adopted in the continuum damage mechanics, (2.7) is as good as (2.6), since in both these equations the parameters should be specified by an appropriate fit to the experimental data, concerning time-to-rupture for various initial stress values. A more detailed account of the physical situation and fluctuation reasons leading to bonds ruptures allows however to get an interpretation of the parameters A_1 and A_2 in (2.7) (especially, for their temperature dependence [4]) — something which is outside the scope of the more formal continuum damage mechanics in the sense of Kachanov [1,2].

The equations (2.4)–(2.6) form the basic system of differential equations for the unknown functions $\lambda(t)$, $\rho(t)$, $c(t)$ of the proposed model of a deteriorating polymer material. The system should be solved under the obvious initial conditions

$$\lambda(0) = 1, \quad \rho(0) = 1, \quad c(0) = 0. \tag{2.8}$$

The analytical solution of the aforementioned system is impossible even for simplest plausible creep functions $K(t,c)$. However, a qualitative picture of the behaviour of the rod can be easily drawn.

Indeed, this behaviour will be decisively determined by the initial stress value σ_0. If σ_0 is very small, the creep phenomenon will be not pronounced strongly and only the "damage" c will evolve achieving a certain critical magnitude, say $c \approx 0.66$, leading to rod's rupture. This situation corresponds obviously to brittle fracture in this case. The peculiar rheological model—metal or hereditary creep type—is clearly not of special importance as it should be since rupture is governed by damage accumulation that follows Kachanov's type law (2.6). As a consequence, it should be then expected that one of the basic facts of Kachanov's damage mechanics, namely the good approximation of dependence of the time-to-rupture versus stress by a linear fit in the log − log coordinates will hold true in the viscoelastic model under study. This will be indeed confirmed in Section 4 for a special and realistic form of the creep function $K(t,c)$ that corresponds to the linear standard body. The peculiarity of the rheology shows up only at higher values of σ_0 when considerable creep deformation evolves and, as a result, the cross-section shrinks down tending to zero; the actual stress increases as a consequence and the rod ruptures due to purely geometrical reasons and fast enough so that there is no time for the damage to attain considerable magnitudes. This situation corresponds obviously to the case of ductile failure. For intermediate values of σ both mechanisms—creep deformation decreasing the cross-section and damage accumulation coexist and interplay and hence a mixed brittle-ductile failure of the rod takes place.

3. Some Facts from Mechanics of Porous Solids

To take into account the influence of the porosity c, treated here as damage parameter, on the effective properties of the rod, we shall invoke some of the simplest relations of mechanics of composite materials. Such relations follow, e.g., from the model of "concentric" spheres as discussed in [5]. This model, as a matter of fact, corresponds to the effective field approach in mechanics of composites [5,6], which in the scalar conductivity case coincides with the well-known Maxwell (or Clausius-Mossoti) relation. The predictions of the model are extremely simple if we assume additionally that the Poisson ratio, ν_0, of the undamaged rod is 0.2. In this case it turns out that $\nu(c) \equiv \nu_0 = 0.2$, independently of c. Hence, from (2.5),

$$\frac{d\lambda}{\lambda} = -0.2\,\frac{d\rho}{\rho}$$

which means that

$$\rho\lambda^5 = 1. \tag{3.1}$$

A somewhat similar relation between ρ and λ exists in the incompressible case, namely, $\rho^2\lambda = 1$. Recall that the latter is essentially used in the Hoff's model of ductile rupture of metals in creep [1–3].

Note that (2.3) and (3.1), when employed in (2.6), yield

$$\dot{c} = B_1 \varepsilon_0^n \lambda^\nu , \quad B_1 = B E_0^n , \quad \nu = 10n, \tag{3.2}$$

thus excluding the function $\rho(t)$. In (3.2) $\varepsilon_0 = \sigma_0 / E_0$ is the initial strain of the rod (at $t = 0$).

In the same concentric shell model at $\nu_0 = 0.2$, the instantaneous Young modulus of the rod is a very simple function of c:

$$E = E(c) = E_0 \frac{1-c}{1+c} \tag{3.3}$$

and similarly for the long-time modulus E_∞. That is why we can assume the creep curves of the undamaged and damaged rod to be proportional

$$K(t,c) = \frac{1+c}{1-c} K(t,0) , \tag{3.4}$$

with a factor, reciprocal to that in (3.3). (The reason is that $K(0,c) = 1/E$, $K(\infty,c) = 1/E_\infty$, where $E = E(c)$ and $E_\infty = E_\infty(c)$ are respectively the instantaneous and long-time elastic moduli of the damaged rod.) The formula (3.4) holds true at $t = 0$ and $t \to \infty$. The detailed analysis of (3.4), performed in [8,9], demonstrates that it provides a very good approximation to the creep curves of a two-phase linear viscoelastic composite and hence it is fully appropriate for the present study which already contains a number of simplifying assumptions.

With (3.3) and (3.4) taken into account, (2.4) becomes

$$\dot{\varepsilon}(t) = \frac{\dot{\lambda}}{\lambda} = \varepsilon_0 \frac{d}{dt} \Big[\lambda^{10} \frac{1+c}{1-c} f(t) \Big] , \tag{3.5}$$

where $f(t) = K(t,0)\, E_0$ is the dimensionless creep curve, $f(0) = 1$, and $\varepsilon_0 = \sigma_0 / E_0$ is the initial strain, defined in (3.2).

A simple differentiation of the right-side of (3.5) yields, taking into account (3.2),

$$\dot{\lambda} = \frac{\varepsilon_0 \lambda^{11} \big[2 B_1 \varepsilon_0^n \lambda^\nu f(t) + (1-c^2) \dot{f}(t) \big]}{(1-c) \big[1 - c - 10 \varepsilon_0 \lambda^{10} (1+c) f(t) \big]} . \tag{3.6}$$

Together with (3.2) we thus get a system of differential equations for $\lambda(t)$ and $c(t)$ that governs the tensile behaviour of the deteriorating viscoelastic rod. The system takes into account the creep characteristics of the latter through the function $f(t)$ and the damage accumulation features through the material parameters B_1 and n. A bit more detailed analysis of this system will be performed in the next Section for a simple and plausible form of the function $f(t)$.

Note that (3.6) makes sense only at $\varepsilon_0 < 0.1$ since at $\varepsilon_0 = 0.1$ the denominator in the right-side of (3.6) vanishes at $t = 0$ which means that at $\sigma_0 = E_0/10$ the rod ruptures instantaneously.

4. Example and Discussion

For the sake of simplicity, assume that the solid part of the rod follows the well-known standard-linear model, so that its creep curve is

$$K_0(t) = K(t,0) = f(t)/E_0, \quad f(t) = 1 + p(1 - e^{-mt}), \tag{4.1}$$

where p and m are positive material parameters; moreover $m = 1/t_*$ and t_* is the retardation time of the model [10]. Introducing the dimensionless time $\tau = t/t_*$, the basic system (3.2), (3.6) becomes

$$\frac{d\lambda}{d\tau} = \frac{\varepsilon_0 \lambda^{11}\left[2\widetilde{B}_1 \varepsilon_0^n \lambda^\nu f(t) + (1-c^2)pe^{-\tau}\right]}{(1-c)\left[1 - c - 10\varepsilon_0 \lambda^{10}(1+c)f(\tau)\right]},$$

$$\frac{dc}{d\tau} = \widetilde{B}_1 \varepsilon_0^n \lambda^\nu, \tag{4.2}$$

where $f(\tau) = 1 + p(1 - e^{-\tau})$ and $\widetilde{B}_1 = B_1 t_* = BE_0^n t_*$ is a dimensionless material parameter.

It is important to point out that to have a reasonable picture of the rod's behaviour we should take $n < 1$. Then at $\varepsilon_0 \ll 1$, the right side of $(4.2)_1$ will be very small also which means that $\lambda = l/l_0$ will remain close to 1. At the same time $\varepsilon_0^n > \varepsilon_0$, and the damage will increase linearly in time: $c \approx \widetilde{B}_1 \varepsilon_0^n t$, as it follows from $(4.2)_2$. The failure occurs when c attains a critical value, say, $c \approx 2/3$, while λ remains close to 1, that is the longitudinal strain is very small. The time-to-rupture, t_R, will be specified then by the relation

$$B\sigma_0^n t_R = 2/3,$$

which shows that $\log \sigma_0$ is a linear function of $\log t_R$. If the kinetic equation of damage (2.7) is adopted, then $\log t_R$ and σ_0 will be interconnected by a linear function, which is often observed in experiments [4].

The aforesaid is demonstrated in Figs. 1 to 3, which correspond to somewhat arbitrarily chosen (for the illustrative purposes) material constants $\widetilde{B}_1 = 0.1$, $n = 0.1$ and $p = 2$ (the latter means that the creep is well pronounced, i.e. $E_0/E_\infty = 3$), varying the initial strain ε_0. In Fig. 1 the functions $\lambda(\tau)$ and $c(\tau)$ are shown for $\varepsilon_0 = 0.0001$. The time-to-rupture is $t_R \approx 11t_*$, corresponding to the moment when the damage c attains the critical value of 2/3. Up to this moment $\lambda = l/l_0$ remains very close to 1 which means that the rod indeed fails due to damage accumulation without showing considerable macroscopic deformation. At $\varepsilon_0 = 0.01$ the situation is entirely different, as is well seen from Fig. 2, where the same functions $\lambda(\tau)$ and $c(\tau)$ are plotted. In this case the function $\lambda(\tau)$ has a vertical asymptote, not shown there, at $\tau \approx 2.6$; at this moment the volume concentration of damage remains low—around 0.2 and hence a typical ductile failure takes place. The intermediate case is shown in Fig. 3, where the initial strain is $\varepsilon_0 = 0.001$ in between those of Figs. 1 and 2. In this

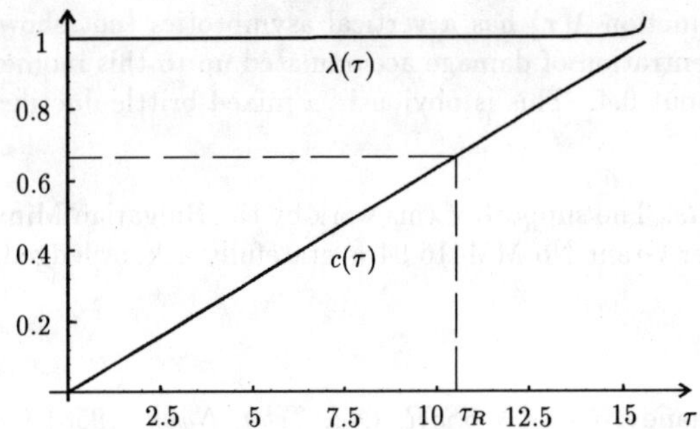

Fig. 1. Longitudinal extension $\lambda = l/l_0$ and damage c as functions of dimensionless time $\tau = t/t_*$; initial strain $\varepsilon_0 = 0.0001$ corresponding to "brittle" rupture.

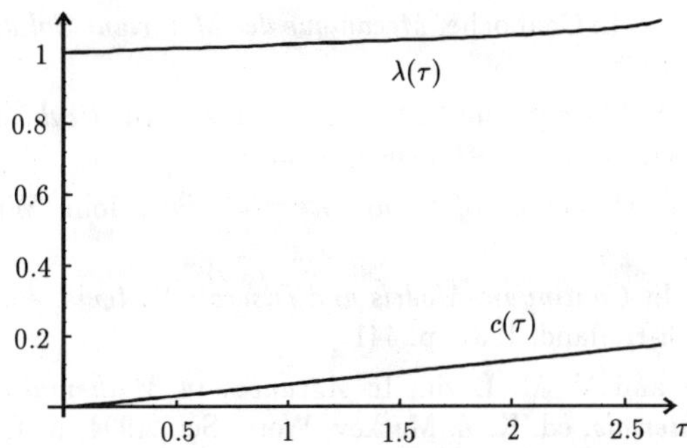

Fig. 2. The same as in Fig. 1 for initial strain $\varepsilon_0 = 0.01$ corresponding to "ductile" rupture.

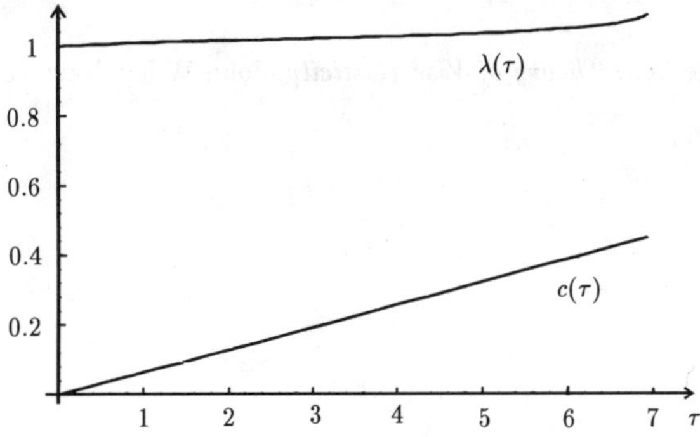

Fig. 3. The same as in Fig. 1 for initial strain $\varepsilon_0 = 0.005$ corresponding to mixed "brittle-ductile" rupture.

case again the function $\lambda(\tau)$ has a vertical asymptotics (not shown) at $\tau \approx 7$, but the volume concentration of damage accumulated up to this moment is already very considerable—about 0.4. This is obviously a mixed brittle-ductile type of failure of the rod.

Acknowledgements. The support of this work by the Bulgarian Ministry of Education and Science under Grant No MM416-94 is gratefully acknowledged.

References

1. L. M. Kachanov, *Izv. AN SSSR, Otd. Tehn. Nauk* (1958) (No 8) 26. (in Russian.)
2. L. M. Kachanov, *Introduction to Continuum Damage Mechanics*, Klüwer Acad. Publ., 1986.
3. J. Lemaitre and J.-L. Chaboche, *Mécanique des Materiaux Solids*, Dunod, Paris, 2-ème edition, 1988.
4. V. R. Regel, A. V. Slutcker and V. V. Tomashevskii, *Kinetical Nature of Strength of Solids*, Nauka, Moscow, 1974. (in Russian.)
5. R. Christensen, *Mechanics of Composite Materials*, John Wiley, New York, 1979.
6. K. Z. Markov, In *Continuum Models and Discrete Systems*, eds. O. Brulin and R. Hsieh, North-Holland, 1981, p. 441.
7. S. K. Kanaun and V. M. Levin, In *Advances in Mathematical Modelling of Composite Materials*, ed. K. Z. Markov, World Sci., 1994, p. 1.
8. K. Z. Markov, *Theor. and Appl. Mech., Bulg. Acad. Sciences*, Year 7 (No 4) (1976) 51.
9. K. Z. Markov, *Theor. Appl. Mechanics, Bulg. Acad. Sci.*, Year 14 (No 4) (1983) 45.
10. R. Christensen, *Theory of Viscoelasticity*, John Wiley, New York, 1979.

Continuum Models and Discrete Systems
Proceedings of 8th International Symposium, June 11-16, 1995, Varna, Bulgaria, ed. K.Z. Markov
© World Scientific Publishing Company, 1996, pp. 623–624

THE MECHANICAL RESPONSE OF CROSS-PLY CERAMIC COMPOSITES

D. ERDMAN and Y. WEITSMAN
Department of Engineering Science and Mechanics, 310 Perkins Hall,
The University of Tennessee, Knoxville, Tennessee 37996-2030, USA

Abstract

This presentation concerns a mechanics model for the multitude of interactive damage processes which occur in fiber reinforced ceramic materials. Unlike polymeric composites, where the fibers serve as the major stiffening agents and load bearing components, ceramic fibers are introduced to enhance the ductility of the otherwise unacceptably brittle ceramic matrices. The improved ductility is achieved by the purposeful inducement of various micro-scale, energy consuming, damage mechanisms which are attributable to the presence of fibers.

In uni-directionally reinforced ceramic composites, loaded parallel to the fiber direction, these mechanisms consist of matrix cracks bridged by the intact, more ductile fibers. The above cracks, which evolve normal to the fibers direction, are accompanied by interfacial slip zones along the fiber/matrix interfaces that emanate from the locations of the aforementioned matrix cracks.

Additional mechanisms include fiber breaks, which occur on the background field caused by multiple matrix cracks and interfacial slips, leading to fiber-break coalescence that results in ultimate failure.

In multi-directionally reinforced ceramic composites the foregoing mechanisms are accompanied by matrix cracking in directions normal or oblique to the fiber orientations, as well as interlaminar slips between distinct plies.

The above mechanisms were analyzed by a mechanics model which accounted for the observed increase in the density of matrix cracks during monotonic loading as well as for the overall ductile-like stress-strain response of the ceramic composite. The model employed a fracture energy criterion for the evolution of matrix cracks and Weibull failure statistics for fiber breaks. The statistical computations were associated with detailed stress fields which incorporated the presence of matrix cracks of increasing density.

The failure of cross-ply laminates is investigated by incorporating the above mechanistic model into the context of a modified shear-lag analysis. That analysis accounts for the presence of interfacial shears between distinct plied and incorporates the non-linear, ductile-like response of the individual plies.

At this time good agreement has been achieved between experimental observations and model predictions for the stress-strain response of both uni-directional and cross-ply ceramic composites. In addition, satisfactory correlation was noted between predicted and observed densities of matrix cracks in the uni-directional case. Additional work is in progress to achieve a similar correlation for cross-ply laminates.

The work on unidirectionally reinforced composites appeared in the articles listed below:

1. Y. Weitsman and H. Zhu, *J. Mech. Physics of Solids* **41** (1993) 351.
2. H. Zhu and Y. Weitsman, *J. Mech. Physics of Solids* **42** (1994) 1601.

Continuum Models and Discrete Systems
Proceedings of 8th International Symposium, June 11-16, 1995, Varna, Bulgaria, ed. K.Z. Markov
© World Scientific Publishing Company, 1996, pp. 625–632

NONLOCAL THEORY FOR MICROMORPHIC MODELS OF DAMAGED MATERIALS

E. INAN
Technical University of Istanbul, Faculty of Science and Letters, Maslak, 80626 Istanbul, Turkey

Abstract. The nonlocal theory of damaged materials is presented in this work by considering damaged solids as micromorphic continuum. Two types of constitutive equations are proposed for brittle and ductile materials. The propagation of plane waves in brittle microisotropic damaged solids is investigated.

1. Introduction

The concept of damage as a single scalar parameter was introduced by Kachanov [1]. Although its very instructive and useful form, this approach has some arbitrary assumptions and cannot be considered as a powerful tool in the analysis of more complicated damage problems. To establish a more realistic damage mechanics, Markov [2] proposed a micromorphic model [3,4] for damaged materials by considering the damage process as a generation and the growth of microdefects. As it is pointed out by Markov there are still some open points in the theory which must be investigated and improved. To overcome the difficulties several attempts have been done but all seem rather artificial. It is agreed by several investigators that the most convincing concept, or rather at least less artificial approach, is the introduction of the nonlocal concept. Micromorphic theories, in principle, are nonlocal theories where the nonlocality is achieved through moment tensor associated with each point of the body. Thus, to obtain a reasonable theory both ideas, micromotions and long range interactions must be employed together. In the following section we will give the results of the formulation of nonlocal theory for micromorphic model of damaged materials following the works given by Eringen in his several papers [5].

2. Fundamental Equations

We consider a damaged thermoelastic solid that responds to the changes in the motion, temperature and micromotion, which is defined as damage, and their gradients at one point. A nonlocal damaged thermoelastic solid is effected by this set of variables at all points of the body. Therefore, we take the ordered set

$$L(\mathbf{x}(\mathbf{X}',t), \mathbf{x}_{,K}(\mathbf{X}',t), \boldsymbol{\alpha}(\mathbf{X}',t), \boldsymbol{\alpha}_{,K}(\mathbf{X}',t), \theta(\mathbf{X}',t), \theta_{,K}(\mathbf{X}',t)), \quad \forall \mathbf{X}' \in V. \tag{2.1}$$

In addition, of course, we have the local variables. According to the state variables given by Eq. (2.1), any response function can be writen as follows

$$\Psi(\mathbf{X},t) = \psi(\mathbf{F}(\cdot), \mathbf{F}_K(\cdot), \mathbf{K}_K(\cdot), \mathbf{K}_{KL}(\cdot), \theta(\cdot), \theta_K(\cdot); \mathbf{x}_{,K}, \boldsymbol{\alpha}, \boldsymbol{\alpha}_{,K}, \theta, \theta_{,K}, \mathbf{X}), \tag{2.2}$$

where $(\cdot)$ denotes the difference functions as

$$\mathbf{F}(\cdot) = \mathbf{x}(\mathbf{X}') - \mathbf{x}(\mathbf{X}); \quad \mathbf{F}_K(\cdot) = \mathbf{x}_{,K}(\mathbf{X}') - \mathbf{x}_{,K}(\mathbf{X}). \tag{2.3}$$

The balance laws, entropy inequality and necessary modifications for nonlocal theory have been given by Eringen [5]. Following these modifications we calculate the time derivative of the internal energy function and substitute it into the entropy inequality to obtain the constitutive equations for damaged bodies. For nonheat conducting materials, the so-obtained results coincide with Eringen's result [5]. Constitutive equations for brittle-damaging materials can be obtained by applying the known precedure. Details are not given here. As it is the case in local micromorphic solid C_{KL}, Ψ_{KL} and Γ_{KLM} may be defined in terms of deformation gradient and damage tensor as

$$C_{KL} = \mathbf{x}_{,K} \cdot \mathbf{x}_{,L}\,, \quad \Psi_{K,L} = \mathbf{x}_{,K} \cdot \boldsymbol{\alpha}_L\,, \quad \Gamma_{KLM} = \mathbf{x}_K \cdot \boldsymbol{\alpha}_{L,M} \tag{2.4}$$

and nonlocal strain measures $B_K(\mathbf{X}')$, $D_{KL}(\mathbf{X}')$, $\Delta_{KL}(\mathbf{X}')$ and $\Lambda_{KLM}(\mathbf{X}')$ as

$$\begin{aligned} B_K(\mathbf{X}') &= \mathbf{x}_{,K} \cdot \mathbf{F}(\mathbf{X}'), \quad D_{KL}(\mathbf{X}') = \mathbf{x}_{,K} \cdot \mathbf{F}_L, \\ \Delta_{KL}(\mathbf{X}') &= \mathbf{x}_{,K} \cdot \mathbf{K}_L(X'), \quad \Lambda_{KLM}(\mathbf{X}') = \mathbf{x}_{,K} \cdot \mathbf{K}_{LM}(X'), \end{aligned} \tag{2.5}$$

we find

$$\begin{aligned} t_{kl} &= {}_E t_{kl} + \int_V [{}_E t'_{kl}]\, dv', \quad s_{kl} = {}_E s_{kl} + \int_V [{}_E s'_{kl}]\, dv', \\ \lambda_{klm} &= {}_E \lambda_{klm} + \int_V [{}_E \lambda'_{klm}] dv', \end{aligned} \tag{2.6}$$

where

$$\begin{aligned} {}_E t_{kl} =& (\lambda + \tau) e_{mm} \delta_{kl} + 2(\mu + \sigma) e_{kl} + \eta \varepsilon_{mm} \delta_{kl} + \nu \varepsilon_{lk} + \kappa \varepsilon_{kl}\,, \\ {}_E s_{kl} =& (\lambda + 2\tau) e_{mm} \delta_{kl} + 2(\mu + 2\sigma) e_{kl} + (2\eta - \tau) \varepsilon_{mm} \delta_{kl} + (\nu + \kappa - \sigma)(\varepsilon_{kl} + \varepsilon_{lk})\,, \\ {}_E \lambda_{klm} =& -\tau_1(\delta_{ml}\gamma_{krr} + \delta_{mk}\gamma_{rrl}) - \tau_2(\delta_{ml}\gamma_{rkr} + \delta_{kl}\gamma_{rrm}) - \tau_3 \delta_{ml}\gamma_{rrk} \\ & - \tau_4 \delta_{mk}\gamma_{lrr} - \tau_5(\delta_{mk}\gamma_{rlr} + \delta_{kl}\gamma_{mrr}) - \tau_9 \gamma_{lkm} - \tau_{10}\gamma_{mlk} - \tau_{11}\gamma_{kml} \\ & - \tau_6 \delta_{kl}\gamma_{rmr} - \tau_7 \gamma_{lmk} - \tau_8(\gamma_{mkl} + \gamma_{klm}) \end{aligned} \tag{2.7}$$

and

$$e_{kl} = \frac{1}{2}(u_{k,l} + u_{l,k}), \quad \varepsilon_{kl} = \varphi_{kl} + u_{k,l}, \quad \gamma_{klm} = -\varphi_{kl,m}. \tag{2.8}$$

All the quantities carrying (′) under the integral sign in Eqs. (2.6) are functions of all the points of the body, $\mathbf{x}' \in V$, and the material coefficients like λ', μ', ..., depend on the distance of the point under consideration to all point of the body, $|\mathbf{x}' - \mathbf{x}|$. If the solid undergoes plastic straining accompanied by damaging, the model for such materials must be modified and plastic straining should also be included among the internal variables remembering that $\mathbf{E}^p$ is subjected to certain additional conditions arising from the yield of the material. Thus, it will be perhaps more convenient to write the array or independent state variables by the addition of $\mathbf{E}^p$ as

$$(\mathbf{E}, \mathbf{E}^p, \boldsymbol{\alpha}, \nabla\boldsymbol{\alpha}, \theta, \mathbf{E}', \mathbf{E}'^p, \boldsymbol{\alpha}', \nabla\alpha', \theta'). \tag{2.9}$$

Following the ideas given by Eringen [6] for nonlocal plasticity, we may say that the entropy inequality must remain form invariant under the following transformation

$$\dot{E}^p_{KL} = \dot{E}^p_{0KL} + \lambda\rho_{KL}\dot{g}, \quad \dot{g} = 0, \quad \dot{\alpha}_{KL} = \dot{\alpha}_{0KL} + \mu\beta_{KL}\dot{f}, \quad \dot{f} = 0\,, \tag{2.10}$$

where in general λ, ρ_{KL}, g, μ, β_{KL}, f are functions of $(\mathbf{E}, \mathbf{E}^p, \boldsymbol{\alpha}, \nabla\boldsymbol{\alpha}, \theta)$ and functionals of $(\mathbf{E}', \mathbf{E}'^p, \boldsymbol{\alpha}', \nabla\alpha', \theta')$. Following physical considerations on flow of solid materials, in plastic region, we shall assume that [6,7] strains and damage at distant points from $\mathbf{X}$ do not affect appreciably the plastic flow and damage evaluation at point $\mathbf{X}$. Therefore we may assume that λ, ρ_{KL}, ..., are functions of local variables only. Then we have two criteria: the yield criterion

$$g(\mathbf{E}, \mathbf{E}^p, \boldsymbol{\alpha}, \nabla\boldsymbol{\alpha}, \theta, \kappa) = 0, \quad \dot{\kappa} = h_{RS}(\mathbf{E}, \mathbf{E}^p, \boldsymbol{\alpha}, \nabla\boldsymbol{\alpha}, \theta)\dot{E}^{(p)}_{RS} \tag{2.11}$$

and the damage criterion

$$f(\mathbf{E}, \mathbf{E}^p, \boldsymbol{\alpha}, \nabla\boldsymbol{\alpha}, \theta, k) = 0, \quad \dot{k} = l_{RS}(\mathbf{E}, \mathbf{E}^p, \boldsymbol{\alpha}, \nabla\boldsymbol{\alpha}, \theta)\dot{\alpha}_{RS}\,, \tag{2.12}$$

where h_{RS} and l_{RS} are second order tensors. For a fixed $\mathbf{E}^p = \mathbf{E}^p_0$, Eq. (2.11) gives

$$\dot{g} = \frac{\partial g}{\partial E_{KL}}\dot{E}_{KL} + \frac{\partial g}{\partial \alpha_{KL}}\dot{\alpha}_{KL} + \frac{\partial g}{\partial \alpha_{KL,M}}\dot{\alpha}_{KL,M} + \frac{\partial g}{\partial \theta}\dot{\theta} = \hat{g}\,; \tag{2.13}$$

as in the local theory Eq. (2.13) takes the form

$$\dot{E}_{KL} = \lambda\rho_{KL}\hat{g}, \quad g = 0\,, \tag{2.14}$$

and we assume Eq. (2.14) is valid when $\hat{g} > 0$ and $g = 0$, otherwise $\mathbf{E}'^p = 0$. This is the loading and unloading criterion used in local theories. During plastic deformation we have

$$\dot{g} = \hat{g} + \frac{\partial g}{\partial E^p_{KL}}\dot{E}^p_{KL} + \frac{\partial g}{\partial \kappa}\dot{\kappa} = 0 \tag{2.15}$$

and we obtain

$$\hat{g} = -H_{KL}\dot{E}^p_{KL}\,, \tag{2.16}$$

where

$$H_{KL} = \frac{\partial g}{\partial E^p_{KL}} + \frac{\partial g}{\partial \kappa} h_{KL} \tag{2.17}$$

and from Eq. (2.14) we find

$$\dot{E}^p_{KL} = (\rho_{KL} H_{MN}/\rho_{RS} H_{RS})\dot{E}^p_{MN}, \quad g = 0, \quad \hat{g} > 0\,. \tag{2.18}$$

Similar calculations can be carried out for damage evolution and the result is

$$\dot{\alpha}^p_{KL} = (\beta_{KL} G_{MN}/\beta_{PQ} G_{PQ})\dot{\alpha}_{MN}, \quad f = 0, \quad \hat{f} > 0\,, \tag{2.19}$$

where

$$G_{MN} = \frac{\partial f}{\partial \alpha_{MN}} + \frac{\partial f}{\partial k} l_{MN}, \quad \hat{f} = f\,. \tag{2.20}$$

According to the order of the theory, several form of expansions may be used to derive the specific forms of the unknown functions g, ρ_{KL}, h_{KL}, f, β_{KL}, l_{MN}. As it is done for the free energy function, a quadratic form for the loading function g can be used to describe a large class of materials including metals. Thus, we write, in the special case of linear dependence on Γ_{KLM},

$$g = g^0 + \frac{1}{2} g^1_{IJKLMN} \Gamma_{IJK} \Gamma_{KLM} - 2\kappa^2\,, \tag{2.21}$$

where g^0 and g^1_{IJKLMN} are isotropic tensor functions of tensor variables $\mathbf{E}$, $\mathbf{E}^p$, $\boldsymbol{\alpha}$ or $\mathbf{E}$, $\mathbf{E}^p$, Ψ. Then, for instance, g^0 may be written as

$$\begin{aligned} g^0 = {} & g^{01}_{KLMN} E_{KL} E_{MN} + g^{02}_{KLMN} E_{KL} E^p_{MN} + g^{03}_{KLMN} E^p_{KL} E^p_{MN} \\ & + g^{04}_{KLMN} E_{KL} \Psi_{MN} + g^{05}_{KLMN} E^p_{KL} \Psi_{MN} + g^{06}_{KLMN} \Psi_{KL} \Psi_{MN}\,. \end{aligned} \tag{2.22}$$

To the same degree of accuracy, linear expansions are written for ρ_{KL} and h_{KL} as

$$\begin{aligned} \rho_{KL} = {} & \rho^1_{KLMN} E_{MN} + \rho^2_{KLMN} E^p_{MN} + \rho^3_{KLMN} \Psi_{MN} + \rho_{KLMNP} \Gamma_{MNP}\,, \\ h_{KL} = {} & h^1_{KLMN} E_{MN} + h^2_{KLMN} E^p_{MN} + h^3_{KLMN} \Psi_{MN} + h_{KLMNP} \Gamma_{MNP}\,. \end{aligned} \tag{2.23}$$

Similar type of expressions may also be given for damage evoluation $\dot{\alpha}_{KL}$. Now the nonlocal formulation for the micromorphic model of damaged materials is completed. As it can be observed from the constitutive equations, there are 18 material coefficents involved in the local case and 18 additional coefficients that appear under the integral signs in its nonlocal version. A more convenient form of Eqs. (2.6) may be given by incorporating local coefficients λ, $\mu,\ldots$, into the nonlocal moduli by expressing them in forms

$$(\lambda', \mu', \kappa', \ldots) = \alpha(|\mathbf{x}' - \mathbf{x}|)\{\lambda, \mu, \kappa, \ldots\}\,. \tag{2.24}$$

There are certain plausible representations for α. One such representation is

$$\alpha(|\mathbf{x}' - \mathbf{x}|, a) = (2\pi\varepsilon^2)^{-1} K_0\left(|\mathbf{x}' - \mathbf{x}|/\varepsilon\right), \tag{2.25}$$

where K_0 is the modified Bessel's function and ε is a quantity related to nonlocality. This function happens to be the Green functions for the infinite plane, satisfying the equation

$$(1 - \varepsilon^2\nabla^2)\alpha = \delta(|\mathbf{x}' - \mathbf{x}|)\,. \tag{2.26}$$

So we may write the field equations as partial differential equations which make it possible to investigate the wave propagation problems for infinite medium. Micromechanical models have great advantages in modelling physical realities with a minimum ambiguity and arbitrariness, but unfortunately they are computationally and mathematically inefficient in practical applications. Even in the linearized form, micromorphic theory is rather complicated in the local theory, only few problems have been solved in the general formulation. It can not be expected to get better solution for its nonlocal version. To cope with this unatractive feature of the theory several further simplifications may be considered.

3. Propagation of the Plane Waves in Damaged Materials

We consider here a kind of isotropy similar to the one known in the classical elasticity which may be called microisotropy [8]. Since strain and microstrain are taken to be independent of each other, their principle directions are in general different and also from the various stress measures. On the basis of the microisotropy, we consider that the principle directions of a particular strain measure would coincide with some suitable stress measure. The selection of the appropriate sets of stress-strain measure is not all together arbitrary. We observe from the field equations that the pertinent stress measures are t_{kl}, relative stress tensor $\sigma_{kl} = t_{kl} - s_{kl}$ and the hyperstress tensor λ_{klm}. To determine the corresponding strain measures suitable to these stresses, we observe the pairing in the energy equation that these three strain measures are macrostrain e_{km}, the microdisplacement gradient φ_{km} (corresponding to damage tensor α_{km}) and microdeformation gradient $\varphi_{mn,k}$ (corresponding to the gradient of damage tensor, $\alpha_{mn,k}$). Accordingly, microisotropy can be defined as follows:

a) The principle directions of the symmetrical part of t_{kl} coincide with the principle directions of macrostrain e_{kl};

b) The principle directions of σ_{km} coincide with those of damage tensor, and finally

c) The principle directions of the symmetric part of hyperstress $\lambda_{k(mn)}$ coincide with the principle directions of the symmetric part of the damage gradient $\gamma_{(mn)k}$.

Since damage is considered as an independent variable than the macrostrain e_{km} or macrodiplacement $u(x,t)$ and expressed as the motion and the growth of the damage, above assumptions seem reasonable. Using above assumptions, the following field equations are obtained for a continuum of infinite extent

$$\begin{aligned}
&\alpha_1 u_{k,kl} + \alpha_2 u_{l,kk} + 2\alpha_3 \varepsilon_{pkl}\phi_{p,k} - (1-\varepsilon^2\nabla^2)\left(\rho\frac{\partial^2 u_l}{\partial t^2}\right) = 0\,, \\
&\beta_1 \phi_{pp,nn}\delta_{kl} + \beta_2 \phi_{(kl),mm} + \alpha_4 \phi_{pp}\delta_{kl} + \alpha_5 \phi_{(kl)} + \rho f_{(kl)} \\
&\quad - (1-\varepsilon^2\nabla^2)\left(\frac{1}{2}\rho j\frac{\partial^2 \phi_{lk}}{\partial t^2}\right) = 0\,, \\
&\beta_3 \phi_{k,kl} + \beta_4 \phi_{l,mm} + 2\alpha_3 \varepsilon_{lmn} u_{n,m} + 4\alpha_3 \phi_l + \rho l_p \\
&\quad - (1-\varepsilon^2\nabla^2)\left(\rho j\frac{\partial^2 \phi_p}{\partial t^2}\right) = 0\,,
\end{aligned} \tag{3.1}$$

where

$$\begin{aligned}
&\alpha_1 = \lambda + \mu + \tau + \sigma - \kappa, \quad \alpha_2 = \mu + \sigma + \kappa, \quad \alpha_3 = \kappa, \quad \phi_p = \frac{1}{2}\varepsilon_{pkm}\phi_{km}\,, \\
&l_p = -\varepsilon_{pmn} f_{mn}, \quad \alpha_4 = \tau, \quad \alpha_5 = 2\sigma, \quad \beta_1 = \tau_3\,, \\
&\beta_2 = \tau_7 + \tau_{10}, \quad \beta_3 = 4(\tau_4 + \tau_9), \quad \beta_4 = -2(\tau_7 + 4\tau_4 + 4\tau_9 - \tau_{10})\,.
\end{aligned} \tag{3.2}$$

To investigate the wave propagation along the x_1-direction in the unbounded medium, we assume that u_i, $\phi_{(ij)}$, ϕ_i are functions of x_1 and t and substitute them into Eqs. (3.1) to find twelve wave equations. Obtained equations are not given here to provide the shortness. Twelve waves propagating in the medium may be clasified into one set of four longitudinal waves, two identical sets of transverse waves, each containing four waves.

We consider now the plane waves advancing along the x_1-direction and write

$$(u_i, \varphi_{(ij)}, \varphi_i) = (u_{i0}, \varphi_{(ij)0}, \varphi_{i0})\exp[i(kx_1 - \omega t)] \tag{3.3}$$

to obtain the following results:

1. Substituting (3.3) into the equation for u_1 gives the phase velocity

$$v_1^2 = \left(\frac{\omega}{k}\right)^2 = \frac{c_\alpha^2}{1-\varepsilon^2 k^2}\,. \tag{3.4}$$

Since the displacement u_1 is in the direction of propagation for the waves traveling with the phase velocity v_1, we designate them longitudinal displacement waves.

For $\varepsilon = 0$, Eq. (3.4) represents the conventional longitudinal waves which is non-dispersive.

2. Second longitudinal wave is given for φ_1 which shows the antisymmetric part of φ_{23}. Substituting Eq. (3.3) into corresponding wave equation, we find

$$[\alpha_3^2(-k^2) - 2\omega_0^2 + \varepsilon^2 k^2 \omega^2 + \omega^2]\varphi_{10} = 0\,, \tag{3.5}$$

which gives

$$v^2 = \frac{\alpha_3^2}{1 - 2(\omega_0/\omega)^2} + \frac{\omega^2}{1 - 2(\omega_0/\omega)^2} e^2\,. \tag{3.6}$$

Eq. (3.6) shows that the speed of propagation depends on the frequency. Hence, these waves are dispersive. Here ω_o is a cut-off frequency which depends on the material constants. This wave is called longitudinal microrotational wave.

3. By the analogy to φ_1 investigated above, $\varphi_{(23)}$ also represents a longitudinal wave. This may be called longitudinal microshear wave.

4. Another longitudinal wave can be obtained for φ_{pp} and the corresponding equation is

$$(3\alpha_1^2 + \alpha_2^2)\frac{\partial^2 \varphi_{pp}}{\partial x_1^2} - [(3c_5^2 + 2c_4^2)/j]\varphi_{pp} + \varepsilon^2 \frac{\partial^4 \varphi_{pp}}{\partial t^2 \partial x_1^2} - \frac{\partial^2 \varphi_{pp}}{\partial t^2} = 0\,. \tag{3.7}$$

Applying the same procedure we find the following dispersion relation

$$v^2 = [(3\alpha_1^2 + \alpha_2^2) - \omega^2\varepsilon^2]/[1 - (\omega_1/\omega)^2] \tag{3.8}$$

which gives another cut-off frequency

$$\omega_c^2 = \omega_1^2 = (3c_5^2 + 2c_4^2)/j\,. \tag{3.9}$$

This wave is called longitudinal dilatational wave.

5. A similar result can be obtained for φ_{11}. We find another cut-off frequency which is

$$\omega_c^2 = \omega_2^2 = 2c_4^2/j\,. \tag{3.10}$$

This is called longitudinal microextensional wave.

6. Next we shall consider the coupled equations given for the transversal displacement u_2 and microrotation φ_3. We call the wave associated with u_2, the transverse displacement wave and the one associated with φ_3 transverse microrotational wave. The velocities of these waves are determined by carrying Eq. (3.3) into the proper equations which gives

$$a(v^2)^2 + bv^2 + c = 0\,, \tag{3.11}$$

where

$$a = 1 - \left(\frac{\omega_0}{\omega}\right)^2, \quad b = -\left[c_p^2 + c_\alpha^2\left\{1 - 2\left(\frac{\omega_0}{\omega}\right)^2\right\} + 4c_3^2\left(\frac{\omega_0}{\omega}\right)^2 + 2\omega^2\varepsilon^2\left(1 - \left(\frac{\omega_0}{\omega}\right)^2\right)\right], \quad c = c_\alpha^2 c_\beta^2 - \omega^2(c_p^2 + c_\alpha^2)\varepsilon^2 + \omega^4\varepsilon^4 . \tag{3.12}$$

Eq. (3.11) is a quadratic equation in v^2 that yields two distinct speeds of propagation. These waves are dispersive and ω_0 is the cut-off frequency. For $\omega = \omega_0$, Eq. (3.11) gives

$$bv^2 + c = 0 , \tag{3.13}$$

which results in

$$v^2 = \frac{c_\alpha^2 c_\beta^2}{c_p^2 + 2c_3^2} - \frac{2\omega_0^2}{c_p^2 + 2c_3^2}\left[\frac{c_\alpha^2 c_\beta^2}{c_p^2 + 2c_3^2} + c_p^2 + 2c_\alpha^2\right]\varepsilon^2 . \tag{3.14}$$

Eq. (3.14) tells us that the inclusion of the nonlocal effects decreases the value of phase velocity.

7. Second group of coupled equations are given for the second transversal displacement u_3 and microrotation φ_3 similar to the problem given in case 5. These are also called transverse displacement and transverse microrotational waves.

8. Repeating the calculations done for φ_{pp} or φ_{11} given above, we obtain similar results for φ_{22} and φ_{33}. They are called transverse microextensional waves.

9. Last group of equations given for $\varphi_{(12)}$ and $\varphi_{(13)}$ are uncoupled equations and represent transverse microshear waves.

As it is pointed out at the beginning, investigation of of plane waves propagation is very important in nonlocal theories for the determination of unkown material constants. It may be possible to match the theoretical results to those found experimentally or through lattice dynamics in the entire Brouillion zone although it is not easy to find experimental evidences to show the effect of microrotations.

References

1. L. M. Kachanov, *Izv. Acad. Nauk. SSSR. Otd. Tekhn.Nauk.* No 8 (1958) 26. (in Russian.)
2. K. Z. Markov, In *Yielding, Damage and Failure of Anisotropic Solids*, ed. J. P. Boechler, Mechanical Engineering Publications, London, 1989, p. 665.
3. A. C. Eringen and S. E. Şuhubi, *Int. J. Engng. Sci.* **2** (1964) 189.
4. S. E. Şuhubi and A. C. Eringen, *Int. J. Engng. Sci.* **2** (1964), 389.
5. A. C. Eringen, *Int. J. Engng. Sci.* **10** (1972) 1.
6. A. C. Eringen, *Int. J. Engng. Sci.* **19** (1981) 1461.
7. S. Xia, G. Li and H. Lee, *Int. Journal of Fracture* **34** (1987) 239.
8. S. L. Koh, *Int. J. Engng. Sci.* **8** (1970) 583.

Continuum Models and Discrete Systems
Proceedings of 8th International Symposium, June 11-16, 1995, Varna, Bulgaria, ed. K.Z. Markov
© World Scientific Publishing Company, 1996, pp. 633–638

NEUTRON INDUCED EMBRITTLEMENT OF WWER1000 REACTOR PRESSURE VESSEL STEELS

Tz. KAMENOVA
Institute for Metal Science, Bulgarian Academy of Sciences,
67 Shipchenski Prohod, Blvd., BG-1574 Sofia, Bulgaria

Abstract. The problem of neutron induced embrittlement of WWER1000 reactor pressure vessel (RPV) steel have been discused. A review of the contemporary knowledge on the mechanisms of radiation embrittlement process and the calculative procedures for determination of the embrittlement rate have been presented. The change of RPV critical temperature of embrittlement during operation of real WWER1000 nuclear power reactor have been evaluated.

1. Introduction

Neutron induced embrittlement is the main ageing process running during exploitation of reactor pressure vessel (RPV) steels which is decisive for the radiation life time extension of nuclear power unit [1.2] The interaction between irradiation defects (vacancies and interstitial atoms) and alloying and impurity elements (P, Cu, As, Sb and Ni) causes formation of dislocation loops, phase precipitation in grain volume and on the grain boundaries. The microstructural changes result in increasing of strength and decreasing plasticity of reactor pressure vessel steel. Parallely a shift (ΔT_{kf}) of the temperature of ductile to brittle transition (T_{kf}) to the higher temperatures is observed. These effects are characteristic for all RPV steels. The rate of embrittlement depends on irradiation condition and on chemical composition—especially on impurity content. The weld metal (WM) is more sensible to neutron induced embrittlement than base metal (BM) of shells.

The low alloyed ferric-pearlitic 15Ch2NMFAA steel is a new grade reactor steel with low level of P and Cu content. The increasing of Ni concentration improves the weldability, increases yield strength and decreases the temperature of ductile to brittle transition (T_{ko}) in comparison to WWER440 steel 15Ch2MFAA. The law describing the rate of neutron embrittlement in 15Ch2MFAA is relatively well established. For nickel containing steel no adequate data are available till now.

The aim of this report is to predict the increase of the WWER1000 RPV critical temperature of embrittlement increase during operation by the means of different models and to assess the radiation life time using input parameters for Unit 5 NPP "Kozloduy."

2. Models and Input Parameters

The control and monitoring of the mechanical properties of irradiated RPV metal are of a great importance for safety operation of the nuclear plant. The prediction of T_{kf} till the end of life-time is one of the methods used for the radiation life time assessment of power unit. In the world practice low alloyed ferric-pearlitic steel with similar composition are used for manufacturing of RPV the methods. Nevertheless the different empirical relation are used for calculation of ΔT_{kf} and T_{kf}.

An empirical formulae for calculation of neutron induced critical temperature of embrittlement change (ΔT_{kf}), valid for 15Ch2MFAA, is suggested in Russian standard (RS) [3]. For the moment because of the lack of data the same formulae is accepted for 15Ch2NMFAA steel also. In this case the T_{kf} is as follows:

$$T_{kf} = T_{ko} + \Delta T_{kf} + \Delta T_{kn} + \Delta T_{kt}, \tag{1}$$

where T_{ko} is the temperature of ductile to brittle transition of non irradiated steel and ΔT_{kf}, ΔT_{kn}, ΔT_{kt} are the shifts of T_{kf} due to fast neutron bombardment, cyclic loading and thermal ageing respectively. For RPV in R.S. the $\Delta T_{kn} = \Delta T_{kt} = 0$ is accepted and ΔT_{kf} is presented by:

$$\Delta T_{kf} = A_f (F/10^{18})^{0.33}\,, \tag{2}$$

where A_f[°C] is a chemical coefficient, F[cm^{-2}] the neutron fluence (E > 0.5 MeV). For the metal in the welding between shells in front of core zone A_f is calculated according to Ermakov [2]:

$$A_f = 800(\mathrm{P} + 0.07\mathrm{Cu})\,; \tag{3}$$

here P and Cu are concentrations of both elements in the steel. In Russian standard [3] the A_f values 20 and 23 are proposed for A_f for the base metal (WM) and for the weld metal (BM) respectively.

The nickel presence in steel is not taken into account in Eq. (3). There exist results supporting synergetic action of Cu and Ni on neutron embrittlement process. Interstitial Ni atoms concentrate on copper precipitates in solid solution or on grain boundaries and also form cluster complexes [4]. Nickel increases the copper-rich precipitates volume fraction. The Ni influence is related to P and Cu concentration and to the magnitude of neutron fluence [5]. The high nickel concentration is most effective when combined with a high copper content [6]. That is why the future verification of validity of Eqs. (2) and (3) for 15Ch2NMFAA is necessary. The data obtained by testing of surveillance specimens from this grade steel show higher A_f value for BM ($A_f = 30$) [7].

In [8] a new empirical formula (Vichkarev–Zvezdin) for 15Ch2NMFAA A_f calculation is derived from experimental data:

$$A_f = 57.5\,(\mathrm{Cu} + 10\mathrm{P}) + 11.3\,. \tag{4}$$

In this case again the Ni concentration is not included in relation also.

In USA and France similar low alloyed ferric-pearlitic steels grade A508 or A533 are used for manufacturing RPV. In these steels the Mn content (< 1.6 %) is higher and Cr content (< 0.7 %) is lower than in the Russian RPV steel. Ni content is up to 1.3 % in weld metal.

In the start of the USA Surveillance Programme (US R.G.1) the phosphorus and copper contents are considered for calculating T_{kf} trend curves [9]:

$$T_{kf}[^{\circ}\mathrm{F}] = T_{ko} + [40 + 1000(\mathrm{Cu} - 0.08) + 5000(\mathrm{P} - 0.008)](F/10^{19})^{0.5}, \qquad (5)$$

with F [cm^{-2}] denoting hereafter the neutron fluence (E > 1MeV).

On the basis of statistical analysis of data, obtained in US RPV surveillance programme, Guethrie [10] proposed the following equation:

$$T_{kf}[^{\circ}\mathrm{F}] = T_{ko} + 48 + [-10 + 470\mathrm{Cu} + 350\mathrm{CuNi}](F/10^{19})^{0.27}. \qquad (6)$$

Later due to the Ni major effect on embrittlement process the Ni concentration is included in T_{kf} calculation and phosphorus influence is neglected (US R.G.2) [10]. This formula is based on results from USA Surveillance Programmes and gives conservative estimation of T_{kf} for low P content (0.01 %):

$$T_{kf}[^{\circ}\mathrm{F}] = T_{ko} + 56 + A_f(F/10^{19})^{[0.28-0.10\log(F/10^{19})]}. \qquad (7)$$

The A_f-values for different concentrations of Cu and Ni are given in Table 1.

New expression (FIZ formulae) based on surveillance results is proposed in France [12], in which the influence of P, Cu and Ni concentration is taken in account:

$$T_{kf}[^{\circ}\mathrm{F}] = T_{ko} + 8 + [24 + 1538(\mathrm{P} - 0.008) + 238(\mathrm{Cu} - 0.08) + 191\mathrm{CuNi}^2](F/10^{19})^{0.35}. \qquad (8)$$

This expression is valid for $0.08 < \mathrm{Cu} < 0.19$ %, $0.008 < \mathrm{P} < 0.021$ %, $\mathrm{Ni} < 1.9$ % and F-neutron fluence (E > 1MeV).

3. Results and Discussion

3.1. Calculation of Input Parameters

In NPP Kozloduy two WWER1000 are in operation. For our calculation the parameters of Unit 5 weld 3 metal are chosen because of higher $T_{ko} = 0^{\circ}$C and high Ni content. The neutron fluence (F) for one effective year (292 full power days) on weld 3 is calculated by design fluence on inner wall of RPV in front of core zone centre divided by factor 1.47.

No sufficient Charpy impact test data from surveillance specimens are available for WWER 1000 weld metal. Only one experimental data $A_f = 30$ for BM is reported in [7]. The coefficients depending on chemical composition calculated for weld 3 using different relation and the reported values are shown in Table 1.

TABLE 1
Standard A_f, experimental A_f and calculated A_f values with composition of weld 3 metal

Ref.	[8]	[3]	[13]	[7]	[9]	[11]	[10]	[12]
A_f, °C	18.2	20	23	30	25	22.8	12.2	42

3.2. Radiation Life-Time Assessment

The calculated A_f and F values are used for prediction of the increasment of ΔT_{kf} during operation time. For weld 3 metal $T_{ko} = 0°C$ so $\Delta T_{kf} = T_{kf}$. The ΔT_{kf} trend curves calculated by Russian standard using different A_f values for 15Ch2NMFAA steel are shown on Fig. 1. Practically the Vishkarev value $A_f = 18.2$ and the standard value $A_f = 20$ gives the same results. At the end of design life time considerable difference (40°C) between T_{kf} calculated with standard and experimental A_f is observed. In the case $A_f = 30$ after 30 effective operation years the radiation life time is over, e.g., when $T_{kf} = 90°C$ [13] is reached.

The ΔT_{kf} trend curves, calculated by relation (2),(5),(6),(7) and (8), are compared on Fig. 2. In the first several operation years the Guethrie and US R.G.2 formulae give more conservative results in comparison to the other formulae. Later the Russian standard method ($A_f = 20$) gives results close to FIZ and US R.G.2. The ΔT_{kf} trend curve, calculated by Russian formulae with experimental $A_f = 30$ shows the highest rate of embrittlement of RPV metal. So this method must to be used for the radiation life time assessment until more accurate data for A_f and embrittlement law are obtained.

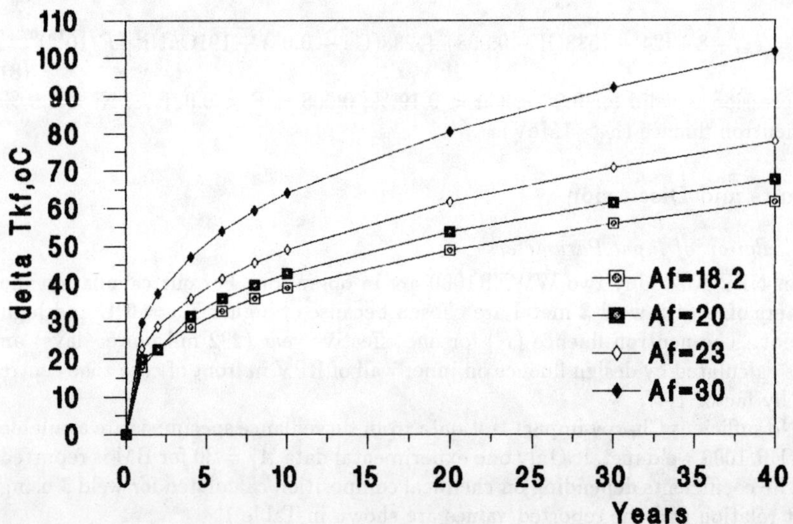

Fig. 1. T_{kf} trend curves, calculated using different chemical factors.

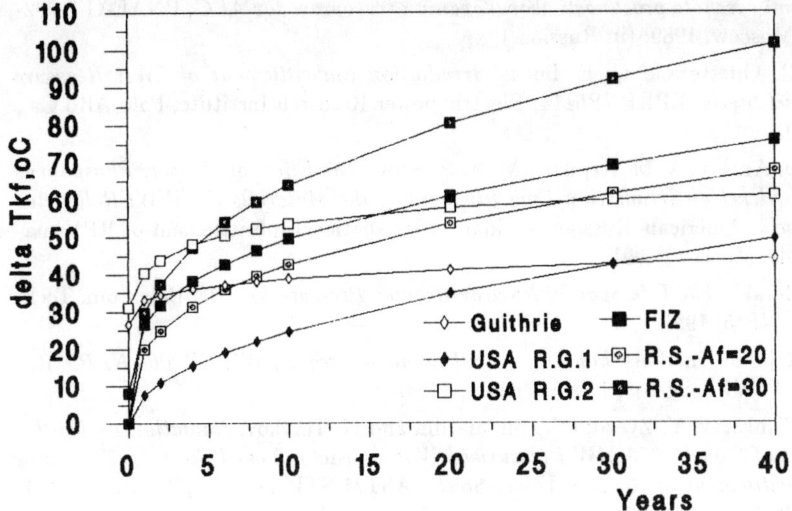

Fig. 2. T_{kf} trend curves, calculated by different methods.

4. Conclusions

1. There are considerable differences in the A_f standard, A_f calculated and A_f experimentaly determinated values.

2. The application of experimental A_f value gives more conservative assessment of embrittlement rate of RPV metal. In this case the radiation life time of Unit 5 will be exhausted before the designed term.

3. The ΔT_{kf} trend curve, calculated by means of Russian standard method and parameters, is close to the curves determinated by French and USA R.G. 2. However, it is much lower than the experimentally found A_f, according to the Russian method.

4. Future experiments are obligatory for clarifying the 15Ch2NMFAA neutron embrittlement law and obtaining more accurate data for A_f.

Acknowledgements. The financial support of the Bulgarian National Foundation for Scientific Research (Contract Th309/93) is gratefully acknowledged.

References

1. L. E. Steel, *Neutron Irradiation Embrittlement of Reactor Pressure Vessel Steels*, Technical Report 163, IAEA, Viena, 1975.
2. N. I. Ermakov and Y. Dragunov, *Some WWER-440 RPV material characteristics, Workshop on Strength of RPV WWER*, Plzen, Chech Republic (1986), p. 23.

3. *Normi raschota prochnosti oborudovania i truboprovodov AEC*, PNAE-G-7-002-86, Moscow, 1989. (in Russian.)
4. G. R. Odette and G. E. Lucas, *Irradiation Embrittlement of LWR Reassure Vessel Steels*, EPRI NP6114, Electric power Research Institute, Palo Alto Ca., 1989.
5. A. D. Amaev, A. M. Krjukov, M. A. Socolov, *The Effect of Copper, Phosphorus and Nickel on Irradiation Embrittlement of the Materials for WWER-Reactor Vessels*, American–Russian seminar on Irradiation Embrittlement of RPV materials, Moccow, 1991.
6. R. Gerald, *The Life Span of Nuclear Reactor Pressure Vessels*, Belgatom, Tech. Rep. NR 5, 1991.
7. A. Krjukov, In *Workshop "Pressure boundary Integrity WWER1000 NPP"*, Report WWER-Sg-103, IAEA Vienna, 1994.
8. O. Vishkarev, Y. Zvezdin, V. Shamardin and G. Tulykov, *Radiation Embrittlement of Soviet 1000-MW Pressurised Water Reactor Vessel Steel*, In *Radiation Embrittlement of Reactor Vessel Steels*, ASTM STP 1170, ASTM Phil., 1993, p. 216.
9. Regulatory Guide 1.99 rev. 1 *Radiation Embrittlement of Reactor Vessel Materials*, U.S. Nuclear Regulatory Commission 1975.
10. G. L. Guithrie, In *LWR Pressure Vessel Surveillance Dosimeter Improvement Program*, NUREG-CR-3391, vol. 2, 1984.
11. Regulatory Guide 1.99 rev. 2 *Radiation Embrittlement of Reactor Vessel Materials*, U.S. Nuclear Regulatory Commission, 1988.
12. C. Rieg, In *Workshop of the European Communities EWGRD-JRC PETTEN*, The Netherlands, 1992, p. 2.
13. J. Brinda, V. Cherny and M. Brmovsky, *Qualification tests Programme of WWER 1000 RPV Materials*, Report Nuclear Research Institute Rez, Chech Republic, 1993.

Continuum Models and Discrete Systems
Proceedings of 8th International Symposium, June 11-16, 1995, Varna, Bulgaria, ed. K.Z. Markov
©World Scientific Publishing Company, 1996, pp. 639–645

FRACTURE MECHANICS OF CONCRETE COMPOSITES

C. OUYANG
Iowa Department of Transportation, 800 Lincoln Way, Ames, Iowa 50010, USA

M. A. TAŞDEMIR
Faculty of Civil Engineering, Istanbul Technical University,
80626 Maslak/Istanbul, Turkey

and

S. P. SHAH
Center for Advanced Cement-Based Materials, Northwestern University,
Evanston, Illinois 60208, USA.

Abstract. Topics discussed in this paper include toughening mechanisms in the fracture process zone of concrete, principles of linear elastic fracture mechanics, various nonlinear fracture models, the determination of material fracture parameters, and mixed mode and mode II fracture.

1. Introduction

This paper summarized a series of studies on application of fracture mechanics to cracking and failure of cement-based materials. The basic concepts of linear elastic fracture mechanics (LEFM) are reviewed. Toughening mechanisms in the fracture process zone of concrete are discussed. The determination of material fracture parameters of concrete are mentioned. Various nonlinear fracture models and mixed mode fracture of concrete are discussed. It is shown that fracture mechanics has now been established as a fundamental approach to describe failure of concrete structures.

2. Linear Elastic Fracture Mechanics

Consider a structure of an elastic-brittle material (such as glass), with a crack of length a. When loading is applied, the structure can supply potential energy (U) at the rate $dU/da = G$ (termed as the strain energy release rate). On the other hand, the crack propagation at the crack tip needs to consume some energy, which is denoted as W at the rate $dW/da = R$ (termed as the fracture resistance). Consequently, the linear elastic fracture mechanics (LEFM) criterion for crack growth is:

$$G = R\,, \tag{1}$$

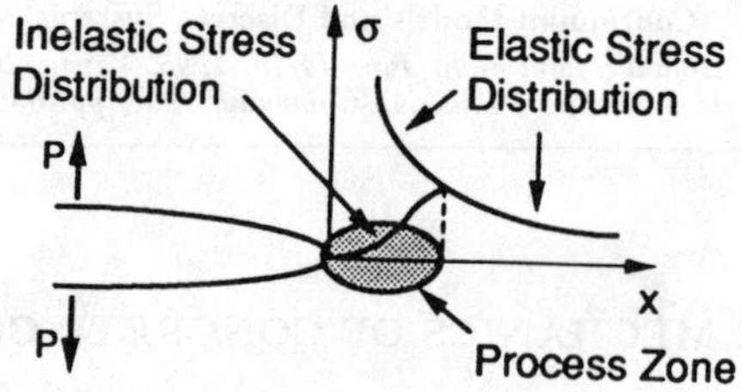

Fig. 1. Stress distribution at a crack-tip and process zone.

R is a material constant for linear elastic brittle materials, whereas G is the function of structural geometry and applied loads. For the plane stress condition, G can be written in terms of the stress intensity factor (K) as

$$G = \frac{K^2}{E}\,, \tag{2}$$

where E is the elastic modulus of the material. Therefore, the criterion (1) can alternatively be written as:

$$K = K_R\,, \tag{3}$$

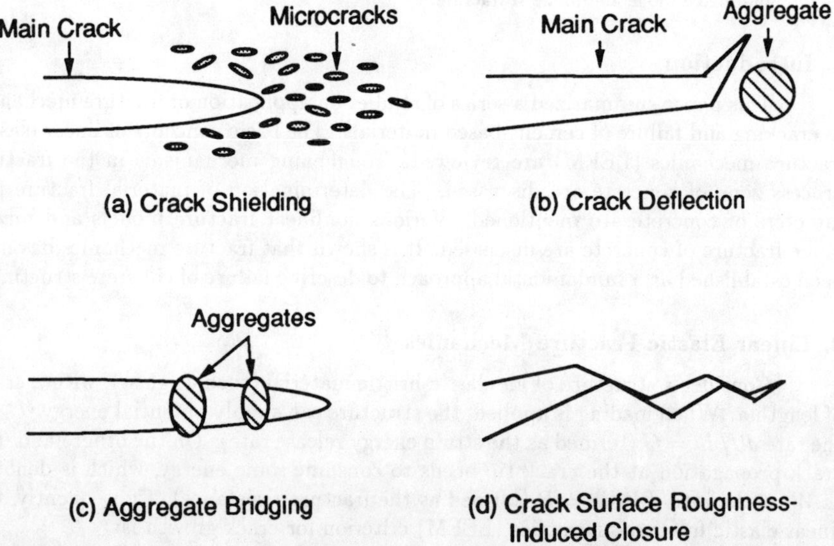

Fig. 2. Toughening mechanisms in concrete.

where $K_R = (ER)^{1/2}$ is termed as the critical stress intensity factor. According to LEFM principles, the stresses and strains at a sharp crack-tip approach infinity.

3. The Fracture Process Zone

Since an inelastic region is always present at the crack-tip in a real material, stresses do not become infinite. This region is often called the fracture process zone. The resulting stresses differ from the elastic field only in this zone as in Fig. 1. The presence of the fracture process zone may be attributed to the inherent material heterogeneity of concrete [1]. Many mechanisms which are responsible for the fracture process zone have been reported. Some of these toughening mechanisms are indicated in Fig. 2. During fracture, the high stress state near the crack-tip causes microcracking at flaws, which result from water-filled pores, air voids acquired during casting, and shrinkage cracks due to the curing process. This phenomenon, known as microcrack- shielding as shown in Fig. 2a, consumes some external energy caused by the applied load [2]. Crack deflection occurs when the path of least resistance is around a relatively strong particle or along a weak interface (Fig. 2b). This mechanism has been studied in detail by Faber and Evans [3]. Other important toughening processes in concrete are grain bridging [4] as shown in Fig. 2c. Bridging occurs when the crack has advanced beyond an aggregate that continues to transmit stresses across the crack until it ruptures or is pulled out. Also, during grain pullout or the opening of a tortuous crack there may be some contact (or interlock) between the faces (Fig. 2d). This causes energy dissipation through friction, and some bridging across the crack.

4. Mode I Fracture

As realized in 1971 by Shah and McGarry [5] LEFM cannot be directly applied due to the presence of a sizeable fracture process zone in these materials. Generally, the cohesive nature of the fracture process zone can be modeled by a traction pressure acting on the crack surface as shown in Fig. 3. The strain energy release rate for mode I crack propagation in concrete can then be expressed as [6]:

$$G = \frac{K_I^2}{E} + \int_0^{\mathrm{CTOD}} \sigma(w)\, dw\,, \tag{4}$$

where K_I is the net stress intensity factor for mode I crack, $\sigma(w)$ is the normal traction pressure which depends on the crack opening displacement (w), and CTOD is the crack tip opening displacement. A crack is assumed to be a line in Eq. (4). In Eq. (4) the Griffith-Irwin energy dissipation mechanism is represented by a non-zero stress intensity factor and the Dugdale-Barenblatt energy dissipation mechanism is represented by the traction term. For simplicity, one may approximately use models only based on a single fracture energy dissipation mechanism, either Griffith-Irwin mechanism by assuming $\sigma(w) = 0$ or Dugdale-Barenblatt mechanism by assuming $K_I = 0$.

Hillerborg *et al.* [7] proposed a fictitious crack model with $K_I = 0$. In their model a critical strain energy release rate (GF), which is defined as the area under softening part of stress-separation curve, and a tensile strength are assumed to be the material properties. A crack is assumed to initiate and propagate when the principal tensile stress reaches the tensile strength.

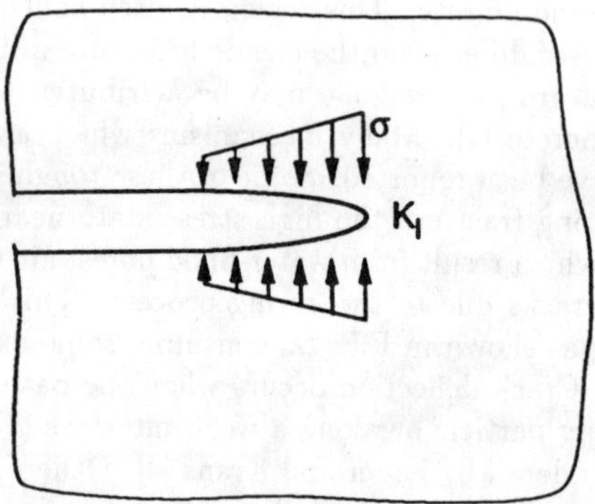

Fig. 3. A cohesive mode I crack subjected to loading.

Both Eq. (4) and the Hillerborg's model require a unique $\sigma(w)$ curve to achieve energy dissipation. Many different shapes of $\sigma(w)$ curves, including linear, bi-linear, polynomial and other smoothly varying functions, have been used. The choice of $\sigma(w)$ function influences prediction of the structural response significantly [8], and the local behavior in the fracture, for example the crack opening, is particularly sensitive to the shape. Experimental determination of $\sigma(w)$ directly from tension tests has been suggested [9], but this is difficult and the results may vary with specimen size and shape. Li *et al.* [10] have proposed a J-integral based method for obtaining the entire $\sigma(w)$ curve. A recent study [11] applied moire interferometry to deduce crack profiles which were used to determine the stress-opening relation in the process zone. Until cohesive fracture parameters, GF, the tensile strength and the form of $\sigma(w)$, are further identified to be geometry and size independent, the cohesive models require some curve fitting.

The fracture process zone in concrete can also be modeled by a single Griffith-Irwin energy mechanism by assuming $\sigma(w) = 0$. Since the effective crack length has been experimentally found to be specimen size and geometry dependent, it cannot directly be used as the fracture criterion. Therefore, an additional quantity should be introduced as the fracture criterion. Most of these effective crack models use two fracture parameters to define the inelastic fracture process. These effective crack models may be represented by Bazant's size effect law [12] and Jenq and Shah's two parameter fracture model [13].

Using the concept of the effective crack, Jenq and Shah [13] proposed the two parameter fracture model as shown in Fig. 4a. In their model, the two independent material fracture properties are the critical stress intensity factor (K^s_{Ic}) and the critical crack tip opening displacement (CTOD_c), which are defined in terms of the effective crack. The fracture criteria for an unstable crack are:

$$K_I = K^s_{\mathrm{Ic}}\,,$$
$$\mathrm{CTOD} = \mathrm{CTOD}_c\,. \qquad (5)$$

In this model, the effective crack exhibits a compliance equal to the unloading compliance of the actual structure.

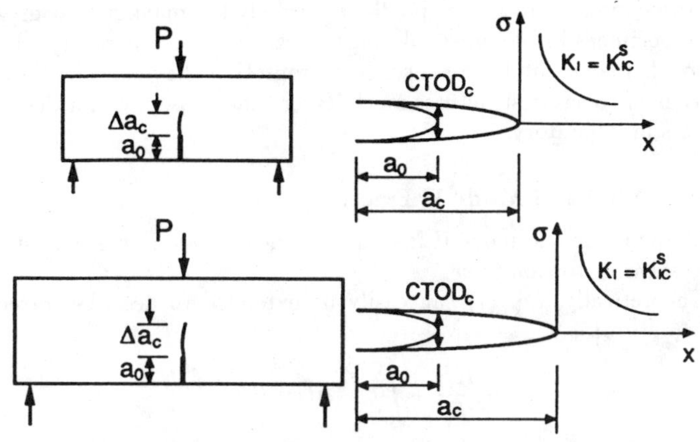

(a) Fracture criteria: $K_I = K^s_{Ic}$ and $\mathrm{CTOD} = \mathrm{CTOD}_c$;

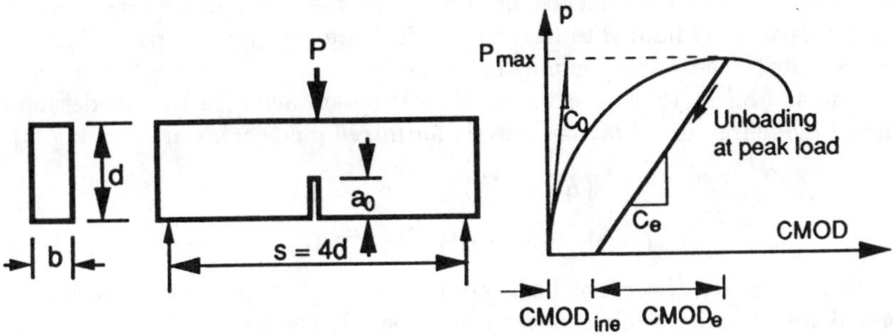

(b) Determination of K_{Ic} and CTOD_c from C_0 and C_e obtained in a notched beam test (C_0 and C_e are the initial compliance and the unloading compliance at the peak load, respectively).

Fig. 4. Two-parameter fracture model.

The procedure for experimentally obtaining the values of K_{Ic}^{s} and $CTOD_c$ is detailed in a 1990 RILEM recommendation. As shown in Fig. 4b, a three-point bend (single-edge notched) beam is to be tested under crack mouth opening displacement (CMOD) control. After the peak, within 95 % of the peak load (P_{max}) the unloading compliance (C_e) is measured. Using C_e and the initial compliance (C_0) along with LEFM equations, the corresponding critical effective crack length (ac) can be determined. Consequently, K_{Ic}^{s} and $CTOD_c$ of the material can be obtained at the critical load P_{max}. This method yields size-independent values for other geometries as well.

Planas and Elices [14] have compared Hillerborg's cohesive model, Jenq and Shah's two parameter fracture model and Bazant's size effect law. They concluded that all three models are able to predict accurately the maximum loading capacity of concrete specimens in the practical range of sizes used in laboratory. However, these three models may exhibit differences in asymptotic behavior when they are used in the analysis of concrete structures much larger than those used for determining their parameters in laboratory.

5. Mixed Mode and Mode II Fracture

Mixed mode and mode II fracture are encountered in concrete structures subjected to shear or torsion force.

Theoretically, Eq. (4) can easily be extended to describe mixed mode and mode II fracture15,

$$G = \frac{K_I^2}{E'} + \frac{K_{II}^2}{E'} + \int_0^{\mathrm{CTOD}} \sigma(w)\, dw + \int_0^{\mathrm{CTSD}} \tau(s)\, ds\,, \tag{6}$$

where K_{II} is the mode II stress intensity factor, $\tau(s)$ is the shear traction force which is the function of crack sliding displacement (s), CTSD is the crack tip sliding displacement. For mode II fracture and mixed mode fracture with the high ratio of K_{II}/K_I, the shear traction term in Eq. (6) may play an important role in fracture processing. Because of the difficulty in formulating the expression for $\tau(s)$, most of studies in this area are limited to mixed mode fracture with the low ratio of K_{II}/K_I, where effect of the shear traction may be negligible.

Jenq and Shah [15] have extended their two parameter fracture model to predict mixed mode fracture. Fracture criteria for mixed mode crack are (see Fig. 5):

$$\left(K_I^2 + K_{II}^2\right)^{1/2} = K_{IC}^{s}\,,$$

$$\left(\mathrm{CTOD}^2 + \mathrm{CTSD}^2\right)^{1/2} = \mathrm{CTOD}_C\,. \tag{7}$$

When the ratio of K_{II}/K_I is not high, Eq. (7) can predict the path of crack propagation and the ultimate loading capacity of the specimen.

6. Summary

A series of studies on application fracture mechanics to failure of concrete have been summarized in this paper. Due to the presence of the sizeable fracture process zone, LEFM criterion can not be directly applied to fracture of concrete. Various nonlinear fracture models proposed to describe fracture of concrete have been presented.

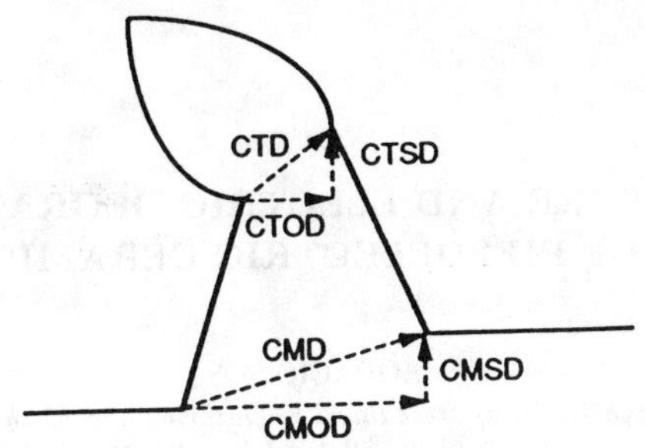

Fig. 5. Extension of two parameter fracture model to mixed mode crack.

References

1. S. Mindess and S. Diamond, *Materials and Structures* **15**(86) (1982) 107.
2. M. Kachanov, In *Fracture Toughness and Fracture Energy of Concrete*, ed. F. H. Wittmann, Elsevier Science, Amsterdam, 1986, p. 3.
3. K. T. Faber and A. G. Evans, *Acta Metall.* **31** (1983) 565.
4. J. G. M. van Mier, *Cement and Concrete Research* **21** (1991) 1.
5. S. P. Shah and F. J. McGarry, *J. Eng. Mech. Div.* **97** (1971) 1663.
6. Y. S. Jenq and S. P. Shah, *Eng. Fracture Mechanics* **21** (1985) 1055.
7. A. Hillerborg, M. Modeer and P. E. Petersson, *Cem. Concr. Res.* **6** (1976) 773.
8. R. E. Roelfstra and F. H. Wittmann, In *Fracture Toughness and Fracture Energy in Concrete*, ed. F. H. Wittmann, Elsevier Science Publishers, Amsterdam, 1986, p. 163.
9. M. Wecharatana, In *Serviceability and Durability of Construction Materials*, ed. B. A. Suprenant, ASCE, New York, 1990, p. 966.
10. V. C. Li, C.-M. Chan and C. K. Y. Leung, *Cem. Concr. Res.* **17** (1987) 441.
11. J.-H. Yon, N. W. Hawkins and A. S. Kobayashi, *J. Eng. Mechanics* **117** (1991) 1595.
12. Z. P. Bazant and M. T. Kazemi, *Int. J. of Fracture* **44** (1990) 111.
13. Y. S. Jenq and S. P. Shah, *J. Eng. Mechanics* **111** (1985) 1227.
14. J. Planas and M. Elices, *Eng. Fracture Mechanics* **35** (1990) 87.
15. Y. S. Jenq and S. P. Shah, *Int. J. of Fracture* **51** (1991) 103.

Continuum Models and Discrete Systems
Proceedings of 8th International Symposium, June 11-16, 1995, Varna, Bulgaria, ed. K.Z. Markov
©World Scientific Publishing Company, 1996, pp. 646-652

FRACTURE AND ELECTRIC DEGRADATION OF PIEZOELECTRIC CERAMICS

HORACIO SOSA
Department of Mechanical Engineering and Mechanics
Drexel University, Philadelphia, PA 19104, USA

Abstract. Issues concerning mechanical and electrical failure of piezoelectric ceramics are discussed. Particular emphasis is given to the roles of material anisotropy, defect orientation and electric boundary conditions.

1. Introduction

Piezoelectric ceramics (PEC) are extensively used in modern technology to convert mechanical energy into electrical energy and vice versa. The importance of this unique characteristic is reflected in the growing number of cutting edge technologies using these ceramics in such applications as electronic packaging, medical and underwater instrumentation, automotive, aerospace and energy distribution.

For decades, studies on piezoelectric materials have been evolving along lines of research as varied as applied physics, electronics, materials science and control theory. Here we are concerned with continuum mechanics views to address a rather new theme in the arena of piezoelectricity, namely mechanical and electrical failure of PEC. Research in this particular field has become quite intense during the last years as a direct result of new demands imposed on piezoelectric ceramics. Indeed, recent applications involve large piezo-ceramic bodies of various shapes and subjected to severe loading conditions. As a consequence, the likelihood of damage and failure of such materials has increased dramatically, and with it the genre of research here addressed.

No attempt will be made to provide a thorough review of the evolution of a topic like fracture of piezo-ceramics. The purpose of this article is to present concisely a number of issues that the author feels of crucial character to understand the underlying mechanisms involved in a phenomenological description of a piezoelectric solid with defects. Hence, short sections with commentaries on boundary conditions, field-defect interactions and material anisotropy are provided. The selected material is intended to be of descriptive nature, avoiding mathematical details except where necessary to emphasise certain results.

2. A Brief on Piezoelectric Ceramics

Piezoelectricity is a process of electromechanical interaction that occurs exclusively in anisotropic solids, thus found in most crystals (for example, quartz and Rochelle salt). The piezoelectric effect, however, can also be induced in originally isotropic ceramics (such as barium titanate and lead titanate zirconate) through a process known as poling [4]. Because this process is crucial to understand the motivations behind the works referred to in this article, a description of such procedure is provided in this section.

The ceramic body is first manufactured into its final shape and size according to traditional methodologies. From a macroscopic point of view such finished body can be regarded as a collection of crystallites (each with its own polar direction) randomly oriented. Since in the average there is no net polarisation the body is considered isotropic. Nevertheless, it can be made piezoelectric if subjected to a high dc electric field (the poling field $\mathbf{E}_\mathrm{P}$) for a certain period of time and at a given temperature (not far below the Curie point). The dipoles are thus aligned in the direction of $\mathbf{E}_\mathrm{P}$, a phenomenon macroscopically characterised by the polarisation vector $\mathbf{P}$. The result is a transversely isotropic ceramic body with its basal (isotropic) plane perpendicular to $\mathbf{P}$, therefore, the material can also be described as an hexagonal crystal of symmetry class 6mm.

Many mathematical analyses of PEC bodies can be performed with respect to a Cartesian co-ordinate system where the positive z axis is chosen in the direction of $\mathbf{P}$. The axes x and y are then chosen to form a right hand triad. It is with respect to such co-ordinate system (or poling direction) that all elastic and electric variables are referred to. In particular, the electric field strength is defined as positive when it has the same sense as the poling field. Under this convention, a cylindrical body made of a poled ferroelectric ceramic subjected to a positive field will elongate, while if subjected to a negative field will shorten.

3. Failure of Piezoelectric Ceramics

Piezoelectric materials can fail both in a mechanical and electrical sense in the presence of defects [3]. Mechanical failure of piezoelectric ceramics is a consequence of their low fracture resistance (or fracture toughness) which, typically, is in the order of 1 MPa $\mathrm{m}^{1/2}$. Therefore, the presence of defects such as micro cracks (which in most cases are induced during poling), inclusions, impurities or voids (as a result of poor mixing of powders or sintering) is clearly detrimental from a structural reliability point of view. This situation has become even worse with recent applications, where the demands for larger ceramic bodies bring associated an increase in the density of defects introduced at the various stages of manufacturing processes.

Defects can also be responsible for what in this article we refer as electric degradation. A typical example is dielectric breakdown, which takes place in the presence of very high electric fields. Because of these fields large numbers of electrons may suddenly be excited to energies within the conduction band, resulting into large currents that may deteriorate the material in an irreversible manner. Breakdown occurs when the electric field exceeds the material's dielectric strength, which for ceramics is in the range of $10^4 - 10^5$ V/mm.

Another example of electric degradation (in our view more serious because it takes place at lower voltage levels) is depoling, a phenomenon that can be manifested in three manners: (1) Thermal depoling takes place under excessive heat. More specifically, at temperatures close to or above the Curie point (whose values can range from 115° to 320° C) the electric dipoles can regain their unaligned state, producing a deterioration of the piezoelectric performance. (2) Electrical depoling is the result of a strong electric field applied to a PEC body in a sense opposite to the original poling voltage. The field strength necessary to cause serious depoling depends on the grade of the material, the duration of application and the temperature, but typically ranges from 500 V/mm to 1000 V/mm when applied statically. Alternating fields have also a depoling effect during the cycles opposite to the poling field. (3) Mechanical depoling can occur under high values of stress, in which case the directional orientation in the ceramic can be loss immediately, which in turn results in a much poorer piezoelectric performance of the material.

This article is devoted to the last two depoling factors, since many applications involve working temperatures well below the Curie point. However, it must be kept in mind that there are situations, in particular in the area of multilayer devices, where thermal stresses tend to contribute to failure or damage of the device.

In passing it is important to note that defects are not the only source of damage. In fact we have shown [15] that mechanical and electric failure can also occur in ideally defect-free PEC. For example, extremely high stresses and electric fields are induced in the neighbourhood of concentrated loads as well as around the tips of embedded or surface electrodes.

Finally, it seems important to comment that most applications of PEC involve time dependent phenomena. However, for purposes of simplification most mathematical analyses of ceramics with defects are constrained to static or quasi-static formulations, assuming that the working frequencies are low. Such approaches are allowed with poled ferroelectric ceramics in view of the fact that they satisfy the conditions of high permittivity and low conductivity necessary for static models to be valid.

4. On Anisotropy and Experimental Results

The interactions of elastic and electric fields with defects in PEC bodies depend strongly on the location of the defect with respect to the poling axis as well as on the nature of the applied loads. This is a consequence of the material symmetry of the ceramic, which is isotropic in planes perpendicular to **P**, and anisotropic in planes containing the polarisation vector.

The role of anisotropy in damage characterisation of PEC has been a theme under debate for many years, in particular from an experimental standpoint. Many early laboratory investigations have been carried out according to concepts of classical fracture mechanics, thus avoiding anisotropy and electro-elastic interactions by invoking that such effects are of negligible nature. A number of works [6,10,11], however, have eluded such a priori assumptions by performing experiments on poled and unpoled ceramics containing cracks. It is not in the spirit of the paper to provide an analysis of such experimental work, but rather to highlight results that have been predicted independently via theoretical models: In [10] it was found that an unpoled (and therefore isotropic) ceramic behaves exactly in the same manner as the isotropic plane of a poled ceramic, as far as fracture toughness is concerned. Moreover, if a crack is considered in the anisotropic plane of the specimen, it will have greater tendency to propagate if placed perpendicularly rather than parallel to **P** [10,11]. It was also found that assuming the crack faces perpendicular to the poling direction, a positive (negative) field induces positive (compressive) stresses, thus enhancing (retarding) crack propagation. In section 6 we show equivalent predictions via an equation derived in closed form. Needless to say that the particular result of crack arrest through an applied voltage has promoted studies in a number of potential applications to mitigate structural damage, among which we can cite the healing of bones. Finally, using poled ceramics it was observed [6] that fields applied perpendicularly to the crack plane always turned the crack in the direction opposite to **P**, this effect being independent of sign and strength of the applied field. The phenomenon of crack skewing is still open to further studies for two main reasons: first, other researchers have been unable to reproduce it and second, analytical results have provided only partial verification [14].

5. On Defect Orientation

In general, defects in materials can be oriented in any arbitrary manner. However, in transversely isotropic ceramics we recognise two particular orientations with respect to the poling axis. To understand these orientations we view the defect as an elliptic cylinder whose generator in the case of a crack becomes its leading edge or front. Two problems can be studied within the realm of a two-dimensional model: (a) the generator (or crack front) is parallel to the poling axis of the ceramic, in which case the elliptic hole is contained in the basal plane; and (b) the generator is perpendicular to the poling axis, in which case the plane where the hole is located is anisotropic. In the first case coupling effects arise due to the interplay of out-of-plane stresses with in-plane electric fields and not from the constitutive equations. In the theory of fracture mechanics this situation corresponds to the so-called mode III or anti-plane problem. Here the resulting system of governing equations is rather simple to solve and we refer the reader to [7-9] for excellent accounts on the physical implications of the results.

The second case, that is the hole in the anisotropic plane, recast in its formulation a fully coupled system of partial differential equations, reflecting elastic-electric interactions and the inherent anisotropy of the material. This geometrical configuration seems also to be more realistic from a practical point of view by virtue of the nature of the loads that are usually involved in most applications. This particular problem appears to have been studied only by the author [13,14,16] through a complex variables approach. From a fracture mechanics point of view modes I and II can be considered in this case, the first one being particularly suitable to be reproduced in the laboratory.

6. On Electric Boundary Conditions

Another issue of importance to be considered in the modelling of PEC bodies with defects is that of the electric boundary conditions to be prescribed at the rim of a hole or faces of a crack. Seeking mathematical simplification it has been common [1,7-9,13,14,17,18] to assume the hole (or crack) filled with gas (typically air or vacuum) and free of tractions and surface charge. This situation corresponds to the case of two dielectric bodies (the piezoelectric medium and the inner portion of the defect) with different dielectric permittivities. In general, at such a boundary the normal component of the induction vector and the electric potential (or, alternatively, the tangential component of the electric field) are continuous. When one of the mediums is air or vacuum, these conditions can be relaxed, and one can merely state that the normal component of the induction in the matter is zero at the boundary (a condition referred to as the "impermeable" condition). It is clear that under such circumstances the originally two-domain boundary-value problem is reduced to finding the electromechanical variables within only one domain: that of matter. Probably the most notable result emanating from the impermeable assumption is the one concerning sharp cracks: It has been found that when the solid is subjected to remote electric load, the electric field is singular at the tip of the crack. However, it has been shown [5] (although under the restriction of isotropic electrostriction) that at the tip of a sharp crack the electric field is large but not singular. In fact, the magnitude of the field is bounded by a material parameter reflecting the permittivities of the material and gas enclosed by the cavity or crack.

A step forward to generalise the aforementioned results was carried out in [2] by considering a transversely isotropic piezo-ceramic subjected to anti-plane deformation. Now, it is tempting to view such a deformation as an impediment to establish more general conclusions with respect to the behaviour of certain variables in the neighbourhood of cracks and holes. In other words, it is not quite clear how the anisotropy of the material may affect some results. Therefore, very recently [16] we have reconsidered the in-plane problem using exact boundary conditions. Several new results are further discussed in the remaining of the section.

To this end, consider an elliptic cavity of axes $2a$ and $2b$ ($a > b$) in x- and z-directions, respectively, embedded in an infinite transversely isotropic piezoelectric body. If the body is subjected to an electric load $\mathbf{E} = E_0\mathbf{e}_z$ (i.e. in the direction of

P), the field inside the cavity can be shown to be given by

$$E_x^c = 0, \quad E_z^c = \frac{(\alpha\gamma + 2c_{11}c_{33}^{-1})\ E_0}{\alpha\gamma + 2\epsilon_0 c_{11}} \tag{1}$$

where $\alpha = b/a$, γ is a material parameter reflecting electromechanical interactions, c_{11} and c_{33} are the "effective" impermittivities of the material in x- and z-directions, respectively and ϵ_0 is the permittivity of the gas enclosed by the hole (typically air, in which case we can assume $\epsilon_0 = 8.85 \times 10^{-12}$ N/m^2).

For most ceramics [12], $\epsilon_0 c_{ii} \sim 10^{-4}$ and $\gamma \sim 1.5-2.0$, thus it is clear that the "exact" and "impermeable" (which is recovered from (1) by setting $\epsilon_0 = 0$) models give virtually the same results even for cavities with ratios $\alpha = 1/100$. However, the results stemming from these models differ drastically in the limiting case of the slit crack: that is, as $\alpha \to 0$, $E_z^c \to E_0/\epsilon_0 c_{33}$. This means that at the crack tip the field is not singular. From a practical point of view this result appears to provide small consolation since fields of only 0.1 V/mm will be enough to degrade at least locally the material. Note that a model based on the impermeable condition yields a singular field when $\alpha \to 0$.

When the body is subjected to a stress $\mathbf{T} = T_0 \mathbf{e}_z \otimes \mathbf{e}_z$ ($T_0 > 0$), a uniform field is induced in the cavity: the first component is here omitted while the second is given by [16]

$$E_z^c = \frac{k_1\ T_0}{\alpha\gamma + 2\epsilon_0 c_{11}} \tag{2}$$

Here the results for the exact and impermeable models follow the same patterns of (1). Moreover, since for typical ceramics $k_1 \sim 10^{-3}$, in the limit of the slit crack the induced field is 10 times the applied stress and depoling will occur for values of T_0 as small as 0.1 MPa.

Finally, let us consider the expression for the normal component of stress in the direction of **P** at the point (a,0) on the elliptic cavity, due to an applied field in z-direction, namely

$$T_{zz}(a, 0) = \frac{k_2\ (c_{33}^{-1} - \epsilon_0)\ E_0}{\alpha\gamma + 2\epsilon_0 c_{11}} \tag{3}$$

where $k_2 \sim 10^8 - 10^9$ is a positive constant. If the piezoelectric solid is in addition subjected to remote forces, the corresponding stresses can be added to this expression according to the principle of superposition. On this point, it is useful to note that the sign of $T_{zz}(a, 0)$ in (3) depends solely on the sign of E_0, since the rest of the expression is positive. Thus, E_0 can increase or reduce the intensity of the normal stress generated by say tensile forces applied in an independent manner.

We note that since $c_{33}^{-1} >> \epsilon_0$, the results given by the exact and impermeable models are quite similar even for very slender holes. However, when $\alpha \to 0$ the impermeable model predicts singular stresses, while the exact model gives stresses of order $\sim 10^4 - 10^6 E_0$. That is, a field $E_0 = 100$ V/m can produce stresses that vary between 1 and 100 MPa. Such a result appears to be serious enough considering the fact that the tensile strength of a typical piezoelectric ceramic is of approximately 80 MPa.

The results given by (1)-(3) clearly show a dramatic qualitative difference regarding the behaviour of elastic and electric fields at the tip of a sharp crack whether the description is done through an exact or an impermeable model. Less clear is, however, the impact of these results from a practical point of view. In fact, how open or closed a crack is seems to be a rather subjective matter. Endless digressions on this issue could be circumvented by invoking Wittgenstein's most celebrated phrase.

Acknowledgements. This work was supported by the National Science Foundation under Grants Nos. MSS-9215296 and INT-9414656. Partial support was also provided by the Ministry of Education of Spain during the author's visit to Instituto de Investigación Tecnológica, Universidad Pontificia Comillas, Madrid.

References

1. W. F. Deeg, *The analysis of dislocation, crack and inclusion problems in piezoelectric solids*, Ph.D. Thesis, Stanford University, CA, 1980.
2. M. Dunn, *Eng. Fracture Mech.* **48** (1994) 25.
3. S. W. Freiman and R. C. Pohanka, *J. Am. Ceram. Soc.* **72** (1989) 2258.
4. B. Jaffe, W. R. Cook and H. Jaffe, *Piezoelectric Ceramics*, Academic Press, New York, 1971.
5. R. M. McMeeking, *ZAMP* **40** (1989) 615.
6. K. D. McHenry and B. G. Koepke, In *Fracture Mechanics of Ceramics,* Vol. 5, ed. by R. C. Bradt et al., Plenum Press, New York, 1983 p. 337.
7. Y. E. Pak, *ASME J. Appl. Mech.* **112** (1990) 647.
8. Y. E. Pak, *Int. J. Fracture* **54** (1992) 79.
9. Y. E. Pak, *Int. J. Solids Structures* **29** (1992) 2403.
10. Y. Pak and A. Tobin, In *Mechanics of Electromagnetic Materials and Structures,* AMD Vol. 161, ed. by J. Lee et al., ASME, New York, 1993.
11. G. G. Pisarenko, V. M. Chushko and S. P. Kovalev, *J. Am. Ceram. Soc.* **68** (1985) 259.
12. Quartz and Silice, Nemours, France, 1995 (private communication).
13. H. Sosa, *Int. J. Solids Structures* **28** (1991) 491.
14. H. Sosa, *Int. J. Solids Structures* **29** (1992) 2613.
15. H. Sosa and M. Castro, *J. Mech. Phys. Solids* **42** (1994) 1105.
16. H. Sosa and N. Khutoryansky, *Int. J. Solids Structures* (submitted).
17. H. A. Sosa and Y. E. Pak, *Int. J. Solids Structures* **26** (1990) 1.
18. Z. Suo, C. M. Kuo, D.M. Barnett and J. R. Willis, *J. Mech. Phys. Solids* **40** (1992) 739.

AUTHOR INDEX

(∗: not personally present)

LIST OF PARTICIPANTS

(with mailing addresses)

Taij ADACHI
Department of Mechanical Engineering
Faculty of Engineering
Kobe University Nada, Kobe 657
JAPAN
e-mail: adachi@mp-1.mech.kobe-u.ac.jp

Abderrazzak AZIRHI
Theoretische Physik
Universitat-Gesamthochschule-Paderborn
Warburger Str. 100
D-33098 Paderborn
GERMANY
e-mail: azi@lagrange.uni-paderborn.de

Angel BALTOV
Institute of Mechanics
Bulgarian Academy of Sciences
Acad. G. Bonchev str., bl. 8
BG-1113 Sofia
BULGARIA

Victor BERDICHEVSKY
Mecanical Engineering Department
College of Engineering
Wayne State University.
Detroit, MI 48202
USA
e-mail: vberd@me1.eng.wayne.edu

David J. BERGMAN
School of Physics and Astronomy
Tel Aviv University
Ramat Aviv
IL-69978 ISRAEL
e-mail: bergman@albert.tau.ac.il

Marcel BERVEILLER
Laboratoire de Physique et Mécanique des Materiaux, ENIM - CNRS
Université de Metz
Ile du Saulcy
F-57045 Metz Cedex 01
FRANCE

Paolo BISCARI
Dipartimento di Matematica
Universita' di Pisa
Via Buonarroti 2
I-56100 PISA
ITALY
e-mail: biscari@vaxsns.sns.it

Michel BORNERT
Laboratoire de Mécanique des Solides
École Polytéchnique
F-91128 Palaiseau Cedex
FRANCE
e-mail: bornert@athena.polytechnique.fr

Jordan BRANKOV
Institute of Mechanics
Bulgarian Academy of Sciences
Acad. G. Bonchev str., bl. 8
BG-1113 Sofia
BULGARIA
e-mail: brankov@bgearn.bitnet

Valeri A. BURYACHENKO
Institüt für Leicht- und Flugzeugbau (E317)
TU-Wien
Gußhausstraße 25-29
A-1040 Wien
AUSTRIA
e-mail: buryach@ilfb03.tuwien.ac.at

Pim BUSSINK
Shell Research
Volmerlaan 8, P.O. Box 60
2280 AB Rijswijk
The NETHERLANDS
e-mail: bussinkp@ksepl.nl

Gianfranco CAPRIZ
Consorzio Pisa Ricerche
Piazza Alessandro d'Ancona 1
I-56127 Pisa
ITALY
e-mail: capriz@iei.pi.cnr.it

André CHRYSOCHOOS
Laboratoire de Mécanique et Génie Civil
Université de Montpellier II
Place Eugene Bavaillon
F-34095 Montpellier Cedex 05
FRANCE
e-mail: chryso@lmgc.univ-montp2.fr

Alexander CHUDNOVSKY
Department of Civil Engineering
University of Illinois at Chicago
P.O.Box 4348
Chicago, IL 60680
USA
e-mail: u14777@uicvm.cc.uic.edu

Bogdan CICHOCKI
Institute of Theoretical Physics
Warsaw University
Hoza 69
PL-00-681 Warsaw
POLAND
e-mail: cichocki@fuw.edu.pl

Bernard COLLET
Université Pierre et Marie Curie (Paris VI)
Modelisation en Mécanique
Tour 66, 4 place Jussieu, boite 162
F-75252 Paris,Cedex 05
FRANCE
e-mail: pouget@frcpn11.in2p3.fr

Cristian DASCALU
Institute of Mathematics of the
Romanian Academy of Sciences
P.O.Box 1-764
70 700 Bucarest
ROMANIA
e-mail: cdascalu@imar.ro

Manuel DE LEÓN
Instituto de Matemàticas y Fisica Fundamental
Consejo Superior de Investigaciones Cientificas
Serrano 123, 28006 Madrid
SPAIN
e-mail: ceeml02@cc.csic.es

Gianpietro DEL PIERO
Istituto di Ingegneria
Via Scandiana 21
I-44100 Ferrara
ITALY
e-mail: luca@ing12.unife.it

Pawel DLUZEWSKI
Institute of Fundamental Technological Research
Polish Academy of Sciences
Swietokrzyska 21
PL-00-049 Warsaw
POLAND
e-mail: pdluzew@ippt.gov.pl

Galja DRAGANOVA
Institute of Chemical Technolgy and Biotechnology
P.O.Box 110
BG-7200 Razgrad, BULGARIA

Phillip M. DUXBURY
Physics and Astronomy Department
Michigan State University
205 Physics-Astronomy Bldg.
East Lansing, MI 48824-1116
USA
e-mail: duxbury@msupa.pa.msu.edu

Saadet ERBAY
Faculty of Science and Letters
Department of Mathematics
Istanbul Technical University
80 626 Maslak/Istanbul
TURKEY
e-mail: feherbay@cc.itu.edu.tr

Marcelo EPSTEIN
University of Calgary
Department of Mechanical Engineering
2500 University Drive N.W.
Calgary, Alberta T2N 1N4
CANADA
e-mail: epstein@enme.ucalgary.ca

François FEUILLEBOIS
École Superieure de Physique et Chimie
Industrielles de la ville de Paris
PMMH, 10 rue Vauquelin
F-75231 Paris Cedex 05
FRANCE
e-mail: feuilleb@pmmh.espci.fr

Werner FRANK
Max-Planck-Institut für Metallforschung
Institut für Physik
Heisenbergstraße 1
D-70568 Stuttgart
GERMANY
e-mail: schemi@physix.mpi-suttgart.mpg.de

Barbara GAMBIN
Institute of Fundamental Technological Research
Polish Academy of Sciences
Swietokrzyska 21
00-049 Warsaw
POLAND
e-mail: bgambin@lksu.ippt.gov.pl

Luiz Carlos GARCIA DE ANDRADE
Universidade do Estado Rio de Janeiro (VERJ)
Instituto de Fisika - Dept de Fisika Teorica
Rua São Francisco Xavier, 524 -- Sala 3001-D
Rio de Janeiro, RJ, Maracanã, Cep: 20550-013
BRAZIL
e-mail: sandraeg@vmesa.uerj.br

Pasquale GIOVINE
Dipartimento di Meccanica dei
Fluidi ed Ingegneria Offshore
Via E.Cuzzocrea, 48
I-89128 Reggio Calabria
ITALY
e-mail: giovine@gauss.dm.unipi.it

Helmut GÜNTHER
Fachhochschule Bielefeld
Fachbereich Elektrotechnik (FB2)
Wilhelm-Bertelsmann-Str. 10, Postfach 2830
D-33602 Bielefeld
GERMANY

Eveline HERVÉ
Laboratoire de Mécanique des Solides
École Polytéchnique
F-91128 Palaiseau Cedex
FRANCE
e-mail: herve@athena.polytechnique.fr

Javor HRISTOV
Faculty of Mathematics and Informatics
"St. Kl. Ohridski" University of Sofia
5 blvd J. Bourchier
BG-1164 Sofia
BULGARIA

Esin INAN
Faculty of Science
Department of Engineering Sciences
Istanbul Technical University
80 626 Maslak/Istanbul
TURKEY
e-mail: feinan@cc.itu.edu.tr

Dominique JEULIN
École des Mines de Paris
Centre de Morphologie Mathématique
35 rue Saint-Honoré
F-77300 Fontainebleau
FRANCE
e-mail:jeulin@cmm.ensmp.fr

Mark L. KACHANOV
Dept of Mechanical Engineering
Tufts University
Medford, MA 02155
USA
e-mail: mkacaho@emerald.tufts.edu

Vratislav KAFKA
Institute of Theoretical and Applied Mechanics
Academy of Sciences of Czech Republic
74 Prosecká
CZ-190 00 Prague 9
CZECH REPUBLIC

V. KAMBOUROVA
Institute of Chemical Technolgy and
Biotechnology
P.O.Box 110
BG-7200 Razgrad
BULGARIA

Sergei KANAUN
Division de Graduados e Investigacion
Instituto Tecnológico y de Estudios Superiores
de Monterrey
Campus Estado de México, Apdo. postal 18
Modulo de Servicio Postal
Atizapán de Zaragoza
MÉXICO 5296
e-mail: kanaoun@servdgi.cem.itesm.mx

Robert KAZANDJIEV
Institute of Mechanics
Bulgarian Academy of Sciences
Acad. G. Bonchev str., bl. 8
BG-1113 Sofia
BULGARIA

Mikhail KOLEV
151, Evl. Georgiev Str.
BG-1504 Sofia
BULGARIA

Arnold M. KOSEVICH
Institute of Low Temperature Physics and
Engineering
Ukrainian Academy of Sciences
47, Lenin Avenue
Kharkov 310164
UKRAINE
e-mail: kosevich@ilt.kharkov.ua

Romuald KOTOWSKI
Institute of Fundamental Technological Research
Polish Academy of Sciences
Swietokrzyska 21
PL-00 049 Warsaw
POLAND
e-mail: rkotow@lksu.ippt.gov.pl

Ekkehardt KRÖNER
Institut für Theoretische und Angewandte Physik
Univesität Stuttgart
Pfaffenwaldring 57
D-70550 Stuttgart
GERMANY
e-mail: doro@itap.physik.uni-stuttgart.de

Ilka LAMBOVA
Faculty of Mathematics and Informatics
"St. Kl. Ohridski" University of Sofia
5 blvd J. Bourchier
BG-1164 Sofia,
BULGARIA
e-mail: ilambova@fmi.uni-sofia.bg

K. C. LE
Universität Bochum
Lehrstuhl für Allgeneine Mechanik
D-44780 Bochum
GERMANY
e-mail: chau.le@ruba.rz.ruhr-uni-bochum.de

Valeri M. LEVIN
Department of Civil Engineering
Petrozavodsk State University
Lenina 33
185640 Petrozavodsk
RUSSIA
e-mail: levin@mainpgu.carelia.ru

Therese LÉVY
Université Pierre et Marie Curie (Paris VI)
Modelisation en Mécanique
Tour 66, 4 Place Jussieu, boite 162
F-75252 Paris Cedex 05
FRANCE
e-mail: lmm@circrp.jussieu.fr

Christian LEXCELLENT
Laboratoire de Mécanique Appliquée
Université de Franche-Comté
Route de Gray, La Boulaie
F-25030 Besancon Cedex
FRANCE

Konstantin Z. MARKOV
Faculty of Mathematics and Informatics
"St. Kl. Ohridski" University of Sofia
5 blvd J. Bourchier
BG-1164 Sofia
BULGARIA
e-mail: kmarkov@fmi.uni-sofia.bg

Tchavdar MARINOV
Department of Mathematics
Technical University Varna
BG-9010 Varna
BULGARIA
e-mail: marinovi@tu-varna.bg

Rositza MARINOVA
Department of Mathematics
Technical University Varna
BG-9010 Varna
BULGARIA
e-mail: marinovi@tu-varna.bg

Jose-Luis MARQUÉS
Universitat Paderborn, Fachbereich Physik
Warburger Str. 100
D-33095 Paderborn
GERMANY
e-mail: fanth1@lagrange.uni-paderborn.de

Milan MICUNOVIC
Masinski Fakultet
ul. Sestre Janjica 6
34000 Kragujevac
YUGOSLAVIA
e-mail: mmicun@nis0.uis.kg.ac.yu

Ivan MIHOVSKY
Faculty of Mathematics and Informatics
"St. Kl. Ohridski" University of Sofia
5 blvd J. Bourchier
BG-1164 Sofia
BULGARIA

Orlin MINCHEV
Institüt für Leicht- und Flugzeugbau (E317)
TU-Wien, Gußhausstraße 25-29
A-1040 Wien
AUSTRIA
permanent address:
Institute of Mechanics
Bulgarian Academy of Sciences
Acad. G. Bonchev str., bl. 8
BG-1113 Sofia
BULGARIA

Donald F. NELSON
Department of Physics
Worcester Polytechnic Institute
100 Institute Road
Worcester, MA 01609
USA
e-mail: dnelson@wpi.wpi.edu

E. PATOOR
Laboratoire de Physique et Mécanique des Matériaux, URA CNRS
Université de Metz
Ile du Saulcy
F-57045 Metz Cedex 01
FRANCE
e-mail: patoor@lpmm.univ-metz.fr

Pedro PONTE CASTANEDA
Mechanical Engineering
University of Pennsylvania
Philadelphia, PA 19104-6284
USA
e-mail: ponte@sol1.lrsm.upenn.edu

Antony POPOV
Faculty of Mathematics and Informatics
"St. Kl. Ohridski" University of Sofia
5 blvd J. Bourchier
BG-1164 Sofia, BULGARIA

Joel POUGET
Université Pierre et Marie Curie (Paris VI)
Modelisation en Mécanique
Tour 66, 4 place Jussieu, boite 162
F-75252 Paris,Cedex 05
FRANCE
e-mail: pouget@frcpn11.in2p3.fr

Dimitar PUSHKAROV
Institute of Solid-State Physics
Bulgarian Academy of Sciences
BG-1184 Sofia
BULGARIA
e-mail: dpushk@bgearn.bitnet

Franz G. RAMMERSTORFER
Institüt für Leicht- und Flugzeugbau (E317)
TU-Wien
Gußhausstraße 25-29
A-1040 Wien, AUSTRIA
e-mail: ra@ilfb03.tuwien.ac.at

Nicolas RIVIER
Laboratoire de Physique Theorique
Institut de Physique
Université Louis Pasteur
3, Rue de l'Université
F 67 084 Strasbourg Cedex
FRANCE
e-mail nick@fresnel.u-strasbg.fr

Krassimir RUSSEW
Institute of Metal Science
Bulgarian Academy of Sciences
67 Shipchensky prohod Str.
BG-1574 Sofia
BULGARIA

Federico J. SABINA
I.I.M.A.S -- Universidad Nacional Autònome de Mexico
Apartado Postal 20-726, Admon. No 20
Delegacion de Alvaro Obregon
01000 Mexico, D.F.
MEXICO
e-mail: fjs@uxmym1.iimas.unam.mx

Tony SACKFIELD
Department of Mathematics and Statistics
Nottingham Trent University
Burton Street
Nottingham NG1 4BU, UK

Igor B. SEVOSTYANOV
Department of theory of elasticity
Faculty of Mathematics and Mechanics
St. Petersburg State University
Bibliotechnaja pl. 2
198904 St. Petersburg
RUSSIA

Yves STEYT
Université Libre de Bruxelles
Centre for Nonlinear Phenomena and Complex Systems ,Campus Plaine U.L.B., C.P. 231
Boulevard du Triomphe
B-1050 Bruxelles
BELGIUM
e-mail: ysteyt@ulb.ac.be

PING SHENG
Department of Physics
University of Science and Technology
Clear Water Bay
Kowloon, HONG KONG
e-mail: phsheng@ usthk.ust.hk

A. J. M. SPENCER
Dept of Theoretical Mechanics
University of Nottingham
Nottingham, NG7 2RD
ENGLAND, U.K.

Liljana STOJANOVA
Institute of Metal Science
Bulgarian Academy of Sciences
67 Shipchensky prohod Str.
BG-1574 Sofia
BULGARIA

Pierre SUQUET
C.N.R.S. -- LMA
31 Chemin Joseph Aiguire
B.P. 71
F-13 402 Marseille
FRANCE
e-mail: suquet@lma.cnrs-mrs.fr

David R. S. TALBOT
Department of Mathematics
Coventry University
Priory Street
Coventry CVI 5FB
ENGLAND, UK
e-mail: mtx008@cck.coventry.ac.uk

Mehmet Ali TASDEMIR
Faculty of Civil Engineering
Division of Engineering Materials
Istanbul Technical University
80 626 Maslak/Istanbul
TURKEY
e-mail: inakkaya@cc.itu.edu.tr

Carmine TRIMARCO
Universita di Pisa
Instituto di Matematiche Applicate
Via Bonanno 25B
I-56126 Pisa
ITALY
e-mail: ming@vm.cnuce.cnr.it

Mikhail TODOROV
Institute of Applied Mathematics and Informatics
Technical University - Sofia
BG-1156 Sofia
BULGARIA

Christopher TRUMAN
Department of Mathematics and Statistics
Nottingham Trent University
Burton Street
Nottingham NG1 4BU
ENGLAND, UK

Jason TURNER
Blackett Laboratory, Imperial College
Prince Consort Road
London SW7 2BZ
ENGLAND, UK
e-mail: j.m.turner@ic.ac.uk

Gazanfer ÜNAL
Faculty of Science
Department of Engineering Sciences
Istanbul Technical University
80 626 Maslak/Istanbul
TURKEY
e-mail: feunal@cc.itu.edu.tr

Epifanio G. VIRGA
Universita di Pisa
Istituto di Scienza delle Costruzioni
Facolta' di Ingegneria
via Diotisalvi, 2
I-56126 Pisa
ITALY

Steven WALLACE
Department of Mathematics and Statistics
Nottingham Trent University
Burton Street
Nottingham NG1 4BU
ENGLAND, UK
e-mail: mat3wallasg@nottingham-trent.ac.uk

Y. Jack WEITSMAN
Department of Engineering Science and Mechanics, 310 Perkins Hall
University of Tennessee
Knoxville, Tennessee 37996-2030
USA
e-mail: weitsman@utkvx.utk.edu

Krzysztof WILMANSKI
Universität-GH-Essen
Fachbereich 120
Bauwesen Institüt für Mechanik
Universitätstraße 15
D-45117 Essen
GERMANY
e-mail: kw110@bauwesen.uni-essen.de

Michael ZAISER
Max-Planck-Institut für Metallforschung
Institut für Physik
Heisenbergstraße 1
D-70568 Stuttgart
GERMANY
e-mail: zaiser@physix.mpi-stuttgart.mpg.de

Zapryan ZAPRYANOV
Faculty of Mathematics and Informatics
"St. Kl. Ohridski" University of Sofia
5 blvd J. Bourchier
BG-1164 Sofia
BULGARIA

Ivanka ZHELEVA
Institute of Chemical Technolgy and Biotechnology
P.O.Box 110
BG-7200 Razgrad
BULGARIA

Henryk ZORSKI
Institute of Fundamental Technological Research
Polish Academy of Sciences
Swietokrzyska 21
PL-00 049 Warsaw
POLAND
e-mail: azachara@ippt.gov.pl

Krassimir D. ZVYATKOV
Faculty of Mathematics and Informatics
"K. Preslavski" University
BG-9700-Schumen
BULGARIA